Quantum Geometry

Fundamental Theories of Physics

An International Book Series on The Fundamental Theories of Physics: Their Clarification, Development and Application

Volume 48

Quantum Geometry

A Framework for Quantum General Relativity

by

Eduard Prugovečki
Department of Mathematics,
University of Toronto,
Toronto, Canada

KLUWER ACADEMIC PUBLISHERS
DORDRECHT / BOSTON / LONDON

Library of Congress Cataloging-in-Publication Data

Prugovečki, Eduard.
Quantum geometry : a framework for quantum general relativity / by Eduard Prugovečki.
p. cm. -- (Fundamental theories of physics ; v. 48)
Includes bibliographical references (p.) and index.
ISBN 0-7923-1640-1 (HB : acid-free paper)
1. Geometric quantization. 2. General relativity (Physics)
I. Title. II. Series.
QC174.17.G46P78 1992
530'.1--dc20 92-2540

ISBN 0-7923-1640-1

Published by Kluwer Academic Publishers,
P.O. Box 17, 3300 AA Dordrecht, The Netherlands.

Kluwer Academic Publishers incorporates
the publishing programmes of
D. Reidel, Martinus Nijhoff, Dr W. Junk and MTP Press.

Sold and distributed in the U.S.A. and Canada
by Kluwer Academic Publishers,
101 Philip Drive, Norwell, MA 02061, U.S.A.

In all other countries, sold and distributed
by Kluwer Academic Publishers Group,
P.O. Box 322, 3300 AH Dordrecht, The Netherlands.

Printed on acid-free paper

Printed in the Netherlands

To the memory of a great physicist of penetrating vision and lasting achievements, whose scientific legacy and uncompromising integrity continue to inspire:

Paul Adrien Maurice Dirac

(1902-1984)

"One must seek a new relativistic quantum mechanics and one's prime concern must be to base it on sound mathematics."

P. A. M. Dirac (1978)

Table of Contents

Note: *The sections marked with an asterisk can be omitted at a first reading.*

Preface

The present monograph provides a systematic and basically self-contained introduction to a mathematical framework capable of incorporating those fundamental physical premises of general relativity and quantum mechanics which are not mutually inconsistent, and which can be therefore retained in the unification of these two fundamental areas of twentieth-century physics. Thus, its underlying thesis is that the equivalence principle of classical general relativity remains true at the quantum level, where it has to be reconciled, however, with the uncertainty principle. As will be discussed in the first as well as in the last chapter, conventional methods based on classical geometries and on single Hilbert space frameworks for quantum mechanics have failed to achieve such a reconciliation. On the other hand, foundational arguments suggest that new types of geometries should be introduced.

The geometries proposed and studied in this monograph are referred to as quantum geometries, since basic quantum principles are incorporated into their structure from the outset. The mathematical tools used in constructing these quantum geometries are drawn from functional analysis and fibre bundle theory, and in particular from Hilbert space theory, group representation theory, and modern formulations of differential geometry. The developed physical concepts have their roots in nonrelativistic and relativistic quantum mechanics in Hilbert space, in classical general relativity and in quantum field theory for massive and gauge fields. However, the principal aim of this monograph is to deal not with specific physical theories, beyond QED and quantum gravity, but rather with general mathematical structures that can serve as frameworks within which such theories can be developed in an epistemologically and mathematically sound manner. On the other hand, we shall demonstrate that the novel features of these frameworks not only clarify some long-standing questions of quantum field theory in curved spacetime and of quantum gravity, but also give rise to some new perspectives on the world of elementary particles.

The essential ideas and techniques of the varied and rich disciplines treated in this monograph are explained in the appropriate sections of its text. In order to deepen the reader's understanding of those more technical aspects which could not be included due to limitations on space, the reader is directed in a carefully guided manner to specific sections of a score of key references, singled out from the list of references provided at the end of this book. It is therefore hoped that despite the advanced nature of the presented material, this monograph will be accessible to most graduate students in physics and in mathematics. Thus, although it is desirable that a student already have some understanding of the mathematical foundations of classical general relativity (cf., e.g., Chapters 1-5 of [W] from amongst the aforementioned key references) and of standard nonrelativistic quantum mechanics (cf., e.g., Chapters 1-4 of [PQ]), that is not absolutely mandatory, since all the basic concepts and results are explained in the text, and for the details which are not covered, instructions are given as to where to find them in the key references. For readers at a more advanced level, detailed references to conference proceedings, lecture notes and contemporary papers published in professional physics and mathematics journals are provided in the notes at the end of each chapter. Consequently, this book can be used also as a reference manual and guide to literature for research in the areas it covers.

From the mathematical point of view, the quantum geometries presented in this monograph are infinite-dimensional fibre bundles associated with principal bundles [C,I] whose structure groups incorporate the Poincaré group – or its covering group ISL(2,C)). The base manifolds of these fibre bundles are Lorentzian manifolds, or their appropriate frame-bundle extensions; whereas, their typical fibres are infinite-dimensional (pseudo-) Hilbert spaces or superspaces. The study of connections on such fibre bundles poses interesting mathematical problems, which appear to have received scant attention thus far.

From the physics point of view, the principal areas of application of the present framework are to quantum field theory in curved spacetime and to quantum gravity. The ensuing methodology is distinct from that of other approaches to these disciplines in that it is derived from *foundational* measurement-theoretical considerations. These considerations have led to a formulation of nonrelativistic and special-relativistic quantum theory on phase space, whose pre-1984 results have been comprehensively summarized in a preceding monograph [P], and whose subsequent developments can be found in some of the more recent works cited in the main text of the present monograph. These developments reflect the possibility of resolving long-standing quantum paradoxes by a careful analysis of old as well as of new quantum-measurement schemes and experimental procedures.

From the point of view of the quantum mechanics on phase space presented in Part I of [P], the present quantum geometries enable an extrapolation of their special-relativistic frameworks to the general-relativistic regime. In this context it should be noted that such an extrapolation had been attempted in Part II of [P], but that it ran into the same main difficulty as the more conventional approaches to quantum field theory in curved spacetime – namely, it did not succeed in properly adapting the equivalence principle of classical general relativity to the quantum regime. The present fibre-theoretical framework has, however, succeeded in that task. In fact, although this framework displays many novel features which are of independent mathematical interest, the achievement of that goal represents its major motivation from the point of view of physics. Consequently, abundant corroborative quotations of well-known authorities in the field are provided, not only as vouchers of the fundamental need for a radical revision of many of the conventional ideas in relativistic quantum theory, but also as a guide to further independent study, that might lead some readers to new ideas of their own.

Central to the application of the present quantum geometry framework to quantum physics is the idea of *geometro-stochastic propagation*, developed by the present author as a mathematically and epistemologically sound extrapolation of the standard path-integral formalism. Fundamentally, this idea proposes a modification of the concept of quantum point particle (which according to the orthodox interpretation displays the behavior of *either* a wave *or* a particle) into that of geometro-stochastic *exciton*, whose microdynamics simultaneously embodies the classical attributes of "waves" as well as of "particles". Thus, geometro-stochastic excitons possess *proper* state vectors belonging to the fibres of the quantum bundles constituting quantum geometries, and are localized in relation to *quantum* Lorentz frames, which take over in the geometro-stochastic framework the role played by their classical counterparts in classical general relativity. These proper state vectors propagate externally (i.e., in the classical spacetimes which constitute the base manifolds of those quantum bundles) along stochastic paths in a manner which is *formally* analogous to that of classical particles in diffusion processes. On the other hand, they are superimposed at each location in the base manifold in the manner which in classical physics is associated only with the behavior of waves. Thus, the wave-particle dichotomy present in the orthodox approach to quantum mechanics is replaced, at the *micro*-level, with a unified physical picture of quantum behavior, which has no counterpart in classical physics. It is only at the *macro*-level that, in complete accord with Bohr's ideas, geometro-stochastic exciton behavior manifests itself, in the context of *certain* experimental arrangements, as that of a "particle"; whereas, in the context of some other experimental arrangements, it manifests itself as that of a "wave". Furthermore, in the nonrelativistic context, the mathematical apparatus of quantum geometry allows the transition to a sharp-point limit, which mathematically corresponds to the transition to proper wave functions which are δ-like. In that limit the conventional quantum mechanics of pointlike particles is recovered, thus demonstrating that geometro-stochastic dynamics is a natural outgrowth of conventional quantum theory (cf. Chapters 3 and 4).

Thus, when taken in conjunction with the idea of geometro-stochastic quantum propagation, the quantum geometry framework provides a viable blueprint for the consistent unification of general relativity and quantum theory, which does not give rise to conflicts with conventional theory in those areas where that theory has received unquestionable experimental support. Moreover, as will be seen in the present monograph, quantum geometries can also incorporate many of the theoretical ideas which are in vogue at the present time. Hence, it can serve as a basic framework within which such ideas can be formulated with the mathematical clarity and rigor required for the understanding of their multifaceted physical implications. It is, therefore, primarily as a *framework* of ideas, rather than as a finished theory, that the results on quantum geometries are presented in this monograph.

In the course of the four decades which followed after the early numerical successes of renormalization theory in quantum electrodynamics, the world of elementary particle physics has witnessed a great variety of constantly changing fashions: in turn, there were vigorous promotions of dispersion relations, Regge poles, current algebras, quark models and QCD, supersymmetry, grand-unification, superstrings (the "Theory Of Everything") – and new fashions are constantly emerging. On the other hand, as will be seen from the numerous quotations provided in the present text, during that same period, the still living grand masters of twentieth century physics repeatedly urged the pursuit of deeper mathematical as well as epistemological analyses. Although their pleas remained mostly unheeded in the world of very rapidly changing fashions in quantum theoretical physics, they kept warning against the prevailing professional complacency in some of the most relevant areas. Most steadfast in his refusal to accept fashionable but fundamentally flawed developments in those areas of quantum theory which he had founded* was P. A. M. Dirac.

---

* A fitting eulogy was provided by Abdus Salam in a recent volume commemorating Dirac's life and work: *"Paul Adrien Maurice Dirac was undoubtedly one of the greatest physicists of this or any other century. In three decisive years – 1925, 1926 and 1927 – with three papers, he laid the foundations, first, of quantum physics, second, of the quantum theory of fields, and, third, of the theory of elementary particles, with his famous equation of the electron. No man except Einstein has had such a decisive influence, in so short a time, on the course of physics in this century. For me, personally, Dirac represented the highest reaches of personal integrity of anyone I have ever known."* (Salam, 1990), p. 262.

When I was a graduate student, I had the privilege of seeing Dirac, and of listening to him, as I attended seminars at which he was present during an extended visit which he made to the Institute of Advanced Studies in Princeton in the early 1960s. By that time he was, of course, already a legendary figure – one of the revered, great physicists of this century, whose professional stature overshadowed that of all the other distinguished physicists who were in regular attendance at those weekly seminars. However, while many of those well-known luminaries liked to impress the audience with comments which confirmed their intellectual brilliance, as a rule Dirac kept his counsel, and limited himself to only an occasional question, which would be pertinent but always unprepossessing. It was only many years later, after I began to read his critical assessments (cf., e.g., pp. 8, 190, 244, 289, 481) of the very foundations of some of the theories presented at those seminars, that I came to appreciate the understated greatness of his genius, and the depth of his commitment to an ideal of *truth* in physical theories, which manifested itself in mathematical beauty – a beauty totally at odds with the ungainliness inherent in the *ad hoc* "working rules" of conventional renormalization theory. Indeed, some of his characteristic comments, such as the one that "people are ... too complacent in accepting a theory which contains basic imperfections" (Dirac, 1978a, p. 20), reveal that, already by that time, he was all too painfully aware that his goals and ideals had become decidedly "unfashionable" in the world of perpetually changing fashions in elementary particle physics.

Considering that Dirac is the universally acknowledged founder of quantum field theory, the very title of his last paper ("The Inadequacies of Quantum Field Theory" – cf. p. 190), bespeaks of lofty professional ideals and ethical standards, maintained throughout his entire life with a steadfastness and uncompromising integrity that has almost no parallel in this century. Although his critical comments might not have reached the fashion-conscious amongst the theoretical and the mathematical physicists, his words and deeds have inspired and provided moral support to me, as I trust it did to many other researchers who share his ideals and basic values with regard to the principles and aims of theoretical physics. It is therefore with genuine reverence and overwhelming spiritual gratitude that I dedicate this work to the memory of Paul A. M. Dirac – a great physicist, as well as a truly great man.

I would like to express my thanks to all those colleagues who have contributed useful comments and information in the course of more than two decades, during which the programme recounted in this and my two preceding monographs ([PQ] and [P]) was gradually developed. Professors James A. Brooke and Wolfgang Drechsler had the opportunity to examine the first draft of many of the chapters in the present monograph, and I hereby thank them for their valuable comments. Special thanks are due to Scott Warlow for his very careful proofreading of the entire first draft, and for his many insightful comments and suggestions. I also wish to thank the editor of the Fundamental Theories of Physics series, Professor Alwyn van der Merwe, for enabling me, for a second time in eight years, to expound the outcome of my research into the foundations of relativity and quantum theory in the cogent form of a basically self-contained monograph published in this series. Last, but certainly not least, I wish to express my gratitude to my wife, Margaret R. Prugovečki, for her assistance with the preparation of the manuscript, and her moral support during what proved to be a protracted and arduous journey into uncharted territories.

EDUARD PRUGOVEČKI
Toronto, October 1991

Chapter 1

Principles and Physical Interpretation of Quantum Geometries

The principal aim of this monograph is to present, in a systematic and self-contained manner, geometric frameworks into which the essential features of general relativity as well as of quantum theory can be incorporated in a consistent fashion. Philosophers of science use the term *physical geometry* (Reichenbach, 1957; Pap, 1962; Lorenzen, 1987) to designate those geometric frameworks that are viewed not only as abstract mathematical disciplines, but also as structures meant to depict the physical world around us. In fact, this type of attitude towards geometry coincides with that adopted by Riemann (1854), Einstein (1916), Weyl (1923, 1949), and others. We therefore find that the geometric frameworks in this monograph are described most aptly by the generic term *quantum geometry*[1].

The quantum geometries presented in this monograph could be described more specifically as being geometro-stochastic frameworks. As such, they are based on the physical and mathematical ideas of a *geometro-stochastic* (GS) formulation of quantum theory presented in a series of recent papers (Prugovečki, 1985-91, Prugovečki and Warlow, 1989), in which geometric techniques meant to reflect the equivalence principle [M,W] at the quantum level were combined with the methods and results of the special relativistic formulation of the stochastic quantum mechanics[2] presented in [P][3]. The outcome of this combination is an extrapolation of the conventional single-Hilbert space framework for quantum theory, which forms the basis of conventional treatments of quantum mechanics (cf., e.g., [PQ]), into a multi-Hilbert space framework, which has the general structure of a (pseudo-) Hilbert or super-Hilbert bundle over a pseudo-Riemannian manifold or supermanifold.

The motivation and most basic features of the quantum geometries resulting from this extrapolation are described in nontechnical language in Sec. 1.3 of this chapter. They are then developed in a mathematically systematic manner in the subsequent chapters by using – and when necessary generalizing – the concepts and techniques of the standard fibre bundle theory that can be found in various textbooks aimed at mathematicians [SC] or at physicists (cf., e.g., [C], [I], or [NT]).

On the level of physics, the GS approach relies on a formulation of quantum propagation in terms of a geometro-stochastic parallel transport in the aforementioned quantum fibre bundles. The most basic features of GS propagation are described in Sec. 1.4. Their examination will reveal that this concept has its roots in the standard path-integral approach to quantum mechanics (cf. [F] or [ST]). However, in the GS approach the fundamental idea of "sum over paths" has been geometrized as well as generalized, so as to be in accord with the basic principles of general relativity.

In this chapter we shall discuss the physical origins of quantum geometries and examine their generic aspects, with particular emphasis on their epistemological aspects[4]. Those readers primarily interested in their mathematical aspects can proceed directly to Chapter 2 – or even to Chapter 3 in case they are already familiar with the fibre bundle formulation of the basic aspects of Lorentzian geometries.

1.1. The Historical Background of Physical Geometries

At the purely epistemological level, the general idea of physical geometry which motivates the considerations in this monograph can be traced back to Greek antiquity. That becomes evident when it is recalled that ancient Greek geometry had its roots in the culture of ancient Babylonia and Egypt, where it arose in response to practical problems encountered in land surveying (Lorenzen, 1987) – as reflected by the original Greek meaning of the term "geometry" (i.e., "earth-measurement"). In keeping with this fact, in the later course of history, some of the ancient Greek philosophers, most notably Aristotle, regarded the concepts of geometry as rooted in the physical world. However, others, most notably Plato, viewed them as a reflection of a world of pure ideas (Maziarz and Greenwood, 1968). Consequently, after Euclid compiled (*c.* 300 B.C.) in his *Elements* the very rich body of theorems, deduced from a few basic definitions and axioms, that was eventually named after him, many came to regard that body of knowledge primarily as an abstract discipline whose value and validity totally transcended the realm of experience. This attitude was epitomized by I. Kant in his *Critique of Pure Reason*, first published in 1781, where he claimed that Euclidean geometry enjoyed an *a priori* status (Kant, 1961). That claim appeared to emerge from a philosophy which provided many deep insights into ancient problems concerning the relationship of mind to matter, and eventually established Kant as "the greatest of modern philosophers" (Russell, 1945, p. 704), but the underlying reason for the claim itself was that, in the words of a contemporary historian of mathematics, Kant's "inability to conceive of another geometry convinced him that there could be no other"[5] (Kline, 1980, p. 76).

The arguments underlying the claim that Euclidean geometry enjoys an a priori status were not seriously questioned by Kant's contemporaries. On the other hand, towards the end of the eighteenth century, Gauss had already discovered non-Euclidean geometries – but never published his results fearing the disapproval of his peers (Kline, 1980). However, he became involved in geodetic goniometry and related professional activities that clearly reflected his empirical approach to geometry (Jammer, 1969; Reich, 1977). For example, he attempted to establish the deviations from 180° in the angles of large triangles formed by the light rays travelling between observational towers on three mountain peaks – but the measured 15" discrepancy was within the observational error bounds. Nevertheless, in 1830 he overcame his qualms and announced his conviction that geometry is *not* an a priori science. However, neither the prestige of Gauss, nor the independent discovery of non-Euclidean geometries by Bolyai and Lobachevski, aroused amongst the mathematicians and physicists of the last century great interest in this subject matter, or in the related question as to whether Euclidean geometry is indeed the physical geometry of space[6].

During the middle of the last century one notable exception to this prevalent lack of interest about the foundations of geometry was provided by Riemann, who publicly expressed his belief in treating geometry as an empirical science. Thus, in his famous 1854 inaugural lecture (attended by Gauss) he stated that "the theorems of geometry cannot be deduced from general notions of quantity, but that those properties which distinguish Space from other conceivably triply extended quantities can only be deduced from experience" (cf. [SC], p. 135). He went even further with this thesis, and asserted that all data supporting a choice of physical geometry "are – like all data – not logically necessary, but only of empirical certainty, [i.e.] they are hypotheses; one can therefore investigate their likelihood, which is [for Euclidean geometry] certainly very great within the bounds of observation, and afterwards decide upon the legitimacy of extending them beyond the bounds of observation, both in the direction of the immeasurably large, and in the direction of the immeasurably small" (cf. [SC], p. 136).

The last part of the above quotation is very pertinent to the fundamental thesis underlying the developments in the present monograph, namely that, whereas classical general relativity denies the legitimacy of extending the bounds of validity of Euclidean geometry "in the direction of the immeasurably large", and requires instead the adoption of Lorentzian geometries, the existence of intrinsic quantum fluctuations at the microlevel removes the bounds of validity not only of Euclidean but also of Lorentzian geometries as we move "in the direction of the immeasurably small" – with all contemporary evidence indicating that the role of "immeasurably small" is played by the Planck length, i.e., that it occurs at approximately 10^{-33} cm (cf., e.g., DeWitt, 1962). We shall therefore argue in favor of replacing, at the microlevel, Lorentzian geometries – which are "classical" in the sense that their empirical interpretation, as provided by Einstein (1905, 1916), was in deterministic terms – with geometries which are "quantum" in the sense that they allow a probabilistic interpretation which embodies, from the outset, Heisenberg's uncertainty principle for position-momentum and time-energy measurements, as well as the limitations imposed on spatio-temporal relationships by Planck's length regarded as an *absolute* lower limit for the operational distinguishability of spacetime points (DeWitt, 1962; Mead, 1964).

Although the replacement of classical with quantum geometries represents a natural step when viewed from such a perspective, it does not represent the only logically acceptable one, any more than the replacement by Einstein of Euclidean with Lorentzian geometries represented the only acceptable alternative. In fact, exactly half a century after Riemann made the earlier quoted statements, H. Poincaré examined at great length in his monograph *Science and Hypothesis* the possibility of adopting non-Euclidean geometries in the description of physical space, only to conclude that "experience ... tells us not which is the truest geometry, but which is the most *convenient*" (cf. Feigle, 1953, p. 180). As the, by general consensus, most outstanding mathematician and mathematical physicist of his age, Poincaré eventually decreed that Euclidean geometry fulfilled such a "convenience criterion" – cf. (Jammer, 1969), pp. 165 and 209.

It is interesting to note that at the very same time when Poincaré arrived at this conclusion, he was also working on the dynamics of the electron. That work led him to the independent discovery of many important results in special relativity (Pais, 1982), including some of the key properties of the inhomogeneous Lorentz group, that was to be later named after him. Hence, with a different outlook he might have made the intellectual leap that eventually led Einstein, via the equivalence principle, to the general theory of relativity. However, whereas Einstein developed the special theory of relativity from operational considerations based on light signals, which provided an interpretation of Lorentz transformations as part of a theory of the (later to be named) Minkowski space, Poincaré's treatment was restricted to the electromagnetic phenomena themselves. Thus, it was left to Einstein to eventually reject totally and decisively Poincaré's conventionalistic point of view, by stating that "the question whether the structure of [the spacetime] continuum is Euclidean, or in accordance with Riemann's general scheme, or otherwise, is . . . a physical question which must be answered by experience, and not a question of mere convention to be selected on practical grounds" – cf. (Feigle, 1953), p. 193.

It has been, however, pointed out by philosophers of science that Poincaré's conventionalism cannot be rejected on either logical or empirical grounds alone since the "propositions of a physical geometry cannot be tested until a method of measurement has been chosen" (Pap, 1962, p. 112). In other words, the tests of the validity of a given choice of *physical* geometry depend crucially on the underlying theory of measurement[7] – in particular on the choice of test bodies, reference frames, measuring devices and signals, as well as on the assumed properties and behavior of all these physical entities. For example, although in his key papers on relativity theory Einstein did not spell out all these details, it is clear from his writings (Einstein, 1905, 1916) that he considered his test particles to be pointlike massive objects, that he viewed his measuring devices to be macroscopic "rigid rods" and "standard clocks", that light signals provided the fundamental means of communication within his measurement-theoretical scheme, and that he envisaged all these objects as behaving in a strictly deterministic manner. Later systematic studies (Kronheimer and Penrose, 1967; Ehlers *et al.*, 1972-73; Woodhouse, 1973) confirmed the feasibility of deriving the geometric structure of general relativity from a consistent set of axioms about the behavior of these physical objects viewed as *classical* objects. In particular, these studies eliminated the assumption about the existence of rigid rods (Marzke and Wheeler, 1964), whose postulated "rigidity" clearly contradicted the relativity principle (Stachel, 1980), and proved not to be required by the operational interpretation of general relativity.

That still left open, however, the question of whether there indeed exist in nature objects which display, either in a strict sense, or at least in an arbitrarily close approximation, the features and behavior postulated by Einstein. Indeed, in the well-known collection of essays honoring Einstein's seventieth birthday, H. Margenau voiced the criticism[8] that "particles may not be regarded as points but as structures of finite size . . . , which threatens the validity of [relativistic] causal description" (cf. Schilpp, 1949, p. 259). In the same publication, K. Menger pointed out that all our empirical knowledge is statistical in nature, so that a thorough geometrization of physics might require "the introduction of probability into the foundations of geometry" (cf. Schilpp, 1949, p. 472). It was, however, M. Born who, after many years of extensive correspondence with Einstein (Born, 1971), categorically stated that "the mathematical concept of a point in a continuum has no direct physical significance" (Born, 1955, p. 3), and arrived at the following two key epistemological conclusions, which we emphasize by means of italics: "*Statements like 'A quantity x has a completely definite value'* (expressed by a real number and represented by a point in the physical continuum) seem to me to *have no physical meaning*. Modern physics has achieved its greatest successes by applying a principle of methodology that *concepts whose application requires distinctions that cannot in principle be observed are meaningless and must be eliminated*" (Born, 1956, p.167).

In this monograph we shall adopt the two italicized passages in the preceding quotation as fundamental methodological principles. As such, we shall refer to them, respectively, as *Born's first maxim* and *Born's second maxim*.

In the very same essay in which he implicitly criticized Poincaré's conventionalism, Einstein admitted "that [his] proposed physical interpretation [of Lorentzian geometry] breaks down when applied immediately to spaces of sub-molecular order of magnitude" (cf. Feigl, 1953, p. 193). In view of this observation, if one tries to extend the domain of validity of relativity theory to the microscopic domain, one has the following clear-cut choices: 1) to consider the choice of geometry at the subatomic level from a conventionalistic point of view and claim, as Poincaré did in the case of Euclidean geometry, that the adoption at the microscopic level of Minkowski and, more generally, Lorentzian geometries is a matter of convenience that transcends empirical verification; 2) to face the fact that, according to generally accepted quantum principles, the behavior of matter at the subatomic level is not in accordance with the deterministic behavior postulated by the operational interpretations of these geometries, and that therefore new types of geometric structures, which incorporate quantum principles, have to be introduced.

If the same fundamental epistemic principles hold true in all of physics, it is only the second alternative that is epistemologically sound. This was, of course, realized by many physicists right after Bohr's old quantum mechanics transcended into the modern quantum mechanics of Heisenberg and Schrödinger, and Born proposed his statistical interpretation for quantum states and observables. Indeed, soon after the publication of the papers by Heisenberg and Schrödinger that founded modern quantum mechanics, papers speculating on the atomistic structure of spacetime began appearing (Latzin, 1927; Pokrowski, 1928; Schames, 1933). Moreover, after the introduction by Heisenberg (1927) of the uncertainty principle, the idea of the quantization of spacetime began to be related to elementary space and time uncertainties (Ruark, 1928; Flint and Richardson, 1928; Fürth, 1929; Landau and Peierls, 1931, Glaser and Sitte, 1934; Flint, 1937). The subsequent suggestion made by Heisenberg (1938, 1943) that there is a fundamental length in nature led Born (1938, 1949) to the idea of a quantum metric operator, and also led other researchers (March, 1937; Markov, 1940) to a statistical concept of metric. The greatest temporary popularity was enjoyed, however, by a notion of discrete quantized spacetime proposed by Snyder (1947), which was for a while pursued by a host of researchers (Yang, 1947; Flint, 1948; Schild, 1948; Hill, 1950; Das, 1960; Gol'fand, 1960, 1963; Kadyshevskii, 1962, 1963). However, that idea seems to have eventually sunk into rather deep oblivion. Indeed, during the last few years several researchers (Bombelli *et al.*, 1987; Finkelstein, 1989; Gudder, 1988a; Isham, 1990) have proposed discrete descriptions for spacetime, but without citing any of the aforementioned earlier work on this subject.

From a foundational point of view, the common weak point of all the above proposals for quantum spacetimes reside in their lack of a clear-cut theory of measurement, that could point the way to reliable and empirically well founded geometric ideas. Such ideas could have either enabled the injection of the fundamental principles of relativistic invariance, causality and locality into these frameworks, or might have alternatively produced compelling principles that would transcend into the aforementioned ones in the macroscopic regime, where classical special and general relativity have received most convincing experimental confirmation. But no such fundamentally new epistemic ideas emerged, since most of the above work on quantum spacetimes was of a formal mathematical nature. Consequently, after the numerical successes of renormalization theory in quantum electrodynamics in the late forties, a conventionalistic attitude began to prevail in the physics community despite the fact that, as we shall see in Sec. 9.6, renormalization theory had removed neither the conceptual nor the technical difficulties encountered by conventional relativistic quantum mechanics and quantum field theory. In fact, as we shall illustrate in the next section with quotations from original sources, the principal founders of quantum mechanics and quantum field theory were very much aware of the source of these fundamental difficulties – and some of them (Dirac, 1973, 1977, 1978; Heisenberg, 1976; Wigner, 1976, 1981) often urged a deeper reconsideration of the foundations of these disciplines. Unfortunately, however, their criticisms and suggestions have remained basically unheeded by the great majority of researchers working in these areas (cf. Chapter 12).

1.2. The Incompatibility of Quantum Theory with Classical Relativistic Geometries

It is universally acknowledged that conventional methods of quantization have not produced an acceptable theory of quantum gravity, and have not succeeded in consistently unifying quantum mechanics with general relativity (cf., e.g., the key articles in MacCallum, 1987, or in Markov *et al*, 1988). However, it is still not usually recognized in the contemporary physics literature that this failure extends to the special relativistic level.

The principal difficulties stem from the imposition of the classical conceptualization of locality in the special relativistic quantum regime, whereby quantum "particles" are envisaged as being capable of "occupying" with "certainty" arbitrarily small regions in Minkowski space. Indeed, if one does not make such an assumption about the very smallest blocks of matter, there is no physical justification for adopting at the microlevel a Lorentzian geometry as the *physical* geometry of spacetime in general, or of a Minkowski geometry in those particular cases where gravitational effects are deemed to be negligible. Furthermore, in the summary of his review article "Interpretation of Quantum Mechanics", E. P. Wigner clearly states that "every attempt to provide a precise definition of a position coordinate [for quantum point particles] stands in direct contradiction with special relativity", and that the answer to the question "whether [conventional] quantum field theory avoids the difficulties mentioned . . . [is] negative also" (cf. [WQ], p. 313).

These statements by the principal pioneer of group theoretical methods in quantum mechanics and leading expert in the quantum theory of measurement might be startling to those who derive their knowledge of these subjects exclusively from the multitude of textbooks on relativistic quantum mechanics and quantum field theory, where these fundamental inconsistencies are usually not mentioned – and if they are, they are glossed over with incorrect 'solutions'. For example, one of the earliest discovered inconsistencies pertains to the indefiniteness of the time-like component $j^0(x)$ of the Klein-Gordon current, which is supposed to provide the probability density for observing a relativistic spin zero particle at some point x in Minkowski space. In standard textbooks on relativistic quantum mechanics, it is usually correctly asserted that only those solutions of the Klein-Gordon equation that correspond to positive energy are acceptable wave functions for single quantum point particles, whereas those corresponding to negative energy are acceptable wave functions for single quantum point antiparticles. This assertion is sometimes followed, however, by the *incorrect* claim that, for such positive-energy wave functions, one has $j^0(x) > 0$ everywhere in Minkowski space, so that therefore "a consistent theory [of localization in Minkowski space] can be developed for a free [relativistic spin-zero] particle" (cf. [SI], p. 56). On the other hand, as *rigorously* proved by Gerlach *et al* (1967), the opposite is true: for *any* positive-energy solution of the Klein-Gordon equation there are points in Minkowski space where $j^0(x) < 0$, in addition to those where $j^0(x) > 0$.

Contrary to claims made in many standard textbooks, in the case of the Dirac current for spin-1/2 particles, the original promise that historically motivated its introduction, namely that of supplying a relativistically covariant probability current for single quantum point particles of spin 1/2, has not been fulfilled either: the Dirac current is positive-definite only because it is related to superpositions of positive and negative energy solutions (cf. Sec. 6.1), which therefore correspond to objects that would simultaneously manifest themselves as both particles and antiparticles (a feature commonly described in literature as *Zitterbewegung*). However, with regard to the issue of single particle (or single antiparticle) localization, "the Klein-Gordon equation is neither better nor worse than the Dirac equation" (Wightman, 1972, p. 98). In fact, it has been rigorously proven in recent years (Thaller, 1984) that *all* positive-energy solutions of the Dirac equation are different from zero at *all* points in Minkowski space. Hence, *none* of these solutions can describe a state of a spin-1/2 quantum *point* particle localized in any finite region in space or in spacetime, since the wave function for such a *localized* state would have to vanish outside that region.

Although not that well-known, the history of the many attempts to bypass the above difficulties with the conventional notion of relativistic point particle localization is long and involved, and we refer the reader interested in it to a review article by Kálnay (1971), to Wigner's aforementioned article ([WQ], pp. 260-314), to Secs. 2.1-2.2 of [P], and to a recent review paper by Hegerfeldt (1989) – all of which contain many additional references. It is interesting to mention, however, that the effort which for a while received most attention was the one by Newton and Wigner (1949). However, it was later discovered (Wightman and Schweber, 1955; Fleming, 1965) that the Newton-Wigner proposal conflicted with Einstein causality, since it led to superluminal relativistic particle propagation. Subsequently, it was established by Hegerfeldt (1974) that this violation of relativistic causality occurs for *any* projector-valued measure that supposedly describes relativistic quantum point particles strictly localized in Minkowski space. Thus, Hegerfeldt showed with total mathematical rigor that, if it is *assumed* that a quantum state can be prepared in any given bounded region of Minkowski space, or in any bounded region of a spacelike hyperplane in Minkowski space, then it will instantaneously propagate to points outside the causal future of that region. This result was later extended to multi-particle relativistic systems (Perez and Wilde, 1977), to very general types of interactions (Hegerfeldt and Ruijsenaars, 1980), and finally even to the case of approximate localizations (Hegerfeldt, 1985), whereby the choice of Minkowski geometry for spacetime is retained, but the wave packet is allowed to display "exponentially bounded tails" outside the chosen finite region.

It is sometimes maintained (Eberhard and Ross, 1989) that conventional quantum field theory resolves the fundamental incompatibility between basic quantum principles and special relativity based on arbitrarily precise localization, on account of the claim that quantum fields can be measured in arbitrarily small regions of Minkowski space, but that such measurements cannot be treated within the context of single-particle or even multi-particle relativistic quantum mechanics due to the phenomenon of pair-creation. However, as pointed out on many occasions by the founders of quantum mechanics and quantum field theory, *all fields*, regardless of whether they are viewed as classical or quantum, *require test bodies for their detection and localization*, so that such suggestions merely lead to a vicious circle. Indeed, as Wigner emphasized in the aforementioned review article, "the best known discussion of the measurement of field strengths, that given by Bohr and Rosenfeld (1933, 1950), postulates an electric test charge with arbitrarily large charge and arbitrarily small size", so that "it does not resolve our problems [with regard to quantum field localization]" (cf. [WQ], p. 313). This fact was actually acknowledged by Bohr and Rosenfeld themselves when they stated in their well-known 1933 article on the subject of quantum field localization that "the fundamental difficulties which confront the consistent utilization of field theory in the atomic theory remain entirely untouched by the present investigation" (cf. [WQ], p. 481), and when they very clearly pointed out that the "consideration of the atomistic constitution of all measuring instruments would be essential for an assessment of the connection between these difficulties and the well-known paradoxes of the measurement problem in relativistic quantum mechanics" (cf. [WQ], p. 482).

Although the above statements by Bohr and Rosenfeld were made in 1933, their substance remained true even after the advent of renormalization theory in quantum electrodynamics in the late 1940s. Indeed, despite the well-known numerical successes of renormalization theory, the following was acknowledged by Schwinger in the preface to a well-known 1958 reprint collection on the subject: "The observational basis of quantum electrodynamics is self-contradictory... . The localization of charge with indefinite precision requires for its realization a coupling with the electromagnetic field that can attain arbitrarily large magnitudes. The resulting appearance of divergences, and contradictions, serves to deny the basic measurement hypothesis. We conclude that *a convergent theory cannot be formulated consistently within the framework of present space-time concepts.*" (Schwinger, 1958, pp. xv-xvi) – emphasis added.

These statements by one of the main founders of renormalization theory are in total agreement with many statements by the acknowledged founder of relativistic quantum mechanics and quantum field theory, P.A.M. Dirac, who, on numerous occasions in the years following the advent of the renormalization programme, expressed his reservations about the very foundations on which that mathematically elaborate structure had been erected (Dirac, 1951, 1962, 1965, 1973, 1977, 1978, 1987). This is perhaps best illustrated by the following quotation: "Our present quantum theory is very good provided we do not try to push it too far – we do not try to apply it to particles with very high energies and we do not try to apply it to very small distances. When we try to push it in these directions, we get equations which do not have sensible solutions ... It is because of these difficulties that I feel that the foundations of quantum mechanics have not yet been correctly established. Working with the present foundations, people have done an awful lot of work in making applications in which they find rules for discarding the infinities. But these rules, even though they may lead to results in agreement with observations, are artificial rules, and I just cannot accept that the present foundations are correct." (Dirac 1978a, p. 20).

Indeed, although the question of mathematical consistency of a physical theory can be regarded as distinct from that of the physical validity of its underlying theory of measurement, the mathematical formalism of conventional quantum field theory in Minkowski space has defied all attempts at a mathematically rigorous and yet physically nontrivial formulation. For example, after more than two decades of protracted efforts by a large and mathematically very skilled group of researchers into the mathematical foundations of conventional quantum field theory, the two main initiators of that program had to acknowledge that "arguments favoring triviality [of ϕ^4 and Yukawa quantum field theory as well as of quantum electrodynamics in 4-dimensional Minkowski space] seem to be stronger, but a definitive answer seems to be out of reach of available methods" (Glimm and Jaffe, 1987, p. 120). However, since in the case of quantum electrodynamics it is difficult to reconcile this "triviality" assessment with the wealth of experimental data indicating the opposite, Feynman's eventually reached conclusion that "it's also possible that [conventional quantum] electrodynamics is not a consistent theory" (Feynman, 1989, p. 199) represents a more likely possibility (cf. also Secs. 7.6, 9.6 and 12.3).

As is well-known, the inclusion of gravity into conventional quantum field theory greatly exacerbates all the fundamental difficulties encountered by that discipline in Minkowski space. The underlying reasons for that become obvious when it is recalled that the application of the Bohr-Rosenfeld methods to the analysis of measurements of quantum fields in the presence of gravity led B. S. DeWitt to the conclusion that "10^{-32} cm constitutes the absolute limit on the applicability of classical concepts, even as modified by [Bohr's] principle of complementarity" (DeWitt, 1962, p. 371). In Sec. 1.3 of a later review article DeWitt (1975) also listed a number of additional epistemic as well as technical difficulties encountered by conventional quantum field theory in curved spacetime (cf. also Sec. 7.2). The essence of these difficulties was very clearly and succinctly summarized in a recent monograph dealing with conventional quantum field theory in curved spacetime as follows (with the italics reproduced as in the original): "*There does not exist a quantum field theory formalism in an arbitrary curved spacetime.* This problem is deep rooted and arises from the fact that standard formalisms of field theory require a preferential slicing in spacetime. In other words, quantum field theory, as we know today, is Lorentz invariant, but is *not* generally covariant. This conceptual problem introduces a new level of observer dependence in quantum theory, the implications of which are not yet well understood." (Narlikar and Padmanabhan, 1986, p. 277).

In view of all these striking foundational difficulties, it is not surprising that much of the surge of interest in string theory in recent years was spurred by the conclusion that the "quantum field theory of point particles seems to be inconsistent in the presence of gravity" (Green *et al.*, 1987, p. 54). However, the advocates of superstring theory have attempted to resolve the foundational difficulties encountered by the renormalization program by abandoning the concept of quantum point particle in favor of that of a superstring *without basing their programme on foundational arguments* in general, or on the quantum theory of measurement in particular. This fact is recognized by many of the leading researchers in the field[9]. Indeed, in the words of one of the original contributors to string theory, "for a theory [of superstrings] that makes the claim of providing a unifying framework for all physical laws, it is the supreme irony that the theory itself appears so disunited" (Kaku, 1988, p. 5). On the other hand, the same author points out that "by contrast, when Einstein first discovered general relativity, he started with physical principles, such as the equivalence principle, and formulated it in the language of general covariance" (Kaku, 1988, p. 6).

The rejection of the concept of point particle in quantum relativity is shared by the geometro-stochastic (GS) approach to the unification of general relativity and quantum theory. However, by contrast with even the most recent developments (Ginsparg, 1989) in superstring theory, which leave "the fundamental physical and geometric principles that lie at [its] foundation . . . still unknown" (Kaku, 1988, p. viii), *the basic strategy of the GS approach is to extrapolate from the outset the equivalence principle to the quantum regime,* and to formulate such an extrapolation from the beginning in a language that reflects the principle of *quantum* general covariance. It turns out that *quantum* geometry provides such a language, and that it is the GS theory of measurement which, in turn, determines what those general features of such quantum geometries should be like.

We therefore turn now to the examination of the fundamental epistemic principles underlying the proposed quantum geometries, emphasizing basic physical ideas rather than mathematical techniques. The mathematical techniques themselves will be presented in detail in Chapters 4 -11. Those readers interested only in the latter can proceed right away to Chapter 2 – or to Chapter 3, in case that they are already familiar with fibre bundle theory.

1.3. Basic Principles of Quantum Geometry

As demonstrated by the historical developments recounted in Sec. 1.1, the geometry of spacetime is not *a priori* given, but it rather depends on the fundamental operational and mathematical features of the adopted theory of measurement of spatio-temporal relationships. The fundamental premise underlying the development of the GS theory of measurement is based on a thesis which in essence can be traced to von Neumann (1932), namely that an epistemologically satisfactory description of nature should not display a dichotomy in its treatment of a "system" and an "apparatus", such as the one in the orthodox approach to the quantum theory of measurement, whereby the former is treated as a quantum object, whereas the latter is viewed as a classical entity. Hence, in the GS approach, test bodies and reference frames are treated exclusively as quantum objects. As such, they are subject to quantum localization laws, which incorporate from the outset the uncertainty relations.

Although the implications of this epistemic attitude are far-reaching, there is actually no conflict on the operational level between the GS approach and the orthodox approach. Indeed, in the orthodox approach, the reason for advocating the treatment of any apparatus as a classical object, despite its generally acknowledged atomic and subatomic structure, was rooted in a positivistic philosophy which asserted that "all non-analytic knowledge is based on experience", and therefore "a sentence makes a cognitively meaningful assertion . . . only if it is either (1) analytic or self-contradictory or (2) capable, at least in principle, of experiential test" (Ayer, 1966, p. 108). Hence, in keeping with such an outlook, Bohr maintained that all meaningful statements of quantum theories should on one hand pertain only to experimental observations, whereas on the other hand, according to him, "by the word 'experiment' we refer to a situation where we can tell others what we have done and what we have learned and that, therefore, the account of the experimental arrangement and of the results of observations must be expressed in unambiguous language with suitable application of the terminology of classical physics" (Bohr, 1961, p. 39).

At the observational level, the GS theory of measurement does not reject such an epistemic attitude, but rather proposes a broader interpretation of the terms "experiential test" and "experimental arrangement" that would allow the replacement in the last of the above quotations of the term "classical physics" with "*quantum* physics". Indeed, the terminology of quantum theory can be made to be as "unambiguous" as that of classical theory once proper care is exercised in defining its basic concepts, stating its fundamental premises, and deducing its principal results. This was exemplified in [PQ], by first formulating the basic axioms of conventional nonrelativistic quantum mechanics in precise Hilbert space terms, and then deducing the main body of the theory found in standard textbooks on quantum mechanics by mathematically rigorous methods, and without any appeal to "the terminology of classical physics", except where comparisons with classical mechanics were drawn. Furthermore, although our *perceptions* of the macroscopic world are more immediate and therefore more easily describable in *everyday* language than those of the microscopic world, the wealth of experimental data on perceptual processes accumulated by psychologists over recent decades indicates that even the most immediate perceptions of the world around us are based on *constructs* in which the data provided by our senses are grouped together into perceptual *models* for various "objects" (including apparatuses!) by a complex process of associations – cf. (Popper, 1976), p. 139. Hence, in the ultimate analysis there is no sharp[10] conceptual demarcation line between those "objects" which we can observe "directly" (provided that we possess normal eyesight!), such as "pointers" and "dials", and those other "objects" which we can observe only with the aid of instruments, such as "quasars" and "Brownian particles" – or, for that matter, "molecules", "atoms", "hadrons" and "leptons". Rather, the fundamental distinctions lie in the *patterns of behavior* of these various "objects", which is radically different at the microlevel from that displayed at the macrolevel, to the extent that in the latter case, the *totality* of manifestations of that behavior cannot be *faithfully* and *exhaustively* described by means of classical concepts.

The orthodox solution to this fundamental difficulty with using classical concepts in the quantum domain was to divide the part of the physical world under observation into "system" and "apparatus", and to apply classical concepts and terminology to the latter part on account of its macroscopic nature at the level of "interfacing" between a human observer and the employed instrumentation. However, although in the nonrelativistic regime this leads to a *mathematically* consistent framework [PQ] as well as to an abundance of phenomenologically satisfactory models, it is seldom recognized that at a *foundational* level such a solution eventually proved to be unsatisfactory, due to a fundamental incompatibility between the language and the concepts of classical theories on one hand, and some of the observational features of quantum phenomena on the other. In fact, it can be argued that the genesis of the various paradoxes (cf. Tarrozzi and van der Merwe, 1988) that plague conventional quantum theory does not lie in the nature of quantum phenomena themselves, but rather in the unwarranted *imposition* of classical language and conceptualizations based on "yes-no" type of statements in a realm which *de facto* does not admit such statements.

The most elementary illustration of this fact is provided by the conventional 'up-or-down' depictions of measurements of spin 1/2 components. Indeed, a theorem by Wigner (1952) shows that any categorical statement to the effect that, say, in the Stern-Gerlach experiment, a measured spin component is *certainly* "up" is, strictly speaking, invalid, and it suggests as more appropriate the idea of *stochastic* spin values (Prugovečki, 1977b; Schroeck, 1982; Busch, 1986), in which a confidence function is attached to spin values even at the theoretical level – cf. (Schroeck, 1991) for a comprehensive review and discussion of stochastic spin measurements. In fact, in principle this observation extends to the measurement of *any* quantum observable. Indeed, Wigner's (1952) theorem was extended by Araki and Yanase (1960) (cf. also Yanase, 1961), who demonstrated that, in Wigner's words, "no precise measurement of any quantity is possible unless [it] commutes with all conserved additive quantities" ([WQ], p. 313). Consequently, as explicitly demonstrated by Busch (1985b), no arbitrarily accurate measurements of position are *in principle* feasible even at the nonrelativistic level – i.e., even if the theorem by Hegerfeldt, mentioned in the preceding section, is not applicable, since it pertains to relativistic quantum theory. As, in the ultimate analysis, any measurement involves evaluations of position coincidences and relationships, this conclusion can be said to possess general epistemological validity.

Consequently, the first principle on which the GS approach to quantum theory is founded incorporates the first of Born's maxims, cited in Sec. 1.1, in the following form.

Principle 1 *(The principle of irreducible indeterminacy).* In GS quantum theory, the measurement outcome of any given physical quantity X is not represented, even *in principle*[11], by a single number x in the spectrum $\sigma(X) \subset \mathbf{R}^1$ of that quantity, but rather it is described by a *stochastic value* (x,χ_x); i.e., it has to be supplemented, even in the (imagined) infinite limit of progressively accurate measurements tending to *optimal* accuracy, by *confidence functions* $\chi_x(y)$, $y \in \sigma(X)$, which reflect *irreducible* indeterminacies by providing the probabilities of observing statistical fluctuations from any measured *base*[12] value x under the most ideal repeatability conditions for any class of experimental arrangements.

Naturally, the above principle flatly contradicts one of the most deeply ingrained *beliefs* in the perfect reproducibility of measurement results in the context of *ideal* measurements – i.e., of von Neumann (1932, 1955) measurements of the first kind. Yet, in the ultimate analysis, all measurements involve *estimates* of position coincidences, and it is a most basic observational fact that no such estimates can be carried out with "absolute precision", even at the classical level. Hence, this belief in turn relies on the *credo* that absolutely accurate values of physical quantities are "possessed" by physical systems, and that, as such, they can be displayed by physical systems at least "in principle", i.e., in the limit of empirically unrealizable *infinite* sequences of increasingly accurate measurements. Thus, this credo is rooted in a naïve realism which surpasses any possibility whatsoever of direct empirical verification. Its roots are therefore ontological, and probably extend all the way to Plato's cosmogony and theory of ideas (Russell, 1945). Indeed, this ontological attitude was already reflected[13] by Euclidean geometry, when it first emerged as an abstract mathematical discipline from the science of measurement of geometric relationships.

On the other hand, the naïve physical realism based on the credo that absolutely accurate values of measurable quantities *exist* as intrinsic properties of all physical systems has underlined the entire development of classical physics. We shall therefore refer to it as the *doctrine of existence of classical reality*. It is this doctrine that underlies the Einstein-Podolsky-Rosen (EPR) criterion of physical reality, which was formulated as follows: "If, without in any way disturbing a system, we can predict with certainty (i.e., with probability equal to unity) the value of a physical quantity, then there exists an element of physical reality corresponding to this physical quantity." (Einstein *et al.*, 1935, p. 777).

With the advent of modern quantum mechanics, indeterminacies were introduced into the very interpretation of its theoretical framework, by irreducibly coupling those "uncertainties" which were found to be present, even in principle, in the measurement of pairs of "incompatible" observables – such as position and momentum. However, Bohr's version of the orthodox approach to quantum theory did not postulate the existence of irreducible uncertainties for single observables. Therefore, Bohr, through his well-known complementarity principle, had to deny all meaning to the question of existence of values for such "observables" in the absence of actual "observations", and to view them exclusively as representing the outcomes of those kinds of interactions between "system" and "apparatus" which, by postulation, constituted an act of measurement (Bohr, 1961). But this epistemic attitude left open the still debated question of whether measurement acts can be described by the quantum theory itself – cf. articles in (Lahti and Mittelstaedt, 1985), (Kostro *et al.*, 1988), and (van der Merwe *et al.*, 1988). In fact, the very question as to what generically constitutes a measurement act was left open to interpretation, so that from the orthodox interpretation it is not even clear at exactly what stage of the "observational process" the famed "reduction of the wave packet" (von Neumann, 1932) takes place.

In the approach to quantum theory advocated by Bohr, the doctrine of existence of classical reality was removed from the main body of physical theory, but it was nevertheless implicitly retained as part of the theory of quantum measurement, which was treated as if it constituted a separate realm of its own. Thus, once a measurement was performed, its outcome was treated as if it were a classical measurement result, which as such could be, at least in principle, regarded as perfectly accurate, and which was only deterministically influenced (if at all!) by any further acts of measurements of the same quantity, so that perfect repeatability could be ensured in principle, despite its infeasibility in practice. Indeed, it is a basic observational fact that statistical fluctuations are present in all measurement results on quantum systems, but they are conventionally incorporated into "error margins" on the "observed values", instead of being treated as integral parts of those values themselves.

The imposition of such a dichotomy between the undisturbed dynamical evolution of the "quantum system" and the measurement of its "observables" by means of "classically behaving" apparatuses did not remove, however, all the conceptual difficulties and inconsistencies (Cini and Serva, 1990; Stuart, 1991) from the orthodox approach – or for that matter from all the other approaches to the quantum theory of measurement that have adopted it. This point was graphically made by Bell in the context of his critical examination of Everett's many-worlds interpretation of quantum mechanics (DeWitt and Graham, 1973) as follows: "In dividing the world into pieces *A* and *B* Everett is indeed following an old convention of abstract quantum measurement theory, that the world does fall neatly into such pieces – instruments and systems. In my opinion this is an unfortunate convention. The real world is made of electrons and protons and so on, and as a result the boundaries of natural objects are fuzzy, and some particles in the boundary can only doubtfully be assigned to either object or environment. I think that fundamental physical theory should be so formulated that such artificial divisions are manifestly inessential." (Bell, 1987b, p. 96).

The arbitrary division of the physical world into "system" and "apparatus" can be also viewed (Bell, 1987b) as the basic source of all the well-known paradoxes that plague the orthodox approach (Schroeck, 1991). However, its retainment, together with all its implications for the quantum theory of measurement (Jammer, 1974; d'Espagnat, 1976), was not unconditionally advocated by some of the other main representatives of the Copenhagen school – such as Born (1956) or Heisenberg (1958). Thus, as we mentioned in Sec. 1.1, Born had questioned the *physical* meaning of the concept of "point in a continuum" (Born, 1955), and in fact had even introduced the idea of a quantum metric operator (Born, 1938, 1949). Similarly, Heisenberg had postulated the existence of a fundamental length in nature already in the developmental stages of quantum theory (Heisenberg, 1938), and, while discussing the internal spatio-temporal structure of elementary particles, had pointed out a "possibility of which neither Kant nor the ancient philosophers could have thought: the word 'dividing' loses its meaning" (Heisenberg, 1976, p. 38). Since all measurements of physical quantities can be reduced, in the ultimate analysis, to measurements of spatio-temporal relationships, any conceptualization of spacetime which negates literal physical meaning to the *mental* procedure of indefinite subdivision of spacetime into smaller and smaller parts implies that there is a residual *irreducible* indeterminacy in the measurement of *all* physical quantities. Hence, within the ontology resulting from such an outlook, the doctrine of existence of classical reality can be maintained only as a convenient *approximation* at a *pragmatic* macroscopic level – but not as a fundamental epistemic principle.

At a deeper measurement-theoretical level, we have already seen that the doctrine of existence of classical reality, according to which classical concepts and classical physics are quite suitable for the *faithful* description of the outcome of spatio-temporal localization measurements of quantum particles, is implicitly contradicted (Busch, 1985) by the Wigner-Araki-Yanase theorem already at the nonrelativistic level – with the Hegerfeldt (1974) theorem merely adding new problems at the relativistic level. Its total removal from quantum theories could be therefore said to reflect the existence of an *all-pervading quantum reality*, which might be picturesquely described as being "unsharp" (Busch, *et al.*, 1989, 1990, 1991) *in relation* to the reality envisaged in classical physics – but which simply *exists* when viewed on its own accord, and as *being* the *truly fundamental reality*.

The issue of completeness of quantum mechanics, which lies at the basis of the still ongoing debate over the EPR paradox (cf. the articles in (Lahti and Mittelstaedt, 1985)), is closely associated with the principle of irreducible indeterminacy. Indeed, the formulation of that paradox takes for granted the doctrine of the existence of a classical reality (cf. [WQ], p. 138). Once that doctrine is dropped, the use of complete sets of observables in quantum mechanics as a means for completely determining, even in principle, the state of a quantum system becomes totally unfeasible. In fact, even if that principle were adopted, a careful mathematical analysis (Prugovečki, 1969b) of the concept of "complete set of observables" reveals its inapplicability for the complete determination of the (Hilbert space) state vector in the realistic context of measurements of "complete sets" of those kinds of observables which possess continuous spectra – such as it is the case with the position or the momentum observables in nonrelativistic quantum mechanics (cf. [PQ], Ch. IV, Sec. 5.1). Therefore, the issue of *informational completeness*[14] of a set of observables has to be then raised (cf. Sec. 3.7 for a more complete discussion): can one, at least in principle, uniquely determine (modulo a phase factor) *any* state vector ψ of a quantum system by means of measurements of any of the well-known families of "complete sets" of observables?

It turns out (Reichenbach, 1948; Gale *et al.*, 1968; Prugovečki, 1977a; Stulpe and Singer, 1990) that the answer to this question is negative not only for measurements of position observables, or of momentum observables, on separate ensembles of identical systems in the same quantum state, but even in the highly idealized situation of their measurement in an ensemble of identical quantum particles which are known to be in the same quantum state, and for which the probability distributions for perfectly accurate position measurements are determined on a subensemble, whereas that of perfectly accurate momentum measurements are determined on the remaining subensemble (so that the uncertainty principle is not violated – cf. also [P], p. 142): even if all those measurements were carried out with *perfect* accuracy, there are quantum states which still would not be uniquely determined by such a measurement procedure. On the other hand, perhaps surprisingly, *any* quantum state of quantum particles *can* be, in principle, pinpointed by means of *simultaneous* measurements of positions and momenta performed with those *finite* accuracies that are commensurate with the uncertainty principle (Ali and Prugovečki, 1977). Practically implementable devices and schemes for such measurements are by now known (Yamamoto and Haus, 1986; Busch and Schroeck, 1989; Busch and Lahti, 1990), so that it is justified even at the technological level to adopt the following principle.

Principle 2 *(The principle of quantum informational completeness).* For any quantum system, there are measurement schemes that satisfy the principle of irreducible indeterminacy (i.e., Principle 1), and which can, at least in principle, completely determine any (pure) quantum state of that system by assigning to it a unique (modulo a phase factor) state vector ψ, which embodies the outcomes of the measurements performed in the vicinity of a *base* spacetime location (cf. Principle 3) $x \in \mathbf{M}$.

As we recall from Sec. 1.1, according to Einstein's views on the role played by geometry in physics, the geometry of spacetime is neither *a priori* given, nor does its choice represent a matter of convenience. Rather, any *physical* geometry should incorporate *all* the *fundamental* limitations on spatio-temporal measurement procedures, namely all those limitations which reflect a dynamics based on the mutual interaction between geometry and matter. Since all limitations on such procedures which are of a quantum nature can be traced to the presence of state vectors providing quantum probability amplitudes that embody irreducible measurement indeterminacies, this implies that the geometry of spacetime should be a *quantum geometry* , in the sense that it should be based on a notion of "point" that incorporates that of state vector. In view of the general phenomenon of "evanescence of the wave packet" under free (or free-fall) conditions, this necessitates the introduction of a privileged class of state vectors **Φ**, called[15] *proper quantum state vectors*. Their presence in a quantum geometry insures that such a geometry obeys Born's maxims (cited in Sec. 1.1), by providing a physical interpretation of the emerging spatio-temporal relationships in which, to paraphrase Born's second maxim, "all distinctions in spacetime localization that cannot even in principle be observed are meaningless, and have been therefore eliminated".

Naturally, the adoption of the above two basic principles of a *quantum* form of realism cannot determine the structure of these geometries in precise mathematical terms any more than the adoption of the doctrine of classical reality could, by itself, determine the mathematical structure of the geometries adopted in classical general relativity. Rather, Born's maxims proved conducive to a chain of extrapolations that began with conventional nonrelativistic quantum theory [PQ], but in which the canonical quantization was replaced by a new method of quantization (Prugovečki, 1978d) that is based on such most fundamental kinematical groups as the Galilei and the Poincaré groups, so that it has a *direct* operational interpretation in terms of procedures carried out with *quantum* test bodies. This quantization method led to the stochastic quantum mechanics (SQM) framework presented in [P] (and reviewed in Chapter 3 of this monograph), which deals effectively with the problems raised by the Hegerfeldt theorem in special relativistic quantum mechanics.

However, SQM did not provide a correct formulation of *general* relativistic quantum theory, since on its own it could not take into account the equivalence principle[16], namely the principle that *in any local Lorentz frame all the nongravitational laws of physics should take on their special relativistic forms*. The conclusion was eventually reached (Prugovečki, 1985, 1987) that, in order to incorporate this most fundamental principle into the quantum geometries for external gravitational fields, such geometries had to be formulated in terms of fibre bundles, in a manner analogous to the way in which classical general relativity (CGR) can be reformulated in modern fibre-theoretical terms (cf. Chapter 2). Indeed, the equivalence principle requires that CGR should be a Lorentz gauge invariant theory. On the other hand, the basic features of quantum propagation, as well as Weyl's (1924, 1952) interpretation of general covariance, require the incorporation into it of infinitesimal spacetime translations that give rise to a general relativity principle (cf., Treder *et al.*, 1980, pp. 14 and 30), which *de facto* extends this Lorentz gauge invariance into Poincaré gauge invariance. Thus, we arrive at the following conclusion (cf. the discussion in Sec. 1.5).

Principle 3 *(The GS quantum general relativity principle)*. In the absence of non-Abelian internal gauge degrees of freedom the mathematical structure of a quantum geometry is that of a fibre bundle **E**, whose structure group **G** *incorporates* the Poincaré group, whose base manifold is a Lorentzian manifold (**M**,g), with metric g and with elements $x \in \mathbf{M}$ that represent *base* spacetime locations in accordance with Principle 1, and whose fibres $\mathbf{F}_x$ are Hilbert spaces (or pseudo-Hilbert spaces, in case of photons) containing *quantum* inertial frames; those quantum frames are informationally complete in $\mathbf{F}_x$ in accordance with Principle 2, and provide the transition amplitudes supplying the confidence functions χ_x describing the fluctuations around those values, as stipulated by Principle 1.

The presence of non-Abelian gauge degrees of freedom – that occur in the treatment of Yang-Mills fields and of quantum gravity – leads to additional technical difficulties, which in conventional quantum field theory are treated by means of Faddeev-Popov ghost fields and well-known BRST techniques [IQ]. The adaptation to the GS framework of geometrized versions of those techniques has proved feasible (Prugovečki, 1988b), and has led to quantum geometries which are represented by superfibre bundles over appropriate extensions of Lorentzian manifolds (Prugovečki, 1989b; Warlow, 1992). We shall postpone the discussion of their mathematical structure until the last two chapters of this monograph, where we introduce the mathematical techniques required for their treatment.

1.4. Physical Features of Geometro-Stochastic Propagation

In any physical theory or model, the choice of dynamics follows the choice of geometry. This remains true even of classical general relativity (CGR), since the generic choice of Lorentzian geometries (as opposed to other non-Euclidean geometries) is dictated[17] by the equivalence principle (Einstein, 1916, §4). Of course, in formulating the initial-value problem, there is a fundamental distinction between CGR and all other physical theories, including special relativity: whereas the choice of geometry in all other physical theories is fixed from the beginning (i.e., *before* the initial problem is solved), in the case of CGR the specific Lorentzian geometry for a given set of initial data on a given three-dimensional manifold σ_0 is generated as part of the solution of that problem – i.e., both the 4-dimensional manifold $\mathbf{M}$ and the Lorentz metric $\boldsymbol{g}$ on it, that constitute the resulting Lorentzian manifold $(\mathbf{M},\boldsymbol{g})$, are obtained by solving Einstein's equation for those initial data [M,W]. This, of course, creates difficult conceptual as well as technical problems in quantum general relativity (QGR). The presentation of the solution of these problems within the GS framework will be therefore postponed until Chapter 11.

The remainder of this monograph will be devoted to the gradual and careful development of ideas and techniques necessary for understanding that solution. This program of gradual extrapolation will be initiated in Chapter 3, within the more familiar setting of nonrelativistic quantum theory. This setting will also provide the basic testing ground for the central idea of geometro-stochastic (GS) propagation. Such GS propagation will be subsequently studied primarily within the context of various general relativistic external-field problems that deal with the propagation of the most standard non-gravitational quantum fields in a given Lorentzian manifold $(\mathbf{M},\boldsymbol{g})$, i.e., as a GS alternative to conventional quantum field theory in curved spacetime [BD]. Although in such studies the action of these non-gravitational fields on the gravitational field itself is neglected, the insights achieved by undertaking them will nevertheless turn out to be of great value in quantizing the gravitational field in Chapter 11 – and thus, eventually, providing a framework capable of taking into account the kind of gravitational reaction whereby the distribution of matter can, in turn, influence the quantum geometry of spacetime.

The concept of GS propagation emerges from the following two fundamental physical ideas of CGR and of nonrelativistic quantum mechanics, respectively:

(A) According to CGR, a massive point particle in free fall propagates from a point $x' \in \mathbf{M}$ to a point $x'' \in \mathbf{M}$ in its chronological future $I^+(x')$ by following a timelike geodesic in a given spacetime Lorentzian manifold $(\mathbf{M},\boldsymbol{g})$ [M,W].

(B) In the path-integral formulation (Feynman, 1948) of nonrelativistic quantum particle propagation, the Feynman propagator is obtained as a path integral

$$K(\mathbf{x}(t'');\mathbf{x}(t')) = \int \exp[iS(\mathbf{x}(t'');\mathbf{x}(t'))]\, \mathcal{D}\mathbf{x}(t) \ , \tag{4.1}$$

which is defined by a "sum" over all paths connecting the initial point $\mathbf{x}' = \mathbf{x}(t')$ at time t' to the final point $\mathbf{x}'' = \mathbf{x}(t'')$ at time $t'' > t'$. This "summing" procedure is formulated (Feynman and Hibbs, 1965, p. 33) by constructing all possible broken polygonal paths, obtained by slicing the section in Newtonian spacetime between $\sigma(t')$ and $\sigma(t'')$ with hyperplanes $\sigma(t_n)$ corresponding to $t' = t_0 < t_1 < \cdots < t_N = t''$ and connecting pairs of points $\mathbf{x}(t_{n-1})$ and $\mathbf{x}(t_n)$ on two neighboring slices $\sigma(t_{n-1})$ and $\sigma(t_n)$ by straight lines, then carrying out an averaging procedure over all such paths, and in the end going to the type of limit encountered in Riemann integration, whereby $\varepsilon = \max(t_n - t_{n-1}) \to +0$. Alternatively, the same result can be arrived at by applying Trotter's product formula [PQ] to the above sequence of time slices, so that (cf. Chapter 1 in [ST])

$$K(\mathbf{x}(t''); \mathbf{x}(t')) = \lim_{\varepsilon \to +0} \int K(\mathbf{x}(t_N); \mathbf{x}(t_{N-1})) \prod_{n=N-1}^{1} K(\mathbf{x}(t_n); \mathbf{x}(t_{n-1}))\, d^3\mathbf{x}(t_n) . \tag{4.2}$$

The basic idea of GS propagation is to combine the essential features of (A) and (B) by foliating the chronological future $I^+(x')$with spacelike hypersurfaces $\sigma(t)$ depending on an abstract parameter[18] t, and to replace in (B) the straight line segments constituting polygonal paths by the geodesics connecting pairs of points $x(t_{n-1})$ and $x(t_n)$ on two neighboring slices $\sigma(t_{n-1})$ and $\sigma(t_n)$, so that a broken worldline for free-fall would result (cf. Sec. 5.4). For *strongly* causal GS propagation (cf. Sec. 5.4) the equivalence principle is then reflected by the requirement that in the limit $\varepsilon \to +0$ the quantum propagation is carried out for infinitesimal durations along *classically* causal geodetic paths. Furthermore, whereas in the integral in (4.2) the nonrelativistic propagators $K(\mathbf{x}(t_n); \mathbf{x}(t_{n-1}))$ are obtained from the time evolution governed by a Hamiltonian operator acting in a single Hilbert space $\mathcal{H}$,

$$\begin{aligned} K(\mathbf{x}(t''); \mathbf{x}(t')) &= K(\mathbf{x}'', t''; \mathbf{x}', t') = \left\langle \mathbf{x}'' \middle| e^{-iH_0(t''-t')} \middle| \mathbf{x}' \right\rangle \\ &= (m/2\pi i t)^{3/2} \exp[i(m/2t)(\mathbf{x}''-\mathbf{x}')^2] , \qquad t = t'' - t' , \end{aligned} \tag{4.3}$$

in the GS approach the Hilbert spaces change from point to point, exactly as in CGR the tangent spaces $T_x\mathbf{M}$ carrying the 4-velocity vectors of free-fall motion change from point to point in $\mathbf{M}$. Consequently, the definition of the corresponding free-fall GS propagators is rooted in the notion of parallel transport from $\mathbf{F}_{x(t')}$ to $\mathbf{F}_{x(t'')}$. Thus parallel transport plays as fundamental a role in the resulting framework for quantum general relativity (QGR) as it plays in CGR. Of course, whereas classical general relativistic propagation is deterministic, and takes place along a unique geodesic for each complete set of initial data (such as those supplied by the specification of a spacetime location $x' \in \mathbf{M}$ and of a 4-velocity $\boldsymbol{v}' \in T_{x'}\mathbf{M}$), strongly causal GS propagation takes place along all possible causal stochastic paths (cf. also Sec. 5.6), so that even if a complete set of initial data is prescribed in $\mathbf{F}_{x'}$, it can reach, with various degrees of probability, any of the base locations $x'' \in I^+(x')$.

In the absence of an external gravitational field, the CGR framework reduces to a special relativistic one upon the introduction of global Lorentz frames $\mathcal{L}$, and of the concomitant identification of all the tangent spaces $T_x\mathbf{M}$ with a single Minkowski space – which can be itself identified with $\mathbf{M}$, or which can be more generally deemed to contain the flat manifold **M**, if that latter manifold displays no unusual global topological features, such as a cylindrical structure that would make it spatially closed. A similar identification $\mathbf{F}_x \equiv \mathbf{H}(\mathcal{L})$ of all the fibres $\mathbf{F}_x$ in the corresponding quantum geometry can be then carried out (cf. Secs. 2.3 and 5.4). Furthermore, since **M** is flat, parallel transport in the quantum fibre bundle representing that geometry becomes path-independent, so that the above GS propagation transcends into a more conventional type of quantum propagation, governed by such standard equations of motion as the Klein-Gordon equation or the Dirac equation. These equations then totally determine the propagation of all state vectors within the single Hilbert space $\mathbf{H}(\mathcal{L})$. In this manner, as we shall see already in Chapters 5 and 6, we shall recover the special relativistic stochastic quantum mechanics presented in [P] – whose principal concepts, aspects and results will be reviewed in Chapter 3.

Although the above described conceptualization of GS propagation is very much in keeping with Feynman's path integral approach, there are a number of fundamental mathematical as well as physical differences.

The most obvious mathematical difference is that the Feynman path integral approach is based on action integrals, such as the one in (4.1), whereas GS propagation is derived by using purely geometric considerations. Hence, as we shall see in subsequent chapters, GS action integrals are not postulated, as it is often the case in conventional path-integral approaches. Rather, they are derived in basically the same manner in which (4.1) can be derived from (4.2) and (4.3) – cf. Chapter 1 in [ST]. The physical differences originate from the fact that in the Feynman approach quantum propagation is envisaged as being that of a *point* particle. Hence, the concept of its "path" is merely heuristic (Exner, 1985, p. 307), since the actual determination of points along the path of a quantum point particle requires an infinite amount of energy (Blokhintsev, 1968; Skagerstam, 1976), and the formulation of its underlying probabilities results in inconsistencies (Sorkin, 1991). Infinities are implicit also in the mathematically heuristic procedures which are employed (Feynman and Hibbs, Chapter 3) in all the actual computations of such path integrals as those in (4.1). Indeed, it is easily seen that not even the integrals in (4.2) exist in a mathematically rigorous sense, on account of the fact that the absolute values of the Feynman propagator in (4.3) are constant, so that the integrations which in (4.2) have to be carried out over $\mathbf{R}^1$ are divergent by basic theorems in the theory of Lebesgue integration (cf. Chapter 2 in [PQ]). The heuristic nature of the Feynman paths is also exemplified in the special relativistic regime by Feynman's (1950) treatment of the propagation of a Klein-Gordon particle, where a Schrödinger equation with respect to a fictitious time variable is introduced, by "blind mathematical analogy" (cf. [ST], p. 226) with the nonrelativistic case, in order to formulate an action-based path integral. Hence the resulting path integration has to be carried out over *all* possible broken paths, regardless of whether those paths are causal or not[19].

On the other hand, the object that propagates in the GS approach is the *proper* state vector of an extended quantum object conceptualized as a *geometro-stochastic exciton*[20], represented by excitations in the quantum fibres $\mathbf{F}_x$ *above* various *base* spacetime locations $x \in \mathbf{M}$. Such proper state vectors incorporate into the GS formalism, from the outset, the principles of irreducible indeterminacy and of informational completeness, formulated in the preceding section. Hence, neither do any divergences occur in the corresponding path integrals (Prugovečki, 1981a), nor do there arise any contradictions with the uncertainty principle if actual probabilities are assigned to families of paths passing through a finite number of specified quantum spacetime locations. Indeed, the probabilities of determining the spatio-temporal relationships of those locations can never be brought arbitrarily close to one, i.e., they cannot lead to practical certainty, due to upper bounds on measurement accuracies which are embedded into the GS framework from the outset.

In the ultimate analysis, the GS approach gives rise to an all-encompassing quantum mechanical conceptualization of *all* parts of an experimental setup, including those macroscopic parts which in the orthodox approach are treated by classical means. For example, from the GS perspective, the Feynman and Hibbs (1965) description of the two-slit experiment is viewed as an *approximation* of a realistic quantum description in which the screen S and the detector D are deemed to consist of a very large number of excitons. From such a microscopic point of view, an exciton E with proper state vector $\mathbf{\Phi}$ propagates along all possible causal stochastic paths from its source O, situated at some base location x', to the base spacetime locations where it encounters the excitons in the screen S; upon interacting there with those screen excitons, it proceeds in the same manner towards the base locations of its eventual collisions with the excitons of the detector. Thus, in mathematical terms, the proper state vector $\mathbf{\Phi}$ remains unaltered in relation to the *local* frames at each x along each stochastic path γ, with the well-known "spreading of the wave packet" occurring only from the perspective of very massive macroscopic frames, which can be introduced in the role of *global* inertial frames if (and only if) the curvature of the base manifold is negligible (cf. Chapter 5). In fact, this stochastic propagation can be best depicted in quantum field theoretical terms, namely as a perpetual process of exciton annihilations at each $x \in \gamma$, and exciton recreation (i.e., *local* vacuum excitation) at a neighboring base location $x + \delta x \in \gamma$. In this GS process the interactions between the exciton E that was emitted by the source O, and the multitude of excitons which, from a fundamentally microscopic point of view constitute the screen or the detector, occur in geometrically local terms (Prugovečki, 1990b). Thus, in the GS approach to quantum propagation, new excitons are produced in the Fock fibres (cf. Chapter 7) $\mathcal{F}_x \supset \mathbf{F}_x$ above each base location $x + \delta x$ under conditions that ensure the local conservation of energy-momentum, and which, moreover, are such that Bose and Fermi statistics are obeyed by bosons and fermions, respectively.

A conceptualization of this GS quantum process, that is closer to the one envisaged by Feynman and Hibbs (1965), emerges under the simplifying assumption that the screen provides a totally impenetrable barrier except at its two slits, so that the source-excitons reaching the base locations of the screen-excitons are annihilated without trace. In that case the contributions to the geometro-stochastic propagator (cf. Chapters 4-6) of a source-exciton E from those stochastic paths which do not pass through those slits equal zero. But even under such simplifying assumptions, a fundamental distinction remains between the GS and the conventional conceptualization of this same quantum process, due to the fact that the GS framework is based on quantum rather than on classical geometries. Hence, in the GS framework the location of the points in the upper slit U cannot be *operationally* distinguished with *absolute* certainty from the location of the points of the lower slit L. Indeed, although the *base* locations x' and x'' of any two points $(x',\mathbf{\Phi}') \in U$ and $(x'',\mathbf{\Phi}'') \in L$ are distinct, these points belong to GS quantum geometries, and as such they are not *absolutely* distinguishable in measurement-theoretical terms, due to the generally nonzero transition probability amplitudes between (the parallel transports to the same base location of) $\mathbf{\Phi}'$ and $\mathbf{\Phi}''$. Of course, the ensuing probabilities are negligible for those base separations of the two slits which are of macroscopic orders of magnitude. Hence, the conventional conceptualization is recovered as an *approximation*, which is sufficiently good for all *practical* purposes. However, from a theoretical point of view, those minuscule transition probabilities are significant since, as discussed in (Prugovečki, 1990b), as well as in

Chapters 5, 7 and 11, they drastically influence the conventional concept of locality and relativistic causality in general, and make inapplicable the "local (anti)commutativity" axiom of conventional quantum field theory in particular.

On the other hand, in complete accord with conventional ideas, as well as with well-known experimental results on neutron interferometry (Badurek *et al.*, 1986; Rauch, 1987, 1988), the propagation of the exciton E proceeds by stochastic parallel transport through *both* slits, until the excitons constituting the detector D are encountered. There, the detection process can be envisaged to take place in accordance with the Machida-Namiki (1980) multi-Hilbert-space theory of measurement (cf. also Murayama, 1990), with the Fock fibres occupied by the excitons in D providing the required Hilbert spaces. Since that theory is illustrated in some review articles (Namiki, 1988) in the context of the Stern-Gerlach experiment, we shall discuss its adaptation to the GS framework in that same context.

A careful analysis (cf., e.g., Bohm, 1986, Sec. XIII.1) of the Stern-Gerlach experiment, from the point of view of conventional nonrelativistic quantum theory, reveals that even if simplifying assumptions are made, the reduction of the wave packet resulting from position observations that are correlated to spin values along a vertical axis cannot be explained by the conventional quantum theory of measurement as part of the dynamics based on the Schrödinger equation for "system-plus-apparatus". Moreover, as the Wigner-Araki-Yanase (1952, 1960) theorem mentioned in Sec. 1.3 confirms, such correlations cannot be 100% (cf. Busch and Schroeck, 1989, §3.1, or Schroeck, 1991, Chapter 2), due to the fact that the supports of the component wave functions in that reduction are never totally disjoint in configuration space. The second of these two problems can be taken care of by introducing (Prugovečki, 1977b; Busch, 1986) the concept of stochastic spin values mentioned in Sec. 1.3, but the first problem cannot be solved in the context of unitary evolution in a single Hilbert space – as indicated by another theorem originally obtained by Wigner (1963), and later refined by Fine (1970) and Shimony (1974). Let us therefore first conceptualize this experiment in realistic terms within the GS quantum framework, which bypasses this last no-go theorem due to its multi-Hilbert-space mathematical structure.

In such a realistic quantum GS depiction, an exciton E propagates geometro-stochastically to the region between the poles of the electromagnet constituting part of the Stern-Gerlach apparatus A, where it interacts with the excitons in the electromagnet through the exchange of photons. After further GS propagation, it eventually collides with the excitons in the detecting devices D. In computational terms, however, such a conceptualization encounters not only the problem of dealing with the enormous number of excitons in A, but also of taking into account all the (generally nonlinear) quantum field theoretical interactions that hold the excitons of A together as part of a *macroscopic* apparatus. Consequently, computationally (but not in principle), a simplifying assumption is required whereby, as in the conventional quantum theory of this experiment (Bohm, 1986), the effects of the excitons of the electromagnet are collectively represented by an external inhomogeneous magnetic field. In the GS approach this simplification can be achieved upon assuming that the effects of the (ever-present) external gravitational field are negligible, so that an identification $\mathbf{F}_x \equiv \mathbf{H}(\mathcal{L})$ of all those single-exciton fibres $\mathbf{F}_x$, in which the propagation of E produces excitations, can be carried out upon the introduction of a (very massive) global Lorentz frame $\mathcal{L}$ (cf. Chapter 5). It is with respect to this global Lorentz frame $\mathcal{L}$ that the aforementioned external magnetic field can be then introduced. It is with respect to that same frame that all the *geometrically local* (Prugovečki, 1990b) excitations occurring in various fibres $\mathbf{F}_x$ can be *collectively* described by a single wave packet $\psi \in \mathbf{H}(\mathcal{L})$ for E, whose propagation is then governed by that external magnetic field.

The second simplification is achieved by following Namiki (1988) and treating the quantum bonding forces which give rise to the macroscopic features of each detector D from the point of view of the quantum statistical mechanics formulated in the GS framework of Chapter 3 in [P]. The reduction of the wave packet then follows from the Machida-Namiki theory, since that theory entails only two basic properties, namely (Namiki, 1988, p. 50): 1) the macroscopic nature of the local system of detectors – which in the GS context are represented by direct sums or integrals of the Fock fibres $\mathcal{F}_x$ constituting Hilbert spaces above the base locations of the excitons in D; 2) the phase shift formula of the S-matrix for collisions of E with a local system of finite size – which in the GS framework follows from the adaptation of conventional quantum scattering theory to the SQM framework carried out in (Prugovečki, 1978b). It should be noted, however, that from the *foundational* GS point of view, the averaging procedure required by the Machida-Namiki theory reflects only "our ignorance about [the] microscopic details of a local system of detectors" (Namiki, 1988, p. 52). On the other hand, as earlier mentioned, those "microscopic details" pertain to generally *nonlinear* aspects of the complicated net of field theoretical interactions that hold the excitons of D together, and govern their interactions with the exciton E playing the role of the "system". Such interactions make Wigner's (1963) theorem inapplicable, since that theorem was derived under the assumption of a linear and deterministic law of quantum evolution. Furthermore, recent measurement-theoretical models (Ghirardi *et al.*, 1986, 1990) confirm that non-linearity in the interaction between a "system" and an "apparatus" can indeed give rise to wave-packet reduction. It can be therefore surmised that a future perfected GS theory of quantum measurement might indeed reduce the problem of such wave packet reduction, in those cases where it occurs at all (Ballentine, 1990), to the status of a "symmetry-breaking problem, such as the Higgs mechanism in modern field theory and/or the phase transitions in critical phenomena" (cf. Namiki, 1988b, pp. 9-10).

For the present, however, the GS conceptualization of the measurement processes is as follows. Both system and apparatus are viewed, at the most fundamental microscopic level, in realistic quantum terms, namely as collections of excitons. If the measurement procedure is decomposed into preparatory and determinative measurements (Prugovečki, 1967), then at the preparatory stage the excitons constituting the system are produced in the fibres above the base locations of various macroscopic configurations of the excitons within the source. Their propagation into the causal future of those base locations proceeds geometro-stochastically, with mutual interactions taking place by the exchange of excitons corresponding to various gauge fields (cf. Chapters 9-11). Thus, from the GS microscopic point of view, there are no global wave packets[21], but only coherent configurations of excited proper quantum states in the Fock fibres of a quantum spacetime described by a given quantum geometry. Hence, in the GS approach to quantum theory, *global wave packets occur only from the macroscopic point of view*, after very massive *global* inertial frames are introduced in those regions of the Lorentzian base manifold where curvature effects are negligible. Moreover, under those circumstances, the macroscopic nature of the apparatus necessitates the description of its quantum states by means of density operators involving averaging procedures over Fock fibres above different base locations. Under the specific conditions of the Machida-Namiki theory, this can give rise to the "reduction of the wave packet". Such a reduction in fact represents a correlation of macroscopic phenomena, reflecting a GS wave function decoherence, conventionally described as "detection".

1.5. The Physical Nature of Geometro-Stochastic Excitons

The principal subject of this monograph is quantum geometry. As espoused in most of this monograph, and further discussed in Sec. 12.5, quantum geometry can be formulated and studied[22] without any specification of the physical nature of GS excitons, in the same manner in which (pseudo-)Riemannian geometry, even as applied to CGR, can be studied without any specification of the physical nature of the test particles or light signals used in deriving the basic predictions of CGR – such as gravitational redshifts, perihelion precessions, etc. (cf., e.g., [W], Chapters 5 and 6). Indeed, the proper quantum metric states introduced in Sec. 3.6 contain a fundamental length ℓ that characterizes a given quantum geometry, whose features could be then viewed as merely reflecting measurement-theoretical limitations (cf. Chapters 3 and 5) – such as those that certainly occur at the Planck-length order of magnitude. Hence, quantum geometries do not require the existence of physical entities which *exactly* "fit" into their points any more than classical geometries require truly pointlike particles that exactly fit into theirs. On the other hand, the attitudes prevailing nowadays in elementary particle physics envisage that there are *truly* elementary building blocks of the universe around us. We could therefore surmise that they manifest themselves as GS excitons that can "fit" into quantum spacetime points. However, given the fluid state of affairs in elementary particle physics, where ideas about what objects are elementary has been changing drastically over the last few decades – with indications existing (Glashow, 1980; Harari, 1983; Weinberg, 1987) that the end of these changes has not been reached – it is preferable to take a detached and objective look at various physical and epistemic possibilities as to the possible nature of such excitons, rather than to automatically identify them with any of those entities which are at the present time regarded as being "elementary".

From a pure mathematical point of view, one could argue in the same manner as those quantum field theorists who claim that the basic constituents of the physical world are merely phenomenological approximations of quantized fields. In the GS context this would imply that only the quantum fields presented in Chapters 7-11 would have to be endowed with physical reality, whereas the single-exciton considerations in Chapters 3-6 would then refer to merely heuristic phenomenology. This attitude is, however, epistemologically as questionable in the GS context as it is in the setting of conventional quantum field theory. Indeed, classical as well as quantum fields are merely mathematical abstractions in the absence of suitable test bodies, which are required (Bohr and Rosenfeld, 1933, 1950) for their detection and localization with the degree of precision allowed on foundational grounds – or at least arbitrarily close to it.

Similarly, one might be tempted to adopt the methodology promoted in some of the very fashionable contemporary quantum cosmology, which postulates that our entire universe was created *ex nihilo* as a quantum fluctuation in some kind of primordial "vacuum", or that it is a "fluctuation" within an abstract superspace of Riemannian 3-geometries (cf. Secs. 11.2, 11.12 and 12.3). However, in that case all boundaries between mathematical abstraction and physical reality are removed, and *any* theoretical idea becomes credible. Despite the present-day ready acceptance of such ideas, concern with epistemic soundness cautions against transposing physical concepts from a domain of discourse where they have a well defined empirical meaning to situations where that meaning is lost (Shimony, 1978; Davies, 1984), and only formal mathematical analogies are retained in support of the argument – as it is the case when quantum tunneling through an actually *existing* potential barrier is extrapolated into an idea of universe creation *ex nihilo*, or when the idea of statistical fluctuations in actually realizable *ensembles* of quantum systems is extrapolated into an idea of a superspace of geometries that describe alternative states of the entire universe.

Heisenberg examined in his very last paper, published posthumously in 1976, the role of epistemological considerations in physics in general, and in high energy physics in particular. He was very blunt as he made the following statements: "I believe that certain erroneous developments in particle theory – and I am afraid that such developments do exist – are caused by a misconception by some physicists that it is possible to avoid philosophical arguments altogether. Starting with poor philosophy, they pose the wrong questions. It is only a slight exaggeration to say that good physics has at times been spoiled by poor philosophy. ... Having witnessed similar mistakes in the development of quantum mechanics fifty years ago, I am in the position to make some suggestions to avoid such errors in the future." (Heisenberg, 1976, p. 32). In the same article he then proceeded to illustrate his point with enlightening occurrences which he witnessed first-hand, and which illustrate how, in the past, demonstrably erroneous conclusions were reached on the basis of considerations "of purely formal mathematical nature" (Heisenberg, 1976, p. 38).

There is no real evidence indicating that Heisenberg's criticisms and suggestions have ever had much impact on developments in particle physics since the time they were made[23]. Nevertheless, in the remainder of this chapter, we shall try and follow his suggestions by outlining a consistent epistemological viewpoint on quantum phenomena (which will be discussed more fully in Chapter 12), as well as by retaining a critically detached judgement vis-à-vis the currently most fashionable models in elementary particle physics and quantum cosmology. Hence, we shall not summarily dismiss alternatives to such models, but rather give thought to questions of their mathematical consistency and epistemological soundness.

Thus, we shall proceed by first presenting a physically straightforward formulation of the idea of GS exciton, conceptualized as an excitation of the *local* vacuum at a base location $x \in \mathbf{M}$ in a quantum spacetime, and then examining the various possibilities for its mathematical implementation – rather than the other way around. Basically, this idea can be traced to Born's (1938, 1949) reciprocity theory and his formulation of a quantum metric operator $D^2 = Q^2 + P^2$. In Born's view, the postulate of such an operator was necessary in order to formulate the reciprocal relationship between the spacetime location and the 4-momentum of quantum matter, vis-à-vis the occurrence in nature of a discrete mass spectrum for elementary particles. Such a physical idea leads, in a natural manner, to a concept of quantum spacetime which displays discrete spacetime degrees of freedom *in addition* to the macroscopically observable four continuous degrees of freedom described by $x \in \mathbf{M}$, in the same manner in which elementary particles display a discrete mass spectrum despite the fact that their 3-momenta $\mathbf{p}$ can assume a continuum of values within $\mathbf{R}^3$.

Indeed, superficially, such a reciprocity idea might seem to suggest that space and time coordinate variables should be "quantized" in the sense that they assume only discrete values. This would lead us, however, to the idea of a discrete spacetime which, as we have seen in Sec. 1.2, has already been unsuccessfully considered in the past[24]. Moreover, not only does such an approach represent a most radical departure from all preceding descriptions of spacetime as a "continuum", but it is not necessitated in the least by the idea of quantization, since quantum theory allows the existence of observables with continuous spectrum, and there is neither experimental evidence nor cogent foundational arguments to the contrary. In fact, foundational arguments based on general covariance indicate the opposite: any "quantization" of spacetime coordinates that would be analogous to the quantization of the total internal energy of a system, resulting in bound states that characterize the transition from the classical to the quantum regime, would produce discrete values for those coordinates. Hence, it would impart to spacetime coordinates an intrinsic physical meaning denied to them by Einstein (1916) in his formulation of CGR, since it would have to single out preferred systems of coordinates and elevate them to the status of "observables".

On the other hand, Born's reciprocity idea can be combined with the first two principles stated in Sec. 1.3, upon reformulating it in terms of spacetime location as well as momentum-energy values that are *fundamentally* indeterminate. Hence, in the immediate vicinity of spacetime locations where curvature effects are negligible, those indeterminacies would be reflected by a discrete spectrum of *proper* state vectors that describe *complementary* indeterminacies of the Heisenberg type, so that $(\Delta q)_n \cdot (\Delta p)_n = h_n/4\pi$, $n = 1,2,\ldots$, where $(\Delta q)_n$ as well as h_n (and therefore also $(\Delta p)_n$) can assume only a discrete set of values; moreover, in case of $(\Delta q)_n$, these fundamental indeterminacies should be bounded from below by the Planck length, and, in the case of h_n , by the Planck constant. As we shall see in the next chapter, after formulating CGR in terms of fibre bundles, this formulation then suggests in a natural manner a "quantization" of general relativity, whereby the above complementary indeterminacies are realized by treating the aforementioned proper state vectors as elements of quantum fibres, in which they generate *quantum* Lorentz frames that take over the role played by classical Lorentz frames in CGR. Thus, from a *macroscopic* point of view, the "quantization of spacetime" becomes a quantization of the *fundamental* indeterminacies in the classical spacetime locations $x \in \mathbf{M}$, and of the corresponding (of necessity mathematically local) 4-momenta $\boldsymbol{p} \in T_x^*\mathbf{M}$ of the excitons which represent the *truly* elementary constituents of matter and radiation.

Consequently, from the GS point of view, the "particle" aspect of micro-phenomena is not an *intrinsic* feature of the GS excitons themselves, but only a *macroscopic* manifestation of their behavior under those very specific experimental arrangements in which part of a system of excitons constitutes a "detecting apparatus". Similarly, their "wave" aspect is yet another macroscopic manifestation exhibited under those 'complementary' experimental arrangements in which subsystems of excitons constitute screens, filters, or some of the other macroscopic devices whose role is *not* to provide *macroscopic* detection. Thus, we arrive at Bohr's (1961) complementarity principle as a principle for macroscopic observations, but not as a reflection of fundamental *quantum* reality: that reality[25] consists of various excitons, and of their unceasing geometro-stochastic propagation that gives rise to the universe around us; "observations" are merely information-gathering procedures that do not change the intrinsic nature of that reality at the microscopic level, but only our *knowledge* about that reality at the macroscopic level, where that information can be encoded into relatively simple classical models. Indeed, in the GS context, a quantum state of the universe consists (cf. Sec. 11.12) of a complex structure of coherently interrelated exciton proper states, which in their totality constitute, at the cosmic level, a GS counterpart of the "wave function of the universe" of Hartle and Hawking (1983) – cf. also (Tipler, 1986: Barrow and Tipler, 1986). However, even when the universe of discourse is restricted to a very limited macroscopic region (such as an experimental set-up), the comparison of theory with experiment necessitates, from a practical point of view, a partial replacement of such a detailed microscopic description by a macroscopic description expressed in simple classical language. That part of the universe of discourse to which such a simplification is applied becomes the "apparatus", and the remainder becomes the "system". When such a *simplification* is undertaken, then in complete accord with Bohr's (1961) epistemology, various experimental arrangements for the collection of data on the same "system" produce *macro*-data which cannot be consistently described in their totality within a *classical picture*[26], since no such single picture can reflect *all* the aspects of the underlying *quantum* reality. *That* reality is in fact holistic, so that different choices of "experimental arrangements" might represent altogether different choices of universes of discourse to which those simplifying assumptions are applied. On the other hand, the GS philosophy is that an underlying *quantum* reality does *exist*, and that it can be consistently described in its *totality* by appropriate use of a suitable *quantum* language – such as that of quantum geometry – albeit not by means of an exclusively classical language.

Thus, from a GS perspective, the wave-particle duality of quantum phenomena is merely a reflection of the use of classical concepts at the macroscopic level of discourse, where state vectors are regarded as mathematical objects external to a background classical geometry, and as such are represented by wave packets defined *within* the manifold describing that geometry. At the microscopic level, however, the GS approach envisages a quantum reality in which the *proper* state vectors $\mathbf{\Phi}$ of spacetime excitons form an integral part of the concept of point in quantum geometries (cf. Sec. 12.5), in terms of which the GS quantum notion of spacetime is formulated. In the context of such geometries, quantum propagation can be viewed *only* as a geometro-stochastic process, whereby a spacetime exciton propagates by stochastic parallel transport along all possible (classical or quantum – cf. Chapter 5) causal stochastic paths available to it – with the total amplitude resulting, at any given point in a quantum spacetime, from such (respectively, strongly or weakly causal) propagation being obtained by the superposition principle. Such a conceptualization of quantum propagation is in keeping with the conventional one in those instances where the latter is applicable, without giving rise to the inconsistencies reviewed in Sec. 1.2, and yet it bypasses those inconsistencies where they occur, thus extending the domain of applicability of the latter. For example, as we have seen in the preceding section, the quantum measurement process, including the phenomenon of wave packet reduction, can be treated as a quantum geometro-stochastic process, which as such constitutes an integral part of a *unique and indivisible quantum reality*.

Of course, it is not necessary to accept the above GS philosophy in order to deal with the present formalism for quantum geometries, any more than it is necessary to accept the orthodox interpretation of quantum theory in order to deal with the conventional formalism of quantum mechanics. On the other hand, in applying quantum geometries to quantum cosmology, there do not appear to be any sensible alternatives for a very basic reason: the entire universe assumes, in that case, the role of the system, so that there is no extra room left for placing the "apparatus". And once an "apparatus" is placed within a quantum spacetime that is supposed to describe our universe, it becomes absurd to envisage that a GS counterpart of the "reduction of the wave function of the universe" occurs each time a given apparatus registers a measurement result; or, in case that a GS alternative to the many-worlds theory (DeWitt and Graham, 1973; Barrow and Tipler, 1986) is adopted, that a change in the entire quantum geometry of spacetime is effected, so that we switch branches in a superspace of universes each time some observation is carried out (cf. Sec. 12.5).

From a mathematical point of view, quantum geometries are represented by well-defined mathematical structures which will be described in detail in Chapters 4-11 of this monograph. In case that the gravitational field is treated semiclassically as an external field, which as such influences the dynamics of quantum objects, but is not in turn influenced by those objects, we shall see from Chapter 3 onwards that the mathematical realization of this idea of quantum geometry is carried out in terms of soldered fibre bundles (Drechsler and

Mayer, 1977; Trautman, 1982) associated with a principal bundle **P(M,G)** over a Lorentzian manifold **M**. The structure group **G** of that principal bundle (called a "gauge group" in some of the physics literature) has to incorporate the Poincaré group, or its covering group ISL(2,**C**), on account of the equivalence principle and the fundamental role played by the generators of spacetime translations in quantum dynamics. In addition, it will also have to incorporate all the internal symmetry groups characteristic of a given model.

The well-known standard model of elementary particle physics (Halzen and Martin, 1984) suggests adopting SU(3)×SU(2)×U(1) as an internal symmetry group, so that photons, gravitons, leptons, quarks and gluons are to be cast in the role of excitons, with the hadrons being viewed as composite objects. However, the still unresolved problem of quark confinement (Nachtmann, 1990), and the possibility of subquark structure, such as a common preon substructure for quarks and leptons (Pati and Salam, 1975, 1984), indicate that the question of "ultimate constituents" is still open for debate, and that other possible alternatives for hadron structure are not to be altogether dismissed within the GS context.

The mathematically, as well as conceptually, simplest of these alternatives is that hadrons are treated as GS excitons, so that their proper state vectors can be derived (Brooke and Prugovečki, 1984; cf. also [P], §4.5) from such a simple model as that based on the aforementioned Born's quantum metric operator. An ensuing mass formula (Prugovečki, 1981b) was found to be in good agreement with experimental data (Brooke and Guz, 1984), so that this alternative cannot be dismissed out of hand. However, given the great gulf between the typical rms radius of hadrons ($\approx 10^{-13}$ cm) and that of leptons, which by comparison appear pointlike – with the experimental upper bound for the latter being set at less than 10^{-16} cm (Barber *et al.*, 1979), but with Planck's length ℓ_P being, on foundational grounds, the most attractive possibility – this first alternative is the furthest away from the theoretical ideas that are prevalent at the present time in elementary particle circles.

The second alternative is to adopt for hadrons various models of relativistic extended objects, such as those based on quantum relativistic oscillators or rotators (Fujimura *et al.*, 1970-71; Feynman *et al.*, 1971; Takabayashi, 1979; Aldinger *et al.*, 1983; Bohm *et al.*, 1988). Mathematically, such models are closely related to those of the first alternative, since in them the role played by Born's quantum metric operator is taken over by various relativistic Hamiltonians which govern collective motion within quantum fibres. These motions reflect internal degrees of freedom whose variables give rise to spectrum generating groups, such as SO(3,2). In this conceptualization, "quark" constituents can be envisaged, but they are confined by the presence of postulated internal potentials. Since similar "interquark potentials" do not partake in the external dynamics between hadrons, from a semiclassical point of view the quark confinement problem is resolved by viewing these constituents as existing in the tangent spaces $T_x\mathbf{M}$, rather than in the classical spacetime continuum **M**. On the other hand, from the point of view of quantum geometry, such models can be viewed purely as a means of generating hadronic proper state vectors.

The third alternative is to start from the differential geometric description of extended hadrons proposed by Drechsler (1975, 1977), in which the de Sitter group SO(4,1) operates as a structural group on fibres which are de Sitter spaces characterized by a radius of curvature $R \approx 10^{-13}$ cm. The geometro-stochastic quantization of such models leads to quantum geometries described by bundles associated with the de Sitter principal bundle **P(M**,SO(4,1)) over Lorentzian spacetime manifolds **M** (Drechsler and Prugovečki, 1991; Drechsler, 1991). This means that the de Sitter group SO(4,1) takes over, in case of hadrons, the role played in the geometro-stochastic propagation of leptons by the Poincaré group, with the latter emerging in the limit $R \to \infty$. Such models could therefore explain the experimentally observed sharp deflections of leptons in their collisions with hadrons without giving rise to confinement problems, since the $R \to \infty$ limit plays only a mathematical role, rather than representing an experimentally feasible procedure.

To approach exciton structure in a manner that comes closer to the ideas which at present still enjoy great popularity amongst elementary particle physicists would require treating them as quantum states of strings or superstrings. Such a treatment can be partly achieved by taking advantage of the formal similarities that some methods of string quantization share with quantum metric operators which incorporate mass-zero states (Prugovečki, 1988a). As outlined in Sec. 12.5, in such an approach string vibrations giving rise to zero-mass excitons could be viewed at the semiclassical level as occurring in certain ten-dimensional subfibres of the twelve-dimensional bundle $T\mathbf{M} \oplus T^*\mathbf{M}$, rather than within a ten-dimensional spacetime continuum $\mathbf{M}^{10}$. Hence, in such a context there is no need for assuming the spontaneous compactification of six dimensions in $\mathbf{M}^{10}$, that is *postulated* in the standard approach to string theory (Green *et al.* 1987; Kaku, 1988).

On the other hand, one of the pioneers of modern string theory has pointed out the following: "One ought to formulate string theory in a much larger space – something like the space of all possible positions of a string. ... The reason we use the language of ten dimensions or four dimensions is because we have so far been forced to talk about string theories in an approximate way" (Green, 1989, p. 130-131). If such an attitude is adopted, then the "much larger stringy space which really has an infinite number of dimensions" envisaged by Green (1989, p. 131) can be indeed provided in the GS context by the infinite-dimensional fibre bundles of the resulting quantum geometries for spacetime, which are both mathematically, as well as physically, the actual arena of the GS method of quantization, presented in this monograph from Chapter 4 onwards.

Clearly, the above alternatives do not exhaust all the possibilities for treating hadron structure and the world of "elementary particles" within the context of the quantum spacetimes described in the present work. In fact, perhaps the most aesthetically pleasing idea is that of a single type of ground exciton state $\boldsymbol{\Phi}$, representing a graviton (which is the most likely candidate for an "ultimate building block" in the universe, and therefore for an exciton), with all remaining exciton states being constructed as soliton states in a nonlinear quantum field theory in quantum spacetime. However, this, as well as other similar ideas, represent thus far only pure speculation. Consequently, in our subsequent studies of quantum geometries, we shall leave the choice of proper exciton states unspecified whenever quantum geometry itself is the principal object of interest. On those occasions when we desire to discuss specific models (as it will be the case in Chapters 9-11), we shall opt for a realization in terms of the ground state of the relativistic harmonic oscillator supplied by a certain type of representation of Born's quantum metric operator D^2 (cf. Sec. 12.5).

Notes to Chapter 1

[1] It should be noted that this term has been used occasionally also by other researchers in quantum gravity to describe ideas which have no direct relationship to the ones studied in the present work.

[2] Not to be confused with various *stochastic mechanics* formulations of quantum theory based on the *classical* theory of stochastic processes (Nelson, 1967, 1986; Guerra, 1981; Smolin, 1986; Namsrai, 1986) – in some of which the full mathematical equivalence with quantum mechanics in Hilbert space is lacking (Wallstrom, 1989) – nor with the *stochastic quantization* method (Damgaard and Hüffel, 1987).

[3] For [M,W] and [P], as well as for all the other references cited by means of capital letters in between square brackets, read the note on *key references* that precedes the list of references at the end of this book, and then search out in that list those references which are headed by that combination of letters.

[4] Sections 1.2-1.3 represent a slightly modified and expanded version of the discussion of these fundamental epistemological aspects that was presented in (Prugovečki, 1990a).

[5] In the concluding chapter of the same publication, Kline sets this statement in its proper epistemological perspective when he points out that "Kant was wrong in insisting upon Euclidean geometry, but his point that man's mind determines how nature behaves is a partial explanation [for why] mathematics works" (Kline, 1980, p. 341) when it is applied to the physical world around us.

[6] It is interesting to compare the attitude towards non-Euclidean geometries that prevailed during the second half of the last century with the analogous attitude towards various proposals for non-Lorentzian spacetime geometries that has been prevalent during the second half of this century. The following quotation from (Kline, 1980, p. 88) summarizes its essence: "Non-Euclidean geometry and its implications about the truth of geometry were accepted gradually by mathematicians [in the course of the last century], but not because the arguments for its applicability were strengthened in any way. Rather the reason was given in the early 1900s by Max Planck, the founder of quantum mechanics: 'A new scientific truth does not triumph by convincing its opponents and making them see light, but rather because its opponents eventually die, and a new generation grows up that is familiar with it'."

[7] Kline (1980) provides a simple illustration of this fact, and concludes that: "Poincaré's philosophy of science has merit. We do try to use the simplest mathematics and alter physical laws if necessary to make our reasoning conform with physical facts. However, the criterion used by mathematicians and scientists today is the simplicity of the *whole* of mathematical and physical theory. And if we must use a non-Euclidean geometry – as Einstein did in his theory of relativity – to produce the simplest combined theory, we do so." (Kline, 1980, p. 344).

[8] In his autobiographical notes preceding those essays, Einstein had, however, already clarified some of his earlier ideas by stating that "the material point . . . can hardly be conceived any more as the basic concept of [relativity] theory", and that "strictly speaking measuring rods and clocks would have to be represented as solutions of the basic equations (*objects consisting of moving atomic configurations*), not, as it were, as theoretically self-sufficient entities" (Schilpp, 1949, pp. 59-61) – emphasis added.

[9] Compare the interviews with some of them carried out by Davies and Brown (1989).

[10] In Wigner's words: "The only difference between the existence of [a] book and the existence of [a] magnetic field in interstellar space is that the usefulness of the concept of the book is much more direct, both for guiding our actions, and for communicating with other people." (Wigner, 1964, p. 251). However, the concept of *proper* state vector, introduced in the next section, leads to a treatment of the "reduction of the wave packet" which does not involve an observer's "consciousness", and is very different from the one advocated by Wigner (1962).

[11] That is, not even in the *imagined* limit of the conventionally postulated type of infinite sequence of increasingly accurate measurement procedures, that underlie all theories of measurement.

[12] Depending on circumstances and the design features of measuring devices, this base value could be the mean, the median, or some other value statistically related to the observed readings.

[13] One of the key postulates of Euclidean geometry is the Archimedean (separability) axiom. This axiom is, however, *not* logically necessary (Lorenzen, 1987, pp. 258 and 270), and in fact it cannot hold in any kind of physical geometry which incorporates the principle of irreducible indeterminacy.

[14] According to the original definition of informational completeness (Prugovečki, 1977a), a set of commuting observables with joint spectral measure [PQ] $E(B)$ is informationally complete if $\mathrm{Tr}(\rho_1 E) = \mathrm{Tr}(\rho_2 E)$ for all values assumed by the spectral measure E implies that the density operators ρ_1 and ρ_2 are equal. This definition can be generalized to POV measures associated with noncommuting observables (cf. [P], Sec. 3.2).

[15] And referred to as *proper wave functions* in [P], and all preceding publications on the GS approach. In this work we shall, however, avoid the use of this term for reasons explained in Note 21 of this chapter.

[16] The formulation of the equivalence principle which we adopt (cf., e.g., [M], p. 386) is sometimes called the *strong* equivalence principle (Carmeli, 1977; Dicke, 1964), to distinguish it from the *weak* equivalence principle (sometimes also called the Galilei-Eötvös principle – cf. Treder *et al.*, 1980, p. 15), which demands merely the equality of inertial and passive gravitational masses. An excellent analysis of

the various possible interpretations of the equivalence principle, as well as of their exact physical and mathematical meaning, is given by Friedmann (1983) – cf. especially Sec. V.4. We shall present, in Sec. 2.7, a mathematically precise formulation of this most fundamental principle of CGR.

17 The equivalence principle, and in fact the validity of all of CGR, was recently questioned in a series of papers by Logunov *et al.* (1986, 1988). However, their arguments, purporting to demonstrate that CGR is of "contradictory character", have no valid foundation, since they are merely resurrecting Einstein's "hole" argument which, together with its implications, are well-known to relativists – cf. Sec. 11.3.

18 There is no absolute or preferred time parameter in CGR, but there have been recent suggestions (cf. 11.2) to introduce a preferred "coordinate time" in quantum gravity in order to give an interpretation to the so-called "wave function of the universe". The GS procedure does not rely on such suggestions – which are not founded on either compelling epistemological arguments, or on any kind of experimental evidence. In fact, GS quantum gravity demonstrates (cf. Chapter 11) that such *ad hoc* assumptions are not in the least necessary to quantum gravity *per se*, which requires only a concept of *proper local* time.

19 Within such a formulation, relativistic causality is ensured from the beginning by the incorporation of the Klein-Gordon equation into the aforementioned Schrödinger equation, containing a fictitious-time parameter τ in addition to the four conventional spacetime variables – cf. Chapter 25 in [ST], or Appendix A in (Feynman, 1950); Sec. 5.7 contains an interpretation of τ within the GS framework.

20 The term *geometrodynamical exciton* was coined by Wheeler (1962, 1967) to describe so-called "spacetime foam" fluctuations, which in his theory represent a manifestation of the physical existence of a superspace of *classical* 3-geometries. This idea has inspired some recent work on quantum topology (Isham, 1988, 1990). To underline the distinction from the present approach, in which the spacetime fluctuations are of a geometro-stochastic origin, we refer to excitations occurring within the fibres of quantum geometries as *geometro-stochastic* excitons (cf. also Sec. 1.5).

21 Since the term "wave function" is often used as a synonym for "wave packet", in the present monograph we do not make use of the term *proper wave function* – although it was used in most of the previous publications on the GS approach to quantum theory. Instead, we shall use exclusively the term *proper state vector*, and thus avoid any possibility of confusion.

22 This would be in accord with the approach advocated by Weyl (1949), when he made the following statement: "A truly realistic mathematics should be conceived, in line with physics, as a branch of theoretical construction of the one real world, and should adopt the same sober and cautious attitude toward hypothetic extension of its foundations as is exhibited by physics." On the other hand, it should be observed that in the GS approach wave packets, as well as proper state vectors, are associated with single systems, rather than with ensembles – the latter being described in each fibre $\mathbf{F}_x$ by statistical or density operators [PQ] representing "mixed states". Thus, in GS theory a "wave packet" is a *geometro-stochastic wave function* resulting from the GS propagation of a proper state vector along all causally allowed geometro-stochastic paths in the base manifold of a quantum bundle – cf. Secs. 4.4, 5.4 and 5.6.

23 Many of these same incisive points were previously made by Heisenberg, and illustrated with examples, in a 1968 talk delivered at ICTP in Trieste, entitled "Theory, Criticism and Philosophy" – reproduced in (Salam, 1990), pp. 85-124. The influences of Einstein and Bohr are there explicitly acknowledged.

24 There have also been some recent suggestions (cf. Isham, 1990, §7) to abandon all differentiable structures in the description of spacetime, and to instead describe spacetime by a finite number of points.

25 In terms of Wigner's classification of *reality*, based on his observation that "excepting immediate sensations and, more generally, the content of my consciousness, everything is a construct", the present use of the term "reality" is related exclusively to Wigner's *universal reality of second kind*, whose relationship to an individual's consciousness constitutes "a continuous spectrum of reality of existence from absolute necessity for life to insignificance" (Wigner, 1964, pp. 254-256).

26 For example, a "particle track" in a Wilson chamber or a bubble chamber does not tell us where an exciton *is* in relation to the walls of that chamber, or other macroscopic points of reference, but merely where an exciton has *produced* a series of macroscopic effects – which we then simplistically call a "particle track". The quantum theory of measurement probabilistically correlates such phenomena to other experimental procedures performed with *macroscopic* objects – such as to the measurement procedures of the relative positioning of the macroscopic "target" and of the "particle accelerator" required to derive the magnitude of scattering angles and of cross-sections in a given scattering experiment.

Chapter 2

The Fibre Bundle Framework for Classical General Relativity

The formulation of general relativity given by Einstein in its final form in 1916 advanced the idea that spacetime should be described by a Lorentzian manifold $(\mathbf{M},g)$ – i.e., by a 4-dimensional manifold $\mathbf{M}$ carrying a Lorentzian metric g – which in the presence of gravitational sources would display non-zero curvature. The mathematical description of such manifolds and of associated tensor structures that was available to Einstein in the second decade of this century was formulated in $\mathbf{M}$-coordinate dependent language, so that its "coordinate-independence" was reflected only by the transformation laws for covariant and contravariant tensor components under changes of coordinates within $\mathbf{M}$. This established amongst physicists a tradition which has been maintained until recently in almost[1] all textbooks on classical general relativity and cosmology, despite the fact that already in the twenties E. Cartan (1923-25) had developed, for general relativity, a more general formulation of connection, curvature, and other basic differential geometric concepts in terms of "moving frames", and that this formulation later transcended into the present-day fibre bundle formulation of differential geometry. On the other hand, ever since the classic work by Kobayashi and Nomizu (1963, 1969), such fibre-theoretical formulations have become, amongst mathematicians, the standard mode for all presentations of differential geometry[2].

The advantages of the fibre-theoretical formulation of the geometry of Lorentzian manifolds are not only formal, but can also clarify fundamental physical principles. The best illustration of this fact is provided by Einstein's original formulation of the *general covariance principle* : "The general laws of nature are to be expressed by equations which hold good for all systems of co-ordinates, that is, are co-variant with any substitutions whatsoever" (Einstein, 1916, §3). However, soon after he advanced this formulation of one of the two basic principles of classical general relativity (CGR), Einstein (1918) had to concur with Kretschmann's (1917) comment that the mathematical formulation of *any* proposed law whatsoever can be written in a form "which holds good for all systems of co-ordinates" – a fact that would have been self-evident from the outset in the fibre-theoretical formulation of CGR. Furthermore, the fibre-theoretical formulation of CGR also makes it clear that the physical essence of Einstein's general covariance principle[3] is embedded into the requirement that all basic laws of CGR should be formulated in terms of geometric objects which can be expressed exclusively in terms of sections of bundles associated with Lorentz or Poincaré principal bundles over given differential manifolds $\mathbf{M}$. In conjunction with the equivalence principle, this leads to the conclusion that CGR can be viewed as being a gauge theory for the Lorentz group, or for the Poincaré[4] group. Such a formulation also makes it clear that statements to the effect that the gauge group of CGR is the diffeomorphism group of **M** are based on a different interpretation of the meaning of the term "gauge group" – cf. Note 30 to this chapter, as well as Secs. 10.2 and 11.3.

The fibre bundle framework of CGR outlined in the present chapter results from the straightforward application of the modern formulation of differential geometry [C,I,NT, SC] to the special case of four-dimensional differential manifolds with Lorentzian metric. Readers already familiar with such formulations, as well as with the basic physical ideas of CGR [M,W], might therefore prefer to proceed directly to Chapter 3.

2.1. Tensor Bundles over Four-Dimensional Differential Manifolds

In CGR a model for a given spacetime is described by specifying a Lorentzian manifold $(\mathbf{M},g)$, from which all other relevant physical and mathematical information can be, in principle, deduced. The specification of such a manifold requires, first of all, the specification of a 4-dimensional differentiable manifold **M**. In the present section we shall therefore deal with those basic geometric objects that can be associated with any such manifold even in the absence of a metric structure, specified by some metric tensor g. We shall concentrate on the 4-dimensional case, which is of primary interest in CGR, but all our definitions and considerations remain actually valid for any finite-dimensional manifold.

By definition [I,N], a set **M** is a *4-dimensional differentiable* (or *differential*) *manifold* if it is a topological space[5] which can be covered with a family of open sets M_α, and if there is a homeomorphism between each M_α and some open subset O_α in $\mathbf{R}^4$,

$$\phi_\alpha:\ x \mapsto (x^0,x^1,x^2,x^3) \in O_\alpha \subset \mathbf{R}^4, \qquad x \in M_\alpha \subset \mathbf{M}, \tag{1.1}$$

which assigns coordinates x^μ, $\mu = 0,1,2,3,$ to each $x \in M_\alpha$ in such a manner that the family of all *charts* (M_α,ϕ_α), called an *atlas*, displays the following basic property: on any of the open subsets $M_\alpha \cap M_\beta$ of **M** in which to each $x \in M_\alpha \cap M_\beta$ is assigned two sets of coordinates, namely those in (1.1), as well as those assigned by some other chart (M_β,ϕ_β) by means of the homeomorphism

$$\phi_\beta:\ x \mapsto (x'^0,x'^1,x'^2,x'^3) \in O_\beta \subset \mathbf{R}^4, \qquad x \in M_\beta \subset \mathbf{M}, \tag{1.2}$$

the *coordinate-transformation maps*

$$\phi_\beta \circ \phi_\alpha^{-1}:\ \phi_\alpha(M_\alpha \cap M_\beta) \to \phi_\beta(M_\alpha \cap M_\beta) \subset O_\beta, \tag{1.3}$$

are *smooth*, i.e., the four functions in the map

$$\phi_\beta \circ \phi_\alpha^{-1}:\ (x^0,x^1,x^2,x^3) \mapsto (x'^0,x'^1,x'^2,x'^3),$$

$$(x^0,x^1,x^2,x^3) \in \phi_\alpha(M_\alpha \cap M_\beta) \subset O_\alpha, \tag{1.4}$$

that express the new coordinates $x'^\nu, \nu = 0,1,2,3,$ in terms of the old coordinates x^μ, $\mu =$ 0,1,2,3, are infinitely differentiable, namely possess partial derivatives of any order[6].

On account of this last differentiability requirement, we can say that a real-valued function $f(y)$ defined on some neighborhood $\mathcal{N}_x$ of x (i.e., on an open set containing x) is *smooth* if it is such that, when coordinates are assigned to $y \in \mathcal{N}_x$ by a given chart (M_α,ϕ_α) with domain $M_\alpha \supset \mathcal{N}_x$, the resulting function of four real variables is smooth. Indeed, by the chain rule, this statement remains true upon the transition, by means of (1.4), to any other chart (M_β,ϕ_β) with $M_\beta \supset \mathcal{N}_x$. Similarly, the statement that a curve $\gamma = \{x(t)|\, a \le t \le b\}$ passing through a point $x \in \mathbf{M}$ is smooth can be defined to mean that, in some chart whose domain contains some neighborhood $\mathcal{N}_x$ of x, the coordinates $x^\mu(t)$, $\mu = 0,1,2,3$, are smooth functions of $t \in [a,b]$ – since again this statement remains true in any other chart whose domain contains some neighborhood of x.

The coordinate changes in (1.4) give rise to corresponding changes in the components of vectors and tensors over **M**. Hence, the explicit formulation of their transformation properties in terms of the derivatives of the coordinate transition functions in (1.3) was deemed to be mandatory in the approach to differential geometry originally adopted by Einstein (1916). That trend was also followed by most textbooks on CGR (cf., e.g., [N], Chapter 6) – and we shall call it the **M**-*coordinate-dependent formulation* of structures associated with a differential manifold. The above definition of smooth functions and curves enables, however, the modern approach to differential geometry [C,I,NT,SC] – to which we shall refer as the **M**-*coordinate-independent formulation* of the same structures.

In this latter formulation a vector X tangent to a smooth curve $\gamma = \{x(t)|\, a \le t \le b\}$ passing through a point $x \in \mathbf{M}$ is defined in a coordinate-independent manner by the following procedure.

Let us choose a parametrization $x(t)$ of γ such that $x(0) = x$. We observe that, in view of the above definition of a smooth real-valued function $f(y)$, $y \in \mathcal{N}_x$, the function $f(x(t))$ is differentiable in $t \in (a,b)$. Consequently, the following map

$$X:\ f \mapsto \left.\frac{df(x(t))}{dt}\right|_{t=0} \in \mathbf{R}^1 , \qquad x(0) = x , \tag{1.5}$$

exists and determines, for each given smooth curve γ, a functional on the family of all smooth functions defined on various neighborhoods of $x \in \mathbf{M}$ by assigning to each such function f a real number. The real-valued functional X in (1.5) is then called the 4-vector tangent to γ at the point $x \in \gamma$. The family $T_x\mathbf{M}$ of all tangent vectors, defined by this coordinate-independent construction for all smooth curves passing through $x \in \mathbf{M}$, is called the *tangent space* of **M** at the point x.

The relationship between this rather abstract definition of tangent vectors and the specification of "contravariant vectors" in the **M**-coordinate-dependent formulation, whereby such vectors are defined in terms of their components X^μ in various vector frames, can be easily established upon choosing in some neighborhood $\mathcal{N}_x$ of x a chart (M_α,ϕ_α) with $M_\alpha \supset \mathcal{N}_x$, and then singling out from the tangent space $T_x\mathbf{M}$ the following set of vectors, which constitute a *vector basis* or *linear frame* in $T_x\mathbf{M}$,

$$\{\partial_\mu \mid \mu = 0,1,2,3\} , \qquad \partial_\mu = \left(\frac{\partial}{\partial y^\mu}\right)_{y=x} :\ f \mapsto \left.\frac{\partial f(y)}{\partial y^\mu}\right|_{y=x} . \tag{1.6}$$

Indeed, upon labelling in (1.5) the points $x(t)$ of the curve γ by means of the coordinates supplied in (1.1) by that chart, and then using the chain rule for the differentiation of $f(x(t))$ with respect to t, we easily see that we can expand any vector $X \in T_x\mathbf{M}$ in terms of this vector basis as follows[7]:

$$X = X^\mu \partial_\mu \in T_x\mathbf{M}, \qquad X^\mu = \left(\frac{dx^\mu}{dt}\right)_{t=0} . \tag{1.7}$$

Hence, $T_x\mathbf{M}$ is indeed a 4-dimensional real vector space, and the coordinates of all the vectors in it are supplied by the derivatives of the above coordinate-functions of the parameter t at points on corresponding smooth curves γ. In particular, if γ is the worldline of a massive particle and t is its proper time, then the components of $X \in T_x\mathbf{M}$ coincide with those of its 4-velocity at $x \in \mathbf{M}$, so that X itself can be identified with the 4-velocity vector $\boldsymbol{v}$, despite the rather abstract nature of the mathematical definition supplied by (1.5).

An $\mathbf{M}$-coordinate-independent definition of "covariant vectors" [N] can be now obtained by introducing the *cotangent space* $T_x{}^*\mathbf{M}$ above x as the algebraic dual of $T_x\mathbf{M}$ – i.e., as consisting of all real-valued linear functionals $\boldsymbol{\omega}$ over $T_x\mathbf{M}$. An equivalent definition of cotangent space can be also given in terms of the family of all smooth real-valued functions defined on some neighborhood $\mathcal{N}_x$ of x, which forms the basis of the definition (1.5), by introducing for each element f in that family the following linear maps:

$$df:\ X \mapsto Xf \in \mathbf{R}^1, \qquad X \in T_x\mathbf{M} . \tag{1.8}$$

For the special case of $f(y) = y^\mu$, $\mu = 0,\dots,3$, where y^μ are the coordinates assigned by (1.1) to all the points $y \in M_\alpha \supset \mathcal{N}_x$, we obtain in accordance with (1.6) and (1.7)

$$d^\mu = \left(dy^\mu\right)_{y=x}:\ X \mapsto \left(Xy^\mu\right)_{y=x} = X^\mu , \qquad X \in T_x\mathbf{M} . \tag{1.9}$$

Hence $\{d^\mu \mid \mu = 0,\dots,3\}$ constitutes a *covector basis* or *linear coframe* in $T_x{}^*\mathbf{M}$. This covector basis is dual to the vector basis $\{\partial_\mu \mid \mu = 0,\dots,3\}$ in $T_x\mathbf{M}$, so that any *covector* $\boldsymbol{\omega}$ can be expanded as follows:

$$\boldsymbol{\omega} = \omega_\mu d^\mu , \qquad d^\mu(\partial_\nu) = \delta^\mu_\nu \;\Rightarrow\; \omega_\mu = \boldsymbol{\omega}(\partial_\mu), \qquad \boldsymbol{\omega} \in T^*_x\mathbf{M} . \tag{1.10}$$

We observe that in modern presentations of classical mechanics (Abraham and Marsden, 1978), the momentum of a classical particle is treated as an element of the cotangent space above a given point. Therefore, in CGR we shall accordingly consider that the 4-momenta $\boldsymbol{p}$ of particles at $x \in \mathbf{M}$ assume values in $T_x{}^*\mathbf{M}$.

The linear space of all *tensors of type* (r,s) above $x \in \mathbf{M}$ can be now defined [I,N] in the following manner,

$$T_x^{r,s}\mathbf{M} = \left[\bigotimes_{m=1}^{r} \left(T_x\mathbf{M}\right)_m\right] \otimes \left[\bigotimes_{n=1}^{s} \left(T_x^*\mathbf{M}\right)_n\right] , \tag{1.11}$$

namely by taking the algebraic tensor product of r copies of $T_x\mathbf{M}$ and s copies of $T_x{}^*\mathbf{M}$. Hence, any (r,s) tensor $T \in T_x^{r,s}\mathbf{M}$ can be expanded in terms of the vector basis elements in (1.6) and of the dual covector basis elements in (1.10):

$$T = T^{\mu_1 \cdots \mu_r}_{\nu_1 \cdots \nu_s} \partial_{\mu_1} \otimes \cdots \otimes \partial_{\mu_r} \otimes d^{\nu_1} \otimes \cdots \otimes d^{\nu_s} \ . \tag{1.12}$$

We note that, in the absence of a metric on $\mathbf{M}$, the well-known operation [M,W] of raising and lowering tensor indices is undefined, so that there is no need to leave vacant the spaces below *contravariant* (i.e., upper) or above *covariant* (i.e., lower) indices.

We can now introduce the following sets:

$$T\mathbf{M} = \bigcup_{x\in\mathbf{M}} T_x\mathbf{M}\,, \qquad T^*\mathbf{M} = \bigcup_{x\in\mathbf{M}} T_x^*\mathbf{M}\,, \qquad T^{r,s}\mathbf{M} = \bigcup_{x\in\mathbf{M}} T_x^{r,s}\mathbf{M} \ . \tag{1.13}$$

We shall show that these sets assume the structure of manifolds if, for each chart (M_α,ϕ_α) in the manifold $\mathbf{M}$, a corresponding chart is constructed in $T\mathbf{M}$, $T^*\mathbf{M}$ and $T^{r,s}\mathbf{M}$, by assigning the coordinates in (1.7), (1.10) and (1.12) to all the elements of, respectively, $T_x\mathbf{M}$, $T_x^*\mathbf{M}$ and $T_x^{r,s}\mathbf{M}$, at all $x \in M_\alpha$. Thus, for example, the chart (M_α,ϕ_α) in $\mathbf{M}$ gives rise to a chart (TM_α,ψ_α) within $T\mathbf{M}$, where TM_α is obtained by taking the union of all tangent spaces with $x \in M_\alpha$, whereas ψ_α is given by

$$X \ \mapsto \ (x^0,x^1,x^2,x^3,X^0,X^1,X^2,X^3) \in O_\alpha \times \mathbf{R}^4\,, \qquad X \in T_x\mathbf{M} \ . \tag{1.14}$$

These manifolds are called, respectively, the *tangent bundle* $T\mathbf{M}$, the *cotangent bundle* $T^*\mathbf{M}$, and the *(r,s) tensor bundle* $T^{r,s}\mathbf{M}$ over the given 4-dimensional differential manifold $\mathbf{M}$. Indeed, in the next section, after introducing the general concept of fibre bundle, we shall see that they all represent instances of associated bundles. On the other hand, it is obvious already at this stage that in general $T^{r,s}\mathbf{M}$ is a manifold of dimension $4 + 4^{r+s}$, and that $T\mathbf{M}$ and $T^*\mathbf{M}$ coincide with the *(1,0)* and the *(0,1)* tensor bundles, respectively, so that they are both 8-dimensional manifolds.

To establish that these bundles are indeed differential manifolds, we have to establish that the present counterparts of the coordinate transformation maps in (1.3) are smooth functions. Since these counterparts consist of the coordinate transformation functions in (1.3), plus the ones for tensor components, we only have to establish the smoothness of the latter.

As a straightforward consequence of (1.6) and (1.9)-(1.10), we find that the coordinate transformation maps in (1.3) give rise to the following change of bases in $T_x\mathbf{M}$ and $T_x{}^*\mathbf{M}$, respectively,

$$\partial'_{\mu'} = L_{\mu'}{}^{\mu}\partial_\mu\,, \qquad \partial'_{\mu'} = \left(\partial\big/\partial y'^{\mu'}\right)_{y=x}, \qquad L_{\mu'}{}^{\mu} = \left(\partial y^{\mu}\big/\partial y'^{\mu'}\right)_{y=x}, \tag{1.15a}$$

$$d'^{\nu'} = L^{\nu'}{}_{\nu}d^{\nu}\,, \qquad d'^{\nu'} = (dy'^{\nu'})_{y=x}\,, \qquad L^{\nu'}{}_{\nu} = \left(\partial y'^{\nu'}\big/\partial y^{\nu}\right)_{y=x} \ . \tag{1.15b}$$

From the duality relation in (1.10), and its counterpart in the primed bases, we immediately obtain:

$$L_{\mu}{}^{\lambda} L^{\nu}{}_{\lambda} = L_{\lambda}{}^{\nu} L^{\lambda}{}_{\mu} = \delta^{\nu}_{\mu} \quad . \tag{1.16}$$

Hence, it immediately follows from (1.12), and from the corresponding expansion in the primed bases defined in (1.15), that the components of the tensor T in (1.12) transform according to the following rule, which is well-known from the **M**-coordinate-dependent formulation of the concept of tensor [N]:

$$T'^{\mu'_1 \dots \mu'_r}_{\nu'_1 \dots \nu'_s} = L^{\mu'_1}{}_{\mu_1} \cdots L^{\mu'_r}{}_{\mu_r} L_{\nu'_1}{}^{\nu_1} \cdots L_{\nu'_s}{}^{\nu_s} T^{\mu_1 \dots \mu_r}_{\nu_1 \dots \nu_s} \quad . \tag{1.17}$$

Since $L_{\mu'}{}^{\mu}$ and $L^{\nu'}{}_{\nu}$ in (1.15) are smooth functions in $M_\alpha \cap M_\beta$, the smoothness of the primed (i.e., transformed) tensor components in (1.17), regarded as functions of the unprimed ones, is thus established. Hence, all the sets in (1.13) indeed become manifolds under the earlier made choice of coordinate charts. They also exhibit a fibre bundle structure, to whose formulation and study we turn next.

2.2. General Linear Frame Bundles over Four-Dimensional Manifolds

A *bundle* is a triple (**E**, π, **M**) consisting of a topological space **E**, called its *total space* or *bundle space*, a subspace of **E** homeomorphic to a given topological space **M**, called its *base space*, and a continuous map π of **E** onto **M**, called its *projection map*. For each $x \in \mathbf{M}$ the set $\pi^{-1}(x)$ of all points in the bundle space **E** mapped by the projection map π into x is called the *fibre* above x. A bundle is called a *fibre bundle* if all its fibres are homeomorphic, or equivalently, if there is a homeomorphism between each fibre $\pi^{-1}(x)$ and a fixed topological space **F**, called the *typical fibre* or *standard fibre* of that fibre bundle. We shall follow the general custom and occasionally denote a fibre bundle with total space **E** by the same symbol **E**, as long as it is clear from the context what are its base manifold and its projection map. We note, however, that in general there might be a great variety of *fibrations* of a given differential manifold **E**, i.e., of choices of **M** and π decomposing the manifold **E** into a base manifold and fibres.

A fibre bundle (**E**, π, **M**, **F**) which is homeomorphic to the (topological direct) product $\mathbf{M}\times\mathbf{F}$ of its base manifold **M** and its typical fibre **F** is called *trivial*. Thus $\mathbf{R}^n$ provides, for any $n > 2$, an example of a manifold which can be fibrated in a variety of ways, corresponding to setting $n = k+m$, into a trivial bundle with $\mathbf{M} = \mathbf{R}^k$ and with $\mathbf{F} = \mathbf{R}^m$. There are also many very elementary examples of bundles which are not trivial, and which share the same total space [C,I,NT,SC].

In this chapter we shall be concerned exclusively with C^∞-*bundles*[8] (**E**, π, **M**), in which case the total space **E** and the base space **M** are required to be differential manifolds, with **M** being diffeomorphic[9] to a submanifold of **E**, and the projection map π is required to be smooth. A C^∞-bundle is called a C^∞-*fibre bundle* if there is a diffeomorphism between each one of its fibres $\pi^{-1}(x)$ and a typical fibre **F**.

In the case of the manifolds with the total spaces (1.13) defined in the last section, the natural manner of fibrating them becomes evident once their definition is considered: they all share the manifold **M** as their base manifold, and, upon adopting the respective choices of $T_x\mathbf{M}$, $T_x^*\mathbf{M}$ and $T_x^{r,s}\mathbf{M}$ as fibres above each $x \in \mathbf{M}$, they obviously become fibre bundles, whose typical fibre equals $\mathbf{R}^4$ in the cases of tangent and cotangent bundles, and it equals $\mathbf{R}^{4(r+s)}$ in the generic case of (r,s)-tensor bundles. However, the manner in which the vector coordinates in (1.7), the covector coordinates in (1.10), and tensor coordinates in (1.12) transform with the change of bases in their fibres identifies them, in addition, as bundles *associated* with the same *principal bundle of linear frames* over **M**.

The total space[10] $GL\mathbf{M}$ of this latter bundle, to which we shall refer as the *general linear frame bundle* over **M**, consists of all linear frames $\hat{e}(x)$ at all points $x \in \mathbf{M}$, where, by definition, a *linear frame* at x is an ordered set of four linearly independent vectors e_i, $i = 0,1,2,3$, which belong to $T_x\mathbf{M}$, so that in some chart (M_α, ϕ_α) over a neighborhood of x,

$$\hat{e}(x) = (e_0, e_1, e_2, e_3) \ , \qquad e_i = \lambda_i{}^\mu(x)\partial_\mu \in T_x\mathbf{M} \ . \tag{2.1}$$

Each linear frame has a dual frame, called a *linear coframe*, which in the case of (2.1) is

$$\hat{\partial}(x) = (\partial_0, \partial_1, \partial_2, \partial_3) \ , \qquad \partial_\mu \in T_x\mathbf{M} \ . \tag{2.2}$$

Clearly, **M** is the base manifold of this general linear frame bundle, and its fibre $\mathbf{L}_x$ above each $x \in \mathbf{M}$ consists of all linear frames in $T_x\mathbf{M}$. Indeed, the set $\mathbf{L}_x$ becomes a manifold if its elements are parametrized by the sixteen real numbers $\lambda_i{}^\mu(x)$, $i,\mu = 0,\ldots,3$, which assume the role of coordinates. All these fibres are then diffeomorphic to the general linear group GL(4,**R**) of all non-singular 4×4 real matrices. Indeed, if we view $\|\lambda_i{}^\mu(x)\|$ as a 4×4 matrix, then there obviously is a one-to-one correspondence between all $\hat{e}(x) \in \mathbf{L}_x$ and all elements of GL(4, **R**), which is smooth in both directions. Hence, the general linear frame bundle $GL\mathbf{M}$ is a fibre bundle with typical fibre equal to GL(4,**R**). Furthermore, an element A of GL(4, **R**) acts in a natural manner, from the right, on any linear frame $\hat{e}(x)$ if the 4-tuple of vectors in (2.1) is viewed as a single-row matrix, i.e.,

$$\hat{e}(x)A = (e'_0, e'_1, e'_2, e'_3) \ , \qquad e'_i = e_j A^j{}_i \ , \qquad \det A \neq 0 \ , \tag{2.3}$$

where $A^i{}_j$ denotes the (ij)-th element in the matrix A. If all the vectors e_i in (2.1) are smooth functions of $x \in M_\alpha$ in the sense that the functions $\lambda_i{}^\mu$ are smooth in M_α, then the elements of the linear frame in (2.3) are obviously smooth on $M_\alpha \times$ GL(4, **R**), i.e., they are smooth functions of the coordinates of $x \in M_\alpha$ as well as of the matrix elements $A^i{}_j$ of A. This establishes that the general linear frame bundle $GL\mathbf{M}$ is a principal bundle with structure group GL(4,**R**), since it conforms to the following general definition of a principal bundle.

A fibre bundle (**P**, $\boldsymbol{\Pi}$, **M**, **G**) is a *principal bundle* over **M**, with projection $\boldsymbol{\Pi}$ and structure group **G**, if the elements g of **G** act on all the elements $\boldsymbol{u}$ of the manifold **P** from the right[11] in such a manner that: 1) the map $(\boldsymbol{u},g) \mapsto \boldsymbol{u} \cdot g$ defining this action from the right is a smooth map[12] from $\mathbf{P} \times \mathbf{G}$ into **P**; 2) $(\boldsymbol{u} \cdot g') \cdot g'' = \boldsymbol{u} \cdot (g'g'')$ for all $\boldsymbol{u} \in \mathbf{P}$ and all $g', g'' \in \mathbf{G}$; 3) all the fibres are invariant under the action of **G**, i.e., each fibre element $\boldsymbol{u} \in \boldsymbol{\Pi}^{-1}(x)$ is mapped into an element $\boldsymbol{u} \cdot g \in \boldsymbol{\Pi}^{-1}(x)$ belonging to the same fibre; 4) the action of the structure group **G** on each fibre $\boldsymbol{\Pi}^{-1}(x)$ is transitive, i.e., all the elements within a given fibre $\boldsymbol{\Pi}^{-1}(x)$ are obtained by the action of all the elements of **G** on any given element of that fibre $\boldsymbol{\Pi}^{-1}(x)$ – so that **G** can be identified with its typical fibre.

To define the concept of a fibre bundle **E** over **M** which is *associated* with a principal bundle **P** over **M**, we shall first introduce the concept of bundle coordinates[13]. This can be achieved by adapting to fibre bundles the method of introducing coordinates for manifolds, that was presented in the previous section. Thus, let us assume that the base manifold **M** can be covered with a family of open subsets $\mathbf{M}_\alpha$ (which as such are submanifolds of **M**, not necessarily related to the sets M_α on which all the charts were defined in the preceding section), and that for each $\mathbf{M}_\alpha$ there is a diffeomorphism, called a *local trivialization map*[14],

$$\boldsymbol{\phi}_\alpha \colon u \mapsto (x,f) \in \mathbf{M}_\alpha \times \mathbf{F} , \qquad u \in \boldsymbol{\pi}^{-1}(\mathbf{M}_\alpha) \subset \mathbf{E} \ , \tag{2.4}$$

which assigns to each element u in any of the fibres of **E** above $\mathbf{M}_\alpha$ the *fibre coordinates* (x,f). In that case the restriction of $\boldsymbol{\phi}_\alpha$ to each fibre $\boldsymbol{\pi}^{-1}(x)$ over any $x \in \mathbf{M}_\alpha$ determines a diffeomorphism between that fibre and the typical fibre **F**, so that, in turn, for each $x \in \mathbf{M}_\alpha \cap \mathbf{M}_\beta$,

$$\boldsymbol{\phi}_\beta \circ \boldsymbol{\phi}_\alpha^{-1} \colon (\mathbf{M}_\alpha \cap \mathbf{M}_\beta) \times \mathbf{F} \to (\mathbf{M}_\alpha \cap \mathbf{M}_\beta) \times \mathbf{F} \ , \tag{2.5}$$

determines a diffeomorphism,

$$\mathbf{g}_{\beta\alpha}(x) \colon \ \mathbf{F} \to \mathbf{F} \ , \qquad x \in \mathbf{M}_\alpha \cap \mathbf{M}_\beta \ , \tag{2.6}$$

of the typical fibre **F** onto itself – i.e., an automorphism of that typical fibre. It is easy to see that these automorphisms, constituting the *transition* or *structure functions* of the fibre bundle **E**, satisfy the following *compatibility relations*:

$$\mathbf{g}_{\alpha\beta}(x) = \left(\mathbf{g}_{\beta\alpha}(x)\right)^{-1} \ , \qquad x \in \mathbf{M}_\alpha \cap \mathbf{M}_\beta \ , \tag{2.7a}$$

$$\mathbf{g}_{\alpha\beta}(x)\mathbf{g}_{\beta\gamma}(x) = \mathbf{g}_{\alpha\gamma}(x) \ , \qquad x \in \mathbf{M}_\alpha \cap \mathbf{M}_\beta \cap \mathbf{M}_\gamma \ . \tag{2.7b}$$

If we consider the above compatibility relations in conjunction with the obvious fact that $\mathbf{g}_{\alpha\alpha}(x)$ represents the identity map of **F** onto itself at each $x \in \mathbf{M}$, we discern the underlying presence of a group structure. This is explicitly realized in the context of the following definition.

A fibre bundle (**E**, $\boldsymbol{\pi}$, **M**, **F**), with typical fibre **F**, is said to be *associated to the principal bundle* (**P**, $\boldsymbol{\Pi}$, **M**, **G**) through the representation U of the group **G** by means of diffeomorphisms $U(g): \mathbf{F} \to \mathbf{F}$, $g \in \mathbf{G}$, acting on the typical fibre of **E** from the left if, given any family of local trivialization maps Φ_α of the principal bundle **P** such that

$$\Phi_\alpha: u \mapsto (x,g) \in \mathbf{M}_\alpha \times \mathbf{G}, \qquad u \in \Pi^{-1}(\mathbf{M}_\alpha) \subset \mathbf{P}, \qquad \mathbf{M} = \bigcup \mathbf{M}_\alpha \quad , \tag{2.8}$$

there is an associated family of local trivialization maps (2.4) of **E** such that[15]

$$\mathbf{g}_{\alpha\beta}(x) = U\big(g_{\alpha\beta}(x)\big) \ , \qquad x \in \mathbf{M}_\alpha \cap \mathbf{M}_\beta , \qquad g_{\alpha\beta}(x) \in \mathbf{G} \quad , \tag{2.9}$$

for all intersecting submanifolds $\mathbf{M}_\alpha$ and $\mathbf{M}_\beta$, where $g_{\alpha\beta}(x)$ is the structure group element which is obtained by restricting the bundle coordinate transition maps

$$\Phi_\alpha \circ \Phi_\beta^{-1}: (\mathbf{M}_\alpha \cap \mathbf{M}_\beta) \times \mathbf{G} \to (\mathbf{M}_\alpha \cap \mathbf{M}_\beta) \times \mathbf{G} \tag{2.10}$$

of the principal bundle **P** to the given $x \in \mathbf{M}_\alpha \cap \mathbf{M}_\beta$.

Naturally, in accordance with the above definition, every principal bundle is associated to itself. Furthermore, if two principal fibre bundles are associated to each other, then they are deemed to be equivalent. For example, the general linear frame bundle $GL\mathbf{M}$ is equivalent to the principal fibre bundle **P(M,GL(4, R))** whose fibres are actually equal (rather than just diffeomorphic) to GL(4, **R**). In general, any two fibre bundles associated with the same principal bundle **P** are called *equivalent* if and only if for any given family of local trivialization maps Φ_α of the principal bundle **P** there are diffeomorphisms $\boldsymbol{\lambda}_\alpha(x)$ of their respective fibres above each $x \in \mathbf{M}_\alpha$, so that their transition functions are related as follows:

$$\mathbf{g}'_{\alpha\beta}(x) = \boldsymbol{\lambda}_\alpha^{-1}(x) \circ \mathbf{g}_{\alpha\beta}(x) \circ \boldsymbol{\lambda}_\beta(x) \ , \qquad x \in \mathbf{M}_\alpha \cap \mathbf{M}_\beta \quad . \tag{2.11}$$

The tangent bundle $T\mathbf{M}$ and the cotangent bundle $T^*\mathbf{M}$ are examples of equivalent (but not identical) bundles associated with the principal bundle $GL\mathbf{M}$.

To see this, let us consider any two local trivializations Φ_α and Φ_β of $GL\mathbf{M}$. Thus, let us say that Φ_α is obtained by constructing, from the coefficients $\lambda_i{}^\mu(x)$ in (2.1), the corresponding matrices $\lambda \in$ GL(4, **R**) at all points $x \in \mathbf{M}_\alpha$, so that we arrive at the map

$$\Phi_\alpha: \hat{e} \mapsto (x,\lambda) \in \mathbf{M}_\alpha \times \mathrm{GL}(4,\mathbf{R}), \qquad \hat{e} = (e_i) \in \Pi^{-1}(\mathbf{M}_\alpha) \subset GL\mathbf{M} \quad , \tag{2.12}$$

where, in accordance with (2.3), $\lambda_i{}^\mu(x)$ is the (μi)-th element of λ; similarly, let us assume that Φ_β is obtained by constructing from the coefficients in the expansion of the same vector frame elements in some other chart,

$$\hat{e}(x) = (e_0, e_1, e_2, e_3) \ , \qquad e_i = \lambda'_i{}^\mu(x)\partial'_\mu \in T_x\mathbf{M} \quad , \tag{2.13}$$

the corresponding matrices $\lambda' \in$ GL(4, **R**) at all points $x \in \mathbf{M}_\beta$, so that

$$\boldsymbol{\Phi}_\beta:\ \hat{e}\mapsto(x,\lambda')\in\mathbf{M}_\beta\times\mathrm{GL}(4,\mathbf{R}),\qquad \hat{e}\in\boldsymbol{\Pi}^{-1}(\mathbf{M}_\beta)\subset GL\mathbf{M}\quad . \tag{2.14}$$

Then these matrices constitute, for any $x\in\mathbf{M}_\alpha\cap\mathbf{M}_\beta$, the fibre coordinates of each given $\hat{e}\in\boldsymbol{\Pi}^{-1}(x)$. Those matrices obviously belong to the typical fibre GL(4, **R**) of $GL\mathbf{M}$. Hence, on account of (1.15)-(1.16), and of the action from the right of GL(4, **R**) upon $GL\mathbf{M}$, the corresponding transition function $g_{\beta\alpha}$ between these two sets of fibre coordinates, which results from (2.10), is given by the maps

$$g_{\beta\alpha}(x):\ \lambda\ \mapsto\ \lambda'=\lambda L^{\mathrm{T}}\ ,\quad L=\left\|L^{\mu'}{}_{\mu}\right\|,\quad L^{\mathrm{T}}=\left\|(L^{\mathrm{T}})_{\mu}{}^{\mu'}\right\|,\quad (L^{\mathrm{T}})_{\mu}{}^{\mu'}=L^{\mu'}{}_{\mu}\ . \tag{2.15}$$

On the other hand, if we expand the same vector X in the bases introduced in (1.6) and (1.15a), respectively, we see that the local trivialization maps

$$\phi_\alpha:\ X\mapsto(x,\mathsf{X})\in\mathbf{M}_\alpha\times\mathbf{R}^4\ ,\quad X\in\boldsymbol{\pi}^{-1}(\mathbf{M}_\alpha)\subset T\mathbf{M}\ ,\quad \mathsf{X}=(X^0,\ldots,X^3)\ , \tag{2.16a}$$

$$\phi_\beta:\ X\mapsto(x,\mathsf{X}')\in\mathbf{M}_\beta\times\mathbf{R}^4\ ,\quad X\in\boldsymbol{\pi}^{-1}(\mathbf{M}_\beta)\subset T\mathbf{M}\ ,\quad \mathsf{X}'=(X'^0,\ldots,X'^3)\ , \tag{2.16b}$$

give rise to the 4-tuples $\mathsf{X},\mathsf{X}'\in\mathbf{R}^4$ representing its respective bundle coordinates within the typical fibre $\mathbf{R}^4$ of the tangent bundle $T\mathbf{M}$. Furthermore, if we treat these bundle coordinate 4-tuples as one-column matrices (so that GL(4, **R**) can act on them from the left), then the vector bundle coordinates transform in accordance with

$$\mathsf{g}_{\beta\alpha}(x)=U\big(g_{\beta\alpha}(x)\big)\ :\ \mathsf{X}\ \mapsto\ \mathsf{X}'=\ L^{-1}\mathsf{X}\ , \tag{2.17}$$

i.e., in accordance with the adjoint representation $U{:}\ A\ \mapsto\ (A^{-1})^{\mathrm{T}}$ of the general linear group GL(4, **R**) considered as a group of transformations acting from the left in $\mathbf{R}^4$.

A totally analogous consideration for the cotangent bundle $T^*\mathbf{M}$ can be carried out by using the covector bases in (1.10) and (1.15b) – which are dual to the vector bases in (1.6) and (1.15a), respectively – and leads to the conclusion that $T^*\mathbf{M}$ is associated to $GL\mathbf{M}$ by the fundamental representation $U{:}\ A\ \mapsto\ A$ of GL(4, **R**). In view of the basic relationships (1.11)-(1.13) leading to the definition of the (r,s)-tensor bundle $T^{r,s}\mathbf{M}$, it then becomes evident that $T^{r,s}\mathbf{M}$ is associated to $GL\mathbf{M}$ by the following tensor representation of GL(4, **R**),

$$U{:}\ A\ \mapsto\ (A^{-1})^{\mathrm{T}}\otimes\cdots\otimes(A^{-1})^{\mathrm{T}}\otimes A\otimes\cdots\otimes A\ , \tag{2.18}$$

in which $(A^{-1})^{\mathrm{T}}$ obviously occurs r-times, and A occurs s-times.

A very useful concept in the theory of fibre bundles is that of section. By definition, a *(local) section* of a fibre bundle **E** is a smooth map $s:x\ \mapsto\ u\in\mathbf{F}_x$ which assigns, to each element x from some open neighborhood of the base manifold **M**, a unique element $u\in\mathbf{E}$ in the fibre above it; a section is called *global*, or a *cross-section* of the bundle, if that neighborhood equals all of **M**, i.e., if it is defined on the entire base manifold **M**.

We note that the concept of cross-section of the tangent bundle over a manifold **M** coincides with that of (global) vector field over **M**, and more generally the concept of cross-section of a tensor bundle over **M** coincides with that of (global) tensor field over **M**.

It is quite easy to establish that a principal bundle which possesses a cross-section is trivial, and that, furthermore, in that case all its associated bundles are also trivial, i.e., they are diffeomorphic to the products $\mathbf{M}\times\mathbf{F}$ of their base manifold **M** and their typical fibres **F** [C,I,NT]. However, as we shall see in the next section, when we formulate Geroch's theorem, in the context of CGR the triviality of such bundles does not imply in the least that either the mathematics or the physics to which they might give rise is in any fundamental sense 'trivial'.

The concept of section of the general linear frame bundle $GL\mathbf{M}$ coincides with Cartan's concept of *moving frame* [SC] in the differential manifold **M**, and leads to a very useful generalization of the concept of vector and tensor field coordinate components, as usually defined in the **M**-coordinate dependent formulations of CGR [N]. Indeed, a moving frame $\hat{\mathbf{e}}(\mathcal{N})$ over some open neighborhood $\mathcal{N}$ in **M** is in general obtained by choosing in (2.1) functions $\lambda_i{}^{\mu}(x)$, $i,\mu = 0,\ldots,3$, which are smooth over $\mathcal{N}$. This gives rise to moving frames in terms of which any vector field in that neighborhood can be expanded. A special case of such moving frames are the *holonomic frames* $\hat{\boldsymbol{\partial}}(\mathcal{N})$ obtained, for neighborhoods $\mathcal{N}$ which lie within the domain of definition of a single chart, by introducing at each $x\in\mathcal{N}$ the linear frame in (2.2), consisting of the vectors which are tangent to the coordinates lines defined by that chart. However, when the manifold **M** is curved, there will be in $\mathcal{N}$ many *nonholonomic frames*, i.e., moving frames $\hat{\mathbf{e}}(\mathcal{N})$ for which there are no charts in $\mathcal{N}$ such that $\lambda_i{}^{\mu}(x) = \delta_i^{\mu}$ at all[16] $x\in\mathcal{N}$. In the case of a curved Lorentzian manifold, examples of such nonholonomic frames are provided by the moving frames whose vectors are orthonormal. In CGR these orthonormal moving frames play a very important role from a mathematical as well as physical point of view. Hence, we turn next to their formulation and study.

2.3. Orthonormal Frame Bundles over Lorentzian Manifolds

A given differential manifold **M** becomes a *Lorentzian manifold* if a metric field $\boldsymbol{g}$ is specified on **M** which is given by a globally defined symmetric $(0,2)$-tensor field of signature $(+1,-1,-1,-1)$. In fibre-theoretical language, $\boldsymbol{g}$ can be described as a global section of the tensor bundle $T^{0,2}\mathbf{M}$, so that in any holonomic coframe corresponding to the chart (1.1)

$$\boldsymbol{g} = g_{\mu\nu}\, \boldsymbol{dx}^{\mu} \otimes \boldsymbol{dx}^{\nu} \ ; \tag{3.1}$$

this cross-section has to have the property that at every $x\in\mathbf{M}$, there is a linear frame (2.1) (which we shall call a *vector tetrad*) that has a dual coframe[17]

$$\hat{\boldsymbol{\theta}}(x) = (\boldsymbol{\theta}^0,\boldsymbol{\theta}^1,\boldsymbol{\theta}^2,\boldsymbol{\theta}^3)\ , \qquad \boldsymbol{\theta}^i = \lambda^i{}_{\mu}(x)\,\boldsymbol{dx}^{\mu} \in T_x^*\mathbf{M}\ , \tag{3.2}$$

$$\lambda_i{}^{\mu}(x)\lambda^j{}_{\mu}(x) = \delta_i^j\ , \qquad \lambda_i{}^{\mu}(x)\lambda^i{}_{\nu}(x) = \delta_{\nu}^{\mu}\ , \qquad i,j,\mu,\nu = 0,1,2,3\ , \tag{3.3}$$

(which we shall call a *covector tetrad*) in terms of which

$$g = \eta_{ij}\, \boldsymbol{\theta}^i \otimes \boldsymbol{\theta}^j \ , \qquad \eta_{ij} = \mathrm{diag}(+1,-1,-1,-1) \ . \tag{3.4}$$

We note that since $\eta_{ij} = 0$ for all $i \neq j$, any Lorentz metric tensor field is implicitly symmetric. Furthermore, in any given chart (1.1), smooth solutions of (3.3) and of

$$g_{\mu\nu}(x)\lambda_i^{\ \mu}(x)\lambda_j^{\ \nu}(x) \ = \ \eta_{ij} \ , \tag{3.5}$$

can be always obtained in some neighborhood of any given point in the domain of definition of that chart. Any such solution supplies, therefore, a family $\{\boldsymbol{\theta}^i\}$ of four orthonormal covector fields in that neighborhood, which we shall call a *vierbein* over that neighborhood – in accordance with common practice[18] in physics literature.

We thus see that vierbeins emerge as special cases of local sections of the general linear coframe bundle $GL^*\mathbf{M}$ – which is equivalent to the linear frame bundle $GL\mathbf{M}$. Such sections are of great importance in CGR, since they are the duals of vector frames whose elements are orthonormal with respect to the given Lorentz metric, and as such can play the role of local Lorentz frames. Indeed, the transformation law in (2.3) relating any two such frames and their coframes, namely

$$\boldsymbol{e}'_i = \Lambda_i^{\ j} \boldsymbol{e}_j \ , \qquad \boldsymbol{\theta}'^i = \Lambda^i_{\ j} \boldsymbol{\theta}^j \ , \qquad i,j = 0,1,2,3 \ , \tag{3.6}$$

has to satisfy the following relationships if the relationships (3.2)-(3.5), as well as their counterparts for the second set of frames, are to hold true:

$$\Lambda_i^{\ k} \eta_{kl} \Lambda_j^{\ l} = \eta_{ij} \ , \qquad \Lambda^i_{\ k} \eta^{kl} \Lambda^j_{\ l} = \eta^{ij} \ , \qquad \eta^{ij} = \eta_{ij} \ . \tag{3.7}$$

This means that the matrices

$$\Lambda := \left\| \Lambda^i_{\ j} \right\| \ , \qquad \left(\Lambda^{\mathrm{T}}\right)^{-1} = \left\| \Lambda_i^{\ j} \right\| = \boldsymbol{\eta} \Lambda \boldsymbol{\eta} \ , \qquad \boldsymbol{\eta} = \left\| \eta_{ij} \right\| \ , \tag{3.8}$$

are elements of the full Lorentz group[19] O(3,1), so that the values assumed by the transition functions in (3.6) for changes of such frames, which are the counterparts of the transition functions in (1.15) for holonomic frames, are in general Lorentz transformations.

We can now consider the subset $OL\mathbf{M}$ of $GL\mathbf{M}$ obtained by removing from $GL\mathbf{M}$ all linear frames (2.1) which are not orthonormal with respect to the given Lorentzian metric (3.1). It is easily seen that $OL\mathbf{M}$ is a subbundle of $GL\mathbf{M}$ in the sense that it is a submanifold of $GL\mathbf{M}$ which satisfies all the conditions in the following definition: a bundle $(\mathbf{E}', \boldsymbol{\pi}', \mathbf{M}')$ is a *subbundle* of the bundle $(\mathbf{E}, \boldsymbol{\pi}, \mathbf{M})$ if $\mathbf{E} \supset \mathbf{E}'$ and $\mathbf{M} \supset \mathbf{M}'$, and if $\boldsymbol{\pi}'$ is equal to the restriction of $\boldsymbol{\pi}$ to $\mathbf{E}'$.

It is evident from the above consideration that $OL\mathbf{M}$ is a principal bundle, and that O(3,1) is its structure group. Since O(3,1) is not a connected Lie group [C,I], generically $OL\mathbf{M}$ is not a connected manifold. However, if the base Lorentzian manifold $\mathbf{M}$ of $OL\mathbf{M}$ is connected, orientable and time-orientable [W] – as we shall always assume, in the sequel, to be the case – we can further restrict our attention to a submanifold *LM* of *OLM* containing only those orthonormal frames $\{e_i \mid i = 0,1,2,3\}$ whose e_0-vectors all point into the "future" direction, and whose "spatial" triples $\{e_a \mid a = 1,2,3\}$ are all right-handed. Hence *LM* is a principal bundle whose structure group is obviously the Lorentz group $SO_0(3,1)$, and which we shall call the *Lorentz frame bundle* over **M**. Clearly, the Lorentz frame bundle is a subbundle of *OLM*, and therefore also of *GLM*.

In general we can say that a principal fibre bundle (**P**, *Π*, **M**) with structure group **G** is *reducible* [20] to a principal fibre bundle (**P'**, *Π'*, **M**) whose structure group **G'** is a Lie subgroup of **G** if there is a family of local trivialization maps of (**P**, *Π*, **M**) whose domains cover **P**, and whose transition functions take their values in **G'**. The above construction of Lorentz frame bundles establishes that the general linear frame bundle *GLM* over an orientable and time-orientable Lorentzian manifold **M** is reducible to *LM*.

We note that if a principal fibre bundle is trivial, then it possesses a cross-section, and therefore it is reducible to a bundle whose structure group consists of only the identity transformation. According to a well-known theorem by Geroch (1968), in the case of non-compact[21] **M** a spin structure[22] can be associated with *LM* if and only if *LM* possesses a global section. Since such a spin structure is necessary for formulating any form of quantum theory for half-integer spin fields, particles or excitons, at first sight it might appear that the fibre bundle formulation of CGR leads in that case to a trivial framework, since all the attention can be restricted to a single global section. This, however, is not the case since, as we shall see in Secs. 2.4 and 2.7, distinct sections of *LM* play a crucial physical role in depicting distinct free-fall conditions, which are described by local Lorentz frames (i.e., inertial orthonormal frames) without which the equivalence principle could not be formulated in CGR. In fact, even in the case where **M** is equal to the Minkowski space, the concept of inertial orthonormal frame is crucial in formulating classical as well as quantum relativistic theories – except that the conventional special relativistic formulation is carried out with respect to global Lorentz frames, rather than local ones.

The transition from local to global Lorentz frames can be, in that case, carried out most naturally in the context of affine frame bundles[23] over **M**. A *general affine frame* $(\boldsymbol{a},\boldsymbol{e}_i)$ in the tangent space $T_x\mathbf{M}$ viewed as an affine space [C,K] is a pair consisting of a vector $\boldsymbol{a} \in T_x\mathbf{M}$ and a linear frame $(\boldsymbol{e}_i)$ in $T_x\mathbf{M}$ – such as the one in (2.1). Since the affine frame $(\boldsymbol{0},\boldsymbol{e}_i)$ can be identified with the linear frame $(\boldsymbol{e}_i)$, it follows that the general affine frame $(\boldsymbol{a},\boldsymbol{e}_i)$ can be viewed as the translation of the linear frame $(\boldsymbol{e}_i)$ from the point of contact of **M** and $T_x\mathbf{M}$, where its origin is situated, to the point $\boldsymbol{a} \in T_x\mathbf{M}$.

The *general affine frame bundle* over **M** is obtained by endowing the family *GAM* of all affine frames in *TM* with a manifold structure in the same manner as that was done for *GLM*, i.e., by constructing for each chart (1.1) of **M** a corresponding chart for *GAM*, which assigns to $(\boldsymbol{a},\boldsymbol{e}_i)$ the holonomic coordinates of the 4-vector $\boldsymbol{a}$, as well as the holonomic coordinates in (2.1) of the linear frame $(\boldsymbol{e}_i)$. The outcome is a principal fibre bundle whose structure group is the general affine group GA(4,**R**) – which equals the semidirect product of the additive group $\mathbf{R}^4$ with GL(4,**R**), i.e., whose group product is given by

$$(b,A)(b',A') = (b + Ab', AA'), \qquad (b,A),(b',A') \in \mathrm{GA}(4,\mathbf{R}) = \mathbf{R}^4 \wedge \mathrm{GL}(4,\mathbf{R}). \quad (3.9)$$

The local trivialization map of *GAM*, which is the counterpart of the one in (2.12) for *GLM*, is therefore given by

$$\tilde{\Phi}_\alpha : (a,e_i) \mapsto (x,a,\Lambda) \in \mathbf{M}_\alpha \times [\mathbf{R}^4 \wedge \mathrm{GL}(4,\mathbf{R})] , \quad (a,e_i) \in \tilde{\Pi}^{-1}(\mathbf{M}_\alpha) \subset GA\mathbf{M}, \tag{3.10}$$

and the elements (b,A) of GA(4,**R**) act from the right on the general element of *GA*$\mathbf{M}$ in (3.10), in accordance with (2.3), (2.15) and (2.17), as follows:

$$(a,e_i)\cdot(b,A) = (a',e_i') , \quad a = a^i e_i , \quad a' = (a^i + b^i)e_i , \quad e_i' = e_k A^k{}_i . \tag{3.11}$$

When **M** is a Lorentzian manifold, then in the same manner in which *GL*$\mathbf{M}$ could be reduced to *L*$\mathbf{M}$, the general affine frame bundle *GA*$\mathbf{M}$ can be reduced to the bundle *P*$\mathbf{M}$ of affine orthonormal frames (a,e_i), which are obtained by translating to $a \in T_x\mathbf{M}$ the origins of the (time and space oriented) linear orthonormal frames (e_i) within the tangent space $T_x\mathbf{M}$ where they are situated. We shall call the bundle *P*$\mathbf{M}$ the *Poincaré frame bundle*, and we shall refer to its elements (a,e_i) as *Poincaré frames*[24]. It is obvious from (3.9)-(3.11) that the structure group of the Poincaré frame bundle *P*$\mathbf{M}$ is the Poincaré group (i.e., the inhomogeneous Lorentz group) $\mathrm{ISO}_0(3,1)$, which can be obtained by restricting the nonsingular 4×4 matrices A in (3.9) to assuming only values $\Lambda \in \mathrm{SO}_0(3,1)$.

If **M** is equal to the Minkowski space of special relativity, then obviously *P*$\mathbf{M}$ is a trivial bundle, and as such it possesses various global sections. A special class of such cross-sections of *P*$\mathbf{M}$ can be identified with the family of global Lorentz frames $\mathcal{L}$ in **M** by the following straightforward construction: each global Lorentz frame (i.e., Minkowski frame) $\mathcal{L} = (e_i(O))$ with origin at $O \in \mathbf{M}$ gives rise to Minkowski coordinates x^i in **M**, which can be used in constructing coordinate charts for *T*$\mathbf{M}$; consequently, all the points in the tangent space $T_x\mathbf{M}$ at each $x \in \mathbf{M}$ can be identified with corresponding points in the Minkowski space **M**, so that in turn we can carry out the following identification,

$$\mathcal{L} \leftrightarrow s(\mathcal{L}) := \left\{ (a(x),e_i(x)) \middle| \, a(x) = -x^i e_i(x) \in T_x\mathbf{M} , \; x = x^i e_i(O) \in \mathbf{M} \right\}, \tag{3.12}$$

where $(e_i(x))$ is the parallel transport (equal in this case to the parallel translation) of the linear frame $(e_i(O))$ to $x \in \mathbf{M}$. This identification of each global Lorentz frame $\mathcal{L}$ with a cross-section $s(\mathcal{L})$ of *P*$\mathbf{M}$ will play an important role in Chapter 5, in establishing that geometro-stochastic quantum mechanics, generically formulated in a curved Lorentzian manifold **M**, reduces to the special relativistic stochastic quantum mechanics in [P] in case that **M** is a Minkowski space.

*2.4. Parallel Transport and Connections in Principal and Associated Bundles

The concept of parallel transport within the tangent bundle *T*$\mathbf{M}$, and the closely related concept of connection governing such parallel transport at the infinitesimal level, are of fundamental importance in CGR. Indeed, Einstein (1916) had postulated[25] from the outset that, in CGR, the motions of all (classical and neutral) massive point particles in free fall within a classical spacetime manifold **M** follow timelike geodesics, i.e., that they are such that their 4-velocity vectors v are parallel transported along their worldlines γ. This postulate, together with its counterpart for light-rays (or for zero-mass particles, such as the photon), whose free-fall motions are assumed to follow null geodesics, is then implicitly used in deriving all the best-known predictions of CGR – such as the cosmological as well as the gravitational red-shifts, the perihelion precessions of planetary motions, the bending of light rays in the gravitational fields of massive bodies, such as the Sun, the time delay of radar signals, etc. [M,W].

The **M**-coordinate dependent formulation of the concept of connection originally adopted by Einstein (1916), and until recent times also adopted by all textbooks on CGR, was expressed in terms of Christoffel symbols directly related, in any given chart, to the metric components appearing in (3.1) – i.e., this formulation of the concept of connection was provided only in terms of the holonomic frames of *GL***M**. On the other hand, the **M**-coordinate independent formulation of the same concept developed by Cartan has led to the concept of Cartan connection in *GL***M**, whose connection coefficients can be expressed with respect to *any* moving frame in *GL***M** [SC] – and in particular with respect to orthonormal (linear or affine) moving frames, which in recent times have turned out to be of particular importance in CGR [W]. Furthermore, this concept of Cartan connection also eventually led to that of Ehresmann connection [SC], which can be formulated for arbitrary principal bundles, and which therefore turns out to be of especial importance in Yang-Mills field theories [I,NT], with which we shall be concerned in Chapter 10.

In the modern approach to the theory of connections, first recounted in textbook form by Kobayashi and Nomizu (1963), the concept of connection is introduced in principal bundles, and then extended into that of parallel transport on those bundles. Once parallel transport is defined in a principal bundle, it is then easily extended to all its associated bundles, where it can be used to define, in a natural manner, covariant differentiation.

To define the concept of connection on a principal bundle ($\mathbf{P}$, Π, $\mathbf{M}$,$\mathbf{G}$) over a 4-dimensional manifold $\mathbf{M}$, and with a structure group $\mathbf{G}$ which is a Lie group of dimension n, we first recall that the total space $\mathbf{P}$ of that principal bundle is itself an (n+4)-dimensional manifold, and that each one of its fibres $\Pi^{-1}(x)$ is an n-dimensional submanifold of $\mathbf{P}$. Hence, the tangent space $T_{\boldsymbol{u}}\mathbf{P}$ of $\mathbf{P}$ at any $\boldsymbol{u} \in \Pi^{-1}(x)$ is an (n+4)-dimensional linear space, and the tangent space $T_{\boldsymbol{u}}\Pi^{-1}(x)$ of $\Pi^{-1}(x)$ at the same point $\boldsymbol{u} \in \Pi^{-1}(x)$ is an n-dimensional linear subspace of $T_{\boldsymbol{u}}\mathbf{P}$ – which is called the *vertical subspace* of $T_{\boldsymbol{u}}\mathbf{P}$, and as such is denoted by $V_{\boldsymbol{u}}\mathbf{P}$. Since a connection governs the parallel transport of frames along smooth paths γ lying within the base manifold $\mathbf{M}$, it is intuitively clear that such transport should take place in $\mathbf{P}$ only along directions specified by vectors $\mathbf{X} \in T_{\boldsymbol{u}}\mathbf{P}$ which *do not* lie within $V_{\boldsymbol{u}}\mathbf{P}$. Furthermore, since the tangent vectors $X \in T_x\mathbf{M}$ to all such paths passing through a given point $x \in \mathbf{M}$ span the entire 4-dimensional space $T_x\mathbf{M}$ tangent to $\mathbf{M}$ at that point, the corresponding vectors $\mathbf{X} \in T_{\boldsymbol{u}}\mathbf{P}$ in whose direction parallel transport can generally take place in $\mathbf{P}$ should span a 4-dimensional subspace $H_{\boldsymbol{u}}\mathbf{P}$ of $T_{\boldsymbol{u}}\mathbf{P}$. We thus arrive at the following general definition of connections in principal bundles over $\mathbf{M}$.

A *connection in a principal bundle* ($\mathbf{P}$, Π, $\mathbf{M}$,$\mathbf{G}$) over the 4-dimensional manifold $\mathbf{M}$ is an assignment, to each point $\boldsymbol{u}$ in the total space $\mathbf{P}$ of that bundle, of a subspace $H_{\boldsymbol{u}}\mathbf{P}$ of $T_{\boldsymbol{u}}\mathbf{P}$, called the *horizontal subspace* of $T_{\boldsymbol{u}}\mathbf{P}$ for that connection, which is such that $H_{\boldsymbol{u}}\mathbf{P}$ depends smoothly[26] on $\boldsymbol{u}$, and for which

$$T_{\boldsymbol{u}}\mathbf{P} = H_{\boldsymbol{u}}\mathbf{P} \oplus V_{\boldsymbol{u}}\mathbf{P}, \qquad \forall \boldsymbol{u} \in \mathbf{P} \ , \tag{4.1a}$$

$$(R_g)_* H_{\boldsymbol{u}}\mathbf{P} = H_{\boldsymbol{u}\cdot g}\mathbf{P}, \qquad \forall g \in \mathbf{G}, \ \forall \boldsymbol{u} \in \mathbf{P} \ , \tag{4.1b}$$

where $(R_g)_*$ denotes the push-forward of the vectors in $H_u\mathbf{P}$ by the right action

$$R_g\colon \boldsymbol{u} \mapsto \boldsymbol{u} \cdot g\,, \qquad \forall g \in \mathbf{G}\,, \quad \forall \boldsymbol{u} \in \mathbf{P}\,, \tag{4.2}$$

of any given group element $g \in \mathbf{G}$ upon the bundle space $\mathbf{P}$.

In the above context, it should be noted that the right action in (4.2) defines a diffeomorphism of **P** onto **P**, and that for any smooth map ϕ from any manifold $\mathbf{M}'$ into any other manifold $\mathbf{M}''$ (including $\mathbf{M}'$ itself) the *push-forward by* ϕ of a vector $X \in T_x\mathbf{M}'$ is a vector[27] $\phi_* X \in T_{\phi(x)}\mathbf{M}''$ defined, in accordance with (1.5), by the map

$$\phi_* X\colon\ f \mapsto \frac{df}{dt}\big(\phi(x(t))\big)\bigg|_{t=0}\,, \qquad x = x(0)\,, \tag{4.3}$$

whereas the *pull-back by* ϕ of a covector $\boldsymbol{\omega}$ in $T^*\mathbf{M}''$ is defined as a covector $\phi^*\boldsymbol{\omega}$ in $T^*\mathbf{M}'$ for which, in accordance with the general definition in Sec. 2.1 of covectors as linear functionals, we have:

$$\phi^*\boldsymbol{\omega}\colon\ X \mapsto \boldsymbol{\omega}(\phi_* X) \in \mathbf{R}^1, \qquad \boldsymbol{\omega} \in T^*_{\phi(x)}\mathbf{M}''\,, \qquad \phi^*\boldsymbol{\omega} \in T^*_x\mathbf{M}'\,. \tag{4.4}$$

In case that ϕ is actually a diffeomorphism, then pull-backs of vectors and push-forwards of covectors can be also defined as follows:

$$\phi^* X := (\phi^{-1})_* X\,, \qquad\qquad \phi_*\boldsymbol{\omega} := (\phi^{-1})^*\boldsymbol{\omega}\,. \tag{4.5}$$

Furthermore, in that case the definition of pull-backs and push-forwards can be extended to arbitrary tensors over the two respective manifolds, by either using (1.11) and applying multi-linear algebra techniques (Abraham and Marsden, 1978, p. 58), or more simply, by applying the operations (4.3)-(4.5) to the vectors and covectors in the expansion (1.12) in a given basis, and then verifying that the outcome is basis-independent on account of (1.17).

In view of the fact that the projection map $\Pi\colon \mathbf{P} \to \mathbf{M}$ is a smooth map of the entire principal bundle space **P** onto the base manifold **M**, then according to (4.3), its push-forward $\Pi_*\colon T\mathbf{P} \to T\mathbf{M}$ is a globally defined map from the entire tangent bundle $T\mathbf{P}$ over that bundle space onto the tangent bundle $T\mathbf{M}$ over its base manifold. If $x = \Pi(\boldsymbol{u})$, then due to the fact that $x = \Pi(\boldsymbol{u}(t))$ for any curve in **P** with $\boldsymbol{u}(t) \in \Pi^{-1}(x)$ and $\boldsymbol{u}(0) = \boldsymbol{u}$, we conclude from (4.3) that Π_* maps all vertical vectors into the zero vector in $T_x\mathbf{M}$, so that

$$\Pi_*\colon \mathbf{X} = X^* + \mathbf{X}^\ddagger \mapsto X = \Pi_* X^* \in T_x\mathbf{M}\,, \qquad \mathbf{X} \in T_u\mathbf{P},\ \ X^* \in H_u\mathbf{P},\ \ \mathbf{X}^\ddagger \in V_u\mathbf{P}, \tag{4.6}$$

in accordance with (4.1a). Hence $\Pi_*\colon H\mathbf{P} \to T\mathbf{M}$ is a one-to-one map which assigns to each vector X^* in the horizontal subspace $H_u\mathbf{P}$ of $T_u\mathbf{P}$ a unique vector $X \in T_x\mathbf{M}$. The vector $X^* \in H_u\mathbf{P}$, tangent to the bundle space, is therefore called the *horizontal lift* of the vector $X \in T_x\mathbf{M}$, tangent to the base manifold.

Thus, for any smooth curve $\gamma = \{x(t) \mid a \le t \le b\}$ starting at $x' = x(a) \in \mathbf{M}$ and ending at $x'' = x(b) \in \mathbf{M}$, each vector $X(t)$ tangent to it will possess horizontal lifts $X^*(t)$

at all points $\boldsymbol{u} \in \boldsymbol{\Pi}^{-1}(x(t))$. If we choose *any* point $\boldsymbol{u}_{x'} \in \boldsymbol{\Pi}^{-1}(x')$ in the fibre above the initial point $x' \in \mathbf{M}$ of γ, then it is not difficult to establish (cf. [I], p. 165, or [SC], p. 363) that there is a unique curve $\gamma^* = \{\boldsymbol{u}(t)|\ a \leq t \leq b\}$ within the principal bundle space **P**, called a *horizontal lift of* γ, which starts at $\boldsymbol{u}_{x'}$ and which at each $t \in [a,b]$ has as a tangent one of the horizontal vectors $\boldsymbol{X}^*(t)$. The value $\boldsymbol{u}_{x''} = \boldsymbol{u}(b)$ assumed by that horizontal lift of γ at its end-point $x'' \in \mathbf{M}$ is called the *parallel transport* of $\boldsymbol{u}_{x'}$ from x' to x'' along γ.

By this straightforward "lifting" procedure we obtain, for each given smooth curve γ within the base manifold **M**, a diffeomorphism

$$\boldsymbol{\tau}_\gamma(x'',x') :\ \boldsymbol{\Pi}^{-1}(x') \to \boldsymbol{\Pi}^{-1}(x'') \quad , \tag{4.7}$$

between the principal bundle fibres above the initial and final points of the curve γ, which represents the parallel transport along γ governed by the connection specified in (4.1).

With the parallel transport operation (4.7) within a principal bundle (**P**, $\boldsymbol{\Pi}$, **M**, **G**) thus specified, the parallel transport along γ from x' to x'' ,

$$\tau_\gamma(x'',x') :\ \boldsymbol{\pi}^{-1}(x') \to \boldsymbol{\pi}^{-1}(x'') \quad , \tag{4.8}$$

within any fibre bundle (**E**, $\boldsymbol{\pi}$, **M**, **F**) associated, in accordance with (2.8)-(2.10), with that principal bundle through the representation U of the group **G**, is given by the maps

$$\tau_\gamma(x'',x') = \phi_\alpha^{-1}(x'') \circ U\big(g_\alpha(\boldsymbol{u}_{x''}) g_\beta(\boldsymbol{u}_{x'})^{-1}\big) \circ \phi_\beta(x') \quad , \tag{4.9a}$$

$$\boldsymbol{\tau}_\gamma(x'',x') :\ \boldsymbol{u}_{x'} \mapsto \boldsymbol{u}_{x''} \quad . \tag{4.9b}$$

For the generic case of $x' \in \mathbf{M}_\alpha$ and $x'' \in \mathbf{M}_\beta$, in (4.9) $g_\alpha(\boldsymbol{u}_{x'}) \in \mathbf{G}$ is the outcome of the trivialization map $\boldsymbol{\Phi}_\alpha$ in (2.8) acting upon $\boldsymbol{u}_{x'}$, and $g_\beta(\boldsymbol{u}_{x''}) \in \mathbf{G}$ is the outcome of a trivialization map $\boldsymbol{\Phi}_\beta$ similarly acting upon $\boldsymbol{u}_{x''}$, whereas $\phi_\alpha(x') : \boldsymbol{\pi}^{-1}(x') \to \mathbf{F}$ is the diffeomorphism induced by (2.4), and the diffeomorphism $\phi_\beta(x'') : \boldsymbol{\pi}^{-1}(x'') \to \mathbf{F}$ is defined in an analogous manner. On first sight, it might appear that this definition depends on the choice of the principal fibre element $\boldsymbol{u}_{x'}$ within the fibre $\boldsymbol{\Pi}^{-1}(x')$ and on the chosen trivialization maps, but that is not actually the case, due to (2.9) and (4.1b).

To see that, and to better understand the geometric meaning of the somewhat abstract definition in (4.9), let us consider the important case of *vector bundles* (**E**, $\boldsymbol{\pi}$, **M**, **F**), i.e., of fibre bundles whose typical fibres are vector spaces. Such bundles are usually associated with principal bundles (**P**, $\boldsymbol{\Pi}$, **M**, **G**) which are *linear frame bundles*, i.e., whose elements $\boldsymbol{u} \in \boldsymbol{\Pi}^{-1}(x)$ either provide vector bases within the fibres $\boldsymbol{\pi}^{-1}(x)$ (such as in the case where $\mathbf{E} = T\mathbf{M}$ and $\mathbf{P} = GL\mathbf{M}$ or $\mathbf{P} = L\mathbf{M}$, and in the case of Yang-Mills bundles [C,I,NT]), or from which bases in $\boldsymbol{\pi}^{-1}(x)$ can be built – such as in the case of (r,s)-tensor bundles with $r+s > 1$, as can be seen[28] from (1.12). In this kind of setting, (4.9) indicates that the parallel transport stipulated in (4.8) can be performed by expanding the vectors in $\boldsymbol{\pi}^{-1}(x')$ in terms of a basis constructed from the elements of $\boldsymbol{u}_{x'}$, and their parallel transport can be effected by simply parallel transporting the elements of that basis. As a consequence of this observation, we see that the parallel transport map in (4.8) provides a linear isomorphism between the fibres $\pi^{-1}(x')$ and $\pi^{-1}(x'')$. For example, if $\mathbf{E} = T\mathbf{M}$ and $\mathbf{P} = GL\mathbf{M}$, then we easily establish, by using (2.12)-(2.17), that (4.9) assumes the following form,

$$\tau_\gamma(x'',x') : \ X = X^i e'_i \ \mapsto \ X^i e''_i \in T_{x''}\mathbf{M} \ , \qquad X \in T_{x'}\mathbf{M} \ , \tag{4.10a}$$

$$\boldsymbol{\tau}_\gamma(x'',x') : (e'_i) \mapsto (e''_i) \ , \tag{4.10b}$$

for any choice of linear frame within $T_{x'}\mathbf{M}$.

In any vector bundle ($\mathbf{E}$, π, $\mathbf{M}$, $\mathbf{F}$), the *covariant derivative* of a vector field Ψ (i.e., of a section Ψ of $\mathbf{E}$, with values $\Psi_x \in \pi^{-1}(x)$) can be defined for any vector $X \in T_x\mathbf{M}$ by choosing a smooth curve γ whose tangent at x equals X, and then taking the following limit, which produces an element of $\pi^{-1}(x)$:

$$\nabla_X \Psi_x = \lim_{t\to 0} \frac{1}{t}\left[\tau_\gamma(x,x(t))\,\Psi_{x(t)} - \Psi_x\right] \ , \qquad x(0) = x \in \mathbf{M} \ , \quad \dot{x}(0) = X \in T_x\mathbf{M}. \tag{4.11}$$

It can be easily shown (cf. [I], p. 173, or [SC], p. 368) that the covariant derivative defined by (4.11) possesses the following fundamental properties at all points $x \in \mathbf{M}$ within the common domain of definition of f, X, Y, Ψ, Ψ' and Ψ'' – where f denotes any smooth function defined in $\mathbf{M}$ (so that $X(f)$ is defined by (1.5)), X and Y are any vector fields in $T\mathbf{M}$, whereas Ψ, Ψ' and Ψ'' are any vector fields in $\mathbf{E}$:

$$\nabla_{X+Y}\Psi = \nabla_X \Psi + \nabla_Y \Psi \ , \tag{4.12a}$$

$$\nabla_X(\Psi' + \Psi'') = \nabla_X \Psi' + \nabla_X \Psi'' , \tag{4.12b}$$

$$\nabla_{fX}\Psi = f\,\nabla_X \Psi \ , \tag{4.12c}$$

$$\nabla_X(f\Psi) = f\,\nabla_X \Psi + X(f)\,\Psi \ . \tag{4.12d}$$

The above four properties indicate that the *covariant differentiation operator* ∇, which maps fields X and Ψ from $T\mathbf{M}$ and $\mathbf{E}$, respectively, into vector fields from $\mathbf{E}$, computed in accordance with (4.11), is a *Koszul connection* [SC] on the vector bundle $\mathbf{E}$ – i.e., that it could have been defined directly as a map that assigns a vector field $\nabla_X\Psi$ in $\mathbf{E}$ to any two vector fields X and Ψ in $T\mathbf{M}$ and $\mathbf{E}$, respectively, in such a manner that it possesses the four properties in (4.12) (cf. [SC], Chapter 6). In fact, some textbooks on CGR (Hawking and Ellis, 1973; Straumann, 1984) define a connection on vector fields assuming values in $\mathbf{E} = T\mathbf{M}$ as being such an operator, and then extend this definition to (r,s)-tensor fields (i.e., to vector fields from $\mathbf{E} = T^{r,s}\mathbf{M}$) by imposing in addition the Leibniz rule and the commutativity of ∇ with contractions – both properties of ∇ which actually follow from (4.11), by taking into account (1.11) and the fact that all these tensor bundles can be associated with one and the same principal bundle, namely with the general linear frame bundle $GL\mathbf{M}$ over $\mathbf{M}$. We shall follow, however, the more general route, whereby connection forms will be next introduced. In turn, these connection forms supply connection coefficients by means of which covariant derivatives can be computed.

*2.5. Connection and Curvature Forms on Principal Bundles

To make the purely geometric formulation of connections presented in the preceding section suitable for computations, we have to supplement it with additional results from the general fibre bundle theory of connections (cf. [K], Chapters II and III). These results are arrived at by associating with each connection (4.1) a connection form $\boldsymbol{\omega} : T\mathbf{P} \to L(\mathbf{G})$, which assigns to each $\mathbf{X} \in T_u\mathbf{P}$ an element $\boldsymbol{\omega}(\mathbf{X})$ from the Lie algebra $L(\mathbf{G})$ of $\mathbf{G}$.

To construct this connection form, let us first recall [C,I] that the Lie algebra $L(\mathbf{G})$ of any Lie group $\mathbf{G}$ consists of all left-invariant vector fields

$$\tilde{\mathbf{X}}_g = (\tilde{L}_g)_* \tilde{\mathbf{X}}, \qquad \tilde{\mathbf{X}} \in T_e\mathbf{G}, \qquad \tilde{L}_g(g') := gg' \in \mathbf{G}, \quad \forall g \in \mathbf{G}, \tag{5.1}$$

on $\mathbf{G}$, and that as a vector space $L(\mathbf{G})$ can be identified with the tangent space $T_e\mathbf{G}$ of $\mathbf{G}$ at the unit element $e \in \mathbf{G}$; upon such an identification, the Lie product in $T_e\mathbf{G}$ is supplied by the Lie bracket of the corresponding fields in $L(\mathbf{G})$, defined as the commutator of the operators representing those fields in accordance with (1.5):

$$[\tilde{\mathbf{X}}, \tilde{\mathbf{Y}}] := [\tilde{\mathbf{X}}_g, \tilde{\mathbf{Y}}_g]_{g=e}, \qquad \tilde{\mathbf{X}}, \tilde{\mathbf{Y}} \in T_e\mathbf{G}. \tag{5.2}$$

In turn, the generic element of $T_e\mathbf{G}$ in (5.1) can be viewed as being the tangent vector at $e \in \mathbf{G}$ to the curve produced by the exponential map[29], i.e., that in accordance with (1.5), it acts as follows upon smooth functions $\tilde{f}$ defined on some neighborhood of $e \in \mathbf{G}$:

$$\tilde{\mathbf{X}}(\tilde{f}) = d\tilde{f}(\exp(\tilde{\mathbf{X}}t))/dt\big|_{t=0}, \qquad \tilde{\mathbf{X}} \in T_e\mathbf{G}. \tag{5.3}$$

Hence, at any $u \in \mathbf{P}$ we can attach, to each such element of $T_e\mathbf{G}$, a unique element of the vertical subspace of $T_u\mathbf{P}$ which acts upon smooth functions, defined on some neighborhood of $u \in \mathbf{P}$, as follows :

$$\hat{\mathbf{X}}_u(\hat{f}) = d\hat{f}(u \cdot \exp(\tilde{\mathbf{X}}t))/dt\big|_{t=0}, \qquad \tilde{\mathbf{X}} \in T_e\mathbf{G}, \quad \hat{\mathbf{X}}_u \in V_u\mathbf{P}. \tag{5.4}$$

By this construction, we have attached to each element (5.1) of the Lie algebra $L(\mathbf{G})$ a unique global section of $T\mathbf{P}$, called the *fundamental field* corresponding to that element. Furthermore, we note that the map of vectors within $T_e\mathbf{G}$ into vectors within $T_u\mathbf{P}$, which is implicitly defined by (5.3) and (5.4),

$$\tilde{\mathbf{X}} \leftrightarrow \hat{\mathbf{X}}_u, \qquad \tilde{\mathbf{X}} \in T_e\mathbf{G}, \quad \hat{\mathbf{X}}_u \in V_u\mathbf{P}, \tag{5.5}$$

provides an isomorphism between $T_e\mathbf{G}$ and the vertical space $V_u\mathbf{P}$ at each $u \in \mathbf{P}$.

On the basis of (4.1a) and of this last observation, the *connection form* $\boldsymbol{\omega}$, viewed as a map $T\mathbf{P} \to L(\mathbf{G})$ corresponding to the connection (4.1), can be unambiguously defined in the following manner,

$$\boldsymbol{\omega}(\overline{\mathbf{X}}+\hat{\mathbf{X}}_u) := \tilde{\mathbf{X}} \in T_e\mathbf{G}\ , \qquad \overline{\mathbf{X}} \in H_u\mathbf{P}, \qquad \hat{\mathbf{X}}_u \in V_u\mathbf{P}\ , \tag{5.6}$$

i.e., as a mapping which takes all the horizontal vectors from $T\mathbf{P}$ into the zero vector in $T_e\mathbf{G}$, and which takes each vertical vector from $T\mathbf{P}$ into the element of $T_e\mathbf{G}$ that produces the fundamental field to which that vertical vector corresponds. As an immediate consequence of (4.1b), we then easily obtain that [C,I,SC]:

$$\boldsymbol{\omega}\big((R_g)_*\mathbf{X}\big) = \mathrm{Ad}_{g^{-1}}\boldsymbol{\omega}(\mathbf{X})\ , \qquad \forall g \in \mathbf{G}, \quad \forall \mathbf{X} \in T_u\mathbf{P}, \quad \forall u \in \mathbf{P}\ , \tag{5.7a}$$

$$\mathrm{Ad}_g := (\tilde{L}_g\tilde{R}_{g^{-1}})_* : \ L(\mathbf{G}) \to L(\mathbf{G})\ , \qquad \tilde{L}_g\tilde{R}_{g^{-1}}: \ g' \mapsto gg'g^{-1}\ . \tag{5.7b}$$

Conversely, if we define an *Ehresmann connection* [SC] as a mapping of $T\mathbf{P}$ onto $L(\mathbf{G})$ which satisfies (5.7), and which maps the elements of each fundamental field into the element of $L(\mathbf{G})$ to which that field belongs, it is then easily seen that to any Ehresmann connection corresponds a unique connection (4.1) such that (5.6) is satisfied, i.e., such that the given Ehresmann connection maps all its horizontal vectors into zero [C,I,K,SC].

Let us now expand the value (5.6) of the connection form in any basis in $T_e\mathbf{G}$, so that

$$\boldsymbol{\omega}(\mathbf{X}) = \omega^a(\mathbf{X})\tilde{\mathbf{Y}}_a\ , \qquad \mathbf{X} \in T_u\mathbf{P}, \qquad \tilde{\mathbf{Y}}_a \in T_e\mathbf{G}, \quad a = 1,\ldots,n\ , \tag{5.8}$$

$$[\tilde{\mathbf{Y}}_a, \tilde{\mathbf{Y}}_b] = C_{ab}{}^c\tilde{\mathbf{Y}}_c\ , \qquad a,b,c = 1,\ldots,n\ , \quad n = \mathit{dim}\,\mathbf{G}\ , \tag{5.9}$$

where $C_{ab}{}^c$ are the structure constants of the Lie algebra $L(\mathbf{G})$. We observe that

$$\omega^a : \ \mathbf{X} = \overline{\mathbf{X}}+\hat{\mathbf{X}}_u \mapsto X^a, \qquad \hat{\mathbf{X}}_u = X^a\hat{\mathbf{Y}}_{a;u} \in V_u\mathbf{P}\ , \qquad \tilde{\mathbf{X}} = X^a\tilde{\mathbf{Y}}_a \in T_e\mathbf{G}\ , \tag{5.10}$$

are global sections of $T^*\mathbf{P}$, which we shall call *Cartan connection forms*, or the *connection one-forms* associated with the chosen basis in the Lie algebra of the structure group $\mathbf{G}$. In this context, it should be noted that if M is any differential manifold, then by definition any smooth function f in M is called a *zero-form*; a *one-form* is by definition a covector field in M, i.e., a section of $T^{0,1}M$; whereas for $k > 1$, a *k-form* $\boldsymbol{\omega}$ is by definition a section of an antisymmetric $(0,k)$-tensor field in M , i.e., a section of $T^{0,k}M$ which in any coordinate chart in M can be expressed, in accordance with (1.9)-(1.12), as follows,

$$\boldsymbol{\omega} = \omega_{\mu_1\cdots\mu_k}\,\mathbf{d}x^{\mu_1}\otimes\cdots\otimes\mathbf{d}x^{\mu_k} = \frac{1}{k!}\omega_{\mu_1\cdots\mu_k}\,\mathbf{d}x^{\mu_1}\wedge\cdots\wedge\mathbf{d}x^{\mu_k}\ , \tag{5.11}$$

and whose coefficients are antisymmetric under the interchange of any two of their indices.

The *exterior product* (or *wedge product*) of a k-form $\boldsymbol{\omega}'$ and an l-form $\boldsymbol{\omega}''$, implicitly used in (5.11), produces a $(k+l)$-form as a consequence of the following definition,

$$\boldsymbol{\omega}'\wedge\boldsymbol{\omega}'' = \frac{1}{k!l!}\sum_\pi(\mathrm{sign}\,\pi)\ \omega'_{\mu_1\cdots\mu_k}\omega''_{\nu_1\cdots\nu_l}\,\mathbf{d}x^{\mu_1}\otimes\cdots\otimes\mathbf{d}x^{\mu_k}\otimes\mathbf{d}x^{\nu_1}\otimes\cdots\otimes\mathbf{d}x^{\nu_l}, \tag{5.12}$$

in which the summation is performed over all permutations of $(1,\ldots,k+l)$. Consequently, the *exterior derivative* of the k-form $\boldsymbol{\omega}$ in (5.11) is a $(k+1)$-form, since it is defined by

$$\mathbf{d}\omega = \frac{1}{k!}\frac{\partial\omega_{\mu_1\cdots\mu_k}}{\partial x^\nu} dx^\nu \wedge dx^{\mu_1} \wedge \cdots \wedge dx^{\mu_k} \ . \tag{5.13}$$

It is easily verified that the definitions in (5.12) as well as in (5.13) are independent of the choice of chart, on account of the $T^{0,k}M$-counterpart of the transformation laws in (1.17).

By applying the definition (5.13) to a k-form $\boldsymbol{\omega}$ defined on the principal bundle space $\mathbf{P}$, we can define its *exterior covariant derivative* as the $(k+1)$-form which is such that

$$\mathbf{D}\omega(\mathbf{X}_1,\ldots,\mathbf{X}_{k+1}) = \mathbf{d}\omega(\overline{\mathbf{X}}_1,\ldots,\overline{\mathbf{X}}_{k+1}) \ , \tag{5.14}$$

for any set $\{\mathbf{X}_1,\ldots,\mathbf{X}_{k+1}\}$ of vector fields in $\mathbf{P}$ whose horizontal components on the right-hand side of (5.14) are defined by the connection (4.1). In particular, we can thus define the exterior derivatives, and therefore also the exterior covariant derivatives, of all the Cartan connection forms in (5.10), which as such are two-forms on the principal bundle space $\mathbf{P}$. Hence, the exterior covariant derivative $\mathbf{D}\boldsymbol{\omega}$ of the connection form in (5.6) equals

$$\boldsymbol{\Omega} := \mathbf{D}\boldsymbol{\omega} = \mathbf{D}\omega^a\tilde{\mathbf{Y}}_a \ : \ T_u\mathbf{P} \otimes T_u\mathbf{P} \ \rightarrow \ T_e\mathbf{G} \ , \tag{5.15}$$

and is an $L(\mathbf{G})$-valued two-form on $\mathbf{P}$, called the *curvature form* of the connection defined by (4.1). It can be proved (cf. [I], p. 175 or [SC], p. 376, as well as Sec. 10.3) that this curvature form satisfies the *Cartan structural equation*

$$\boldsymbol{\Omega}(\mathbf{X},\mathbf{Y}) = \mathbf{d}\boldsymbol{\omega}(\mathbf{X},\mathbf{Y}) + [\boldsymbol{\omega}(\mathbf{X}),\boldsymbol{\omega}(\mathbf{Y})] \ , \qquad \mathbf{X},\mathbf{Y} \in T_u\mathbf{P} \ , \tag{5.16a}$$

which gives rise to a system of equations relating the components of the curvature form to the Cartan connection forms in (5.10):

$$\Omega^a = \mathbf{d}\omega^a + \tfrac{1}{2}C_{bc}{}^a\omega^b \wedge \omega^c \ , \qquad \boldsymbol{\Omega} = \Omega^a\tilde{\mathbf{Y}}_a \ . \tag{5.16b}$$

This system is well-known in the context of Yang-Mills field theories [NT] and in the theory of BRST quantization (Baulieu, 1985). Upon computing, in accordance with (5.13)-(5.14), the exterior covariant derivatives of their right-hand sides, we immediately find that they vanish (cf. [C], p. 375), so that we obtain the following *Bianchi identities*,

$$\mathbf{D}\Omega^a = 0 \ , \qquad a = 1,\ldots,n \ , \tag{5.16c}$$

which are equivalent to the statement that $\mathbf{D}\boldsymbol{\Omega} = 0$.

Let us now apply the concept of connection form to the computation of the covariant derivative (4.11) of a vector field Ψ, which assumes its values Ψ_x within the total space $\mathbf{E}$ of a vector bundle $(\mathbf{E}, \pi, \mathbf{M}, \mathbf{F})$, associated with some principal bundle $(\mathbf{P}, \Pi, \mathbf{M}, \mathbf{G})$ through the representation U of its structure group $\mathbf{G}$. For the sake of simplicity, we shall assume that ($\mathbf{P}$, $\boldsymbol{\Pi}$, $\mathbf{M}$, $\mathbf{G}$) is a linear frame bundle. In general, its elements $\boldsymbol{u} \in \mathbf{P}$ then supply *associated frames* $\{\boldsymbol{\Phi}_\rho(\boldsymbol{u})\}$ within all the fibres $\boldsymbol{\pi}^{-1}(x)$ of $\mathbf{E}$, so that

$$\boldsymbol{\Psi}_x = \Psi^\rho_{x;\boldsymbol{u}} \boldsymbol{\Phi}_\rho(\boldsymbol{u}) \in \boldsymbol{\pi}^{-1}(x) \ , \qquad \boldsymbol{u} \in \boldsymbol{\Pi}^{-1}(x) \ , \tag{5.17}$$

where generically ρ is a multi-component index – as is the case, for example, in (1.12), or in the case of tensors associated with Yang-Mills fields [NT].

To express ∇ in terms of the Cartan connection forms in (5.10), we choose a *moving frame* (Cartan, 1935) in the given linear frame bundle, i.e., a section

$$s: \ x \mapsto \ \boldsymbol{u} \in \boldsymbol{\Pi}^{-1}(x) \ , \qquad x \in \mathbf{M}^s \subset \mathbf{M} \ , \tag{5.18}$$

whose domain $\mathbf{M}^s$ contains a segment of the curve $\gamma = \{x(t) | \ 0 \le t \le b\}$ in (4.11), including its initial point $x = x(0)$. We can then insert in (4.11)

$$\boldsymbol{\tau}_\gamma(x,x(t))\boldsymbol{\Psi}_{x(t)} = \Psi^\rho_{x(t);\boldsymbol{u}(t)} \boldsymbol{\Phi}_\rho\big(\boldsymbol{\tau}_\gamma(x,x(t))\boldsymbol{u}(t)\big) , \qquad s: \ x(t) \mapsto \ \boldsymbol{u}(t) \ . \tag{5.19}$$

Comparing with (5.3)-(5.6) and (5.8), it is easy to see that

$$\lim_{t\to 0} \frac{1}{t}\Big[\boldsymbol{\Phi}_\rho\big(\boldsymbol{\tau}_\gamma(x,x(t))\boldsymbol{u}(t)\big) - \boldsymbol{\Phi}_\rho(\boldsymbol{u})\Big] = \boldsymbol{\omega}^a(\mathbf{X})\hat{\mathbf{A}}_{a;\boldsymbol{u}}\boldsymbol{\Phi}_\rho(\boldsymbol{u}) , \quad \boldsymbol{u} = \boldsymbol{u}(0) \in \boldsymbol{\Pi}^{-1}(x) \ , \tag{5.20}$$

where $\hat{\mathbf{A}}_{a;\boldsymbol{u}}$ are generators of the representation U_x of $\mathbf{G}$ (implicitly depending on s – cf. (5.24)) acting within the fibre $\boldsymbol{\pi}^{-1}(x)$ so that, in accordance with (5.5) and (5.10),

$$\exp(\hat{\mathbf{A}}_{a;\boldsymbol{u}}t)\boldsymbol{\Phi}_\rho(\boldsymbol{u}) \ = \ \boldsymbol{\Phi}_\rho(\boldsymbol{u}\cdot \exp(\tilde{\mathbf{Y}}_a t)) , \qquad t \in \mathbf{R}^1 . \tag{5.21}$$

In (5.20) $\mathbf{X} \in T_{\boldsymbol{u}}\mathbf{P}$ is the tangent to the curve $\gamma^s = \{\boldsymbol{u}(t) | \ 0 \ \le t \ \le b\}$ in $\mathbf{P}$ that passes through $\boldsymbol{u} \in \mathbf{P}$, and is determined by γ and the chosen section s in (5.18) in accordance with (5.19), so that $\mathbf{X} = s_* X$, where X is the tangent to γ at x. Consequently, by using the same procedure as in deriving, in ordinary calculus, the formula for the derivative of the product of two functions (cf. (2.17) in Sec. 5.2 for a similar procedure), upon introducing a chart as in (1.1) and taking advantage of (1.6), we obtain that

$$\nabla_X \boldsymbol{\Psi}_x = [\partial_X + \boldsymbol{\omega}^a(s_* X)\hat{\mathbf{A}}_{a;\boldsymbol{u}}]\boldsymbol{\Psi}_x \ , \qquad \partial_X \boldsymbol{\Psi}_x = \big[X^\mu \partial_\mu\big(\Psi^\rho_{x;\boldsymbol{u}}\big)\big]\boldsymbol{\Phi}_\rho(\boldsymbol{u}) \ . \tag{5.22}$$

Since the above formula is valid for all tangent vectors $X \in T_x\mathbf{M}$ at all points $x \in \mathbf{M}$ in the domain of definition of the vector field $\boldsymbol{\Psi}$, we can write in abbreviated form

$$\nabla\boldsymbol{\Psi} = [d + A^s]\boldsymbol{\Psi} , \qquad d = dx^\mu \partial_\mu \ , \quad A^s = \boldsymbol{\omega}^a_s \hat{\mathbf{A}}_{a;\boldsymbol{u}} \ , \quad \boldsymbol{\omega}^a_s(X) = \boldsymbol{\omega}^a(s_* X) \ . \tag{5.23}$$

The operator-valued form A^s is defined on the restriction of the tangent bundle $T\mathbf{M}$ to the domain M^s of the chosen section s . It assumes there values within the Lie algebras

$L(U_x)$ of the group representations U_x of **G** that act in the various fibres of **E**, and which implicitly depend on the choice of section *s* that in general supplies a local trivialization map (2.4) (cf. also the soldering maps in (6.5) below, as well as in (4.3.2a) within Chapter 4):

$$U_x(g) = \phi_\alpha^{-1}(x) \circ U(g) \circ \phi_\alpha(x) : \ \pi^{-1}(x) \to \pi^{-1}(x) \ , \tag{5.24a}$$

$$U(g) : \ \mathbf{F} \to \mathbf{F} \ , \qquad g \in \mathbf{G} \ . \tag{5.24b}$$

This form is called, in physics literature, the *gauge potential* [C,NT] corresponding to the choice (5.18) of section of the principal bundle of frames, and the section itself is in turn called a *gauge choice*. In this same context, the structure group **G** is then often[30] called the "gauge group" of the vector bundle (**E**, π, **M**, **F**), and the transition from one gauge potential to another (cf. Secs. 10.2 and 11.3 for more comprehensive discussions)

$$\mathbf{A}^s \to \mathbf{A}^{s'} \ , \qquad s' : \ x \mapsto u' \in \Pi^{-1}(x) \ , \qquad x \in \mathbf{M}^{s'} \ , \tag{5.25}$$

resulting from a change of choice of section of the frame bundle (i.e., of a gauge choice) is referred to as a (local) *gauge transformation*. It is easy to check[31] that, on account of (5.7),

$$\mathbf{A}^{s'} = U(g^{-1})\, dU(g) + U(g^{-1})\, \mathbf{A}^s\, U(g) \ , \qquad dU(g) = \frac{\partial U(g)}{\partial g^a} dg^a \ , \tag{5.26a}$$

$$g : \ x \mapsto g_x \in \mathbf{G} \ , \qquad u'_x = u_x \cdot g_x \ , \qquad x \in \mathbf{M}^s \cap \mathbf{M}^{s'} \ , \tag{5.26b}$$

within the common domain of definition $\mathbf{M}^s \cap \mathbf{M}^{s'}$ of any two gauge potentials.

2.6. Levi-Civita Connections and Riemannian Curvature Tensors

In applying to CGR the fibre-theoretical framework for treating connections, which was discussed in the preceding two sections, it has to be first noted that if $\mathbf{G} = \mathrm{GL}(4,\mathbf{R})$ then $T_e\mathbf{G}$ can be identified with the linear space $M(4,\mathbf{R})$ of all 4×4 real matrices, and that the Lie bracket of the elements of $L(\mathbf{G})$ can be then identified with the commutator bracket for those matrices. Furthermore, by using in (5.7b) the definition (4.3) of push-forwards of vectors by any map acting on a manifold – which in this case equals GL(4,**R**) – we easily see that for a linear connection form $\boldsymbol{\omega}$ on *GL***M**, (5.7a) assumes the following form,

$$\boldsymbol{\omega}\big((R_A)_* \mathbf{X}\big) = A^{-1} \boldsymbol{\omega}(\mathbf{X}) A \ , \qquad \forall A \in \mathrm{GL}(4,\mathbf{R}) \ , \quad \forall \mathbf{X} \in T_u\mathbf{P} \ , \quad \forall u \in \mathbf{P} \ , \tag{6.1}$$

since both $\boldsymbol{\omega}(\mathbf{X})$ as well as A are represented by 4×4 real matrices.

In the present context, the most natural choices of bases for the sixteen-dimensional Lie algebra $g\ell(4,\mathbf{R})$ of GL(4,**R**) are provided by the holonomic frames

$$\left\{ \tilde{\mathbf{Y}}_i^j \,\middle|\, i,j = 0,1,2,3 \right\} , \qquad \tilde{\mathbf{Y}}_i^j = \partial / \partial a_i^j \in T_I \mathrm{GL}(4,\mathbf{R}) \ , \quad A = \left\| a_i^j \right\| \in \mathrm{GL}(4,\mathbf{R}) \ , \tag{6.2}$$

obtained by adopting as coordinates in a neighborhood of the unit matrix $I \in \mathrm{GL}(4,\mathbf{R})$ the components[32] of the matrices A from that neighborhood. In these holonomic frames the expansion (5.8) assumes the form

$$\boldsymbol{\omega}(\mathbf{X}) = \omega_i^j(\mathbf{X})\tilde{\mathbf{Y}}_j^i\,, \qquad \mathbf{X} \in T_u GL\mathbf{M}\,, \qquad i,j = 0,1,2,3\ . \tag{6.3}$$

The general expression (5.22) for covariant derivatives can be therefore transcribed, in a very straightforward manner, to the tangent bundle $T\mathbf{M}$ and to the cotangent bundle $T^*\mathbf{M}$, as well as to any other tensor bundle $T^{r,s}\mathbf{M}$.

Especially important is the $\mathbf{R}^4$-valued *canonical form* of $GL\mathbf{M}$, defined by

$$\boldsymbol{\theta} : \mathbf{X} \mapsto \boldsymbol{\sigma}_u(\Pi_* \mathbf{X}) \in \mathbf{R}^4\,, \qquad \mathbf{X} \in T_u GL\mathbf{M}\,, \quad u = (e_0,\dots,e_3) \in GL\mathbf{M}\,, \tag{6.4}$$

where $\boldsymbol{\sigma}_u : T_x\mathbf{M} \to \mathbf{R}^4$ is the *soldering map*[33] of $T_x\mathbf{M}$ determined by the frame $\boldsymbol{u}$:

$$\boldsymbol{\sigma}_u : X \mapsto (\theta^0(X),\dots,\theta^3(X)) \in \mathbf{R}^4\,, \qquad \theta^i(e_j) = \delta^i_j\,, \quad X \in T_x\mathbf{M}\,, \ x = \Pi u\ . \tag{6.5}$$

Its exterior covariant derivative $\mathbf{D}\boldsymbol{\theta}$ – defined in accordance with (5.14) – is known as the *torsion form* of the linear connection $\boldsymbol{\omega}$, and it satisfies the structural equation [C,I,NT,SC]

$$\boldsymbol{\Theta}(\mathbf{X},\mathbf{Y}) = \mathbf{d}\boldsymbol{\theta}(\mathbf{X},\mathbf{Y}) + \boldsymbol{\omega}(\mathbf{X})\boldsymbol{\theta}(\mathbf{Y}) - \boldsymbol{\omega}(\mathbf{Y})\boldsymbol{\theta}(\mathbf{X})\,, \qquad \boldsymbol{\Theta} := \mathbf{D}\boldsymbol{\theta}\,, \quad \mathbf{X},\mathbf{Y} \in T_u\mathbf{P}\ . \tag{6.6}$$

This equation, together with the structural equation (5.16a), which holds for any kind of curvature form, constitutes the set of *Cartan structural equations for a linear connection form* $\boldsymbol{\omega}$. These equations are often written in abbreviated form, either in terms of the Lie bracket within the Lie algebra of $\mathrm{GA}(4,\mathbf{R})$ [C,I], or in terms of the wedge product [SC]:

$$\boldsymbol{\Theta} = \mathbf{d}\boldsymbol{\theta} + [\boldsymbol{\omega},\boldsymbol{\theta}] = \mathbf{d}\boldsymbol{\theta} + \boldsymbol{\omega}\wedge\boldsymbol{\theta}\,, \qquad \boldsymbol{\Omega} = \mathbf{d}\boldsymbol{\omega} + [\boldsymbol{\omega},\boldsymbol{\omega}] = \mathbf{d}\boldsymbol{\omega} + \boldsymbol{\omega}\wedge\boldsymbol{\omega}\ . \tag{6.7}$$

They immediately yield the two well-known *Bianchi identities for a linear connection form*,

$$\mathbf{D}\boldsymbol{\Theta} = \boldsymbol{\Omega}\wedge\boldsymbol{\theta}\,, \qquad \mathbf{D}\boldsymbol{\Omega} = 0\,, \tag{6.8}$$

where the wedge product can be computed in an arbitrary holonomic basis in accordance with (5.12). We note that the second Bianchi identity for a linear connection form is equivalent to the set of Bianchi identities (5.16c), which hold true for all connection forms.

We can now use the soldering map introduced in (6.5), as well as its inverse, to define the following operators,

$$\mathbf{T} : (X,Y) \mapsto \boldsymbol{\sigma}_u^{-1}\big(\boldsymbol{\Theta}(\overline{\mathbf{X}},\overline{\mathbf{Y}})\big)\,, \qquad \mathbf{R} : (X,Y,Z) \mapsto \boldsymbol{\sigma}_u^{-1}\big(\boldsymbol{\Omega}(\overline{\mathbf{X}},\overline{\mathbf{Y}})\boldsymbol{\sigma}_u(Z)\big)\,, \tag{6.9a}$$

$$X = \Pi_*\overline{\mathbf{X}}\,, \ Y = \Pi_*\overline{\mathbf{Y}}\,, \ Z \in T_x\mathbf{M}\,, \qquad \overline{\mathbf{X}},\overline{\mathbf{Y}} \in H_u\mathbf{P}\ . \tag{6.9b}$$

It can be easily verified (cf. [SC], pp. 379-381) that these *torsion* and *curvature operators* are independent of the choice of $\boldsymbol{u} \in \Pi^{-1}(x)$, and that in fact, for any choice of $GL\mathbf{M}$-section

(5.18) and for any vector fields X,Y,Z on $\mathbf{M}$, they can be expressed in the following manner in terms of the covariant derivatives defined by (5.22), and specialized to the case of the vector bundle $T\mathbf{M}$:

$$\mathbf{T}(X,Y) = \nabla_X Y - \nabla_Y X - [X,Y] \quad , \tag{6.10}$$

$$\mathbf{R}(X,Y)Z = \nabla_X \nabla_Y Z - \nabla_Y \nabla_X Z - \nabla_{[X,Y]} Z \quad . \tag{6.11}$$

Hence, for each such choice of section of $GL\mathbf{M}$, i.e., for any choice of *Cartan gauge*, we can introduce the *torsion tensor* and the *curvature tensor*, defined as $(1,2)$-tensors and $(1,3)$- tensors, respectively, by the following formulae:

$$\hat{T} = T^i{}_{jk}\, e_i \otimes \theta^j \otimes \theta^k \;, \qquad T^i{}_{jk} = \theta^i\big(\mathbf{T}(e_j,e_k)\big) \;, \tag{6.12}$$

$$\hat{R} = R^i{}_{jkl}\, e_i \otimes \theta^j \otimes \theta^k \otimes \theta^l \;, \qquad R^i{}_{jkl} = \theta^i\big(\mathbf{R}(e_k,e_l)e_j\big) \;. \tag{6.13}$$

We have seen in Sec. 2.3 that in any orientable and time orientable Lorentzian manifold $\mathbf{M}$, the general linear frame bundle $GL\mathbf{M}$ reduces to the Lorentz frame bundle $L\mathbf{M}$, whose elements $\boldsymbol{u}$ are orthonormal frames in which the metric tensor $\boldsymbol{g}$ assumes the form (3.5). Naturally, all the above definitions and results can be reduced to such Lorentz frame bundles by means of pull-backs $s^*\boldsymbol{\omega}$ of the connection forms in (6.3) to those sections of $GL\mathbf{M}$ which consist of orthonormal frames. However, there is an infinity of connections (cf. Sec. 10.2) which can be introduced in a $GL\mathbf{M}$ bundle and, despite the presence of a metric tensor, most of them will not preserve under parallel transport the orthonormality of the axes of a Lorentz frame. Hence, amongst all these connections, a special role is played by those which are compatible with the Lorentzian metric $\boldsymbol{g}$ in the following general sense: a connection in the linear frame bundle $GL\mathbf{M}$ is *compatible with a metric* $\boldsymbol{g}$ on its base manifold $\mathbf{M}$ if, for all smooth curves γ in $\mathbf{M}$, the parallel transport operator (4.8), defined in the tangent bundle $T\mathbf{M}$ by (4.10), is an isometry, i.e., if we have

$$\big\langle \tau_\gamma(x'',x')X, \tau_\gamma(x'',x')Y \big\rangle_{x''} = \langle X,Y \rangle_{x'} \,, \qquad \forall X,Y \in T_{x'}\mathbf{M} \;, \tag{6.14a}$$

$$\langle X,Y \rangle_x := \boldsymbol{g}(X,Y) \,, \qquad X,Y \in T_x\mathbf{M} \;. \tag{6.14b}$$

In view of the definition (4.11) of covariant derivatives, it is obvious that a connection in the linear frame bundle $GL\mathbf{M}$ is compatible with a metric $\boldsymbol{g}$ if and only if $\nabla \boldsymbol{g} \equiv \mathbf{0}$.

Taking into consideration that the Lie algebra $so(3,1)$ of $SO_0(3,1)$ is a six-dimensional subalgebra of $gl(4,\mathbf{R})$, it is convenient to express the covariant derivatives for connections compatible with the metric in $\mathbf{M}$ by means of the following pull-backs $s^*\omega_{ij}$ of the Cartan connection forms in (6.3) to sections s of $L\mathbf{M}$,

$$\omega^s_{ij}(X) = \omega_{ij}(s_* X) \,, \qquad \omega_{ij} := \eta_{ik}\omega^k_j = -\omega_{ji} \quad ; \tag{6.15a}$$

these pull-backs correspond to the following expansion of the connection form $\boldsymbol{\omega}$ on $L\mathbf{M}$ in terms of an $so(3,1)$-basis of infinitesimal generators of $SO_0(3,1)$:

$$\omega(\mathbf{X}) = \tfrac{1}{2}\omega_{ij}(\mathbf{X})\tilde{\mathbf{Y}}^{ij}, \qquad \mathbf{X} \in T_u LM \ . \tag{6.15b}$$

Indeed, the antisymmetry of the connection 1-forms in (6.15a) represents a necessary as well as a sufficient condition for a connection to be compatible with a given metric – as can be easily seen by setting $\boldsymbol{Y} = \boldsymbol{e}_i$ and $\boldsymbol{Z} = \boldsymbol{e}_j$ in the later derived equation (6.22). In the context of the subsequently defined and studied Levi-Civita connections, these connection forms are also called *Ricci rotation one-forms* since, on account of (3.6), $\boldsymbol{\omega}_{ij}(\boldsymbol{s}_*\boldsymbol{X})$ can be viewed, for "infinitesimal" vectors $\boldsymbol{X}$, as being the components of Lorentz transformations describing the change in the orientation of the axes of Lorentz frames that are parallel transported by an "infinitesimal" amount in the direction of $\boldsymbol{X}$.

In view of (6.15), for connections compatible with a metric, the expression (5.22) for the covariant derivative of a given vector field $\boldsymbol{\Psi}$, which assumes its values $\boldsymbol{\Psi}_x$ within the total space **E** of some vector bundle (**E**, $\boldsymbol{\pi}$, **M**, **F**) associated with (L**M**, $\boldsymbol{\Pi}$, **M**, $SO_0(3,1)$) through the representation U of the Lorentz group $SO_0(3,1)$, can be expressed as follows:

$$\nabla_X \boldsymbol{\Psi}_x = [\partial_X + \tfrac{1}{2}\omega^s_{ij}(X)\hat{\mathbf{M}}^{ij}_u]\boldsymbol{\Psi}_x \ , \qquad \hat{\mathbf{M}}^{ij}_u = -\hat{\mathbf{M}}^{ji}_u \ , \tag{6.16a}$$

$$\exp(\hat{\mathbf{M}}^{ij}_u t)\boldsymbol{\Phi}_\rho(\boldsymbol{u}) = \boldsymbol{\Phi}_\rho(\boldsymbol{u}\cdot\exp(\tilde{\mathbf{Y}}^{ij}t)) \ , \qquad i,j = 0,1,2,3 \ . \tag{6.16b}$$

In (6.16), $\hat{\mathbf{M}}^{ij}_u$ are the infinitesimal generators of Lorentz transformations describing the aforementioned "infinitesimal rotations" of the (ij)-axes of Lorentz frames.

According to the general theory of affine connections (cf. [K], Sec. III.3), to every linear connection on GL**M** corresponds a unique *affine connection form* $\tilde{\boldsymbol{\omega}}$ on the general affine frame bundle GA**M** described in Sec. 2.3, which is such that

$$\gamma^*\tilde{\boldsymbol{\omega}} = \boldsymbol{\omega} + \boldsymbol{\theta}, \qquad \gamma : \boldsymbol{u} \mapsto (\mathbf{0},\boldsymbol{u}) \in GA\mathbf{M} \ , \qquad \boldsymbol{u} \in GL\mathbf{M} \ . \tag{6.17}$$

In particular, this observation can be used to extend connections, that are compatible with a Lorentzian metric, from the Lorentz frame bundle L**M** to the Poincaré frame bundle P**M**. Hence, in order to compute the covariant derivative of a vector field $\boldsymbol{\Psi}$, which in general assumes its values $\boldsymbol{\Psi}_x$ within the total space **E** of some vector bundle associated with the principal bundle (P**M**, $\boldsymbol{\Pi}$, **M**, $ISO_0(3,1)$) through the representation U of the Poincaré group $ISO_0(3,1)$, we shall proceed as we did in (5.18) by first selecting a section of P**M**:

$$\boldsymbol{s} : x \mapsto (\boldsymbol{a},\boldsymbol{u}) \in \boldsymbol{\Pi}^{-1}(x) \subset P\mathbf{M} \ , \qquad x \in \mathbf{M}^s \subset \mathbf{M} \ . \tag{6.18}$$

We can then easily adapt to P**M** the arguments that have led to (5.22), thus arriving at the following expression for the required covariant derivative (Drechsler, 1977b, 1984),

$$\tilde{\nabla}_X \boldsymbol{\Psi}_x = [\partial_X + \tilde{\boldsymbol{\theta}}^i(X)\hat{\mathbf{P}}_{i;u} + \tfrac{1}{2}\tilde{\boldsymbol{\omega}}_{ij}(X)\hat{\mathbf{M}}^{ij}_u]\boldsymbol{\Psi}_x \ , \tag{6.19a}$$

$$\partial_X \boldsymbol{\Psi}_x = \left[X^\mu \partial_\mu\left(\Psi^\rho_{x;a,u}\right)\right]\boldsymbol{\Phi}_\rho(\boldsymbol{a},\boldsymbol{u}) \ , \qquad \tilde{\boldsymbol{\theta}}^i_{a,u}(X) = \boldsymbol{\theta}^i_u(X) + (\nabla_X \boldsymbol{a})^i \ . \tag{6.19b}$$

in which, for the sake of simplicity in notation, we have dropped, in the *soldering forms*[33] $\tilde{\boldsymbol{\theta}}^i$ as well as in $\tilde{\boldsymbol{\omega}}_{ij}$, the indices indicating their dependence on the choice of section $\boldsymbol{s}$ of

*GA*M. In (6.19b) $\nabla_X \boldsymbol{a}$ denotes the covariant derivative of the vector field $\boldsymbol{a}$ assuming values within *T*M, so that $(\nabla_X \boldsymbol{a})^i$ is the i-th component of that field in relation to the axes of the linear frame $\boldsymbol{u}$; moreover, in complete analogy with (5.21) and (6.16b), we have set

$$\exp(\hat{\mathbf{P}}_{i;\boldsymbol{u}}t)\boldsymbol{\Phi}_\rho(\boldsymbol{a},\boldsymbol{u}) = \boldsymbol{\Phi}_\rho(\boldsymbol{a}\cdot\exp(\tilde{\mathbf{Y}}_i t),\boldsymbol{u}) \ , \qquad i = 0,1,2,3 \ , \tag{6.20a}$$

$$\exp(\hat{\mathbf{M}}^{ij}_{\boldsymbol{u}}t)\boldsymbol{\Phi}_\rho(\boldsymbol{a},\boldsymbol{u}) = \boldsymbol{\Phi}_\rho(\boldsymbol{a},\boldsymbol{u}\cdot\exp(\tilde{\mathbf{Y}}^{ij}t)) \ , \qquad i,j = 0,1,2,3 \ . \tag{6.20b}$$

We note that, in the derivation of (6.16) and (6.19), the connection was assumed to be compatible with the metric $\boldsymbol{g}$, but not necessarily torsion-free. Lorentzian manifolds which carry connections that are compatible with the metric, but for which the the torsion form in (6.6) is not equal to zero, are used in physics to describe *Riemann-Cartan spacetimes* (Hehl *et al.*, 1973; Drechsler, 1984), so that we shall call them *Riemann-Cartan connections*. An infinity of Riemann-Cartan connections can be introduced in *L*M despite the presence of a metric tensor, and they can be characterized by the various values assumed by the components of the torsion[34] tensors in (6.12). Classical general relativity is, however, predicated on the assumption that the connection of spacetime is not only compatible with the metric, but that in addition its torsion tensor identically equals zero[35].

A connection in the linear frame bundle *GL*M is called a *Levi-Civita connection* (or a *Riemannian connection*) on the, in general, (pseudo-)Riemannian[36] manifold (M,$\boldsymbol{g}$), if in addition to being compatible with the given (pseudo-)Riemannian metric $\boldsymbol{g}$ on M, it is also torsion-free – i.e., it is such that $\boldsymbol{\Theta} = \mathbf{0}$, so that the torsion tensor is identically equal to zero everywhere in M. The *fundamental lemma of Riemannian geometry* then states that each (pseudo-)Riemannian manifold (M, $\boldsymbol{g}$) possesses a unique Levi-Civita connection.

The proof of this fundamental lemma turns out to be quite straightforward (cf. [C], p. 308; [SC], p. 256). In its simplest version, it is based on the observation that, by (6.10),

$$\mathbf{T}(X,Y) = \mathbf{0} \quad \Leftrightarrow \quad \nabla_X Y - \nabla_Y X = [X,Y] \ , \tag{6.21}$$

and that $\nabla \boldsymbol{g} \equiv \mathbf{0}$ implies the *Ricci identity*

$$X(\boldsymbol{g}(Y,Z)) = \boldsymbol{g}(\nabla_X Y,Z) + \boldsymbol{g}(Y,\nabla_X Z) \tag{6.22}$$

for any *T*M-vector fields X,Y and Z defined within some common domain in M. Hence, upon using (6.21) in (6.22), and then algebraically combining the outcome with its two counterparts obtained by cyclically permuting X,Y and Z, we arrive at the following result,

$$\begin{aligned} \boldsymbol{g}(\overline{\nabla}_X Y,Z) = {} & \tfrac{1}{2}\left[X(\boldsymbol{g}(Y,Z)) + Y(\boldsymbol{g}(X,Z)) - Z(\boldsymbol{g}(X,Y))\right] \\ & + \tfrac{1}{2}\left[\boldsymbol{g}(Z,[X,Y]) - \boldsymbol{g}(X,[Y,Z]) - \boldsymbol{g}(Y,[X,Z])\right] \ , \end{aligned} \tag{6.23}$$

for the case where $\overline{\nabla}_X$ is the covariant derivative that corresponds to a Levi-Civita connection. Thus, upon inserting $X = \boldsymbol{e}_i$, $Y = \boldsymbol{e}_j$ and $Z = \boldsymbol{e}_k$ into (6.23), where $\boldsymbol{u} = (\boldsymbol{e}_0,\dots,\boldsymbol{e}_3)$ are the linear frames in the chosen section (5.18) of *GL*M, we obtain expressions which uniquely relate the metric components g_{ij} to the connection coefficients in the chosen Cartan gauge:

$$\Gamma^i{}_{jk} := \theta^i(\overline{\nabla}_j e_k) = g^{il} g(e_l, \overline{\nabla}_j e_k) , \qquad \overline{\nabla}_j := \overline{\nabla}_{e_j} , \tag{6.24a}$$

$$g^{ij} g_{jk} = \delta^i{}_k , \qquad g_{jk} = g(e_j, e_k) . \tag{6.24b}$$

On the other hand, from the general properties (4.12) of covariant derivatives, we obtain

$$\overline{\nabla}_i e_j = \Gamma^k{}_{ij} e_k , \qquad \overline{\nabla}_i \theta^j = -\Gamma^j{}_{ik} \theta^k , \tag{6.25}$$

so that according to (5.22) and (6.3), the connection coefficients in (6.24a) determine the gauge potentials of a Levi-Civita connection in $GL\mathbf{M}$. This means that the Levi-Civita connection on $\mathbf{M}$ is completely determined by the choice of metric g.

The curvature tensor (6.13) corresponding to a Levi-Civita connection is called the *Riemann curvature tensor* of the manifold on which that connection is defined. The components of a Riemann curvature tensor can be easily computed in any moving frame from the connection coefficients of the Levi-Civita connection, by using (6.11) and (6.25), as well as the general properties (4.12) of covariant derivatives. The result is:

$$R^i{}_{jkl} = e_k(\Gamma^i{}_{lj}) - e_l(\Gamma^i{}_{kj}) + \Gamma^i{}_{km}\Gamma^m{}_{lj} - \Gamma^i{}_{lm}\Gamma^m{}_{kj} - \theta^i(\overline{\nabla}_{[e_k, e_l]} e_j) . \tag{6.26}$$

The significance of the Riemann curvature tensor for physics, as well as for differential geometry, emerges from the following fundamental theorem of Riemannian geometry (cf. [C], p. 310): *A Riemannian or pseudo-Riemannian manifold is locally flat if and only if its Riemann curvature tensor is identically equal to zero.*

For Lorentzian manifolds $(\mathbf{M}, g)$, the above theorem signifies that $\mathbf{R} \equiv \mathbf{0}$ if and only if $(\mathbf{M}, g)$ is locally isometric either to the entire Minkowski space, or to a 4-dimensional submanifold of the Minkowski space. In this context, it should be recalled that two Lorentzian manifolds $(\mathbf{M}', g')$ and $(\mathbf{M}'', g'')$ are *isometric* if there is a diffeomorphism ϕ : $\mathbf{M}' \to \mathbf{M}''$ of one onto the other which preserves the metric, i.e., which is such that $g'' = \phi_* g'$. In CGR, isometric Lorentzian manifolds are deemed to describe physically indistinguishable spacetimes (cf. [W], as well as Chapter 11).

2.7. Einstein Field Equations and Principles of General Relativity

The main feature that distinguishes CGR from all other physical theories that preceded it, is that in CGR the energy and momentum contained in matter and nongravitational radiation is coupled to the geometric structure of spacetime. Hence, the principal geometric features of the spacetimes of CGR theories are not adopted on *a priori* grounds – as it is the case in Newtonian and other physical theories. Rather, they reflect the dynamics of that interaction.

This central feature of CGR is achieved by assuming that any possible form of distribution of matter and nongravitational radiation in our universe can be described by a *stress-energy tensor field* $\overline{T}$, and that $\overline{T}$ is coupled to the metric field g of a Lorentzian manifold $(\mathbf{M}, g)$ that describes spacetime in classical terms. In CGR this coupling is embedded into the *Einstein equation*, which, in the Planck units[37] and the (+1,–1,–1,–1) signature that we use throughout this monograph, assumes the following form:

$$\overline{\boldsymbol{R}} - \tfrac{1}{2} \boldsymbol{g} R = -\kappa \overline{\boldsymbol{T}} \ , \qquad \kappa = 8\pi G = 8\pi \ . \tag{7.1}$$

The left-hand side of this equation is known as the *Einstein tensor*. In addition to the metric tensor $\boldsymbol{g}$, it contains the *Ricci tensor* and the *Riemann curvature scalar*, respectively defined in terms of the Riemann curvature tensor in (6.13) as follows :

$$\overline{\boldsymbol{R}} = R_{ij} \boldsymbol{\theta}^i \otimes \boldsymbol{\theta}^j \ , \qquad R_{ij} = R^k{}_{ikj} \ , \tag{7.2a}$$

$$R = R^i{}_i \ , \qquad R^i{}_j := g^{ik} R_{kj} \ . \tag{7.2b}$$

In terms of the components of these tensors with respect to any moving frame, i.e., in any Cartan gauge, (7.1) assumes the familiar form of the Einstein field equations:

$$R_{ij} - \tfrac{1}{2} g_{ij} R = -8\pi T_{ij} \ . \tag{7.3}$$

The fact that the above equations intertwine the stress-energy tensor and the metric tensor into a net of mutual interactions is best seen from (6.23), (6.24) and (6.26). The traditional way of dealing with this system of nonlinear equations is to express them in a holonomic frame – for which we shall employ from now on exclusively Greek indices. Thus, in any such holonomic frame, (6.23) and (6.24) supply the following expression for the connection coefficients in (6.24a) (known in CGR as *Christoffel symbols*):

$$\Gamma^\kappa{}_{\mu\nu} = \tfrac{1}{2} g^{\kappa\lambda} \left(g_{\lambda\mu,\nu} + g_{\lambda\nu,\mu} - g_{\mu\nu,\lambda} \right) \ , \qquad g_{\lambda\mu,\nu} := \partial g_{\lambda\mu} / \partial x^\nu \ . \tag{7.4}$$

This is due to the fact that the Lie bracket of any pair of holonomic frame elements vanishes, so that when (6.23) is used in (6.24a) the second of the square brackets in (6.23) becomes equal to zero. For the same reason the last term in (6.26) also equals zero, so that

$$R^\kappa{}_{\lambda\mu\nu} = \Gamma^\kappa{}_{\lambda\nu,\mu} - \Gamma^\kappa{}_{\lambda\mu,\nu} + \Gamma^\kappa{}_{\mu\rho} \Gamma^\rho{}_{\lambda\nu} - \Gamma^\kappa{}_{\nu\rho} \Gamma^\rho{}_{\lambda\mu} \ , \qquad \Gamma^\kappa{}_{\lambda\mu,\nu} := \partial \Gamma^\kappa{}_{\lambda\mu} / \partial x^\nu \ . \tag{7.5}$$

On the other hand, in any vierbein gauge the (generally nonholonomic) frames of that gauge are orthonormal, so that (3.4) holds, and therefore it is the first of the square brackets in (6.23) that vanishes when it is used in (6.24a). Therefore in a vierbein gauge, generally supplied by a section (5.18) given by means of the vierbein fields in (3.3), the equation (6.23) leads to the following expressions relating the connection coefficients in (6.24a) (which in CGR are known as the *Ricci rotation coefficients* [M,W]), to the Christoffel symbols in (7.4) as well as to the vierbein fields $\lambda_i{}^\mu$ in (3.3):

$$\Gamma^i{}_{jk} = \lambda^i{}_\mu \lambda_j{}^\nu \left(\lambda_k{}^\kappa \Gamma^\mu{}_{\nu\kappa} + \partial_\nu \lambda_k{}^\mu \right) \ . \tag{7.6}$$

We note that in a vierbein gauge the last term in (6.26) does not vanish. However, the 256 components of the Riemann tensor are related by the following identities (cf., e.g., [C], p. 309):

$$R_{ijkl} = -R_{ijlk} = -R_{jikl} = R_{klij} \ , \qquad R_{ijkl} := \eta_{im} R^m{}_{jkl} \ , \tag{7.7a}$$

$$R_{ijkl} + R_{iklj} + R_{iljk} = 0 \ , \tag{7.7b}$$

$$R_{ijkl;m} + R_{ijlm;k} + R_{ijmk;l} = 0 \ , \qquad R_{ijkl;m} := \overline{\nabla}_m R_{ijkl} \ . \tag{7.7c}$$

The first of the equalities in (7.7a), as well as the equalities in (7.7b) and (7.7c), can be easily derived in an arbitrary Cartan gauge – the last two of these equations being obtained by rewriting the Bianchi identities (6.8) in terms of the components of the Riemann tensor, for which therefore $\mathbf{\Theta} = \mathbf{0}$ in the first of those identities. The remaining equalities in (7.7a) are most easily derived in a vierbein gauge [C], by taking advantage of the antisymmetry $\omega_{ij} = -\omega_{ji}$ of the Ricci connection coefficients in (6.14).

We note that by contracting the second Bianchi identity (7.7c), we get, in any Cartan gauge, the following set of four identities (cf., e.g., Straumann, 1984, p. 60):

$$\overline{\nabla}_i \left(R^i{}_j - \tfrac{1}{2} \delta^i{}_j R \right) = 0 \ , \qquad R^i{}_j := g^{ik} R_{kj} \ . \tag{7.8}$$

According to (7.3), these identities imply a conservation law for the stress-energy tensor:

$$\overline{\nabla}^i T_{ij} = 0 \ , \qquad \overline{\nabla}^i := g^{ij} \overline{\nabla}_j \ . \tag{7.9}$$

In view of the nonlinearity of the Einstein field equations in (7.3), as well as the fact that some of them are constraint equations, the solution of the generic initial-value problem in CGR represents a difficult problem [W], whose discussion we shall postpone until Chapter 11. Fortunately, it proved possible already in the early days of CGR to construct explicitly for vacuum spacetime regions, where therefore $T_{ij} = 0$, nontrivial solutions – i.e., solutions which did not describe merely flat spacetime regions, that by the last theorem cited in the preceding section would be isometric to regions of Minkowski space. The best known of these solutions, as well as the earliest, was the Schwarzschild (1916) solution, obtained by requiring that the metric tensor be static as well as spherically symmetric [M,N,W]. This solution proved suitable for describing such features of planetary motion as perihelion precession. Furthermore, cosmological models suitable for the description of the large scale features of our universe were also obtained already in the early days of CGR by adopting very simple forms of the stress-energy tensor – such as those for a perfect fluid, whose only parameters are its density and its proper 4-velocity [M,N,W]. Indeed, upon making simplifying assumptions about the metric tensor, which reduce the number of its independent components by the imposition of various symmetries, one can solve in those cases the Einstein equations (7.3) for the yet undetermined metric components [M,N,W]. These models (amongst which the Robertson-Walker ones are the best-known since they represent viable candidates for contemporary cosmological models) provide, together with the Schwarzschild solution, most of the fundamental predictions of CGR that have been thus far successfully confirmed by astronomical observations or by experiments.

As we mentioned at the beginning of Sec. 2.4, all these fundamental predictions are based on the *geodesic postulate*[25], which can be stated as follows: *all neutral and massive point particles in free fall within a classical spacetime manifold* **M** *follow timelike*[38] *geodesics, whereas all zero-mass particles in free fall follow null geodesics.* In other words, if $\boldsymbol{v}$ denotes the 4-velocity vector along one of their worldlines $\gamma = \{x(\tau) | \tau \in \mathbf{R}^1\}$, then there is a choice of parameter τ (namely the *affine parameter* of the geodesic, which in the case of massive particles represents their proper time [C,M,W]) for which

$$\overline{\nabla}_{v}\dot{x} = 0 \ , \qquad v = \dot{x} := \dot{x}^{\mu}\partial_{\mu} \ , \qquad \dot{x}^{\mu} := dx^{\mu}/d\tau \ . \tag{7.10a}$$

In any coordinate chart whose domain contains some segment of γ, (7.10a) yields the well-known form of the geodesic equation for that segment:

$$\ddot{x}^{\lambda} + \Gamma^{\lambda}_{\mu\nu}\dot{x}^{\mu}\dot{x}^{\nu} = 0 \ , \qquad \ddot{x}^{\lambda} := d^2x^{\lambda}/d\tau^2 \ . \tag{7.10b}$$

On the other hand, in a vierbein gauge, we can expand the 4-velocity vector with respect to the orthonormal frames of the L**M**-section (5.18) corresponding that gauge, to obtain from (7.10a) the system of equations

$$\dot{v}^{i} + \Gamma^{i}{}_{jk}v^{j}v^{k} = 0 \ , \qquad \dot{v}^{i} := dv^{i}/d\tau \ , \qquad v^{i} = \lambda^{i}{}_{\mu}\dot{x}^{\mu} \ , \tag{7.10c}$$

for the geodetic segment which lies within the common domain of the holonomic and nonholonomic frames. Thus, the formulation of (7.10c) in terms of coordinates involves then a system of eight first-order equations, rather than the four second-order equations in (7.10b), but it is nevertheless more suitable for the formulation of the equivalence principle and for the description of free-fall phenomena – as will be shown next.

We shall say that a Cartan gauge, given by a section s of the general linear frame bundle GL**M**, is *adapted to a smooth curve* γ if $\Gamma^{i}{}_{jk} = 0$ at all $x \in \gamma$ which lie in the domain of definition $\mathbf{M}^s$ of that gauge. In particular, a *vierbein gauge adapted to* γ corresponds to a section s of the Lorentz frame bundle L**M** for which $\Gamma^{i}{}_{jk} = 0$ at all $x \in \gamma$ which lie in $\mathbf{M}^s$. Such vierbein gauges can be obtained by restricting an arbitrary section s' of L**M** to a 3-dimensional hypersurface intersected by γ at one point x', constructing a smooth flow [C] containing γ that consists of curves which intersect that hypersurface at other points in a neighborhood of x', and then parallel transporting all the frames at those points along the curves of the given flow. Since the Levi-Civita connection is compatible with the metric of the Lorentzian manifold describing spacetime in CGR, such parallel transport will preserve the mutual orthonormality of the axes in all these frames, so that a section s of L**M** will be obtained that has the above required property.

We shall call a section s of the general linear frame bundle GL**M**, that is adapted to a timelike geodesic γ, an *inertial moving frame* along the geodesic γ, if in the set

$$s_{\gamma} = \{(e_0(x),\dots,e_3(x)) | x \in \gamma \cap \mathbf{M}^s\} \subset s \quad , \tag{7.11}$$

all the frame elements $e_0(x)$, $x \in \gamma$, are equal to the tangent vectors to that geodesic for a suitable choice of affine parameter τ. If, in particular, all the frames in s_γ are orthonormal, we shall call s an *inertial Lorentz moving frame* for γ. Indeed, in that case the restriction s_γ of s has the following properties,

$$g_{ij}(x) = g(e_i(x), e_j(x)) = \eta_{ij} \ , \qquad e_0(x) = \dot{x} \ , \qquad \forall x \in \gamma \cap \mathbf{M}^s \ , \tag{7.12a}$$

$$\Gamma^i{}_{jk}(x) = g^{il} g\left(e_l(x), \overline{\nabla}_{e_j(x)} e_k(x)\right) = 0 \ , \qquad \forall x \in \gamma \cap \mathbf{M}^s \ , \tag{7.12b}$$

so that (7.11) depicts a Lorentz frame that is in free-fall, and which moves in such a manner that an observer situated at its origin has the geodesic γ as his worldline.

Let us introduce now the *normal coordinates associated with an inertial Lorentz moving frame* by the following construction: we set $x^0(\tau) = \tau$ at all points along γ, whereupon at each point $x \in \gamma$ we draw all the spacelike geodesics with tangents

$$X = X^\alpha e_\alpha(x) \in T_x\mathbf{M} \ , \qquad \alpha = 1,2,3, \qquad x \in \gamma \ , \tag{7.13}$$

in affine parametrization; then, in some neighborhood of $x(\tau) \in \gamma$, we assign to each point y which lies on one of these geodesics at unit parametric distance from $x(\tau)$ the coordinates (τ, X^1, X^2, X^3). These coordinates will be therefore defined within some tube around γ. Moreover, at each $x \in \gamma$, we shall have $\partial_0 = e_0(x)$ and $\partial_\alpha = e_\alpha(x)$, $\alpha = 1,2,3$, so that they determine a "local Lorentz frame" in the sense defined in §8.6 of [M].

Consequently, the CGR equivalence principle can be now stated in the following mathematically precise form (cf. Norton, 1989, for a historical analysis of Einstein's original formulation, and Sec. 1.3 for the form adopted in [M]): *For any inertial Lorentz moving frame in free-fall along some timelike geodesic γ, all the nongravitational laws of physics, expressed in the normal coordinates associated with that inertial frame, should at each point along γ equal, up to first-order terms in those coordinates, their special relativistic counterparts expressed in the tensor coordinates associated with the respective Lorentz frames.*

The above seemingly unnecessarily complicated reformulation of the equivalence principle is motivated by M. Friedman's (1983) careful analysis of the various shades of meaning of this fundamental principle. Indeed, that analysis concludes with the following observations: "Standard formulations of the principle of equivalence characteristically obscure [the] crucial distinction between first-order laws and second-order laws by blurring the distinction between 'infinitesimal' laws, holding at a single point, and local laws, holding on a neighborhood of a point. ... What the principle of equivalence says, then, is that special relativity and general relativity have the same 'infinitesimal' structure, not that they have the same local structure." (Friedman, 1983, p. 202).

Because of the presence of second-order laws, such as those involving curvature-dependent terms, the Lorentz and Poincaré gauge invariance of CGR cannot be inferred from the equivalence principle alone. Furthermore, as we pointed out in the second of the introductory paragraphs of this chapter, Einstein's original formulation of the general covariance principle had no physical content whatever, although in the special relativistic context the covariance of physical laws under (global) Lorentz and Poincaré transformations is an experimentally verifiable feature of those laws. The fundamental reason for this difference is that the Minkowski coordinates of special relativity have an operational meaning, whereas those in general relativity are merely labels – namely, as we have seen in Sec. 2.1, they are only a part of the definition of charts that provide a topological space with the structure of a differential manifold.

This suggests an operational reinterpretation of the general covariance principle (cf. [P], Sec. 4.1), whereby operational significance is imparted to the Riemann normal coordinates [M] associated with an inertial Lorentz moving frame. We can say that a neighborhood of some point $x \in \mathbf{M}$ displays, to an observer in free fall along a geodesic γ, a spacetime region of maximum linear dimension $l \in \mathbf{R}^1$ if l is the supremum, over all the points in that neighborhood, of the Riemann normal coordinate values in the inertial Lorentz moving frames with respect to which that observer is at rest. Furthermore, we can then adopt the supremum R of the Gaussian curvature radii corresponding, in Riemann normal coordinates, to all $\tau =$ const. hypersurfaces within that neighborhood as a basic measure of the maximum "curvature radius" measurable by that observer over that neighborhood – e.g., by the geodesic deviation method (cf. [M], Box 14.1). We can then expect that, for any two observers O_i, $i = 1,2$, who are in free fall, and whose geodetic worldlines intersect at a point $x \in \mathbf{M}$, the outcomes of their respective measurements expressed in Riemann normal coordinates display Lorentz and Poincaré covariance, up to the n-th order of magnitude in the maximum linear dimensions l_i of the neighborhood of x over which their joint measurements are performed, if the ratios l_i/R_i, $i = 1,2$, are of at most $n+1$ order of magnitude.

These considerations are, in principle, experimentally implementable, and are in fact close to actual methods of verifying Lorentz and Poincaré covariance in the real world around us, where gravitational effects are ever-present, but can be regarded as being so "small" as to be ordinarily negligible in experiments involving atomic or subatomic phenomena. The above formulation then provides a mathematically clear-cut criterion as to the required degree of experimental accuracy which would indeed render such general-relativistic effects "negligible" over given regions of spacetime within which a given set of observations is carried out. If it is then taken for granted that – given the typical size of these regions and the Gaussian curvature radii prevailing under terrestrial conditions – Lorentz and Poincaré covariance have received more than adequate experimental confirmation, they suggest that the mathematical embodiment of the above described operational principle of general covariance consists of imposing Lorentz and Poincaré gauge invariance in all general relativistic theories.

In this context, it should be noted that in CGR, Lorentz and Poincaré gauge invariance are *operationally* indistinguishable – although some authors of CGR texts give preference either to one or to the other on ideological grounds. Indeed, the basic *directly* observable covariant quantities of CGR, such as velocities, momenta and accelerations, are insensitive to affine structure, since their components transform under a Lorentz transformation Λ in the same manner as they do under any of the associated Poincaré transformations (a,Λ). On the other hand, we have seen in Sec. 2.5 that to each linear connection corresponds an affine connection[23]. However, the additional connection coefficients of that affine connection, namely the soldering forms in (6.19), are of significance only in the quantum context, since quantum state vectors are not insensitive to affine structure – as witnessed by all the representations of the Poincaré group introduced in the next chapter.

Thus, it is only the Levi-Civita connection in $L\mathbf{M}$ that is of significance[4] to CGR. Its extension to $P\mathbf{M}$ is obtained when we remove from (6.19b) the canonical forms $\boldsymbol{\theta}^i$. We then arrive at a special case of a generalized affine connection in the sense of Kobayashi and Nomizu (1963), which coincides with the Levi-Civita connection on those sections of

PM for which $\boldsymbol{a} \equiv \mathbf{0}$, so that they are also sections of *LM*. It is this connection that will play a central role in formulating the concept of GS propagation within the quantum bundles introduced and studied from Chapter 5 onwards.

Notes to Chapter 2

1 Such textbooks as (Adler *et al.*, 1975), (Weinberg, 1972) and [N] provide typical examples; [M] and [W], as well as (Hawking and Ellis, 1973) and (Straumann, 1987) are partial exceptions, whereas (Göckeler and Schücker, 1987) represents a total break with the old tradition.

2 The introductory text [SC] presents in Chapter 5 the **M**-coordinate-based formulation of connections that had emerged from the classic works of Gauss, Riemann, Levi-Civita and other leading geometers of the last century, and then gradually introduces the reader to the modern formulation.

3 In [M] this principle is ignored on account of Kretchmann's remarks [M, p. 431]. According to [W] this principle stipulates that "there are no preferred vector fields or preferred bases of vector fields" [W, p. 57]. On the other hand, this feature of CGR results from the imposition of Lorentz gauge invariance. In Sec. 11.3 general covariance will be related to the diffeomorphism group viewed as a 'gauge group'.

4 Cf., e.g., (Trautman, 1981), p. 306. As discussed in Secs. 2.3, 2.6 and 2.7, the Lorentz gauge symmetry can be always extended into Poincaré gauge symmetry for the tensor bundles that occur in CGR.

5 A *topological space* is a set which contains a family of subsets, which are called *open* sets, to which the set itself as well as the empty set belong, and which are such that the union of any family of open sets, as well as the intersection of any *finite* number of open sets, is again an open set. A map f from one topological space into another is continuous if, for every open set O_2 in the second space, the set $O_1 = f^{-1}(O_2)$ in the first space (which consists of all points in it mapped into O_2) is also open. A *homeomorphism* between two topological spaces is a one-to-one map of one onto the other (i.e., a *bijection*) which is continuous, and whose inverse is also continuous (cf. Sec. 1C in [C] for further details).

6 This definition corresponds to a C^∞-manifold. In CGR it could be considerably relaxed (Hawking and Ellis, 1973) to that of various C^n-manifolds. The latter, however, do not contribute substantially new structures and results in CGR.

7 Here and throughout the remainder of this monograph, we shall use the *Einstein convention* of summation over all values $\mu = 0,...,3$ in each pair of repeated contravariant and covariant indices.

8 In a general bundle **E** and **M** are topological spaces, and the projection map $\boldsymbol{\pi}$ is only continuous [C,I].

9 A *diffeomorphism* between two manifolds is a one-to-one smooth map of one of them onto the other, whose inverse is also smooth (where, in accordance with Note 6, we restrict our attention to C^∞-manifolds). It is therefore natural to identify **M** with a particular submanifold of **E** to which it is diffeomorphic although, strictly speaking, as sets the two are generally distinct. In this context, note that a subset **N** of any given m-dimensional manifold **E** is an n-dimensional *submanifold* of **E** if every point $x \in \mathbf{N}$ is in the domain of a chart (M,ϕ) of **E** which is such that $\phi(x) = (x^1,...,x^n,y^1,...y^{m-n})$, where $(y^1,...y^{m-n})$ is a fixed element of $\mathbf{R}^{m-n}$; it then follows that an atlas of charts (M,ϕ) of **E** gives rise to an atlas of charts $(\mathbf{N}\cap M,\psi)$ of **N** if we set $\psi(x) = (x^1,...,x^n)$ for all $x \in \mathbf{N}\cap M$. (For the mathematically precise definition of other basic concepts of differential geometry, we direct the reader to [C], [I] or [SC]; however, in general, an intuitive grasp of all those concepts which we have not defined either in the main text, or in the notes, should suffice for the understanding of the all basic aspects of the subsequent arguments and proofs.)

10 Kobayashi and Nomizu (1963), as well as other authors, denote the (general) linear frame bundle by *L*(**M**), rather than by *GLM*. We shall use, however, the symbol *LM* to denote the *orthonormal-frame* bundle of local Lorentz frames, which is introduced in the next section.

11 Note that, in case that g is a matrix that acts by matrix multiplication on the elements of $\mathbf{R}^n$, for its action from the right those elements have to be viewed as one-row matrices, whereas for its action from the left they have to be viewed as one-column matrices, so that one mode of such action can be related to the other by taking the transposes of the matrices in question.

12 In the considered case of C^∞-bundles the structure group **G** is required to be a Lie group. As such, **G** is a differential manifold, so that $\mathbf{P}\times\mathbf{G}$ is defined as the product of two manifolds [C,I].

13 There are other approaches to the definition of associated bundles which do not assign prominence to bundle coordinates (cf. the **G**-product definition in Sec. 4.3). However, the present approach, which closely follows that in [C], is the most natural one in those cases where a manifold structure is available in **E** prior to a bundle structure – as is the case with the bundles treated in this chapter.

14 In case that **E** does not have an a priori manifold structure, but it is constructed, as is the case with the quantum bundles first introduced in Chapter 4, by taking a **G**-product of a principal frame bundle and a standard fibre **F**, then the local trivialization maps (2.4) will be derived from *soldering maps* (cf. Secs. 4.3 and 5.1), thus imparting to **E** a bundle structure by (implicitly or explicitly) soldering quantum frames in those fibres to the elements of principal frame bundles to which they thus become associated.

15 Note that *left action* gives rise to group multiplication in an order which is the reversal of that for *right action*, so that representations by diffeomorphisms (which some authors call "realizations" – cf. [C], p. 162) acting from the right and from the left satisfy two distinct composition laws, that are given, respectively, by $U(g)\circ U(g') = U(g'g)$ and $U(g)\circ U(g') = U(gg')$.

16 This can be always achieved at each single point x separately, but not so that the same would hold true in an entire neighborhood of that chosen point if the Riemann curvature tensor is not zero at x.

17 Note that the second set of equalities in (3.3) follows from the duality conditions in (1.10), that relate holonomic frames to their dual frames, and similarly the first set follows from the counterparts of these relationships, which in turn relate tetrads to their respective duals.

18 The terminology varies somewhat from author to author, with the term *tetrad* used as an alternative to *vierbein* (Göckeler and Schücker, 1987), rather than as the value assumed by a vierbein at a given point.

19 Throughout this monograph, *Lorentz group* means the *restricted* Lorentz group, i.e., the proper orthochronous Lorentz group $SO_0(3,1)$. If SO(3,1) denotes the group of matrices in O(3,1) which have a determinant one (so that neither time inversions nor space reflections are included in it), then $SO_0(3,1)$ is its largest subgroup which does not incorporate any spacetime reflections – cf. (Cornwell, 1984), p. 66.

20 In some textbooks [C] the definition of bundle reducibility is formulated in terms of bundle embeddings, so that the present statements would represent a necessary and sufficient condition for reducibility.

21 A topological space is *compact* if every covering of that space with open sets has a finite subcovering [C,W]. However, compact spacetimes exhibit acausal structure, and hence are suspect on physical grounds – cf. (Hawking and Ellis, 1973), p. 189, or (Geroch and Horowitz, 1979), p. 242. Robertson-Walker spacetimes (cf. Chapter 5 in [W]) provide examples of important cosmological models which are not compact, and which at the present time are deemed to provide a good description of the large-scale structure of our universe.

22 Spin structures will be defined and studied in Chapter 6, while formulating the quantum theory for GS spin-1/2 excitons. Their detailed treatment at the classical level can be found in Sec. 13.2 of [W], whereas a brief treatment is given in [C], pp. 415-418. Avis and Isham (1980) have proposed a generalization of this concept.

23 In treating affine frame bundles, and in the next section affine connections, we adopt the terminology in Chapter III, Sec. 3 of [K] – where all the relevant proofs can be also found. It should be noted that, whereas in this terminology an affine connection is a connection on the affine frame bundle $G A\mathbf{M}$, some authors ([N]; Friedmann, 1983; Straumann, 1984) refer by this name to a Koszul connection [SC] defined on a tangent bundle, as well as on tensor bundles in general.

24 This terminology is most suitable, but it is not standard. In fact, in all past publications on GS theory (Prugovečki, 1987-90) we followed Drechsler (1982, 1984), by using $L_A\mathbf{M}$ to denote the Poincaré frame bundle $P\mathbf{M}$, and referring to it as the affine Lorentz frame bundle over **M**.

25 In subsequent years, Einstein and Grommer (1927), as well as many others, attempted to derive this *geodesic* (or *geodetic*) *postulate* from the CGR equations of motion by treating a massive point particle as a limiting case of a continuous distribution of mass in mutual interaction with gravitational fields. However, the claims as to their success "are based on formal, so far not rigorously justifiable approximation methods" (Ehlers, 1987, p. 65). The review article by Havas (1989) provides many details of the rich history of these various attempts, some of which preceded the well-known work by Einstein and Grommer (1927).

26 In the sense that in some neighborhood $\mathcal{N}$ of each point in the total space **P** of the principal bundle there are four linearly independent vector fields $\mathbf{X}_0(u),\ldots,\mathbf{X}_3(u) \in T_u\mathbf{P}$, $u \in \mathcal{N}$, which span the horizontal subspaces $H_u\mathbf{P}$ above $\mathcal{N}$.

27 If an "infinitesimal" vector X is visualized as an oriented line segment joining the point $x \in \mathbf{M}$ to an "infinitesimally close" point $x + dx \in \mathbf{M}$, then its push-forward by ϕ can be visualized as the oriented line segment joining $\phi(x) \in \mathbf{M}$ to $\phi(x + dx) \in \mathbf{M}$ – i.e., as the outcome of "pushing forward", by means of the mapping ϕ, all the points of such an "infinitesimal" line segment within the manifold **M**.

28 Note that once the parallel transport of a frame, such as the one in (2.1), is defined, that of its coframe is also defined: the parallel transported coframe is simply the dual of the parallel transported frame.

29 The *exponential map* is well-defined for any vector in the tangent space T_eG of any Lie group [C,I]. If the elements of that group are matrices, then so are the elements of T_e**G**, and the exponential map can be computed by inserting those matrices into the Taylor expansion of the exponential function at zero.

30 This terminology is sometimes used for Yang-Mills fields (cf. [C], p. 402, or [NT], p. 178), and some authors apply it also to CGR (Drechsler and Mayer, 1977; Drechsler, 1984). However, some other authors mean by a gauge group a certain subgroup of the group of automorphisms of the principal bundle (cf. [I], pp. 127-129; [BG], p. 46). The two definitions are obviously closely related, but they are not equivalent (cf. Sec. 10.2), so that Trautman (1980, 1981) talks of gauge transformations of the first kind and of the second kind, respectively. We shall explain this terminology in Secs. 10.2 and 11.3, when it becomes actually required. In the meantime we shall avoid the usage of the term "gauge group", and employ instead the term "structure group", until all the relevant distinctions are clearly drawn.

31 These relations are usually derived in the context of classical Yang-Mills theories for which $U(g) = g$ (cf. [C], pp. 402-407, [I], p. 161, or [NT], p. 178).

32 The specification of these components by means of upper indices set directly above lower indices is adopted from [K], since it facilitates the denotation of multiplication from the right as well as from the left, without the need of transposing indices (cf. Notes 11 and 15). This type of notation is also employed in [C], as well as in the lecture notes by Drechsler and Mayer (1977), whose notation we shall often follow in this section.

33 Cf. [K], p. 118. The canonical form in (6.4) is sometimes also called the "soldering form" of **M** (cf. [C], p. 376), but following Drechsler (1984), we shall reserve the latter name for the form in (6.19a), which equals the canonical form $\boldsymbol{\theta}$ plus the covariant derivative of the affine vector field $\boldsymbol{a}$, which constitutes part of any section (6.18) of the general affine frame bundle *GA***M**. On the other hand, the soldering map (6.5) is a special case of the soldering maps defined in Eq. (3.2a) of Chapter 4, whereby the fibres of vector bundles are "soldered" to the respective fibres of their principal frame bundles. However, the term "soldering map" is not at all common in literature, albeit the inverses $(\sigma_u)^{-1}$ of soldering maps are routinely used when the fibres of *T***M** are (implicitly) soldered to those of *GL***M**. In that instance the *linear map* $(\sigma_u)^{-1}$ is (somewhat confusingly) denoted by the same letter u that is employed to denote the *linear frame* $\boldsymbol{u}$ to which it corresponds – cf. [SC], pp. 345 and 365, or [K], p. 56.

34 Torsion, which is viewed as the antisymmetric part of a connection compatible with a metric on a manifold, was first introduced by Cartan (1923, 1924), who believed that it led to a possible generalization of CGR (Debever, 1979), in which the torsion of spacetime would be connected to an intrinsic angular momentum of matter. Many researchers subsequently contributed to the investigation of the ensuing Riemann-Cartan spacetimes (cf Hehl *et al.*, 1976, Sec. I.D). Their formulation by means of the modern differential geometry techniques reviewed in this chapter is due to Trautman (1970, 1972, 1973, 1975).

35 In Riemann-Cartan spacetimes the presence of non-zero torsion leads to a type of free-fall behavior of test particles (Adamowicz and Trautman, 1975) for which there is no experimental evidence in the classical context, since matter does not display intrinsic angular momentum at the macroscopic level, so that classically "the very notion of a spinning test particle [is] obscure" (Hehl *et al.*, 1976, Sec. V.D3). However, the possibility remains open that torsion might play a role for some quantum particles with spin, and in fact, the consideration of Riemann-Cartan spacetimes in the quantum regime has been strongly advocated in a series of papers by W. Drechsler (1977b-1985, 1988, 1990).

36 As a generalization of the definition of a Lorentzian manifold given in Sec. 2.3, $(\mathbf{M}, g)$ is a *pseudo-Riemannian manifold* if the metric field g that is specified on **M** has signature $(+1,...,+1,-1,...,-1)$; $(\mathbf{M}, g)$ is called a *Riemannian manifold* if the signature of g is $(+1,...,+1)$.

37 The Planck natural units are those in which $\hbar = c = G = 1$, and in which therefore the Planck length is also equal to one – cf. Appendix F in [W] for further details and *cgs* equivalents.

38 Recall that a smooth curve in a Lorentzian manifold is timelike, null or spacelike if all its tangent vectors X are timelike ($g(X,X) > 0$), null ($g(X,X) = 0$) or spacelike ($g(X,X) < 0$), respectively.

Chapter 3

Stochastic Quantum Mechanics on Phase Space

We have seen in the previous chapter that the concept of locality is of fundamental significance in CGR: without it, the concept of tangent bundle, and more generally of tensor bundles over a manifold **M**, would be impossible to define, so that neither the Einstein's field equations nor the geodesic postulate for free-fall motion could be formulated in their well-known form (cf. Sec. 2.7). However, as we pointed out in Sec. 1.2, at the quantum level, the conventional concept of locality of pointlike particles encounters insurmountable difficulties already in the special relativistic regime, due to Hegerfeldt's (1974) no-go theorem and to foundational problems in the orthodox quantum theory of measurement.

The framework reviewed in this chapter has emerged from an effort to resolve these difficulties by retaining Einstein's principle of special relativistic invariance, as well as Heisenberg's uncertainty principle, but giving up the concept of *sharp* localization – i.e., localization with perfect precision – in favor of a more realistic concept of localization with finite precision, which is of a stochastic nature. This stochastic concept of localization is based on the notion of *stochastic value* introduced in Sec. 1.3 in the course of formulating the principle of irreducible indeterminacy – i.e., Principle 1 in that section. It therefore became natural to describe the resulting theoretical framework as *stochastic quantum mechanics* (SQM), on account of the stochastic aspects embedded in the concept of the *stochastic values* which quantum observables assume in that framework [P]. However, on account of the fundamental role that phase space plays in it (as it will become apparent already in Sec. 3.2), it is more suitable to use for this framework the fuller descriptive title of *stochastic quantum mechanics on phase space* , and thus differentiate it from other unrelated or only remotely related frameworks, which also use the descriptive labels of "stochastic quantization", "stochastic mechanics", or "stochastic quantum mechanics"[1].

In this chapter we shall first review the basic properties of probability measures for the outcome of measurements of compatible observables in ordinary quantum mechanics (QM) from the point of view of the systems of imprimitivity [BR] formulated and later introduced in quantum theory by Mackey (1951-53, 1968). Then, upon presenting the systems of imprimitivity most pertinent to the localization problem in nonrelativistic as well as in relativistic QM, we shall formulate for both these cases their SQM counterparts. These counterparts constitute generalized systems of imprimitivity, which are also known under the shorter, and from the point of view of physics and geometry, more suitable name of systems of covariance. As it will become clear in the context of this presentation, the systems of covariance of SQM are related to measurements in which the values of the pertinent observables are measured with finite precision, whereas the systems of imprimitivity of QM are related to "measurements" by which they are measured with (purportedly) infinite precision. We shall therefore discuss the sharp-point limit of nonrelativistic SQM, which, as the name suggests, corresponds to the ***imagined*** operational procedure of indefinitely increasing the measurement precision until perfectly accurate (i.e., "sharp") values are obtained. The fact that, at the mathematical level, the very same procedure leads in relativistic SQM to divergences, provides an indication of the source of the earlier mentioned difficulties encountered by the concept of sharp localization in relativistic QM, and it will be therefore assigned the appropriate amount of attention in this chapter. Furthermore, as we explained in purely physical terms in Sec. 1.4, the entire idea of geometro-stochastic (GS) propagation rests on path integrals in which, in accordance with the geodesic postulate, free-fall conditions are reproduced by the parallel transport along geodesics. In the flat spacetime case such propagation coincides with the conventional mode of quantum propagation, expressed in the SQM context. The fact that this is indeed the case will be confirmed by comparing in the next two chapters the path integrals for GS propagators with the path integrals for SQM propagators. Therefore, in preparation for that task, we shall recount in Sec. 3.6 of this chapter the basic properties of SQM path integrals.

For the sake of simplicity, in the present chapter we shall deal primarily with the case of massive particles of zero spin. Pertinent results for non-zero spin and/or zero mass will be discussed in later chapters, when the occasion requires it. Furthermore, guides to the original literature on these subjects will be provided in all instances. More detailed reviews of SQM can be found in [P], as well as in an extensive review article by S. T. Ali (1985).

3.1. Nonrelativistic Systems of Imprimitivity

The concept of system of imprimitivity originally arose (Mackey, 1951-53) in the theory of induced representations of locally compact groups [BR]. It involves two basic ingredients.

The first of these ingredients is a unitary representation U of a locally compact group $\mathbf{G}$ that acts as a continuous group of transformations on some locally compact topological space $\mathcal{M}$ [BR]. In general, the values assumed by U are those of unitary operators $U(g)$ acting on a Hilbert space $\mathcal{H}$. A good example of such a representation, taken from the nonrelativistic quantum mechanics of a single spin-zero particle, is the following.

The Hilbert space $\mathcal{H}$ is the space $L^2(\mathbf{R}^3)$, which consists[2] of all the complex-valued functions $\hat{\psi}(\mathbf{x})$, $\mathbf{x} \in \mathbf{R}^3$, that are square-integrable on $\mathbf{R}^3$ with respect to the Lebesgue measure [PQ], and which carries the inner product

$$\langle \hat{\psi}_1 | \hat{\psi}_2 \rangle = \int_{\mathbf{R}^3} \hat{\psi}_1^*(\mathbf{x}) \hat{\psi}_2(\mathbf{x})\, d^3\mathbf{x} \ . \tag{1.1}$$

The group $\mathbf{G}$ is the Euclidean group E(3), whose elements $(\mathbf{a}, R)$ act in $\mathbf{R}^3$ as follows:

$$(\mathbf{a},R):\ \mathbf{x} \mapsto \mathbf{x}' = \mathbf{a} + R\mathbf{x} \ , \qquad \mathbf{a} \in \mathbf{R}^3 , \qquad R \in \mathrm{SO}(3) \ . \tag{1.2}$$

The corresponding well-known unitary representation is given by

$$\hat{U}(\mathbf{a},R):\quad \hat{\psi}(\mathbf{x}) \mapsto \hat{\psi}'(\mathbf{x}) = \hat{\psi}(R^{-1}(\mathbf{x}-\mathbf{a}))\ . \tag{1.3}$$

The second ingredient of a system of imprimitivity is a *projector-valued measure E*, commonly called also a *PV-measure*, on the Borel sets[3] B of the topological space $\mathcal{M}$, and acting in the Hilbert space $\mathcal{H}$. Thus, by definition, the map

$$E:\ B \mapsto E(B)\ ,\qquad E(B) = E^*(B) = E^2(B)\ , \tag{1.4}$$

assigns an orthogonal projection operator to each Borel set B in $\mathcal{M}$ in such a manner that $E(\mathcal{M})$ equals the identity operator $\mathbf{1}$ in $\mathcal{H}$, and that for any countable family of disjoint Borel sets from $\mathcal{M}$ we have

$$E(\cup_{i=1}^{\infty} B_i) = \sum_{i=1}^{\infty} E(B_i) := \mathrm{s}\text{-}\lim_{n\to\infty} \sum_{i=1}^{n} E(B_i)\,,\quad B_i \cap B_j = \varnothing\,,\quad i \neq j\ . \tag{1.5}$$

The strong limit in (1.5) signifies that the convergence of the above infinite series is to be interpreted by applying the partial sum in (1.5) to any vector from $\mathcal{H}$, and then taking the limit in the norm of $\mathcal{H}$ for the resulting sequence of vectors (cf. [PQ], Sec. III-5.3). An example of a PV-measure for the nonrelativistic quantum mechanics of a single spin-zero particle is provided by the spectral measure $E^{\mathbf{Q}}$ corresponding to position operators [PQ], whose generic element multiplies each wave function with the characteristic function[4] of a Borel set from $\mathbf{R}^3$, namely

$$E^{\mathbf{Q}}(B):\quad \hat{\psi}(\mathbf{x}) \ \mapsto\ \chi_B(\mathbf{x})\hat{\psi}(\mathbf{x})\ . \tag{1.6}$$

Hence, according to the orthodox interpretation of QM, the resulting expectation value

$$\langle \hat{\psi} | E^{\mathbf{Q}}(B)\hat{\psi}\rangle \ =\ \int_B |\hat{\psi}(\mathbf{x})|^2\, d^3\mathbf{x} \tag{1.7}$$

represents the probability of obtaining for a sharp (i.e., perfectly accurate) position measurement of a quantum particle under observation an outcome within B, provided that particle is in a quantum state represented by the normalized wave function $\hat{\psi}$.

We can now define a *system of imprimitivity* as a pair (U,E), consisting of a unitary representation U and of a PV-measure E of the above described type, with these two mathematical entities interrelated in such a manner that

$$U(g)E(B)U^{-1}(g) = E(g\cdot B),\qquad g\cdot B = \{g\cdot x | x \in B\} \subset \mathcal{M}\ ,\quad g \in \mathbf{G}\ , \tag{1.8}$$

when the elements of the group $\mathbf{G}$ act on $\mathcal{M}$ from the left.

The significance of this mathematical concept for QM emerges from the easily verifiable fact that the Euclidean group E(3) defined by (1.3) constitutes, in combination with the spectral measure $E^{\mathbf{Q}}$ defined in (1.6), a system of imprimitivity, whose property (1.8) is equivalent to the statement that

$$\langle\hat{\psi}|E^{Q}(B)\hat{\psi}\rangle = \langle\hat{\psi}'|E^{Q}(\mathbf{a}+R\cdot B)\hat{\psi}'\rangle \quad . \tag{1.9}$$

In view of (1.7), this property expresses the covariance of QM probabilities for position measurement under coordinate transformations corresponding to changes of orthonormal affine frames in Euclidean 3-space. Such transformations can be carried out operationally, by performing translations and rotations of inertial frames under the assumption that such frames obey the laws of Newtonian classical mechanics, i.e., that they are sufficiently massive to be described in classical terms, in concordance with the orthodox QM theory of measurement. For this reason, Wightman (1962) and Mackey (1968) have argued in favor of adopting systems of imprimitivity as the basic mathematical tools for formulating the concept of particle localizability in QM.

Indeed, systems of imprimitivity can also describe the covariance features of nonrelativistic QM probabilities for the outcome of measurements of other fundamental observables, such as 3-momenta, angular momenta, spins, etc. For a single particle, the transition from the configuration representation, used in (1.3) and (1.6), to the momentum representation is effected by taking the Fourier transform[5]

$$U_{\mathrm{F}} : \hat{\psi} \mapsto \tilde{\psi} \, , \qquad \tilde{\psi}(\mathbf{k}) = (2\pi)^{-3/2}\int \exp(-i\mathbf{k}\cdot\mathbf{x})\hat{\psi}(\mathbf{x})d^{3}\mathbf{x} \quad . \tag{1.10}$$

Then the corresponding nonrelativistic system of imprimitivity,

$$\tilde{U}(\mathbf{a},R) = U_{\mathrm{F}}\hat{U}(\mathbf{a},R)U_{\mathrm{F}}^{-1} \, , \qquad E^{\mathrm{P}}(B) : \tilde{\psi}(\mathbf{k}) \mapsto \chi_{B}(\mathbf{k})\tilde{\psi}(\mathbf{k}) \, , \tag{1.11}$$

reflects the covariance properties of QM probabilities for momentum measurements under Euclidean transformations – as it will become evident from the relation (1.23), that will be derived in the more general context of Galilean transformations, to whose study we turn next.

A general *Galilean transformation* in Newtonian spacetime, represented by $\mathbf{R}^4$, can be expressed in terms of a linear map of $\mathbf{R}^4$ onto $\mathbf{R}^4$,

$$(b,\mathbf{a},\mathbf{v},R) : \; (\mathbf{x},t) \mapsto (\mathbf{x}' = \mathbf{a} + \mathbf{v}t + R\mathbf{x}, \, t' = t + b) \quad , \tag{1.12}$$

that represents a transformation of Galilean inertial coordinates (cf. Sec. 4.1). It generalizes (1.2) by involving time translations in the amount b, as well as velocity boosts to the 3-velocity $\mathbf{v}$, in addition to spatial translations and rotations. Hence a Galilean transformation reflects the general operational procedure of changing very massive (and therefore classically describable) inertial frames in accordance with the laws of Newtonian classical mechanics, as well as generally changing the zero-point of clocks attached to such frames – which in Newtonian physics supposedly keep track of absolute time. Upon performing two successive general Galilean transformations in accordance with (1.12), it is easily verified by straightforward algebra that the family of all Galilean transformations (1.12) constitutes a group, known as the *Galilei group* $\mathcal{G}$, with the following group multiplication law:

$$(b',\mathbf{a}',\mathbf{v}',R')\cdot(b,\mathbf{a},\mathbf{v},R) = (b'+b, \, \mathbf{a}'+R'\mathbf{a}+b\mathbf{v}', \, \mathbf{v}'+R'\mathbf{v}, \, R'R) \quad . \tag{1.13}$$

To express the effect of the Galilei transformation (1.12) on the wave functions of free spin-zero quantum particles, let us work in the Schrödinger picture, in which the time evolution [PQ] specified by

$$\hat{\psi}_t = \exp(-iH_0 t)\hat{\psi} \ , \qquad H_0 = (-1/2m)\Delta \ , \tag{1.14}$$

provides the general solution of the free Schrödinger equation:

$$i\partial_t \hat{\psi}(\mathbf{x},t) = (-1/2m)\Delta\hat{\psi}(\mathbf{x},t) \ , \qquad \hat{\psi}(\mathbf{x},t) = \hat{\psi}_t(\mathbf{x}) \ . \tag{1.15}$$

As pointed out by Bargmann (1954), under generic Galilean transformations this equation is left invariant by the ray (i.e., projective) representation

$$\begin{aligned} \hat{U}(b,\mathbf{a},\mathbf{v},R) \ : \ \ & \hat{\psi}(\mathbf{x},t) \mapsto \hat{\psi}'(\mathbf{x},t) \\ & = \exp\left[i\left(-\tfrac{1}{2}m\mathbf{v}^2(t-b) + m\mathbf{v}\cdot(\mathbf{x}-\mathbf{a})\right)\right] \hat{\psi}(R^{-1}[\mathbf{x}-\mathbf{a}-\mathbf{v}(t-b)],t-b) \ , \end{aligned} \tag{1.16}$$

rather than by the vector representation that is a straightforward extrapolation of (1.3), namely the representation that produces the right-hand side of (1.16) with the exponential phase factor removed. The acceptability of such representations in QM follows from Wigner's theorem on ray representations (cf. Sec. 3.7), which in turn reflects the fact that in QM it is not the probability amplitudes provided by wave functions that are directly observable, but it is rather the corresponding probability densities that are directly measurable.

We can easily verify by explicit computation that

$$\begin{aligned} & \hat{U}((b',\mathbf{a}',\mathbf{v}',R')\cdot(b,\mathbf{a},\mathbf{v},R)) = \\ & \quad = \exp\left\{-im\left[\tfrac{1}{2}\mathbf{v}'^2\, b + \mathbf{v}'\cdot R'\mathbf{a}\right]\right\} \hat{U}(b',\mathbf{a}',\mathbf{v}',R')\hat{U}(b,\mathbf{a},\mathbf{v},R) \ . \end{aligned} \tag{1.17}$$

Consequently, (1.16) can be also viewed as an ordinary vector representation of the central extension of the Galilei group [BR], whose elements $(\theta,b,\mathbf{a},\mathbf{v},R)$ include the phase θ, so that the phase factor in (1.17) can be incorporated into their group multiplication law:

$$\begin{aligned} & (\theta',b',\mathbf{a}',\mathbf{v}',R')\cdot(\theta,b,\mathbf{a},\mathbf{v},R) \\ & \quad = \left(\theta'+\theta+m\left(\tfrac{1}{2}\mathbf{v}'^2\, b + \mathbf{v}'\cdot R'\mathbf{a}\right),\ b'+b,\ \mathbf{a}'+R'\mathbf{a}+b\mathbf{v}',\ \mathbf{v}'+R'\mathbf{v},\ R'R\right) \ . \end{aligned} \tag{1.18}$$

We note that the restriction of this extension to elements with $b' = b = 0$ and $R' = R = I$ coincides with the well-known Weyl-Heisenberg group [BR], which provides the mathematical foundation of the canonical commutation relations in QM, and plays a crucial role in the derivation of von Neumann's theorem on the uniqueness, modulo unitary transformations, of irreducible representations of these relations for a finite number of degrees of freedom [PQ].

The ray representation in (1.16) constitutes, in combination with the spectral measure in (1.6), the fundamental system of imprimitivity describing sharp localizability properties of spin-zero point particles in nonrelativistic QM, since they lead to the following covariance feature under Galilei transformations,

$$\left\langle \hat{\psi}_t \middle| E^{\mathrm{Q}}(B)\hat{\psi}_t \right\rangle = \left\langle \hat{\psi}'_{t+b} \middle| E^{\mathrm{Q}}(\mathbf{a}+\mathbf{v}t+R\cdot B)\hat{\psi}'_{t+b} \right\rangle \tag{1.19}$$

for the QM probabilities for finding the particle spatially located with respect to the original inertial frame within the Borel set B at the instant t .

The corresponding system of imprimitivity for spin $s>0$ is obtained, upon retaining (1.6), but replacing (1.16) with

$$\hat{U}_s(b,\mathbf{a},\mathbf{v},R)\ :\ \hat{\psi}(\mathbf{x},t)\ \mapsto\ \hat{\psi}'(\mathbf{x},t)$$
$$=\exp\left[i\left(-\tfrac{1}{2}m\mathbf{v}^2(t-b)+m\mathbf{v}\cdot(\mathbf{x}-\mathbf{a})\right)\right]D^s(A)\,\hat{\psi}(R^{-1}[\mathbf{x}-\mathbf{a}-\mathbf{v}(t-b)],t-b), \tag{1.20a}$$
$$\hat{\psi}(\mathbf{x},t)\in\ell^2(2s+1)\ ,\qquad D^s(A)\in \mathrm{GL}(2s+1,\mathrm{C})\ ,\qquad A\in \mathrm{SU}(2)\ , \tag{1.20b}$$

where in general D^s is a representation of the covering group SU(2) of SO(3), so that $\pm A$ are the elements of SU(2) corresponding to $R\in$ SO(3) [BR]. The systematic derivation and classification of all vector, as well as ray, representations of the Galilei group, carried out by Lévy-Leblond (1963, 1971), reveals that the systems of imprimitivity based on (1.6) and (1.20) are, modulo unitary equivalence, the only QM systems of imprimitivity describing the localizability properties of nonrelativistic particles in Newtonian spacetime.

The corresponding momentum systems of imprimitivity can be obtained by using Fourier transforms, as in (1.11). Indeed, we easily derive from (1.10) that

$$\tilde{U}_s(b,\mathbf{a},\mathbf{v},R)\ :\ \tilde{\psi}(\mathbf{k},t)\mapsto \tilde{\psi}'(\mathbf{k},t)$$
$$=\exp\left\{i\left[\tfrac{1}{2}m\mathbf{v}^2(t-b)-\mathbf{k}\cdot(\mathbf{a}+\mathbf{v}(t-b))\right]\right\}D^s(A)\,\tilde{\psi}(R^{-1}[\mathbf{k}-m\mathbf{v}],t-b)\ . \tag{1.21}$$

Hence, in full accordance with the fact that, under a general change of inertial frame, (1.12) is replaced, in the case of classical nonrelativistic momentum components, by

$$(b,\mathbf{a},\mathbf{v},R):\ (\mathbf{k},t)\mapsto(\mathbf{k}'=m\mathbf{v}+R\mathbf{k},\ t'=t+b)\ , \tag{1.22}$$

we obtain the following covariance properties,

$$\left\langle \tilde{\psi}_t \middle| E^{\mathrm{P}}(B)\tilde{\psi}_t \right\rangle = \left\langle \tilde{\psi}'_{t+b} \middle| E^{\mathrm{P}}(m\mathbf{v}+R\cdot B)\tilde{\psi}'_{t+b} \right\rangle,\quad \tilde{\psi}_t(\mathbf{k})=\exp(-i\mathbf{k}^2t/2m)\tilde{\psi}(\mathbf{k}), \tag{1.23}$$

of QM probabilities for (sharp) momentum measurement outcomes.

We observe that the probability measure in (1.23) is actually time-independent, in accordance with the fact that the momentum of a free particle is conserved. Naturally, in the presence of interactions with external sources, that will no longer generally be the case, since momentum is then not conserved.

3.2. Nonrelativistic Systems of Covariance

Already in the 1950s, researchers into the foundations of QM, such as Ludwig (1953, 1958, 1983, 1985), found out that the concept of PV-measure is not only too restrictive in the context of the quantum theory of measurement, but that in view of the fact that measurements of position or momentum can be carried out only with a finite precision, their description suggests the replacement of the orthogonal projection operators, used in computing such expectation values as those in (1.9), (1.19) and (1.23), by positive operators[6]. This leads to the idea of replacing, in the quantum theory of measurement, PV-measures by *POV measures*, i.e., *positive-operator-valued measures* for which the properties of PV measures in (1.4) and (1.5) are made less stringent, by requiring only that

$$E(B) = E^*(B) \geq 0 , \qquad B \subset \mathcal{M} , \tag{2.1a}$$

$$E(B_1 \cup B_2 \cup \cdots) = E(B_1) + E(B_2) + \cdots , \qquad B_i \cap B_j = \varnothing , \quad i \neq j . \tag{2.1b}$$

The above two conditions, namely positivity and σ-additivity, obviously represent necessary and sufficient conditions on a family of operators for having expectation values that produce, in general, *relative* or *conditional* probability measures[7]. Hence, in principle, a physical interpretation can be assigned also to suitable classes of POV measures which are not projector-valued.

A POV measure is said to be *normalized* if $E(\mathcal{M}) = \mathbf{1}$. We note that even if a POV measure is not normalized, its expectation values can be still interpreted as representing conditional probabilities.

A *system of covariance* is defined as a pair (U, E), consisting of a unitary representation U of a group $\mathbf{G}$, that acts from the left on a locally compact topological space $\mathcal{M}$, and of a POV-measure E on the Borel sets of $\mathcal{M}$, which are interrelated in such a manner that

$$U(g)E(B)U^{-1}(g) = E(g \cdot B), \qquad g \in \mathbf{G} , \quad B \subset \mathcal{M} . \tag{2.2}$$

Hence a system of imprimitivity is a special case of a system of covariance. In fact, systems of covariance whose POV measures are not a projector-valued but are normalized are also referred to as "generalized systems of imprimitivity" (Holevo, 1985, 1987).

Non-normalizable systems of covariance in Newtonian spacetime can be easily obtained from those of the preceding section by considering measurements which are of finite duration instead of being instantaneous, i.e., which have to be mathematically described as taking place over regions in Newtonian spacetime. For a given inertial frame and a given clock for measuring the absolute time of Newtonian physics, Newtonian spacetime can be identified with $\mathbf{R}^4$, so that its Borel sets B can be taken to be those of $\mathbf{R}^4$. The relative probabilities for the outcomes of the aforementioned measurements can be then obtained as the expectation values of the following Bochner integrals[8],

$$\overline{E}^{\mathrm{Q}}(B) = \int_{-\infty}^{+\infty} E^{\mathrm{Q}}(B_t)\,dt , \qquad B_t = \{\mathbf{x} | (\mathbf{x}, t) \in B \subset \mathbf{R}^4\} \subset \mathbf{R}^3 . \tag{2.3}$$

It is easily verified that the above operators define a non-normalizable POV measure on the Borel sets of $\mathbf{R}^4$ which, in combination with the representation in (1.16) or in (1.20), constitutes a system of covariance.

In the 1960s, a variety of proofs were given (Urbanik, 1961; Varadarajan, 1962; Cohen, 1966; Gudder, 1968) which followed in the footsteps of von Neumann's (1932) proof of the "impossibility" of the simultaneous measurement of "incompatible" QM observables. However, it was independently realized by several researchers (She and Hefner, 1966; Prugovečki, 1966-67; Park and Margenau, 1968) that the problem of measurement of "incompatible" QM observables, such as position and momentum, assumes new aspects when the possibility that outcomes of their simultaneous measurement which are not sharp is considered – so that no conflict with Heisenberg's uncertainty principle arises, and the aforementioned proofs become irrelevant. Hence, She and Hefner (1966), Davies and Lewis (1969), Holevo (1972) and Prugovečki (1966, 1976) were able to propose probability measures describing the outcome of such measurements, without giving rise to any inconsistencies with the orthodox theory of quantum measurement, since the latter dealt exclusively with the highly idealized (and practically unrealizable) case of perfectly accurate measurement outcomes. Of course, no uniqueness of these probability measures for "unsharp" measurement outcomes could be expected to hold for a given system, since those outcomes are bound to depend as much on the employed test particles or apparatuses as on the system itself. However, the conjunction of "system" *and* "apparatus" could be expected to produce unique POV measures[9], whose expectation values for the quantum states in which the "system" was prepared would produce the required probability measures. The fact that such POV measures belonged to systems of covariance for the Galilei group was subsequently established, and it led to a new method of quantization, which was based exclusively on group-theoretical methods (Prugovečki, 1978d) – cf. Sec. 11.4.

In the nonrelativistic context, the central idea of this purely group-theoretical method of quantization is to consider ray representations of the Galilei group on phase space, rather than on the configuration or on the momentum space alone – as was done in the previous section. Hence, let us introduce the Hilbert space $L^2(\mathbf{R}^6)$ which consists of all the complex-valued functions $\psi(\mathbf{q},\mathbf{p})$ in the variables $\mathbf{q},\mathbf{p} \in \mathbf{R}^3$ that are square-integrable on $\mathbf{R}^6$ with respect to the Lebesgue measure, and which carries the inner product

$$\langle \psi_1 | \psi_2 \rangle = \int_{\mathbf{R}^6} \psi_1^*(\mathbf{q},\mathbf{p})\psi_2(\mathbf{q},\mathbf{p})\, d^3\mathbf{q}\, d^3\mathbf{p} \ . \tag{2.4}$$

In this space we can then consider, by analogy with (1.14) and (1.15), the time evolution

$$\psi_t = \exp(-iH_0 t)\psi \ , \qquad H_0 = (-1/2m)\Delta_{\mathbf{q}} \ , \tag{2.5}$$

which provides general solutions of the Schrödinger equation in the $\mathbf{q}$-variables:

$$i\partial_t \psi(\mathbf{q},\mathbf{p},t) = (-1/2m)\Delta_{\mathbf{q}}\psi(\mathbf{q},\mathbf{p},t) \ , \qquad \psi(\mathbf{q},\mathbf{p},t) = \psi_t(\mathbf{q},\mathbf{p}) \ . \tag{2.6}$$

We can then introduce in $L^2(\mathbf{R}^6)$ the following ray representation of the Galilei group:

$$U(b,\mathbf{a},\mathbf{v},R):\ \psi(\mathbf{q},\mathbf{p},t) \mapsto \psi'(\mathbf{q},\mathbf{p},t) = \exp\left[i\left(-\tfrac{1}{2}m\mathbf{v}^2(t-b) + m\mathbf{v}\cdot(\mathbf{q}-\mathbf{a})\right)\right]$$
$$\times\ \psi(R^{-1}[\mathbf{q}-\mathbf{a}-\mathbf{v}(t-b)], R^{-1}[\mathbf{p}-m\mathbf{v}], t-b) \ . \tag{2.7}$$

It is easily checked (Prugovečki, 1978d) that this representation is unitary, and that it satisfies the same group multiplication law as the one in (1.17), namely:

$$U((b',\mathbf{a}',\mathbf{v}',R')\cdot(b,\mathbf{a},\mathbf{v},R)) =$$
$$= \exp\left\{-im\left[\tfrac{1}{2}\mathbf{v}'^2 b + \mathbf{v}'\cdot R'\mathbf{a}\right]\right\} U(b',\mathbf{a}',\mathbf{v}',R')U(b,\mathbf{a},\mathbf{v},R) \quad . \tag{2.8}$$

However, as opposed to (1.16), the representation in (2.7) is highly reducible. Its complete spectral analysis (Ali and Prugovečki, 1986) reveals the existence of a decomposition

$$U(b,\mathbf{a},\mathbf{v},R) = \bigoplus_{n=1}^{\infty} \bigoplus_{j=1}^{\infty} U_{nj}(b,\mathbf{a},\mathbf{v},R) \tag{2.9}$$

into irreducible components for each choice of a set of functions e_n, $n = 1,2,\ldots$, for which

$$\int_0^{+\infty} e^*_{n'}(\kappa)e_{n''}(\kappa)\ \kappa^2 d\kappa = \delta_{n'n''} \quad , \tag{2.10}$$

so that they constitute an orthonormal basis in the Hilbert space with the inner product defined in accordance with the left-hand side of (2.10).

The phase space representation in (2.7) displays a number of mathematically interesting properties – whose basic statements and proofs can be found in (Prugovečki, 1978d), and whose systematic harmonic analysis is presented in (Ali and Prugovečki, 1986). We shall, however, mention only those which turn out to be of direct physical relevance.

The (nj)-th irreducible representation in (2.9) is unitarily equivalent to the configuration space representation in (1.20) corresponding the integral spin value s (i.e., $s = j = 0,1,2,\ldots$), and therefore also to the momentum space counterpart of that representation in (1.21). If $\mathbf{P}_{nj}$ denotes the orthogonal projection operator of $L^2(\mathbf{R}^6)$ onto the subspace carrying that (nj)-th irreducible representation, so that

$$U_{nj} = \mathbf{P}_{nj}U = W_{nj}\tilde{U}W_{nj}^{-1} \quad , \qquad W_{nj} : \ell^2(2j+1)\otimes L^2(\mathbf{R}^3) \to \mathbf{P}_{nj}L^2(\mathbf{R}^6) \ , \tag{2.11}$$

expresses its unitary equivalence to its momentum space counterpart, then the unitary operator W_{nj} supplying that unitary equivalence is an integral operator. The action of this operator on the momentum space wave functions appearing in (1.21) results into

$$(\mathbf{P}_{nj}\psi)(\mathbf{q},\mathbf{p}) = (2\pi)^{-3/2}\int_{\mathbf{R}^3} \exp(i\mathbf{q}\cdot\mathbf{k})e_n(|\mathbf{p}-\mathbf{k}|)Y_j^l((\mathbf{p}-\mathbf{k})/|\mathbf{p}-\mathbf{k}|)\tilde{\psi}_l(\mathbf{k})d^3\mathbf{k} \ , \tag{2.12}$$

where the summation over the repeated internal angular momentum indices $l = -j,\ldots, +j$ is implicitly understood.

Each one of the above irreducible subspaces is characterized by a unique rotationally invariant *resolution generator*, that is by a unique vector ξ with the property that

$$(\mathbf{P}_\xi\psi)(\mathbf{q},\mathbf{p}) = \langle \xi_{\mathbf{q},\mathbf{p}}|\psi\rangle \ , \qquad \xi_{\mathbf{q},\mathbf{p}} = U(0,\mathbf{q},\mathbf{p}/m,I)\xi \ , \quad \psi \in L^2(\mathbf{R}^6) \ , \tag{2.13}$$

and therefore the orthogonal projection operator $\mathbf{P}_\xi$ onto that subspace can be written in the form of a Bochner integral:

$$\mathbf{P}_\xi = \int_{\mathbf{R}^6} \left|\xi_{\mathbf{q},\mathbf{p}}\right\rangle d^3\mathbf{q}\, d^3\mathbf{p} \left\langle \xi_{\mathbf{q},\mathbf{p}}\right| \ , \qquad \mathbf{P}_\xi = \mathbf{P}_\xi^* = \mathbf{P}_\xi^2 \ . \tag{2.14}$$

Equations (2.13) and (2.14) lead to the conclusion that each one of these subspaces is a reproducing kernel Hilbert space, so that

$$(\mathbf{P}_\xi \psi)(\mathbf{q},\mathbf{p}) = \int_{\mathbf{R}^6} \left\langle \xi_{\mathbf{q},\mathbf{p}} \middle| \xi_{\mathbf{q}',\mathbf{p}'} \right\rangle \psi(\mathbf{q}',\mathbf{p}')\, d^3\mathbf{q}'\, d^3\mathbf{p}' . \tag{2.15}$$

We can now consider, in each one of the above irreducible subspaces $\mathbf{P}_\xi L^2(\mathbf{R}^6)$, a corresponding system of covariance given by

$$U_\xi(b,\mathbf{a},\mathbf{v},R) = \mathbf{P}_\xi U(b,\mathbf{a},\mathbf{v},R) \ , \qquad E_\xi(B) = \int_B \left|\xi_{\mathbf{q},\mathbf{p}}\right\rangle d^3\mathbf{q}\, d^3\mathbf{p} \left\langle \xi_{\mathbf{q},\mathbf{p}}\right| \ . \tag{2.16}$$

From the point of view of the physical interpretation of this kind of system of covariance, most significant are the following *marginality properties* of the expectation values of the wave functions from that irreducible subspace:

$$\left\langle \psi \middle| E_\xi(\hat{B}\times\mathbf{R}^3)\psi \right\rangle = \int_{\hat{B}} d^3\mathbf{q} \int_{\mathbf{R}^3} d^3\mathbf{x}\, \hat{\chi}^l_{\mathbf{q}}(\mathbf{x}) \left|\hat{\psi}_l(\mathbf{x})\right|^2, \quad \hat{\chi}^l_{\mathbf{q}}(\mathbf{x}) = (2\pi)^3 \left|\hat{\xi}^l(\mathbf{x}-\mathbf{q})\right|^2, \tag{2.17}$$

$$\left\langle \psi \middle| E_\xi(\mathbf{R}^3\times\tilde{B})\psi \right\rangle = \int_{\tilde{B}} d^3\mathbf{p} \int_{\mathbf{R}^3} d^3\mathbf{k}\, \tilde{\chi}^l_{\mathbf{p}}(\mathbf{k}) \left|\tilde{\psi}_l(\mathbf{k})\right|^2, \quad \tilde{\chi}^l_{\mathbf{p}}(\mathbf{k}) = (2\pi)^3 \left|\tilde{\xi}^l(\mathbf{k}-\mathbf{p})\right|^2. \tag{2.18}$$

Indeed, the respective right-hand sides of the above marginal measures involve exclusively the configuration and momentum representatives of the phase-space wave functions $\mathbf{P}_\xi \psi$ and ξ, which are obtained by applying to them the inverse of the maps W_{nj} in (2.11). Thus, if $\xi = \xi_{nj} \in \mathbf{P}_{nj} L^2(\mathbf{R}^6)$, then it follows from (2.12) that

$$\tilde{\xi}^l_{nj}(\mathbf{k}) = (2\pi)^{-3/2} e_n(|\mathbf{k}|) Y^l_j(\mathbf{k}/|\mathbf{k}|) \ . \tag{2.19}$$

According to a well-known theorem by Wigner (1932, 1979), *there are no positive definite phase space distribution functions*, associated with either pure or mixed quantum states, *that have as their marginal distributions* (obtained by integrating them in $\mathbf{q} \in \mathbf{R}^3$ and in $\mathbf{p} \in \mathbf{R}^3$, respectively) *the quantum probability densities for the sharp position and sharp momentum measurements* of the preceding section. Hence no such phase space distribution functions can be interpreted as probability distributions for the simultaneous measurement of *sharp* position and of *sharp* momentum. However, the phase-space distribution functions, determined by the wave functions in (2.13), give rise to the following positive definite probability measures on the Borel sets of phase space,

$$\left\langle \psi \middle| E_\xi(B)\psi \right\rangle = \int_B \left|(\mathbf{P}_\xi \psi)(\mathbf{q},\mathbf{p})\right|^2 d^3\mathbf{q}\, d^3\mathbf{p} \ , \tag{2.20}$$

since they bypass the aforementioned 'no-go' theorem by Wigner by virtue of the fact that in their marginal distributions in (2.17) and (2.18), respectively, the quantum probability densities for sharp position and for sharp momentum measurements are smeared out with normalized *confidence functions* (cf. Principle 1 in Sec. 1.3) which are not δ-like. Hence these distribution functions can be interpreted as probability densities for the simultaneous measurement of the spread-out stochastic values

$$(\mathbf{q},\hat{\chi}^{l}_{\mathbf{q}})\times(\mathbf{p},\tilde{\chi}^{l}_{\mathbf{p}})\in\Gamma_{\xi}\ \leftrightarrow\ \mathbf{R}^{6} \tag{2.21}$$

of position and momentum, so that the corresponding probability measures in (2.20) provide probabilities for observing stochastic values within the Borel sets of a given *stochastic phase space* Γ_{ξ} , consisting of all the stochastic values in (2.21) that are obtained as $\mathbf{q}$ and $\mathbf{p}$ vary over $\mathbf{R}^3$.

The implicit presence in these probabilities of the generalized coherent state vectors[10] $\xi_{\mathbf{q},\mathbf{p}}$, which according to (2.13) are obtained by boosting the resolution generator ξ to the 3-velocity $\mathbf{v} = \mathbf{p}/m$, and spatially translating it by the amount $\mathbf{q}$, suggests that ξ should be interpreted as the state vector of an *extended* (rather than pointlike) quantum test body[11] whose center-of-mass is located at the origin of a classical inertial frame. Consequently, $\xi_{\mathbf{q},\mathbf{p}}$ is the state vector of a duplicate of such a test body, submitted to the aforementioned boost and spatial translation operations, in order to be used in the measurement of the mean stochastic position $\mathbf{q}$ and of the mean stochastic 3-momentum $\mathbf{p}$ of a quantum particle in the state ψ. In the case of $j > 0$, a simultaneous measurement of spin components is also implicitly performed by correlating the l-th component of spin of the system particle to the l-th component of the internal angular momentum of the extended test particle[12].The existence of the systems of covariance in (2.16) then supplies the covariance features under Galilei transformations of the corresponding probability measures in (2.20),

$$\left\langle \psi_t \middle| E_{\xi}(\hat{B}\times\tilde{B})\psi_t \right\rangle = \left\langle \psi'_{t+b} \middle| E_{\xi}\big((\mathbf{a}+\mathbf{v}t+R\cdot\hat{B})\times(m\mathbf{v}+R\cdot\tilde{B})\big)\psi'_{t+b} \right\rangle \ , \tag{2.22}$$

which are the counterparts of those in (1.19) and (1.23).

The consistency of the above interpretation with not only the orthodox interpretation of QM, but also with the fundamental aspects of classical and quantum statistical mechanics, has been confirmed by a variety of studies (Prugovečki, 1976-78; Ali and Prugovečki, 1977) – which are reviewed in Chapters 1 and 3 of [P]. Additional studies of measurement-theoretical procedures for the simultaneous measurement of position and momentum, as well as of other "incompatible" observables (Busch, 1985-86, Schroeck, 1981-82), have led to a great variety of recent results, reviewed in (Schroeck, 1991), which demonstrate the practical implementability as well as the theoretical usefulness of a generalized theory of quantum measurement based on systems of covariance. However, the inconsistencies of the relativistic QM theory of measurement, mentioned in Sec. 1.2, indicate that a more radical departure from conventional concepts is required in the relativistic regime. This need for a further extrapolation of conventional QM is underlined by the fact that, whereas the non-relativistic SQM probabilities obtained from (2.20) merge into their conventional counterparts of the preceding section in the sharp-point limits which we shall describe in Sec. 3.5, those limits do not exist in the relativistic case. Therefore, we turn next to the description of the most prominent features of the systems of imprimitivity of relativistic QM, which underline their physical and mathematical limitations, and then compare them with the corresponding systems of covariance in relativistic SQM, which do not share those limitations.

In Sec. 3.7 this will point the way to a geometro-stochastic framework capable of consistently unifying quantum mechanics and relativity theory at their most general level.

3.3. Relativistic Systems of Imprimitivity

In the relativistic regime, the role of Galilei group is taken over by the (restricted) *Poincaré group*, which can be identified, as in Sec. 2.4, with $ISO_0(3,1)$. Its elements incorporate spacetime translations $a \in \mathbf{R}^4$, and act on 4-tuples x of Minkowski coordinates a follows,

$$(a,\Lambda) \;:\; x \mapsto x' = a + \Lambda x \;, \qquad x \in \mathbf{R}^4 \;, \tag{3.1}$$

so that they obey the following group multiplication law (cf. (2.3.9)):

$$(a',\Lambda')(a,\Lambda) = (a' + \Lambda' a,\, \Lambda'\Lambda) \;, \qquad a', a \in \mathbf{R}^4 \;, \qquad \Lambda', \Lambda \in SO_0(3,1) \;. \tag{3.2}$$

In the special relativistic context the above coordinate transformation law replaces the transformation law in (1.12), so that, at least on the surface, the extension to the relativistic regime of the considerations in Sec. 3.1 appears very straightforward. Thus, according to all standard treatments of relativistic QM [BL, IQ, SI], a wave functions $\phi(x)$, purportedly representing the state vectors in the configuration representation of a single relativistic particle of rest mass m and zero spin, can be regarded as being an element of the pre-Hilbert space (extendable by Cauchy completion – cf. [PQ], p. 31) with inner product

$$\langle \hat{\varphi}_1 | \hat{\varphi}_2 \rangle \;=\; i \int_{\mathbf{R}^3} \hat{\varphi}_1^*(x) \overleftrightarrow{\partial}_0 \hat{\varphi}_2(x)\, d^3\mathbf{x} \;, \tag{3.3}$$

whose positive-definiteness is ensured if we restrict ourselves to positive-energy solutions

$$\hat{\varphi}_{x^0} = \exp(-iP_0 x^0)\hat{\varphi} \;, \qquad P_0 = \left(\mathbf{P}^2 + m^2\right)^{1/2} , \qquad \mathbf{P}^2 = -\Delta \;, \tag{3.4}$$

of the Klein-Gordon equation,

$$\left(\partial_\mu \partial^\mu + m^2\right)\hat{\varphi}(x) = 0 \;, \qquad \hat{\varphi}(x^0, \mathbf{x}) = \hat{\varphi}_{x^0}(\mathbf{x}) \;. \tag{3.5}$$

The inner product (3.3) replaces (1.1), and can be also written in the following manifestly covariant form,

$$\langle \hat{\varphi}_1 | \hat{\varphi}_2 \rangle \;=\; i \int_\sigma \hat{\varphi}_1^*(x) \overleftrightarrow{\partial}_\mu \hat{\varphi}_2(x)\, d\sigma^\mu(x) \;, \qquad d\sigma^\mu(x) = n^\mu(x) d\sigma(x) \;, \tag{3.6a}$$

$$\hat{\varphi}_1^*(x) \overleftrightarrow{\partial}_\mu \hat{\varphi}_2(x) := \hat{\varphi}_1^*(x)\left(\partial \hat{\varphi}_2(x)/\partial x^\mu\right) - \left(\partial \hat{\varphi}_1^*(x)/\partial x^\mu\right)\hat{\varphi}_2(x) \;, \tag{3.6b}$$

where n^μ, $\mu = 0,...,3$, are the components, with respect to any global Lorentz frame, of the future-pointing unit vector normal to the maximal spacelike hypersurface σ over which the integration is carried out. The following maps,

$$\hat{U}(a,\Lambda) : \quad \hat{\varphi}(x) \mapsto \hat{\varphi}'(x) = \hat{\varphi}(\Lambda^{-1}(x-a)) \, , \tag{3.7}$$

provide a unitary irreducible representation of the Poincaré group. We note that

$$\hat{U}(a',\Lambda')\hat{U}(a,\Lambda) = \hat{U}(a'+\Lambda' a, \Lambda'\Lambda) \, , \tag{3.8}$$

so that, as opposed to their nonrelativistic counterparts in (1.16), these maps define an ordinary vector representation – in agreement with the fact that the Poincaré group has no nontrivial unitary ray representations (Bargmann, 1947, 1952; Inönü and Wigner, 1952).

It would appear that the stage is set for introducing a relativistic system of imprimitivity, similar to the nonrelativistic system of imprimitivity based on the PV measure defined by (1.6) and on the representation defined by (1.16), so that the probability density that supplants the one in (1.7) would be equal to $|\hat{\varphi}(x)|^2$. However, such a system of imprimitivity is not physically acceptable. Indeed, it is easily verified that in any Lorentz frame the resulting total probability of finding the particle anywhere in space at a given time would not be then conserved, since the integral of this density over the hyperplanes corresponding to different instants with respect to that frame is not time-independent[13].

As mentioned in Sec. 1.2, it is sometimes asserted (cf. [SI], p. 56) that the timelike component $j^0(x)$ of the Klein-Gordon current

$$j^\mu(x) = (i/2m)\,\hat{\varphi}^*(x)\,\overleftrightarrow{\partial}^\mu\,\hat{\varphi}(x) \, , \qquad \overleftrightarrow{\partial}^\mu = \eta^{\mu\nu}\overleftrightarrow{\partial}_\nu \, , \tag{3.9}$$

provides the desired probability density in configuration space. However, this cannot actually be the case, since Gerlach *et al.* (1967) have rigorously proved that, for any positive-energy solution (3.4) of the Klein-Gordon equation, there are some points in Minkowski space where $j^0(x)$ is positive, as well as other points where $j^0(x)$ is negative, so that the positive-definiteness required of any probability density is violated. As a matter of fact, Hegerfeldt's (1974) theorem, discussed in Sec. 1.2, demonstrates that the search for systems of imprimitivity consistent with a notion of a relativistically invariant *sharp* localizability of quantum particles is bound to be fruitless. Hence, we turn instead to the formulation and interpretation of systems of imprimitivity for momentum measurements.

In the relativistic regime it is customary to express the momentum wave functions in the Heisenberg picture [PQ], so that they are time-independent. By analogy with (1.10), and with this stipulation in mind, the transition to the momentum representation can be effected by a mapping which takes the positive-energy solutions (3.4) of the Klein-Gordon equation into the time-independent momentum wave functions defined by

$$\tilde{\varphi}(k) = 2k^0(2\pi)^{-3/2}\int_{x^0=0}\exp(ik\cdot x)\hat{\varphi}(x)d^3\mathbf{x} \, , \quad k = (k^0,\mathbf{k}) \, , \; k^0 = (\mathbf{k}^2+m^2)^{1/2} . \tag{3.10}$$

Since this map is basically given by a Fourier transform, its inverse is provided by

$$\hat{\varphi}(x) = (2\pi)^{-3/2} \int_{k^0>0} \exp(-ik\cdot x)\tilde{\varphi}(k)\, d\Omega_m(k) \ , \tag{3.11a}$$

$$d\Omega_m(k) = \delta(k^2 - m^2) d^4k \ , \tag{3.11b}$$

where (3.11b) gives rise to a Lorentz-invariant measure on the forward mass hyperboloid

$$V_m^+ = \left\{ k \middle| k^2 := k\cdot k = m^2, \ k^0 > 0 \right\} \subset \mathbf{R}^4 \ . \tag{3.12}$$

Hence, it is easily verified that (3.10) and (3.11) provide unitary maps between the Hilbert space of positive-energy solutions (3.4) of the Klein-Gordon equation, with inner product (3.3), and the Hilbert space of functions on the forward mass hyperboloid (3.12), that are square-integrable with respect to this invariant measure, and which carries the inner product

$$\langle \tilde{\varphi}_1 | \tilde{\varphi}_2 \rangle = \int_{k^0>0} \tilde{\varphi}_1^*(k)\tilde{\varphi}_2(k)\, d\Omega_m(k) = \int_{V_m^+} \tilde{\varphi}_1^*(k)\tilde{\varphi}_2(k) \frac{d^3\mathbf{k}}{2\sqrt{\mathbf{k}^2 + m^2}} \ . \tag{3.13}$$

The momentum space counterparts of the 'configuration space' wave function maps in (3.7), which describe their behavior under Poincaré transformations, can be derived in the same manner as their nonrelativistic counterparts in (1.21), with the following result:

$$\tilde{U}(a,\Lambda) \ : \ \tilde{\varphi}(k) \mapsto \tilde{\varphi}'(k) = \exp(ia\cdot k)\tilde{\varphi}(\Lambda^{-1}k) \ . \tag{3.14}$$

Hence, if we introduce the PV measure

$$\tilde{E}(B) : \tilde{\varphi}(k) \ \mapsto \ \chi_B(k)\tilde{\varphi}(k) \ , \qquad \tilde{\varphi} \in L^2(V_m^+, d\Omega_m) \ , \quad B \subset V_m^+ \ , \tag{3.15}$$

then (3.14) and (3.15) give rise to a system of imprimitivity for momentum measurements.

Thus, in full accordance with the fact that under a general change of inertial frame the classical relativistic momentum components transform in accordance with

$$(a,\Lambda) \ : \ k \mapsto k' = \Lambda k \ , \qquad k \in V_m^+ \ , \tag{3.16}$$

we obtain the desired covariance properties,

$$\langle \tilde{\varphi} | \tilde{E}(B)\tilde{\varphi} \rangle = \int_B \tilde{\varphi}_1^*(k)\tilde{\varphi}_2(k)\, d\Omega_m(k) = \langle \tilde{\varphi}' | \tilde{E}(\Lambda B)\tilde{\varphi}' \rangle \ , \tag{3.17}$$

for the QM probabilities of sharp momentum measurement outcomes. In fact, (3.17) represents the relativistic counterpart of (1.23), since it can be easily checked that, if the 3-velocity $\mathbf{v}$ of the relative motion of two inertial frames has a magnitude which is small in relation to the speed of light, then (3.17) can be well approximated by (1.23).

The systems of imprimitivity for sharp momentum measurement outcomes of massive particles with spin $s > 0$ are rather similar, and can be obtained by combining the PV measures with elements given by the linear maps

$$\tilde{E}_s(B):\tilde{\varphi}(k)\mapsto\chi_B(k)\tilde{\varphi}(k)\,,\qquad\tilde{\varphi}\in\ell^2(2s+1)\otimes L^2(V_m^+,d\Omega_m)\,,\ B\subset V_m^+\,,\quad(3.18)$$

with the following unitary representation,

$$\tilde{U}_s(a,\Lambda):\ \tilde{\varphi}(k)\mapsto\tilde{\varphi}'(k)=\exp(ia\cdot k)D^s(R_{k,\Lambda})\tilde{\varphi}(\Lambda^{-1}k)\ ,\qquad(3.19a)$$

$$R_{k,\Lambda}=\Lambda_k^{-1}\Lambda\Lambda_{\Lambda^{-1}k}\in SO(3)\subset SO_0(3,1)\ ,\qquad(3.19b)$$

where Λ_k denotes the Lorentz boost that imparts the 4-momentum k to a classical particle which has rest-mass m, and which was originally at rest. The SO(3) element in (3.19b) lies in the subgroup of the Lorentz group that describes the spatial rotations of global Lorentz frames; it is called a *Wigner rotation,* since it was introduced by Wigner (1939) in the context of his well-known original derivation and classification of all unitary irreducible representations of the Poincaré group.

The modern derivation and classification of all these representations for the Poincaré group, as well as for its universal covering group ISL(2,C), is based on Mackey's (1951-53) theory of induced representations, in which systems of imprimitivity play a central role from a purely mathematical point of view (cf. §2 in Chapter 17 of [BR]). This systematic derivation reveals that, modulo unitary equivalence, the representations in (3.19) are the only irreducible unitary representations corresponding to all the possible spin values s = 0,1/2,1,3/2,2,5/2,... , and to all the possible rest masses $m > 0$ (where, in case of half-integer spin, Λ_k has to be replaced by a corresponding $A_k \in SU(2)$). For example, the familiar Dirac representation for s = 1/2 can be obtained from its s =1/2 counterpart in (3.19) by means of a Foldy-Wouthuysen transformation – cf. Sec. 6.1.

3.4. Relativistic Systems of Covariance

In the preceding section we have seen that, whereas there exists a close parallelism between the derivation of systems of imprimitivity for measurements of sharp quantum particle momentum in the nonrelativistic and the relativistic regime, that parallelism breaks down in the case of sharp position measurements. In fact, whereas in the nonrelativistic regime probability measures for sharp position measurements were suggested by Born (1926) only one year after the birth of modern quantum mechanics, in the relativistic regime their discovery has defied the efforts of several generations of researchers, until it was eventually proven by Hegerfeldt's (1974) theorem that all such efforts have been in vain[14], if an interpretation consistent with Einstein causality were desired. It is therefore remarkable that, in the context of systems of covariance based on POV measures over phase space, it turned out (Prugovečki, 1978c,d) to be possible to maintain that parallelism in all its essential physical aspects, and that the complete harmonic analysis (Ali and Prugovečki, 1986) of the relativistic counterparts of the phase space representations of Sec. 3.2 revealed many very close mathematical analogies between the nonrelativistic and relativistic SQM framework[15].

Let us therefore proceed as in Sec. 3.2, and introduce, as the relativistic counterpart of the Hilbert space $L^2(\mathbf{R}^6)$ in that section, the Hilbert space $L^2(\Sigma_m)$ which consists of all the complex-valued functions $\varphi(q,p)$ in the variables $q = (0,\mathbf{q})$ and $p = (p^0,\mathbf{p})$, with $\mathbf{q},\mathbf{p}\in\mathbf{R}^3$ and $p^0 = (\mathbf{p}^2+m^2)^{1/2}$, and which carries the inner product

$$\langle \varphi_1 | \varphi_2 \rangle = \int_{q^0=0} \varphi_1^*(q,p)\varphi_2(q,p)\, d^3\mathbf{q}\, d^3\mathbf{p} \ . \tag{4.1}$$

In this space we can consider, by analogy with (3.4) and (3.5), the time evolution

$$\varphi_{q^0} = \exp(-iP_0 q^0)\varphi \ , \qquad P_0 = \left(\mathbf{P}^2 + m^2\right)^{1/2} , \quad \mathbf{P}^2 = -\Delta_{\mathbf{q}} \ , \tag{4.2}$$

which provides positive-energy solutions of the Klein-Gordon equation in the q-variables:

$$\left(\partial_\mu \partial^\mu + m^2\right)\varphi(q,p) = 0 \ , \qquad \varphi(q,p) = \varphi_{q0}((0,\mathbf{q}),p) \ , \qquad \partial_\mu = \partial/\partial q^\mu \ . \tag{4.3}$$

In the Hilbert space $L^2(\Sigma_m)$, with inner product (4.1), we can then introduce the following representation of the Poincaré group:

$$U(a,\Lambda) \ : \ \varphi(q,p) \mapsto \varphi'(q,p) = \varphi(\Lambda^{-1}(q-a), \Lambda^{-1}p) \ . \tag{4.4}$$

It is easily verified that this representation satisfies the same group multiplication law,

$$U(a',\Lambda')U(a,\Lambda) = U(a' + \Lambda' a, \Lambda'\Lambda) \ , \tag{4.5}$$

as the representations of the Poincaré group studied in the preceding section. However, as opposed to those representations, the representation defined by (4.4) is highly reducible – as was the case with its nonrelativistic counterpart defined in (2.7). Its spectral analysis can be carried out in basically the same manner as in the latter case (Ali and Prugovečki, 1986). The main difference is that, for any choice of a set of functions e_n, $n = 1,2,\ldots$, such that

$$\int_m^{+\infty} e_{n'}^*(\kappa) e_{n''}(\kappa)\ \kappa(\kappa^2 - m^2)^{1/2} d\kappa = \delta_{n'n''} \ , \tag{4.6}$$

and such that this set constitutes an orthonormal basis in the Hilbert space with the inner product defined by the left-hand side of (4.6), that analysis reveals the existence of a linear operator Θ, which is unbounded, but has a bounded inverse, and which is such that

$$U(a,\Lambda) = \bigoplus_{n=1}^{\infty} \bigoplus_{j=1}^{\infty} \Theta U_{nj}(a,\Lambda)\Theta^{-1} \ , \tag{4.7}$$

where U_{nj} are irreducible unitary representations, while their Θ-transforms are not unitary.

The relativistic phase space[16] representation defined by (4.4) displays many mathematical properties which are very similar to those of its nonrelativistic counterpart in (2.7), despite the fact that (4.4) defines an ordinary vector representation, whereas (2.7) determines a ray representation which is not equivalent to a vector one. We shall mention, without proof, only those properties which are of direct physical interest – and direct the reader interested in further details to (Prugovečki, 1978c,d) and to (Ali and Prugovečki, 1986) for the proofs and for a more complete exposition (cf. also Secs. 2.3-2.7 in [P]).

First of all, we observe that the inner product in (4.1) can be written in the following manifestly covariant form,

$$\langle\varphi_1|\varphi_2\rangle = \int_{\Sigma_m} \varphi_1^*(q,p)\varphi_2(q,p)\,d\Sigma_m(q,p) \ , \qquad \Sigma_m = \sigma_0 \times V_m^+ \ , \tag{4.8a}$$

$$d\Sigma_m(q,p) = 2p_\mu\delta(p^2-m^2)d\sigma^\mu(q)d^4p = 2p_\mu d\sigma^\mu(q)d\Omega_m(p) \ , \tag{4.8b}$$

where $d\Sigma_m$ is the unique, modulo a positive multiplicative constant, covariant measure on the hypersurface Σ_m within the relativistic phase space of particles of rest mass m (Ehlers, 1971). In (4.8), that constant was adjusted so that, in any global Lorentz frame with respect to which σ_0 is a hyperplane corresponding to $q^0 = 0$ (or more generally to constant q^0), the element of measure $d\Sigma_m(q,p)$ equals $d^3\mathbf{q}d^3\mathbf{p}$, and (4.8a) therefore assumes the form (4.1). It should be also noted that if Σ_m is identified (cf. [P], pp. 85-86) with the homogenous space $ISO_0(3,1)/(\mathcal{T}\times SO(3))$ of left cosets [BR] with respect to the direct product $\mathcal{T}\times SO(3)$ of the time-translation subgroup $\mathcal{T}$ with the spatial rotation subgroup SO(3) of the Poincaré group, then $d\Sigma_m$ gives rise to a left invariant measure on that space. On the other hand, if Σ_m is identified with the homogenous space $(\mathcal{T}\times SO(3))\backslash ISO_0(3,1)$ of right cosets, then it is instead $d\Sigma_m/p^0$ that gives rise to a right invariant measure on that space.

The (nj)-th irreducible representation in (4.7) is unitarily equivalent to the momentum space representation in (3.19) that corresponds to $s = j = 0,1,2,\ldots$. Let $\mathbf{P}_{nj}$ denote the orthogonal projection operator which maps the Hilbert space $L^2(\Sigma_m)$, carrying the representation in (4.4), onto the subspace carrying that (nj)-th irreducible representation, so that

$$U_{nj} = \mathbf{P}_{nj}U = W_{nj}\tilde{U}W_{nj}^{-1} \ , \qquad W_{nj}: \ \ell^2(2j+1)\otimes L^2(V_m^+, d\Omega_m) \to \mathbf{P}_{nj}L^2(\Sigma_m) \ , \tag{4.9}$$

expresses its unitary equivalence to its momentum space counterpart. Then the unitary operator W_{nj}, supplying that unitary equivalence, is an integral operator whose action on the momentum space wave functions appearing in (3.18) results in

$$(\mathbf{P}_{nj}\varphi)(q,p) = \left(m/4\pi^3\right)^{1/2}\int_{V_m^+} \exp(-iq\cdot k)e_n(p\cdot k/m)Y_j^l\left(R_{k,\Lambda_p}\mathbf{p}\right)\tilde{\varphi}_l(k)\,d\Omega_m(k), \tag{4.10}$$

where, as was the case in (2.12), the summation over the repeated internal angular momentum indices $l = -j,\ldots,+j$ is implicitly understood.

Each one of the above irreducible subspaces is characterized by a unique rotationally invariant *resolution generator* η, namely by a unique vector belonging to that subspace and having the property that for all vectors φ from $L^2(\Sigma_m)$

$$(\mathbf{P}_\eta\varphi)(q,p) = \langle\eta_{q,p}|\varphi\rangle \ , \qquad \eta_{q,p} = U(q,\Lambda_p)\eta \ . \tag{4.11}$$

Consequently, the orthogonal projection operator $\mathbf{P}_\eta$ onto that subspace can be written in the form of the following Bochner integral:

$$\mathbf{P}_\eta = \int_{\Sigma_m} |\eta_{q,p}\rangle d\Sigma_m(q,p)\langle\eta_{q,p}| \ , \qquad \mathbf{P}_\eta = \mathbf{P}_\eta^* = \mathbf{P}_\eta^2 \ . \tag{4.12}$$

As in the nonrelativistic case, equations (4.11) and (4.12) then lead to the conclusion that each one of these subspaces is a reproducing kernel Hilbert space, so that

$$(\mathbf{P}_\eta \varphi)(q,p) = \int_{\Sigma_m} \langle \eta_{q,p} | \eta_{q',p'} \rangle \, \varphi(q',p') \, d\Sigma_m(q',p') \; . \tag{4.13}$$

We note that, up to this point, the analogies between the nonrelativistic relations (2.11)-(2.15) and their present relativistic counterparts were so thorough that the corresponding considerations of Sec. 3.2 could be transferred almost verbatim to the present section. The first significant difference between these two cases emerges from the fact that, whereas Galilei transformations preserve simultaneity, Poincaré transformations generically do not. Consequently, the relativistic POV measure in each of the irreducible subspaces $\mathbf{P}_\eta L^2(\Sigma_m)$, which constitutes, together with the irreducible representation induced by (4.4) in $\mathbf{P}_\eta L^2(\Sigma_m)$, the system of covariance

$$U_\eta(a,\Lambda) = \mathbf{P}_\eta U(a,\Lambda) \; , \qquad \bar{E}_\eta(B) = \int_{-\infty}^{+\infty} E_\eta(B_{q^0}) \, dq^0 \; , \tag{4.14a}$$

$$E_\eta(B_{q^0}) = \int_{B_{q^0}} |\eta_{q,p}\rangle \, d\Sigma_m(q,p) \langle \eta_{q,p}| \; , \tag{4.14b}$$

$$B_{q^0} = \{(\mathbf{q},p) \,|\, (q,p) \in B \subset \mathbf{R}^4 \times V_m^+\} \subset \Sigma_{m,q^0} \; , \tag{4.14c}$$

is of necessity a measure over Borel sets in the *relativistic* phase space, which involves the time variable in such a manner that it cannot be extricated from the space variables – as it was possible, for example, in the case of the POV measures in (2.3).

On the other hand, since $dq^0 d\Sigma_m$ is not an invariant measure element in relativistic phase space, it is not immediately obvious that the pair in (4.14a) indeed constitutes a system of covariance, i.e., that it satisfies (2.2). To see that this is actually the case, we observe that in each of the irreducible subspaces $\mathbf{P}_\eta L^2(\Sigma_m)$ (but not on all of $L^2(\Sigma_m)$) the inner product in (4.8) can be written in an alternative form (cf. Sec. 2.6 in [P]), which is analogous to the one in (3.6a), namely

$$\langle \varphi_1 | \varphi_2 \rangle \;=\; iZ_{\eta,m}^{-1} \int_{\Sigma_m} \varphi_1^*(q,p) \overleftrightarrow{\partial}_\mu \varphi_2(q,p) \, d\sigma^\mu(q) d\Omega_m(p) \; , \tag{4.15a}$$

$$Z_{\eta,m} = 2m \int_{V_m^+} |e(p^0)|^2 \, d\Omega_m(p) \; , \tag{4.15b}$$

$$\varphi_1^*(q,p) \overleftrightarrow{\partial}_\mu \varphi_2(q,p) := \varphi_1^*(q,p) \big(\partial \varphi_2(q,p)/\partial q^\mu\big) - \big(\partial \varphi_1^*(q,p)/\partial q^\mu\big) \varphi_2(q,p) \; . \tag{4.15c}$$

This alternative form involves the renormalization constant $Z_{\eta,m}$, which depends on the resolution generator η via its momentum space representative e, generically assigned to it by (4.10) – i.e., on the proper state vector of the test particle used in the measurement of stochastic position and momentum, to which the system of covariance determined by (4.14) will be shortly related. Hence, the POV measure in (4.14a) can be expressed in the form

$$\bar{E}_\eta(B) = \; iZ_{\eta,m}^{-1} \int_{-\infty}^{+\infty} dq^0 \int_{B_{q^0}} |\eta_{q,p}\rangle \overleftrightarrow{\partial}_0 \langle \eta_{q,p}| \, d^3\mathbf{q} \, d\Omega_m(p)$$

$$= iZ_{\eta,m}^{-1} \int_B \left|\eta_{q,p}\right\rangle n^\mu \overleftrightarrow{\partial}_\mu \left\langle \eta_{q,p}\right| d^4q d\Omega_m(p) , \tag{4.16}$$

which involves the invariant measure element $d^4qd\Omega_m(p)$ on relativistic phase space. It therefore now becomes straightforward to deduce from (4.4) and (4.11) that the pair in (4.14a) satisfies the relationship

$$U_\eta(a,\Lambda)\overline{E}_\eta(B)U_\eta^{-1}(a,\Lambda) = \overline{E}_\eta(a+\Lambda B) \ , \tag{4.17}$$

which is a special case of (2.2). This establishes that we are indeed dealing with a Poincaré system of covariance.

All these results suggest that, by analogy with the nonrelativistic case, η should be interpreted in the relativistic case as the state vector of an extended quantum test body at the origin of a global Lorentz frame, that $\eta_{q,p}$ should be interpreted as the state vector of a duplicate of such a test body, that was submitted to Lorentz boost and spacetime translation operations in order to be used in the measurement of the mean spacetime stochastic location q and mean stochastic 4-momentum p of a quantum particle in the state φ, and that the expectation values for φ of the POV measure in (4.14a) should provide the relative probabilities of such measurement outcomes. Hence,

$$\left\langle \varphi \middle| E_\eta(B_{q^0})\varphi \right\rangle = \int_{B_{q^0}} \left|(\mathbf{P}_\eta\varphi)(q,p)\right|^2 d\Sigma_m(q,p) , \qquad B_{q^0} \subset \Sigma_{m,q^0} \ , \tag{4.18}$$

then represents the probability of obtaining such measurement results within the Borel set B_{q^0}, if those measurements were carried out at a given instant q^0 with respect to the chosen global Lorentz frame.

In the present relativistic context, we cannot deduce this interpretation for unsharp measurements from a corresponding interpretation for sharp measurements – as we did in Sec. 3.2, by means of the marginality relations (2.17) and (2.18). Indeed, as we have seen in Sec. 3.3, in relativistic quantum mechanics there is no consistent theory of sharp spacetime localization for quantum particles. It is, however, interesting to note that the W_η-unitary transforms of the operators of multiplication by q^a, $a = 1,2,3$, are equal to the Newton-Wigner (1949) operators in any Lorentz frame in which Σ_m corresponds to a hypersurface of constant q^0, and that similarly the W_η-unitary transforms of the operators of multiplication by p^a, $a = 1,2,3$, are equal, modulo a positive constant, to the 3-momentum operators (cf. [P], pp. 98-99). This provides indirect evidence that a concept of unsharp spacetime localization, based on the systems of covariance in (4.14), can be consistently maintained, so that the violation of relativistic causality by the Newton-Wigner operators becomes merely a reflection of the fact that the idealized concept of perfectly sharp quantum particle spacetime localization is inconsistent in the relativistic regime. The basic support for the aforementioned interpretation of (4.18) then comes from the existence of the conserved and covariant probability currents described in the next section, as well as from the existence of covariant relativistic propagators for proper state vectors, derived in Sec. 3.6 – which share with their conventional counterparts key features, that are of direct experimental significance. A probability interpretation of (4.18), that is also consistent with the basic tenets of *general* relativistic theory, will be given in Sec. 5.5.

3.5. Probability Currents and Sharp-Point Limits

The existence of the unitary maps W_{nj} in (2.11) and (4.9) demonstrates that the frameworks of both nonrelativistic and relativistic SQM are unitarily equivalent to the respective QM frameworks. Consequently, in purely mathematical terms, every single statement and result can be translated from the latter into the former framework, and vice versa. That means that the differences between these frameworks emerge only at the physical level. We can compare this situation with that of the configuration and momentum representations of nonrelativistic QM: they are unitarily equivalent, and yet they each display their own advantages vis-à-vis the type of physical results and measurement procedures that are being considered – depending on whether sharp position or sharp momentum measurements are of primary interest. Similarly, the nonrelativistic SQM framework described in Sec. 3.2 displays its own advantages when nonsharp measurements of these quantities are of primary concern – as it can be the case in statistical mechanics, especially when studying Brownian motion, the Boltzman equation, etc. (cf. [P], Chapter 3). On the other hand, in the relativistic regime, it is only the SQM framework that is based on notions of relativistic systems of covariance that can be consistently related to the localization of relativistic quantum particle in spacetime. It is therefore of importance to understand in which sense nonrelativistic SQM quantities of direct physical significance merge into their nonrelativistic QM counterparts during a steady increase in measurement precision, with a sharp-point limit of perfect precision as its theoretically conceivable (but practically unachievable) goal, as well as why the same procedure is not mathematically feasible in the relativistic regime – so that no theory of sharp relativistic quantum particle localization can emerge in this manner.

The quantities of most direct physical significance in QM are the configuration and momentum space densities. In the case of a single particle of zero spin[17] they are given by

$$\hat{\rho}(\mathbf{x},t) = \left|\hat{\psi}(\mathbf{x},t)\right|^2 , \qquad \tilde{\rho}(\mathbf{k},t) = \left|\tilde{\psi}(\mathbf{k},t)\right|^2 , \tag{5.1}$$

in accordance with (1.7) and (1.10)-(1.11). By (2.17)-(2.18) and (2.20), we have

$$\rho_\xi(\mathbf{q},t) = \int_{\mathbf{R}^3} d^3\mathbf{p}\left|\psi(\mathbf{q},\mathbf{p},t)\right|^2 = \int_{\mathbf{R}^3} d^3\mathbf{x}\,\hat{\chi}_{\mathbf{q}}(\mathbf{x})\left|\hat{\psi}(\mathbf{x},t)\right|^2, \qquad \psi \in \mathbf{P}_\xi L^2(\mathbf{R}^6) , \tag{5.2}$$

$$\tilde{\rho}_\xi(\mathbf{p},t) = \int_{\mathbf{R}^3} d^3\mathbf{q}\left|\psi(\mathbf{q},\mathbf{p},t)\right|^2 = \int_{\mathbf{R}^3} d^3\mathbf{k}\,\tilde{\chi}_{\mathbf{p}}(\mathbf{k})\left|\tilde{\psi}(\mathbf{k},t)\right|^2, \qquad \psi = W_\xi \tilde{\psi} , \tag{5.3}$$

even when potential interactions govern the time evolution. We can then immediately see that, under suitable conditions[18] on the configuration space wave functions, we shall have

$$\rho_\xi(\mathbf{x},t) \to \hat{\rho}(\mathbf{x},t), \qquad \hat{\chi}_0(\mathbf{x}) = (2\pi)^3\left|\hat{\xi}(\mathbf{x})\right|^2 \to \delta^3(\mathbf{x}) , \tag{5.4}$$

$$\tilde{\rho}_\xi(\mathbf{k},t) \to \tilde{\rho}(\mathbf{k},t), \qquad \tilde{\chi}_0(\mathbf{k}) = (2\pi)^3\left|\tilde{\xi}(\mathbf{k})\right|^2 \to \delta^3(\mathbf{k}) . \tag{5.5}$$

The type of limit in (5.4), in which extended test particles possessing increasingly better spatially localized proper state vectors are employed until perfectly accurate localization is (conceptually) achieved, will be called a *sharp-point limit*; whereas the type of limit in (5.5), in which extended test particles with proper state vectors that are increasingly better localized in momentum space are adopted in order to (conceptually) arrive at perfectly accurate measured momentum values, will be called a *sharp-momentum limit.*

It is remarkable that the sharp-point limit of nonrelativistic SQM exists not only for the stochastic configuration space probability densities – where, in view of (2.17), its equality to the conventional probability density in (5.1) is not at all surprising – but that a similar result holds for the respective probability currents (cf. Sec. 1.8 in [P]), i.e.,

$$\mathbf{j}_\xi(\mathbf{x},t) \to \mathbf{j}(\mathbf{x},t), \qquad \hat{\chi}_0(\mathbf{x}) = (2\pi)^3 \left|\hat{\xi}(\mathbf{x})\right|^2 \to \delta^3(\mathbf{x}) \ , \tag{5.6}$$

under the single stipulation that the configuration space representatives of the proper state vectors that appear in (5.4) are real functions. Furthermore, the probability current

$$\mathbf{j}(\mathbf{x},t) = (2im)^{-1}\, \hat{\psi}^*(\mathbf{x},t)\, \overleftrightarrow{\nabla} \hat{\psi}(\mathbf{x},t) \ , \tag{5.7}$$

is routinely derived in QM (cf., e.g., [Messiah, 1962]) only on the basis of the requirement that the continuity equation,

$$\partial_t \hat{\rho}(\mathbf{x},t) + \nabla \cdot \mathbf{j}(\mathbf{x},t) = 0 \ , \tag{5.8}$$

be satisfied, whereas the definition of the SQM current

$$\mathbf{j}_\xi(\mathbf{q},t) = \int_{\mathbf{R}^3} \frac{\mathbf{p}}{m} \left|\psi(\mathbf{q},\mathbf{p},t)\right|^2 d^3\mathbf{p} \ , \qquad \psi \in \mathbf{P}_\xi L^2(\mathbf{R}^6) \ , \tag{5.9}$$

is obtained in the same physically well-grounded manner as in the case of its counterpart in classical statistical mechanics (Balescu, 1975) – i.e., from the interpretation of $|\psi(\mathbf{q},\mathbf{p},t)|^2$ in (5.2) as a probability density in phase space, and from the interpretation of $\mathbf{p}/m$ as a stochastic 3-velocity. Consequently, no close relationship could be expected on *a priori* grounds between these two types of currents. Nevertheless, the stochastic configuration space probability density in (5.2) and the current in (5.9) satisfy the continuity equation

$$\partial_t \rho_\xi(\mathbf{q},t) + \nabla \cdot \mathbf{j}_\xi(\mathbf{q},t) = 0 \ , \tag{5.10}$$

and the sharp-point limit (5.6) holds true for any state vector ψ which in the configuration space can be represented by a once continuously differentiable wave function.

In the free relativistic regime, the counterpart of the stochastic momentum density in (5.3) is q^0-independent, and has the following sharp-momentum limit (cf. [P], p. 95),

$$\tilde{\rho}_\eta(p) = \int_{\mathbf{R}^3} \left|\varphi((q^0,\mathbf{q}),p)\right|^2 d^3\mathbf{q} \ \to \ (2p^0)^{-1} |\tilde{\varphi}(p)|^2, \qquad \varphi = W_\eta \tilde{\varphi} \ , \tag{5.11}$$

as it could be expected on the basis of (3.13) and (3.17). However, the corresponding probability density

$$\rho_\eta(q) = \int_{p\in V_m^+} |\varphi(q,p)|^2\, d^3\mathbf{p} \ , \qquad \varphi \in \mathbf{P}_\eta L^2(\Sigma_m) \ , \tag{5.12}$$

in stochastic configuration space, which is the relativistic counterpart of the one in (5.2), does not display the existence of a sharp-point limit. This is totally in accordance with the fact that relativistic QM does not possess a relativistically covariant *probability* density for particle localization in Minkowski space.

On the other hand, the probability density in (5.12) equals, modulo the factor m^{-1}, the timelike component of the probability current

$$j_\eta^\mu(q) = 2\int_{p^0>0} \frac{p^\mu}{m}|\varphi(q,p)|^2\, d\Omega_m(p) = \int_{V_m^+} \frac{p^\mu}{mp^0}|\varphi(q,p)|^2\, d^3\mathbf{p} \ , \tag{5.13}$$

which is obviously the relativistic counterpart of the probability current in (5.9). This current (studied in greater detail in [P], Sec. 2.8) is relativistically covariant:

$$j_\eta^\mu(q) \mapsto j_\eta^\mu(q') = \Lambda^\mu{}_\nu j_\eta^\nu(q) \ , \qquad q' = a + \Lambda q \ . \tag{5.14}$$

It is also conserved, provided that the function $e(p\cdot k/m)$, associated by (4.10) with the proper state vector η in (4.11), is real (Ali *et al.*, 1981, 1988). Hence, in that case we have

$$\partial_\mu j_\eta^\mu(q) = 0 \ , \qquad \partial_\mu = \partial/\partial q^\mu \ . \tag{5.15}$$

This ensures [P] that the q-integration for the inner product in (4.8a) can be performed over arbitrary spacelike Cauchy [W] hypersurfaces σ, and not just over spacelike hyperplanes.

The relativistic SQM probability current has no relationship to the Klein-Gordon current in (3.9) – which, as we recall, is not a probability current at all, but rather it is usually described as a "charge current". However, it is interesting to note that, in relativistic SQM, there is a formally analogous "charge current", namely

$$J_\eta^\mu(q) = i(mZ_{\eta,m})^{-1} \int_{p^0>0} \varphi^*(q,p)\overleftrightarrow{\partial}_\mu \varphi(q,p)\, d\Omega_m(p) \ . \tag{5.16}$$

Its existence is closely related to the existence of the form (4.15a) of the inner product in (4.8a). This "charge current" is relativistically covariant and conserved (cf. [P], Sec. 2.8):

$$J_\eta^\mu(q) \mapsto J_\eta^\mu(q') = \Lambda^\mu{}_\nu J_\eta^\nu(q) \ , \qquad \partial_\mu J_\eta^\mu(q) = 0 \ . \tag{5.17}$$

On the other hand, despite its formal similarity to the Klein-Gordon current in (3.9), in the sharp-point limit the current in (5.16) does not merge into the one in (3.9), except in the asymptotic limit $m \to \infty$ of infinitely massive particles (cf. [P], p. 115) – i.e., in the limit where the physical behavior of a quantum particle begins to resemble classical behavior.

Thus, at the physical level, relativistic SQM displays features which are not present in the conventional relativistic QM reviewed in Sec. 3.3. These features demonstrate that a consistent probabilistic interpretation of quantum particle spacetime localization is feasible within the SQM framework, despite the fact that it is not feasible within conventional QM.

In agreement with the various observations cited in Sec. 1.2, in the relativistic quantum regime this interpretation necessitates the abandoning of the notion of perfectly sharp localization. It should be noted, however, that in the SQM framework reviewed in this chapter classical geometries are still retained. In fact, it is only the theory of quantum measurement that is first extrapolated in the nonrelativistic SQM regime, so as to accommodate the notion of extended test particle, and subsequently the extrapolated nonrelativistic SQM theory is transferred to the relativistic regime, so as to arrive at *bona fide* relativistically covariant probability measures and probability currents for such measurements.

This means that SQM still allows for the possibility of *arbitrarily accurate* spacetime localization, and only precludes the possibility of taking the sharp-point limit that would actually lead to *infinitely* accurate spacetime localization[19]. However, the various analyses (DeWitt, 1962; Mead, 1964; Blokhintsev, 1973) of those aspects of the quantum theory of measurement, which take into account the ever-present and all-pervasive gravitational phenomena, indicate that even this rather limited SQM idealization cannot be maintained in the unification of quantum theory and general relativity, since Planck's time and Planck's length impose an absolute lower bound on the accuracy of spacetime localization. Consequently, a reformulation of quantum theory which goes beyond merely replacing pointlike test particles with extended ones is required, in which the classical concept of pointlike localization is totally removed from quantum theory, by embedding this lower bound into the very geometry of spacetime. To carry out this task in the subsequent chapters, we first have to express the concept of quantum propagation exclusively in terms of the proper state vectors of extended quantum objects. We therefore turn now to this task.

3.6. Path Integrals in Stochastic Quantum Mechanics

In the nonrelativistic context, the Feynman path integral in (1.4.1) leads to expressions for propagators of pointlike quantum particles. These propagators can be heuristically[20] interpreted as providing the probability amplitudes for a particle to reach the spatial location $\mathbf{x}(t'')$ at time t'' if it has started from the location $\mathbf{x}(t')$ at time t' (Feynman and Hibbs, 1965). On the other hand, the expression (1.4.3) for the free propagator can be also written in terms of the representation of the Galilei group in (1.16) in the form

$$K(\mathbf{x}'',t'';\mathbf{x}',t') = \langle \mathbf{x}'',t''|\mathbf{x}',t'\rangle \ , \qquad t''>t' \ , \tag{6.1a}$$

$$|\mathbf{x}',t'\rangle = \hat{U}(t',\mathbf{x}',\mathbf{0},I)|\mathbf{0},0\rangle \ , \qquad |\mathbf{x}'',t''\rangle = \hat{U}(t'',\mathbf{x}'',\mathbf{0},I)|\mathbf{0},0\rangle \ , \tag{6.1b}$$

which provides its operational interpretation in geometric terms, namely as a transition amplitude between the initial state vector $|\mathbf{x}',t'\rangle$ and the final state vector $|\mathbf{x}'',t''\rangle$, obtained by translating the state vector $|\mathbf{0},0\rangle$, situated at time zero at the origin of an inertial frame in a Newtonian spacetime, to the spacetime locations $(\mathbf{x}',t')$ and $(\mathbf{x}'',t'')$, respectively.

With this geometric interpretation in mind, we can immediately extrapolate the definition of the nonrelativistic propagator (6.1) to the proper state vectors of extended particles:

$$K_\xi(\mathbf{q}'',\mathbf{p}'',t'';\mathbf{q}',\mathbf{p}',t') = \left\langle \xi_{\mathbf{q}'',\mathbf{p}'',t''} \middle| \xi_{\mathbf{q}',\mathbf{p}',t'} \right\rangle \ , \tag{6.2a}$$

$$\xi_{\mathbf{q}',\mathbf{p}',t'} = U_\xi(t',\mathbf{q}',\mathbf{p}'/m,I)\xi \ , \qquad \xi_{\mathbf{q}'',\mathbf{p}'',t''} = U_\xi(t'',\mathbf{q}'',\mathbf{p}''/m,I)\xi \ . \tag{6.2b}$$

It then turns out (Prugovečki, 1981a) that the main physical features of Feynman's (1948) path integral formulation of quantum mechanics are retained if (6.1) is replaced by (6.2). At the same time this extrapolation removes the difficulties originating from the fact that (6.1a) is neither a bona fide inner product, nor does it determine a bona fide probability density. Indeed, (6.2a) determines a bona fide probability density, on account of the fact that

$$(2\pi)^3 \int_{\mathbf{R}^6} \left| K_\xi(\mathbf{q}'',\mathbf{p}'',t'';\mathbf{q}',\mathbf{p}',t') \right|^2 d^3\mathbf{q}''\, d^3\mathbf{p}'' = 1 \ , \tag{6.3}$$

As a matter of fact, in accordance with the interpretation discussed in Sec. 3.2, this density provides, for an extended particle which at time t' is at the stochastic location $\mathbf{q}'$ and has stochastic 3-momentum $\mathbf{p}'$, the probabilities of its reaching at the time t'' various stochastic locations $\mathbf{q}''$, and displaying there various stochastic 3-momenta $\mathbf{p}''$. Furthermore, on account of (2.14), the *SQM propagator* in (6.2a) also possesses the following properties,

$$\begin{aligned} K_\xi(\mathbf{q}'',\mathbf{p}'',t'';\mathbf{q}',\mathbf{p}',t') &= K_\xi^*(\mathbf{q}',\mathbf{p}',t';\mathbf{q}'',\mathbf{p}'',t'') \\ &= \int_{\mathbf{R}^6} K_\xi(\mathbf{q}'',\mathbf{p}'',t'';\mathbf{q},\mathbf{p},t) K_\xi(\mathbf{q},\mathbf{p},t;\mathbf{q}',\mathbf{p}',t')\, d^3\mathbf{q} d^3\mathbf{p} \ , \end{aligned} \tag{6.4}$$

which characterize also Feynman propagators. Indeed, the following counterparts of (6.4),

$$K(\mathbf{x}'',t'';\mathbf{x}',t') = K^*(\mathbf{x}',t';\mathbf{x}'',t'') = \int_{\mathbf{R}^3} K(\mathbf{x}'',t'';\mathbf{x},t)K(\mathbf{x},t;\mathbf{x}',t')\, d^3\mathbf{x} \ , \tag{6.5}$$

are satisfied by (6.1a) – but only at a mathematically formal level, since we see from (1.4.3) that the integral in (6.5) cannot exist in the Lebesgue sense.

On the other hand, (6.5) can be treated as the sharp-point of (6.4), by adopting as a proper state vector the ground state of the nonrelativistic harmonic oscillator (Messiah, 1962). In fact, in the configuration representation this ground state is given by

$$\hat{\xi}^{(\ell)}(\mathbf{x}) = (8\pi^3\ell^2)^{-3/4} \exp(-\mathbf{x}^2/4\ell^2) \ , \qquad \ell > 0 \ , \tag{6.6}$$

where $\sqrt{3}\ell$ can be interpreted as the rms radius of an extended particle[21]. By combining (1.16), (2.11) and (2.13), we get

$$\hat{\xi}^{(\ell)}_{\mathbf{q},\mathbf{p}}(\mathbf{x}) = (8\pi^3\ell^2)^{-3/4} \exp\left\{-[(\mathbf{x}-\mathbf{q})/2\ell]^2 + i\mathbf{p}\cdot(\mathbf{x}-\mathbf{q})\right\} \ . \tag{6.7}$$

Hence, upon inserting (6.7) into the formula for the Fourier transform in (1.1), we can explicitly compute the corresponding momentum space wave functions, namely we get

$$\tilde{\xi}^{(\ell)}_{\mathbf{q},\mathbf{p}}(\mathbf{k},t) = (\ell^2/2\pi^3)^{3/4} \exp\left\{-\ell^2(\mathbf{k}-\mathbf{p})^2 - i\mathbf{k}\cdot\mathbf{q} - i\mathbf{k}^2 t/2m\right\} \ , \tag{6.8}$$

by using standard formulae on Gaussian integrals (cf. [ST], p. 20). The same formulae can be then also used to compute the inner product in (6.2a), with the following end result:

$$K_{\xi(\ell)}\left(\mathbf{q}'',\mathbf{p}'',t'';\mathbf{q}',\mathbf{p}',t'\right) = (2\pi)^{-3}(\ell^2/\ell_{m,t}^2)^{3/2}\exp\left\{-[(\mathbf{q}''-\mathbf{q}')^2/8\ell_{m,t}^2]\right\}$$
$$\times\exp\left\{-(\ell^2/2\ell_{m,t}^2)\left[\ell^2(\mathbf{p}''-\mathbf{p}')^2 - i(\mathbf{q}''-\mathbf{q}')\cdot(\mathbf{p}''+\mathbf{p}') + (it/2m)(\mathbf{p}''^2+\mathbf{p}'^2)\right]\right\}, \quad (6.9a)$$

$$\ell_{m,t}^2 = \ell^2 + (i/4m)t \ , \qquad t = t''-t' \in \mathbf{R}^1 \ . \quad (6.9b)$$

Consequently, by comparing (6.9) with (1.4.3), we see that

$$(\pi/2\ell^2)^{3/2}K_{\xi(\ell)}(\mathbf{q}'',\mathbf{p}'',t'';\mathbf{q}',\mathbf{p}',t') \xrightarrow[\ell\to+0]{} K(\mathbf{q}'',t'';\mathbf{q}',t') \ , \quad (6.10)$$

so that, indeed, in the sharp-point limit the Feynman propagator in (6.1) emerges as a renormalized limit of the SQM propagator in (6.2) – where, for $t' = t''$, that limit has to be interpreted in the sense of distributions, i.e., of generalized functions (Gel'fand and Shilov, 1964), since it equals a δ-function. On the other hand, the fact that the renormalization constant in (6.10) becomes infinite in that limit emerges as one of the root causes of the difficulties with Feynman propagators and path integrals, mentioned in Note 20.

The fundamental importance of the reproducibility properties, presented in (6.5) for the Feynman propagator in (6.1), lies in the fact that, in the presence of generically time-dependent potential interactions,

$$H(t) = H_0 + H_I(t) \ , \qquad H_0 = \mathbf{P}^2/2m \ , \qquad H_I(t) = V(\mathbf{Q},t) \ , \quad (6.11)$$

so that they lead to path integral formulae in (1.4.1) for the Feynman propagators

$$\hat{K}(\mathbf{x}'',t'';\mathbf{x}',t') = \left\langle \mathbf{x}'' \middle| T\int_{t'}^{t''} e^{-iH(t)}dt \middle| \mathbf{x}' \right\rangle \ . \quad (6.12)$$

This is demonstrated in Chapter 1 of [ST] by using the Trotter product formula. However, the application of this formula in this context is at the mathematical level only heuristic, since the bra and ket vectors in (6.1) are not elements of the Hilbert space $L^2(\mathbf{R}^3)$.

On the other hand, the vectors appearing in the expression (6.2a) for the SQM propagator are bona fide elements of the Hilbert space $L^2(\mathbf{R}^6)$. Hence, we can legitimately use (2.14) in between each consecutive pair of operators in the second of the following inner products

$$K_\xi(\mathbf{q}'',\mathbf{p}'',t'';\mathbf{q}',\mathbf{p}',t') = \left\langle \xi_{\mathbf{q}'',\mathbf{p}''} \middle| e^{-iH_0 t}\xi_{\mathbf{q}',\mathbf{p}'} \right\rangle$$
$$= \left\langle \xi_{\mathbf{q}'',\mathbf{p}''} \middle| \prod_{n=N}^{1} e^{-iH_0(t_n-t_{n-1})}\xi_{\mathbf{q}',\mathbf{p}'} \right\rangle \ . \quad (6.13)$$

Thus, upon setting $\mathbf{q}' = \mathbf{q}_0$, $\mathbf{p}' = \mathbf{p}_0$, $\mathbf{q}'' = \mathbf{q}_N$ and $\mathbf{p}'' = \mathbf{p}_N$, we obtain

$$K_\xi(\mathbf{q}'',\mathbf{p}'',t'';\mathbf{q}',\mathbf{p}',t') = \lim_{\varepsilon\to+0}\int K_\xi(\mathbf{q}_N,\mathbf{p}_N,t_N;\mathbf{q}_{N-1},\mathbf{p}_{N-1},t_{N-1})$$
$$\times\prod_{n=N-1}^{1} K_\xi(\mathbf{q}_n,\mathbf{p}_n,t_n;\mathbf{q}_{n-1},\mathbf{p}_{n-1},t_{n-1})\,d^3\mathbf{q}_n d^3\mathbf{p}_n \quad (6.14)$$

as a counterpart of (1.4.2), in which all the integrals exist in a bona fide Lebesgue sense, and in which the above limit exists pointwise as $\varepsilon = \max(t_n - t_{n-1}) \to +0$. Furthermore, for Hamiltonian operators $H(t)$ acting in $\mathbf{P}_\xi L^2(\mathbf{R}^6)$, and generically involving time-dependent potential interactions, so that they are obtained by setting in (6.12) (cf. [P], Sec. 1.7)

$$P_A = -i\partial/\partial q^A\ , \qquad Q^A = q^A + i\partial/\partial p^A\ , \qquad A = 1,2,3\ , \quad (6.15)$$

we can introduce, as a counterpart of (6.12), the nonrelativistic SQM propagator

$$\hat{K}_\xi(\mathbf{q}'',\mathbf{p}'',t'';\mathbf{q}',\mathbf{p}',t') = \left\langle \xi_{\mathbf{q}'',\mathbf{p}''}\middle| T\left(\int_{t'}^{t''} e^{-iH(t)}\,dt\right)\xi_{\mathbf{q}',\mathbf{p}'}\right\rangle\ . \quad (6.16)$$

The application of the Trotter product formula in the following generalized form,

$$\hat{K}_\xi(\mathbf{q}'',\mathbf{p}'',t'';\mathbf{q}',\mathbf{p}',t') = \lim_{\varepsilon\to+0}\left\langle \xi_{\mathbf{q}'',\mathbf{p}''}\middle|\prod_{n=N}^{1} e^{-iH_I(t_n)(t_n-t_{n-1})}e^{-iH_0(t_n-t_{n-1})}\xi_{\mathbf{q}',\mathbf{p}'}\right\rangle\ , \quad (6.17)$$

is again mathematically legitimate. Hence, upon using (2.14) in between the elements of each consecutive pair of operators in (6.11), we obtain the expression

$$\hat{K}_\xi(\mathbf{q}'',\mathbf{p}'',t'';\mathbf{q}',\mathbf{p}',t') = \lim_{\varepsilon\to+0}\int d^3\mathbf{q}'_N d^3\mathbf{p}'_N \prod_{n=N-1}^{1} d^3\mathbf{q}'_n d^3\mathbf{p}'_n d^3\mathbf{q}_n d^3\mathbf{p}_n$$
$$\times\prod_{n=N}^{1}\left\langle \xi_{\mathbf{q}_n,\mathbf{p}_n}\middle| e^{-iH_I(t_n)(t_n-t_{n-1})}\xi_{\mathbf{q}'_n,\mathbf{p}'_n}\right\rangle K_\xi(\mathbf{q}'_n,\mathbf{p}'_n,t_n;\mathbf{q}_{n-1},\mathbf{p}_{n-1},t_{n-1})\ . \quad (6.18)$$

where we have set, as before, $\mathbf{q}' = \mathbf{q}_0$, $\mathbf{p}' = \mathbf{p}_0$, $\mathbf{q}'' = \mathbf{q}_N$ and $\mathbf{p}'' = \mathbf{p}_N$. From (6.18), a path integral formula in stochastic phase space can be derived (Prugovečki, 1981a), which on the formal level is quite analogous to the one in (1.4.1) – cf. also Secs. 1.9-1.11 in [P].

In the relativistic regime, Feynman propagators are normally defined [IQ,SI] for particle propagation in conjunction with antiparticle propagation, on account of the pair creation and annihilation processes that play a crucial role in relativistic quantum field theory, and require that 4-momentum values on the backward mass hyperboloid be assigned to antiparticles. For scalar quantum fields, describing particles and antiparticles of spin zero and rest mass $m > 0$, the Feynman propagator is a generalized function defined by[22]

$$K_F(x''-x') = -\frac{2i}{(2\pi)^4}\lim_{\varepsilon\to+0}\int d^4k\,\frac{\exp[-ik\cdot(x''-x')]}{k^2 - m^2 + i\varepsilon}\ . \quad (6.19)$$

We shall be dealing with corresponding particle-antiparticle GS propagators in later chapters on quantum field theory (cf. Sec. 9.6), but if we concentrate for the time being on a single relativistic quantum particle propagating forward in time, then (6.19) yields

$$K_F(x''-x') = 2(2\pi)^{-3}\int_{k^0>0} \exp[ik\cdot(x'-x'')]\,d\Omega_m(k)\,, \qquad x''^0 > x'^0 \;. \tag{6.20}$$

On the other hand, the relativistic propagator for proper state vectors of free stochastically extended particles can be defined by analogy with the nonrelativistic one in (6.2), i.e., as

$$K_\eta(q'',p'';q',p') \;=\; \left\langle \eta_{q'',p''} \middle| \eta_{q',p'} \right\rangle \;, \tag{6.21a}$$

$$\eta_{q',p'} = U_\eta(q',\Lambda_{p'})\eta \;, \qquad\qquad \eta_{q'',p''} = U_\eta(q'',\Lambda_{p''})\eta \;. \tag{6.21b}$$

Consequently, upon specializing (4.10) to the case of $j = 0$, and upon taking advantage of the unitarity of the W_η map in (4.9) in order to express the inner product in (6.21a) in the momentum representation, we can write (6.21a) in the following form:

$$K_\eta(q'',p'';q',p') \\ = \left(m/4\pi^3\right)\int_{k^0>0} \exp[ik\cdot(q'-q'')]\; e^*(p''\cdot k/m)e(p'\cdot k/m)\,d\Omega_m(k)\;. \tag{6.22}$$

It turns out (Prugovečki, 1981a) that all the main physical features of the nonrelativistic SQM path integral formulation of quantum mechanics are retained if relativistic propagation is based on the free SQM propagator in (6.22). In particular,

$$\frac{(2\pi)^3}{2m}\int_{\Sigma_m} \left|K_\eta(q'',p'';q',p')\right|^2 d\Sigma_m(q'',p'') \;=\; 1 \;, \tag{6.23}$$

so that, in accordance with the interpretation discussed in Sec. 3.4, the propagator in (6.22) provides, for an extended particle which at the stochastic spacetime location q' is known to have the stochastic 4-momentum p', the probabilities of its reaching the various stochastic locations q'' along a spacelike hypersurface σ, and of displaying there various stochastic 4-momenta p''. On account of (4.12), the SQM propagator in (6.22) also has the properties

$$K_\eta(q'',p'';q',p') \;=\; K^*_\eta(q',p';q'',p'') \\ = \int_{\Sigma_m} K_\eta(q'',p'';q,p)K_\eta(q,p;q',p')\,d\Sigma_m(q,p) \;, \tag{6.24}$$

where the integration can be performed along any phase-space hypersurface $\Sigma_m = \sigma\times V^+_m$, for which σ a spacelike Cauchy surface in Minkowski space. More generally, we can insert any time-ordered family of such hypersurfaces between (q',p') and (q'',p''), so that, upon iterating (6.24), we arrive at the following relativistic counterpart of (6.14):

$$K_\eta(q'',p'';q',p')$$
$$= \lim_{N\to\infty} \int K_\eta(q_N,p_N;q_{N-1},p_{N-1}) \prod_{n=N-1}^{1} K_\eta(q_n,p_n;q_{n-1},p_{n-1})\, d\Sigma_m(q_n,p_n) \ . \quad (6.25)$$

Furthermore, a similar technique can be used in deriving (Prugovečki, 1981a) relativistic counterparts of (6.17) for the propagation of extended quantum particles in the presence of external electromagnetic fields – cf. also [P], Sec. 2.10.

The role of the proper state vector represented by the ground state of a nonrelativistic harmonic oscillator is naturally taken over in the relativistic context by the proper state vector represented by the ground state of a relativistic harmonic oscillator (Fujimura *et al.*, 1970; Feynman *et al.*, 1971). Such a proper state vector can be, however, also interpreted (Prugovečki, 1981b) as the ground state of Born's (1949) quantum metric operator. In the momentum representation this *quantum metric ground state* is represented by the function

$$\tilde{\eta}^{(\ell)}(k) = (m/4\pi^3)^{1/2} e^{(\ell)}(k^0) = \tilde{Z}_{\ell,m}^{-1/2} \exp(-\ell m k^0) \ , \quad \tilde{Z}_{\ell,m} = 8\pi^4 m^3 \ell^{-1} \mathrm{K}_2(2\ell m^2), \quad (6.26)$$

in terms of which (6.22) assumes the form[23]

$$K_{\eta^{(\ell)}}(q'',p'';q',p') = \tilde{Z}_{\ell,m}^{-1} \int_{k^0>0} \exp\{k\cdot[i(q'-q'') - \ell(p'+p'')]\}\, d\Omega_m(k) \ . \quad (6.27)$$

The integration in (6.27) can be carried out explicitly (cf. [P], pp. 119-121), so that the propagator for the ground state of the relativistic harmonic oscillator can be expressed in terms of the modified Bessel functions K_1 and K_2 (with K_2 entering via (6.26)) as follows:

$$K_{\eta^{(\ell)}}(q'',p'';q',p') = \frac{2\pi m\ \mathrm{K}_1\left(m\sqrt{-(q+i\ell p)^\mu (q+i\ell p)_\mu}\right)}{\tilde{Z}_{\ell,m}\sqrt{-(q+i\ell p)^\mu (q+i\ell p)_\mu}} \ , \quad (6.28a)$$

$$q = q'-q'' \ , \qquad p = p'+p'' \ , \qquad q',q'' \in \mathbf{R}^4 \ , \quad p',p'' \in \mathbf{V}_m^+ \ . \quad (6.28b)$$

By comparing (6.27) with (6.20), we see that

$$2(2\pi)^{-3} \tilde{Z}_{\ell,m} K_{\eta^{(\ell)}}(q'',p'';q',p') \xrightarrow[\ell\to+0]{} K_F(q''-q') \ , \qquad q''^0 > q'^0 \ . \quad (6.29)$$

Since a totally analogous relation holds for the corresponding antiparticle propagators (cf. [P], Sec. 5.1), we arrive at the conclusion that in the *formal* sharp-point limit of the type (6.29), the relativistic Feynman propagator in (6.19) emerges as a renormalized limit of SQM particle-antiparticle propagators. In that limit, the renormalization constant in (6.26) becomes infinite. Nevertheless, this observation indicates that, at the purely computational level, the S-matrix elements of quantum field theories formulated in the SQM framework (cf. Chapter 5 in [P]) can be made to match, to an arbitrary degree of accuracy, their conventional counterparts. Indeed, computational practice is based on a Dyson perturbation series for the S-matrix [SI,IQ], whose convergence, however, has never been established[24]. Rather, partial sums are computed up to a given order – which is normally very low (usually no higher than fourth) on account of computational complexities. This is achieved by summing up Feynman diagrams, whose numerical values are computed on the basis of two-point functions that represent Feynman propagators, such as the one in (6.19), but which are usually expressed in the momentum representation. To each such Feynman diagram therefore corresponds a topologically homeomorphic SQM diagram, whose numerical value converges, upon appropriate renormalization, to its Feynman counterpart in the above described type of formal sharp-point limit.

Viewed from such a purely pragmatic and computational perspective, the sharp-point limit procedure can only add in quantum field theory yet another renormalization scheme to an already rich list of such schemes [IQ]. However, as Dirac has pointed out on numerous occasions, in relativistic quantum field theory "the renormalization idea would be sensible only if applied with finite renormalization factors, not with infinite ones, [and consequently] *one must seek a new relativistic quantum mechanics and one's prime concern must be to base it on sound mathematics*" (Dirac, 1978b, pp. 5-6) – emphasis added.

With this, as well as other, foundational concerns in mind, in the next chapter we shall turn to a further extrapolation of conventional nonrelativistic quantum mechanics. In turn, this nonrelativistic extrapolation will lead in Chapters 5-10 to a formulation of relativistic quantum mechanics and of quantum field theory that will ultimately provide the framework for a mathematically consistent unification of relativity and quantum theory described in Chapter 11. However, as shown in the next three sections, geometric aspects which are essential to this unification can be found already in the SQM framework.

3.7. Quantum Frames and Quantum Informational Completeness

The bridge between the SQM framework, described in this chapter, and the GS framework, whose various facets will be described in the remaining chapters of this monograph, can be established via the concept of quantum frame[25]. The operational aspect of this concept had actually provided the basic impetus for the original formulation (Prugovečki, 1978c,d) of the SQM method of quantization: in operational terms, a quantum frame is obtained by subjecting identical duplicates of a quantum test body to all the basic kinematical procedures (i.e., the operations of spacetime translation, spatial rotation and boost) in order to obtain an array of kinematically correlated microdetectors, which can be then used for the spatio-temporal localization of quantum systems. In mathematical terms, a quantum frame is obtained by taking the proper state vector of such a test body, which marks the origin of that frame, and subjecting it to the entire group **G** of transformations that describes those kinematical operations, i.e., to suitable representations of the Galilei and of the Poincaré group in the nonrelativistic and the relativistic regime, respectively. If the adopted proper state vector Φ^u is a *bona fide* element of a Hilbert space $\mathcal{H}$ (rather than an object describing a perfectly localized quantum point test particle, such as a δ-function, which lies within an extension[26] of $\mathcal{H}$), the outcome of this construction can be one of the families of vectors that supplied some of the continuous resolutions of the identity in Secs. 3.2 and 3.4.

From a general mathematical point of view, an SQM *quantum frame* is a family Q($\boldsymbol{u}$) of generalized coherent states, in a Hilbert space $\mathcal{H}$, which is uniquely assigned to a global classical frame of reference $\boldsymbol{u}$ in a flat (nonrelativistic or relativistic) spacetime,

$$u \leftrightarrow Q(u) = \left\{ \Phi^u_\zeta = U(g_\zeta)\Phi^u \in \mathcal{H} \mid \zeta \in \mathcal{M} \right\} , \qquad u \in X , \tag{7.1}$$

provided that u belongs to a family X of such classical frames upon which a kinematical group $\mathbf{G}$ of transformations g acts transitively and effectively[27] from the right, and in such a manner that the following conditions are satisfied:

1) The *extended phase space* $\mathcal{M}$ of parametric values ζ indexing the generalized coherent states of a quantum frame in (7.1) is a locally compact space equal to the Cartesian product $\mathbf{R}^4 \times \mathcal{Y}$ of the set $\mathbf{R}^4$, consisting of all the coordinate 4-tuples q, with respect to the frame u, belonging to the points in that flat spacetime (such as Newtonian or Minkowski spacetime), with the set $\mathcal{Y}$ of all the additional components of ζ, representing coordinate values, with respect to coordinate axes of the same frame, of other observable quantities – such as 3-momentum $\mathbf{p}$, or 4-momentum p (or, alternatively, 3-velocity $\mathbf{v}$, or 4-velocity v), sharp spin s with respect to a given axis in u (or, alternatively, stochastic spin s), etc.

2) For any quantum state vector Ψ, and for any quantum frame in (7.1), the *extended phase space distribution functions*

$$\rho_u(\Psi;\zeta) = \langle \Psi | \Phi^u_\zeta \rangle \langle \Phi^u_\zeta | \Psi \rangle , \qquad \Psi \in \mathcal{H} , \quad \|\Psi\| = 1 , \tag{7.2}$$

represent probability densities, in general with respect to that frame and its dual (cf. Sec. 3.9), that correspond to the expectation values $\langle \Psi | E(B) \Psi \rangle$ of a POV measure $E(B)$ (cf. Sec. 3.2) over the Borel sets B of the locally compact space $\mathcal{M} = \mathbf{R}^4 \times \mathcal{Y}$.

3) The POV measure $E(B)$ is one of the ingredients of a system of covariance for the group $\mathbf{G}$, whose other component consists of the unitary representation U of $\mathbf{G}$ by means of which the generalized coherent states in (7.1) were constructed, so that, for each change $u \mapsto u \cdot g$ of classical frames, the effect of the replacement of (7.1) with

$$u \cdot g \leftrightarrow Q(u \cdot g) = \left\{ \Phi^{u \cdot g}_\zeta \mid \zeta \in \mathbf{R}^4 \times \mathcal{Y} \right\} , \qquad g \in \mathbf{G} , \tag{7.3}$$

gives rise to the following transformation law of the probability amplitudes in (7.2):

$$\Psi_{u \cdot g}(\zeta) = \left(U(g^{-1}) \Psi_u \right)(\zeta) , \qquad \forall g \in \mathbf{G} , \qquad \forall u \in X . \tag{7.4}$$

The fact that quantum frames can be used as a basic tool for the geometrization of quantum mechanics, which was originally discussed in (Prugovečki, 1982b), relies on three additional features which they ordinarily possess – and which were established in the nonrelativistic context in (Prugovečki, 1977a, 1978a,b): A) The informational completeness (cf. Sec. 1.3 as well as Note 14 to Chapter 1) of the POV measures to which the quantum frames give rise in those instances, namely the fact that the probability densities in (7.2) uniquely determine a (unit) Hilbert ray[28]

$$\tilde{\Psi} = \left\{ c\Psi / \|\Psi\| \, \middle| \, |c| = 1, c \in \mathbf{C}^1, \Psi \neq \mathbf{0}, \Psi \in \mathcal{H} \right\} \in \mathcal{P} , \tag{7.5}$$

which in turn uniquely corresponds to a pure quantum state described by state vectors in $\mathcal{H}$.

B) The possibility of approximating such frames with their classical counterparts by suitable and operationally feasible choices of some of their basic parameters, such as the mass of their constituents. C) The transition of the quantum into the classical phase space mode of description, when such an approximation is carried out; in other words, as a whole, the family of *all theoretically* possible quantum frames exhibits a gradual transition from the classical to the quantum regime, although otherwise that family might contain even a preponderance of elements whose behavior in no way resembles classical behavior.

For the sake of simplicity, let us illustrate the above features with the case of the spin-0 nonrelativistic quantum frames, which implicitly occur in (2.13)-(2.22) when we choose $j=0$ in (2.12) and (2.19). We then have

$$\Phi_\zeta^u = U_\xi(g_\zeta)\Phi^u , \qquad \zeta = (t,\mathbf{q},\mathbf{v}) \in \mathbf{R}^7 , \quad g_\zeta \in \mathcal{G} , \quad \Phi^u = W_{n0}\xi , \quad \xi = \xi_{n0} , \tag{7.6}$$

where g_ζ is the Galilei transformation in classical phase space defined by (1.12) and (1.22) upon setting $R=I$. The informational completeness of such frames (cf. Theorem 1 in [P], p. 138) can be expressed as the existence of a one-to-one mapping

$$\tilde{\Psi} \leftrightarrow \rho_u , \qquad \forall \tilde{\Psi} \in \mathcal{P} \cong \mathcal{H}/S^1 , \qquad \mathcal{H} = \mathbf{P}_\xi L^2(\mathbf{R}^6) = W_{n0}L^2(\mathbf{R}^3) , \tag{7.7}$$

between unit rays in the Hilbert space $\mathcal{H}$ in (7.7), defined in accordance with (7.5), and the probability densities

$$\rho_u(\Psi;\zeta) = |\Psi_u(\zeta)|^2 , \qquad \Psi_u(\zeta) = \langle \Phi_\zeta^u | \Psi \rangle , \quad \Psi \in \mathcal{H} , \quad \|\Psi\| = 1 , \tag{7.8}$$

on the stochastic phase space Γ_ξ defined in (2.21). Consequently, as it is the case with classical frames, the outcomes of measurements[29] of localization with respect to such quantum frames can, in principle, unambiguously determine *any* quantum state.

The fact that, in the limit of infinitely massive quantum test particles, the kinematical behavior of such quantum frames approximates that of classical frames, follows from the the results on the formulation of statistical quantum mechanics on stochastic phase spaces, reviewed in Chapter 3 of [P], whose principal conclusions are as follows.

1) When quantum mechanics, or, more generally, quantum statistical mechanics, is formulated by means of density distributions on stochastic phase space, such as those in (7.8), then the von Neumann equation for density operators transcends into a Liouville equation for those densities. That equation has the appearance of the Liouville equation for classical phase space distributions, but contains additional terms that depend on the mass and the spread of the stochastically extended quantum test particles. In the absence of external fields, i.e., in the case of inertial propagation, the quantum Liouville equation merges into its classical counterpart in the limit where the test particles which constitute the quantum frame become infinitely massive. This implies that, for example, in the case of quantum test particles with proper state vectors given by (6.6), the time-dependence of the stochastic phase space distribution in (7.8) can be explicitly computed[30], with the result

$$\rho_u(\Psi;t,\mathbf{q},\mathbf{p}/m) \approx \rho_u(\Psi;0,\mathbf{q}-\mathbf{v}t,\mathbf{p}/m) , \qquad \mathbf{v} = \mathbf{p}/m , \qquad |t| \ll \ell^2 m/\hbar , \tag{7.9}$$

so that the aforementioned approximately classical behavior is explicitly revealed. We note that, although in this monograph we otherwise use Planck natural units, in which $\hbar = 1$, the dependence of the range of validity of this approximation upon $\hbar$ is explicitly displayed, in order to emphasize that such a comparison of classical to quantum behavior intrinsically involves Planck's constant $h = 2\pi\hbar$ – which is not a physically dimensionless number.

2) Classical and quantum statistical mechanics can be formulated in a common master Liouville space, which coincides with the Hilbert-Schmidt class [PQ] over the Hilbert space $\mathcal{H}$ in (7.7). This mathematically unified approach to classical and quantum statistical mechanics significantly extends the scope of the "Koopmanism" of modern approaches to classical mechanics (Abraham and Marsden, 1978), since classical as well as quantum observables are represented by superoperators[31] acting in that common master Liouville space. In particular, that is true of the classical Hamiltonian superoperator, which emerges as the first term in an expansion in powers of $\hbar$ and ℓ of the corresponding quantum Hamiltonian superoperator, when the latter is expressed with respect to a given massive quantum frame (cf. [P], Secs. 3.4-3.6). Hence, any geometric features of the quantum formalism can be transferred, by using these correspondences, from the quantum to the classical regime – cf. Sec. 3.9.

The geometric features of the quantum formalism at its most general level emerge from the dichotomy between its Hilbert space origins, which are necessarily linear on account of the superposition principle, and its physical interpretation, which is based on the complex projective space[32] $\mathcal{P}$ in (7.5), namely on an infinite-dimensional complex manifold which is *not* linear, but it is a Kähler manifold [KN], i.e., it is a complex manifold with a Hermitian metric and a connection compatible with that metric. We shall refer to the space $\mathcal{P}$, consisting of all unit rays in $\mathcal{H}$, as the *projective Hilbert space* associated with $\mathcal{H}$.

We shall describe that metric and connection, as well as their basic properties, in the next section. In the finite-dimensional case, this metric is commonly known under the name of the Fubini-Study metric [KN] – and we shall retain this name in the infinite-dimensional case. The essential observation is that a (Hermitian) metric on a Kähler manifold gives rise to a distance function between any two of its elements. That distance can be defined as in the Riemannian case (cf. [K], p. 157), namely as the infimum of the lengths (in the Hermitian metric of that manifold) of all piecewise smooth curves joining those two points.

As we shall see in the next section, the *Fubini-Study distance* between two unit rays is

$$d(\bar{\Psi}_1, \bar{\Psi}_2) = 2 \arccos |\langle \Psi_1 | \Psi_2 \rangle| , \qquad \Psi_i \in \bar{\Psi}_i \subset \mathcal{H} , \quad \|\Psi_i\| = 1 , \quad i = 1,2 , \tag{7.10}$$

so that their Fubini-Study distance is zero only if they coincide; moreover, we also have

$$0 \le d(\bar{\Psi}_1, \bar{\Psi}_2) = d(\bar{\Psi}_2, \bar{\Psi}_1) \le d(\bar{\Psi}_1, \bar{\Psi}_3) + d(\bar{\Psi}_3, \bar{\Psi}_2) , \qquad \forall \bar{\Psi}_1, \bar{\Psi}_2, \bar{\Psi}_3 \in \mathcal{P}, \tag{7.11}$$

on general grounds (cf. [K], p. 157). We observe that there is a one to one map between the Fubini-Study distance, defined within the projective Hilbert space $\mathcal{P}$, and the most fundamental quantity in the theory of quantum measurement, namely the *transition probability*

$$|\langle \Psi_1 | \Psi_2 \rangle|^2 = \cos^2\left(\tfrac{1}{2} d(\bar{\Psi}_1, \bar{\Psi}_2)\right) \le 1 , \qquad \|\Psi_1\| = \|\Psi_2\| = 1 , \tag{7.12}$$

between any two quantum states identifiable with rays from that space.

The appearance of the projective Hilbert space $\mathcal{P}$ in the physics literature can be traced to *Wigner's theorem on ray transformations*. This fundamental theorem is usually formulated in terms of the transition probabilities in (7.12) (cf. [BR], p. 398), but it can be now rephrased in purely geometric terms as follows: to any Fubini-Study isometric map

$$\bar{U} : \mathcal{P} \to \mathcal{P} , \qquad d(\bar{\Psi}_1, \bar{\Psi}_2) = d(\bar{U}\bar{\Psi}_1, \bar{U}\bar{\Psi}_2) , \qquad \forall \bar{\Psi}_1, \bar{\Psi}_2 \in \mathcal{P}, \tag{7.13}$$

on the projective Hilbert space $\mathcal{P}$, corresponds an operator U in $\mathcal{H}$, for which

$$U\Psi \in \bar{U}\bar{\Psi} \subset \mathcal{H} , \qquad \forall \Psi \in \mathcal{H} - \{\mathbf{0}\} ; \tag{7.14}$$

this operator is either unitary or antiunitary, and it is unique modulo a phase factor. Naturally, it is this very fundamental theorem that provides the physical motivation for the introduction of such ray representations as those in (1.16), (1.20), (1.21), (2.7) and (2.11).

In the case of informationally complete quantum frames, there is a one-to-one correspondence – such as the one in (7.7) – between the rays identifiable with quantum states, and the probability distribution functions over the extended stochastic phase spaces identifiable with the index set $\mathcal{M}$ in (7.1). Thus, the Fubini-Study metric also provides a metric within the space of these distribution functions. Furthermore, we can now see that the values assumed by these distribution functions possess a geometric meaning. For the distribution functions in (7.8), that meaning follows from (7.12):

$$\rho_u(\Psi;\zeta) = \cos^2\left(\tfrac{1}{2} d(\bar{\Phi}^u_\zeta, \bar{\Psi})\right) , \qquad \Psi \in \mathcal{H} , \quad \|\Psi\| = 1 . \tag{7.15}$$

Consequently, we can now provide a characterization of informational completeness which is of a purely geometric nature: *a quantum frame is informationally complete if and only if the (directly measurable) Fubini-Study distance of any given quantum state to all the elements of that frame uniquely determines that state.*

We note that a quantum frame also provides a *continuous resolution of the identity* operator $\mathbf{1}$ in the Hilbert space $\mathcal{H}$ in which it acts, i.e.,

$$\int_\Sigma \left|\tilde{\Phi}^u_\zeta\right\rangle d\Sigma(\zeta) \left\langle \Phi^u_\zeta\right| = \mathbf{1} , \qquad \Sigma \subset \mathcal{M} = \mathbf{R}^4 \times \mathcal{Y} \tag{7.16}$$

– where examples[33] of measures of integration $d\Sigma$ over hypersurfaces Σ in $\mathcal{M}$ are provided by (2.14), as well as by (4.8b) and (4.12). This last property is, of course, very similar to the well-known *completeness property*[34] of an orthonormal basis in $\mathcal{H}$, namely [PQ]

$$\sum_{\alpha=1}^{\infty} |w_\alpha\rangle\langle w_\alpha| = \mathbf{1} , \qquad \langle w_\alpha | w_\beta \rangle = \delta_{\alpha\beta} , \qquad \alpha,\beta = 1,2,\ldots . \tag{7.17}$$

However, the Bochner integral in (7.16) is generically weakly convergent, whereas the infinite sum of projectors in (7.17) is strongly convergent. On the other hand, quantum frames as well as orthonormal bases are *total sets*, i.e, their linear spans are dense in $\mathcal{H}$.

*3.8. Kähler Metrics and Connections in Hopf Bundles and Line Bundles

Investigations of the geometric significance of the Berry (1984) phase factors that appear during adiabatic evolutions of quantum systems (Simon, 1983; Wilczek and Zee, 1984) led to the realization that the adiabatic approximation was not actually required in their formulation (Aharonov and Anandan, 1987; Page, 1987), that their geometric features were retained in the classical limit (Hannay, 1985; Anandan, 1988), and that, in fact, the underlying physical phenomenon was of a purely geometric origin (Anandan and Pines, 1989; Anandan and Aharonov, 1990). In view of the observations made in the preceding section, it is not surprising that this origin is of a measurement-theoretical nature, and that the various experiments (Tycko, 1987; Suter *et al.*, 1987) that can be interpreted (Anandan and Aharonov, 1988) as measurements of geometric angles and phases, give rise to techniques of localization with respect to various quantum frames.

To derive the Fubini-Study metric that underlies these considerations, as well as the notion of distance (7.10) between two rays in the space of rays $\mathcal{P}$ associated with a Hilbert space $\mathcal{H}$, let us first consider $\mathcal{H}$ from the point of view of a flat Kähler manifold with a Hermitian metric G and a fundamental 2-form W equal to, respectively (cf. [KN], p. 159),

$$G = \delta_{\alpha\beta}\, d\bar{z}^{\alpha} \otimes dz^{\beta} \quad , \qquad W = i\delta_{\alpha\beta}\, d\bar{z}^{\alpha} \wedge dz^{\beta} \ . \tag{8.1}$$

Upon the adoption in $\mathcal{H}$ of the orthonormal basis in (7.17), the complex one-forms in (8.1) are defined by the following linear and antilinear functionals, respectively,

$$dz^{\alpha} : f \mapsto \langle w_{\alpha} | f \rangle = z^{\alpha} \ , \quad d\bar{z}^{\alpha} : f \mapsto \langle f | w_{\alpha} \rangle = \bar{z}^{\alpha}, \quad f \in \mathcal{H} \ , \quad \alpha = 1,2,\ldots \ . \tag{8.2}$$

We see that the above linear forms constitute a basis in the dual $\mathcal{H}^*$ of $\mathcal{H}$. Hence, their definition is analogous to the definition of the one-forms in (2.1.9), which belong to the cotangent space of a real finite-dimensional manifold **M**. On the other hand, $\mathcal{H}$ is a linear complex infinite-dimensional manifold, so that its cotangent spaces can be identified with its topological dual $\mathcal{H}^*$. Furthermore, $\mathcal{H}$ possesses an inner product, so that by the Riesz theorem (cf. [PQ], p. 184) it can be identified with that dual – as always done in practice.

Due to Parseval's relation in $\mathcal{H}$ (cf. [PQ], p. 38) we have,

$$G(f,g) = \delta_{\alpha\beta}\, d\bar{z}^{\alpha}(f)\, dz^{\beta}(g) = \sum_{\alpha=1}^{\infty} \langle f | w_{\alpha} \rangle \langle w_{\alpha} | g \rangle = \langle f | g \rangle \ , \tag{8.3a}$$

$$W(f,g) = i\delta_{\alpha\beta}\left(d\bar{z}^{\alpha}(f)\, dz^{\beta}(g) - d\bar{z}^{\alpha}(g)\, dz^{\beta}(f) \right) = -2 Im \langle f | g \rangle \ . \tag{8.3b}$$

Consequently, the Hermitian metric in (8.1) basically represents a geometrized manner of writing the inner product of $\mathcal{H}$. As we shall see shortly, such geometrized versions of inner products have the advantage of being extendible to manifolds which are not linear.

To uncover the significance of the fundamental form in (8.1), let us take note of the fact that if **M** is any finite-dimensional manifold, then its cotangent bundle T^***M** possesses a natural symplectic structure (cf. [KN], p. 165), reflected by the presence of a sympletic 2-form $\boldsymbol{\omega} = d\boldsymbol{\theta}$ derivable from a canonical 1-form $\boldsymbol{\theta}$. As it is well known, in classical physics this fact can be used in the interpretation of $T^*\mathbf{M}$ as a classical phase space $\mathcal{M}$ (Abraham and Marsden, 1978), with the particular case of $\mathbf{M} = \mathbf{R}^3$ and $\mathcal{M} = \Gamma$ emerging already in the elementary Newtonian mechanics of a single classical particle.

This construction can be easily generalized to the present infinite-dimensional case by introducing the following two real linear submanifolds of $\mathcal{H}$,

$$\Re(\{w_\alpha\}) = \left\{ q^\alpha w_\alpha \middle| q^1, q^2, \ldots \in \mathbf{R}^1,\ q^\alpha w_\alpha := \sqrt{2}\sum_{\alpha=1}^{\infty} Re\langle w_\alpha | f\rangle w_\alpha,\ f \in \mathcal{H} \right\}, \quad (8.4a)$$

$$\Im(\{w_\alpha\}) = \left\{ \sum_{\alpha=1}^{\infty} p_\alpha w_\alpha \in \mathcal{H} \,\middle|\, p_\alpha = \sqrt{2} Im\langle w_\alpha | f\rangle,\ \alpha = 1,2,\ldots,\ f \in \mathcal{H} \right\}, \quad (8.4b)$$

and then carrying out the identifications

$$\mathcal{H} \leftrightarrow \Re(\{w_\alpha\}) \oplus \Im(\{w_\alpha\}) \leftrightarrow T^*\Re(\{w_\alpha\}) \equiv \mathcal{H}(J), \quad (8.5)$$

which implicitly correspond to the assignment of the following complex structure[35],

$$J(\{w_\alpha\}) : (q^\alpha w_\alpha) \oplus \sum_{\alpha=1}^{\infty} p_\alpha w_\alpha \to \left(-\sum_{\alpha=1}^{\infty} p_\alpha w_\alpha\right) \oplus (q^\alpha w_\alpha), \quad (8.6)$$

to the corresponding real Hilbert space $\mathcal{H}(J)$. We can then rewrite the fundamental 2-form in $\mathcal{H}(J)$ in terms of the real-valued coordinates within its two real linear subspaces in (8.4),

$$W = dp_\alpha \wedge dq^\alpha = dT, \qquad T = p_\alpha\, dq^\alpha. \quad (8.7)$$

Thus, its similarity with symplectic forms on classical phase spaces (cf. [Abraham and Marsden, 1978], p. 179) becomes apparent. Hence, for a given *complex structure J*, the Hilbert space $\mathcal{H}$ can be identified with an infinite-dimensional symplectic linear space $\mathcal{H}(J)$, with *Kähler metric G* and *symplectic form W*. In turn, this real symplectic manifold has a complexification $\mathcal{H}^c(J)$ with corresponding Kähler metric and fundamental form (cf. (8.8b)), so that $\mathcal{H}$ can be identified with a subspace of $\mathcal{H}^c(J)$, and (cf. [KN], p. 155)

$$G = 2\tilde{G}_{\overline{\alpha}\beta}\, d\overline{\zeta}^\alpha \otimes d\zeta^\beta, \qquad W = 2i\tilde{G}_{\overline{\alpha}\beta}\, d\overline{\zeta}^\alpha \wedge d\zeta^\beta, \quad (8.8a)$$

$$\tilde{W}(\tilde{f},\tilde{g}) = \tilde{G}(\tilde{f},\tilde{J}\tilde{g}),\ \tilde{G}_{\overline{\alpha}\beta} = \tilde{G}_{\beta\overline{\alpha}} = \delta_{\alpha\beta},\ \tilde{G}_{\alpha\beta} = \tilde{G}_{\overline{\alpha}\overline{\beta}} = 0,\ \alpha,\beta,\overline{\alpha},\overline{\beta} = 1,2,\ldots. \quad (8.8b)$$

We thus recover (8.1) from (8.8) by setting $\zeta^\alpha = 2^{-1/2} z^\alpha$ – which is in complete accordance with (8.2) and (8.4). Furthermore, we can exploit this construction in deriving a number of additional geometric structures, by taking advantage of the results in Sec. IX.5 of [KN].

To arrive at a better understanding of this profusion of real and complex geometric structures, let us remove from the Hilbert space $\mathcal{H}$ the zero vector $\mathbf{0}$, and regard the outcome $\mathcal{H}^\dagger$ as a fibre bundle over the projective space $\mathcal{P}$, with projection map

$$\Pi : \Psi \mapsto \bar{\Psi} \in \mathcal{P}, \qquad \Psi \in \mathcal{H}^\dagger \equiv \mathcal{H} - \{\mathbf{0}\}, \quad (8.9)$$

and structure group $U(1)$, i.e., as the *line bundle* $(\mathcal{H}^\dagger, U(1), \mathcal{P})$ with total space $\mathcal{H}^\dagger$ and base space $\mathcal{P}$ (cf. Sec. 2.2). The outcome of this construction is a trivial fibre bundle, but the principal bundle $\mathcal{S} = \mathcal{P} \times U(1)$ with which it is associated, namely

$$(\mathcal{S}, \Pi, \mathcal{P}, U(1)) \ , \qquad \mathcal{S} = \{\Psi \| \|\Psi\| = 1\} \subset \mathcal{H} \ , \quad U(1) = \{e^{i\theta} | \theta \in \mathbf{R}^1\} \ , \quad (8.10)$$

can be viewed[36] as an infinite-dimensional *Hopf bundle* (cf. [KN], p. 137), which is itself endowed with a Hermitian metric. In turn, we shall see that this latter metric basically results from the invariant metric of $U(1)$ regarded as a Lie group (which coincides with the familiar Riemannian metric of the unit circle S^1 in $\mathbf{R}^2$), and from the Fubini-Study metric in the projective Hilbert space $\mathcal{P}$, namely the metric that gives rise to the distance function defined in (7.10).

The Fubini-Study metric is usually expressed in homogeneous coordinate charts, which in the present infinite-dimensional case are given for $\rho = 1,2,\ldots$ by

$$\{w_\rho^\alpha = z^\alpha / z^\rho | \alpha = 1,2,\ldots,\rho-1,\rho+1,\ldots\} \leftrightarrow \vec{\Psi} \in \vec{O}_\rho \ , \quad (8.11a)$$

$$\Psi \in O_\rho = \{U w_\rho | U^* U = U U^* = \mathbf{1}\} \ , \quad (8.11b)$$

and are induced by the canonical coordinates z^α in $\mathcal{H}$, which were defined in (8.2). Thus, upon omitting, for the sake of notational simplicity, the dependence of these coordinates upon the index ρ, we can express the *Fubini-Study metric* as follows,

$$\vec{G} = 2\vec{G}_{\overline{\alpha}\beta}\, d\overline{w}^\alpha \otimes dw^\beta, \ \ \tfrac{1}{2}\vec{G}_{\overline{\alpha}\beta} = \left(1 + w_\gamma \overline{w}^\gamma\right)^{-1} \delta_{\alpha\beta} - \left(1 + w_\gamma \overline{w}^\gamma\right)^{-2} w_\alpha \overline{w}_\beta \ , \quad (8.12a)$$

$$w_\alpha = z_\alpha / z_\rho \ , \quad \overline{w}_\alpha = \overline{z}_\alpha / \overline{z}_\rho \ , \quad z_\alpha = \delta_{\alpha\beta} z^\beta \ , \quad \alpha = 1,2,\ldots,\rho-1,\rho+1,\ldots \ , \quad (8.12b)$$

in accordance with (8.8a) and with Theorem 7.8 on p. 160 of [KN] – whose generalization to the present infinite-dimensional case is straightforward. If we then introduce within the open submanifold O_ρ of $\mathcal{H}^\dagger$, defined in (8.11b), the corresponding coordinates

$$(r = \|\Psi\|, \theta, w^1, \ldots, w^{\rho-1}, w^{\rho+1}, \ldots) \leftrightarrow \Psi \in O_\rho \ , \quad re^{i\theta} = z^\rho \left(1 + w_\gamma \overline{w}^\gamma\right)^{1/2} , \quad (8.13)$$

then the Kähler metric on the line bundle $\mathcal{H}^\dagger$, which was expressed in (8.1) and (8.8a) in the canonical coordinates z^α, can be expressed in these new coordinates by computing the coordinate transformation maps in (2.1.15) for the present change of coordinates. The outcome is given by (cf. [Page, 1987] for the finite-dimensional case)

$$G = dr \otimes dr + r^2 \left[(d\theta - \tilde{T}) \otimes (d\theta - \tilde{T}) + \tfrac{1}{2} \vec{G}_{\overline{\alpha}\beta}\, d\overline{w}^\alpha \otimes dw^\beta \right] \ , \quad (8.14a)$$

$$\tilde{T} = \tfrac{i}{2} \left(1 + w_\gamma \overline{w}^\gamma\right)^{-1} \left(\overline{w}_\alpha\, dw^\alpha - w_\alpha\, d\overline{w}^\alpha\right) \ , \quad (8.14b)$$

$$\tilde{W} = d\tilde{T} = i\left(1 + w_\gamma \overline{w}^\gamma\right)^{-2}\left[w_\alpha \overline{w}_\beta - \left(1 + w_\gamma \overline{w}^\gamma\right)\delta_{\alpha\beta}\right] d\overline{w}^\alpha \wedge dw^\beta \quad , \tag{8.14c}$$

where, modulo the factor 1/2, (8.14b) and (8.14c) are equal to the respective pull-backs to $\mathcal{P}$ of the (canonical) 1-form and of the (fundamental) 2-form in (8.7).

It now becomes a computationally routine matter to adapt to the Hopf manifold $\mathcal{S}$, whose metric is obviously given by the expression between the square brackets in (8.14a), the results derived in Sec. IX.5 of [KN] for the case of *Kähler connections*, i.e., for torsion-free complex connections ∇ in Kähler manifolds, which leave invariant their metrics – and therefore also their complex structures J. By analogy with the fundamental lemma of Riemannian geometry, which singles out Levi-Civita connections, such connections are unique in the case of finite-dimensional Kähler manifolds. This fact actually follows from the counterpart of (2.6.23) for a Kähler metric – cf. Theorem 4.8 on p. 152 of [KN].

This result turns out to remain true in the case of infinite-dimensional Kähler manifolds (Freed, 1985). However, it should be noted that for such manifolds covariant derivatives are only densely [PQ] defined as linear operators that act in tangent spaces, which are infinite-dimensional Hilbert spaces. In the case of the Hopf bundle $\mathcal{S}$, this bundle as well as all its tangent spaces can be isometrically imbedded [K,KN] into the flat manifold $\mathcal{H}$, in accordance with the expression between the square brackets (8.14a), which supplies the inner product in its tangent spaces $T_\Psi \mathcal{S}$, $\Psi \in \mathcal{S}$. As a Hilbert space, $\mathcal{H}$ then supplies the required dense domains on which the counterpart of (2.6.23) stays valid.

The same general observations hold true of the projective Hilbert space $\mathcal{P}$, which serves as base manifold for the Hopf bundle $\mathcal{S}$. Hence, the Fubini-Study distance between any two points in its base manifold $\mathcal{P}$ is obtained [K,KN] by minimizing the arc length

$$L(\bar{C}) = \int_a^b \sqrt{\bar{G}(\dot{\bar{\Psi}}(t), \dot{\bar{\Psi}}(t))}\, dt \ , \qquad \bar{C} = \left\{\bar{\Psi}(t) \middle| a \le t \le b\right\} \subset \mathcal{P} \ , \tag{8.15}$$

between those two points, and as such it gives rise to the expression in (7.10).

To see that, we first note that for the Kähler connection in $\mathcal{S}$ there are geodesics

$$C = \left\{\Psi(t) \middle| a \le t \le b\right\} \subset \mathcal{S} \subset \mathcal{H} \ , \qquad \bar{\Psi}(t) = \Pi(\Psi(t)) \in \bar{C} \ , \tag{8.16}$$

that minimize the arc length between any of its two given points. Consequently, their tangents vectors have to satisfy the subsidiary condition

$$d\theta(\dot{\Psi}(t)) = \tilde{T}(\Pi_* \dot{\Psi}(t)) \quad , \qquad a \le t \le b \ , \tag{8.17}$$

since this is obviously a sufficient as well as necessary condition for obtaining a zero contribution to this arc length from the first term in the square bracket in (8.14a). In that case, the arc length τ between its initial point and any other of its points becomes equal to the Fubini-Study distance between the projections in $\mathcal{P}$ of those two points,

$$\tau = \int_a^t \sqrt{\bar{G}(\dot{\bar{\Psi}}(t'), \dot{\bar{\Psi}}(t'))}\, dt' = \int_a^t \sqrt{G(\dot{\Psi}(t'), \dot{\Psi}(t'))}\, dt' \ , \qquad \Psi(t') \in C \subset \mathcal{S} \ , \tag{8.18}$$

and can be used as a special choice of affine parameter for that geodesic – cf. (2.7.10).

In general, due to (8.17), the horizontal lift C (cf. Sec. 2.4) of a smooth curve in $\mathcal{P}$ to the principal bundle $\mathcal{S}$ corresponds to the connection form (cf. Sec. 2.5),

$$\boldsymbol{\omega}: \mathbf{X} \mapsto \langle \Psi | \mathbf{X} \rangle \in L(U(1)) \cong T_e U(1) \ , \qquad \mathbf{X} \in T_\Psi \mathcal{S} \ . \tag{8.19}$$

Indeed, since $\langle \Psi(t) | \Psi(t) \rangle = 1$ along C, we obtain by differentiation with respect to the parameter t that $2\mathrm{Re}\langle \Psi(t) | \mathbf{X} \rangle = 0$, provided that $\mathbf{X}$ denotes the vector tangent to C at any given point $\Psi(t) \in C$. Hence, on account of (8.3b), (8.7) and (8.17), the 1-form which assumes the values $-\mathrm{Im}\langle \Psi(t) | \mathbf{X} \rangle = i\langle \Psi(t) | \mathbf{X} \rangle$ supplies the gauge potential for the resulting parallel transport, generically defined for any bundle associated with a given principal frame bundle in accordance with (2.5.26a). In turn, (7.10) then follows[37] from (8.18).

According to (8.13), for any choice of section s of the Hopf principal bundle $\mathcal{S}$, we can express any fibre of $\mathcal{H}^\dagger$ in the form

$$\Pi^{-1}(\tilde{\Psi}) = \left\{ \xi \hat{\Psi} \middle| \ \xi = re^{i\theta}, \ r > 0, \ \theta \in \mathbf{R}^1, \ \hat{\Psi} \in s \right\} \subset \mathcal{H}^\dagger \ . \tag{8.20}$$

Hence, it follows from (2.5.22) and (8.19) that the covariant derivative for the Kähler connection ∇ on the line bundle $\mathcal{H}^\dagger$ can be expressed along any smooth curve in $\mathcal{P}$ as follows:

$$\nabla_X \Psi = \dot{\xi} \hat{\Psi} + \left\langle \hat{\Psi} \middle| \dot{\hat{\Psi}} \right\rangle \Psi , \quad X = \dot{\tilde{\Psi}} \in T_{\tilde{\Psi}} \mathcal{P}, \ \hat{\Psi} \in C \subset s, \ e^{\hat{\mathbf{A}}t} = e^{it} \in U(1) \ . \tag{8.21}$$

For any connection in any principal fibre bundle ($\mathbf{P}$, $\boldsymbol{\Pi}$, $\mathbf{M}$,$\mathbf{G}$) (cf. Sec. 2.2), the *holonomy group* $H(x)$ of that connection at a point x in that base manifold $\mathbf{M}$ is, by definition, the group of all isomorphisms of the fibre $\boldsymbol{\Pi}^{-1}(x)$, obtained by carrying out the parallel transport of that fibre $\boldsymbol{\Pi}^{-1}(x)$ along all possible closed piecewise smooth curves in $\mathbf{M}$ that begin and end at x (cf. [K], p. 71, or [C], p. 386). In general, the holonomy group is a subgroup of the structure group $\mathbf{G}$. In the case of the Hopf bundle $\mathcal{S}$, the elements of the holonomy group for the Kähler connection ∇ are obviously elements of $U(1)$. As such, each of them is a complex number, that can be expressed by means of a real phase β.

Thus, the parallel transport along a given piecewise smooth path C in $\mathcal{S}$, whose projection $\boldsymbol{\Pi}(C)$ in the base manifold $\mathcal{P}$ is closed, gives rise to the particular holonomy group element $\exp[i\beta(C)]$ of $U(1)$, whose phase number is called the *geometric phase* of that path. According to (8.17), (8.19) and (8.21), the geometric phase of such a path C, defined as in (8.16), has the following value (cf. also [Page, 1987]):

$$\beta(C) = i\int_C \langle \Psi(t) | d\Psi(t) \rangle = \int_a^b \tilde{T}\left(\dot{\tilde{\Psi}}(t) \right) dt \ , \qquad \nabla_{\dot{\tilde{\Psi}}} \Psi \equiv \mathbf{0} \ , \quad \tilde{\Psi}(a) = \tilde{\Psi}(b) \ . \tag{8.22}$$

On account of (8.14c), we can use under certain conditions Stokes' theorem[38] to express the geometric phase of C as an integral over a surface in $\mathcal{P}$ that is enclosed by $\boldsymbol{\Pi}(C)$:

$$\beta(C) = \oint_{\tilde{C}} \tilde{T} = \int_S \tilde{W} \ , \qquad \tilde{W} = \boldsymbol{d}\tilde{T} \ . \tag{8.23}$$

Let us consider now any isometric flow in the projective Hilbert space $\mathcal{P}$. By the general definition of a flow in any manifold (cf. Abraham and Marsden, 1978, p. 61), such a flow is given by a continuous family of transformations

$$\bar{U}(t_2,t_1) \, : \, \bar{\Psi}(t_1) \mapsto \bar{\Psi}(t_2) \in \mathcal{P} \ , \qquad \bar{\Psi}(t_1) \in \mathcal{P} \ , \tag{8.24}$$

$$\bar{U}(t_3,t_1) = \bar{U}(t_3,t_2) \circ \bar{U}(t_2,t_1) \ , \qquad a \le t_1 \le t_2 \le t_3 \le b \ , \tag{8.25}$$

of $\mathcal{P}$ onto itself, which are equal to the identity transformation whenever $t_1 = t_2$. The fact that the flow is isometric means that it preserves the Fubini-Study distance in $\mathcal{P}$. Hence, according to the reformulation of Wigner's (1959) theorem on ray transformations, presented in the preceding section, each one of the isometric transformations of such a flow gives rise to a corresponding operator $U(t_2,t_1)$ in $\mathcal{H}$, which is either unitary or antiunitary, and is unique modulo a complex factor of unit absolute value. In view of the continuity property in t of such a flow, the possibility of antiunitarity is ruled out, since the product of two antiunitary operators is a unitary operator.

If we now assume that for some Hilbert space state vector $\Psi(a)$ the curve

$$C = \left\{ \Psi(t) = U(t,a)\Psi(a) \middle| a \le t \le b \right\} \subset \mathcal{H}^{\dagger}, \quad \Psi(b) = e^{i\theta(C)}\Psi(a), \ \theta(C) \in \mathbf{R}^1, \tag{8.26}$$

is piecewise smooth and its end points belong, as indicated above, to the same fibre in the line bundle $\mathcal{H}^{\dagger}$, then its projection

$$\bar{C} = \left\{ \bar{\Psi}(t) = \Pi(\Psi(t)) \middle| a \le t \le b \right\} \subset \mathcal{P} , \qquad \bar{\Psi}(b) = \bar{\Psi}(a) \ , \tag{8.27}$$

in $\mathcal{P}$ is also a piecewise smooth curve. Moreover, since that projection is closed, there is a geometric phase (8.22) associated with C regardless of whether or not $\Psi(t)$ satisfies a Schrödinger equation. If, however, the particular vector-valued function $\Psi(t)$ in (8.26) does satisfy a Schrödinger equation, which in general incorporates a time-dependent Hamiltonian $H(t)$, then a straightforward argument by Aharonov and Anandan (1987) (originally meant to establish the existence of a geometric phase), supplies the formula

$$\theta(C) = \beta(C) - \int_a^b \langle \Psi(t) | H(t) \Psi(t) \rangle dt \ . \tag{8.28}$$

The expressions in (8.23) for the geometric part of the above total phase θ for that Schrödinger evolution are equivalent to those previously obtained by Samuel and Bhandari (1988), as well as by Anandan and Aharanov (1988, 1990), without the use of Wigner's theorem. However, those derivations were based on the assumption that the underlying evolution governed by a Schrödinger equation applies *globally* on $\mathcal{H}$. That poses subtle domain questions (cf. [PQ], Ch. IV, Sec. 7.9; or Yosida, 1974, Ch. XIV, Sec. 4) that restrict their validity at a mathematically rigorous level to the case of finite-dimensional Hilbert spaces. Those problems are averted by the above purely geometric approach, which requires no assumptions about the domains of Hamiltonians for quantum evolutions.

3.9. Quantum Frames and Quantum Metrics in Typical Quantum Fibres

In all the subsequent chapters, we shall be dealing with quantum bundles in which a particular reproducing kernel Hilbert space $\mathcal{H}$ plays the role of typical fibre **F**, and which could be deemed to be associated to principal bundles of quantum frames whose general mathematical features are specified by (7.1)-(7.4). Each fibre $\mathbf{F}_x$ of such a quantum bundle over a point x in its base manifold **M** is thereby *soldered* to that base manifold (cf. Sec. 4.3 or Sec. 5.1). Consequently, much of the geometric structure in tangent and other tensor bundles over that same base manifold **M** has its counterpart in those quantum fibres. In particular, that is true of the Hermitian sesquilinear [PQ] forms G which are the counterparts of those in (8.1). We shall therefore refer to such a Hermitian sesquilinear form G in $\mathbf{F}_x$ as the *quantum metric* within the quantum fibre $\mathbf{F}_x$ (cf. Sec. 5.2).

We can formulate already at the present stage some of the geometric aspects of these quantum bundle structures within any Hilbert space $\mathcal{H}$ that could play the role of typical quantum fibre **F**. Hence, let us introduce the following continuous linear functionals in **F**,

$$\tilde{\Phi}^{\zeta} : \Psi \mapsto \langle \Phi^{u}_{\zeta} | \Psi \rangle = \Psi^{\zeta} , \qquad \Psi \in \mathcal{H} \equiv \mathbf{F} , \tag{9.1}$$

which therefore belong to the dual $\mathbf{F}^*$ of the Hilbert space **F**. Borrowing the terminology of Chapter 2, and in particular that of Sec. 2.1, we shall refer to

$$\tilde{\mathrm{Q}}(u) = \left\{ \tilde{\Phi}^{\zeta} \in \mathcal{H}^* \equiv \mathbf{F}^* \,\middle|\, \zeta \in \mathcal{M} = \mathbf{R}^4 \times \mathcal{V} \right\} \tag{9.2}$$

as a *coframe of the quantum frame* $\mathrm{Q}(\boldsymbol{u})$.

We can now extend the Einstein convention in (2.1), defined there for summations over pairs of contravariant and covariant discrete indices with equal values, to integrations over continuous indices. Thus, we shall write, in accordance with (7.16),

$$\Psi = \Psi^{\zeta} \Phi_{\zeta} := \int_{\Sigma} d\Sigma(\zeta)\, \Psi^{\zeta} \Phi^{u}_{\zeta} , \qquad \forall \Psi \in \mathcal{H} \equiv \mathbf{F} , \tag{9.3}$$

where, as we have seen already in Secs. 3.2 and 3.4, the outcome of the above integration is independent of the choice of the hypersurface Σ within the extended phase space $\mathcal{M}$.

Let us use asterisks to denote the antilinear functional defined by taking the complex conjugate of the image of the map in (9.1). Then, instead of expressing the quantum metric in **F** in terms of the 1-forms in (8.2), namely as we did in (8.1), we can express it in the form

$$G = \Phi_{\zeta_1 \zeta_2} (\tilde{\Phi}^{\zeta_1})^* \otimes \tilde{\Phi}^{\zeta_2} , \qquad \Phi_{\zeta_1 \zeta_2} := \Phi_{\zeta_2}(\zeta_1) , \tag{9.4}$$

with respect to a quantum frame $\mathrm{Q}(\boldsymbol{u})$. Indeed, in accordance with (9.3), the Hermitian metric G assumes in the dual of $\mathbf{F} \otimes \mathbf{F}$ the same value as in (8.3a):

$$G(\Psi_1 \otimes \Psi_2) = \langle \Psi_1 | \Psi_2 \rangle , \qquad \Psi_1, \Psi_2 \in \mathbf{F} . \tag{9.5}$$

We thus see that the analogies between classical and quantum frames of reference are very striking. These analogies are, however, not merely formal, but they extend to the physical level. Indeed, as we discussed in Secs. 3.2 and 3.4-3.6, the *coordinate wave function* of a state vector Ψ, assigned to each such state vector in the typical fibre **F** by the map in (9.1), has a physical meaning as probability amplitude that gives rise to the probability density in (7.2).

On the other hand, note should be taken of the fact that the same is *not* generally true of the components of the same vector with respect to an orthonormal basis in **F**, such as the one in (7.17). This is an established measurement-theoretical fact, although it runs counter to the still wide-spread popular *belief* that, in the absence of superselection rules, any orthonormal basis in a Hilbert space $\mathcal{H}$ of quantum state vectors can be *somehow* assigned to a complete set of 'observables'.

A rationalization of that belief can be arrived at only *if* the general *mathematical* features (cf. [PQ], Ch. IV, Sec. 5) of complete sets of commuting self-adjoint operators are combined with von Neumann's (1932) *postulate* that *all* self-adjoint operators in the Hilbert space $\mathcal{H}$ of a quantum system, subject to no superselection rules, represent observables. However, as Wigner (1963) pointed out, even in nonrelativistic quantum mechanics it is *not known* how to assign measurement procedures even to (appropriately symmetrized) products of *both* the position and the momentum operators – not to mention the prevailing multitude of all the other self-adjoint operators. Moreover, in general, procedures for the arbitrarily precise measurement of most self-adjoint operators *cannot exist even in principle*[39], since "there are serious limitations on the measurability of an arbitrary quantity. They blur the mathematical elegance of von Neumann's original postulate that all self-adjoint operators are measurable." (Wigner, 1981, p. 298).

At the physical level, informationally complete (Prugovečki, 1977b; Busch *et al.*, 1987) quantum frames obviate these problems, since any quantum state can be, in principle, uniquely determined by measuring its Fubini-Study distance from all the elements of such a quantum frame. Hence, there is no need for von Neumann's postulate that, in the absence of superselection rules, any self-adjoint operator in the Hilbert space $\mathcal{H}$ of a quantum system represents an observable: according to (7.15), the Fubini-Study distance of a quantum state to a quantum frame is directly related to a probability density, whose measurement can unambiguously determine *any* quantum state. On the other hand, the same is *never* true of orthonormal frames – a fact which represented a major motivation for von Neumann's postulate. For example, even in two-dimensional spaces for spin 1/2 quantum observables, the measurement of the transition *probabilities* to the "spin-up" and "spin-down" eigenstates do not in general uniquely determine the quantum spin state. It was for that reason, as well as on account of the measurement-theoretical fact that, as established by Wigner (1952), as well as by Araki and Yanase (1960), no arbitrarily precise measurements of those probabilities for *sharp* spin values are in principle feasible, that *stochastic* spin values were originally introduced (Prugovečki, 1977a), as spin coordinate values measured in relation to a *quantum* spin frame. Later work by Schroeck (1982a) and Busch (1986, 1988) established the measurement-theoretical implementability of this concept – which was later extended to the relativistic regime by Brooke and Schroeck (1989).

At the mathematical level, quantum frames can provide solutions to the domain problems mentioned at the end of the preceding section. Indeed, the linear manifolds spanned by such frames can supply a common dense core [PQ] for all the values $H(t)$ assumed by a generally time-dependent Hamiltonian. In fact, as can be seen from (6.15), that core is then shared by other fundamental observables, such as position, momentum, etc. Furthermore, generally speaking, the resolution generators Φ^u for such frames, which physically play the role of their origins, are uniquely determined (cf. [P], Secs. 1.7 and 2.7) by the adopted representation of the fundamental kinematical group in $\mathcal{H}$ – namely of the Galilei group or the Poincaré group in the nonrelativistic or the relativistic context, respectively. Consequently, when viewed as physical states of quantum objects, the elements of such frames are, in principle, uniquely determined by measurement procedures as Hilbert space vectors, and not just as rays. We can therefore say that the elements of quantum frames represent quantum states which provide the *reference phases*, in relation to which the *relative* geometric phases of nearby (in the Fubini-Study metric) state vectors can be measured and compared in the course of a certain quantum evolution. Indeed, any of the thus far envisaged (Anandan and Aharonov, 1988) measurement procedures of geometric phases for cyclic quantum evolutions can be, in principle, adapted to supply procedures for the measurement of such relative phases for quantum state vectors, which can be expanded in terms of coordinate wave functions assigned to that vector by the map in (9.1).

In this monograph we shall be concerned primarily with those quantum frames which are inertial, i.e., whose propagation in time is governed by a free Hamiltonian H_0. When the constituents of such frames are sufficiently massive in relation to the quantum systems for whose localization[40] they are employed, their behavior approaches that of classical inertial frames – as can be seen from (7.9).

On the other hand, we can also entertain the possibility of non-inertial frames that are acted upon by forces which, in the nonrelativistic regime, give rise to time-dependent Hamiltonians $H(t)$. In such cases, the SQM wave functions representing those frames are derivable from the nonrelativistic SQM propagator in (6.15) – and again an approximately classical behavior can be discerned under certain physical conditions.

To illustrate this point, let us consider a quantum frame which at $t = 0$ coincides with the one defined by (2.13) and (7.6) for the spin zero case. In that case

$$\Phi^u_\zeta = U(0,\mathbf{q},\mathbf{v},I)\Phi^u, \qquad \Phi^u = \xi \in L^2(\mathbf{R}^6), \quad \zeta = (0,\mathbf{q},\mathbf{v}) \;, \tag{9.6}$$

where, according to (2.12) and (2.19), the proper state vector

$$\Phi^u(\mathbf{q},\mathbf{p}) = \int_{\mathbf{R}^3} \exp(i\mathbf{q}\cdot\mathbf{k})\tilde{\xi}(\mathbf{p}-\mathbf{k})\tilde{\xi}(\mathbf{k})\,d^3\mathbf{k} \;, \tag{9.7}$$

corresponds to the quantum object marking its origin, and the representation U in (9.6) is the one specified in (2.7). In the Schrödinger picture, the coordinate wave function of its constituent which at $t = 0$ was located at the stochastic mean position $\mathbf{q}(0)$, and displayed there the stochastic mean velocity $\mathbf{v}(0) = \mathbf{p}(0)/m$, evolves with the passage of time into

$$\Phi^u_{t,\mathbf{q}(0),\mathbf{v}(0)}(\zeta) = \left\langle \Phi^u_\zeta \middle| U(t,0)\Phi^u_{0,\mathbf{q}(0),\mathbf{v}(0)} \right\rangle \;, \qquad U(t,0) = T\left(\int_0^t e^{-iH(t')}\,dt'\right) . \tag{9.8}$$

We note that the above coordinate wave function is indeed a special case of (6.16).

Under the mild conditions on the proper state vector Φ^u of the quantum frame origin (such as those that make the approximation in (7.9) valid), and under additional relatively mild conditions on the potential in (6.11), the solution of the following equation (cf. [P], Eq. (4.23), Sec. 3.4),

$$i\frac{\partial}{\partial t}\Phi^{u}_{\mathbf{q}(t),\mathbf{v}(t)} = \left\{\left[\frac{\mathbf{p}^2}{2m} + V(\mathbf{q},t)\right] - i\left[\frac{p_A}{m}\frac{\partial}{\partial q^A} - \frac{\partial V}{\partial q^A}\frac{\partial}{\partial p_A}\right]\right\}\Phi^{u}_{\mathbf{q}(t),\mathbf{v}(t)} \ , \tag{9.9}$$

provides, modulo a phase factor[41], a good approximation to the wave function in (9.8):

$$\Phi^{u}_{t,\mathbf{q}(0),\mathbf{v}(0)}(\mathbf{q},\mathbf{p}/m) \approx \exp[i(\mathbf{q}\cdot\mathbf{p} - \mathbf{q}(t)\cdot\mathbf{p}(t))]\Phi^{u}_{\mathbf{q}(t),\mathbf{v}(t)}(\mathbf{q},\mathbf{p}/m) \ . \tag{9.10}$$

We recognize in the second term between square brackets in (9.10) the Liouville operator for evolution in classical statistical mechanics. Hence, modulo a phase factor supplied by the classical Hamiltonian between the first set of square brackets in (9.9), the solution of (9.9) for the approximate coordinate wave functions in (9.10) coincides with that provided by the solutions $(\mathbf{q}(t),\mathbf{p}(t))$ of the corresponding *classical* problem formulated in the phase space $\Gamma = \mathbf{R}^6$. Moreover, that classical Hamiltonian compensates for the last term present in (8.28). For stochastic phase space wave functions which are sufficiently sharply peaked around the classical values $(\mathbf{q}(t),\mathbf{p}(t))$, we easily obtain from the multiplication law in (2.8), or by a direct computation of (9.8) based on (2.12), the following approximate formula,

$$\left\langle \Phi^{u}_{\mathbf{q}(t),\mathbf{v}(t)} \middle| d\Phi^{u}_{\mathbf{q}(t),\mathbf{v}(t)}/dt \right\rangle = -i\mathbf{p}(t)\cdot\dot{\mathbf{q}}(t)\left\|\Phi^{u}_{\mathbf{q}(t),\mathbf{v}(t)}\right\|^2 \ . \tag{9.11}$$

Hence, when the quantum evolution in (9.6) gives rise to a closed path in $\mathcal{P}$, this supplies the following classical approximation of the geometric phase for that evolution[42]:

$$\beta(C) \approx 2\oint_{\gamma} p_A\, dq^A = 2\int_{S_\gamma} dp_A \wedge dq^A \ , \qquad A = 1,2,3 \ . \tag{9.12}$$

We observe that in (9.12) γ is the closed path in the *classical* phase space $\Gamma = T^*\mathbf{R}^3$. This path corresponds to the closed path in (8.27) in the case where $\Psi(a)$ is the element of the quantum frame labelled by $(0,\mathbf{q}(a),\mathbf{p}(a))$, in accordance with (9.6). Furthermore, in (9.12) S_γ is a surface enclosed by γ in the phase space $\Gamma = \mathbf{R}^6$. Hence, in view of (8.7), the analogy between the quantum case and its classical approximation is complete for the geometric phase of such nonrelativistic quantum frames.

Nonrelativistic quantum frames play a most fundamental role in the formulation of geometro[43]-stochastic (GS) quantization in Newton-Cartan spacetimes, presented in the next chapter. The main purpose of this formulation is to provide a test case of the main ideas of the GS method of quantization in general, and of the validity of the formulation of GS propagation as an embodiment of the equivalence principle, in particular. Readers not interested in these aspects can, therefore, skip the next chapter, and later consult it only on those occasions when in subsequent chapters reference is made to some of its principal results.

Notes to Chapter 3

1 Cf. Note 2 to Chapter 1, and Sec. 5.6. Thus far few connections have been established (Hajra and Bandyopadhyay, 1991) between the various "stochastic" formulations of quantum mechanics and quantum field theory – of which the Parisi-Wu (1981) is the best known in elementary particle physics circles.

2 Strictly speaking, the elements of $L^2(\mathbf{R}^3)$ are not the square-integrable functions $\psi(\mathbf{x})$ themselves, but rather equivalence classes consisting of such functions that are equal almost everywhere in $\mathbf{R}^3$ with respect to the Lebesgue measure (cf. [PQ], Sec. 4.2 in Chapter II). Hence the value of $\psi(\mathbf{x})$ can be changed at will on a set of Lebesgue measure zero (such as on a finite or countably infinite set) without changing the quantum state vector represented by that wave function.

3 The family of *Borel sets* on a locally compact topological space equals the Boolean σ-algebra generated by the open sets of that topological space – i.e., heuristically speaking, it consists of sets obtained by taking countable unions and intersections of open sets, and then indefinitely repeating these operations. In the case of $\mathbf{R}^n$ the family of Borel sets can be obtained by applying the same operations to the family of all intervals in $\mathbf{R}^n$ – cf. Sec. 1.3 in Chapter II of [PQ].

4 Recall that, by definition, the *characteristic function* of any given subset of a set S is defined everywhere in S by assigning 1 to points within that subset, and 0 to points outside that subset.

5 For functions from $L^2(\mathbf{R}^3)$ which are square-integrable but not integrable, the Lebesgue integral in (1.10) does not exist, so that the Fourier transform has to be extended into the Fourier-Plancherel transform, which involves limiting the integration to bounded Borel sets in $\mathbf{R}^3$, and then taking a limit-in-the-mean [PQ]. For the sake of simplicity in notation, in this monograph we shall not explicitly display this limit-in-the-mean, but rather tacitly assume its presence whenever required.

6 A nonzero operator E is called *positive* (symbolically written $E \geq 0$) if $\langle\psi|E\psi\rangle \geq 0$ for all vectors ψ in its domain of definition within a given Hilbert space. Ludwig (1983, 1985), as well as some of his followers (Kraus, 1983), call positive operators, associated with QM measurement procedures, *effects*, so that in their terminology a POV measure could be called an effect-valued measure.

7 In conventional probability theory, *conditional probability* measures $P_A(B) = P(B|A) := P(B\cap A)/P(A)$ are defined in terms of normalized probability measures P for all measurable subsets B of a measurable set A – cf., e.g., Sec. 33 in (Billingley, 1979). However, QM probabilities are normalized over space for each instant in time, so that in dealing with probability measures over sets in spacetime, there are no finite probabilities over all of spacetime. Hence for A compact in spacetime, or for a spacetime segment which is finite in time (and given, in the special relativistic regime, by all events between two given instants, determined with respect to a given Lorentz frame), we shall refer to $P_A(B)$ as a *relative probability* measure. Indeed, it obviously does not make sense to add the relative probabilities in two such disjoint spacetime segments, so as to say, for example, that the probability of observing a particle in that union equals the sum of probabilities in the constituent spacetime segments. Hence for relative probability measures, as opposed to conditional probability measures, $P_A(B)$ might not represent a probability on the family of *all* the measurable subsets B of a given measurable set A, but rather only for a subfamily determined within the context of such operational considerations as the above ones.

8 It follows from the definition of Bochner integrals (cf. Appendix 3.7 in Chapter V of [PQ]) that the expectation value of the operator on the left-hand side of (2.3) is equal to the Lebesgue integral of the expectation values of the projectors on the right-hand side.

9 These independently discovered and rediscovered nonrelativistic POV measures have led to a wealth of new ideas and results in various branches of physics and mathematics that previously seemed unrelated – as exemplified in a variety of monographs, such as those by Davies (1976), by Helstrom (1976), and by Holevo (1982) – in addition to [P]. The short review article by Busch *et al.* (1989), written with the professed goal of making the physics community at large aware of the existence of these results, and therefore hopefully preventing the wasteful duplication of effort invested in their constant rediscovery by new researchers entering the field, provides an extensive bibliography on the subject – and yet it cannot help omitting a significant number of important references – cf., e.g., (Martens and de Muinck, 1990) for some additional recent references.

10 The notion of *coherent states* can be traced to the work of Schrödinger (1926), who studied them as nonspreading wave packets for quantum oscillators; von Neumann (1932), who employed them to partition phase space into cells; Bloch and Nordsieck (1937), who used them to treat the "infrared

catastrophy"; and Schwinger (1953), who formally expounded their properties. However, their name, as well as the realization of their significance for quantum optics, is due to the well-known work of Glauber (1963) on the subject. The generalization of this notion to homogeneous (coset) spaces for locally compact groups was first formulated by Barut and Girardello (1972), as well as by Perelomov (1972). The latter subsequently published an extensive monograph on the subject (Perelomov, 1986), which systematically deals with the derivation of coherent states for a great variety of groups.

11 For position measurements down to the molecular and atomic order of magnitude, the role of such a test body could be played, e.g, by a hydrogen atom; at the subatomic level, hadrons can play the role of extended test bodies in position measurements down to the 10^{-13} cm range. Of course, as discussed in Sec. 1.5, one of the central theses in this monograph is that all elementary quantum objects are extended (albeit not necessarily in the stringlike sense of Green *et al.*, 1987), and that the Planck length provides a lower bound on the linear dimensions of that extension, so that pointlike localization loses its physical *raison d'être*. That this might be the case was surmised more than half-a-century ago by Born (1938, 1949), Dirac (1949), Heisenberg (1938, 1943), and other pioneers and founders of quantum mechanics. It is due primarily to the post-World War II historical factors described in Chapter 12 that their clear-cut admonishments, such as the following one by Dirac, did not succeed to steer quantum physics on the right course much earlier, despite the initial failures in that direction mentioned in Sec. 1.1: "Present-day atomic theories involve the assumption of localizability, which is sufficient but it is *very likely to be too stringent*. The assumption requires that the theory shall be built up in terms of dynamical variables that are each localized at some point in space-time, two variables localized at two points lying outside each other's light-cones being assumed to have zero P.b. [i.e., Poisson bracket]. *A less drastic assumption may be adequate*, e.g., that there is a fundamental length λ such that the P.b. of two dynamical variables must vanish if they are localized at two points whose separation is space-like and greater than λ, but need not vanish if it is less than λ." (Dirac, 1949, p. 399) – emphasis added.

12 For the sake of simplicity and clarity of exposition, in the present brief review of the salient features of nonrelativistic SQM we have concentrated on the most straightforward choice of phase space ray representation for the Galilei group, namely the one in (2.7). This representation can obviously describe only the case where the quantum *system* particle as well as the quantum *test* particle have the same (integer-valued) spin as well as the same mass. However, there are phase space ray representations of the Galilei group that correspond to any given combination of spins and masses for 'test' particles and 'system' particles – cf. (Ali and Prugovečki, 1986).

13 Note also that this density does not correspond to the joint spectral measure of the three operators of multiplication by x^a, $a = 1,2,3$, since these operators are not symmetric (i.e., Hermitian) with respect to the inner product in (3.3). Stueckelberg (1941) has therefore proposed a new form of quantum mechanics, in which the wave functions would be square-integrable in time as well as in space, i. e., they would belong to $L^2(\mathbf{R}^4)$. This formulation was extended by Horwitz and Piron (1973) to the many-body case, and has received a considerable amount of treatment up to the present time (Horwitz, 1984; Saad *et al.*, 1989). However, it requires the introduction of an additional "historical time" parameter τ, so that in this and other respects it represents a radical departure from standard physical and epistemological ideas on quantum mechanics and classical spacetime structure.

14 Cf. Kálnay (1971) and Hegerfeldt (1989) for review articles, which can serve as a (by no means exhaustive) guide to the literature on this subject. The extent to which many of the key results reviewed in these articles have remained unknown even to experts is illustrated by the fact that Hegerfeldt's (1974) theorem is not even mentioned in the most recently published lecture notes on localizability in quantum mechanics (Bacry, 1988), and that in Chapter 20 of the 1986 revised edition of [BR], as well as in the most recent textbook on relativistic quantum mechanics (Greiner, 1990), the Newton-Wigner (1949) operators are still presented as "relativistic position operators", despite Wigner's own clarification of this misconception (cf. [WQ], pp. 310-313). It can be, however, deemed encouraging for the future of this subject that some textbooks on conventional relativistic QM, such as the one published in the well-known Landau series, point out already in the introduction that "there is as yet no logically consistent and complete relativistic quantum theory" (Berestetskii *et al.*, 1982, p. 4). Hence, at least *their* readers are not left with very basic and elementary misconceptions about the foundations of this subject.

15 More recent results by Ali *et al.* (1989-90) and by De Bièvre (1989) have indeed established the existence of an underlying general theory of coherent states for certain types of locally compact groups. This was expected on the basis of the striking analogies between the Galilei and the Poincaré cases, originally

established in (Prugovečki, 1978c,d), and further studied by Ali and Prugovečki (1986). Their theory provides a generalization of Perelomov's (1972, 1986) treatment of an already generalized notion of coherent states. The need for such a generalization emerged from the limitations imposed by *physical* considerations on the, to a large extent, parallel treatment of coherent states for the Galilei and Poincaré group – originally presented in (Prugovečki, 1978c,d) for the spin-0 states, obtained upon setting $j = 0$ in (2.12) and (4.10), respectively. Indeed, comparison with Perelomov's (1986) prescription reveals that the states in (2.12) are generalized coherent states in the sense of Perelomov, but for the *isochronous* Galilei subgroup, rather than for the full Galilei group. In SQM this restriction to the isochronous case of $t =$ const. slices of Newtonian spacetime is required on physical grounds. As a matter of fact, if Perelomov's construction is applied to the entire Galilei group, it results in unphysical wave functions, i.e., wave functions which are square integrable in the time as well as in the space variables, and as such can describe only quantum objects which materialize out of nowhere, and which, with the passage of time, rescind back into nowhere, even when left alone. On the other hand, due to lack of absolute simultaneity in special relativity, there is no isochronous subgroup of the Poincaré group. Hence, the states in (4.10) are of necessity generalized coherent states for the entire Poincaré group, but *not* in the sense of Perelomov, since the action of time translations, resulting from the representation in (4.4), upon the resolution generator is related neither to a character of the subgroup of time translations, nor to an isotropy group for the quantum state represented by that resolution generator. The technical details of such constructions, as well as the prospects for further applications and generalization, are discussed from a general group-theoretical point of view by Ali and Antoine (1989), and by Antoine (1990). Particular derivations of this kind of generalized coherent states for the de Sitter group in two and in four spacetime dimensions can be found in (Ali *et al.*, 1990) and in (Drechsler and Prugovečki, 1991), respectively.

16 Classical *relativistic phase space* is defined (Ehlers, 1971) in the CGR context as a submanifold of the cotangent bundle $T^*\mathbf{M}$ obtained by restricting the domain of the covectors in each cotangent space to the forward mass hyperboloids (3.12) for each of the rest masses of the considered particles. In the special-relativistic context, a classical relativistic phase space can be therefore obviously identified with the direct topological product of Minkowski space and those forward mass hyperboloids, so that in any given global Lorentz frame its elements are labeled by the 8-tuples (q, p) that appear in (4.3) and (4.4). The measure of integration for the inner product in (4.1), expressed in manifestly covariant form in (4.8b), is the measure generally adopted in such classical relativistic phase spaces (Ehlers, 1971), but it has the effect of making the representation in (4.4) globally non-unitary on $L^2(\mathbf{R}^6)$ – although *all* its irreducible components (4.7) *are* unitary. This mathematical inconvenience can be, however, easily rectified by adding an extra relativistically invariant term to the inner product in (4.1), and then extending $L^2(\mathbf{R}^6)$ by constructing an equipped Hilbert space (Prugovečki, 1973). The corresponding extension of the representation in (4.4) is then unitary – cf. (Ali and Antoine, 1989), p. 35.

17 All the results in this section were originally derived for the spin zero case (Prugovečki, 1978a; Ali *et al.*, 1981), but were later generalized to the case of arbitrary spin values (Prugovečki, 1980; Ali *et al.*, 1988, Brooke and Schroeck, 1989). However, for the sake of simplicity we shall review in this section only the case of zero spin.

18 Recall from Note 2 that $\rho(\mathbf{x},t)$ is not, in general, uniquely defined for all $\mathbf{x} \in \mathbf{R}^3$, but rather its values can be changed at will on sets of Lebesgue measure zero. On the other hand, the stochastic phase space wave functions are always continuous, so that $\rho_\xi(\mathbf{q},t)$ are always uniquely defined for all $\mathbf{q} \in \mathbf{R}^3$. However, in order for their sharp-point limits in (5.4) to exist, the configuration space wave function also has to be continuous. Similarly, for the sharp momentum limits in (5.5) to exist, the momentum space wave function has to be continuous – as it will be the case if the corresponding configuration space wave functions are in addition Lebesgue integrable over $\mathbf{R}^3$. The set of vectors in $L^2(\mathbf{R}^3)$ which are continuous, as well as Lebesgue integrable over $\mathbf{R}^3$, is dense in the Hilbert space $L^2(\mathbf{R}^3)$, but it is not equal to it (cf. [PQ], Sec. 5.4 of Chapter II).

19 The results by Hegerfeldt (1985) on approximate localization "with exponentially bounded tails" indicate that, if we assume that the probability density in (5.12) is exponentially bounded in $|\mathbf{q}|$ at $q^0 = 0$ in some Lorentz frame, then it does not stay exponentially bounded at any later time (as it should, in accordance with Einstein causality), despite the established fact that asymptotic causality is satisfied as $q^0 \to \infty$ (Greenwood and Prugovečki, 1984). If this result would turn out to be true also for less restrictive modes of approximate localization (Kosinski and Maslanka, 1990), that would mean that the GS mode of propagation introduced in the general relativistic context in Sec. 5.4 is, even in the case of

flat spacetime, the only mode of quantum propagation in *strict* abidance with Einstein causality – i.e., that the geometro-stochastic framework has to be used also in the special relativistic regime in order to satisfy strict Einstein causality, and at the same time retain a relativistically consistent notion of quantum particle localization. On the other hand, in Sec. 5.7 we shall discuss the possibility that at the quantum level the requirements of strict Einstein causality could be relaxed when one is not dealing with pointlike quantum particles.

20 This interpretation is merely heuristic at the mathematical as well as at the physical level: mathematically, since the time evolution operator in (1.4.3) is applied to such objects as the ket vectors $|\mathbf{x}(t')\rangle$, which do not belong to the Hilbert space $L^2(\mathbf{R}^3)$; physically, since the square of the absolute value of the propagator in (1.4.3) does not yield a bona fide probability density, because it is not integrable in $\mathbf{x} \in \mathbf{R}^3$. The first difficulty can be removed by the use of rigged or equipped Hilbert spaces (Antoine, 1969; Prugovečki, 1973), but the second one remains, and can be dealt with only by regarding the propagator of a point quantum particle as a renormalized limit (involving an infinite renormalization) of propagators for state vectors in $L^2(\mathbf{R}^3)$.

21 The adoption of the wave function (6.6) – which modulo the normalization factor $(2\pi)^{-3/2}$ (cf. (2.19)) equals the ground state of the nonrelativistic harmonic oscillator – as a model for fundamental extended particles "within" which classical spacetime concepts lose their meaning, was first proposed by Landé (1939). The discovery by Hofstadter of the extended nature of hadrons was based on form factor measurements in scattering experiments, which generally provided values around 10^{-13} cm for their rms radii. Later on, with the advent of quark models for hadrons, nonrelativistic harmonic oscillator wave functions were used in formulating static quark-antiquark models of mesons (Close, 1969). The ground states of the nonrelativistic harmonic oscillators were also shown to provide state vectors for *optimal* localization in stochastic phase space (Ali and Prugovečki, 1977a), and to play a central role in models for the actual simultaneous measurement of the stochastic positions and momenta of quantum particles (Busch, 1985a).

22 The choice of factor in front of the integral in (6.19) is largely a matter of convention, since in the relativistic regime the Feynman propagator is always defined on the basis of vacuum expectation values of time-ordered products of field operators, i.e., by means of a two-point function, rather than on the basis of the propagation of Klein-Gordon particles. The definition in (6.19) is the one given in [SI], pp. 442-444, and it is, by a $2i$ factor, distinct from the one given in [IQ], p. 124. As mentioned in Note 18 to Chapter 1, Feynman's (1950) treatment of propagation of Klein-Gordon particles was carried out by means of a Schrödinger equation involving a fictitious time parameter, "somewhat analogous to proper time" (Feynman's, 1950, p. 453). That fictitious time parameter was, however, introduced primarily "by blind mathematical analogy" ([ST], p. 226) with the path-integration method in the nonrelativistic case, rather than on physical grounds that could impart to that formal mathematical procedure any kind of operational meaning. However, in Sec. 5.7 we shall discuss the possibility of imparting a physical meaning to this parameter.

23 The value of the normalization constant in (6.26) can be deduced form the reproducibility property in (6.24). The computation has been explicitly carried out in [P], p. 120, for the case where, for dimensional reasons, one of the factors used in the exponential is slightly different from the one in (6.26) . In this context, recall that throughout this monograph Planck dimensionless units are used, in which the Planck length has the value one, so that it does not explicitly make its appearance as one of the factors in the exponential function in (6.26).

24 In fact, the consensus is that in conventional relativistic quantum field theory this series is merely an asymptotic series (Dyson, 1952). In the context of the numerologically most successful model, namely QED, Schwinger describes the situation as follows: "If one excludes the consideration of bound states, it is possible to expand the elements of a scattering matrix in powers of the coupling constant, and examine the effect of charge and mass renormalization, ... It appeared that, for any process, the coefficient of each power in the renormalized coupling constant was completely finite (Dyson, 1949). *This highly satisfactory result did not mean, however, that the act of renormalization had, in itself, produced a more correct theory* [emphasis added]. The convergence of the power series is not established, and the series doubtless has the significance of an asymptotic expansion." (Schwinger, 1958, pp. xiii-xiv). This last statement means that, even if one *assumes* that an S-matrix itself actually exists, the partial sums of the renormalized expansion do not converge to the actual values of S-matrix elements. On the other hand, as it was mentioned in Sec. 1.2, the *existence of the S-matrix has never been*

established for any nontrivial four-dimensional model in conventional quantum field theory. On the contrary, protracted efforts in that direction have led to the conclusion that "arguments favoring triviality [of such models] seem to be stronger" (Glimm and Jaffe, 1987, p. 120), even in the case of QED.

25 This concept was first introduced in (Prugovečki, 1981d), after it was realized that, from an operational point of view, in the special relativistic regime the distribution in (5.12) cannot be interpreted as a probability density for stochastic values with respect to *classical* frames of reference – as is the case with its nonrelativistic counterpart in (5.2). Indeed, if that were the case, then a special relativistic counterpart of (5.4) would hold – but the lack of absolute simultaneity in special relativity makes that impossible. The operational interpretation of quantum frames was later elaborated in Sec. 4.3 of [P], in the context of extending their operational significance to the general relativistic regime (cf. Sec. 11.5 for a recapitulation). From the observations made in (Prugovečki, 1982b) it is very clear, however, that the same basic operational interpretation applies in the nonrelativistic regime – as explicitly demonstrated also by Aharanov and Kaufherr (1984). The general treatment of these, and related geometric issues, in the present and in the next section coincides with the one in (Prugovečki, 1991c).

26 Such extensions were studied in a quantum mechanical context by Antoine (1969), Prugovečki (1973), and many others, and supply a mathematically rigorous interpretation of the Dirac (1930) formalism.

27 A group G of transformations $g: \mathcal{X} \to \mathcal{X}$ acts on a set $\mathcal{X}$ *effectively* if $g(x) = x$ for all $x \in \mathcal{X}$ implies that $g = e$ (i.e., that g equals the identity transformation), it acts *freely* if $g(x) \neq x$ except when $g = e$, and it acts *transitively* if for any $x, x' \in \mathcal{X}$ there is a $g \in \mathbf{G}$ such that $g(x) = x'$; the action of G on $\mathcal{X}$ is *from the left* if $(g'g'')(x) = g'(g''(x))$ for all $x \in \mathcal{X}$ and all $g', g'' \in \mathbf{G}$, and it is *from the right* if $(g'g'')(x) = g''(g'(x))$ for all $x \in \mathcal{X}$ and all $g', g'' \in \mathbf{G}$ (cf. [C], p. 153). In accordance with general conventions in modern differential geometry, which we implicitly adopted already in Chapter 2 (cf. Notes 11 and 15 to Chapter 2), we consider that transformations act from the right on the set $\mathcal{X}$ when $\mathcal{X}$ consists of frames. The concept of *generalized coherent state* is discussed at length by Perelomov (1986), but a further generalization of Perelomov's method of construction is required (cf. Note 15) in order to cope with the situation in relativistic SQM – such as was the case in Sec. 3.4.

28 Cf. Chapter 13, §2 in [BR], or Chapter IV, §2.1 in [PQ], for a discussion of the *fundamental* significance of unit Hilbert rays to the formulation and interpretation of quantum theory in general. The concept of *ray* was introduced in quantum theory already by Dirac (1930).

29 Experimental designs for such measurements can be deduced from the marginality properties (2.17)-(2.18), and are the subject of a considerable amount of literature (Busch, 1985, 1987; Yamamoto and Haus, 1986; Busch and Schroeck, 1989; Busch and Lahti, 1990; Schroeck and Foulis, 1990; etc.), which is systematically reviewed by Busch *et al.* (1991) and by Schroeck (1991). It should be noted that orthonormal bases, in any Hilbert space of dimension larger than one, constitute complete systems of vectors [PQ], but are *not* informationally complete: even if the transition probabilities with respect to all the elements of an orthonormal basis can be actually measured, they uniquely determine a quantum state *only* if that state is represented by an element of that orthonormal basis.

30 Cf. Sec. 3 of (Prugovečki, 1978a). Similar conclusions were more recently reached in (Aharonov and Kaufherr, 1984) and in (Anandan and Aharonov, 1988). In fact, this latter work also employs the term "quantum frame", and discusses its implementability in practice. However, this work derives some of the mathematical features of quantum frames by geometric phase arguments that are similar to the ones employed in the next section, rather than by the primarily analytic methods employed in the original work on the subject (Prugovečki, 1977a, 1978a,b).

31 In fact, the SQM method of quantization emerged from the observation that in the nonrelativistic context this method assigns quantum counterparts to classical observables which are functions of position and momentum variables with respect to a given classical inertial frame (Prugovečki, 1978d). More recently, this led to the conclusion that various choices of resolution generators (i.e., of proper quantum state vectors marking the origin of a quantum frame) leads, during the ensuing transition to the quantum regime, to various orderings of those classical variables (Ali and Doebner, 1990).

32 Cf. [KN], p. 134. In general, Chapter IX of [KN] deals with finite-dimensional complex manifolds, but many of its basic definitions and results can be generalized to the case of infinite-dimensional complex Banach manifolds (Lang, 1972), defined in Chapter VII of [C] by extending the concept of chart and atlas, presented in Sec. 2.1, to that case. That can be accomplished in a very straightforward manner by replacing in (2.1.1) and (2.1.2) $\mathbf{R}^4$ with any (real or complex, finite or infinite-dimensional) Banach

space (i.e., complete normed space) **B**, and requiring that the corresponding coordinate transformation maps in (2.1.3) are smooth in the sense of being infinitely many times differentiable in accordance with the following general definition (cf. [C], p. 71; or [Lang, 1972], p. 6): a map F from an open set $\boldsymbol{O}$ in **B** into **B** is ***differentiable*** at a point $f \in \boldsymbol{O}$ if there is a continuous (in the norm $\|\cdot\|$ of **B**) linear map $DF(f)$ (called the ***Fréchet differential*** of F at f) of **B** into **B** for which $F(f+h) - F(f) = DF(f)h + R(h)$, where $\|R(h)\|/\|h\| \to 0$ as $\|h\| \to 0$. Many of the mathematical features of infinite-dimensional Kähler manifolds, that are our main concern in the remainder of this chapter, are derived in (Freed, 1985, 1988).

33 Cf. also (6.3.2a) for the case of Dirac quantum frames, but take note of the fact that the set Σ in (6.3.2a) does not coincide with Σ in (7.16), since it does not incorporate the spin variables s. The incorporation of those variables results in a dual Dirac frame into which the summation over r is incorporated, and for which the summation over s is part of the integration with respect to the resulting measure.

34 This similarity has resulted in the usage of the term "overcomplete family" to describe continuous resolutions of the identity, on account of the fact that their elements are not generically linearly independent in a *topological* (Hilbert space) sense, i.e., that a given element of a quantum frame can be approximated arbitrarily well in the norm topology of that Hilbert space by linear combinations of other elements – or, equivalently, that the orthogonality of a vector to certain characteristic proper subsets of a family of coherent states implies that vector must equal the zero vector. This is not the case with orthonormal bases – which, on the other hand, are never *informationally* complete. We shall, however, avoid the use of the term "overcomplete family", since it can be sometimes misleading. Thus, elements of quantum frames can be *algebraically* linearly independent, i.e., none of their elements might equal *finite* linear combinations of some other elements. Furthermore, at the physical level, they correspond to kinematically distinguishable operations, namely operations which display distinct physical outcomes when applied to the adopted quantum test bodies. For example, if a proper state vector of such a test body displays rotational symmetry, then spatial rotation operations do *not* produce physically distinguishable states – and correspondingly they do not give rise to distinct quantum frame elements. On the other hand, certain quantum frames, such as those whose resolution generators are equal to ground harmonic oscillator eigenstates in (6.6), can be *informationally* overcomplete – even to the extent that the Fubini-Study distance of a quantum state to all the frame elements in any given (however small) neighborhood of the origin of those frames can uniquely determine that state. This feature is due to well-known analyticity properties in the variables $z^A = (2\ell)^{-1}q^A + i\ell p^A$ (cf. [P], p. 32) of the corresponding Fock-Bargmann representation (Perelomov, 1986) of coordinate wave functions with respect to such quantum frames. However, such a feature is highly desirable in the curved spacetime context of subsequent chapters, since it implies that localization measurements performed with respect to such frames, over neighborhoods of a point $x \in \mathbf{M}$ in the base manifold of a quantum bundle which are so small that curvature effects can be neglected (cf. Sec. 2.7), suffice in determining quantum states in the fibres above those points – cf. Sec. 12.5.

35 A *complex structure* on a *real* vector space $\mathcal{V}$ is a linear one-to-one map J of the vector space $\mathcal{V}$ onto $\mathcal{V}$ which is such that $J^2 = -1$, where 1 is the identity operator in that vector space (cf. [K], p. 114).

36 This principal bundle also results when Perelomov's (1972) prescription for constructing generalized coherent states is applied to the group $\mathbf{G} = U(\mathcal{H})$, consisting of all unitary operators in $\mathcal{H}$. To see that, let us identify each element of $U(\mathcal{H})$ with the orthonormal basis in $\mathcal{H}$ that is obtained when that unitary operator acts on the orthonormal basis singled out in (7.17). Let $U(\mathcal{H},w_1)$ denote the stability [BR] subgroup of $U(\mathcal{H})$ that leaves unchanged the element w_1 in that preferred basis. Since $U(1)$ leaves invariant the unit ray corresponding to w_1, the direct product $\mathbf{H} = U(\mathcal{H},w_1)\otimes U(1)$ plays in Perelomov's (1972,1986) construction the role of isotropy group for the quantum state (i.e., unit ray) represented by the state vector w_1. According to that construction, $U(\mathcal{H})$ emerges in an obvious manner as a principal fibre bundle with typical fibre $U(\mathcal{H},w_1)\otimes U(1)$, and base space $U(\mathcal{H})/U(\mathcal{H},w_1)\otimes U(1)$. In turn, this principal fibre bundle can be identified with the bundle in (8.10) by first identifying each unit vector Ψ in $\mathcal{H}$ with the element of the coset space $U(\mathcal{H})/U(\mathcal{H},\Psi)$ whose members map w_1 into Ψ – where $U(\mathcal{H},\Psi)$ denotes the isotropy subgroup of Ψ, so that $U(\mathcal{H})/U(\mathcal{H},\Psi)$ is isomorphic to $U(\mathcal{H})/U(\mathcal{H},w_1)$, since $U(\mathcal{H},\Psi)$ and $U(\mathcal{H},w_1)$ are isomorphic. The base space of this principal bundle can be then identified in an obvious manner with the projective Hilbert space $\mathcal{P}$ by identifying the unit ray corresponding to each unit vector Ψ with $U(\mathcal{H})/U(\mathcal{H},\Psi)\otimes U(1)$, and consequently also with the homogeneous space

$U(\mathcal{H})/U(\mathcal{H},w_1)\otimes U(1)$. This construction can be used to motivate the introduction of the homogeneous coordinates in (8.11a) for the case of $\rho = 1$. Since analogous identifications can be carried out for all the remaining w_ρ, $\rho = 2,3...$, the introduction of all the other charts in (8.11) can be viewed from a similar perspective. In these mathematical considerations, note should be taken of the fact that, according to a theorem by Kuiper (1965), the infinite-dimensional group $U(\mathcal{H})$ is contractible to a point.

37 The detailed argument runs as follows (cf. Anandan and Aharonov, 1990, p. 1698): For the self-adjoint operator A whose matrix elements in the orthonormal basis in (7.17) are all equal to zero, with the exception of $\langle w_1|Aw_2\rangle = \langle w_2|Aw_1\rangle = 1$, it is easily computed that $\exp(iAt) = (\cos t)\mathbf{1} + i(\sin t)A$ by expanding the exponential in a power series. Consequently, $C = \{\Psi(t) = \exp(iAt)w_1 \mid t \in \mathbf{R}^1\}$ is a smooth curve in $\mathcal{S}$, and since $\boldsymbol{\omega}(d\Psi(t)/dt) = \langle\Psi(t)|d\Psi(t)/dt\rangle = \langle w_1|iAw_1\rangle = 0$, it must coincide with the horizontal lift of $\Pi(C)$ (cf. Sec. 2.5). On account of $(4\langle d\Psi(t)/dt|d\Psi(t)/dt\rangle)^{1/2}dt = 2(\langle iA\Psi(t)|iA\Psi(t)\rangle)^{1/2}dt = 2dt$ being, in accordance with (8.12a) and (8.14a), the length element along C, we get from (8.18) that $\tau = 2t$. On the other hand, for any two given unit rays in (7.10), we can choose representatives ψ_1 and ψ_2 as well as an othonormal basis in $\mathcal{H}$ which is such that $\psi_1 = w_1$ and $\psi_2 = (\cos t)w_1 + i(\sin t)w_2$ for some $t \in \mathbf{R}^1$. Hence, we have $\cos t = \langle\psi_1|\psi_2\rangle$, and since $\psi_2 = \exp(iAt)\psi_1$, we can join those representatives by means of the corresponding horizontal lift C. This means that the corresponding value of τ is the Fubini-Study distance between the two given unit rays in $\mathcal{P}$, so that $\cos \tau/2 = \langle\psi_1|\psi_2\rangle = |\langle\Psi_1|\Psi_2\rangle|$ for *any* two unit vector representatives Ψ_1 and Ψ_2 of those two unit rays.

38 Cf. [C], p. 216, or (Lang, 1972), Ch. IX, or (Abraham *et al.*, 1988), §7.2. However, since in all these references, which otherwise deal with infinite-dimensional manifolds (the last two doing that from the outset), this fundamental theorem is nevertheless stated and proved only for the finite-dimensional case, the validity of the expression involving the surface integral in (8.23) can be deemed established only when the path C is entirely contained within a finite-dimensional subspace of $\mathcal{H}$.

39 Cf. also Sec. 7.8 and Chapter 12 for further discussions of this *fact*.

40 In the nonrelativistic context, the first detailed studies of such measurement-theoretical procedures have been apparently carried out independently by Aharonov and Kaufherr (1984), and by Busch (1985) – cf. also Notes 29 and 30.

41 This phase factor corresponds to a particular choice of stochastic phase space gauge, and it is essential when comparing the quantum to the classical regime, since such comparisons are intrinsically gauge-dependent – cf. (Prugovečki, 1978b).

42 This result can be compared with the one presented on p. 7 of (Anandan, 1990) for the case, *de facto*, of quantum frames with the configuration space resolution generators (i.e., proper state vectors) defined in (6.6). However, the present general argument establishes its validity even for the cases of nonrelativistic quantum frames whose constituents do not necessarily possess proper state vectors of optimal extension in phase space.

43 In addition to the extensive use of geometric techniques, the method of *geometro-stochastic quantization* and the method of *geometric quantization* (Śniatycki, 1980; Woodhouse, 1980; Hurt, 1983), developed by Kostant (1970) and Souriau (1970), share several common *mathematical* features in their first stages, namely in the nonrelativistic SQM stage for the former, and in the prequatization stage for the latter. Indeed, in those stages both methods make use of L^2 spaces over classical phase space (cf. Sec. 3.2). Moreover, it has been shown (Ali and Emch, 1986) that the choice of resolution generator in the former is equivalent to the choice of polarization of phase space in the latter. The method of quantization proposed by Berezin (1975) further bridges the technical gap between the Kostant-Souriau geometric quantization method and the non-relativistic SQM quantization method outlined in this chapter, since in its application of geometric techniques it makes extensive use of coherent states. Berezin's method is discussed in detail in Chapter 16 of (Perelomov, 1986), with particular emphasis on its applicability to general homogeneous Kähler manifolds. Hence, in the nonrelativistic regime the Kostant-Souriau as well as the Berezin method of quantization are more general than the SQM method outlined in Sec. 3.2. However, their methodologies do not extend to the special relativistic regime, due to their emphasis on the existence of quantum counterparts of classical algebras of observables (cf. Note 31), which, on account of the difficulties with sharp localization in relativistic quantum theory, presents such approaches with severe foundational problems. This is even more true in the general relativistic regime, with which most of the subsequent chapters will be dealing.

Chapter 4

Nonrelativistic Newton-Cartan Quantum Geometries

The basic physical principles and ideas of geometro-stochastic (GS) quantum theory were described in Secs. 1.3-1.5, and their implementation within a general relativistic context will be carried out in the next and subsequent chapters. However, as a basic testing ground for these principles and ideas in general, and of the central concept of GS propagation in particular, we shall choose in this chapter the more familiar, as well as experimentally more extensively investigated, territory of nonrelativistic quantum theory. In this realm the concept of sharp localization gives rise to no foundational inconsistencies at the theoretical level, and the experimental confirmation of the conventional theory is indubitable and conclusive. Thus, we shall demonstrate that in the nonrelativistic context the proposed GS framework merges in the sharp-point limit (cf. Sec. 3.5) into a framework that is equivalent to conventional nonrelativistic quantum mechanics in the presence of an external Newtonian gravitational field. In this manner, we shall reach the assurance that the GS framework gives rise to no inconsistencies with that part of quantum theory that has already received experimental confirmation under the empirical limitations stipulated by the imposition of the nonrelativistic regime. Hence, the motivation for the investigations in this chapter is, in this last respect, analogous to the motivation for considering the linearized gravity approximation of the classical theory of general relativity (CGR): the existence of such a weak-gravity limit, in which classical general relativistic solutions merge into their Newtonian counterparts, has provided, at the inception[1] of general relativity, the necessary assurance that the CGR framework does not give rise to any conflict with the wealth of observational data supporting Newtonian theory, despite the total dissimilarities in the mathematical structures of these two frameworks for describing gravitational phenomena.

The work by Cartan (1923, 1924) brought these two frameworks for classical gravity considerably closer by showing that the Newtonian framework can be recast in a fully geometrized form, which came to be known as Newton-Cartan spacetime[2]. Furthermore, Trautman (1963, 1966) and Havas (1964) were later able to show that Newton's gravitational field equations could be recast in a form involving the curvature tensor of a suitable connection (cf. Sec. 4.1), which bears a close similarity to Einstein's field equations (cf. Sec. 2.7). Hence, the transition from general relativity to Newton-Cartan theory could be eventually carefully worked out (Künzle, 1972, 1976) with full mathematical rigor. It therefore became natural to formulate a generalized type of Schrödinger equation on Newton-Cartan spacetimes (Kuchař, 1980; Duval and Künzle, 1984; Duval *et al.*, 1985) in order to geometrically describe nonrelativistic quantum particles in free fall in an external

Newtonian gravitational field, and thus better understand the role that the equivalence principle plays in the formulation of relativistic quantum theories in curved spacetime.

The Newton-Cartan quantum geometries described in this chapter are based on the opposite type of approach to this same problem (Prugovečki, 1987a; De Bièvre, 1989a): the equivalence principle is assigned the central role of guiding principle in formulating a concept of GS propagation of quantum particles in free fall in an external Newtonian gravitational field. The fact that such propagation is, in the sharp-point limit, in agreement with the conventional theory (usually based on the Schrödinger equation for such a field) is then deduced as a consequence of fundamental theoretical significance: *this agreement provides direct evidence for the validity of the equivalence principle at the quantum level, and, by implication, also of the idea of GS propagation as a mathematical embodiment of that fundamental principle.* Consequently, in the next chapter the idea of GS propagation can be extrapolated to the relativistic regime with the total assurance that, *in the weak-gravity nonrelativistic sharp-point limit, the resulting geometro-stochastic framework does not give rise to conflicts with the observational data supporting conventional theory.* As this is the main purpose of the present chapter, readers not interested in such a verification can skip it.

We shall first review in Sec. 4.1 the most basic ideas and results of the classical Newton-Cartan theory of spacetime, using the fibre-theoretical language and tools of Chapter 2. We shall then adapt these classical ideas to the quantum framework by constructing in Sec. 4.2 Bargmann frame bundles, and in Sec. 4.3 quantum fibre bundles associated with these principal frame bundles, whose typical fibres are the SQM Hilbert spaces described in Sec. 3.2. Upon formulating in Sec. 4.4 parallel transport and deriving GS propagators within these quantum bundles, we demonstrate that in the sharp-point limit GS propagation indeed merges into conventional propagation governed by a Schrödinger equation.

*4.1. Classical Newton-Cartan Geometries

The concept of Newton-Cartan spacetime can be formulated with the greatest degree of generality as part of the theory of Galilean manifolds (Dombrowski and Horneffer, 1964; Künzle, 1972). By definition, a *Galilean manifold* is a 4-dimensional manifold **M** which carries a *Galilei structure* provided by a contravariant degenerate *space metric* $\boldsymbol{\gamma}$ of signature (0;1,1,1), as well as a covariant degenerate *time metric* $\boldsymbol{\beta}$ of signature (1;0,0,0), with the two metrics interrelated in such a manner that

$$\gamma^{\mu\nu}\beta_\nu = 0 \ , \qquad \boldsymbol{\beta} = \beta_\mu \, dx^\mu \ , \qquad \boldsymbol{\gamma} = \gamma^{\mu\nu}\partial_\mu \otimes \partial_\nu \ , \qquad \mu,\nu = 0,1,2,3 \ . \quad (1.1)$$

More specifically, the presence of a space metric and a time metric with the aforementioned signatures means that, in the tangent space $T_x\mathbf{M}$ of each $x \in \mathbf{M}$, there are special types of linear frames $\{\boldsymbol{e}_i \mid i = 0,1,2,3\}$, called *Galilei frames*, which are such that

$$\boldsymbol{\beta} = \boldsymbol{\theta}^0 \ , \qquad \boldsymbol{\gamma} = \delta^{AB}\, \boldsymbol{e}_A \otimes \boldsymbol{e}_B \ , \qquad A,B = 1,2,3 \ , \qquad (1.2)$$

where $(\boldsymbol{\theta}^i)$ denote the duals of these frames. In turn, this implies that there are sections s of the general linear frame bundle $GL\mathbf{M}$ – i.e., moving frames defining Cartan gauges (cf.

Sec. 2.6) – whose domains cover **M**, and in which (1.2) is satisfied, so that we can call them *Galilei moving frames*.

We observe that the Galilei moving frames are similar to the vierbeins in CGR (cf. Sec. 2.3) – albeit they do not diagonalize a non-degenerate 4-dimensional metric. On the other hand, the presence of the time metric $\boldsymbol{\beta}$ can be related to the absolute time of Newtonian physics. Indeed, all vectors $\boldsymbol{X}$ in each tangent space $T_x\mathbf{M}$ can be classified[3] as being *timelike* if $\boldsymbol{\beta}(\boldsymbol{X}) \neq 0$ and *spacelike* if $\boldsymbol{\beta}(\boldsymbol{X}) = 0$. Furthermore, if the maximal hypersurfaces σ_t in **M**, whose tangent vectors are spacelike at each one of their points, foliate **M**, then they can be indexed by a parameter t determining a scalar field globally defined on **M** and such that $\boldsymbol{dt} = \boldsymbol{\beta}$. Hence, the three-dimensional manifolds σ_t take over the role that absolute space plays in Newtonian mechanics, and t takes over the role of absolute time. In each of these space manifolds σ_t, the space metric $\boldsymbol{\gamma}$ in **M** induces a Riemannian metric, which, upon the later introduction of a suitable connection, can be used to determine the (absolute) spatial distances between its points, which are the hallmark of the Newtonian theory of absolute space and time. Thus, indeed, the time metric operationally corresponds to measurements with standard clocks, whereas the space metric operationally corresponds to measurements with rigid rods – whose existence does not contradict the principles of Newtonian mechanics, as it does those of relativistic physics (Stachel, 1980).

With these interpretations in mind, it is not difficult to establish that if we introduce the *Galilei frame bundle* $H\mathbf{M}$ as the subbundle of $GL\mathbf{M}$ consisting of Galilei frames, then $H\mathbf{M}$ becomes a principal frame bundle whose structure group is the homogeneous Galilei group, i.e., the group of Galilei transformations with zero spacetime translations, which acts from the right on the Galilei frames in $H\mathbf{M}$ as follows (Duval and Künzle, 1977):

$$(0,\mathbf{0},\mathrm{v},R) : (\boldsymbol{e}_0,\boldsymbol{e}_A) \mapsto (\boldsymbol{e}_0' = \boldsymbol{e}_0 + \boldsymbol{e}_A v^A, \boldsymbol{e}_A' = \boldsymbol{e}_B R^B{}_A), \quad \mathrm{v} \in \mathbf{R}^3, \ R \in \mathrm{SO}(3). \tag{1.3}$$

Since (1.3) can be obviously written in the alternative form (Duval *et al.*, 1985),

$$\boldsymbol{e}_i' = \boldsymbol{e}_j A^j{}_i(\mathrm{v},R) , \qquad A(\mathrm{v},R) = \begin{Vmatrix} 1 & \mathbf{0} \\ \mathbf{v} & R \end{Vmatrix} \in \mathrm{GL}(4,\mathbf{R}) , \tag{1.4}$$

in which the homogeneous Galilei transformation $(0,\mathbf{0},\mathrm{v},R)$ has been identified with a 4×4 matrix $A \in \mathrm{GL}(4,\mathbf{R})$, we see that the homogeneous Galilei group can be identified with a subgroup $\mathrm{H}(4,\mathbf{R})$ of $\mathrm{GL}(4,\mathbf{R})$, and that in the presence of a Galilei structure the general linear frame bundle $GL\mathbf{M}$ is reducible (cf. Sec. 2.3) to the Galilei frame bundle $H\mathbf{M}$ that has $\mathrm{H}(4,\mathbf{R})$ as its structure group (Künzle, 1972).

A *symmetric Galilei connection* on a Galilean manifold $(\mathbf{M},\boldsymbol{\beta},\boldsymbol{\gamma})$ can be now defined as a torsion-free (cf. Sec. 2.6) connection on the Galilei frame bundle $H\mathbf{M}$ that is compatible (cf. Sec. 2.6) with the space metric $\boldsymbol{\gamma}$ and the time metric $\boldsymbol{\beta}$ (Künzle, 1972; Duval *et al.*, 1985). More specifically, this definition means that if ∇_X denotes the corresponding covariant derivative of a Galilei connection (cf. Sec. 2.5), then at any $x \in \mathbf{M}$,

$$\nabla_X \boldsymbol{\beta} = \mathbf{0} = \nabla_X \boldsymbol{\gamma} , \qquad \nabla_X Y - \nabla_Y X = [X,Y] , \qquad \forall X,Y \in T_x\mathbf{M} . \tag{1.5}$$

Since $\boldsymbol{\beta}$ and $\boldsymbol{\gamma}$ are degenerate metrics, the fundamental lemma of Riemannian geometry does not apply, so that there can be many symmetric Galilei connections on a

M endowed with a Galilei structure. However, amongst Galilei connections, a distinguished role is played by *Newtonian connections*, defined as those symmetric Galilei connections whose curvature tensors satisfy the following type of constraints (Cartan, 1923, 1924):

$$\gamma^{il} R^{j}{}_{klm} = \gamma^{jl} R^{i}{}_{klm} \ , \qquad i,j,k,m = 0,1,2,3 \ . \tag{1.6}$$

The curvature tensor components in (1.6) are given by (2.6.26), with $\overline{\nabla}$ denoting the operator of covariant differentiation for that Newtonian connection in any moving frame.

A Galilean manifold $(\mathbf{M},\boldsymbol{\beta},\boldsymbol{\gamma})$ with a Newtonian connection is called a *Newton-Cartan manifold.* Newton-Cartan spacetimes can be constructed as Newton-Cartan manifolds in which, for a prescribed mass density ρ, the following counterparts of the Einstein field equations (2.7.3) are satisfied:

$$R_{ij} = 4\pi\rho\beta_i\beta_j \ , \qquad R_{ij} = R^{k}{}_{ikj} \ . \tag{1.7}$$

The essential observation is that for the Newtonian connection on a Newton-Cartan manifold there is a scalar field ϕ, defined globally on **M**, as well as a connection that is locally flat (i.e., whose curvature tensor is zero at all $x \in \mathbf{M}$), and which is such that the curvature tensor components of the Newtonian connection can be expressed in the form[4]

$$R^{i}{}_{jkl} = 2\beta_j \gamma^{im} (\phi_{;m;k}\beta_l - \phi_{;m;l}\beta_k) \ , \tag{1.8}$$

where the semicolons in front of subscript indices indicate covariant differentiation with respect to that flat connection. If we introduce the Cartan connection coefficients

$$\Gamma^{i}{}_{jk} := \theta^i\left(\overline{\nabla}_j e_k\right) \ , \qquad \hat{\Gamma}^{i}{}_{jk} := \theta^i\left(\hat{\nabla}_j e_k\right) \ , \tag{1.9}$$

for the Newton-Cartan connection and for the flat connection, respectively (cf. the first of the equations in (2.6.24a)), then the two are related as follows (Friedman, 1983, p. 100):

$$\Gamma^{i}{}_{jk} = \hat{\Gamma}^{i}{}_{jk} + \gamma^{il}\phi_{;l}\beta_j\beta_k \ . \tag{1.10}$$

As a consequence, *Galilean inertial coordinates* with $x^0 = t$ can be introduced in **M** in which the connection coefficients of the flat connection vanish identically, so that (1.10) assumes a very simple form:

$$\Gamma^{\lambda}{}_{\mu\nu} = (1 - \delta_0^\lambda)\delta_\mu^0\delta_\nu^0\, \partial\phi/\partial x^\lambda \ , \qquad \beta_\mu = \delta_\mu^0 \ , \qquad \gamma^{\mu\nu} = \mathrm{diag}(0,1,1,1) \ . \tag{1.11}$$

In these coordinates the scalar field ϕ plays the role of gravitational potential, since in them the field equations (1.7) assume the form of a Poisson equation (Künzle, 1972):

$$\Delta_{\mathbf{x}}\phi(\mathbf{x},t) = 4\pi\rho(\mathbf{x},t) \ , \qquad \mathbf{x} \in \mathbf{R}^3 \ , \qquad t = x^0 \in \mathbf{R}^1 \ . \tag{1.12}$$

In this manner the conventional Newtonian classical mechanics for particles moving in the external gravitational field described by that potential is basically recovered. However, it has to be mentioned that since the Newton-Cartan theory is fundamentally a local theory of spacetime, its local field equations are not capable of singling out a unique Newtonian potential and a corresponding unique class of Galilean inertial coordinates, except if additional global assumptions are made (Friedman, 1983, Sec. III.4). Therefore, whenever we later compare the GS propagation in a Newton-Cartan spacetime with the conventional quantum propagation in a Newtonian spacetime identifiable with $\mathbf{R}^3 \times \mathbf{R}^1$ (cf. Sec. 4.4), we shall implicitly assume that such conditions have been imposed, so that a unique Newtonian potential is available for the Schrödinger equation governing that conventional form of quantum propagation.

In a Newton-Cartan spacetime, a classical test particle is in free fall if and only if its motion takes place along timelike geodesics, i.e., along curves that satisfy the geodesic equation for the given Newtonian connection in any of the forms presented in (2.7.10), and whose tangent vectors are everywhere timelike. Therefore, the analogy with the situation in CGR is very close, so that we can, almost verbatim, transfer to the present situation all the key concepts discussed in the last part of Sec. 2.7.

Thus, we shall say that a Cartan gauge, given by a section s of the general linear frame bundle $GL\mathbf{M}$, is *adapted to a smooth curve* γ if the connection coefficients $\Gamma^i{}_{jk}$ of the Newton-Cartan connection vanish at all $x \in \gamma$ which lie in the domain of definition $\mathbf{M}^s$ of that gauge. In particular, a *Galilei gauge adapted to* γ corresponds to a section s of the Galilei frame bundle $H\mathbf{M}$ for which $\Gamma^i{}_{jk} = 0$ at all $x \in \gamma$ which lie in $\mathbf{M}^s$. As was the case with the corresponding vierbein gauges in Sec. 2.7, such adapted Galilei gauges can be obtained by restricting an arbitrary section s' of $H\mathbf{M}$ to a 3-dimensional hypersurface intersected by γ at one point x', constructing a smooth flow [C] containing γ and depicted by curves which similarly intersect that hypersurface at other points in a neighborhood of x', and then parallel transporting all the frames at those points along the curves of the given flow. Since the Newtonian connection is compatible with both the space and the time metric of a Newton-Cartan spacetime, the parallel transports of Galilei frames will also be Galilei frames, so that a section s of $H\mathbf{M}$ that has the above required property will be obtained in this manner.

A section s of the general linear frame bundle $GL\mathbf{M}$ that is adapted to a timelike geodesic γ in a Newton-Cartan spacetime will be called an *inertial moving frame* in free-fall along the geodesic γ if, for a suitable choice of affine parameter τ, all the frame elements $e_0(x)$ in its restriction

$$s_\gamma = \{(e_0(x),\ldots,e_3(x)) \,|\, x \in \gamma \cap \mathbf{M}^s\} \subset s \tag{1.13}$$

to γ are equal to the tangent vectors to that geodesic. If, in particular, all the frames in s_γ are Galilei frames, we shall call s an *inertial Galilei moving frame* for the geodesic γ, since

$$\Gamma^i{}_{jk}(x) = \boldsymbol{\theta}^i\left(\overline{\nabla}_{e_j(x)} e_k(x)\right) = 0 \ , \qquad \forall x \in \gamma \cap \mathbf{M}^s \ , \tag{1.14}$$

so that (1.13) depicts a Galilei frame that is in free-fall, and which moves in such a manner that an observer situated at its origin has the geodesic γ as his worldline.

These *local* inertial Galilei moving frames should be distinguished from the *global Galilei frames* with respect to which the *inertial Galilean coordinates* occurring in (1.12) are defined, and which become the Newtonian spacetime coordinates used in Sec. 3.1 if **M** is identifiable with $\mathbf{R}^3\times\mathbf{R}^1$ (cf. Sec. 4.4). These global Galilei frames are inertial with respect to the flat connection rather than with respect to the Newton-Cartan connection. As we see from (1.11), these two connections coincide if and only if the potential ϕ is equal everywhere to a constant, i.e., if there is no Newtonian gravitational field. Given the fact that gravitational fields are ever-present, since they are long-range and cannot be screened, we see that, strictly speaking, the pre-CGR concept of inertial frame that underlies Newtonian mechanics, as well as special relativity, refers to a physically fictional situation.

*4.2. Newton-Cartan Connections in Bargmann Frame Bundles

In order to construct a quantum Newton-Cartan bundle with typical fibre equal to one of the Hilbert spaces $\mathbf{P}_\xi L^2(\mathbf{R}^6)$ described in Sec. 3.2, we have to extend the Galilei frame bundle $H\mathbf{M}$, since these spaces are reproducing kernel Hilbert spaces related to ray representations of the full Galilei group, rather than to representations of its subgroup consisting of only homogeneous Galilei transformations. Hence, we have to first extend $H\mathbf{M}$ into a bundle $G\mathbf{M}$ consisting of the affine Galilei frames $(\boldsymbol{a},\boldsymbol{e}_i)$. These frames are the nonrelativistic counterparts of the Poincaré frames in Sec. 2.3. All the general results pertaining to affine frame bundles that were discussed in Sec. 2.3, as well as in Sec. 2.6, naturally apply also to $G\mathbf{M}$. In particular, in accordance with (2.6.17), any Newton-Cartan connection defined on some Galilei frame bundle $H\mathbf{M}$ can be extended in a unique manner to $G\mathbf{M}$.

We recall now that ray representations of the Galilei group, such as the ones defined by (3.1.16) or (3.2.7), can be regarded as ordinary representations of a central extension of the Galilei group. We note that there is such a central extension for each value of $m > 0$, but that the substitution $\theta \to \theta' = m\,\theta$ provides an isomorphism between the one for any given mass $m > 0$ and the one for $m = 1$. The central extension of the Galilei group for $m = 1$ is known also under the name of *Bargmann group*.

As can be seen from (3.1.18), the general element $(\theta,b,\mathbf{a},\mathbf{v},R)$ of the Bargmann group consists of the general element of the Galilei group to which a phase $\theta \in \mathbf{R}^1$ is added, and its group multiplication law is given by

$$(\theta',b',\mathbf{b}',\mathbf{v}',R')\cdot(\theta,b,\mathbf{b},\mathbf{v},R)$$
$$= \left(\theta'+\theta+\left(\tfrac{1}{2}\mathbf{v}'^2\,b+\mathbf{v}'\cdot R'\,\mathbf{b}\right),\, b'+b,\, \mathbf{b}'+R'\mathbf{b}+b\mathbf{v}',\, \mathbf{v}'+R'\mathbf{v},\, R'R\right) . \qquad (2.1a)$$

Hence, a *Bargmann frame* $(\chi,\boldsymbol{a},\boldsymbol{e}_i)$ is correspondingly obtained by adding to an affine Galilei frame $(\boldsymbol{a},\boldsymbol{e}_i)$ a vector χ from the domain $\mathbf{R}^1$ in which the θ-variable assumes its values – which is therefore merely a real number[5]. The *Bargmann frame bundle* $B\mathbf{M}$ over $\mathbf{M}$ can be now defined as the principal bundle whose general element is a Bargmann frame over some $x \in \mathbf{M}$, and whose structure group is the Bargmann group.

The action, from the right, of the Bargmann group on Bargmann frames $(\chi,\boldsymbol{a},\boldsymbol{e}_i) \in B\mathbf{M}$ is given by (Duval and Künzle, 1977):

$$(\theta,\boldsymbol{b},\mathbf{v},R) : (\chi,\boldsymbol{a},\boldsymbol{e}_0,\boldsymbol{e}_A) \mapsto (\chi',\boldsymbol{a}'=\boldsymbol{a}+\boldsymbol{e}_i\,b^i,\boldsymbol{e}'_0=\boldsymbol{e}_0+\boldsymbol{e}_A\,v^A,\boldsymbol{e}'_A=\boldsymbol{e}_B\,R^B{}_A) , \qquad (2.1b)$$

$$\chi' = \chi + \theta - v_A b^A (1 - b^0/2) + v_B a^B + v_A v^A a^0/2 \ , \tag{2.1c}$$

$$v_A := \delta_{AB} v^B \ , \qquad A,B = 1,2,3 \ , \qquad i = 0,1,2,3 \ . \tag{2.1d}$$

Consequently, the Newton-Cartan connection that was previously extended to GM, can be now further extended into a connection on $B\mathbf{M}$, with a connection form $\boldsymbol{\omega}$ that can be expressed, in accordance with (2.5.8), as

$$\boldsymbol{\omega} = \tilde{\boldsymbol{\omega}}^i \tilde{\mathbf{f}}_i + \boldsymbol{\omega}^A \tilde{\mathbf{F}}_A + \tfrac{1}{2} \boldsymbol{\omega}^{AB} \tilde{\mathbf{F}}_{AB} + \boldsymbol{\omega}^* \tilde{\mathbf{f}}_* \ , \tag{2.2}$$

where $\{\tilde{\mathbf{f}}_i, \tilde{\mathbf{F}}_A, \tilde{\mathbf{F}}_{AB}, \tilde{\mathbf{f}}_*\}$ constitutes a basis in the Lie algebra of the Bargmann group[6]. At any given point $\boldsymbol{u} \in B\mathbf{M}$, represented in some given chart in $\mathbf{M}$ by a Bargmann frame with

$$\boldsymbol{a} = a^\mu \partial/\partial x^\mu \ , \qquad \boldsymbol{e}_i = e_i^\mu \partial/\partial x^\mu \ , \qquad i = 0,1,2,3 \ , \tag{2.3}$$

the corresponding Cartan connection forms that occur in (2.2), and are generically defined in (2.5.10), can be suitably expressed as follows (Duval and Künzle, 1977),

$$\tilde{\boldsymbol{\omega}}^i = \boldsymbol{\theta}_\lambda^i [d a^\lambda + (\Gamma^\lambda_{\ \mu\nu} a^\nu + \delta_\mu^\lambda) d x^\mu] \ , \qquad \boldsymbol{\theta}^i = \boldsymbol{\theta}_\mu^i \, d x^\mu \ , \tag{2.4a}$$

$$\boldsymbol{\omega}^A = \boldsymbol{\theta}_\lambda^A (d e_0^\lambda + \Gamma^\lambda_{\ \mu\nu} e_0^\nu \, d x^\mu) \ , \tag{2.4b}$$

$$\boldsymbol{\omega}^{AB} = \boldsymbol{\theta}_\lambda^A \delta^{BC} (d e_C^\lambda + \Gamma^\lambda_{\ \mu\nu} e_C^\nu \, d x^\mu) \ , \tag{2.4c}$$

$$\boldsymbol{\omega}^* = d\chi + \delta_{AB} \boldsymbol{\omega}^A a^B - \gamma_{\mu;\boldsymbol{u}} \, d x^\mu + \phi \, d x^0 \ , \tag{2.4d}$$

where the connection coefficients $\Gamma^\lambda_{\ \mu\nu}$ are computed from the conditions (1.5), and $\gamma_{\mu;\boldsymbol{u}}$ is related to the space and time metric components. According to (1.11), in inertial Galilean coordinates only the coefficients $\Gamma^A_{\ 00} = \partial\phi/\partial x^A$ are different from zero. Furthermore, in such coordinates, $\gamma_{\mu;\boldsymbol{u}}$ in (2.4d) assumes the form

$$\gamma_{\mu;\boldsymbol{u}} = \delta_{\mu i} e_0^i - \delta_\mu^0 e_0^0 - \tfrac{1}{2} \delta_\mu^0 e_0^A \delta_{AB} e_0^B \ . \tag{2.5}$$

With the Newton-Cartan connection form $\boldsymbol{\omega}$ on the Bargmann frame bundle $B\mathbf{M}$ thus specified, we can compute the parallel transport

$$\boldsymbol{\tau}_\gamma(x'', x') : \ \boldsymbol{u}_{x'} = (\chi(x'), \boldsymbol{a}(x'), \boldsymbol{e}_i(x')) \mapsto \boldsymbol{u}_{x''} = (\chi(x''), \boldsymbol{a}(x''), \boldsymbol{e}_i(x'')) \tag{2.6}$$

of Bargmann frames along any smooth curve $\gamma = \{x(t) | \ a \le t \le b\}$, starting at $x' = x(a) \in \mathbf{M}$ and ending at $x'' = x(b) \in \mathbf{M}$, by computing the horizontal lift $\gamma^* = \{\boldsymbol{u}(t) | \ a \le t \le b\}$ of γ described in Sec. 2.4. This computation can be carried out by imposing, at all $t \in [a,b]$, the condition $\boldsymbol{\omega}(\boldsymbol{X}^*(t)) = \boldsymbol{0}$ on the tangent vectors $\boldsymbol{X}^*(t)$ of γ^* – thus ensuring that these vectors are the horizontal lifts of the vectors $\boldsymbol{X}(t)$ tangent to the curve γ.

In particular, this computation can be carried out for the case where γ is a timelike geodesic in $\mathbf{M}$, whose inertial Galilean coordinates therefore satisfy the geodesic equation

$$\ddot{x}^\lambda + \Gamma^\lambda_{00}(\dot{x}^0)^2 = 0 \ , \qquad \Gamma^A_{00} = \partial\phi/\partial x^A \ , \qquad \Gamma^0_{00} = 0 \ , \tag{2.7}$$

obtained by using (1.11) in (2.7.10b). In that case a section s of BM containing the frames

$$u_{x(\tau)} = (\chi(x(\tau)), a(x(\tau)), e_i(x(\tau))) = \tau_\gamma(x(\tau), x(0))\, u_{x(0)} \ , \qquad \tau \in [0,\infty) \ , \tag{2.8}$$

describes an *inertial Bargmann frame* in free fall along γ that is analogous to an inertial Galilean frame. Using (2.3)-(2.5) and (2.7) we find that

$$a(x(\tau)) = [a^\mu(x(0)) - \tau \dot{x}^\mu(\tau)]\partial_\mu \ , \qquad \partial_\mu = \partial/\partial x^\mu \ , \tag{2.9a}$$

$$e_0(x(\tau)) = [e_0^\mu(x(0)) + (\dot{x}^A(\tau) - \dot{x}^A(0))\delta^\mu_A/\dot{x}^0(\tau)]\partial_\mu \ , \tag{2.9b}$$

$$e_A(x(\tau)) = e_A^\mu(x(0))\partial_\mu \ , \tag{2.9c}$$

$$\chi(x(\tau)) = \chi(x(0)) + \int_0^{x^0(\tau)} [\tfrac{1}{2}\dot{\mathbf{x}}^2(t) - \phi(\mathbf{x}(t),t)]\, dt - \tfrac{1}{2}\tau \dot{x}^A(0)\delta_{AB}\dot{x}^B(0)/\dot{x}^0(\tau) \ . \tag{2.9d}$$

Comparison with (2.1) shows that

$$u_{x(\tau)} = (\chi(x(0)),\, a^\mu(x(0))\partial_\mu,\, e_i^\mu(x(0))\partial_\mu)\cdot(\theta_\tau, b_\tau^0, \mathbf{b}_\tau, \mathbf{v}_\tau, I) \ , \quad \tau \in [0,\infty) \ , \tag{2.10a}$$

$$\theta_\tau = \int_0^{x^0(\tau)} [\tfrac{1}{2}\dot{\mathbf{x}}^2(t) - \phi(\mathbf{x}(t),t)]dt - \tfrac{1}{2}\tau \dot{x}^A(0)\delta_{AB}\dot{x}^B(0)/\dot{x}^0(\tau) \in \mathbf{R}^1 \ , \tag{2.10b}$$

$$b_\tau = -\tau\dot{x}(\tau) \in \mathbf{R}^4 \ , \qquad \mathbf{v}_\tau = [\dot{\mathbf{x}}(\tau) - \dot{\mathbf{x}}(0)]/\dot{x}^0(\tau) \in \mathbf{R}^3 \ , \tag{2.10c}$$

where it should be noted that the integral in (2.10b) equals, for mass $m = 1$, the classical action that is used in Feynman path integration [ST]. This fact indicates the possibility of a geometric interpretation of the Feynman path integral method, which will eventually emerge from the considerations in the remainder of this chapter.

*4.3. Quantum Newton-Cartan Bundles

We can now construct a vector bundle $\mathbf{E}$ with typical fibre $\mathbf{F}$ equal to the family of all solutions (3.2.5) of the free Schrödinger equation which belong to one of the Hilbert spaces $\mathbf{P}_\xi L^2(\mathbf{R}^6)$ described in Sec. 3.2,

$$\mathbf{F} = \left\{ \Psi \,\middle|\, \Psi(\chi, q, \mathbf{v}) := e^{i\chi}\psi_{q^0}(\mathbf{q}, m\mathbf{v}) \ , \ \psi_{q^0} = \exp(-iH_0 q^0)\psi_0 \in \mathbf{P}_\xi L^2(\mathbf{R}^6) \right\} \ , \tag{3.1}$$

and with structure group equal to the Bargmann group, i.e., to the central extension of the Galilei group. This construction can be achieved by taking the $\mathbf{G}$-product [I,K] of the Bargmann frame bundle BM with the typical fibre $\mathbf{F}$, where in the present case the role of $\mathbf{G}$ is played by the Bargmann group.

In general, for any given principal bundle ($\mathbf{P}$, Π, $\mathbf{M}$,$\mathbf{G}$) and given topological space $\mathbf{F}$ on which the group $\mathbf{G}$ acts from the left as a topological group of transformations [BR, I], their $\mathbf{G}$-*product* $\mathbf{P}\times_{\mathbf{G}}\mathbf{F}$ is defined as follows ([K], p. 54): On the topological product $\mathbf{P}\times\mathbf{F}$, the elements g of the group $\mathbf{G}$ act from the right by mapping each $(u,\Psi)\in\mathbf{P}\times\mathbf{F}$ into $(u,\Psi)\cdot g := (u\cdot g, g^{-1}\cdot\Psi)\in\mathbf{P}\times\mathbf{F}$; the set $\mathbf{P}\times_{\mathbf{G}}\mathbf{F}$ is then defined as the family of equivalence classes $\boldsymbol{\Psi}$ of elements in $\mathbf{P}\times\mathbf{F}$, whereby two such elements (u',Ψ'), $(u'',\Psi'')\in\mathbf{P}\times\mathbf{F}$ are deemed equivalent if there is some $g\in\mathbf{G}$ such that $(u'',\Psi'') = (u',\Psi')\cdot g$. The map which takes each $(u,\Psi)\in\mathbf{P}\times\mathbf{F}$ into $x=\Pi(u)\in\mathbf{M}$ then induces a projection map $\pi:\mathbf{P}\times_{\mathbf{G}}\mathbf{F}\to\mathbf{M}$, which is continuous if $\mathbf{P}\times_{\mathbf{G}}\mathbf{F}$ is endowed with the natural topology induced in it by the topological product of $\mathbf{P}$ and $\mathbf{F}$. Furthermore, if $\mathbf{F}$ is a manifold, and if $\mathbf{G}$ acts on it from the left as a Lie group of transformations [I], then a differentiable structure can be introduced in $\mathbf{P}\times_{\mathbf{G}}\mathbf{F}$ by first noting that each $x\in\mathbf{M}$ has a neighborhood $\mathcal{N}$ for which the set $\Pi^{-1}(\mathcal{N})$ is homeomorphic to $\mathcal{N}\times\mathbf{G}$, and then imposing the requirement that the corresponding $\pi^{-1}(\mathcal{N})$ be an open submanifold of $\mathbf{P}\times_{\mathbf{G}}\mathbf{F}$ that is diffeomorphic to $\mathcal{N}\times\mathbf{F}$. It then follows that ($\mathbf{P}\times_{\mathbf{G}}\mathbf{F}$,$\pi$, $\mathbf{M}$) is a fibre bundle associated with the principal bundle ($\mathbf{P}$, Π, $\mathbf{M}$,$\mathbf{G}$), and that the natural projection[7] $\mathbf{P}\times\mathbf{F}\to\mathbf{P}\times_{\mathbf{G}}\mathbf{F}$ produces diffeomorphisms

$$\sigma_x^u : \boldsymbol{\Psi}\mapsto\Psi\in\mathbf{F}\ ,\qquad u\in\Pi^{-1}(x)\ ,\qquad \boldsymbol{\Psi}\in\mathbf{F}_x=\pi^{-1}(x)\ ,\tag{3.2a}$$

between the fibres $\pi^{-1}(x)$ of that associated bundle and its standard fibre $\mathbf{F}$.

We can now define the *Newton-Cartan quantum bundle* $\mathbf{E}$ as being equal to the $\mathbf{G}$-product $(B\mathbf{M})\times_{\mathbf{G}}\mathbf{F}$, where $\mathbf{F}$ is given by (3.1). In the context of such quantum bundles the maps in (3.2a) will be called *generalized soldering maps*[8], since they can be used to solder Bargmann frames to their corresponding quantum counterparts – cf. (3.7). These soldering maps can be alternatively expressed as maps between wave function amplitudes,

$$\sigma_x^u : \boldsymbol{\Psi}(\boldsymbol{\zeta})\mapsto\Psi(\zeta)\ ,\qquad u=(\chi,a,e_i)\in\Pi^{-1}(x)\subset B\mathbf{M}\ ,\tag{3.2b}$$

$$\boldsymbol{\zeta}=(\chi,a+q^i e_i,v^A e_A)\in\mathbf{R}^1\times T_x\mathbf{M}\times T_x\mathbf{M}\ ,\qquad \zeta=(\chi,q,\mathbf{v})\in\mathbf{R}^8\ ,\tag{3.2c}$$

where the complex number $\boldsymbol{\Psi}(\boldsymbol{\zeta})$ can be regarded as the frame-independent value of $\boldsymbol{\Psi}\in\mathbf{F}_x$ at any given $\boldsymbol{\zeta}\in\mathbf{R}^1\times T_x\mathbf{M}\times T_x\mathbf{M}$ with $v^0=0$, to which the wave function amplitude $\Psi(\zeta)$ is then assigned from the standard fibre. We observe that, in the transition from one choice of section of $B\mathbf{M}$ to another, these wave functions are related in the following manner,

$$\sigma_x^{u'}\circ(\sigma_x^u)^{-1} : \Psi\mapsto\Psi'=U_\xi(\theta(x),b^0(x),\mathbf{b}(x),\mathbf{v}(x),R(x))\Psi\ ,\tag{3.3a}$$

$$(\chi(x),a(x),e_i(x))=(\chi'(x),a'(x),e_i'(x))\cdot(\theta(x),b^0(x),\mathbf{b}(x),\mathbf{v}(x),R(x))\ ,\tag{3.3b}$$

where the representation in (3.3a) is the extension of the ray representation in (3.2.16) of the Galilei group into the following representation of the Bargmann group,

$$U_\xi(\theta,b^0,\mathbf{b},\mathbf{v},R)=\exp(im\theta)\mathbf{P}_\xi U(b^0,\mathbf{b},\mathbf{v},R)\ :\quad \Psi(\chi,q^0,\mathbf{q},\mathbf{p})$$

$$\mapsto\ \Psi'(\chi,q^0,\mathbf{q},\mathbf{p})=\Psi(\chi',q^0-b^0,R^{-1}[\mathbf{q}-\mathbf{b}-\mathbf{v}(q^0-b^0)],R^{-1}[\mathbf{p}-m\mathbf{v}]),\tag{3.4a}$$

$$\chi' = \chi + m(\theta - \tfrac{1}{2}\mathbf{v}^2(q^0 - b^0) + \mathbf{v}\cdot(\mathbf{q} - \mathbf{b})) \ , \tag{3.4b}$$

obtained in accordance with (3.1.18) and (3.2.7). Thus, the fibres $\mathbf{F}_x$ of $\mathbf{E}$ consist, at any $x \in \mathbf{M}$, of the equivalence classes of such wave functions constructed in accordance with (3.2). Hence, for each choice of section s of the Bargmann frame bundle $B\mathbf{M}$, the soldering maps in (3.2) provide the local trivialization maps generically defined in (2.2.4), so that (3.3a) gives rise to (2.2.6). Thus, the Newton-Cartan quantum bundle $\mathbf{E}$ indeed has the Bargmann group, defined in the preceding section, as its structure group, and it is also associated to the principle bundle $B\mathbf{M}$ in the sense of the definition in Sec. 2.2.

The fibres $\mathbf{F}_x$ of $\mathbf{E}$ are Hilbert spaces with inner products defined as follows,

$$\langle \boldsymbol{\Psi}_1 | \boldsymbol{\Psi}_2 \rangle = \int_{\Sigma_m} \Psi_1^*(\zeta)\Psi_2(\zeta)\, d\Sigma_m(\zeta) \ , \qquad \Psi_1 = \boldsymbol{\sigma}_x^u \boldsymbol{\Psi}_1 \ , \quad \Psi_2 = \boldsymbol{\sigma}_x^u \boldsymbol{\Psi}_2 \ , \tag{3.5}$$

where the integration can be carried out along any of the following surfaces,

$$\Sigma_m = \left\{ \zeta \,|\, \zeta = (\chi, q, \mathbf{v}), \quad \chi, q^0 = \text{const.} \right\} \ , \qquad d\Sigma_m(\zeta) = m^3 d^3\mathbf{q}\, d^3\mathbf{v} \ , \tag{3.6}$$

with respect to the above measure. Those elements $\boldsymbol{\Psi}$ of each fibre which are normalized with respect to the above inner product will be called *local state vectors*.

For each choice of Bargmann frame $\boldsymbol{u} \in B\mathbf{M}$, the family

$$\left\{ \boldsymbol{\Phi}_\zeta^u = (\boldsymbol{\sigma}_x^u)^{-1} U_\xi(\zeta, I)\xi \,\middle|\, \zeta = (\chi, q^0, \mathbf{q}, \mathbf{v}) \in \mathbf{R}^8 \right\} \ , \qquad \boldsymbol{u} \in \Pi^{-1}(x) \subset B\mathbf{M} \ , \tag{3.7}$$

of generalized coherent states within $\mathbf{F}_x$ is called a *quantum frame* at $x \in \mathbf{M}$ (cf. Sec. 3.7). According to (3.2.13)-(3.2.15), and in view of the unitarity of the soldering maps, any vector in $\mathbf{F}_x$ can be expanded in terms of the elements of such a quantum frame as follows:

$$\boldsymbol{\Psi} = \int_{\Sigma_m} d\Sigma_m(\zeta)\Psi_x^u(\zeta)\boldsymbol{\Phi}_\zeta^u \ , \qquad \boldsymbol{\Psi} \in \mathbf{F}_x \ , \tag{3.8a}$$

$$\Psi_x^u(\zeta) = \langle \boldsymbol{\Phi}_\zeta^u | \boldsymbol{\Psi} \rangle \ , \qquad \Psi_x^u = \boldsymbol{\sigma}_x^u \boldsymbol{\Psi} \in \mathbf{F} \ . \tag{3.8b}$$

The functions in (3.8b) will be therefore referred to as the *coordinate wave functions* (cf. Sec. 3.9) of the local vector $\boldsymbol{\Psi}$ with respect to the given quantum frame (3.7). It should be noted that on account of the properties of reproducing kernel Hilbert spaces which underlie the basic structure of the fibres $\mathbf{F}_x$, all coordinate wave functions are continuous functions in the index variables $\zeta \in \mathbf{R}^8$. As a matter of fact, for such choices of proper state vector as the optimal ones in (3.6.6), they are also smooth functions of ζ.

The soldering of the quantum frames to the Bargmann frames in $B\mathbf{M}$, which is implicit in (3.7), supplies a one-to-one map between $B\mathbf{M}$ and the set $Q\mathbf{M}$ of all such quantum frames, which is such that $Q\mathbf{M}$ assumes in a natural manner the structure of a *quantum frame bundle* that is isomorphic [C,I] to $B\mathbf{M}$. Hence parallel transport can be defined first within $Q\mathbf{M}$ on the basis of (2.6),

$$\tau_\gamma(x'',x') \ : \quad \boldsymbol{\Phi}_\zeta^{u_{x'}} \quad \mapsto \quad \boldsymbol{\Phi}_\zeta^{u_{x''}} \ , \qquad\qquad \forall\zeta \in \mathbf{R}^8 \ , \tag{3.9a}$$

and then extended to the quantum bundle **E** in accordance with (3.8):

$$\tau_\gamma(x'',x') \ : \quad \boldsymbol{\Psi}' = \int d\Sigma_m(\zeta)\Psi(\zeta)\boldsymbol{\Phi}_\zeta^{u_{x'}} \quad \mapsto \quad \boldsymbol{\Psi}'' = \int d\Sigma_m(\zeta)\Psi(\zeta)\boldsymbol{\Phi}_\zeta^{u_{x''}} \ . \tag{3.9b}$$

This definition of parallel transport of local quantum state vectors $\boldsymbol{\Psi}$ is easily seen to be frame independent on account of the Bargmann gauge covariance of the entire framework.

For any given open domain D on which a section s of the Bargmann bundle $B\mathbf{M}$ is defined, a *section of the Newton-Cartan quantum bundle* **E** is defined as a map that assigns to each $x \in D$ a vector $\boldsymbol{\Psi}_x \in \mathbf{F}_x$ in such a manner that the coordinate wave functions in (3.8b) that correspond to $\boldsymbol{u} = s(x)$ are smooth functions of $x \in D$ for each fixed choice of the index variables $\zeta \in \mathbf{R}^8$. This definition is obviously Bargmann-gauge independent, i.e., independent of the choice of section s of the Bargmann bundle $B\mathbf{M}$, or, equivalently, of the choice of section over D of the quantum frame bundle $Q\mathbf{M}$.

The *covariant derivative* of such sections $\{\boldsymbol{\Psi}_x \mid x \in D\}$ of **E** can be now defined, in accordance with (2.4.11), by the following strong limit[9]:

$$\nabla_X \boldsymbol{\Psi}_x = \text{s-}\lim_{t\to 0} \frac{1}{t}\left[\tau_\gamma(x,x(t))\boldsymbol{\Psi}_{x(t)} - \boldsymbol{\Psi}_x\right] \ , \quad x(0) = x \in \mathbf{M} \ , \ \dot{x}(0) = X \in T_x\mathbf{M} \ . \tag{3.10}$$

By specializing the general arguments leading to (2.5.22) and (2.5.23), we obtain upon using (2.2) that, for any given choice of section s of the Bargmann bundle $B\mathbf{M}$,

$$\nabla_X = \partial_X + \tilde{\omega}^j(s_*X)\hat{\mathbf{f}}_{j;u} + \omega^A(s_*X)\hat{\mathbf{F}}_{A;u} + \tfrac{1}{2}\omega^{AB}(s_*X)\hat{\mathbf{F}}_{AB;u} + \omega^*(s_*X)\hat{\mathbf{f}}_* \ . \tag{3.11}$$

The operators on the right-hand side of (3.11) are directly related to the self-adjoint infinitesimal generators of the unitary representation

$$U_{x;u}(\theta,b,\mathbf{a},\mathbf{v},R) = (\sigma_x^u)^{-1}U_\xi(\theta,b,\mathbf{a},\mathbf{v},R)\sigma_x^u \ , \qquad\qquad u = s(x) \ , \tag{3.12}$$

of the Bargmann group, which correspond, in accordance with (2.5.21), to a basis in the 11-dimensional Lie algebra of the Bargmann group.

From the physical point of view, the natural choices for these infinitesimal generators are the generators of spacetime translations, Galilean velocity boosts, spatial rotations and phase changes. These generators can be easily expressed as differential operators (cf. (3.6.15)) by using the chain rule in the partial differentiation of the wave function in (3.4), that results from the action of the representation of the Bargmann group, with respect to the parameters θ, b^i and v^A, as well as with respect to the angles of rotation in the (AB)-planes:

$$\hat{\mathbf{f}}_{j;u} = -iP_{j;u} = -\partial/\partial q^j \ , \quad \hat{\mathbf{F}}_{A;u} = imQ_{A;u} = im(q^A + i\partial/\partial p^A) - q^0\,\partial/\partial q^A \ , \tag{3.13a}$$

$$\hat{\mathbf{F}}_{AB;u} = -iJ_{AB;u} = -i(Q_{A;u}P_{B;u} - Q_{B;u}P_{A;u}) \ , \qquad \hat{\mathbf{f}}_* = imf_* = m\,\partial/\partial\chi \ . \tag{3.13b}$$

Naturally, in case of (3.13), all of these generators are computed in relation to the local Bargmann frame $\boldsymbol{u}$, rather than in relation to global frames in flat Newtonian spacetime.

In the abbreviated notation introduced in (2.5.23), the operator of covariant differentiation (i.e., the Koszul connection) assumes the form

$$\nabla = \boldsymbol{d} + i(-\tilde{\omega}_S^j P_{j;u} + m\omega_S^A Q_{A;u} - \tfrac{1}{2}\omega_S^{AB} J_{AB;u} + m\omega_S^* f_*) \ . \tag{3.14}$$

The infinitesimal generators in (3.14) are self-adjoint but, in general, unbounded operators, whose domains of definition are therefore dense in $\mathbf{F}_x$, but cannot coincide with $\mathbf{F}_x$ on account of the Hellinger-Toeplitz theorem [PQ]. Consequently, the strong limit in (3.10) does not exist for an arbitrary section. Rather, a sufficient condition for the existence of that limit, at all points in the domain of definition D of some given section $\{\Psi_x\}$, is that Ψ_x belong to a common dense core[10] of all the (self-adjoint) infinitesimal generators in (3.13) whose coefficients in (3.11) do not vanish.

It follows from (3.8) and (3.9a) that the operator for parallel transport in (3.9b) is unitary for any smooth curve γ. In other words, the parallel transport in the Newton-Cartan quantum bundle $\mathbf{E}$ is *compatible with the metric* determined by the inner product (3.5) in the fibres of $\mathbf{E}$. From this fact, or alternatively from the self-adjointness of the infinitesimal generators in (3.13), we can deduce that, within the common domain of definition of any two sections of $\mathbf{E}$, we can write down an equation which is analogous in outward appearance to the one in (2.6.22):

$$\left\langle \nabla_X \Psi_x \middle| \Psi'_x \right\rangle + \left\langle \Psi_x \middle| \nabla_X \Psi'_x \right\rangle = X\left\langle \Psi_x \middle| \Psi'_x \right\rangle \ . \tag{3.15}$$

However, it should be noted that, whereas (2.6.22) holds at all points in the common domain of definition of two sections of TM, (3.15) holds only at those points x in the common domain of definition of the two sections of $\mathbf{E}$ where the above two covariant derivatives are defined.

In this manner, we have established on the Newton-Cartan quantum bundle $\mathbf{E}$ a quantum geometry incorporating a notion of parallel transport which shares all the basic features of its classical counterpart in Sec. 4.1. We now turn to the task of formulating a consistent quantum dynamics within the context of such geometries.

*4.4. Geometro-Stochastic Propagation in Quantum Newton-Cartan Bundles

The quantum Newton-Cartan geometry introduced in the preceding section admits a notion of local quantum state vector Ψ for which there is no counterpart in either conventional QM or in SQM. Naturally, in case that, as mentioned in Sec. 4.1, a unique Newtonian potential defined on a corresponding unique class of Galilean inertial coordinates has been singled out on the basis of additional global assumptions, and if the resulting Newtonian spacetime with flat connection is identifiable with $\mathbf{R}^3 \times \mathbf{R}^1$, then a bridge can be built between GS quantum theory and SQM formulated on the same differential manifold $\mathbf{M}$.

Let us first make the above notion of identification more precise by comparing the salient features of the two classical spacetimes, which in that case exist side by side.

Under the above stipulated global conditions, the Newton-Cartan spacetime and the Newtonian spacetime both share the same differential manifold **M**, which can be foliated in both cases into the same equal-time hypersurfaces:

$$\mathbf{M} = \bigcup\nolimits_{t=-\infty}^{+\infty} \sigma_t \quad , \qquad \sigma_t = \left\{ x \,\middle|\, (x^1, x^2, x^3) = \mathbf{x},\ x^0 = t \right\} \leftrightarrow \mathbf{R}^3 \ . \tag{4.1}$$

Furthermore, the diffeomorphisms indicated in (4.1) for each $t \in \mathbf{R}^1$ can be constructed in both cases by envisaging *coherent flows* of classical test particles. By definition, such a flow is obtained by locating a point test particle at each point $\mathbf{x}$ of an initial-data hypersurface σ_0, imparting to all these particles the same 3-velocity (in the sense that the resulting velocity field in σ_0 has zero covariant derivatives, for the flat connection, in any direction tangential to σ_0), and then letting all these test particles move in free fall along the geodesics of the connection in the respective spacetime. In this manner coordinates can be assigned to points $x \in \mathbf{M}$, by first assigning to all $x \in \sigma_0$ a set $\mathbf{x}$ of coordinates that are Euclidean (in the sense that they are Riemannian normal coordinates with respect to the space metric $\boldsymbol{\gamma}$, and that their coordinate lines are geodesics with respect to the flat connection), and then using the coherent flow to assign corresponding coordinates to points $x \in \sigma_t$ for various choices of $t \in \mathbf{R}^1$. In this manner, for each choice of constant 3-velocity field $\mathbf{v}$ and of Euclidean coordinates along σ_0, a system of coordinates will result in **M**, which in the case of a New-tonian spacetime are the *inertial Galilean coordinates* used in Secs. 3.1-3.2 and 3.5-3.6. In the case of a Newton-Cartan spacetime, they were first mentioned towards the end of Sec. 4.1, and will be henceforth called *inertial Newton-Cartan coordinates*.

In the case of a Newtonian spacetime, the connection is flat regardless of whether or not gravitational sources are present, and the inertial Galilean coordinates, resulting from coherent flows for the same choice of Euclidean coordinates along σ_0 but distinct choices of 3-velocity $\mathbf{v}$, are related by the globally defined linear Galilean transformations in (3.1.12). In the case of a Newton-Cartan spacetime, the connection is not flat in the presence of gravitational sources, and although the resulting parallel transports preserve the spatial distance determined by the space metric $\boldsymbol{\gamma}$, they do not preserve the Euclidean spatial distance defined by means of the inertial Galilean coordinates. Hence, although both types of flows can give rise to diffeomorphisms between **M** and $\mathbf{R}^3 \times \mathbf{R}^1$, these diffeomorphisms are not identical, and it is only the first one that can be viewed as giving rise to an *identification* of **M** and $\mathbf{R}^3 \times \mathbf{R}^1$ in all the pertinent geometric aspects.

Physically, the first type of coherent flow takes place along geodesics that are straight lines, and as such could correspond to *actual* classical motion of isolated nonrelativistic test particles only in the total absence of gravitational forces. When gravitational sources are present, as is the case in the real world around us, it is only the second type of coherent flow that can take place in the form of free-fall motion along the generally curved geodesics of the given Newton-Cartan connection.

A corresponding type of motion of local quantum state vectors in a quantum Newton-Cartan bundle **E** could be envisaged as given by

$$\boldsymbol{\Psi}_{x(\tau)} = \tau_\gamma(x(\tau), x(0))\, \boldsymbol{\Psi}_{x(0)} \quad , \qquad \tau \in [0, \infty) \ , \tag{4.2}$$

if γ is a timelike geodesic in **M**. However, instead of defining such a motion in purely geometric terms, we can also view it as being governed by a propagator for parallel transport, whose definition we shall now provide.

In keeping with the earlier comparison of classical propagation in the Newtonian and Newton-Cartan cases, and in accordance with (3.6.2), the *propagator for parallel transport* along γ can be defined, for an arbitrary smooth curve $\gamma = \{x(t) | \, a \leq t \leq b\}$ starting at $x' = x(a) \in \mathbf{M}$ and ending at $x'' = x(b) \in \mathbf{M}$, and in any choice of Bargmann gauge, as follows:

$$K_\gamma(x'',\zeta'';x',\zeta') = \left\langle \Phi_{\zeta''}^{u(x'')} \middle| \tau_\gamma(x'',x')\Phi_{\zeta'}^{u(x')} \right\rangle \; . \tag{4.3}$$

It can be then immediately deduced from (3.7) that

$$\Psi_{x(\tau)}(\zeta) = \int K_\gamma\big(x(\tau),\zeta;x(a),\zeta'\big)\, \Psi_{x(a)}(\zeta')d\Sigma_m(\zeta') \; , \tag{4.4}$$

where the integration can be carried out along any of the hypersurfaces (3.6), that lie within $\mathbf{R}^1 \times T_{x'}\mathbf{M} \times T_{x'}\mathbf{M}$. Furthermore, upon setting $\tau_\gamma(x'', x') = \tau_\gamma(x'', x)\, \tau_\gamma(x, x')$ in (4.3), and then performing at x similar integrations, we can write, in analogy with (3.6.4),

$$\begin{aligned} K_\gamma(x'',\zeta'';x',\zeta') &= K_\gamma^*(x',\zeta';x'',\zeta'') \\ &= \int K_\gamma(x'',\zeta'';x,\zeta)K_\gamma(x,\zeta;x',\zeta')\,d\Sigma_m(\zeta) \; . \end{aligned} \tag{4.5a}$$

Consequently, by iterating (4.5a), we obtain a counterpart of (3.6.14):

$$K_\gamma(x'',\zeta'';x',\zeta') = \lim_{\varepsilon \to +0} \int K_\gamma(x_N,\zeta_N;x_{N-1},\zeta_{N-1}) \times \prod_{n=N-1}^{1} K_\gamma(x_n,\zeta_n;x_{n-1},\zeta_{n-1})\,d\Sigma_m(\zeta_n) \; . \tag{4.5b}$$

Despite these formal mathematical similarities between the SQM propagators of Sec. 3.6 and the propagators for parallel transport defined in (4.3), the latter cannot possess a *direct* physical significance, as was the case with the former. Indeed, according to the heuristic interpretation (Feynman and Hibbs, 1965, pp. 31-38) of the Feynman propagators in (3.6.1a), they do not emerge from the propagation of the initial state in (3.6.1b) along a single smooth path, but rather along all possible broken paths consisting of straight line segments – i.e., in geometric language, along all possible broken paths built from geodetic segments belonging to timelike geodesics of the flat connection in a Newtonian spacetime. On account of (3.6.14), this interpretation can be also retained for SQM propagators, except that in this latter case it is the proper state vector of an extended quantum object, rather than of a point particle (cf. Hartle and Kuchař, 1984), that propagates in this manner. On the other hand, the propagator in (4.3) relates to parallel transport along a single smooth path, and as a consequence the integrations in (4.5) do not involve hypersurfaces embedded in the spacetime manifold $\mathbf{M}$, but rather hypersurfaces embedded in the tangent spaces $T_x\mathbf{M}$ at points along the smooth curve γ.

Let us therefore adapt the basic physical, as well as mathematical, features of the expression (3.6.18) for the general nonrelativistic SQM propagator in the presence of external non-gravitational fields to the present situation of quantum propagation within a

Newton-Cartan bundle, by slicing the section in Newtonian spacetime between two hypersurfaces $\sigma_{t'}$ and $\sigma_{t''}$ in (4.1) containing the points x' and x'', respectively, with hypersurfaces σ_{t_n} corresponding to $t' = t_0 < t_1 < \cdots < t_N = t''$, and then connecting pairs of points x_{n-1} and x_n on two successive slices $\sigma_{t_{n-1}}$ and σ_{t_n} by the timelike geodesics of the corresponding Newton-Cartan spacetime, rather than by the straight lines which represent the timelike geodesics of the coexisting flat Newtonian spacetime. By carrying out an averaging procedure over all such broken geodetic paths, and in the end going to the limit $\varepsilon = \max(t_n - t_{n-1}) \to +0$, we arrive at the following *geometro-stochastic propagator* from $x' \in \mathbf{M}$ to the future Newton-Cartan base location $x'' \in \mathbf{M}$,

$$\hat{\boldsymbol{K}}(x'',\zeta'';x',\zeta') = \lim_{\varepsilon\to+0}\int d\Sigma_m(\zeta'_N)\prod_{n=N-1}^{1} d^3x_n\, d\Sigma_m(\zeta_n)d\Sigma_m(\zeta'_n)$$

$$\times\prod_{n=N}^{1}\delta^4(q_n + a(x_n))\left\langle \boldsymbol{\Phi}^{\boldsymbol{u}(x_n)}_{\zeta_n}\right| e^{-iH_I(x_n)(t_n - t_{n-1})}\left.\boldsymbol{\Phi}^{\boldsymbol{u}(x_n)}_{\zeta'_n}\right\rangle K(x_n,\zeta'_n;x_{n-1},\zeta_{n-1}) . \quad (4.6)$$

In this expression, $K(x_n,\zeta_n;x_{n-1},\zeta_{n-1})$ denotes the propagator (4.3) for parallel transport along the timelike geodesic $\gamma(x_{n-1},x_n)$ connecting the points x_{n-1} and x_n on the two consecutive slices $\sigma_{t_{n-1}}$ and σ_{t_n}, the integration in x_n is carried out over the entire hypersurface σ_{t_n} , and $H_I(x_n)$ is the interaction Hamiltonian with external nongravitational sources expressed in the interaction picture [PQ] in relation to the affine Galilei frame at x_n, whereas the presence of

$$\delta^4(q_n + a(x_n)) \ , \qquad \zeta_n = (\chi(x_n), a(x_n) + q^i_n e_i(x_n), v^A_n\, e_A(x_n)) \ , \quad (4.7)$$

ensures that such an interaction takes place at the point of contact between the manifold **M** and the tangent space $T_x\mathbf{M}$. This fact becomes very evident if we perform the integrations in the q_n variables, introduce inertial Galilei coordinates, and write (4.6) in operator notation:

$$\hat{\boldsymbol{K}}(x'',\zeta'';x',\zeta') = \lim_{\varepsilon\to+0}\int d\Sigma_m(\zeta'_N)\prod_{n=N-1}^{1} d^3\mathbf{x}_n\, d^3\mathbf{p}_n\, d\Sigma_m(\zeta'_n)$$

$$\times\prod_{n=N}^{1}\left\langle \boldsymbol{\Phi}^{\boldsymbol{u}(x_n)}_{\hat{\zeta}_n}\right| e^{-iH_I(x_n)(t_n - t_{n-1})}\left.\boldsymbol{\Phi}^{\boldsymbol{u}(x_n)}_{\zeta'_n}\right\rangle\left\langle \boldsymbol{\Phi}^{\boldsymbol{u}(x_n)}_{\zeta'_n}\right|\boldsymbol{\tau}_{\gamma(x_{n-1},x_n)}(x_n,x_{n-1})\left.\boldsymbol{\Phi}^{\boldsymbol{u}(x_{n-1})}_{\hat{\zeta}_{n-1}}\right\rangle, \quad (4.8a)$$

$$\hat{\zeta}_0 = \zeta' \ , \qquad \hat{\zeta}_N = \zeta'' \ , \qquad p^A_n = m v^A_n \ , \qquad n = 1,2,\ldots,N-1 \ , \quad (4.8b)$$

$$\hat{\zeta}_n = (\chi(x_n), -a^i(x_n)e_i(x_n), v^A_n\, e_A(x_n)) \ , \qquad n = 1,2,\ldots,N-1 \ . \quad (4.8c)$$

In the absence of an external gravitational field parallel transport in $Q\mathbf{M}$ becomes path independent. Hence, upon comparing (4.6) with (3.6.18) rewritten in the following form,

$$\hat{K}_\xi(\mathbf{x}'',\mathbf{p}'',t'';\mathbf{x}',\mathbf{p}',t') = \lim_{\varepsilon\to+0}\int d^3\mathbf{q}'_N d^3\mathbf{p}'_N \prod_{n=N-1}^{1} d^3\mathbf{q}'_n d^3\mathbf{p}'_n d^3\mathbf{x}_n d^3\mathbf{p}_n$$

$$\times\prod_{n=N}^{1}\left\langle \xi_{\mathbf{x}_n,\mathbf{p}_n,t_n}\right| e^{-i\hat{H}_I(t_n)(t_n - t_{n-1})}\left.\xi_{\mathbf{q}'_n,\mathbf{p}'_n,t_n}\right\rangle K_\xi(\mathbf{q}'_n,\mathbf{p}'_n,t_n;\mathbf{x}_{n-1},\mathbf{p}_{n-1},t_{n-1}) , \quad (4.9a)$$

$$\hat{H}_I(t) = \exp(iH_0t)H_I(t)\exp(-iH_0t) \ , \qquad t \in \mathbf{R}^1 \ , \tag{4.9b}$$

in which the interaction Hamiltonian is expressed in the interaction picture, we immediately see that, in view of (3.1)-(3.2), the two propagators coincide in this case modulo the phase factor $\exp[i(\chi(t_n)-\chi(t_{n-1}))]$.

On the other hand, we observe that the physical roles of the **x**-variables and the q^A-variables are now different. Indeed, in the SQM approach the **q**-variables were inertial Galilei stochastic coordinates which marked the mean position of the stochastic values in (3.2.21), and the $(\mathbf{x},t)$-variables marked the points of a classical Newtonian spacetime. In the present GS approach the q^A-variables are internal gauge variables related to a choice of section of the Bargmann bundle, and therefore also of the affine Galilei bundle. As such, they no longer label points on hypersurfaces σ_t in the base manifold **M**, but rather points in the tangent space at some $x \in \sigma_t$. It is only via the exponential map defined by the Newton-Cartan connection that they can be uniquely mapped into points in some neighborhood of a base location $x \in \sigma_t$ in a manner that approximately preserves the spatial metric relationships in the corresponding neighborhood of the point of contact between the tangent space $T_x\sigma_t$ and σ_t. In this manner, the q^A-variables can be related to *geometro-stochastic fluctuations* around the base location $x \in \mathbf{M}$ reflecting the fact that the points of a quantum Newton-Cartan spacetime are *not* the points of a finite-dimensional manifold, but rather of an infinite-dimensional manifold, which in addition to classical degrees of freedom also incorporates quantum degrees of freedom directly in its geometry.

The presence of additional ordinary *quantum fluctuations* emerges from the natural interpretation of the *GS wave function* $\{\boldsymbol{\Psi}_{x(t)} \mid x(t) \in \sigma_t\}$ obtained, in a quantum Newton-Cartan spacetime, by the propagation of a local state vector $\boldsymbol{\Psi}_{x(0)}$ from a base location $x(0) \in \sigma_0$ to hypersurfaces σ_t in its future, where the coordinate wave functions will be

$$\Psi_{x(t)}(\zeta) = \int \hat{\boldsymbol{K}}(x(t),\zeta;x(0),\zeta')\,\Psi_{x(0)}(\zeta')d\Sigma_m(\zeta') \ . \tag{4.10}$$

Comparing with the flat Newtonian spacetime situation, we see that

$$\psi(\chi,x(t),\mathbf{p}) = \left\langle \boldsymbol{\Phi}^{u(x(t))}_{\hat{\zeta}(x(t))} \middle| \boldsymbol{\Psi}_{x(t)} \right\rangle, \quad \hat{\zeta}(x(t)) = (\chi, -a^i(x(t)), v^A) \ , \quad \mathbf{p} = m\mathbf{v} \ , \tag{4.11}$$

should be interpreted as a conditional probability amplitude, i.e., that

$$\int_B |\psi(\chi,x(t),\mathbf{p})|^2\, d^3\mathbf{x}\, d^3\mathbf{p} \ , \qquad x(t) \leftrightarrow (\mathbf{x},t) \ , \qquad B \subset \mathbf{R}^6 \ , \tag{4.12}$$

represents the conditional probability that, provided a local state $\boldsymbol{\Psi}_{x(0)}$ were prepared at $x(0) \in \sigma_0$, then the quantum object of mass m and zero spin[11] originally in that state would be found, by measurements of stochastic position and momentum performed at the global Galilean time t with test particles of mass m and proper state vectors ξ, to display phase space values with inertial Galilei coordinates within the Borel subset B.

It now remains to investigate how the GS propagator in (4.6) compares with the SQM propagator in (4.9a) for an interaction Hamiltonian containing the gravitational potential ϕ in addition to a potential V from non-gravitational sources. In the inertial Galilei coordinates x^A, $A = 1,2,3$, which in Sec. 3.6 were incorporated into the triple $\mathbf{q} \in \mathbf{R}^3$, such an interaction is given by

$$H_I(t) = V(\mathbf{Q},t) + m\,\phi(\mathbf{Q},t) \ , \qquad Q^A = x^A + i\partial/\partial p^A \ , \quad A = 1,2,3 \ . \tag{4.13}$$

In view of the fact that the quantum Newton-Cartan framework is Galilei gauge invariant, we can carry out the comparison in any choice of section of the Galilei frame bundle $H\mathbf{M}$. Hence, let us choose the holonomic frame $\{\boldsymbol{\partial}_\mu \mid \mu = 0,1,2,3\}$, which corresponds to partial derivatives with respect to the inertial Galilei variables in some global Galilei inertial frame. This frame can be extended into a holonomic Bargmann frame with $a^\lambda \equiv 0$, $\boldsymbol{e}_\mu \equiv \boldsymbol{\partial}_\mu$ and $\chi \equiv 0$, in which the Newton-Cartan connection coefficients in (3.14) are therefore given, in accordance with (1.11) and (2.4), by

$$\tilde{\omega}^i_s = dx^i \ , \qquad \omega^A_s = (\partial\phi/\partial x^A)\, dx^0 \ , \qquad \omega^{AB}_s = 0 \ , \qquad \omega^*_s = \phi\, dx^0 \ . \tag{4.14}$$

Let us now consider the parallel transports along the timelike geodesic $\gamma(x_{n-1},x_n)$ of the $n-1$ quantum frame in (4.6),

$$\boldsymbol{\Phi}_{n-1}(x_n(\tau)) = \boldsymbol{\tau}_{\gamma(x_{n-1},x_n)}(x_n(\tau),x_{n-1})\boldsymbol{\Phi}^{u(x_{n-1})}_{\hat{\zeta}_{n-1}} \ . \tag{4.15}$$

If τ denotes the affine parameter for that geodesic, then the covariant derivative of (4.15) is equal to zero everywhere along $\gamma(x_{n-1},x_n)$. Hence, from (3.14) and (4.14), we get

$$\partial_\tau \boldsymbol{\Phi}_{n-1} = i(-\dot{x}^\mu_n P_{\mu;u} + m(\partial\phi/\partial x^A)_{x_n}\dot{x}^0_n Q_{A;u} + m\phi(x_n)\dot{x}^0_n)\,\boldsymbol{\Phi}_{n-1} \ , \quad \partial_\tau = \partial/\partial\tau, \tag{4.16}$$

upon setting $q^0 = 0$ with respect to the frame $\boldsymbol{u} = \boldsymbol{u}(x_n(\tau))$ that belongs to the chosen section of the Bargmann frame bundle $B\mathbf{M}$ at a given $x_n(\tau) \in \gamma(x_{n-1},x_n)$. In the limit $\varepsilon = \max(t_n - t_{n-1}) \to +0$, we can therefore insert in (4.8a)

$$\begin{aligned}\boldsymbol{\tau}_{\gamma(x_{n-1},x_n)}(x_n,x_{n-1})\boldsymbol{\Phi}^{u(x_{n-1})}_{\hat{\zeta}_{n-1}} &= \exp[i(x^A_n - x^A_{n-1})P_A(x_n) \\ -i(H_0(x_n)+m\phi(x_n)+m(\partial\phi/\partial x^A)Q_A(x_n))&(t_n - t_{n-1}) + O((t_n - t_{n-1})^2)]\boldsymbol{\Phi}^{u(x_{n-1})}_{\hat{\zeta}_{n-1}},\end{aligned} \tag{4.17}$$

where all the infinitesimal generators are computed at the end point of each geodetic arc:

$$H_0(x_n) = -P_{0;u(x_n)} \ , \qquad P_A(x_n) = P_{A;u(x_n)} \ , \qquad Q_A(x_n) = Q_{A;u(x_n)} \ . \tag{4.18}$$

In performing now the comparison between (4.8a) and (4.9a), the following correspondences should be taken into account,

$$H_0(x_n) \leftrightarrow H_0 \ , \quad P_A(x_n) \leftrightarrow P^A = -i\partial/\partial x^A \ , \quad Q_A(x_n) \leftrightarrow Q^A - x^A_n \ , \tag{4.19a}$$

$$\exp[i(x_n^A - x_{n-1}^A)P_A(x_n)]\Phi_{\zeta_{n-1}}^{u(x_n)} \leftrightarrow \xi_{\mathbf{x}_{n-1},\mathbf{p}_{n-1},t_{n-1}} , \qquad \Phi_{\zeta_n'}^{u(x_n)} \leftrightarrow \xi_{\mathbf{q}_n',\mathbf{p}_n',t_n} . \quad (4.19b)$$

We note that the first of the identifications in (4.19b) involves a space translation, in view of the fact that the inertial Galilei coordinates x^A are taken with respect to a fixed global Galilei frame, whereas the chosen section of the Bargmann bundle contains only local inertial Galilei frames, whose origins are in all cases at the points of contact of the respective tangent spaces $T_x\mathbf{M}$ to the base manifold $\mathbf{M}$.

Assuming that the classical gravitational potential is given by an analytic function, we can expand its quantum counterpart into a power series around each point $x_n \in \mathbf{M}$, so that

$$\phi(\mathbf{Q},t) = \phi(x_n) + \phi_{,A}(x_n)(Q^A - x_n^A) + \cdots , \qquad \phi_{,A}(x_n) = (\partial\phi/\partial x^A)_{x_n} . \quad (4.20)$$

Consequently, the GS propagator in (4.8) corresponds to an SQM propagator in (4.9) with a truncated gravitational potential, obtained by neglecting in (4.20) quadratic and higher terms. Such an approximation is obviously satisfactory for gravitational forces which, for each $x_n \in \sigma_{t_n}$, do not vary significantly over spatial neighborhoods that are so small that the presence of the proper state vector in the fibre above that base location gives rise to geo-metro-stochastic spatial fluctuations that are comparable to their size.

For example, let us consider the optimally localized (in phase space) proper state vectors that correspond to (3.6.6). In accordance with (3.6.9), for any such choice of proper state vector, the elements of the quantum frame in (3.7) can be written out in the following explicit form:

$$\Phi_{\ell;\zeta}^{u(x)}(\zeta') = (2\pi)^{-3}(\ell^2/\ell_{m,\bar{q}^0}^2)^{3/2} \exp\left\{i(\chi-\chi') - [(\mathbf{q}-\mathbf{q}')^2/8\ell_{m,\bar{q}^0}^2]\right\}$$
$$\times \exp\left\{-(\ell^2/2\ell_{m,\bar{q}^0}^2)\left[\ell^2(\mathbf{p}-\mathbf{p}')^2 - i(\mathbf{q}-\mathbf{q}')\cdot(\mathbf{p}+\mathbf{p}') - (i\bar{q}^0/2m)(\mathbf{p}^2+\mathbf{p}'^2)\right]\right\}, \quad (4.21a)$$

$$\bar{q}^0 = q'^0 - q^0 , \quad \zeta = (\chi, q^\mu \partial_\mu(x), p^A \partial_A(x)/m) , \quad \zeta' = (\chi', q'^\mu, p'^A/m) . \quad (4.21b)$$

It is obvious from (4.21a) that the aforementioned approximation holds good if the external gravitational force has negligible variations over regions whose linear dimensions are of the same order as ℓ . This is indeed the case under normal observational circumstances if ℓ is of the same order of magnitude as the Planck length – as we shall assume to be the case in the remainder of this monograph. Moreover, upon performing on the free-fall propagators within (4.8a) the same type of renormalization as in taking the sharp-point limit in (3.6.10), as well as renormalizations compensating for the lack of stochastic momentum variables of integration in that limit, we obtain

$$(\pi/2\ell^2)^{3/2}\hat{K}_{\xi(\ell)}(x'',\zeta'';x',\zeta') \xrightarrow[\ell\to+0]{} \exp[i(\chi''-\chi')]\hat{K}(\mathbf{x}'',t'';\mathbf{x}',t') , \quad (4.22)$$

where the propagator on the right-hand side is the Feynman propagator in (3.6.12) with the interaction Hamiltonian in (4.13).

Thus, we can indeed claim that in the sharp-point limit nonrelativistic GS propagation coincides with conventional quantum propagation. This conclusion can be also reached more directly by formulating[12] the Feynman propagation for point particles in geometro-stochastic terms by means of the formulae (2.9)-(2.10) for the free-fall parallel transport of Bargmann frames, and then comparing the expressions (4.6)-(4.8) for the GS propagator with the thus geometrically reformulated path integral for the propagator in (3.6.12).

Notes to Chapter 4

1 As authoritatively recounted by Pais (1982), investigations of the agreement in weak gravitational fields between Newtonian theory and gravitational models that eventually were to become part of the CGR framework can be traced to a series of papers published by Einstein already in the ten-year period preceding his seminal 1916 paper. A central role in these early considerations was played by attempts to derive quantitative estimates of the perihelion precession [M,W] of the planet Mercury by supplying corrections to Newtonian theory based on the idea that gravitational interaction is transmitted with the finite speed of light, rather than instantaneously. The possibility of such an explanation for Mercury's perihelion precession can be traced to P. Gerber (1898), to whom E. Mach accredits that "from the perihelion motion of Mercury, forty-one seconds in a century, [he] finds the velocity of propagation [of gravitational interaction] to be the same as that of light" (Mach, 1907, p. 535).

2 As reviewed by Havas (1964), spacetime formulations of Newtonian mechanics were provided already by P. Frank (1909) and H. Weyl (1923) before Cartan's work. A condensed comparison between the Newton-Cartan and the Einstein classical theory of gravity can be found in Box 12.3 on p. 297 of [M].

3 There are no null vectors in the Newton-Cartan theory for obvious physical as well as mathematical reasons. However, Galilean manifolds, as well as corresponding Lorentzian manifolds, can be simultaneously embedded in 5-dimensional pseudo-Riemannian manifolds (Künzle and Duval, 1986), and in that manner they can be studied side by side. These 5-dimensional counterparts of Galilean manifolds are *Bargmann manifolds* whose fifth degree of freedom can be represented by coordinates x^5 corresponding to the phase variable θ in the central extension of the Galilei group (cf. (3.1.18)). The same interpretation can be also retained for the corresponding 5-dimensional counterparts of Lorentzian manifolds. For that reason, in the original formulation of Newton-Cartan quantum geometries (Prugovečki, 1987a), Bargmann rather than Galilei manifolds were used as base manifolds.

4 This result was first derived by Trautman (1963), who also imposed the conditions $\beta_i R^j{}_{klm} = \beta_k R^j{}_{ilm}$, which turned out to be redundant (Künzle, 1972, p. 352).

5 Whereas $\boldsymbol{a}$ and $\boldsymbol{e}_i$ assume values in the tangent spaces of $\mathbf{M}$, this is not the case with χ . However, as mentioned in Note 3, $\mathbf{M}$ can be embedded in a 5-dimensional Bargmann manifold $\mathbf{M}^5$, where there are charts for which the role of χ is taken over by vectors $\boldsymbol{e}_5 = \partial/\partial x^5$ that belong to the 5-dimensional tangent spaces of $\mathbf{M}^5$.

6 From a purely mathematical point of view, any basis in the present 11-dimensional Lie algebra could be chosen. The indexing as well as the choice of bases is, however, conditioned in the present context by physical considerations, which assign a physical meaning to the Lie algebra basis elements in (2.2), as will become clear in the next section – cf. (3.11)-(3.14).

7 This is the map which takes the pair $(\boldsymbol{u},\Psi) \in \mathbf{P} \times \mathbf{F}$ into the equivalence class $\boldsymbol{\Psi} \in \mathbf{P} \times_{\mathbf{G}} \mathbf{F}$ containing it – cf. Proposition 5.4 on p. 55 of (Kobayashi and Nomizu, 1963). Note that this reference deals with finite-dimensional manifolds and bundles, but that this result, as well as other results and definitions in it, remain valid in the case of infinite-dimensional manifolds and bundles (cf. Chapter VII in [C]), such as the quantum bundles in this monograph. Note also that in this reference, as well as in some other references (cf., e.g., Sec. 3.3 in [I]), the very *definition* of an *associated bundle* with typical fibre $\mathbf{F}$ and structure group $\mathbf{G}$ is that of a $\mathbf{G}$-product $\mathbf{P} \times_{\mathbf{G}} \mathbf{F}$, where $\mathbf{P}$ is a principal bundle with the same structure group $\mathbf{G}$.

8 The term "soldering map" is not usually used for the mapping in (3.2a), but it proved convenient in adapting the more specialized concept of "soldered bundle" in (Kobayashi, 1956, 1957; Drechsler, 1977-1990) to the quantum regime, by viewing the infinite-dimensional quantum bundles studied in this

monograph as "soldered" to the (tangent spaces of the) base manifold by such maps. In the previous papers dealing with quantum bundles (Prugovečki, 1987-89; Prugovečki and Warlow, 1989) the soldering maps were associated with sections s of the principal bundle **P**, rather than with the individual elements in it. The same pattern of construction could be followed in the present case by soldering the quantum frames directly to the Bargmann frames of the BM bundle in accordance with (3.7), and subsequently defining the wave functions Ψ in (3.2b), by means of (3.8b), as components of fibre coordinates. Such a method of construction will be actually presented from Chapter 6 onwards, once the *quantum* frame bundles become sufficiently familiar to be regarded as *the* fundamental principal bundles with which various other quantum bundles are associated, rather than as some auxiliary mathematical objects. We give here, as well as in Sec. 5.1, an alternative formulation for pedagogical reasons, and also because it is more in keeping with the definition in standard physics literature [C,I] of associated bundles as G-products $\mathbf{P} \times_G \mathbf{F}$, where **P** is a principal bundle consisting of classical rather than quantum frames. However, the former point of view actually becomes mandatory in the GS quantization of non-Abelian gauge fields, where the use of classical frame bundles becomes insufficient, and Grassmannian gauge degrees of freedom of quantum origin have to be introduced.

9 In all the cases treated in Chapter 2 the vector bundles had finite-dimensional fibres, so that the limits in (2.4.11), as well as (2.5.20), could be unambiguously defined as in vector calculus, i.e., by taking the corresponding limits for vector components in an arbitrary vector basis. In the present case the fibres are infinite-dimensional Hilbert spaces, so that a number of inequivalent topologies are of importance, of which the weak topology and the strong (or norm) topology [PQ] play the most significant roles. In a particular context, the choice of topology is conditioned by fundamental existence theorems. In the case of (3.10), the central theorem is a theorem which bears no specific name in the literature (in [PQ], p. 288, it is denoted as Theorem 3.1), but which is very closely related to the well-known Stone's theorem. This central theorem states that if A is a self-adjoint operator, then the vector-valued function $\exp(iAt)\psi$ will be differentiable in the strong sense at $t = 0$ (and therefore also at all other $t \in \mathbf{R}^1$) if and only if the vector ψ belongs to the domain of definition of A – which in case of unbounded A is only dense in the given Hilbert space. This theorem therefore ensures the existence of the strong limit in (3.10) under the subsequently discussed conditions.

10 A *core* of a self-adjoint operator A is a domain of essential self-adjointness, i.e., a linear space to which the restriction of A has a *unique* self-adjoint extension – which, of course, is A itself (cf. [PQ], p. 366, for a more general definition that applies to any closed operator). The most suitable cores for the covariant derivative operators in (3.11) are obtained by taking the linear spans of all elements of quantum frames belonging to some section of the quantum frame bundle QM, whose domain of definition contains the points where the covariant derivative is being computed – cf. Theorem A.2 in (Prugovečki and Warlow, 1989b).

11 As mentioned in Note 12 to Chapter 3, the case of an arbitrary choice of spin and mass can be treated by a suitable choice of SQM system of covariance.

12 This formulation and verification of (4.22) was explicitly carried out at the formal level by De Bièvre (1989a), who constructed a quantum Newton-Cartan bundle with $L^2(\mathbf{R}^3)$ as typical fibre, and (implicitly) used quantum frames $\Phi_{x;\mathbf{k}}(\chi,q^0,\mathbf{q})$ built from the plane waves $\Phi_\mathbf{k}(\mathbf{q}) = (2\pi)^{-3/2}\exp(i\mathbf{k}\cdot\mathbf{q})$ – cf. Eqs. (3.6*b,c*) on p. 735 of (De Bièvre, 1989a). Such a construction reflects the fact that in QM textbooks, a plane wave $\Phi_\mathbf{k}(\mathbf{x})$ is routinely interpreted as providing the "probability amplitudes" for observing a quantum point particle of 3-momentum $\mathbf{k}$ at the spatial location $\mathbf{x}$ in relation to a given inertial frame (which is conventionally always envisaged as being macroscopic and behaving in a totally classical manner). Of course, plane waves do not belong to $L^2(\mathbf{R}^3)$, but the resulting framework can be made mathematically more rigorous by using typical fibres which are rigged or equipped Hilbert spaces (Antoine, 1969; Prugovečki, 1973). However, the long-standing difficulties with the precise mathematical meaning of Feynman path integrals still remain even under those circumstances. Moreover, physical difficulties stemming from the uncertainty principle are also present, since the elements of such bundles, by their very definition, represent sharply localized states – and yet some of those elements are given in the form of plane waves which purportedly correspond to sharp values of 3-momentum. On the other hand, all these difficulties can be removed by regarding plane waves as representing sharp-point limits of quantum states giving rise to POV measures associated with systems of covariance in $L^2(\mathbf{R}^3)$ – namely of states represented by wave functions belonging to $L^2(\mathbf{R}^3)$, such as the ones in (3.6.7).

Chapter 5

Relativistic Klein-Gordon Quantum Geometries

In this chapter we shall adapt to the relativistic regime the construction of the nonrelativistic quantum bundles presented in the preceding chapter. The central idea in this adaptation is to cast in the role of standard fibres for quantum bundles the Hilbert spaces that carry the systems of covariance for the Poincaré group described in Sec. 3.4. The main reason for this choice of typical fibres is that, for physical as well as mathematical reasons, it is not possible to *consistently* combine the fundamental principles of general relativity with a fibre-theoretical framework based on the adoption of standard fibres that are equal to the Hilbert spaces of the conventional special relativistic quantum mechanics, and which carry the systems of imprimitivity for the Poincaré group described in Sec. 3.3.

Indeed, at the physical level, the equivalence principle and the concept of locality are fundamental to general relativity. Consequently, a fibre-theoretical framework for quantum general relativity has to be based on fibres containing *local* state vectors, as was the case even in the nonrelativistic context of the Newton-Cartan geometries studied in the preceding chapter. In case such local state vectors represent the quantum states of point particles, the existence of such states would entail the sharp localization of those particles at the points of the base spacetime manifold **M**. On the other hand, basic kinematical considerations, as well as basic quantum measurement schemes, require the adoption of fibres which consist of wave functions in the momentum representation. However, for a quantum point particle which is already known to be located at a given point $x \in \mathbf{M}$, the existence of local quantum states, represented by wave functions of *arbitrarily* narrow spreads in momentum space in a classical spacetime, is in obvious violation of the Heisenberg uncertainty principle[1].

At the mathematical level, the adoption of standard fibres equal to the Hilbert spaces of the conventional special relativistic quantum mechanics would entail the casting of the plane waves $\Phi_{x;k}(q)$ in the role of quantum frames[2] at each $x \in \mathbf{M}$. Already in the non-relativistic regime this gives rise to the mathematical difficulties, mentioned in Sec. 3.6, with path integrals for propagators of such quantum states. However, those difficulties can be ignored at the formal computational level; moreover, under certain physically stringent conditions, rigorous mathematical variations of those path integration methods have been devised[3]. On the other hand, in the special relativistic regime, the adoption of plane waves as the basic vehicle of quantum propagation leads to the well-known divergences of conventional quantum field theory, for which no effective cures have been found in the realistic case of 4-dimensional Minkowski space – albeit the renormalization programme has provided algorithms for computing numerical results. Furthermore, the presence of spacetime curvature gives rise to additional difficulties with formal renormalization schemes already in the semi-classical regime of quantum field theory in curved spacetime[4], and leads to an unending regressive sequence of renormalizations in the case of perturbative expansions in quantum gravity, thus making the renormalization idea totally ineffective in that context even as a purely computational tool[5].

For all of the above reasons, as well as the additional ones which were discussed at some length in Chapter 1, we shall not adopt as standard fibres of the quantum bundles in this chapter the Hilbert spaces of Sec. 3.3, but rather those of Sec. 3.4, with their corresponding irreducible representations of the Poincaré group and related systems of covariance. Of course, at first sight it might be thought that such a choice of fibres violates the general relativistic concept of locality, since we have seen in Chapter 3 that the quantum states in those Hilbert spaces are not sharply localizable in Minkowski space. However, from the point of view of its physical interpretation, the geometry of Minkowski space is a classical geometry. Hence, if we do not adhere to the dogma that the geometry of Nature is given *a priori* in the form of a *classical* geometry[6], then we can tackle the task of searching for alternative *quantum* geometries, in which fibre-theoretical adaptations of such quantum states are local by virtue of the mathematical structure and of the physical interpretation of those geometries.

The first step in that search was already undertaken in the preceding chapter in the nonrelativistic regime. However, in that context, such a step was not mandatory either from the mathematical or from the physical point of view. On the other hand, the mathematical as well as the physical ideas introduced in that chapter can be adapted to the relativistic regime, and, as we shall see in the remainder of this monograph, they lead to a host of relativistic quantum geometries, which cover the entire range of relativistic models which have thus far proven to be of importance in quantum physics. These geometries will be presented in the subsequent chapters in order of their increasing level of mathematical sophistication, as well as of their overall physical significance.

The Klein-Gordon quantum geometries studied in this chapter are the simplest both physically as well as mathematically. In introducing them, we follow the pattern set out in the preceding chapter. Thus, in Sec. 5.1 we define Klein-Gordon quantum bundles. In Sec. 5.2 we formulate parallel transport and connections in such bundles, and in Sec. 5.3 we study related geometric concepts. In Sec. 5.4 we introduce the concept of GS propagation, and in Sec. 5.5 we deal with its physical interpretation. In Secs. 5.6-5.7 we discuss alternative formulations of GS propagation, and their physical implications.

5.1. Klein-Gordon Quantum Bundles

We shall construct in this section a Klein-Gordon quantum bundle **E** over a given Lorentzian manifold **M** by the method developed in Sec. 4.3 in the context of quantum Newton-Cartan bundles. To underline the similarities, in this and the next section we shall follow the same order and method of exposition as in Sec. 4.3. On the other hand, this will eventually also bring out in stronger relief the differences between the nonrelativistic and relativistic case, due to the presence of the light-cone structure in the latter context.

We start by first choosing as standard fibre **F** the family of all positive-energy solutions (3.4.2) of the free Klein-Gordon equation which belong to one of the Hilbert spaces $\mathbf{P}_\eta L^2(\Sigma_m)$ described in Sec. 3.4,

$$\mathbf{F} = \left\{ \Psi \,\middle|\, \Psi(q,v) := \psi_{q^0}((0,\mathbf{q}),mv) \ , \ \ \psi_{q^0} = \exp(-iP_0 q^0)\psi_0 \in \mathbf{P}_\eta L^2(\Sigma_m) \right\} . \quad (1.1)$$

The above Hilbert space **F** can be also described, in keeping with the specialization of (3.4.10) to the spin-zero case, as consisting of all functions

$$\Psi(q,v) = Z_{f,m}^{-1} \int_{u^0>0} \exp[-imq\cdot u] f(u\cdot v) \tilde{\Psi}(u)\, d\Omega(u) \ , \quad (1.2)$$

obtained, for a yet unspecified renormalization constant $Z_{f,m}$, as $\tilde{\Psi}$ varies over all functions that are square-integrable, along the forward 4-velocity hyperboloid V^+, with respect to the following Lorentz-invariant measure on it,

$$d\Omega(u) = \delta(u^2 - 1)\, d^4u \ , \qquad\qquad V^+ = \left\{ u \,\middle|\, u^2 := u^i u_i = 1 \ , \ \ u^0 > 0 \right\} . \quad (1.3)$$

The fixed function f in (1.2), which characterizes the typical fibre **F** as well as the system of covariance in it, will be called the *quantum spacetime form factor* of the Klein-Gordon quantum bundle which we are constructing. We note that the set of all available quantum spacetime form factors f stands in one-to-one correspondence with the set of all resolution generators η of stochastic phase space systems of covariance, which can be obtained for the spin-zero case from the SQM framework formulated in Sec. 3.4.

We define now the *Klein-Gordon quantum bundle* **E** with typical fibre **F** as a bundle associated with the Poincaré frame bundle *PM* described in Sec. 2.3, by setting it equal to the **G**-product $(PM) \times_{\mathbf{G}} \mathbf{F}$ of the Hilbert space in (1.1) with the principal bundle *PM*. As was the case in Sec. 4.3, the natural projection $(PM) \times \mathbf{F} \to (PM) \times_{\mathbf{G}} \mathbf{F}$ gives rise to the *generalized soldering maps*

$$\sigma_x^u : \ \boldsymbol{\Psi} \mapsto \Psi \in \mathbf{F} \ , \qquad u \in \Pi^{-1}(x) \subset PM \ , \qquad \boldsymbol{\Psi} \in \mathbf{F}_x \subset \mathbf{E} \ , \quad (1.4)$$

which can be also expressed as maps between wave function amplitudes,

$$\sigma_x^u : \ \boldsymbol{\Psi}(\boldsymbol{\zeta}) \mapsto \Psi(\zeta) \ , \qquad u = (a, e_i) \in \Pi^{-1}(x) \ , \quad (1.5a)$$

$$\boldsymbol{\zeta} = (a + q^i e_i, v^i e_i) \in T_x\mathbf{M} \times V_x^+ \subset T_x\mathbf{M} \times T_x\mathbf{M} \ , \quad \zeta = (q,v) \in \mathbf{R}^4 \times V^+ \subset \mathbf{R}^8 . \quad (1.5b)$$

For any given value of $\boldsymbol{\zeta}$ in (1.5b), the complex number $\boldsymbol{\Psi}(\boldsymbol{\zeta})$ in (1.5a) can be regarded as the frame-independent value of $\boldsymbol{\Psi} \in \mathbf{F}_x$, to which the soldering map assigns the *coordinate wave function amplitude* $\Psi(\zeta)$ belonging to an element Ψ in the standard fibre **F**. In the transition from one choice of section of *PM* to another, these wave functions are related by the adaptation to 4-velocity variables v,

$$U_\eta(b,\Lambda) \ : \ \Psi(q,v) \mapsto \Psi'(q,v) = \Psi(\Lambda^{-1}(q-b), \Lambda^{-1}v) \ , \qquad \Psi \in \mathbf{F} \ , \quad (1.6)$$

of the Poincaré transformations defined by (3.4.14a) in combination with (3.4.4). Consequently, these transition maps can be expressed as follows,

$$\sigma_x^{u'} \circ (\sigma_x^u)^{-1} : \Psi \mapsto \Psi' = U_\eta(b(x), \Lambda(x))\Psi \ , \tag{1.7a}$$

$$(\boldsymbol{a}(x), \boldsymbol{e}_i(x)) = (\boldsymbol{a}'(x), \boldsymbol{e}'_i(x)) \cdot (b(x), \Lambda(x)) \ , \tag{1.7b}$$

so that they underline the fact that the fibres $\mathbf{F}_x$ of $\mathbf{E}$ consist at each $x \in \mathbf{M}$ of equivalence classes of all the wave functions resulting from all possible changes of Poincaré gauge.

For each choice of section s of the Poincaré frame bundle $P\mathbf{M}$, the generalized soldering maps in (1.4) provide the local trivialization maps

$$\phi^s : \boldsymbol{\Psi} \mapsto (x, \sigma_x^u \boldsymbol{\Psi}) \in \mathbf{M}^s \times \mathbf{F} \ , \qquad \boldsymbol{\Psi} \in \mathbf{F}_x \subset \pi^{-1}(\mathbf{M}^s) \ , \qquad u = s(x) \ , \tag{1.8}$$

generically defined in (2.2.4). Hence, the Klein-Gordon quantum bundle $\mathbf{E}$ indeed emerges as a fibre bundle associated with the principal bundle $P\mathbf{M}$ in the sense of the definition in Sec. 2.2.

The fibres $\mathbf{F}_x$ of $\mathbf{E}$ are Hilbert spaces which carry the inner products

$$\langle \boldsymbol{\Psi}_1 | \boldsymbol{\Psi}_2 \rangle = \int_\Sigma \Psi_1^*(\zeta) \Psi_2(\zeta) \, d\Sigma(\zeta) \ , \qquad \Psi_1 = \sigma_x^u \boldsymbol{\Psi}_1 \ , \quad \Psi_2 = \sigma_x^u \boldsymbol{\Psi}_2 \ , \tag{1.9}$$

where the integration is to be performed with respect to the measure

$$d\Sigma(\zeta) = 2v_\mu \, \delta(v^2 - 1) \, d\sigma^\mu(q) d^4v = 2v_\mu \, d\sigma^\mu(q) d\Omega(v) \ , \tag{1.10}$$

along any of the following surfaces

$$\Sigma = \left\{ \zeta \middle| \zeta = (q, v) \, , \quad q^0 = \text{const.}, \quad v \in V^+ \right\} \, . \tag{1.11}$$

The renormalization constant $Z_{f,m}$ in (1.2) can be now fixed so that

$$\langle \boldsymbol{\Psi}_1 | \boldsymbol{\Psi}_2 \rangle = \int_{u^0 > 0} \tilde{\Psi}_1^*(u) \, \tilde{\Psi}_2(u) \, d\Omega(u) \ . \tag{1.12}$$

In view of (3.4.15a), the inner product in (1.9) can be also written in the form

$$\langle \boldsymbol{\Psi}_1 | \boldsymbol{\Psi}_2 \rangle = i\hat{Z}_{f,m} \int_\Sigma \Psi_1^*(q,v) \overleftrightarrow{\partial}_\mu \Psi_2(q,v) \, d\sigma^\mu(q) d\Omega(v) \ , \tag{1.13}$$

with the new renormalization constant $\hat{Z}_{f,m}$ adjusted to the use of 4-velocity variables in such a manner that (1.12) remains true. A straightforward computation then yields (cf. [P], Sec. 2.6):

$$\hat{Z}_{f,m}^{-1} = (2\pi)^3 (m Z_{f,m})^{-2} \int_{V_m^+} \left| f(v^0) \right|^2 d\Omega(v) \ . \tag{1.14}$$

On account of the Lorentz covariance of the framework, the integration in (1.9), or in (1.13), can be also carried out along the Lorentz transforms of any of the surfaces in

(1.11). Furthermore, in view of the existence of the conserved currents in (3.5.13) and (3.5.17) for choices of quantum spacetime form factors f that are real, the hyperplane component of these surfaces can be deformed into an arbitrary spacelike Cauchy surface. Hence, *throughout the remainder of this monograph we shall assume that the quantum spacetime form factors* f *are real functions*. Although, as functions of v^0, it is sufficient that they be defined only on the interval $[1,+\infty)$, in order to be able to later define their off-shell derivatives, quantum-frame congruence, sharp-point limits, etc., we shall assume that they are defined for all values within $[0,+\infty)$. All these conditions are certainly satisfied for the quantum spacetime form factors in Eq. (5.5) of Sec. 5.5, that correspond to the adoption of GS exciton proper state vectors which are eigenstates of Born's quantum metric operator (cf. [P], p. 204) for any positive-valued choice of the fundamental length ℓ.

The elements $\boldsymbol{\Psi}$ of each fibre which are normalized with respect to the above inner product will be called *local Klein-Gordon state vectors* . Furthermore, for each choice of Poincaré frame $\boldsymbol{u} \in P\mathbf{M}$, the family

$$\left\{ \boldsymbol{\Phi}_\zeta^{\boldsymbol{u}} = (\boldsymbol{\sigma}_x^{\boldsymbol{u}})^{-1} U_\eta(q,\Lambda_v)\eta \,\middle|\, \zeta = (q,v) \in \mathbf{R}^4 \times V^+ \right\} , \qquad \boldsymbol{u} \in \boldsymbol{\Pi}^{-1}(x) \subset P\mathbf{M} \, , \quad (1.15)$$

of generalized coherent states within $\mathbf{F}_x$ will be called a *Klein-Gordon quantum frame* at $x \in \mathbf{M}$. Indeed, according to (3.4.11)-(3.4.13), and in view of the unitarity of the generalized soldering maps embedded in the definition of the inner product in (1.9), we can expand an arbitrary vector $\boldsymbol{\Psi}$ in $\mathbf{F}_x$ as follows:

$$\boldsymbol{\Psi} = \int_\Sigma d\Sigma(\zeta)\, \Psi_x^{\boldsymbol{u}}(\zeta) \boldsymbol{\Phi}_\zeta^{\boldsymbol{u}} \, , \qquad \boldsymbol{\Psi} \in \mathbf{F}_x \, , \quad (1.16a)$$

$$\Psi_x^{\boldsymbol{u}}(\zeta) = \left\langle \boldsymbol{\Phi}_\zeta^{\boldsymbol{u}} \middle| \boldsymbol{\Psi} \right\rangle , \qquad \Psi_x^{\boldsymbol{u}} = \boldsymbol{\sigma}_x^{\boldsymbol{u}}\, \boldsymbol{\Psi} \in \mathbf{F} \, . \quad (1.16b)$$

Hence, each Klein-Gordon quantum frame at x provides a resolution of the identity $\mathbf{1}_x$ in the fibre $\mathbf{F}_x$ above x:

$$\int_\Sigma \left| \boldsymbol{\Phi}_\zeta^{\boldsymbol{u}} \right\rangle d\Sigma(\zeta) \left\langle \boldsymbol{\Phi}_\zeta^{\boldsymbol{u}} \right| = \mathbf{1}_x \, , \qquad \mathbf{1}_x : \boldsymbol{\Psi} \mapsto \boldsymbol{\Psi} , \qquad \forall \boldsymbol{\Psi} \in \mathbf{F}_x \, . \quad (1.17)$$

We note that, for a given quantum spacetime form factor f, the quantum frames above the same base point x , corresponding to all the possible choices of rest mass $m > 0$,

$$\boldsymbol{\Phi}_{f,m;q,v}^{\boldsymbol{u}(x)}(q',v') = Z_{f,m}^{-2} \int_{u^0>0} \exp[im(q-q')\cdot u] f(u\cdot v') f(u\cdot v)\, d\Omega(u) \quad (1.18)$$

are congruent, in the sense that they can be all obtained from the *standard quantum frame* that corresponds to unit mass in Planck units (i.e., to the Planck mass [W]),

$$\boldsymbol{\Phi}_{f;q,v}^{\boldsymbol{u}(x)}(q',v') = Z_f^{-2} \int_{u^0>0} \exp[i(q-q')\cdot u] f(u\cdot v') f(u\cdot v)\, d\Omega(u) \, , \quad (1.19)$$

by the following rescaling[7] procedure:

$$\Phi^{u(x)}_{f,m;q,v}(q',v') = (Z_f/Z_{f,m})^2 \Phi^{u(x)}_{f;mq,v}(mq',v') \ , \qquad (Z_f/Z_{f,m})^2 = m^3 \ . \tag{1.20}$$

The functions in (1.16b) will be called the *coordinate wave functions* (cf. Sec. 3.9) of the local vector Ψ with respect to the Klein-Gordon quantum frame in (1.15). These coordinate wave functions are always continuous functions in the index variable ζ, and for such quantum frames as the standard ground-exciton Klein-Gordon frames that will be introduced in Sec. 5.5, they are actually smooth functions of ζ (cf. (5.8)).

The soldering of the Klein-Gordon quantum frames to the Poincaré frames in *P*M, implicitly carried out in (1.15), supplies a one-to-one map (cf. (3.7.1))

$$\rho \ : \ u \leftrightarrow \ \mathbf{Q}(u) = \left\{ \Phi^u_\zeta \mid \zeta \in \mathbf{R}^4 \times V^+ \right\} \ , \qquad u \in P\mathbf{M} \ , \tag{1.21}$$

between the Poincaré frame bundle *P*M and the *Klein-Gordon quantum frame bundle*

$$K\mathbf{M} \ = \ \{ \mathbf{Q}(u) | u \in P\mathbf{M} \} = \bigcup\nolimits_{x \in \mathbf{M}} \{ \mathbf{Q}(u) | u \in \Pi^{-1}(x) \} \ . \tag{1.22}$$

This bijective map is a bundle isomorphism [C,I], in the sense that it obviously preserves the fibre structure, and that we have:

$$u \cdot (b, \Lambda) \ \leftrightarrow \ \left\{ \Phi^{u \cdot (b,\Lambda)}_\zeta \mid \zeta \in \mathbf{R}^4 \times V^+ \right\} \ , \qquad \forall \, (b, \Lambda) \in \mathrm{ISO}_0(3,1) \ . \tag{1.23}$$

The existence of this bundle isomorphism will be used in the next section to define parallel transport within the Klein-Gordon quantum bundle **E**.

5.2. Parallel Transport in Klein-Gordon Bundles

The relationship of the quantum frame bundle *K*M defined in (1.22) to the Klein-Gordon quantum bundle **E** is very similar to the relationship of the Poincaré frame bundle *P*M to the tangent bundle *T*M. This analogy can be underlined by a suitable choice of notation.

By using the Lorentzian metric $\boldsymbol{g}$ on **M**, we can expand any tangent vectors $\boldsymbol{X}$ at $x \in \mathbf{M}$ in any Poincaré frame $(\boldsymbol{a},\boldsymbol{e}_i)$ above the same point x as follows:

$$\boldsymbol{X} = \boldsymbol{a} + X^i \boldsymbol{e}_i \ , \qquad X^i = \boldsymbol{\theta}^i(\boldsymbol{X} - \boldsymbol{a}) \ = \eta^{ij} \, \boldsymbol{g}(\boldsymbol{e}_j, \boldsymbol{X} - \boldsymbol{a}) \ . \tag{2.1}$$

Let us therefore introduce the following linear functionals on $\mathbf{F}_x$ (cf. Sec. 3.8):

$$\tilde{\Phi}^\zeta : \ \Psi \mapsto \langle \Phi_\zeta | \Psi \rangle = \Psi^\zeta \ , \qquad \Phi_\zeta := \Phi^u_\zeta \ . \tag{2.2}$$

Since they are continuous, they belong to the dual $\mathbf{F}^*_x$ of the Hilbert space $\mathbf{F}_x$ Customarily, $\mathbf{F}^*_x$ is identified with $\mathbf{F}_x$ by using the Riesz theorem [PQ]. However, we shall view $\mathbf{F}^*_x$ as an entity that is separate from $\mathbf{F}_x$, and therefore view the linear functionals in (2.2) as constituting a *coframe of the quantum frame* $\{\boldsymbol{\Phi}_\zeta\}$. In that case, by extending the Einstein convention from summation to integration over continuous indices, in the same manner as we did in (8.6) in the case of any typical fibre **F**, we can write in each $\mathbf{F}_x$,

$$\Psi^\zeta \boldsymbol{\Phi}_\zeta := \int_\Sigma d\Sigma(\zeta)\, \Psi^\zeta \boldsymbol{\Phi}_\zeta^u \quad , \tag{2.3}$$

where the integration is independent of the choice of Σ, so that in analogy[8] with (2.1):

$$\boldsymbol{\Psi} = \Psi^\zeta \boldsymbol{\Phi}_\zeta \quad , \qquad \Psi^\zeta = \tilde{\boldsymbol{\Phi}}^\zeta(\boldsymbol{\Psi}) = \langle \boldsymbol{\Phi}_\zeta | \boldsymbol{\Psi} \rangle \quad . \tag{2.4}$$

This also demonstrates that the inner products in the quantum fibres, defined by (1.9), indeed play a role very analogous to that played by the Lorentzian metric tensor g vis-à-vis the tangent bundle $T\mathbf{M}$ – and tensor bundles in general. In fact, while constructing Fock bundles in the quantum field theoretical context of Chapter 7, we shall have to deal with *Klein-Gordon (r,s)-tensor bundles,*

$$\mathbf{E}^{r,s} = (P\mathbf{M}) \times_{\mathbf{G}} \mathbf{F}^{r,s} \quad , \qquad \mathbf{G} = \mathrm{ISO}_0(3,1) \quad , \tag{2.5}$$

whose typical fibres are Hilbert tensor products (cf. [PQ], Sec. II-6.5) of r copies of the fibre **F** in (1.1), and of s copies of its dual $\mathbf{F}^*$. They are, therefore, the analogues of the finite-dimensional tensor bundles $T^{r,s}\mathbf{M}$, and their fibres can be identified in a natural manner with the Hilbert tensor products[9]

$$\mathbf{F}_x^{r,s} = (\mathbf{F}_x)^{\otimes r} \otimes (\mathbf{F}_x^*)^{\otimes s} = \mathbf{F}_x \otimes \cdots \otimes \mathbf{F}_x \otimes \mathbf{F}_x^* \otimes \cdots \otimes \mathbf{F}_x^* \quad . \tag{2.6}$$

Hence, we can generalize (1.16) and (2.3) as follows:

$$\boldsymbol{\Psi} = \Psi^{\zeta_1 \cdots \zeta_r}{}_{\zeta'_1 \cdots \zeta'_s}\, \boldsymbol{\Phi}_{\zeta_1} \otimes \cdots \otimes \boldsymbol{\Phi}_{\zeta_r} \otimes \tilde{\boldsymbol{\Phi}}^{\zeta'_1} \otimes \cdots \otimes \tilde{\boldsymbol{\Phi}}^{\zeta'_s} \quad , \qquad \boldsymbol{\Psi} \in \mathbf{F}_x^{r,s} \quad , \tag{2.7a}$$

$$\Psi^{\zeta_1 \cdots \zeta_r}{}_{\zeta'_1 \cdots \zeta'_s} = \langle \boldsymbol{\Phi}_{\zeta_1} \otimes \cdots \otimes \boldsymbol{\Phi}_{\zeta_r} \otimes \tilde{\boldsymbol{\Phi}}^{\zeta'_1} \otimes \cdots \otimes \tilde{\boldsymbol{\Phi}}^{\zeta'_s} | \boldsymbol{\Psi} \rangle \quad , \qquad \boldsymbol{\Psi} \in \mathbf{F}^{r,s} \quad . \tag{2.7b}$$

Following the same procedure as in Sec. 4.3, we define a *section of the Klein-Gordon quantum bundle* **E** on any given open domain D on which a section $\boldsymbol{s}$ of the Poincaré frame bundle $P\mathbf{M}$ is defined as a map that assigns to each $x \in D$ a vector $\boldsymbol{\Psi}_x \in \mathbf{F}_x$ in such a manner that the coordinate wave functions

$$\Psi_x^\zeta = \tilde{\boldsymbol{\Phi}}_x^\zeta(\boldsymbol{\Psi}) := \langle \boldsymbol{\Phi}_\zeta^{u(x)} | \boldsymbol{\Psi}_x \rangle \quad , \qquad \boldsymbol{s} : x \mapsto \boldsymbol{u}(x) \quad , \tag{2.8}$$

are smooth functions of $x \in D$ for each fixed choice of the index variables ζ. On account of the Poincaré covariance of the Klein-Gordon quantum bundle, this definition is independent of the adopted Poincaré gauge, i.e., of the choice of section $\boldsymbol{s}$ of the Poincaré frame bundle. Furthermore, it can be extended to arbitrary regions D, which might not lie within the domain of a single section of the Poincaré frame bundle, by covering those regions with overlapping open sets, such that each one of those sets lies within the domain of a single section s of $P\mathbf{M}$. In particular, it can be extended to the case of a *global section* of $\mathbf{E}$, whose domain of definition D equals all of $\mathbf{M}$.

The above definitions extend in an obvious manner to sections of arbitrary Klein-Gordon tensor bundles $\mathbf{E}^{r,s}$. By analogy with (3.8.7), we can therefore define a *quantum metric* on $\mathbf{M}$ as a section G of $\mathbf{E}^{0,2}$, which in any quantum frame $\{\boldsymbol{\Phi}_\zeta\}$ has the form

$$G = \Phi_{\zeta_1\zeta_2}(\tilde{\boldsymbol{\Phi}}^{\zeta_1})^* \otimes \tilde{\boldsymbol{\Phi}}^{\zeta_2} , \qquad \Phi_{\zeta_1\zeta_2} := \Phi_{\zeta_2}(\zeta_1) . \tag{2.9}$$

The quantum metric therefore generically assumes values in the dual of $\mathbf{F}_x \otimes \mathbf{F}_x$ (which is identifiable with the Hilbert-Schmidt class over $\mathbf{F}_x$), and is such that

$$G(\boldsymbol{\Psi}_1 \otimes \boldsymbol{\Psi}_2) = \langle \boldsymbol{\Psi}_1 | \boldsymbol{\Psi}_2 \rangle , \qquad \boldsymbol{\Psi}_1, \boldsymbol{\Psi}_2 \in \mathbf{F}_x , \tag{2.10}$$

in complete analogy with the corresponding relation (3.8.8) in the typical fibre $\mathbf{F}$.

Connections on the Poincaré frame bundle $P\mathbf{M}$ that are compatible with a Lorentzian metric $\boldsymbol{g}$ were discussed in Sec. 2.6. We referred there to such connections as Riemann-Cartan connections. We shall regard the affine extension to $P\mathbf{M}$ of the Levi-Civita connection on a Lorentzian manifold[10] as a special case of such connections, which corresponds to the case of zero-torsion form for the pull-back to $L\mathbf{M}$ defined in (2.6.17).

For any choice of Riemann-Cartan connection on $P\mathbf{M}$, a corresponding parallel transport along any piecewise smooth curve γ, connecting $x' \in \mathbf{M}$ to $x'' \in \mathbf{M}$, can be defined first within $P\mathbf{M}$ by the "horizontal lift" procedure described in Sec. 2.4,

$$\tilde{\tau}_\gamma(x'',x') : \quad u_{x'} = (a', e'_i) \ \mapsto \ u_{x''} = (a'', e''_i) , \tag{2.11}$$

and then extended to $T\mathbf{M}$,

$$\tilde{\tau}_\gamma(x'',x') : \quad X' = a' + X^i e'_i \ \mapsto \ X'' = a'' + X^i e''_i , \tag{2.12}$$

as well as to $T^{r,s}\mathbf{M}$ bundles in general. In a totally analogous manner we can take the parallel transport within $K\mathbf{M}$,

$$\tau_\gamma(x'',x') : \quad \boldsymbol{\Phi}_\zeta^{u_{x'}} \ \mapsto \ \boldsymbol{\Phi}_\zeta^{u_{x''}} , \tag{2.13}$$

and extend it, in accordance with the general types of expansions in (2.4) and (2.7), to the Klein-Gordon quantum bundle $\mathbf{E}$,

$$\tau_\gamma(x'',x') : \quad \boldsymbol{\Psi}' = \psi^\zeta \boldsymbol{\Phi}_\zeta^{u_{x'}} \ \mapsto \ \boldsymbol{\Psi}'' = \psi^\zeta \boldsymbol{\Phi}_\zeta^{u_{x''}} , \tag{2.14}$$

as well as to $\mathbf{E}_{r,s}$ bundles in general:

$$\psi^{\zeta_1 \cdots \zeta_r}{}_{\zeta'_1 \cdots \zeta'_s} \boldsymbol{\Phi}_{\zeta_1}^{u_{x'}} \otimes \cdots \otimes \tilde{\boldsymbol{\Phi}}^{u_{x'};\zeta'_s} \ \mapsto \ \psi^{\zeta_1 \cdots \zeta_r}{}_{\zeta'_1 \cdots \zeta'_s} \boldsymbol{\Phi}_{\zeta_1}^{u_{x''}} \otimes \cdots \otimes \tilde{\boldsymbol{\Phi}}^{u_{x''};\zeta'_s} . \tag{2.15}$$

Naturally, this definition of parallel transport within the Klein-Gordon quantum bundle, as well as within Klein-Gordon tensor bundles in general, is frame independent on account of the Poincaré covariance of all those bundles.

Given the compatibility of a Riemann-Cartan connection with the Lorentzian metric, it follows from (2.13) that the operator for parallel transport in (2.14) is unitary for any smooth curve γ. This means that the parallel transport in the Klein-Gordon quantum bundle **E** is *compatible with the quantum metric* **G** in (2.9) and (2.10).

The *covariant derivative* of a section $\{\boldsymbol{\Psi}_x \mid x \in D\}$ of **E** can be now defined as in (4.3.10), i.e., by the strong limit[11]

$$\nabla_X \boldsymbol{\Psi}_x = \operatorname{s-lim}_{t\to 0} \frac{1}{t}\left[\tau_\gamma(x,x(t))\boldsymbol{\Psi}_{x(t)} - \boldsymbol{\Psi}_x\right] , \quad x(0) = x \in \mathbf{M} ,\ \dot{x}(0) = X \in T_x\mathbf{M}. \tag{2.16}$$

This definition can be also applied to sections of arbitrary Klein-Gordon tensor bundles, but we shall postpone its study in that context until Chapter 7.

To express (2.16) in a form analogous to (2.6.19), we choose any section s of the Poincaré frame bundle $P\mathbf{M}$, and insert

$$\begin{aligned} &\boldsymbol{\sigma}_x^s\left[\tau_\gamma(x,x(t))\boldsymbol{\Psi}_{x(t)} - \boldsymbol{\Psi}_x\right] \\ &\quad = \left(\boldsymbol{\sigma}_{x(t)}^s \boldsymbol{\Psi}_{x(t)} - \boldsymbol{\sigma}_x^s \boldsymbol{\Psi}_x\right) + \left[U_\eta\left(b_\gamma^s(x,x(t)), \Lambda_\gamma^s(x,x(t))\right) - \mathbf{1}\right]\boldsymbol{\sigma}_{x(t)}^s \boldsymbol{\Psi}_{x(t)} , \end{aligned} \tag{2.17a}$$

$$\tilde{\boldsymbol{\tau}}_\gamma(x,x(t))s(x(t)) = s(x)\cdot\left(b_\gamma^s(x,x(t)), \Lambda_\gamma^s(x,x(t))\right) , \tag{2.17b}$$

in (2.16), noting that

$$\partial_X \boldsymbol{\Psi}_x := (\boldsymbol{\sigma}_x^s)^{-1} \operatorname{s-lim}_{t\to 0} \frac{1}{t}\left(\boldsymbol{\sigma}_{x(t)}^s \boldsymbol{\Psi}_{x(t)} - \boldsymbol{\sigma}_x^s \boldsymbol{\Psi}_x\right) = [X^\mu \partial_\mu(\Psi_x^\zeta)]\, \boldsymbol{\Phi}_\zeta^{u(x)} \tag{2.18}$$

exists due to the smoothness property of any section of **E**. We therefore obtain that, for the given choice of section s of the Poincaré frame bundle $P\mathbf{M}$,

$$\nabla_X \boldsymbol{\Psi}_x = [\partial_X + \tilde{\boldsymbol{\theta}}^i(X)\hat{\mathbf{P}}_{i;\boldsymbol{u}} + \tfrac{1}{2}\tilde{\boldsymbol{\omega}}_{ij}(X)\hat{\mathbf{M}}_{\boldsymbol{u}}^{ij}]\, \boldsymbol{\Psi}_x , \tag{2.19}$$

where the operators on the right-hand side of (2.19) are the infinitesimal generators of the unitary representation

$$U_{x;\boldsymbol{u}}(b,\Lambda) = (\boldsymbol{\sigma}_x^s)^{-1} U_\eta(b,\Lambda)\boldsymbol{\sigma}_x^s , \qquad \boldsymbol{u} = s(x) , \tag{2.20}$$

of the Poincaré group. These generators are obtained, in accordance with (2.6.20), from a basis of the Lie algebra *iso*(3,1) of $\mathrm{ISO}_0(3,1)$. For the Poincaré frame $(\boldsymbol{a},\boldsymbol{e}_i)$ assigned by the chosen section s at some $x \in \mathbf{M}^s$, it is natural to parametrize the spacetime translations in (1.6) by means of the components b^i of b, the spatial rotations around each of the axes $\boldsymbol{e}_i$, $i = 1,2,3$, by their rotation angles, and the Lorentz boosts in the direction of those same axes by the parameter that imparts to them the formal appearance of Lorentz "rotations", so that, for example, in the case of the Lorentz boosts in the direction of e_3 we have

$$(a, e_0, e_1, e_2, e_3)\cdot(0, \Lambda_\vartheta^{(3)}) = (a, e'_0(\vartheta), e_1, e_2, e'_3(\vartheta)) \quad , \tag{2.21a}$$

$$e'_0(\vartheta) = e_0 \cosh\vartheta + e_3 \sinh\vartheta \quad , \qquad e'_3(\vartheta) = e_0 \sinh\vartheta + e_3 \cosh\vartheta \quad , \tag{2.21b}$$

$$q^0(\vartheta)' = q^0 \cosh\vartheta - q^3 \sinh\vartheta \quad , \qquad q^3(\vartheta)' = q^3 \cosh\vartheta - q^0 \sinh\vartheta \quad . \tag{2.21c}$$

From (1.2) and (1.6), we can then immediately deduce that the infinitesimal generators of spacetime translations and spatial rotations are given in the present spin-zero context by[12] (cf. [P], Sec. 2.7),

$$\hat{\mathbf{P}}_{j;u} = -iP_{j;u} \quad , \qquad \hat{\mathbf{M}}_u^{jk} = iM_u^{jk} = i(Q_u^j P_u^k - Q_u^k P_u^j) \quad , \tag{2.22a}$$

$$P_{j;u} = i\partial/\partial q^j , \qquad Q_u^j = q^j - i\partial/\partial p_j \quad , \qquad p_j = m\eta_{jk} v^k \quad . \tag{2.22b}$$

We note, however, that in order to preserve for the infinitesimal generators of Lorentz boosts the same formal appearance as for those of spatial rotations, we have to extend the wave functions in (1.2) off the 4-velocity shell V^+. The following extension,

$$\Psi(q,v) = Z_{f,m}^{-1} \int_{u^0>0} \exp[-imq\cdot u] f(1-\tfrac{1}{2}(u-v)^2)\tilde{\Psi}(u)\, d\Omega(u) \quad , \tag{2.23}$$

obviously coincides for all $v \in V^+$ with the expression in (1.2), and leads to Q-operators for which (2.22a) holds true for all ten generators of the representation (2.20) of the Poincaré group[13] – as can be verified in the representative case of Lorentz boosts in the direction of e_3 by substituting into (2.23) in place of q the expression for q' that follows from (2.21c), in conjunction with a similar substitution for v', and then perform the partial differentiation with respect to the parameter ϑ.

The operator of covariant differentiation (i.e., the Koszul connection on **E**), generically defined by (2.5.23), assumes now the form

$$\nabla = \mathbf{d} - i\tilde{\boldsymbol{\theta}}^i P_{i;u} + \tfrac{i}{2}\tilde{\boldsymbol{\omega}}_{jk} M_u^{jk} \quad , \qquad \mathbf{d} = \boldsymbol{\theta}^i \partial_i \quad , \qquad \partial_i := \partial_{e_i} \quad . \tag{2.24}$$

Since the representations in (2.20) are unitary, their infinitesimal generators in (2.22) are self-adjoint operators despite the fact that the Q-operators in (2.22b) involve off-shell differentiation. On the other hand, these infinitesimal generators are unbounded operators, whose domains of definition are therefore dense in $\mathbf{F}_x$, but cannot coincide with $\mathbf{F}_x$ on account of the Hellinger-Toeplitz theorem [PQ]. Consequently, in a given Poincaré gauge specified by some section s of PM, the strong limit in (2.16) does not exist for an arbitrary section of **E**. Rather, it exists if and only if the elements of the section $\{\Psi_x \mid x \in D\}$ belong to the domain of the self-adjoint operator densely defined by the following strong limits[14]:

$$\left(\tilde{\boldsymbol{\theta}}^i(X)P_{i;u} - \tfrac{1}{2}\tilde{\boldsymbol{\omega}}_{jk}(X)M_u^{jk}\right)\Psi = \operatorname*{s-lim}_{t\to 0} \frac{i}{t}\left[U_{x;u}\left(b_\gamma^s(x,x(t)), \Lambda_\gamma^s(x,x(t))\right) - \mathbf{1}_x\right]\Psi \,. \tag{2.25}$$

Hence, from the compatibility of the quantum connection with the quantum metric G, we deduce that for any two sections of **E** we can write down an equation which is analogous in outward appearance to the one in (2.6.22), namely

$$\left\langle \nabla_X \boldsymbol{\Psi}_x \middle| \boldsymbol{\Psi}'_x \right\rangle + \left\langle \boldsymbol{\Psi}_x \middle| \nabla_X \boldsymbol{\Psi}'_x \right\rangle = X \left\langle \boldsymbol{\Psi}_x \middle| \boldsymbol{\Psi}'_x \right\rangle \ , \tag{2.26}$$

but holds only at those points x in the common domain of definition of the two sections of **E** where the above two covariant derivatives are defined.

To deal with this domain problem for covariant derivatives of vector fields in **E**, we introduce the concept of *coherent section* of **E**: by definition, the section $\{\boldsymbol{\Psi}_x \mid x \in D\}$ is coherent for some choice s of Poincaré gauge, if it can be obtained at all $x \in D$ by setting $\boldsymbol{\Psi}_x = \boldsymbol{\Phi}_{x;\zeta}$ for some fixed value of ζ, where $\{\boldsymbol{\Phi}_{x;\zeta}\} = \boldsymbol{\rho}(s)$ is the section of K**M** corresponding to s. Coherent sections are in the domain of definition of covariant derivatives for any choice of Poincaré gauge (cf. Prugovečki and Warlow, 1989b, Theorem A.2). Furthermore, in the particular Poincaré gauge s with respect to which they are coherent, the term in (2.18) vanishes, and (2.19) assumes the following form:

$$\nabla_X \boldsymbol{\Phi}_{x;\zeta} = i[-\tilde{\boldsymbol{\theta}}^j(X) P_{j;u} + \tfrac{1}{2} \tilde{\boldsymbol{\omega}}_{jk}(X) M_u^{jk}] \boldsymbol{\Phi}_{x;\zeta} \ . \tag{2.27}$$

To deal now with the domain problem for covariant derivatives of arbitrary sections of the Klein-Gordon bundle **E**, we transfer to Poincaré gauges the concept of a Cartan gauge adapted to a smooth curve, which was defined in Sec. 2.7 . Thus, we shall say that a Poincaré gauge, given by a section s of the Poincaré frame bundle GL**M**, is *adapted to a smooth curve* γ if the connection one-forms in (2.24) vanish at all $x \in \gamma$ which lie in the domain of definition $\mathbf{M}^s$ of that gauge. If X is tangent to a curve γ with respect to which s is adapted, then it can be proven (cf. Prugovečki and Warlow, 1989b, Theorem A.4) that the corresponding covariant derivative in the direction X exists for any section $\boldsymbol{\Psi}$ of **E**, since it is given by

$$\nabla_X \boldsymbol{\Psi}_x = \partial_X \boldsymbol{\Psi}_x + \boldsymbol{\Psi}_x^\zeta \nabla_X \boldsymbol{\Phi}_{x;\zeta} \ , \qquad \partial_X \boldsymbol{\Psi}_x = [X^\mu \partial_\mu (\Psi_x^\zeta)] \boldsymbol{\Phi}_{x;\zeta} \ , \tag{2.28}$$

where, for each fixed value of ζ, the section $\{\boldsymbol{\Phi}_{x;\zeta}\}$ is coherent in the Poincaré gauge adapted to γ. Hence, we can always find Poincaré gauges in which such basic relationships as (2.26) have a well-defined meaning, since for any given smooth curve γ we can always construct a section of P**M** adapted to it by the parallel transport method described in Sec. 2.7.

In this manner we have established that for any Riemann-Cartan connection in P**M**, defined as in Sec. 2.6, we can introduce in any given Klein-Gordon quantum bundle **E** a notion of parallel transport, as well as a general theory of quantum connections compatible with the quantum metric in (2.9). These parallel transports in **E**, and the quantum connections related to them, share all the basic features of their classical counterparts in T**M**. In basing on them the concept of GS propagation arrived at by the considerations in the next two sections, we shall assume for physical reasons that the connection adopted in P**M** is the affine extension of the Levi-Civita connection. However, from a mathematical point of view, most definitions and results in the next couple of sections apply equally well to the generic case of Riemann-Cartan connections.

*5.3. Quantum Torsion and the Klein-Gordon Quantum Connection

In the preceding section we defined a quantum metric G on the Klein-Gordon quantum bundle $\mathbf{E}$, as well as a family of quantum connections ∇ in $\mathbf{E}$ corresponding to various choices of Riemann-Cartan connections in PM. In this context, a *quantum connection* can be viewed as a Koszul connection [SC], namely as an operator for covariant differentiation on $\mathbf{E}$, that assigns a vector field $\nabla_X \Psi$ in $\mathbf{E}$ to any two vector fields X and Ψ in TM and $\mathbf{E}$, respectively – and does that in such a manner that the four properties in (2.4.12) are satisfied for any vector field in $\mathbf{E}$ (i.e., for any section of $\mathbf{E}$) assuming values within its domain of definition. Thus, by analogy with the definition of connections in some textbooks on CGR (Hawking and Ellis, 1973; Straumann, 1984), we could extend this definition of quantum connection to Klein-Gordon (r,s)-tensor fields assuming values in $\mathbf{E}^{r,s}$ by imposing the Leibniz rule and the commutativity of ∇ with contractions. Given the fact that we already have a concept of parallel transport in $\mathbf{E}^{r,s}$, based on (2.15), the outcome of such an approach is identical with the more direct definition[15] of the quantum connection ∇ on $\mathbf{E}^{r,s}$ obtained by setting $\nabla_X \Psi$ equal to (2.16) for generic sections of $\mathbf{E}^{r,s}$. However, quantum counterparts of basic classical geometric concepts can be then introduced.

For quantum connections, we can define a *quantum torsion field* by analogy with (2.6.10), namely as an operator-valued form which, for any given section[16] s of the Lorentz frame bundle LM, assigns to any two vector fields X and Y from TM a linear operator acting on sections Ψ of $\mathbf{E}$, as well as on sections of $\mathbf{E}^{r,s}$ in general, as follows,

$$\hat{T}(X,Y):\ \Psi \mapsto \hat{\nabla}_X \hat{\partial}_Y \Psi - \hat{\nabla}_Y \hat{\partial}_X \Psi - (\partial_X \hat{\partial}_Y - \partial_Y \hat{\partial}_X)\Psi\,, \qquad \hat{\nabla} = d + \tfrac{i}{2}\tilde{\omega}_{jk} M_u^{jk}\,, \tag{3.1}$$

provided, of course, that the values assumed by Ψ are within the common (dense) domain of all the operators occurring in the above mapping. In (3.1) the operators ∂_X and ∂_Y are defined as in (2.18), whereas their hatted counterparts act in a similar manner on the internal gauge variables, i.e., in the case of sections of $\mathbf{E}$ they act in the following manner:

$$\partial_X \Psi = [X^\mu \partial_\mu(\Psi^\zeta)]\,\Phi_\zeta\,, \qquad \partial_\mu(\Psi^\zeta)\big|_x = \partial \Psi_x^\zeta / \partial x^\mu\,, \qquad \Psi = \Psi^\zeta \Phi_\zeta\,, \tag{3.2}$$

$$\hat{\partial}_X \Psi = [X^i \hat{\partial}_i(\Psi^\zeta)]\,\Phi_\zeta\,, \qquad \hat{\partial}_i(\Psi^\zeta)\big|_x = \partial \Psi_x^\zeta / \partial q^i\,, \qquad \zeta = (q,v)\,. \tag{3.3}$$

We can now ask the question as to which of the quantum connections in $\mathbf{E}$, as well as in $\mathbf{E}^{r,s}$ in general, possess quantum torsion operators which identically equal zero, so that

$$\hat{\nabla}_X \hat{\partial}_Y \Psi - \hat{\nabla}_Y \hat{\partial}_X \Psi \equiv (\partial_X \hat{\partial}_Y - \partial_Y \hat{\partial}_X)\Psi\,. \tag{3.4}$$

Upon comparing (2.19) with the covariant derivative (2.6.19) for sections of arbitrary finite-dimensional vector bundles associated with the Poincaré frame bundle PM, it can be expected that the answer will be arrived at by considering the affine extension to $P\mathbf{M}$ of the linear Levi-Civita connection, whose connection coefficients were computed in (2.7.4) and (2.7.6). Hence, we surmise that this affine Levi-Civita connection is the only Riemann-Cartan connection which leads to a quantum connection that is torsion-free – and which we therefore will call the *Klein-Gordon quantum connection.* However, in view of the technicalities entailed in dealing with infinite-dimensional fibre bundles, such as the domain problems for covariant derivatives, as well as the fact that their fibres are complex rather than real vector spaces, this result cannot be claimed to follow directly[17] from the fundamental lemma of Riemannian geometry. Rather, it warrants subsequent more careful consideration, which will be based on introducing a quantum counterpart of the notion of torsion form $\boldsymbol{\Theta}$ of a linear[18] connection in $GL\mathbf{M}$, that was defined in Sec. 2.6.

The existence of the bundle isomorphism $\boldsymbol{\rho}$ in (1.21) enables us to transfer the basic concepts of connection form and of canonical form, defined in Secs 2.5 and 2.6, respectively, from the Lorentz frame bundle $L\mathbf{M}$ to the subbundle $K_L\mathbf{M} = \boldsymbol{\rho}(L\mathbf{M})$ of the Klein-Gordon quantum frame bundle $K\mathbf{M}$. Thus, let us introduce in the typical fibre $\mathbf{F}$ of the Klein-Gordon quantum bundle the (densely defined) operators,

$$\hat{\partial}_X : \Psi \mapsto X^i \, \partial\Psi / \partial q^i \, , \qquad X \in \mathbf{R}^4 \, , \qquad \Psi \in \mathbf{F} \, . \tag{3.5}$$

We can then rewrite the soldering map (2.6.5), expressed in Sec. 2.6 as a map from vectors in $T\mathbf{M}$ to vectors within the typical fibre $\mathbf{R}^4$ of $T\mathbf{M}$, as a corresponding map,

$$\hat{\boldsymbol{\sigma}}_{\boldsymbol{u}} : (\hat{\partial}_{\boldsymbol{X}} \tilde{\boldsymbol{\Phi}}^\zeta)\boldsymbol{\Phi}_\zeta \quad \mapsto \quad (\hat{\partial}_X \tilde{\boldsymbol{\Phi}}^\zeta)\boldsymbol{\Phi}_\zeta \, , \tag{3.6a}$$

$$\boldsymbol{u} = (\boldsymbol{e}_0, \ldots, \boldsymbol{e}_3) \in L\mathbf{M} \, , \qquad \boldsymbol{X} = X^i \boldsymbol{e}_i \in T_x\mathbf{M} \, , \qquad x = \boldsymbol{\Pi} \boldsymbol{u} \, , \tag{3.6b}$$

from operators acting within the fibres of $\mathbf{E}$ into operators acting within its typical fibre $\mathbf{F}$. Consequently, we can now define a *quantum canonical form* of $\mathbf{M}$ as a map from the tangent bundle $TK_L\mathbf{M}$ into operators acting within the typical fibre $\mathbf{F}$ of the Klein-Gordon quantum bundle (cf. (2.4.5)):

$$\hat{\boldsymbol{\theta}} : \hat{\mathbf{X}} \mapsto (\hat{\partial}_X \tilde{\boldsymbol{\Phi}}^\zeta)\boldsymbol{\Phi}_\zeta \, , \qquad \hat{\mathbf{X}} = \boldsymbol{\rho}_* \mathbf{X} \in T_{\boldsymbol{\rho}(\boldsymbol{u})} K_L \mathbf{M} \, , \quad X = \boldsymbol{\sigma}_{\boldsymbol{u}}(\boldsymbol{\Pi}_* \mathbf{X}) \, . \tag{3.7}$$

Similarly, to any (linear) Riemann-Cartan connection form $\boldsymbol{\omega}$ in $L\mathbf{M}$ we can assign a corresponding *quantum connection form* in $K_L\mathbf{M}$ by setting (cf. (2.4.5)):

$$\hat{\boldsymbol{\omega}} : \hat{\mathbf{X}} \mapsto \tfrac{1}{2} \boldsymbol{\omega}_{ij}(\boldsymbol{\rho}^* \hat{\mathbf{X}}) \hat{\mathbf{M}}^{ij}_{\boldsymbol{u}} \, , \qquad \hat{\mathbf{X}} \in T_{\boldsymbol{\rho}(\boldsymbol{u})} K_L \mathbf{M} \, . \tag{3.8}$$

We can now define, by analogy with (2.6.6), the following *quantum torsion form,*

$$\hat{\boldsymbol{\Theta}} : (\hat{\mathbf{X}}, \hat{\mathbf{Y}}) \mapsto \mathbf{d}\hat{\boldsymbol{\theta}}(\hat{\mathbf{X}}, \hat{\mathbf{Y}}) + \hat{\boldsymbol{\omega}}(\hat{\mathbf{X}})\hat{\boldsymbol{\theta}}(\hat{\mathbf{Y}}) - \hat{\boldsymbol{\omega}}(\hat{\mathbf{Y}})\hat{\boldsymbol{\theta}}(\hat{\mathbf{X}}) \, , \qquad \hat{\mathbf{X}}, \hat{\mathbf{Y}} \in T_{\boldsymbol{\rho}(\boldsymbol{u})} K_L \mathbf{M} \, , \tag{3.9}$$

which assigns an operator acting within the typical fibre $\mathbf{F}$ to each pair of vectors tangent to $K_L\mathbf{M}$. It is easy to check that the quantum torsion field in (3.1) is related to the above quantum torsion form in a manner analogous to that in (2.6.9), namely that

$$\hat{T} : (X,Y) \mapsto (\sigma_x^u)^{-1}\left(\hat{\Theta}(\hat{\hat{X}},\hat{\hat{Y}})\right) , \qquad \hat{\hat{X}},\hat{\hat{Y}} \in T_{p(u)}K_L\mathbf{M} . \tag{3.10}$$

It is clear from the above construction that the quantum torsion form in (3.9) is zero if and only if its classical counterpart in (2.6.6) equals zero. In turn, the quantum torsion field in (3.1) is zero if and only if the quantum torsion form in (3.9) is zero. Hence, we have arrived at the desired conclusion: the Klein-Gordon quantum connection has zero quantum torsion, and is the only quantum connection which has this property.

For any quantum connection form in (3.8), we can define its quantum curvature form by analogy with (2.5.16a):

$$\hat{\Omega}(\hat{\mathbf{X}},\hat{\mathbf{Y}}) = \mathbf{d}\hat{\omega}(\hat{\mathbf{X}},\hat{\mathbf{Y}}) + [\hat{\omega}(\hat{\mathbf{X}}),\hat{\omega}(\hat{\mathbf{Y}})] , \qquad \hat{\mathbf{X}},\hat{\mathbf{Y}} \in T_{p(u)}K_L\mathbf{M} . \tag{3.11}$$

Similarly, for any quantum connection ∇ we can define, by analogy with (2.6.11), the *quantum curvature field* as follows:

$$\hat{R}(X,Y) : \quad \Psi \mapsto \nabla_X\nabla_Y\Psi - \nabla_Y\nabla_X\Psi - \nabla_{[X,Y]}\Psi . \tag{3.12}$$

It should be noted, however, that the quantum curvature field in (3.12) is not, in general, the one that emerges from the quantum curvature form in (3.11). Rather, that role is played by the operator between the round brackets in the following decomposition:

$$\hat{R}(X,Y) = (\hat{\nabla}_X\hat{\nabla}_Y - \hat{\nabla}_Y\hat{\nabla}_X - \hat{\nabla}_{[X,Y]}) + \hat{S}(X,Y) . \tag{3.13}$$

By comparing the respective expressions for quantum curvature forms, we can establish that the two quantum curvatures in (3.13) are equal if and only if the torsion tensor equals zero, as is the case also in the similar relationship between linear connections in *GLM* and their affine counterparts in *GAM* – cf. Proposition 3.4 in [K], p. 130. In the present context, this singles out again the Klein-Gordon connection as the one and only quantum connection for which these two quantum curvature fields are equal. Thus, the Klein-Gordon quantum connection displays many desired geometric features in a unique manner. Consequently, in the remainder of this chapter we shall give preference to this unique quantum connection of zero quantum torsion, and deal with it exclusively.

5.4. Geometro-Stochastic Propagation in Klein-Gordon Quantum Bundles

In this section we shall formulate the GS propagation of local quantum states in the Klein-Gordon quantum bundle **E**, based on the parallel transport of those states,

$$\Psi_{x(\tau)} = \tau_\gamma(x(\tau),x(0))\,\Psi_{x(0)} , \qquad \Psi_{x(\tau)} \in \mathbf{F}_{x(\tau)} , \quad x(\tau) \in \gamma , \quad \tau \in [0,\infty) , \tag{4.1}$$

that is governed by the Klein-Gordon quantum connection. It should be noted that all the subsequent considerations can be very easily transferred to arbitrary Klein-Gordon tensor bundles. In fact, this task will be implicitly carried out in Chapter 7, in the context of Klein-

Gordon-Fock bundles.

For the Klein-Gordon quantum connection, as well as for any other quantum connection in **E**, the *propagator for parallel transport* can be defined along any smooth curve $\gamma = \{x(t)| \, a \leq t \leq b\}$ joining any two points $x' = x(a) \in \mathbf{M}$ to $x'' = x(b) \in \mathbf{M}$, and with respect to any choice of Poincaré gauge. Its definition,

$$K_\gamma(x'',\zeta'';x',\zeta') = \left\langle \Phi^{u(x'')}_{\zeta''} \middle| \tau_\gamma(x'',x')\Phi^{u(x')}_{\zeta'} \right\rangle , \qquad \zeta',\zeta'' \in \mathbf{R}^4 \times V^+ , \tag{4.2}$$

is totally analogous to the corresponding definition in (4.4.2) for Newton-Cartan quantum connections.

Upon inserting (1.17) between the operator for parallel transport and the local state vector on the right-hand side of (4.1), and then taking (1.16) into consideration, it can be immediately deduced that

$$\Psi_{x(\tau)}(\zeta) = \int K_\gamma(x(\tau),\zeta;x(0),\zeta') \, \Psi_{x(0)}(\zeta') d\Sigma(\zeta') , \tag{4.3}$$

where the integration can be carried out along any of the hypersurfaces described in Sec. 5.1, in the paragraph following equation (1.14). Furthermore, upon inserting $\tau_\gamma(x'', x') = \tau_\gamma(x'', x) \, \tau_\gamma(x , x')$ into the right-hand side of the inner product in (4.2), and performing a similar integration at $x \in \gamma$, we arrive at the following relativistic counterpart of (4.4.5a):

$$\begin{aligned} K_\gamma(x'',\zeta'';x',\zeta') &= K^*_\gamma(x',\zeta';x'',\zeta'') \\ &= \int K_\gamma(x'',\zeta'';x,\zeta) K_\gamma(x,\zeta;x',\zeta') d\Sigma(\zeta) . \end{aligned} \tag{4.4}$$

Consequently, by iteration we can also obtain the relativistic counterpart of (4.4.5b):

$$\begin{aligned} K_\gamma(x'',\zeta'';x',\zeta') = \lim_{\varepsilon\to+0} \int & K_\gamma(x_N,\zeta_N;x_{N-1},\zeta_{N-1}) \\ & \times \prod_{n=N-1}^{1} K_\gamma(x_n,\zeta_n;x_{n-1},\zeta_{n-1}) \; d\Sigma(\zeta_n) . \end{aligned} \tag{4.5}$$

The idea is now, in essence, to proceed by analogy with the definition of the GS propagation in Newton-Cartan quantum bundles, presented in Sec. 4.4, and to therefore extrapolate this last result into a definition of GS propagation in the Klein-Gordon quantum bundle **E**. Hence, pursuing that analogy, we surmise that GS propagation should take place along broken paths consisting of arcs of timelike geodesics of the Levi-Civita connection in **M**, and that along each such broken path, it should be carried out by the parallel transport in (4.1), which is governed by the Klein-Gordon quantum connection defined in the preceding section. However, at this point we encounter the first physical difference between the present relativistic situation, and its nonrelativistic counterpart treated in Chapter 4, which has fundamental geometric implications: whereas in Newton-Cartan manifolds we were able to foliate in (4.4.1) the base manifold in a unique manner, due to the existence of a time metric $\boldsymbol{\beta}$ that led to a global time parameter t which could play the role of absolute time, this is no longer[19] the case with the Lorentzian base manifold **M** of the present Klein-Gordon quantum bundles.

On the other hand, there are good arguments[20] supporting the thesis that in CGR all the physically realistic classical models of spacetime must be described by globally hyperbolic manifolds **M** (cf. [M,W] or Sec. 11.1). Hence, although there is no absolute or preferred time parameter in CGR, any globally hyperbolic spacetime manifold can be viewed from a geometrodynamic perspective, namely as an evolution of 3-geometries

$$\mathbf{S} = \left\{\left(\mathbf{M}_t^{(3)}, g_t^{(3)}\right) \middle| \; t \in \mathbf{I} \subset \mathbf{R}^1\right\} \; . \tag{4.6}$$

The above set **S** then describes a family of diffeomorphic 3-dimensional Riemannian manifolds, whose metrics evolve smoothly in such a manner that, taken in conjunction, they give rise to the metric $\boldsymbol{g}$ of a Lorentzian manifold **M** (cf. Sec. 11.1 for further details).

From a more traditional CGR point of view, the 3-geometries in (4.6) are obtained as foliations of an already given 4-dimensional classical spacetime manifold **M**,

$$\mathbf{M} = \bigcup\nolimits_{t \in \mathbf{I}} \sigma_t \quad , \qquad \sigma_t = \left\{x \middle| (x^1, x^2, x^3) = \mathbf{x},\; x^0 = t\right\} \leftrightarrow \mathbf{M}_t^{(3)} \quad , \tag{4.7}$$

into diffeomorphic Cauchy surfaces, which we shall call *reference surfaces*, and which in general are maximal spacelike hypersurfaces in **M**. The coordinates in (4.6) emerge in a natural manner in the context of solving the initial-value problem in CGR (cf. Sec. 11.1; [M], Ch. 21; [W], Ch. 10) by the ADM method (Arnowitt, Deser and Misner, 1962). As a result, t indeed can play the role of a "global time". However, in CGR there is no single "global time", but rather an infinity of "global times", that correspond to an infinity of physically as well as mathematically feasible and yet distinct foliations of the form (4.7).

From a geometrodynamic perspective [M], a classical spacetime can be alternatively viewed as an equivalence class of *geometrodynamic evolutions* of the form (4.6), which provide *maximal* Cauchy developments (cf. Thm. 10.2.2 in [W]), in the sense that they cannot be isometrically embedded into some larger Lorentzian manifold, so as to become one of its proper subsets; the equivalence relation in this equivalence class can be defined by the requirement that two geometrodynamic evolutions are deemed equivalent if the Lorentzian manifolds to which they give rise are isometric (cf. Kuchař, 1976, 1977).

Mathematically, the coordinates in (4.7) can be obtained by following the integral curves of the vector field $\boldsymbol{\partial}_t$ dual to $\boldsymbol{dt}$ [W]. These timelike integral curves can be operationally interpreted as representing the worldlines of classical point particles. However, generically a classical point particle that follows such a worldline might not be in free fall, but rather it might have to be subjected to constant accelerations and decelerations to keep it on the prescribed worldline, which operationally require the imposition of some outside force field to maintain such a state of motion. On the other hand, we can always choose foliations and coordinate systems that correspond to coherent flows of classical test particles, which are operationally defined to a large extent as in the nonrelativistic context of Newton-Cartan spacetimes, except that, because of the lack of a flat connection in a generic Lorentzian manifold **M**, we cannot mathematically or operationally define the meaning of a 4-velocity field that is constant along an initial-data Cauchy surface σ_0 . Hence, in the present CGR context, a *coherent flow* of classical test particles is operationally defined by locating a point test particle at each point in an initial-data Cauchy surface σ_0, imparting to these test particles a 4-velocity such that the resulting 4-velocity field is orthogonal to σ_0, and then allowing all these test particles to move in free fall along the geodesics of the Levi-Civita connection in the considered spacetime. In this manner we can operationally introduce *synchronous (Gaussian normal) coordinates* (cf. [M], p. 717; or [W], p. 42), by first covering an initial-data Cauchy surface σ_0 with charts that assign to all $x \in \sigma_0$ one or more sets $\mathbf{x}$ of coordinates, and then using the coherent flow whose flow lines are orthogonal to σ_0 to assign corresponding coordinates to points $x \in \sigma_t$ for various choices of $t \in \mathbf{R}^1$, where t is the proper time of the classical test particles in that flow. The synchronous coordinates that result from this procedure provide a system of globally defined coordinates, that mathematically come closest[21] to the inertial Newton-Cartan coordinates defined in Sec. 4.4, and can be used as a basis in the comparison of GS propagation within the Klein-Gordon and Newton-Cartan bundles describing the same weak gravitational field from a relativistic and nonrelativistic perspective, respectively.

In adapting the basic physical and mathematical features of the GS propagation from Newton-Cartan to Klein-Gordon quantum bundles, we do not require, however, the adoption of any particular coordinate system. Rather, in defining free-fall GS propagation between two points x' and x'' in the base manifold of a Klein-Gordon bundle $\mathbf{E}$, we only require the foliation, provided by (4.6) and (4.7), of the segment in that base Lorentzian manifold situated between two hypersurfaces $\sigma_{t'}$ and $\sigma_{t''}$ containing the points x' and x'', respectively. With that foliation given, we can follow the pattern set in Sec. 4.4, by introducing hypersurfaces σ_{t_n}, $n = 0,\ldots,N$, corresponding to $t' = t_0 < t_1 < \cdots < t_N = t''$, and then connecting pairs of points x_{n-1} and x_n on two successive slices $\sigma_{t_{n-1}}$ and σ_{t_n} by the timelike geodesics of the Levi-Civita connection in $\mathbf{M}$.

However, at this point we encounter the second fundamental physical difference between the present relativistic situation, and its geometrized nonrelativistic counterpart studied in Chapter 4: the total absence of a light-cone structure in Newton-Cartan manifolds, as compared with its existence in Lorentzian manifolds. Thus, albeit in both the nonrelativistic and the relativistic regimes GS propagation is to take place along all possible timelike geodesics of the respective Newton-Cartan and Lorentzian manifolds, in the former case the tangents $X \in T_{x_n}\mathbf{M}$ at any point $x_n \in \sigma_{t_n}$ cover the entire tangent space $T_{x_n}\mathbf{M}$ (which is isomorphic to $\mathbf{R}^4$), except for the subspace of spacelike directions tangential to σ_{t_n} (which is isomorphic to $\mathbf{R}^3$, and therefore of lower dimensionality), whereas in the latter case they cover only the interior of the light cone at x_n ; hence, the subset of spacelike directions, which are inaccessible to any relativistically causal propagation, constitutes a submanifold of $T_{x_n}\mathbf{M}$ of the same dimensionality as $T_{x_n}\mathbf{M}$ itself.

The existence of the light-cone structure, in conjunction with the geodesic postulate, according to which free-fall relativistic propagation can take place only along timelike geodesics, has profound implications for free-fall GS propagation.

Indeed, in the nonrelativistic Newton-Cartan theories it is possible that any point $x_{n+1} \in \sigma_{t_{n+1}}$ can be reached by free-fall propagation from some given point $x_n \in \sigma_{t_n}$, and conversely, a point $x_n \in \sigma_{t_n}$ can receive contributions from all points $x_{n-1} \in \sigma_{t_{n-1}}$. That means that in a geometrically local nonrelativistic quantum theory, such as the one in Chapter 4, it is physically meaningful to insist on "conservation of probability", since any quantum state represented by some local quantum state vector $\Psi \in \mathbf{F}_{x_0}$, $x_0 \in \sigma_0$, can propagate instantaneously to all points along a hypersurface σ_{t_1} which lies infinitesimally close into its (nonrelativistic, and therefore absolute) future. The fact that in Newton-Cartan spacetime manifolds there is a flat connection, side-by-side with the Newton-Cartan connection, enabled us to make, in Sec. 4.4, the transition from GS propagation to a time-evolution governed by a Schrödinger equation, with a potential that incorporated a Newtonian gravitational potential. In that manner the *global* unitarity of that evolution was ensured. Hence, once a local quantum state had propagated all along σ_{t_1}, so that the quantum object in that state *could* be detected, with total probability one, *anywhere* along σ_{t_1}, then the probability of detecting it *anywhere* along σ_{t_2} at a later instant t_2 remained one. Thus, the *global probability of detection anywhere in space* was conserved, in accordance with the conventional interpretation of the QM concept of "conservation of probability".

On the other hand, in the present general relativistic regime, a local quantum state vector $\Psi \in \mathbf{F}_{x_0}$, $x_0 \in \sigma_0$, can propagate in a *strictly causal* manner (cf. Sec. 5.7) only to those points along a hypersurface σ_{t_1} which lie within the chronological future[22] $I^+(x_0)$, i.e., only to the points $x_1 \in \sigma_{t_1} \cap I^+(x_0)$; from each $x_1 \in \sigma_{t_1} \cap I^+(x_0)$ it can then propagate only to points $x_2 \in \sigma_{t_2} \cap I^+(x_1)$, etc. Hence, GS propagation that is *strictly* causal can reach only points which lie within the chronological future $I^+(x_0)$ of the initial base point x_0. Consequently, for such strictly causal relativistic GS propagation the concept of "conservation of probability", in the sense of the conservation of *global* probability of detection anywhere in "space", is not physically meaningful, since not all points along any of the hypersurfaces in that foliation are accessible to propagation of a locally prepared state of an object (cf. also the discussion in Sec. 7.3). In fact, this last remark applies equally well to the process of observation and detection by any physically realizable family of local "observers", since it would literally take an (uncountable) infinity of such "observers" to monitor all the points along the hypersurfaces of any one of the foliations in (4.7). Furthermore, even if we allow for the existence, in principle, of such a family of "observers", the outcomes of their "detection" activities would still be inconclusive from the point of view of *global* probability conservation. Indeed, on one hand, according to general relativistic principles, there does not exist the possibility of instantaneous exchange of information between them; on the other hand, in a curved spacetime it is impossible to carry out repetitions of preparatory and observational procedures under the *identical global conditions* . However, the possibility of such (systematically unbiased) repetitions is utterly essential to the concept of the statistical interpretation of conventional quantum mechanics.

It is in this last respect that the quantum mechanical *general* relativistic regime is different, at the deepest epistemological level, not only from the nonrelativistic regime, but also from the *special* relativistic regime. Indeed, in the special relativistic regime, the structure of spacetime is fixed by the adopted geometry of Minkowski space, which is not only globally homogeneous and isotropic [W], but also translationally invariant. Hence, any physical phenomenon that can be produced and observed in a finite region in Minkowski space, can be faithfully reproduced and observed under spatio-temporally identical conditions in any other region which can be obtained by spacetime translation from the first region. This possibility, however, is completely lacking in curved spacetime. Therefore, whereas it made sense to transfer the SQM concept of *globally* extended test bodies from the nonrelativistic regime of Sec. 3.2, to the relativistic regime of Sec. 3.4, that concept becomes physically meaningless in the general relativistic context[23]. In fact, any special relativistic physical concept that is based on *global* probabilistic notions (in the sense that it involves probability measures over spacelike hyperplanes, or more generally, over maximal spacelike hypersurfaces in Minkowski space) is bound to lead to paradoxes[24] when it is transferred in its original form to Cauchy surfaces in curved classical spacetimes. Indeed, close epistemological analysis often reveals that during the transferal process such a concept might have actually lost its physical meaning[25].

The following definition of the *geometro-stochastic propagator* in a Klein-Gordon quantum bundle, from the base location $x' \in \mathbf{M}$ to some future base location $x'' \in I^+(x')$ in the base Lorentzian manifold **M**,

$$K(x'',\zeta'';x',\zeta') = \lim_{\varepsilon\to+0} \int K(x_N,\zeta_N;x_{N-1},\zeta_{N-1})$$
$$\times \prod_{n=N-1}^{1} d\sigma(x_n)\, d\Sigma(\zeta_n)\, K(x_n,\zeta_n;x_{n-1},\zeta_{n-1}) \,, \tag{4.8}$$

bypasses these difficulties by, first of all, carrying out an averaging procedure only over relativistically causal broken geodetic paths (i.e., paths consisting of segments of future-directed timelike geodesics), prior to taking the limit $\varepsilon = \max(t_n - t_{n-1}) \to +0$; consequently, the integration in x_n is carried out in (4.8) not over the entire hypersurface σ_{t_n}, but rather only over $\sigma_{t_n} \cap I^+(x_{n-1}) \cap I^-(x'')$. Second, but equally essential, in (4.8) $K(x_n,\zeta_n;x_{n-1},\zeta_{n-1})$ denotes the propagator (4.2) for parallel transport along the timelike geodesic $\gamma(x_{n-1},x_n)$, connecting the points x_{n-1} and x_n on the two consecutive slices $\sigma_{t_{n-1}}$ and σ_{t_n}, renormalized by division with the area of $\sigma_{t_n} \cap I^+(x_{n-1}) \cap I^-(x'')$.

This renormalization of the propagator for local parallel transport replaces the δ-function in (4.4.7). Indeed, in weak gravitational fields we can introduce in the differential manifold **M** the linearized theory of gravity as a bridge between general relativity and the Newtonian or Newton-Cartan theory (cf. [M], Sec. 18.1; or [W], Sec. 4.4). In that case, similarly to the situation in the Newton-Cartan theory, the linearized theory supplies a flat connection, which is present side-by-side with the Levi-Civita connection in **M**, as well as a Minkowski metric with which that flat connection is compatible. Thus, there are preferential foliations of **M** into hyperplanes with respect to that coexisting Minkowski metric, for which we can carry out an identification of the tangent space $T_{x_n}(\sigma_{t_n})$ at each given $x_n \in \sigma_{t_n}$ with σ_{t_n} itself. Consequently, after such an identification has been carried out, the roles of the integration in $x_n \in \sigma_{t_n}$ and $q_n \in T_{x_n}(\sigma_{t_n})$ can be interchanged, in view of the dependence of the propagators on only the difference in the Minkowski coordinates of $x_n \in \sigma_{t_n}$ and of $q_n \in T_{x_n}(\sigma_{t_n})$. The resulting integration in q_n takes place effectively over the surface $\sigma_{t_n} \cap I^+(x_{n-1}) \cap I^-(x'')$, and on account of the continuity properties of SQM propagators, division by the area of $\sigma_{t_n} \cap I^+(x_{n-1}) \cap I^-(x'')$ produces in the limit $\varepsilon = \max(t_n - t_{n-1}) \to +0$ the effect of a δ-function at what was the point of contact between σ_{t_n} and $T_{x_n}(\sigma_{t_n})$ prior to their identification.

When an external gravitational field is absent in **M**, parallel transport in *K***M**, as well as in the Klein-Gordon quantum bundle **E**, becomes path independent. Consequently, the performance in equation (4.8) of the integrations in $d\Sigma(\zeta_n)$ at fixed choices of $x_n \in \sigma_{t_n}$, $n = 1,\ldots,N-1$, merely reproduces the formula (4.5), but with a renormalization constant on its right-hand side, and for a piecewise smooth rather than for a smooth curve. That renormalization constant gets cancelled, however, upon performing the integrations with respect to all the surface measures $d\sigma(x_n)$.

This means that, in the special relativistic regime, and upon adopting sections s of the Poincaré frame bundle $P\mathbf{M}$ which represent global Lorentz frames (cf. Sec. 2.3), the present concept of GS propagation faithfully reproduces, within the chronological future of each point in $\mathbf{M}$, the probability amplitudes of the relativistic SQM propagators. Indeed, in such Poincaré gauges, the identification (2.3.12) between s and a global Lorentz frame $\mathcal{L}$ can be carried out, and it leads to an identification of all the fibres $\mathbf{F}_x$ with the typical fibre $\mathbf{F}$, so that the SQM framework of Sec. 3.4 is recovered.

5.5. The Physical Interpretation of the GS Klein-Gordon Framework

The agreement between the SQM approach and the special relativistic regime of the GS approach to quantum propagation is essential in endowing both these mathematical frameworks with a consistent physical interpretation. This point becomes very clear when we compare the situation in the relativistic regime with that in the nonrelativistic regime.

Indeed, as we have seen in Sec. 3.1, in the nonrelativistic quantum regime there is a theory of sharp point particle localization, which *can* be deemed to be physically consistent *if* we accept all the basic physical premises of nonrelativistic physics. Therefore, on account of the marginality properties in (3.2.17), in the nonrelativistic SQM approach the $\mathbf{q}$-variables *could* be consistently interpreted as representing inertial Galilei stochastic coordinates, which marked the mean position of the stochastic values in (3.2.21), whereas the $(\mathbf{x},t)$-variables *could* be interpreted as coordinates marking the points of a classical Newtonian spacetime. In that context, an operationally deterministic interpretation could be assigned to the $(\mathbf{x},t)$-variables by means of coherent flows of classical test particles, for which t played the role of absolute time variable.

In the quantum special relativistic regime these possibilities are lacking for two fundamental reasons: 1) Hegerfeldt's (1974) theorem precludes the possibility of a theory of sharp point particle localization, which is physically consistent with the basic physical premises of relativistic physics, so that there is no counterpart of (3.2.17); 2) the lack of a frame independent concept of simultaneity prohibits the assignment of a special status to the time variable, so that there cannot be a physically consistent counterpart of the stochastic value in (3.2.21) for position measurements, in which the time variable does not appear. In this last respect, even if there were such stochastic values in a given global Lorentz frame, in the transition to a boosted global Lorentz frame the space stochastic fluctuations would become *spacetime* stochastic fluctuations, in a manner in which the space and time degrees of freedom would be inseparably intertwined. Thus, in the relativistic SQM framework, neither the $\mathbf{q}$-variables nor the $(\mathbf{x},t)$-variables can play, in a physically consistent manner, the role which they played in the nonrelativistic SQM framework of Sec. 3.2.

On the other hand, we observe that in the present GS approach the roles of the $(\mathbf{x},t)$-variables and of the q^i-variables are physically and mathematically quite different than they were in Sec. 3.2: the $(\mathbf{x},t)$-variables have lost their direct physical significance as operationally defined coordinates, and are merely abstract mathematical coordinates, with t merely labelling the 3-geometries of a geometrodynamic evolution depicted in (4.6), or the hypersurfaces in a foliation depicted in (4.7), and with $\mathbf{x}$ labelling points x in these 3-dimensional manifolds, in the purely mathematical sense of being assigned to each $x \in \sigma_t$ by charts (cf. Sec. 2.1); the q^i-variables are internal Poincaré gauge variables, related to a choice of section of the Poincaré bundle $P\mathbf{M}$, so that they no longer label points on hypersurfaces σ_t in the base manifold $\mathbf{M}$, but rather points in the tangent space at various $x \in \sigma_t$. It is only via the exponential map, determined by the Levi-Civita connection in $\mathbf{M}$, that the values which these gauge variables assume can be uniquely mapped into Riemann normal coordinates [M] of points in certain neighborhoods $\mathcal{N}_x$ of a base location $x \in \mathbf{M}$, in a manner which approximately preserves the spatio-temporal metric relationships between sufficiently small neighborhoods $\mathcal{N}_x$, and corresponding neighborhoods of the point of contact between the tangent space $T_x\mathbf{M}$ and $\mathbf{M}$, in the following sense (cf. [M], §11.6): if $l \in \mathbf{R}^1$ denotes the maximum linear dimension of a neighborhood $\mathcal{N}_x$, and R the maximum curvature radius within it (cf. Sec. 2.7), then the aforementioned spatio-temporal relationships are preserved up to terms of the second order in l/R in the Taylor expansion of the Lorentzian metric at x.

By using the exponential map on points in $T_x\mathbf{M}$, labelled by the q^i-variables, we can introduce the relative probability densities[26]

$$\rho_x^{\boldsymbol{u}}(\exp_x \boldsymbol{q}, v) = \left|\boldsymbol{\Phi}^{\boldsymbol{u}}_{(-a^i,\hat{v})}(q,v)\right|^2 , \qquad \boldsymbol{u} = (\boldsymbol{a}, \boldsymbol{e}_i) \in \Pi^{-1}(x) , \tag{5.1a}$$

$$\boldsymbol{q} = \boldsymbol{a} + q^i \boldsymbol{e}_i \in T_x\mathbf{M} , \qquad v \in V^+ \subset \mathbf{R}^4 , \qquad \hat{v} = (1,0,0,0) \in \mathbf{R}^4 , \tag{5.1b}$$

and thus provide quantitative measures for the *geometro-stochastic fluctuations* of the quantum metric $\boldsymbol{G}$ in (2.9) in the aforementioned neighborhoods $\mathcal{N}_x$. These GS quantum metric fluctuations at each base location $x \in \mathbf{M}$ reflect the fact that the points of a Klein-Gordon quantum bundle are *not* the points of a finite-dimensional classical spacetime manifold, but rather those of an infinite-dimensional quantum spacetime manifold, which incorporates directly in its geometry quantum degrees of freedom, in addition to those degrees of freedom which are present also in classical spacetimes. These local GS metric fluctuations should be therefore distinguished from the quantum fluctuations around globally computed mean values of wave functions, which are present in all quantum theories.

The presence in the GS framework of this nonlocal type of quantum fluctuations emerges when we follow the propagation of a local state vector $\boldsymbol{\Psi}_{x(0)}$, from a base location $x(0) \in \sigma_0$ to hypersurfaces σ_t in its future, which is governed by the GS propagator in (4.8). The coordinate wave functions

$$\Psi_{x(t)}(\zeta) = \int K(x(t),\zeta;x(0),\zeta')\, \Psi_{x(0)}(\zeta')\,d\Sigma(\zeta') \tag{5.2}$$

then give rise to *GS wave functions* $\{\boldsymbol{\Psi}_{x(t)} \,|\, x(t) \in \sigma_t \cap I^+(x(0))\}$. More generally, any physically feasible distribution $\{\boldsymbol{\Psi}_{x(0)} \,|\, x(t) \in \sigma_0\}$, obtained by the superposition principle from GS wave functions on any chosen initial-data hypersurface σ_0, will give rise to *GS wave packets* $\{\boldsymbol{\Psi}_{x(t)} \,|\, x(t) \in \sigma_t\}$ for $t > 0$.

By the extrapolation of the treatment of Newton-Cartan quantum bundles, the values which a GS wave function $\{\boldsymbol{\Psi}_{x(t)} \,|\, x(t) \in \sigma_t \cap I^+(x(0))\}$ assumes at the points of contact,

$$\psi(x(t),v) = \left\langle \boldsymbol{\Phi}^{\boldsymbol{u}(x(t))}_{\hat{\zeta}(x(t))} \middle| \boldsymbol{\Psi}_{x(t)} \right\rangle , \qquad \hat{\zeta}(x(t)) = (-\,a^i(x(t)), v^i) , \tag{5.3}$$

are to be interpreted as relative probability amplitudes. Thus

$$\int_B |\psi(x,v)|^2 \, d\Sigma(x,v) \ , \qquad B \subset \bigcup_{x\in\sigma_t} V_x^+ \subset TM \ , \tag{5.4}$$

represents the conditional probability that, if a local state $\Psi_{x(0)}$ of a quantum object of rest mass m and zero spin were prepared at $x(0) \in \sigma_0$, then that quantum object would be found by measurements of stochastic position and momentum, performed along σ_t with test bodies which are GS excitons of rest mass[27] m and proper state vectors η, to be located within the Borel subset B in general relativistic stochastic phase space.

We note that in the present GS regime we can, in principle, distinguish operationally between quantum fluctuations related to the spread-out nature of GS wave functions, and local GS metric fluctuations: the former are restricted to the chronological future $I^+(x)$ of a base point x where a preparatory measurement procedure has been carried out, whereas the latter can display contributions outside the causal future $J^+(x)$, since they are related to fundamental uncertainties in the metric relationships of a quantum spacetime, and are directly related to the fundamental length that enters the quantum spacetime form factor f.

It is important, as well as instructive, to examine the above physical interpretations for the special case of Klein-Gordon quantum frames corresponding to spin zero GS excitons of rest mass m, whose proper state vectors are the ground states of Born's (1938, 1949) quantum metric operator. Their explicit mathematical forms can be derived for the *fundamental quantum spacetime form factor* (cf. Sec. 12.5 for a discussion of its significance, as well as of the physical reasons and justification for such a choice)

$$f_\ell(u\cdot v) = \exp(-\ell\, u\cdot v) \ , \qquad \ell > 0 \ , \tag{5.5}$$

by adapting the expressions in (3.6.27)-(3.6.28) to the 4-velocity dependent measure in (1.10). This leads to the following renormalization factors:

$$Z_{\ell,m} = (4\pi^3 m^{-3} \int d^3\mathrm{v} \left| f_\ell(v^0)\right|^2)^{1/2} = (8\pi^4 \mathrm{K}_2(2\ell)/\ell m^3)^{1/2} \ . \tag{5.6}$$

The corresponding standard quantum frame elements,

$$\Phi_{\ell;q,v}^{u(x)}(q',v') = Z_\ell^{-2} \int_{u^0>0} \exp\{[i(q-q') - \ell(v'+v)]\cdot u\} \, d\Omega(u) \ , \tag{5.7}$$

can be therefore computed as in (3.6.28). We thus obtain the following explicit formula for *standard ground-exciton Klein-Gordon quantum frames*:

$$\Phi_{\ell;q,v}^{u(x)}(q',v') = \frac{2\pi\, \mathrm{K}_1\left(\sqrt{-(\bar q + i\ell\bar v)^\mu(\bar q + i\ell\bar v)_\mu}\right)}{Z_\ell^2 \sqrt{-(\bar q + i\ell\bar v)^\mu(\bar q + i\ell\bar v)_\mu}} \ , \tag{5.8a}$$

$$\bar q = q - q' \ , \qquad \bar v = v + v' \ , \qquad q, q' \in \mathbf{R}^4 \ , \qquad v, v' \in V^+ \ . \tag{5.8b}$$

These quantum frames are the relativistic counterparts of those in (4.4.21).

For the choice of fundamental quantum spacetime form factor given in (5.5), the integration in (1.14) can be easily carried out, with the following result:

$$\hat{Z}_{\ell,m} = \mathrm{K}_2(2\ell)/m\mathrm{K}_1(2\ell) = (2m)^{-1}\, d\ln[\ell/\mathrm{K}_1(2\ell)]/d\ell \ . \tag{5.9}$$

It is interesting to note the behavior of both renormalization constants in the sharp-point limit:

$$Z_{\ell,m} = (\ell^3 m^3/4\pi^4)^{-1/2} + O(\ell^{-1/2}) \xrightarrow[\ell\to+0]{} +\infty \ , \tag{5.10a}$$

$$\hat{Z}_{\ell,m} = (\ell m)^{-1} + O(1) \xrightarrow[\ell\to+0]{} +\infty \ . \tag{5.10b}$$

Thus, we see that upon performing "infinite renormalizations" to (5.7), in the special relativistic regime the GS propagator will merge into the free Feynman propagator for spin zero particles – with the antiparticle contribution coming from the backward 4-velocity hyperboloid (cf. Chapter 7).

*5.6. GS Propagation in Klein-Gordon Bundles and Quantum Diffusions

The concept of GS propagation in Klein-Gordon quantum bundles, formulated in Sec. 5.4, relied on the equivalence principle to provide the guidance needed in making the transition from the special relativistic regime treated in Sec. 3.4 to the present general relativistic regime. However, the equivalence principle cannot generally guarantee the uniqueness of this transition (Friedman, 1983, pp. 200-204). It is, therefore, worthwhile to investigate alternative formulations of the concept of GS propagation, which are capable of providing useful mathematical techniques as well as additional insights into this concept. Therefore, in this section we shall investigate the possibility of using the Itô-Dynkin method in the formulation of GS propagation.

The concept of GS propagation formulated in Sec. 5.4 emphasized the geometric aspect, whereas the stochastic aspect was manifested primarily by the "random" manner in which all possible broken causal paths were chosen for the parallel transports whose superpositions gave rise to a GS propagator between two base locations. This procedure is strongly reminiscent of the Itô-Dynkin treatment of stochastic parallel transport, that subsequently inspired the development of stochastic differential geometry methods (Daletskii, 1983; Belopolskaya and Dalecky, 1990). Indeed, at first Itô (1962) formulated the concept of Brownian motion in curved Riemannian manifolds, and used it to define stochastic parallel transport in certain tensor bundles over such manifolds. However, subsequently Dynkin (1968), as well as Itô (1975) himself, showed that no metric structure was required for such a formulation, and that a concept of stochastic parallel transport could be defined, by means of stochastic equations, in tensor bundles over any finite-dimensional manifold **M** with a connection. Furthermore, they also showed that such parallel transport can be alternatively formulated as a limit-in-probability of ordinary parallel transports that took place along broken paths consisting of geodesic arcs.

The formal analogy between the Itô-Dynkin procedure and the one underlying the definition of the GS propagator in (4.8) is unmistakable, but there are also essential differences. First of all, the Itô-Dynkin procedure requires an external universal time parameter t that assumes values within the interval $\mathbf{R}^+=[0,+\infty)$, and which, therefore, is not incorporated into the geometric structure of the manifold $\mathbf{M}$ itself. Hence, no notion of relativistic causality can be *directly* incorporated into their framework. Second, the Itô-Dynkin procedure deals with a well-defined probability measure over all continuous paths in the manifold $\mathbf{M}$, which is derived from the Wiener measure in the typical fibre of the tangent bundle $T\mathbf{M}$, and which is therefore not ordinarily present in the quantum regime.

Nevertheless, the existence of deeper roots for the formal analogy between GS propagation and diffusion processes in curved manifolds is suggested by the Feynman (1950) treatment of path integrals for the Klein-Gordon equation by means of a Schrödinger equation with a fictitious-time parameter, by the Parisi-Wu (1981) stochastic quantization method (Rivers, 1987; Damgaard and Hüffel, 1987), which uses a fictitious-time parameter related to "white noise" sources, and by the fundamental role which Wiener processes play in Euclidean quantum field theory[28] (Glimm and Jaffe, 1987). We shall, therefore, base our present investigation of the possibility of a diffusion-based formulation of GS propagation on the introduction of an auxiliary parameter τ related to diffusion processes in the base manifold $\mathbf{M}$ of the Klein-Gordon quantum bundle $\mathbf{E}$.

We begin by observing that the derivation of the irreducible systems of covariance in (3.4.14) was based on Hilbert spaces of positive-energy solutions (3.4.2) of the Klein-Gordon equations in (3.4.3). Those equations can be, however, extracted also from the following heat equations,

$$\left(\partial_\tau + \tfrac{1}{2}\partial_a\partial^a\right) f_\tau(q_v,v) = 0 \ , \qquad q_v \in \mathbf{R}^4 \ , \quad \partial_a = \partial/\partial q_v^a = -\partial^a \ , \quad a = 1,2,3,4 \ , \tag{6.1}$$

by considering solutions in $\tau \in \mathbf{R}^+=[0,+\infty)$ for which there are analytic continuations to pure imaginary values it, $t \geq 0$, of τ, and for the 4-th component of q_v, so that[29]

$$\varphi_m(q_v^0,\mathbf{q}_v,v) = \int_0^{+\infty} \exp(-im^2 t/2)\, f_{it}(\mathbf{q}_v, iq_v^0, v)\, dt \ , \qquad q_v^0 \in \mathbf{R}^+, \tag{6.2}$$

is well-defined. Such analytic continuations are routinely employed in the transition from the Euclidean to the Minkowski regime in quantum theory (cf. [I], Sec. 3-1-5; [ST], Chapter 9; Rivers, 1987, Chapter 6). Thus, by such standard techniques, we arrive at solutions of the following Klein-Gordon equations,

$$\left(\partial_\mu\partial^\mu + m^2\right)\varphi_m(q_v,v) = 0 \ , \qquad \partial_\mu = \partial/\partial q_v^\mu = \eta_{\mu\nu}\partial^\nu \ , \quad \mu,\nu = 0,1,2,3 \ . \tag{6.3}$$

These solutions are the analytic continuations of the solutions of the heat equation in (6.1) that are subjected to a subsidiary condition involving the rest mass m in such a manner that $m^2/2$ plays the role of eigenvalue for a formal "proper time operator" $i\partial_t$:

$$\left(-i\partial_t + \tfrac{1}{2}\partial_a\partial^a\right) f_{it,m}(q_v,v) = 0 \ , \qquad 2i\partial_t f_{it,m} = m^2 f_{it,m} \ . \tag{6.4}$$

After these solutions of the Klein-Gordon equation in (6.3) are obtained, they can be cast into equivalence classes, modulo Lorentz boosts Λ_v to 4-velocities v, by setting:

$$\varphi_m(\Lambda_v q, v) = \varphi_m(q, \hat{v}) \ , \qquad \forall q = q_{\hat{v}} \in \mathbf{R}^+ \times \mathbf{R}^3, \qquad \forall v \in V^+ \subset \mathbf{R}^4 \ . \quad (6.5)$$

We can now reinterpret the diffusion equation for $\hat{v} = (1,0,0,0) \in \mathbf{R}^4$ in (6.1) as a Fokker-Planck equation[30] for Brownian motion in $\mathbf{R}^4$ of a "phlogiston" – to borrow a terminology employed by Rivers (1987). Its solutions for initial conditions $f_0(q)$ at $\tau = 0$ that are continuous and Lebesgue-integrable in $\mathbf{R}^4$ can be then expressed in the form

$$f_\tau(q,\hat{v}) = E_{0,q}[f_0(q_{\tau,\hat{v}})] := \int_{\mathbf{R}^4} f_0(q')p(\tau,q;0,q')d^4q' \ , \qquad \hat{v} = (1,0,0,0) \ , \quad (6.6)$$

where p denotes the transition density of a Wiener process in $\mathbf{R}^4$ (Arnold, 1974) :

$$p(\tau,q;\tau',q') = (2\pi(\tau-\tau'))^{-2}\exp\left[-|q-q'|^2/2(\tau-\tau')\right] \ , \quad (6.7a)$$

$$\tau > \tau' \in \mathbf{R}^1 \ , \qquad q,q' \in \mathbf{R}^4 \ , \qquad |q-q'|^2 = \sum_{a=1}^4 \left(q^a - q'^a\right)^2 \ . \quad (6.7b)$$

Similarly, in accordance with (6.5), for each $v \in V^+$ we can interpret

$$f_\tau(q_v,v) = E_{0,q_v}[f_0(q_{\tau,v})] := \int_{\mathbf{R}^4} f_0(q',v)p(\tau,q_v;0,q')d^4q' \ , \quad (6.8a)$$

$$f_0(q',v) = f_0(\Lambda_v^{-1}q',\hat{v}) \ , \qquad v = \Lambda_v\hat{v} \in V^+ \subset \mathbf{R}^4 \ , \quad (6.8b)$$

also as a Wiener process that runs with respect to the global Lorentz frame

$$\left\{e_a^{(v)} = \Lambda_v e_a \middle| a = 4,1,2,3\right\} \ , \qquad e_4^{(v)} = v \ , \quad (6.9a)$$

$$e_4 = \hat{v} = (1,0,0,0) \ , \quad e_1 = (0,1,0,0) \ , \quad e_2 = (0,0,1,0) \ , \quad e_3 = (0,0,0,1) \ , \quad (6.9b)$$

which is viewed, however, as an orthonormal frame with respect to a Euclidean metric in $\mathbf{R}^4$, so that the formula (6.7) still applies, with the q-components replaced by the q_v-components with respect to these frames.

We have thus associated with the family of all global Lorentz frames in (6.9) a family $\{q_{\tau,v} \mid v \in V^+\}$ of Wiener processes $q_{\tau,v}$ for "phlogistons", all running simultaneously in $\mathbf{R}^4$ and satisfying at $\tau = 0$ initial conditions that are interrelated by (6.8b). Of course, these interrelationships do not remain true for $\tau > 0$, since this family of Wiener processes is obviously left invariant by the group SO(4), rather then by $SO_0(3,1)$. This indicates that τ should be viewed as being related to the internal "proper time" of "phlogistons" with respect to the v-dependent frame in (6.9a), rather than to the externally observable Minkowski time measured with respect to the global frame in (6.9b).

This externally observable Minkowski time coordinate with respect to the frames in (6.9b), which are interpreted as global Lorentz frames in Minkowski space, is the one denoted by q^0, and it is introduced in (6.2) by performing in (6.7) and (6.8) analytic continuation to pure imaginary values of $q^4 = iq^0$. Indeed, such a procedure results in the analytically continued Lorentz covariant families of functions

$$\hat{f}_{it}(q,v) = \hat{E}_{0,q_v}[f_0(q_{it,v})] \quad , \qquad t,q^0 \in \mathbf{R}^+, \tag{6.10}$$

which are the SQM counterparts of the functions containing a fictitious-time parameter in the Feynman (1950) path integral approach to the propagation of Klein-Gordon particles. However, the expectation values in (6.10), and their subsequent counterparts incorporating connection-dependent terms, represent the outcome of mathematically well-defined functional integrations, which can be performed by the Itô method (Arnold, 1974), or in accordance with the procedure for computing Feynman-Kac integrals for quantum propagators (Kac, 1959; Glimm and Jaffe, 1987), rather than by Feynman's (1948) purely formal method of path integration.

The transition from the SQM special relativistic regime, in which wave functions are represented by the elements Ψ of the typical fibre $\mathbf{F}$ of the considered Klein-Gordon quantum bundle $\mathbf{E}$, to the Euclidean regime in which the Wiener processes of (6.6)-(6.9) operate, can be effected in an unambiguous manner for those quantum spacetime form factors f for which the off-velocity shell extensions in (2.23) are well-defined – such as is the case with all those derived (Brooke and Prugovečki, 1984) from Born's quantum metric operator. Indeed, in that case the auxiliary wave functions

$$\hat{\Psi}(q,v) = Z_{f,m}^{-1}\int_{m-\Delta m}^{m+\Delta m} d\hat{m}\int_{\mathbf{R}^3}\exp\left[-iq\cdot\hat{k}\right]f(1-\tfrac{1}{2}(v-\hat{k}/\hat{m})^2)\tilde{\Psi}(\hat{k}/\hat{m})\,d^3\mathbf{k}/ik^4 \tag{6.11}$$

resulting from analytic extensions to the following k-values,

$$k \rightarrow \hat{k} = (-ik^4,\mathbf{k}) \quad , \qquad k^4 = \sqrt{\mathbf{k}^2+\hat{m}^2} \quad , \qquad \mathbf{k}\in\mathbf{R}^3 \ , \tag{6.12}$$

and integrations over mass-values from a fixed small interval around the considered rest mass value m, are well-defined for the dense set of vectors in the linear span of the coherent states constituting the quantum frame $\{\Phi_\zeta\}$ in $\mathbf{F}$ that is associated with the canonical frame in (6.9b). We shall denote by $\mathbf{F}^0$ the linear space of the corresponding Euclidean wave functions specified by the following four-dimensional Fourier transforms,

$$\hat{f}_0(q_v,v) = -iZ_{f,m}^{-1}\int \exp[iq_v\cdot k]\,f(1-\tfrac{1}{2}(v-\hat{k}/\hat{m})^2)\tilde{\Psi}(\hat{k}/\hat{m})d^4k \ , \tag{6.13}$$

in which the integration extends over the volume in $\mathbf{R}^4$ covered by the mass hyperboloids over which the integration in (6.11) is performed. Hence these Euclidean wave functions are the analytic extensions of the ones in (6.11) to pure imaginary values of q^0, so that

$$\hat{f}_0(q_v,v) = \hat{\Psi}(q,v) \quad , \qquad \hat{f}_0 \in \mathbf{F}^s \ , \qquad q_v^4 = -iq^0 \in \mathbf{R}^+, \tag{6.14}$$

in view of the fact that the inner product in (6.13) is Euclidean. Given now an initial Euclidean wave function in (6.14), the construction based on (6.6)-(6.9) leads to a Euclidean wave function $\Psi_\tau(q,v)$, which can be related to that initial wave function in (6.13) by means of (6.2), or, alternatively, by the following direct-integral decomposition:

$$\hat{\Psi}_{it} = \int_{\mathbf{R}^+}^{\oplus} \hat{\Psi}_{it,\hat{m}^2}\, d\hat{m}^2 \quad , \qquad \hat{\Psi}_{it,\hat{m}^2}(q,v) = \exp(-i\hat{m}^2 t/2)\, \hat{\Psi}(q,v) \quad . \tag{6.15}$$

This follows easily from the observation that, in accordance with (6.6),

$$\hat{f}_\tau(q_v,v) = -iZ_{f,m}^{-1} \int \exp\left[iq_v \cdot k - k^2\tau/2\right] f(1-\tfrac{1}{2}(v-\hat{k}/\hat{m})^2)\, \tilde{\Psi}(\hat{k}/\hat{m}) d^4k \quad . \tag{6.16}$$

We see that in the special relativistic regime it makes no difference which one of the Wiener processes $q_{\tau,v}$ is used, since they all eventually lead to the same quantum propagation in Minkowski space despite the fact that these diffusion processes display SO(4) invariance, rather than SO(3,1) invariance. Naturally, the key to recovering relativistic invariance lies in the choice of initial conditions, which belong to Hilbert spaces that carry relativistic systems of covariance (cf. Sec. 3.4), and to the analytic continuation method, which converts Euclidean into Minkowski metric relationships for wave functions in $\mathbf{R}^4$.

We can now extend these considerations to a curved Lorentzian manifold by using the Itô-Dynkin method. The most straightforward way to adapt this method to the present situation is to transfer the Brownian motions from $\mathbf{R}^4$ to the Lorentzian manifold $\mathbf{M}$, by using the vierbein fields described in Sec. 2.3, and by setting up for an atlas of coordinate charts (cf. Sec. 2.1) the stochastic differential equations

$$dx^\mu_{\tau,v} = \lambda_a{}^\mu(x_{\tau,v},v)\, dq^a_{\tau,v} - \tfrac{1}{2}\hat{g}_s^{\kappa\nu}(x_{\tau,v})\Gamma^\mu{}_{\kappa\nu}(x_{\tau,v})d\tau \quad , \qquad \hat{g}_s^{\kappa\nu} = \sum_{a=1}^4 \lambda_a{}^\kappa \lambda_a{}^\nu \, , \tag{6.17a}$$

$$e_a(x) = \lambda_a{}^\mu(x)\partial_\mu \quad , \qquad \lambda_a{}^\mu(x,v) = (\Lambda_v)_a{}^b \lambda_b{}^\mu(x) \quad , \qquad x^\mu_{0,v} = x^\mu \quad . \tag{6.17b}$$

The solutions of the above equations obviously describe interrelated diffusion processes in $\mathbf{M}$. In fact, if we view the basic diffusion process

$$dx^\mu(\tau) = \lambda_a{}^\mu(x(\tau))\, dq^a(\tau) - \tfrac{1}{2}\hat{g}_s^{\kappa\nu}(x(\tau))\Gamma^\mu{}_{\kappa\nu}(x(\tau))d\tau \quad , \qquad x^\mu(0) = x^\mu \quad , \tag{6.18a}$$

as being related to a section s of the Poincaré frame bundle $P\mathbf{M}$, whose elements are

$$s(x) = \left\{(a(x), e_i(x)) \middle| e_i(x) = \lambda_i{}^\mu(x)\partial_\mu \, , \; i = 0,1,2,3\right\} \quad , \qquad x \in \mathbf{M}^s \quad , \tag{6.18b}$$

then, for each $v \in V^+$, the process $x_{\tau,v}$ can be viewed as being related to the section

$$s^{(v)}(x) = \left\{(a(x), e_i^{(v)}(x)) \middle| e_i^{(v)}(x) = \lambda_i{}^\mu(x,v)\partial_\mu \, , \; i = 0,1,2,3\right\} \quad . \tag{6.19}$$

These diffusion processes give rise to a stochastic parallel transport that is compatible with the Lorentzian metric in $\mathbf{M}$ if we choose the connection coefficients in (6.17a) and (6.18a) to be those of a connection compatible with that metric – in particular, if we set them equal to the Christoffel symbols for the Levi-Civita connection (cf. (2.7.4)), as we shall do in the sequel. Indeed, according to the Itô-Dynkin method, such a stochastic parallel transport can be obtained from the deterministic parallel transport, defined in Secs. 2.4-2.6, as follows: all stochastic paths are first approximated with piecewise smooth curves γ which are constructed from geodesic arcs $\gamma(x(\tau_{n-1}),x(\tau_n))$ of the (in the present case) Levi-Civita connection in the Lorentz manifold **M**, then the geometric type of parallel transport is carried out along all these curves γ, and, in the end, a limit-in-probability with respect to the probability measure of a diffusion process defined by (6.17) is shown to exist as $\varepsilon = \max(\tau_n - \tau_{n-1}) \to +0$. Thus, the stochastic aspect of the definition manifests itself only in taking this last limit. The central role of the diffusion process is to enable the extension of the geometric concept of parallel transport, which requires curves that are piecewise smooth (namely continuous, and consisting of a finite number of sections which are C^1-smooth, at the very least), to paths which are continuous, but do not possess a tangent at any of their points. It is this latter type of paths that particles in all diffusion processes, for which Brownian motion is the prototype, follow with a probability equal to one (Arnold, 1974) – so that there is a zero probability for their following a smooth path.

On the other hand, it is rather obvious that the stochastic parallel transport based on (6.17) and (6.18) is devoid of all *direct* physical meaning in the relativistic context. Indeed, in the original papers by Itô (1962, 1975) and Dynkin (1968), the parameter τ operationally played the physical role of an external time variable, and the stochastic propagation proceeded indiscriminately in all possible directions of the manifold **M**, which in general was not required to possess any kind of metric. However, in the context of (6.16), (6.17) and (6.18), τ is an abstract parameter – and by referring to it by any such popular labels as "fictitious time", "intrinsic time", "parameter time", or "historical time" for "phlogistons", we obviously cannot endow it with actual physical meaning. We shall therefore view the above Itô-Dynkin method as merely providing the first step in the construction of solutions of the general relativistic Klein-Gordon equation in **E**,

$$\left(g^{\mu\nu}\nabla_\mu\nabla_\nu + m^2\right)\boldsymbol{\Psi}_{x,m^2} = 0 \ , \qquad \nabla_\mu = \nabla_{\partial_\mu} \ , \qquad \partial_\mu = \partial/\partial x^\mu \ . \tag{6.20}$$

The next step in this construction is to build, for each choice of section s of the Poincaré frame bundle P**M**, an auxiliary Euclidean quantum bundle $\mathbf{E}^s = \mathbf{P}(\mathbf{M}^s,\mathbf{G}) \times_{\mathbf{G}} \mathbf{F}^s$. This auxiliary bundle is associated with the principal subbundle $\mathbf{P}(\mathbf{M}^s,\mathbf{G})$ of L**M**, which has as structure group **G** the group SO(3), regarded as a subgroup of $SO_0(3,1)$ that leaves invariant the timelike elements of the tetrads in (6.18b). Its standard fibre $\mathbf{F}^s$ is a Hilbert space equal to the completion of the earlier described space $\mathbf{F}^0$ of Euclidean wave functions in (6.13) with respect to the inner product of the momentum space counterparts of those Euclidean wave functions, i.e., $\mathbf{F}^0$ is identifiable with the Hilbert space $L^2(\Omega,d^4k)$, where Ω is the region in $\mathbf{R}^4$ over which the k-integration is carried out in (6.13). Consequently, the typical fibre **F** of **E** can be embedded in $\mathbf{F}^s$, on the basis of (6.13)-(6.14). On the other hand, due to the constitution of $\mathbf{F}^s$, we can allow the elements of this Euclidean bundle $\mathbf{E}^s$ to propagate in accordance with Itô-Dynkin (1962, 1968) types of stochastic equations. Thus, we set up in the typical fibre $\mathbf{F}^s$ the following stochastic differential equations[31],

$$d\hat{\Psi}^\zeta_{\tau,x} = (\Gamma_\mu \hat{\Psi}^\zeta_{\tau,x})dx^\mu(\tau) + \left((\partial_\nu\Gamma_\mu)\hat{\Psi}^\zeta_{\tau,x} + \Gamma_\mu\Gamma_\nu\hat{\Psi}^\zeta_{\tau,x}\right)\hat{g}^{\mu\nu}_s d\tau \ , \tag{6.21a}$$

for Euclidean state vector coordinate functions, where

$$\Gamma_\mu = \partial_\mu - \nabla_\mu = i\lambda^j{}_\mu P_{j;u} - \tfrac{i}{2}\tilde{\omega}_{jk}(\partial_\mu) M_u^{jk} \ , \qquad \hat{g}_s^{\mu\nu} = \sum_{a=1}^{4} \lambda_a{}^\mu \lambda_a{}^\nu \ , \quad (6.21\text{b})$$

with Γ_μ obtained by inserting $X = \partial_\mu$ in (2.19). We note that the action of the operators Γ_μ on parallel transports of Euclidean state vectors with the wave function coordinates in (6.14), carried out along piecewise smooth curves γ constructed from geodesic arcs, is well-defined if we start at a given x from initial conditions $\Psi_{0,x} \in \mathbf{F}_x$ that are in their domain. Hence, by the Itô-Dynkin method we can construct the operators

$$T_{s,\gamma}(x(\tau),x): \ \mathbf{F}_x^s \to \mathbf{F}_{x(\tau)}^s \ , \qquad (6.22)$$

for stochastic parallel transport from x to $x(\tau)$ along the stochastic paths γ resulting from (6.18). Thus, for any initially prescribed field $\Psi_0 = \{\Psi_{0,x} \mid x \in \mathbf{M}\}$, that has support within $\mathbf{M}^s$, the GS Euclidean wave functions defined by the expectation values

$$\hat{\Psi}_{\tau,x} = \hat{E}_{0,x}[T_{s,\gamma}^{-1}(x(\tau),x)\hat{\Psi}_{0,x(\tau)}] \ , \qquad (6.23)$$

can be computed by performing Itô types of integration (Arnold, 1974) over all the stochastic paths joining x and $x(\tau)$ which lie within $\mathbf{M}^s$. Furthermore, by using the same line of argument as in (Dynkin, 1968), we easily establish that these GS Euclidean wave functions satisfy a heat equation in the base manifold $\mathbf{M}^s$ of the bundle $\mathbf{E}^s$:

$$\partial_\tau \hat{\Psi}_{\tau,x} = \tfrac{1}{2}\hat{g}_s^{\mu\nu}\nabla_\mu \nabla_\nu \hat{\Psi}_{\tau,x} \ , \qquad \hat{g}_s^{\mu\nu} = \sum_{a=1}^{4} \lambda_a{}^\mu \lambda_a{}^\nu \ . \qquad (6.24)$$

We observe that, up to this point, the above procedure is only ISO(3), rather than $ISO_0(3,1)$, gauge invariant. This is witnessed also by the presence of the Riemannian metric in (6.17), which also reflects an underlying ISO(4) gauge invariance.

To recover Poincaré gauge invariance, the transition to the relativistic regime is first carried out by analytic continuations which are the reversals of the earlier described ones that effected the transition from the Minkowski to the Euclidean regime. Hence, these analytic continuations take us from the real and positive values of τ and q^4 to the corresponding pure imaginary values it and iq^0, and are also extended via the vierbein fields to the Lorentzian metric in $\mathbf{M}^s$, so that (6.24) transcends into a general-relativistic counterpart of the Stueckelberg (1941) equation:

$$i\partial_t \hat{\Psi}_{it,x} = \tfrac{1}{2} g_s^{\mu\nu}\nabla_\mu \nabla_\nu \hat{\Psi}_{it,x} \ , \qquad g_s^{\mu\nu} = \eta^{ij}\lambda_i{}^\mu \lambda_j{}^\nu \ . \qquad (6.25)$$

We then impose in the auxiliary Euclidean quantum bundle $\mathbf{E}^s$ the same subsidiary conditions as in (6.4). This is tantamount to executing in a smooth manner across all its fibres direct integral decompositions similar to those in (6.15),

$$\hat{\Psi}_{it,x} = \int_{\mathbf{R}^+}^{\oplus} \hat{\Psi}_{it,x,\hat{m}^2}\, d\hat{m}^2 \ , \qquad \hat{\Psi}^{\zeta}_{it,x,\hat{m}^2} = \exp(-i\hat{m}^2 t/2)\, \hat{\Psi}^{\zeta}_{x,\hat{m}^2} \ , \qquad (6.26)$$

so that (6.23), in combination with (6.24), yields the general-relativistic Klein-Gordon equation in (6.20).

In the next chapter we shall have to require that a global section s of $L\mathbf{M}$ exists, in order to secure the existence on $\mathbf{M}$ of a spin structure for the formulation of quantum fermion fields. Hence, under those circumstances the recovery of Poincaré covariance can be carried out globally on $\mathbf{M}$.

In this manner we have arrived at a formulation of the propagation of GS excitons in terms of a stochastic parallel transport of the Itô-Dynkin type, which is based on diffusion processes. Hence, such propagation indeed deserves the name[32] of "quantum diffusion". However, although solutions of the Klein-Gordon equations (6.20) are thus obtained, and a Poincaré covariant mode of propagation results, this construction obviously does not incorporate relativistic causality in a form that is acceptable in CGR (cf. [W], Ch. 8), since the propagation proceeds along all possible stochastic paths, and it therefore does not totally preclude tachyonic behavior. This is a feature which the present procedure shares with all the other treatments of quantum propagation based on stochastic procedures in the Euclidean regime, that were mentioned at the beginning of this section. It therefore warrants closer scrutiny and discussion – which we shall provide in the next section.

5.7. Relativistic Causality and Quantum Stochasticity

The concept of quantum stochasticity, which underlies the formulation of GS propagation in Sec. 5.4 as well as in Sec. 5.6, is at the *epistemological* level very different from the concept of classical stochasticity that underlies Wiener as well as other classical diffusion processes, despite many existing mathematical analogies. Indeed, classical stochasticity is based on the conception that each single classical particle in a large ensemble of identical particles follows a single stochastic path, so that the probability measures, such as Wiener measures, which one associates with families of stochastic paths, merely reflect the observer's inability to predict which specific path a given particle is going to follow. Hence, classical stochasticity *intrinsically* deals with ensembles of particles. In contradistinction, the GS propagators formulated in either Sec. 5.4 or in Sec. 5.6 govern the behavior of single GS excitons in the same manner in which Feynman propagators govern the behavior of single quantum point particles: the GS propagator of a *single* exciton is a "sum" over all possible stochastic paths, where that "sum" does not reflect an averaging over an ensemble, but rather a genuine dynamic feature of quantum propagation. Hence, GS *quantum stochasticity* reflects the fact that the *quantum propagation of each single GS exciton takes place simultaneously over all possible stochastic paths*[33], and is obtained by taking suitable "sums" of propagators for parallel transport over all such *possible* paths – with the fundamental difference between the methods used in Sec. 5.4 and Sec. 5.6, respectively, lying in the choice of what is a "possible" or "allowed" path, as the subsequent discussion will make it clear.

Thus, geometro-stochastic propagation is stochastic by virtue of the nature of the paths along which it proceeds, and by its macroscopic manifestations in the context of measurement processes (cf. Sec. 1.4), rather than on the basis of associating with those paths an intrinsic probability measure that could remain independent of the nature of the measurement process – as is the case in the context of classical stochastic processes. This is therefore a totally different conceptualization of stochasticity than in Nelson's (1967, 1986) stochastic mechanics, which unsuccessfully[34] attempted to reduce quantum propagation to a stochastic dynamics of classical point particles, whose behavior is influenced by the presence of a hypothetical background field. From that point of view, quantum dynamics describes only ensembles of quantum particles whose time evolution is governed by a classical diffusion process, and whose collective behavior proceeds along a family of stochastic paths, with various probabilities being assigned to each measurable family of paths by some suitable diffusion model related to the Schrödinger equation (Guerra, 1981; Nelson 1967, 1986). In contradistinction, *no single GS exciton propagates over a single path*. This enables GS excitons to simultaneously embody the classical attributes of "waves" as well as of "particles", thus removing the dichotomy displayed by the orthodox interpretation of quantum mechanics (cf. Sec. 1.3): the exciton proper state vectors are represented by *internal* wave functions (i.e., wave functions belonging to quantum fibres $\mathbf{F}_x$), which, however, propagate externally (i.e., in the base manifold **M**) along stochastic paths in the manner of classical particles in diffusion processes, but which, on the other hand, are superimposed at each location in the base manifold in the manner which in classical physics is associated only with the behavior of waves.

Hence, the *mathematical* similarities between classical and quantum notions of stochasticity are somewhat deceptive even when the stochastic approach of Sec. 5.6 is adopted: in the classical cases exemplified by (6.6) and (6.23) one deals with probabilistic *mean values*, whereas in the quantum case resulting *after* the analytic continuation is carried out in the transition from the Euclidean-and-classical to the relativistic-and-quantum regime, one deals with a genuine *superposition of probability amplitudes* rather than with a mean value. Nevertheless, the fact that the same family of stochastic paths is used in both cases raises the question whether relativistic causality is satisfied in the quantum context – i.e., after such an analytic continuation is carried out. Indeed, in Sec. 5.6 the same family of stochastic paths is used for the fictional "phlogiston" propagation, as well as for the actual GS "exciton" propagation, so that, after the transition from the Euclidean and classical to the relativistic and quantum regime, many of the stochastic paths allowed in the exciton propagation will be noncausal – in the sense that their approximations by broken geodetic paths would consist of arcs of noncausal geodesics in the Lorentzian manifold **M**.

To answer the question concerning the effect of this on relativistic causality, let us first recall that in the context of classical fields, the universally accepted [M,W] concept of general relativistic causality is that of *Einstein causality*, according to which no classical field in a classical spacetime described by a Lorentzian manifold **M** can propagate outside the causal future $J^+(x)$ of any point $x \in \mathbf{M}$. Thus, if we view in the corresponding special relativistic context the Klein-Gordon equation in (3.3.5) as an equation for a classical field $\varphi(x)$ in the Minkowski space **M**, then the only relativistically invariant solutions that satisfy Einstein causality for initial data given along a maximal spacelike hypersurface σ are

$$\varphi(x) = \int_\sigma \Delta_R(x-x')\overleftrightarrow{\partial}'_\mu \,\varphi(x')\,d\sigma^\mu(x') \ , \qquad \partial'_\mu = \partial/\partial x'_\mu \ , \tag{7.1}$$

where $\Delta_R(x-x')$ is the retarded Green function for the Klein-Gordon equation, i.e., the well-known fundamental solution of the Klein-Gordon equation in Minkowski space [SI]:

$$\Delta_R(x-x') = -\theta(x_0 - x'_0)\Delta(x-x') \ , \qquad \Delta(x) = \Delta^{(+)}(x) + \Delta^{(-)}(x) \ , \tag{7.2}$$

$$\Delta^{(\pm)}(x-x') = \mp i \int_{k^0>0} \exp[\pm i(x'-x)\cdot k] d\Omega_m(k) \ , \qquad x,x' \in \mathbf{R}^4 \ . \tag{7.3}$$

On the other hand, the positive-energy solutions of the Klein-Gordon equation that were introduced in (3.3.4) are expressed in terms of a propagator $i\Delta^{(+)}(x-x')$ (cf. [SI], p. 169),

$$\hat{\varphi}(x) = -\int_\sigma \Delta^{(+)}(x-x') \overleftrightarrow{\partial}'_\mu \hat{\varphi}(x') d\sigma^\mu(x') \ , \qquad \partial'_\mu = \partial/\partial x'_\mu \ , \tag{7.4}$$

which does not vanish outside the causal future $J^+(x')$ of any point $x' \in \mathbf{M}$, but rather decreases exponentially in spacelike directions (cf. [IQ], p. Sec. 1-3-1; [SI], Sec. 7c).

This last fact is in keeping with Hegerfeldt's (1974) theorem as to acausality of quantum particle propagation in conventional relativistic quantum mechanics. However, it is not usually taken as a cause for concern, since the variable x is not deemed to be related to a quantum particle position, for the reasons discussed in Sec. 3.3 and in Note 13 to Chapter 3. Moreover, according to the conventional formulation of "locality" for quantum fields (Wightman and Gårding, 1964; Streater and Wightman, 1964), the input from Einstein causality takes place only in the form of "local (anti)commutativity", i.e., via the properties of the commutators for the boson quantum fields, and of the anticommutators for the fermion quantum fields. Thus, those (anti)commutators *supposedly* obey Einstein causality in the sense that they vanish outside the light cone[35]. For example, for a neutral free scalar quantum field $\phi(x)$ we have,

$$[\phi(x), \phi(x')] = i\Delta(x-x') \ , \qquad \phi(x) = \phi^*(x) = \phi^{(+)}(x) + \phi^{(-)}(x) \ , \tag{7.5}$$

where the above Green function $\Delta(x-x')$, defined in (7.2), is indeed "causal" in that sense.

This conventional formulation of "causality" in terms of "local (anti)commutativity" came, however, under more careful scrutiny (Wightman, 1971, 1973) after the discovery by Velo and Zwanziger (1969) that retarded Green functions are not always Einstein-causal even in the presence of external fields, but rather that non-perturbative "solutions to familiar spin equations propagate acausally, whereas the perturbative solution is causal in every order" (Velo and Zwanziger, 1971, p. 9). This means that some "local" quantum fields

$$\psi(x) = \psi^{\mathrm{in}}(x) - \int_{\mathbf{M}} S_R(x,x';B) B(x') \psi(x') d^4x' \ , \tag{7.6}$$

that obey standard relativistically invariant Dirac-type[36] equations [BR]

$$[\beta^\mu \partial_\mu + m + B(x)]\psi(x) = 0 \ , \qquad B(x) = -ie\beta^\mu A_\mu(x) \ , \tag{7.7}$$

with the minimal coupling to an external electromagnetic field specified in (7.7), as well as other types of couplings (Wightman, 1971, 1973), do *not* obey Einstein causality since the corresponding retarded Green functions, obtained as the solutions of

$$[\beta^\mu \partial_\mu + m + B(x)] S_R(x,x';B) = \delta^4(x-x') I \ , \qquad \partial_\mu = \partial/\partial x^\mu \ , \tag{7.8}$$

$$S_R(x,x';B) = 0 \quad , \qquad x_0 - x_0' < 0 \quad , \tag{7.9}$$

do not vanish when x and x' are spacelike separated; rather, they merely obey "weak retardedness", in the sense that as we proceed to infinity in some spacelike direction, they approach zero faster than some power of $(x-x')^2$. In Wightman's words, "this is a kind of instability of relativistic wave equations not anticipated by earlier work by physicists: *the particles described by the wave equation move faster than light in regions where the external perturbation is nonvanishing.*" (Wightman, 1973, p. 454).

The question can be therefore raised whether the original concept of Einstein causality should be retained at all in the quantum regime, once the *classical* concept of locality is relinquished in quantum relativistic theory. Indeed, one could argue that the metric tensor should be treated in the same manner in which all classical potentials corresponding to nongravitational forces are treated in quantum theory, so that "tunnel effects" can take place, whereby a quantum particle "tunnels" through those spacetime regions which are absolutely forbidden in the classical regime[37]. If the probability amplitudes for such "acausal" propagation vanish sufficiently rapidly as one moves away from the light cone – as is indeed the case with the SQM propagators in (3.6.28) (Greenwood and Prugovečki, 1984) – then no macroscopically observable conflict with Einstein causality would arise under normal circumstances, i.e., in the presence of weak gravitational fields, where the linearized gravity approximation can be employed (cf. Sec. 7.2).

To adapt this point of view to the case of the quantum diffusions of the preceding section, let us consider the Stueckelberg (1941) equation with minimal coupling to an external electromagnetic field,

$$i\partial_\tau \varphi_\tau(x) = \frac{1}{2M}\eta^{\mu\nu}(\partial_\mu - ieA_\mu(x))(\partial_\nu - ieA_\nu(x))\varphi_\tau(x) \quad , \tag{7.10}$$

which is the Minkowski space counterpart of (6.25). Some of its advocates (Horwitz and Piron, 1973; Horwitz 1984, 1992; Horwitz *et al.*, 1988; Saad *et al.*, 1989) view this equation as governing the propagation along various worldlines of a relativistic quantum particle displaying an unsharp mass spectrum – cf. (6.15). In so doing, they allow the possibility of quantum propagation along worldlines which are in part spacelike, while they undergo the transition from the "forward" to a "backward" orientation in Minkowski space (cf., e.g., Arshansky *et al.*, 1983, p. 1172), that is required by them to represent the transition from particle propagation to antiparticle propagation that leads to pair annihilation.

This gives rise, however, to the question as to the meaning of the parameter τ in (7.10) – which in that context is dubbed "historical time" (Horwitz, 1984, 1992). It turns out (Horwitz *et al.*, 1988, 1991) that as long as one restricts oneself to timelike worldlines, one can interpret τ as the proper time of the center of the wave packet $\varphi_\tau(x)$ representing a quantum particle in this approach[38]. However, that interpretation obviously falls by the wayside when one deals with null or with spacelike worldlines.

On the other hand, in the stochastic approach to the "phlogiston" propagation formulated in Sec. 5.6, the role of τ can be played by the affine parameter of the broken geodesics that approximate the stochastic paths in accordance with the Itô (1962) and Dynkin (1968) limit-in-probability theorems. Modulo a multiplicative factor and an additive term [C,SC], the affine parameter is well defined (cf. (2.7.10)) for any geodesic of any connection in any manifold **M**. Hence, in the present context it is well defined, modulo those two constants, upon the adoption of the Levi-Civita connection in the Lorentzian manifold **M**. In turn, those constants can be fixed by the requirement that, for a particular choice of initial-data reference surface σ_0 in the geometrodynamic evolution in (4.6), τ should coincide with proper time for those timelike geodesics that are normal to σ_0. The continuity properties of the paths for diffusion processes then uniquely determine the values of τ along all other paths in such a manner that τ is proportional (but not necessarily equal) to proper time along all those piecewise smooth paths which are timelike.

With these possibilities in mind, we are left in the present quantum geometry framework with two distinct alternatives for implementing quantum propagation: 1) The primarily geometric approach of Sec. 5.4, which abides by *strict relativistic causality*, in the sense that it allows propagation only along stochastic paths that can be approximated by broken paths consisting exclusively of arcs of *timelike* geodesics, and which we can therefore describe as representing *strictly causal GS propagation.* 2) The quantum diffusion approach of Sec. 5.6, which also allows propagation along stochastic paths which cannot be approximated by causal classical broken paths, but in which, in the present case of a single spin-zero exciton, the propagation is governed by the Klein-Gordon equation in (6.20); due to the superposition principle, propagation along classical acausal paths will produce negligible amplitudes, so that we shall describe as *weakly causal GS propagation.*

In this context we note that the strictly causal geometro-stochastic propagation cannot be governed by the Klein-Gordon equation, since the single-particle propagator $i\Delta^{(+)}(x-x')$ violates strict Einstein causality already in the quantum point particle case based on (3.3.5) and (7.4). On the other hand, in the second approach, the question remains as to how to formulate a form of relativistic causality which reflects the fact that effect cannot precede cause, i.e., that some absolute form of time-ordering is ingrained in the framework.

A complete answer to that question will not emerge until the GS approach to quantum gravity is fully developed in Chapter 11. However, a partial answer can be provided even at this stage if the geometrodynamic point of view advocated by Wheeler (1962, 1968) is adopted. According to that view, in reality the very "object which is central to all of classical general relativity, the four-dimensional spacetime geometry, simply does not exist, except in a classical approximation" ([M], p. 1183). In other words: "The dynamic object is not spacetime. It is space. The configuration of space changes with time." ([M], p. 1181).

And indeed, *classical* general relativity postulates the description of *all* of spacetime by a *single* four-dimensional Lorentzian manifold **M**, which despite its dynamical nature reflected by the Einstein field equations in (2.7.3), already *exists* in its entirety – *while in reality the geometrodynamic evolutions embodied in (4.7) are still taking place*, and relativistic observers are *actively partaking* in those evolutions. Thus, the CGR point of view is epistemologically compatible only with a *strict global determinism* that envisages all of physical spacetime as *predetermined in its entirety* at the very instant of birth of our universe. Since matter and geometry interact, that means that the evolutions of all matter fields, as well as of all non-gravitational radiation fields, also had to be determined, in their minutest details, at the instant of creation of the universe. Consequently, all the fundamental quantum epistemological notions which are based on the *intervention* of observers in the quantum measurement process, and which are therefore fundamentally stochastic in nature, have no place at all in such a conceptualization. In fact, the very act of "measurement", as opposed to mere "observation" in the sense of a passive registration of physical events, is *not consistent* with such a deterministic point of view, which postulates the actual existence *in toto* of a unique classical spacetime, whose geometry is nevertheless dynamically influenced by all states of motion of the matter present in it, including that which in part constitutes "observers" (cf. Sec. 11.12 for a fuller discussion of all these foundational issues).

On the other hand, according to the full-fledged theory of *quantum* spacetime, that will emerge in Chapter 11, such a quantum spacetime is to be mathematically described by a *superfibre bundle* constituting an infinite-dimensional supermanifold. Such a supermanifold can be foliated in an infinity of ways, thus arriving at various geometrodynamic evolutions (4.6) describing a universe in the process of evolution, in which matter can dynamically and *creatively* interact with geometry[39]. This, however, can give rise to various base Lorentzian manifolds $(\mathbf{M}, g)$, which can play the role of "classical spacetimes" in quantum bundles in which various classical gravitational fields provide mere approximations of a *quantum* gravitational field. However, as we shall first see in Chapter 7, and then elaborate in subsequent chapters, the fact that these external gravitational fields represent *approximations*, rather than accurate reflections of physical reality, taken in conjunction with the local nature of geometro-stochastic propagation, avoids the need of introducing an epistemologically questionable (Ballentine, 1973; Shimony, 1978) interpretation of such a *geometro-stochastic* quantum framework in terms of the many-world "scenario" advocated by Everett (1957), and more recently by Barrow and Tipler (1986).

The existence of geometrodynamic evolutions of the form (4.6) provides the required basis for the formulation of weak relativistic causality: geometro-stochastic propagation is *weakly causal with respect to a given geometrodynamic evolution* (4.6) if it takes place by stochastic parallel transport only along paths which are the limits, in the Itô-Dynkin sense, of broken paths consisting of arcs of geodesics that join points lying on *successive* reference surfaces in the geometrodynamic evolution provided by (4.6). Thus, propagation along certain spacelike geodesic arcs is allowed by such a weak causality criterion. On the other hand, the time-reversal of such propagation is ruled out, thus providing the assurance that cause always precedes effect.

On the other hand, in the quantum framework based on the Stueckelberg equation (7.10) the particle propagation along portions of worldlines that are oriented "backward" in (Minkowski) time is *formally* allowed, but it is interpreted as actually representing antiparticle propagation that takes place "forward" in time (Arshansky *et al.*, 1983). The same formal interpretation is provided in the context of Feynman's (1950) path-integral approach to the Klein-Gordon equation (cf. [ST], p. 227). However, strictly speaking, there is no room for antiparticles in the context of a single particle theory. Hence, such a "solution" to the causality problem has to be actually interpreted in the context of many-body quantum field theoretical models treated in later chapters.

We shall therefore postpone the further discussion of this point until we start dealing in Chapter 7 with quantum field theoretical models. Let us, however, emphasize that the present formulation of Klein-Gordon quantum geometries does allow the formulation of at least two modes of quantum propagation, which are very distinct in their physical implications: the one of Sec. 5.4 obeys *strict* relativistic causality, whereas the one in Sec. 5.6 definitely does not. The same, of course, remains true of the quantum geometries formulated in subsequent chapters. Epistemologically one should give preference to the strongly causal GS propagation, but, as it will be seen in Chapter 7, weakly causal propagation has the advantage of being *mathematically* closer to the formulations of path integrals in conventional quantum field theory. In the ultimate analysis, it is an experimentally decidable proposition whether *strict* Einstein causality is satisfied at the *micro*level. However, thus far no truly decisive experimental tests seem to have been carried out.

Notes to Chapter 5

1 The corresponding situation in the nonrelativistic regime was discussed in Note 12 to Chapter 4. In the general relativistic regime the situation is exacerbated by the fact that there are no constant universal time hypersurfaces analogous to the absolute time hypersurfaces in Newton-Cartan spacetimes. Consequently, it cannot be argued that a plane wave $\Phi_{\mathbf{k}}(\mathbf{x}) = (2\pi)^{-3/2}\exp(i\mathbf{k}\cdot\mathbf{x})$ is able to represent the quantum state of a point particle of sharp 3-momentum **k**, whose relative probability density $|\Phi_{\mathbf{k}}(\mathbf{x})|^2$ of detection at various locations **x** at the universal time $t = 0$ is constant, so that a degenerate form of the Heisenberg uncertainty principle is at least formally satisfied.

2 See the nonrelativistic definition (4.3.7) of such frames, its realization in (4.4.21), as well as Note 12 of Chapter 4, and then compare with (3.3.10) and (3.6.20), as well as with the relativistic quantum frames in Sec. 5.1.

3 Cf. (Gel'fand and Yaglom, 1960), (Gudder, 1979, 1988a) and (Exner, 1985), as well as the key references cited therein, such as the work of S. Albeverio and R. Høegh-Krohn (1976), C. DeWitt-Morette *et al.* (1979), etc. The physically stringent conditions to which we refer involve such assumptions as the request that the potentials be represented by bounded functions – which are not satisfied by realistic QM models in molecular, atomic or nuclear physics.

4 Cf., e.g., the comments on the trace-anomaly of quantum fields in curved spacetime on p. 220 of (Birrell and Davies, 1986), or on p. 213 of (Fulling, 1989), as well as the concluding section of the second of these two references.

5 As pointed out by Dirac, the canonical quantization of gravity "involves quantities of the form $\delta(x-x')$ $\delta_{,\tau}(x-x')$", but "there does not exist any general method for handling such quadratic quantities in the δ-function, free of inconsistencies" (Dirac, 1968, p. 543). Whether the Feynman "sum over histories" might produce, however, at least a computationally feasible algorithm is still unknown – cf., e.g., the article by B. De Witt in (Markov et al., 1988), p. 94-124.

6 This is an implicit part of the *doctrine of existence of classical reality*, discussed in Sec. 1.3, and dismissed there on account of its lack of epistemological soundness, reflected in the conceptual "paradoxes" as well as in the mathematical inconsistencies to which it gives rise in the quantum relativistic regime.

7 The rescaling defining this congruence is similar to the rescaling $\theta \to \theta' = m\theta$, which we defined at the beginning of Sec. 4.2, and which was originally introduced by Lévy-Leblond (1971) in order to set up an isomorphism between all the central extensions of the Galilei group for various $m > 0$, and the central extension for $m = 1$, namely the Bargmann group, that emerges as the *standard* central extension. In fact, we could have also introduced a similar congruence relationship between all quantum Newton-Cartan frames – the sole reason for not doing that being to avoid too many repetitive definitions.

8 Naturally, this analogy would become even more complete if we would restrict ourselves, in the classical context, to the Lorentz frame bundle LM – as can indeed be done in the context of CGR. In the quantum context, however, spacetime translational gauge degrees of freedom are embedded into the framework on account of the structure of quantum theory at its most fundamental level, and the notation in (2.3) reflects that fact, by including them in the values assumed by the ζ-indices.

9 Such fibres would be obtained directly if the definition of the quantum tensor bundle in (2.5) were given by taking corresponding Whitney products (cf., e.g., Osborn, 1982) of **E** and $\mathbf{E}^*$ bundles. That would have made the analogy with (2.1.11) and (2.1.12) more complete. On the other hand, the tensor bundles $T^{r,s}\mathbf{M}$ of Sec. 2.1 could have been defined as **G**-products of $GL\mathbf{M}$ with their respective standard fibres $\mathbf{F} = \mathbf{R}^r \times (\mathbf{R}^s)^*$, where $\mathbf{G} = \mathrm{GL}(4,\mathbf{R})$; a reduction of $GL\mathbf{M}$ to $L\mathbf{M}$, in conjunction with setting $\mathbf{G} = \mathrm{SO}_0(3,1)$, could have been then carried out in the presence of a Lorentzian metric. It should be also noted that all the definitions concerning quantum tensor bundles presented in this section, as well as the extension in (2.3) and (2.7) of the Einstein convention, could have been formulated in the context of Newton-Cartan quantum bundles. However, in that context, there is no deeper physical motivation for them.

10 As defined in Sec. 2.3, a Lorentzian manifold is a manifold on which a Lorentzian metric field g is defined. In CGR, some authors automatically include in this definition also the Levi-Civita connection, since historically connections were indeed originally *derived* from a metric. Throughout this monograph, we shall, however, treat connections and metrics as two independent concepts, which become related only when additional assumptions are introduced – of which their mutual compatibility is a most natural and significant feature physically as well as mathematically.

11 The remarks in Note 9 to Chapter 4 remain valid. In particular, as discussed shortly, the covariant derivative in (2.16) is defined only on a dense set of vectors Ψ_x .

12 Similar expressions are routinely provided (cf., e.g., [BL], p. 146) for the infinitesimal generators of the Poincaré group in the "configuration"-space representation treated in Sec. 3.3, in which the Q^i-operators are represented by operators of multiplication with the coordinate variable x^i . It should be noted, however, that such operators are not well-defined *in* the Hilbert space of solutions $\varphi(x)$ of the Klein-Gordon equation, since in general $x^i\varphi(x)$ is not again a solution of the Klein-Gordon equation. Hence, the situation is similar to the one in (2.22), where the problem is solved in (2.23) by extending the wave functions off the 4-velocity shell (cf. also Note 13).

13 The rigorous proof of this statement is quite straightforward (cf. Prugovečki, 1985, Appendix B), once care has been taken with justifying differentiation off the 4-velocity shell V^+ by introducing the wave function extensions specified in (2.23). Although (1.2) and (2.23) coincide on V^+, that is no longer true off the 4-velocity shell, since there the respective arguments of the quantum spacetime form factor f no longer have the same value, and the use of the integrand in (1.2) to define an extension does not lead to the appearance of the desired Q-operators in the Lorentz boost generators.

14 The domain of definition of this operator is always dense, since it contains the dense linear span $C_{u(x)}$ of all the generalized coherent states that belong to the Klein-Gordon quantum frame corresponding to the Poincaré frame $u(x)$ from the adopted section s of $P\mathbf{M}$. On account of (1.7), (1.15) and (2.20), $C_{u(x)}$ is actually $u(x)$-independent (and therefore also s-independent), so that it can be appropriately denoted by C_x. This linear set C_x , which is dense in the quantum fibre $\mathbf{F}_x$, is a core [PQ] for the covariant differentiation operator defined by (2.25) – cf. Theorem A.2 in (Prugovečki and Warlow, 1989b).

15 The proof of the equivalence of the definition of a Koszul connection (called an "affine connection" by Abraham and Marsden (1978), by Hawking and Ellis (1973), as well as by Straumann (1984)) to that of a connection on the tangent bundle $T\mathbf{M}$ viewed as a bundle associated to the general linear frame bundle $GL\mathbf{M}$ (or to the Lorentz frame bundle $L\mathbf{M}$, when a Lorentzian metric is present in $\mathbf{M}$), is not a mathematically trivial enterprise – as can be seen from Propositions 9 and 10 on pp. 368-371 of [SC]. In this context, the contents of the present section transfer the basic ideas and techniques presented in [SC] to the quantum bundle $\mathbf{E}$, viewed as a bundle associated to the (principal) Klein-Gordon frame bundle $K\mathbf{M}$ in (1.22), which is isomorphic to the Poincaré frame bundle $P\mathbf{M}$.

16 We implicitly identify here, as well as in the sequel, $L\mathbf{M}$ with the subbundle $\gamma(L\mathbf{M})$ of $P\mathbf{M}$, where γ is the map defined in (2.6.17). We restrict ourselves to sections of $L\mathbf{M}$, since the concept of torsion was defined in Sec. 2.6 only for linear connections – as is generally the case in the literature [C,I,SC,NT].

17 A direct conclusion could be reached if in (3.1) the hatted derivatives could be replaced by their unhatted counterparts. Such a replacement would lead, however, to an inconsistent result, namely that, for a connection with zero quantum torsion, the covariant derivatives in (2.16) do not depend on the values assumed by the connection coefficients. This (incorrect) conclusion can be reached by using the resulting (incorrect) expression for quantum torsion in (2.26), in the same manner in which (2.6.21) was used in (2.6.22), and then deriving for $\langle\nabla_X\Psi'|\Psi''\rangle$ a counterpart of (2.6.23), with X,Y and Z replaced within all the inner products by ∂_X ,∂_Y and ∂_Z , respectively. The aforementioned inconsistency would then be arrived at by choosing coherent sections $\{\Phi_\zeta\}$ of $K_L\mathbf{M}$, and some coordinate chart whose domain has a nonempty intersection with that of $\{\Phi_\zeta\}$, and then setting $\Psi' = x^\mu\Phi_{\zeta'}$, $\Psi'' = x^\mu\Phi_\zeta$ and $Y = Z = \partial_\mu$ in $\langle\nabla_X\Psi'|\Psi''\rangle$.

18 Note that the definition of the torsion form $\mathbf{\Theta}$ remains unchanged when the connection and canonical forms $\boldsymbol{\omega}$ and $\boldsymbol{\theta}$ are restricted to the Lorentz frame bundle $L\mathbf{M}$.

19 As mentioned in Note 18 to Chapter 1, there has been a recent suggestion (Unruh, 1989) to introduce a preferred "coordinate time" in quantum gravity, and to treat "time" in the general relativistic context in the same manner as in the Schrödinger equation of nonrelativistic QM. This suggestion is, however, counter to the spirit of CGR, as exemplified in all of Einstein's (1905-1949) writings. In fact, the existence of a preferred "coordinate time" as a physically observable time would contradict the general covariance principle. Of course, at first sight it might appear that such a preferred "coordinate time" can be extracted from the various cosmological models [N] prevailing at the *present time* – which is an historical, anthropomorphic and terrestrial time, and therefore very emphatically *local* ! On the other hand, it has to be recalled that the time-parameter in the Robertson-Walker metric used in these *large-*

scale cosmological models describes synchronous hypersurfaces in a universe in which matter has been smoothed out over supra-galactic-cluster distances, so that all "local" fluctuations in structure, which characterize the *actual* state of our universe even from a classical point of view, have been totally removed. Hence, the preferred "coordinate time" in these models is very much a *mathematical* coordinate time, in the sense of the coordinate charts defined in Sec. 2.1, rather than being a physical coordinate time, in the operational sense of being observable and measurable with any kind of *real* clocks – as is the case with the time coordinates of Minkowski space.

20 Cf. (Penrose, 1979), as well as Secs. 8.3 and 12.1 in [W]. Even if one does not accept such arguments, one can always split a classical spacetime manifold into domains of dependence of closed achronal sets, and apply the subsequently described geometrodynamic approach to all these domains. On the other hand, there are additional good arguments (cf. Wheeler, 1962, 1968, as well as [M], Chapter 43) in favor of accepting the geometrodynamic point of view as the more fundamental one in a quantum context.

21 The Riemann normal coordinates defined in a neighborhood of a point by means of the exponential map would come even closer if they could be globally defined, but that is not generally possible in the presence of curvature. Even the Gaussian normal coordinates generally develop coordinate singularities as one follows them for increasing values of t from the initial-data hypersurface σ_0, since the worldlines of some of the particles in the coherent flow will eventually intersect, but the locii of these intersections constitutes a manifold of lower dimensionality than that of **M** (Lifshitz and Khalatnikov, 1963). Hence, the Borel subset of **M** on which the synchronous flow coordinates are not defined is of Riemannian measure zero, and therefore of no consequence in the context of GS propagation.

22 Recall that the *chronological future* $I^+(x_0)$ of a point x_0 in a time orientable Lorentzian manifold **M** consists of all points $x \in \mathbf{M}$ which can be connected with x_0 by means of a future-directed timelike smooth curve – i.e., a curve whose tangent at any point lies within the interior of the future light cone at that point [M,W]. Similarly, the *causal future* $J^+(x_0)$ of the same point x_0 consists of all points $x \in \mathbf{M}$ which can be connected with x_0 by means of a future-directed causal smooth curve – i.e., a curve whose tangent at any point lies within the interior of the future light cone at that point, or on the future light cone itself. The definitions of the *chronological past* $I^-(x_0)$ and of the *causal past* $J^-(x_0)$ of the point x_0 are obtained by replacing in the above respective definitions the term "future" with the term "past". Note that if $x \in J^+(x_0)$ but $x \notin I^+(x_0)$ then x and x_0 can be connected only by null causal curves (cf. [W], p. 191), and the same holds true if $x \in J^-(x_0)$ but $x \notin I^-(x_0)$, so that massive particles cannot propagate between such points.

23 A misguided attempt of this nature (Prugovečki, 1981c) led to many of the difficulties encountered by conventional quantum field theory in curved spacetime – such as the need for global timelike Killing fields, and of restricting the foliations in (4.7) to those with hypersurfaces orthogonal to those fields. The only major weakness that it *did* manage to avoid was the paradoxical phenomenon of spontaneous particle creation *ex nihilo* – cf. the next note.

24 For example, the transference of the conventional special relativistic concept of the quantum propagation based on the Klein-Gordon equation to non-static curved spacetimes leads to the "prediction" of spontaneous particle creation *ex nihilo* (cf. [BD], Sec. 3.2) – i.e., not as part of the energy-conserving and well established phenomenon of *pair* creation (cf. Sec. 7.1).

25 In the case of the "explanation" of particle creation *ex nihilo* by transferring a special relativistic analysis involving accelerated observers to the inertial observers in a curved spacetime, the difficulty that emerges is that "as far as Minkowski space is concerned, the [Minkowski] vacuum is a strong candidate for the 'correct' or 'physical' vacuum, ... [whereas] when gravitational fields are present, inertial observers become free-falling observers, and in general no two free-falling detectors will agree on a choice of vacuum" ([BD], p. 55). This and other related points, as well as the solutions to these foundational problems offered by the GS approach based on quantum geometries, will be discussed in Secs. 7.2 and 7.3.

26 Fluctuations in timelike directions can be observed and measured by an adaptation (Prugovečki, 1982) of a method that Wigner (1972) had suggested in the context of defining and observing time-energy uncertainty relations.

27 The case of distinct rest masses for the system and for the test body can be treated by using the congruence relationships that give rise to the rescalings in (1.20), whereas the case of distinct spins can be treated by using corresponding SQM systems of covariance – such as the ones introduced in the next chapter for spin 1/2. It should be noted that in all these considerations the roles of "system" and "test

body" are interchangeable, as befits a purely quantum theory of measurement, which does not draw a sharp demarcation line between "system" and "apparatus".

28 There have also been attempts by Dohrn and Guerra (1978), and by Guerra and Ruggiero (1978), to adapt Feynman's (1950) idea to Nelson's (1967, 1986) formulation of stochastic mechanics, but they did not lead to a relativistically covariant theory. Furthermore, as recently pointed out by Wallstrom (1989), Nelson's stochastic mechanics displays even more fundamental difficulties, that seem to preclude its agreement with quantum mechanics at its most fundamental level, since it gives rise to states which possess no quantum mechanical counterparts.

29 The system of equations (6.1) and (6.2) represent a Euclidean counterpart of the system of equations in Appendix A of (Feynman, 1950) (cf. also [ST], Chapter 25). They lead to the treatment (Prugovečki, 1991b) of GS propagation reviewed in this section, which is in some respects also close to the Hudson-Parthasarathy (1984, 1986) treatment of quantum diffusions.

30 Also known as the Kolmogorov forward equation – cf. (Arnold, 1974), Sec. 2.6. The first five chapters of this standard textbook on classical stochastic processes by Arnold (1974) contain all the theory required for a mathematically thorough understanding of the use of this subject in the present chapter, but such a level of mathematical insight is not essential for a good understanding at the physics level. A review of all the required basic mathematical ideas can be found in Chapter 6 of (Namsrai, 1986). A good and reasonably self-contained guide to the basic theory of the Fokker-Planck equation, written from the physicist's point of view, can be found in Chapter VIII of the textbook by van Kampen (1981).

31 The original papers by Itô (1962, 1975) and by Dynkin (1968) dealt only with the case of finite-dimensional vector bundles. However, as recounted by Belopolskaya and Dalecky (1990), their techniques have been extended to the infinite-dimensional case. For the sake of simplicity of exposition, in this section we shall ignore the more technical aspects entailed by such an extrapolation, and direct instead the interested reader to Chapters 4-6 of (Belopolskaya and Dalecky, 1990).

32 Another concept of quantum diffusions, based on Fock space methods, was developed and systematically studied in the course of the past decade by Hudson and Parthasarathy (1984, 1986) and their collaborators (Applebaum, 1988; Applebaum and Hudson, 1984; Hudson and Robinson, 1988, 1989). Their method requires second-quantization techniques, whose possible significance for the GS framework described in Chapter 7 have been briefly described in Sec. 4 of (Prugovečki, 1991a).

33 Thus, the concept of GS propagation is decidedly at odds with the so-called statistical interpretation (cf. (Jammer, 1974), p. 468) of quantum mechanics (Prugovečki, 1967; Ballentine, 1970), according to which the quantum mechanical state vector pertains to ensembles, rather than to single particles.

34 As explicitly stated by Nelson, already his "interpretation [of quantum interference in the two-slit experiment] flatly contradicts the customary discussion of interference" presented in (Feynman and Hibbs, 1965), since according to him "(I) each particle must go through either the top slit or the bottom slit, and (II) the probability of arrival at a given point is the sum of two parts, [i.e.] the probability of arrival coming through the top slit and the probability of arrival coming through the bottom slit." (Nelson, 1985, pp. 92-93). It is therefore not surprising that, as recently proven by Wallstrom (1989), even at the most common nonrelativistic level Nelson's stochastic mechanics *is not equivalent* in its predictions to nonrelativistic quantum mechanics.

35 On the other hand, the Feynman propagator $i\Delta_F(x) = \theta(x^0)i\Delta^{(+)}(x) - \theta(-x^0)i\Delta^{(-)}(x)$, which is sometimes described as "causal", and which serves as a basis of perturbational quantum field theoretical computations, does not vanish outside the light cone, so that it is not Einstein-causal. The fact that it proved impossible (Glimm and Jaffe, 1987, p. 120) to establish the actual mathematical existence of any physically *nontrivial* relativistically invariant model of interacting quantum fields based on this notion of "locality" in (four-dimensional) Minkowski space indicates that this formulation of micro-locality is *physically* inconsistent (cf. Sec. 1.2 as well as Chapter 7), as can be expected to be the case on the basis of the epistemological considerations presented in Chapter 1 and in Chapter 12.

36 The prototype of these Bargmann-Wigner (1948) equations is the Dirac equation in (6.1.13), which corresponds to $\beta^\mu = -i\gamma^\mu$ in (7.7). The breakdown of strict Einstein causality has been established, however, only for those equations of the form (7.7) which correspond to quantum particles with spin values equal to one, or larger than one, interacting with external fields.

37 This possibility was also proposed by S. W. Hawking (1977) as an explanation for his well-known derivation of particle emission from a black hole (Hawking, 1975a; Wald, 1975). He states in the abstract of his review paper (Hawking, 1977) the following: "Black holes are often defined as areas from

which nothing, not even light, can escape. There is good reason to believe, however, that particles can get out of them by 'tunneling'." In Sec. 7.6 we shall point out that, in fact, if black hole radiation is actually observed, that would represent conclusive proof that at the quantum level *strict* Einstein causality is violated in nature.

38 The Stueckelberg-Piron-Horwitz framework does not associate a quantum particle with a fixed rest mass, but rather with an entire spectrum of rest masses that range around some mean value. In this, as well as other respects, it departs from generally accepted ideas as to what constitutes a quantum particle. However, in its latest developments, reviewed in (Horwitz, 1992), it is pointed out that this theory can lead to chaotic motion, so that "the notion of localizability may lose meaning even in the classical sense, ... [which] can have the effect of stabilizing the mass of the system, suggesting the restoration of a good mass shell approximation when this microcausality is relaxed."

39 As it will be seen in Chapter 11, this *fundamental* irreversibility of geometro-stochastic quantum evolution is in agreement with long-standing conjectures by Prigogine (1980, 1989) about the role of irreversible dynamics in nature and its relationship to non-locality (in the conventional sense), as well as with the conjectures of Penrose (1987), who views irreversibility as essential to any consistent integration of quantum mechanics with general relativity, and emphatically summarizes his views as follows: "*the true quantum gravity must be a time-asymmetrical theory*" (Penrose, 1987, p.37).

Chapter 6

Relativistic Dirac Quantum Geometries

In the present chapter we shall apply the physical ideas and mathematical techniques of the preceding chapter to quantum bundles associated with a Dirac quantum frame bundle $D\mathbf{M}$. These are principal bundles isomorphic to the principal bundle whose typical fibre equals the covering group [BR] ISL(2,C) of the Poincaré group $ISO_0(3,1)$. We shall concentrate on the case of quantum bundles whose standard fibres carry irreducible representations of ISL(2,C) for rest mass $m > 0$ and spin-1/2, but the same procedures can be applied with equal ease to the case of arbitrary spin.

In view of the significance which the Dirac equation enjoys in standard relativistic treatments of massive spin-1/2 quantum particles, we shall adopt standard fibres which carry systems of covariance based on Dirac-type bispinor representations of ISL(2,C). We shall, however, derive these representations from Wigner-type representations of ISL(2,C) for spin-1/2 massive relativistic quantum particles, rather than derive it from relativistic invariance requirements for the Dirac equation in configuration space, as is customary in standard textbooks on relativistic quantum theory (Berestetskii *et al.*, 1982; Bjorken and Drell, 1964; [IQ,SI]). As we shall see, such an approach has decided mathematical advantages when constructing Dirac quantum geometries, and yet it does not affect in the slightest the physical content of the framework, since the original motivation for introducing the Dirac equation in configuration space did not live up to its expectations.

Indeed, it is well-known that the Dirac equation was introduced by Dirac (1928) in the hope of resolving the problems with the localization of relativistic quantum point particles, to which the Klein-Gordon equation gave rise in view of the indefinite nature of the timelike component of the Klein-Gordon current in (3.3.9). At the time it appeared that the ensuing Dirac current in (1.17) did provide the solution to this problem, since its timelike component is evidently positive-definite. The first warning signs indicating that might not be the case, however, appeared when Breit (1928) discovered that the velocity operators associated with the "position" operators for Dirac bispinor wave functions had $\pm c$ as their only eigenvalues, thus paradoxically assigning a velocity equal to the velocity of light to a massive particle. This discovery was soon followed by the well-known paradox of Klein (1929), and Schrödinger subsequently discovered *Zitterbewegung*, namely the presence within the Dirac current of rapidly oscillating cross-terms between positive-energy (i.e., particle) and negative-energy (i.e., antiparticle) solutions of the Dirac equation.

In an independent mathematical investigation, Wigner (1939) derived and classified all unitary irreducible representations of ISL(2,C), thus uncovering the unitary irreducible representations for spin-1/2 massive relativistic quantum particles in (1.5) and (1.7), which are in outward appearance totally similar to those in (3.3.14) for the Klein-Gordon case. An operator that explicitly established the unitary equivalence between the Dirac-type and Wigner-type representations was constructed by Foldy and Wouthuysen (1950). This implicitly indicated that "the Klein-Gordon equation is neither better nor worse than the Dirac equation" (Wightman, 1972, p. 98) with regard to the issue of relativistic quantum point particle localization, since any inconsistency with particle localization occuring in one of these two contexts is mapped by means of these operators into a corresponding inconsistency in the other context. An explicit demonstration of this fact was provided by Thaller (1984), who rigorously proved that *all* positive-energy solutions of the Dirac equation are different from zero at *all* points in Minkowski space, so that *none* of these solutions can describe a state of a spin-1/2 *point* particle localized in a finite region in spacetime.

Although Dirac's treatment of spin-1/2 massive relativistic quantum particles has not provided the solution to the basic problem of relativistic quantum point particle localization that constituted its original motivation, the significance of the Dirac equation has remained unaffected on account of the fact that as a linear relativistically invariant equation it supplies the prototype for all other similar equations for higher spins [BR], and that the second-quantized form of its current plays a key role in quantum electrodynamics.

In Sec. 6.1 we shall briefly review the most salient features of the Wigner and of the Dirac types of wave functions for spin-1/2 particles and antiparticles. After introducing in Sec. 6.2 the corresponding phase space wave functions, we shall formulate in Sec. 6.3 the concept of Dirac quantum frame. The ensuing concept of Dirac quantum frame bundle will then be used in Sec. 6.4 to define and study parallel transport in Dirac quantum bundles associated with those quantum frame bundles.

*6.1. Spinorial Wave Functions in Wigner and Dirac Representations

The concept of spinor, originally introduced by Cartan (1913, 1966), can be best understood in terms of its group-theoretical significance. To see that, we first note that Minkowski space can be identified with the pair $(\mathbf{R}^4,\eta)$ consisting of the 4-dimensional real vector space $\mathbf{R}^4$ and the Minkowski metric η, i.e., the symmetric bilinear form $q'\cdot q'' = \eta_{\mu\nu}q'^{\mu}q''^{\nu}$, $q', q'' \in \mathbf{R}^4$, defined by (2.3.4) and (2.3.8). The Lorentz group can be then identified with the group O(3,1) of matrices Λ that leave this bilinear form invariant, so that they satisfy (2.3.7). Similarly, any *spinor space* W (cf. [W], Sec. 13.1) can be identified with the pair $(\mathbf{C}^2,\boldsymbol{\varepsilon})$ consisting of the 2-dimensional complex vector space $\mathbf{C}^2$ and the antisymmetric bilinear form

$$w'\bullet w'' = \varepsilon_{AB}w'^{A}w''^{B}, \qquad \varepsilon_{00} = \varepsilon_{11} = 0 \ , \quad \varepsilon_{01} = -\varepsilon_{10} = 1 \ , \quad A,B = 0,1 \ . \qquad (1.1)$$

A necessary and sufficient condition for a linear transformation $w \mapsto Aw$, given by the matrix A, to leave this bilinear form invariant is obviously that

$$\varepsilon_{AB} = \varepsilon_{CD}\,A^{C}{}_{A}\,A^{D}{}_{B} \qquad \Leftrightarrow \qquad \det A = 1 \ . \qquad (1.2)$$

Hence the group of 2×2 complex matrices A which leave this bilinear form invariant coincides with SL(2,C).

If W^* denotes the dual of $W = (\mathbf{C}^2, \boldsymbol{\varepsilon})$, then an identification of the Minkowski space $(\mathbf{R}^4, \boldsymbol{\eta})$ with $W^* \otimes W$ can be carried out by means of the map [BL,W]

$$q \mapsto \underset{\sim}{q} = 2^{-1/2} q^\mu \sigma_\mu = \frac{1}{\sqrt{2}} \begin{pmatrix} q^0 + q^3 & q^1 - iq^2 \\ q^1 + iq^2 & q^0 - q^3 \end{pmatrix} , \tag{1.3a}$$

$$\sigma_0 = \begin{pmatrix} 1 & 0 \\ 0 & 1 \end{pmatrix} , \quad \sigma_1 = \begin{pmatrix} 0 & 1 \\ 1 & 0 \end{pmatrix} , \quad \sigma_2 = \begin{pmatrix} 0 & -i \\ i & 0 \end{pmatrix} , \quad \sigma_3 = \begin{pmatrix} 1 & 0 \\ 0 & -1 \end{pmatrix} , \tag{1.3b}$$

where σ_1, σ_2 and σ_3 are the well-known Pauli matrices [BR,BL]. In turn, this identification enables the introduction of the two-to-one covering map ISL(2,C) $\to$ $\mathrm{ISO}_0(3,1)$, whereby $\Lambda(\pm A)$ is the element of $\mathrm{SO}_0(3,1)$ assigned to $\pm A \in \mathrm{SL}(2,\mathrm{C})$ by the well-known requirement[1] that for all $q \in \mathbf{R}^4$

$$A \underset{\sim}{q} A^* = 2^{-1/2} \Lambda^\mu{}_\nu q^\nu \sigma_\mu = \underset{\sim}{\Lambda q} \qquad \Leftrightarrow \qquad \Lambda^\mu{}_\nu = \tfrac{1}{2} \mathrm{Tr}\left(\sigma_\mu A \sigma_\nu A^*\right) . \tag{1.3c}$$

The unitary irreducible representation of ISL(2,C) for spin-1/2 relativistic quantum particles of rest mass m derived by Wigner (1939) acts on spinorial wave functions in the momentum representation. These spinor-valued wave functions can be deemed to assign to each 4-momentum value $k \in V_m$ a one-column complex matrix $\tilde{w}^{(+)}(k)$ if $k^0 > 0$, so that the inner product of the Hilbert space to which they belong can be written in a form analogous to that in (3.3.13):

$$\left(\tilde{w}_1^{(+)} \middle| \tilde{w}_2^{(+)}\right) = \int_{k^0 > 0} \tilde{w}_1^{(+)*}(k) \tilde{w}_2^{(+)}(k)\, d\Omega_m(k) . \tag{1.4}$$

The unitary irreducible representation of ISL(2,C) for spin-1/2 relativistic quantum particles originally derived by Wigner (1939) can be obtained by the general method of induced representations (cf. [BR], Ch.17, §2.D), and is of the form

$$\tilde{\tilde{U}}^{(+)}(a,A) \; : \; \tilde{w}^{(+)}(k) \mapsto \tilde{w}^{(+)\prime}(k) = \exp(ia \cdot k) D_+^{1/2}(A,k) \tilde{w}^{(+)}(\Lambda^{-1}k) , \tag{1.5a}$$

$$D_+^{1/2}(A,k) = (k^\nu \sigma_\nu)^{-1/2} A (A^{-1} k^\nu \sigma_\nu A^{*-1})^{1/2} , \qquad A \in \mathrm{SL}(2,\mathrm{C}) , \tag{1.5b}$$

where Λ equals the earlier described $\Lambda(\pm A) \in \mathrm{SO}_0(3,1)$.

The corresponding unitary irreducible representation of ISL(2,C) for spin-1/2 relativistic quantum antiparticles acts in the Hilbert space of spinor-valued wave functions on the backward mass hyperboloid, which carries the inner product

$$\left(\tilde{w}_1^{(-)} \middle| \tilde{w}_2^{(-)}\right) = \int_{k^0 < 0} \tilde{w}_1^{(-)*}(k) \tilde{w}_2^{(-)}(k)\, d\Omega_m(k) . \tag{1.6}$$

The action of its elements $\tilde{\tilde{U}}^{(-)}(a,A)$ on these antiparticle wave functions is obtained by the complex conjugation of that in (1.5).

The direct sum of these two representations provides the Wigner-type representation,

$$\tilde{\tilde{U}}(a,A): \quad \tilde{w}(k) = \tilde{w}^{(+)}(k) \oplus \tilde{w}^{(-)}(-k) \mapsto \exp(\pm ia \cdot k) D^{1/2}(A,k) \tilde{w}(\Lambda^{-1}k) \ , \tag{1.7a}$$

$$D^{1/2}(A,k) = D_{+}^{1/2}(A,k) \oplus D_{-}^{1/2}(A,k) \ , \qquad D_{-}^{1/2}(A,k) = \overline{D}_{+}^{1/2}(A,k) \ , \tag{1.7b}$$

for spin-1/2 massive particles and antiparticles. This representation acts on the direct sum of the above introduced Hilbert spaces for spin-1/2 particles and antiparticles, which can be identified with the Hilbert space of bispinor-valued functions whose values are one-column matrices with four components.

From the well-known Foldy-Wouthuysen (1950) transformation it can be deduced that the transition

$$\tilde{U}_{FW}: \ \tilde{w}^{(\pm)} \mapsto \tilde{u}^{(\pm)} \ , \qquad (\gamma^{\mu} k_{\mu} \mp m)\tilde{u}^{(\pm)}(\pm k) = 0 \ , \quad k \in V_m^+ \ , \tag{1.8}$$

to the Dirac-type bispinor wave functions can be achieved by multiplying the Wigner bispinor wave function, viewed as a one-column matrix with four elements, with a 4×4 matrix-valued function of the 4-momentum k:

$$\tilde{u}(k) = k_0^{-1}[m/2(k_0 + m)]^{1/2}(k_0 + m - \vec{\gamma} \cdot \vec{k})\tilde{w}(k) \ , \qquad k = (k_0, \vec{k}) \in V_m^+ \ , \tag{1.9a}$$

$$(\gamma^{\mu} k_{\mu} - m)\tilde{u}(k) = 0 \ , \qquad \tilde{u}(k) = \tilde{u}^{(+)}(k) \oplus \tilde{u}^{(-)}(-k) \ , \tag{1.9b}$$

$$\gamma^0 = \beta = \begin{pmatrix} \sigma_0 & 0 \\ 0 & -\sigma_0 \end{pmatrix}, \qquad \vec{\gamma} = \begin{pmatrix} 0 & \vec{\sigma} \\ -\vec{\sigma} & 0 \end{pmatrix} . \tag{1.9c}$$

For the particular choice of Dirac matrices in (1.9c), the above Dirac momentum-space wave functions transform[2] in accordance with the following representation of ISL(2,C):

$$\tilde{U}(a,A) \ : \ \tilde{u}(k) \mapsto \tilde{u}'(k) = \exp(ia \cdot k) S(A) \tilde{u}(\Lambda^{-1}k) \ , \quad A \in \mathrm{SL}(2,\mathbf{C}) \ , \tag{1.10a}$$

$$S(A) = S_0 \begin{pmatrix} A & 0 \\ 0 & A^{*-1} \end{pmatrix} S_0^{-1} \ , \qquad S_0 = S_0^{\mathrm{T}} = S_0^{-1} = \frac{1}{\sqrt{2}} \begin{pmatrix} \sigma_0 & \sigma_0 \\ \sigma_0 & -\sigma_0 \end{pmatrix} . \tag{1.10b}$$

The inner product under which the map in (1.10a) is unitary is given by (cf. [BL], p. 188):

$$\langle \tilde{u}_1 | \tilde{u}_2 \rangle \ = \ \int_{k^0 > 0} \tilde{u}_1{}^*(k) \gamma^0 \left(\gamma^{\mu} k_{\mu} / m \right) \tilde{u}_2(k) \, d\Omega_m(k) \ . \tag{1.11a}$$

In view of (1.8), for the positive-energy and negative-energy Dirac wave functions, respectively, we have

$$\left\langle \tilde{u}_1^{(\pm)} \middle| \tilde{u}_2^{(\pm)} \right\rangle \ = \ \pm \int_{V_m^{\pm}} \tilde{u}_1^{(\pm)*}(k) \gamma^0 \tilde{u}_2^{(\pm)}(k) \, d\Omega_m(k) \ = \ \left(\tilde{w}_1^{(\pm)} \middle| \tilde{w}_2^{(\pm)} \right) . \tag{1.11b}$$

Upon performing the transition to the so-called configuration representation, whose elements are represented by the wave functions

$$\hat{\psi}(x) = (2\pi)^{-3/2} \int_{k^0>0} \exp(-ix^\mu k_\mu) \tilde{u}(k)\, d\Omega_m(k) \ , \tag{1.12}$$

we can easily establish [BL] that these wave functions satisfy the Dirac equation in its most familiar form:

$$(i\gamma^\mu \partial_\mu - m)\hat{\psi}(x) = 0 \ , \qquad \partial_\mu = \partial/\partial x^\mu \ . \tag{1.13}$$

The propagation of the positive-energy and negative-energy solutions of (1.13) can be expressed by means of propagators which are the respective counterparts of the ones in (5.7.3) (cf. [SI], Sec. 8b):

$$\hat{\psi}^{(\pm)}(x) = -i\int_\sigma S^{(\pm)}(x-x')\gamma^\mu \hat{\psi}^{(\pm)}(x')\, d\sigma_\mu(x') \ , \tag{1.14a}$$

$$S^{(\pm)}(x-x') = -(i\gamma^\mu \partial_\mu + m)\Delta^{(\pm)}(x-x') \ . \tag{1.14b}$$

Furthermore, for these Dirac bispinor wave functions the transformation law in (1.10a) gives rise to the representation

$$\hat{U}(a,A) \ : \ \hat{\psi}(x) \mapsto \hat{\psi}'(x) = S(A)\,\hat{\psi}(\Lambda^{-1}(x-a)) \ , \qquad A \in \mathrm{SL}(2,\mathbb{C}) \ . \tag{1.15}$$

It is evident, however, that the no-go theorem of Hegerfeldt (1974) applies to the Wigner-type representation derived from (1.5), and in view of the unitary equivalence of that representation to the Dirac-type representations in (1.10) and (1.15), it applies equally well to the latter. Hence, (1.15) cannot be associated with a system of imprimitivity (cf. Sec. 3.1) for the *sharp* localization in Minkowski space of spin-1/2 quantum particles.

The inner product in (1.11) can be expressed in terms of the Dirac wave functions in (1.12) as follows,

$$\langle \hat{\psi}_1 | \hat{\psi}_2 \rangle = 2m\int_{x^0=0} \hat{\psi}_1{}^*(x)\hat{\psi}_2(x)\, d^3\vec{x} = 2m\int \hat{\bar{\psi}}_1(x)\gamma^\mu \hat{\psi}_2(x)\, d\sigma_\mu(x) \ , \tag{1.16}$$

where the second of the above integrations can be carried out over any spacelike hyperplane. More generally, due to the conservation of the Dirac current,

$$j^\mu(x) = \hat{\bar{\psi}}(x)\gamma^\mu \hat{\psi}(x) \ , \qquad \hat{\bar{\psi}}(x) = \hat{\psi}^*(x)\gamma^0 \ , \qquad \partial_\mu j^\mu = 0 \ , \tag{1.17}$$

it easily follows that this integration can be carried out along any maximal spacelike hypersurface σ in Minkowski space.

We note that the Dirac current in (1.17) is positive definite, i.e., that $j^0(x) \geq 0$ at all points in Minkowski space. However, on account of the difficulties mentioned in the introduction to this chapter, this current does not supply a consistent solution to the problem of *sharp* localization in Minkowski space of spin-1/2 quantum particles. In particular, the result by Thaller (1984), who rigorously proved that *all* positive-energy solutions of the

Dirac equation are different from zero at *all* points in Minkowski space, demonstrates that these solutions do not describe a spin-1/2 *point* particle localizable in any bounded region in spacetime. This conclusion is consistent with the fact that albeit the Green function

$$S(x-x') = S^{(+)}(x-x') + S^{(-)}(x-x') \tag{1.18}$$

obeys Einstein causality, the Dirac particle and Dirac antiparticle propagators in (1.14b) do not obey it, since their counterparts in (5.7.3) do not vanish outside the causal future $J^+(x')$ of any point $x' \in \mathbf{M}$, but merely decrease exponentially in spacelike directions.

6.2. Standard Fibres for Dirac Quantum Bundles

The impossibility of associating with the representation in (1.15) systems of imprimitivity that would interrelate, in accordance with (3.1.8), the restriction of that representation to the irreducible subspace of wave functions for spin-1/2 quantum particles to PV-measures that would leave that subspace invariant, suggests the need for introducing phase space representations of ISL(2,C) which do not describe quantum point particles, and for which, therefore, the problem of localization is formulated in the context of systems of covariance based on POV-measures which are not projector-valued (cf. Sec. 3.2). In Sec. 6.4 we shall relate these phase space representations of ISL(2,**C**) to the structure groups of Dirac quantum bundles, and treat their carrier Hilbert spaces as standard fibres for those bundles. In the present and in the next section we shall describe the most essential features of these representations and their carrier spaces, omitting most proofs[3], since they are not essential to the understanding of the subsequent applications in Sec. 6.4, and in Chapter 8.

The construction of a Dirac typical fibre **F**, as well as of the representation $U(a,A)$ of ISL(2,**C**) which it carries, proceeds in a most natural manner if we start with a Wigner-type phase space representation in the Hilbert space given by the direct sum

$$L^2(\Sigma_m^+) \oplus L^2(\Sigma_m^+) \oplus L^2(\Sigma_m^-) \oplus L^2(\Sigma_m^-) \ . \tag{2.1}$$

The first two components of the direct sum in (2.1) equal the Hilbert spaces of wave functions defined by (3.4.1) and (3.4.2), whereas the last two components coincide with the corresponding Hilbert space obtained by the time-reversal of (3.4.2). Hence, the elements of the Hilbert space in (2.1) can be represented by one-column wave functions with four components, whose time-evolution can be expressed as follows,

$$w_{q^0} = \exp(-i\overline{P}_0 q^0) w \ , \qquad \overline{P}_0 = \beta\left(\overline{\mathbf{P}}^2 + m^2\right)^{1/2} , \quad \overline{\mathbf{P}} = \beta \mathbf{P} \beta \ , \tag{2.2}$$

in terms of the Dirac β-matrix defined in (1.9c). Since, in view of (1.7), we have to deal simultaneously with 4-momentum values on the forward as well as on the backward mass hyperboloid, it is convenient to regard these wave functions as functions of the 4-velocity v, rather than of the 4-momentum p.

In the Hilbert space defined in (2.1) we can consider the following representation of ISL(2,C), suggested by (1.7):

$$\overline{U}(a,A): \ w(q,v) \mapsto w'(q,v) = D^{1/2}(A,\overline{P})w(\Lambda^{-1}(q-a),\Lambda^{-1}v) \ . \tag{2.3}$$

The spectral analysis of the above representation can be carried out along the general lines described in Sec. 3.4. It can be easily deduced from (3.4.7) and (3.4.10) that such a decomposition displays a spin spectrum that incorporates all half-integer spin values, so that the present construction, taken in conjunction with that in Sec. 3.4, supplies phase space wave functions for all possible spin values. We shall concentrate, however, on the case of spin-1/2, which is of the most fundamental importance in elementary particle physics.

The unitary irreducible representations for spin-1/2, that are obtained by the above procedure, correspond to various choices of resolution generators η that emerge from (3.4.6). Each one of these representations is unitarily equivalent to that in (1.7). However, we shall not reproduce here the formulae[4] explicitly exhibiting that equivalence, since in later applications we shall concentrate exclusively on the Dirac-type of phase space wave functions.

In accordance with (1.8) and (1.9a), the transition to these Dirac-type of phase space wave functions can be effected by means of the operator

$$U_{FW} = P_0^{-1}[m/2(P_0+m)]^{1/2}(P_0+m-\vec{\gamma}\cdot\vec{P}) \ , \qquad P=(P_0,\vec{P}) \ , \ P_\mu = i\partial/\partial q^\mu \ , \tag{2.4}$$

for each choice of resolution generator η . These wave functions satisfy the following SQM counterpart of the Dirac equation in (1.13),

$$(i\gamma^\mu\partial_\mu - m)\Psi(q,v) = 0 \ , \qquad \partial_\mu = \partial/\partial q^\mu \ . \tag{2.5}$$

Their transformation properties under the simultaneous changes of spinor and vector frames are analogous to those of the transformations in (1.15):

$$U_\eta(a,A) = U_{FW}\overline{U}(a,A)\mathbf{P}_\eta U_{FW}^{-1} \ : \ \Psi(q,v) \mapsto S(A)\,\Psi(\Lambda^{-1}(q-a),\Lambda^{-1}v) \ . \tag{2.6}$$

The infinitesimal generators of the above representation, which are the counterparts of those in (5.2.22) for the spin-zero case, can be easily computed[5] from (1.10b) and (2.6):

$$\hat{\mathbf{P}}_\mu = -iP_\mu \ , \qquad \hat{\mathbf{M}}^{\mu\nu} = iM^{\mu\nu} = i(Q^\mu P^\nu - Q^\nu P^\mu) + iS^{\mu\nu} \ , \tag{2.7a}$$

$$P_\mu = i\partial/\partial q^\mu , \qquad Q^\mu = q^\mu - (i/m)\partial/\partial v_\mu \ , \qquad S^{\mu\nu} = \tfrac{i}{4}\left[\gamma^\mu,\gamma^\nu\right] \ . \tag{2.7b}$$

They are seen to contain the internal spin terms $S^{\mu\nu}$in addition to the external angular momentum terms that occur in (5.2.22a).

The inner product of the Hilbert space in (2.1) assumes, after the transition to the Dirac phase space representation determined by (2.4), a form analogous to that of the inner product in (1.11a):

$$\langle\Psi_1|\Psi_2\rangle = \int_{q^0=0} \Psi_1^*(q,v)\gamma^0\left(\gamma^\mu P_\mu/m\right)\Psi_2(q,v)\, d^3\mathbf{q}d^3\mathbf{v} \ . \tag{2.8a}$$

Moreover, it can be also written in the following manifestly covariant form:

$$\langle \Psi_1 | \Psi_2 \rangle = \int_\Sigma \overline{\Psi}_1(q,v)\big(\gamma^\mu P_\mu/m\big)\Psi_2(q,v)\, d\Sigma(q,v) \ , \qquad \overline{\Psi}_1 = \Psi_1^* \gamma^0 \ . \tag{2.8b}$$

We shall adopt as standard fibres **F** for the Dirac quantum bundles defined in Sec. 6.4 those Hilbert spaces with the above inner product, which consist of all those phase space Dirac-type wave functions that result from choices of resolution generators η which, in general, correspond to the quantum spacetime form factors f adopted in (5.1.2). By later incorporating the fundamental quantum spacetime form factor f_ℓ in (5.5.5) into the treatment of gauge fields, we shall achieve a unified treatment of all elementary quantum objects. In this approach the (in the classical sense) non-pointlike nature of these objects becomes a fundamental feature of the quantum geometries that can accommodate them.

In the same manner in which the inner product in (5.1.9) can be expressed in the form (5.1.13), which is similar to the familiar form (3.3.6a) of the inner product for Klein-Gordon particles, so the inner product (2.8) can be expressed in the following alternative form, which is similar to the inner product in (1.16):

$$\langle \Psi_1 | \Psi_2 \rangle = 2m\hat{Z}_{f,m} \int_\Sigma \overline{\Psi}_1(q,v)\gamma^\mu \Psi_2(q,v)\, d\sigma_\mu(q) d\Omega(v) \ . \tag{2.9}$$

Furthermore, this form of the inner product corresponds to a covariant current which is conserved on account of (2.5), and which is in fact analogous[6] to the one in (1.17):

$$\hat{J}^\mu_\eta(q) = \int_{v^0>0} \overline{\Psi}(q,v)\gamma^\mu \Psi(q,v)\, d\hat{\Omega}(v) \ , \quad d\hat{\Omega}(v) = 2m\hat{Z}_{f,m} d\Omega(v) \ . \tag{2.10}$$

Hence, the spacelike hyperplane for the q-integration in (2.9) can be deformed into an arbitrary maximal spacelike hypersurface, without changing the value of the integral determining the inner product in **F**.

*6.3. Dirac Quantum Frame Bundles

As a Hilbert space with inner product defined by (2.8) or (2.9), any Dirac standard fibre **F** can be decomposed into a direct sum of two of its subspaces

$$\mathbf{F} = \mathbf{F}^{(+)} \oplus \mathbf{F}^{(-)} \ , \qquad \mathbf{F}^{(\pm)} = \mathbf{P}_f^{(\pm)} \mathbf{F} \ , \tag{3.1}$$

consisting of all the positive energy and the negative energy state vectors, respectively. This can be most conveniently achieved by using the following representations

$$\mathbf{P}_f^{(\pm)} = \pm \sum\nolimits_{r,s=1}^{4} \int \big| \Phi_{q,v,r}^{(\pm)} \big\rangle \gamma_{rs}^0 \, d\Sigma(q,v) \big\langle \Phi_{q,v,s}^{(\pm)} \big| \ , \tag{3.2a}$$

$$\Phi_{q,v,r}^{(\pm)} = U_\eta(q,\Lambda_v) \sum\nolimits_{s=1}^{4} S_{sr}^*(\Lambda_v) \Phi_{f,m,s}^{(\pm)} = U_\eta(q,\Lambda_v) \Phi_{f,m,s}^{(\pm)} \ , \tag{3.2b}$$

$$\Phi^{(\pm)}_{f,m,r;s}(q,v) = \frac{\pm 1}{2m}\left[\left(m \pm i\gamma^{\mu}\partial/\partial q^{\mu}\right)\gamma^{0}\right]_{sr}$$
$$\times \int_{u^0>0} \exp(-imq\cdot u) f(v\cdot u) f(u^0) d\Omega(u) \ , \tag{3.2c}$$

for the respective orthogonal projectors onto those subspaces[7]. The generalized coherent states in (3.2b) play the role of *Dirac quantum frames* (cf. Sec. 3.7), since in the present context they take over the role played in (5.1.15) by the Klein-Gordon quantum frames $\{U_{\eta}(q,\Lambda_{v})\eta\}$ in the typical Klein-Gordon quantum fibre sharing the same quantum space-time form factor f. We note that the action of the second representation in (3.2b), namely of $U_{\eta}(q,\Lambda_{v})$, actually coincides with that of the representation in (5.1.6) when the latter is applied to each component of the Dirac state vector representing the resolution generators of the continuous resolutions of the identity in $\mathbf{F}^{(\pm)}$.

The wave functions in the positive energy and negative energy subspaces $\mathbf{F}^{(\pm)}$ satisfy the respective Dirac equations[8]

$$(i\gamma^{\mu}\partial_{\mu} \mp m)\Psi^{(\pm)}(q,v) = 0 \ , \qquad \Psi^{(\pm)} = \mathbf{P}^{(\pm)}_{f}\,\Psi \ , \tag{3.3}$$

so that for them the inner product in (2.8) can be also expressed as follows:

$$\left\langle \Psi^{(\pm)}_{1} \middle| \Psi^{(\pm)}_{2} \right\rangle = \pm \int \overline{\Psi}^{(\pm)}_{1}(q,v)\Psi^{(\pm)}_{2}(q,v)\, d\Sigma(q,v) \ . \tag{3.4}$$

The action of the projectors onto these subspaces on the elements of the Dirac typical fibre $\mathbf{F}$ can be expressed in the form

$$(\mathbf{P}^{(\pm)}_{f}\Psi)(q,v) = \mp i \int S^{(\pm)}(q,v;q',v')\,\Psi(q',v')\, d\Sigma(q',v') \ , \tag{3.5a}$$

which is analogous to that encountered in textbooks on conventional quantum field theory [BL,IQ,SI]. On account of the alternative form (2.9) of the inner product in $\mathbf{F}$, it can be also expressed in the even more strikingly similar form

$$(\mathbf{P}^{(\pm)}_{f}\Psi)(q,v) = -i \int S^{(\pm)}(q,v;q',v')\gamma^{\mu}\,\Psi(q',v')\, d\sigma_{\mu}(q')d\hat{\Omega}(v') \ , \tag{3.5b}$$

by noting that the integral kernels in (3.5a) satisfy the following reproducibility relations in the chosen phase space variables:

$$S^{(\pm)}(q',v';q'',v'') = -i \int S^{(\pm)}(q',v';q,v)\gamma^{\mu} S^{(\pm)}(q,v;q'',v'')\, d\sigma_{\mu}(q)d\hat{\Omega}(v) \ . \tag{3.6}$$

These kernels can be constructed as follows,

$$S^{(\pm)}(q,v;q',v') = -(1/2m)(m + i\gamma^{\mu}\partial/\partial q^{\mu})\Delta^{(\pm)}(q,v;q',v') \ , \tag{3.7}$$

from those for the spin-zero case, i.e., from the positive-energy reproducing kernels in (5.1.18), and from their negative-energy counterparts, expressed in a notational convention designed to parallel[9] the one used in conventional theory (cf. (5.7.3) and (1.14)):

$$\Delta^{(\pm)}(q,v;q',v') = \mp iZ_{f,m}^{-2}\int_{u^0>0}\exp[\pm im(q'-q)\cdot u]f(v\cdot u)f(v'\cdot u)d\Omega(u) \quad . \tag{3.8}$$

Comparing (3.7) and (3.8) with (3.2b) and (3.2c), we see that

$$\Phi_{q,v,r}^{(\pm)}(q',v',r') = i[S^{(\pm)}(q,v;q',v')\gamma^0]_{r'r} \quad . \tag{3.9}$$

We shall now construct a quantum Dirac frame bundle which is isomorphic to the principal bundle

$$SF\mathbf{M} = \bigcup_{x\in\mathbf{M}}\{(\underset{\sim}{a}(x),\xi_A(x))|a(x)\in T_x\mathbf{M},\ \xi_A(x)\in S_x\mathbf{M},\ A=0,1\} \tag{3.10}$$

of affine spin frames, on which ISL(2,C) acts from the right in accordance with (2.3.11) and (1.3). This can be achieved in the most straightforward manner by first setting its typical fibre equal to the set

$$\mathbf{D} = \bigcup_{(a,A)\in \mathrm{ISL}(2,\mathrm{C})}\left(\mathbf{U}^{(+)}(a,A),\mathbf{U}^{(-)}(a,A)\right) \ , \tag{3.11a}$$

$$\mathbf{U}^{(\pm)}(a,A) = \left\{U_\eta(a,A)\Phi_{f,m,r}^{(\pm)}\middle|\ r=1,2,3,4\right\} \ , \tag{3.11b}$$

consisting of all Dirac quantum frames corresponding to a given choice of quantum space-time form factor. We then define the *quantum Dirac frame bundle* for that form factor as the **G**-product (cf. Sec. 4.3):

$$DF\mathbf{M} = SF\mathbf{M} \times_{\mathbf{G}} \mathbf{D} \ , \qquad \mathbf{G} = \mathrm{ISL}(2,\mathrm{C}) \ . \tag{3.12}$$

According to a well-known theorem by Geroch (1968), for a non-compact (orientable and time-orientable) Lorentzian manifold **M**, the existence of the spin structure[10] embodied in the affine spin-frame bundle in (3.10) is possible if and only if the Lorentz frame bundle $L\mathbf{M}$ is trivial, i.e., it possesses a global section s_0. On the other hand, non-compactness is a consequence of the physically reasonable requirement that **M** should not possess closed timelike curves, since their existence would provide worldlines that obviously violate causality. Hence, on physical grounds we have to allow for the existence of such a global section s_0. It is then easy to conceptualize the quantum Dirac frame bundle $DF\mathbf{M}$ in (3.12) without any direct reference to the affine spin frame bundle $SF\mathbf{M}$ in (3.10): we associate with each Lorentz frame $u_0(x)\in s_0$ a copy of a Dirac quantum frame whose elements are defined in (3.2b), and then construct the fibre of $DF\mathbf{M}$ above $x\in\mathbf{M}$ by applying to that Dirac quantum frame all the transformations $U_\eta(a,A)$, in accordance with the procedure implicit in (3.11b).

We shall therefore give precedence in the sequel to quantum frame bundles over their "classical" counterparts. In fact, whereas it can be argued that linear frames whose elements are vectors in Minkowski space, or belong to the spaces tangent to a curved classical space-time manifold **M**, are physically realizable in terms of "classical" macroscopic structures, that is certainly no longer the case with spin frames[11]. Hence, on physical as well as mathematical grounds, it is advantageous to conceive of all the subsequent constructions of quantum bundles directly in terms of the quantum frame bundles to which they are associated, rather than to resort to constructions based on more traditional principal frame bundles, as we did in Chapters 4 and 5 (cf. Note 8 to Chapter 4).

*6.4. Dirac Quantum Bundles

In accordance with the point of view advocated in the last paragraph of the preceding section, we define now a *Dirac quantum bundle* $D\mathbf{M}$, with a typical fibre $\mathbf{F}$ defined in Sec. 6.2, as the $\mathbf{G}$-product (cf. Sec. 4.3)

$$D\mathbf{M} = DF\mathbf{M} \times_{\mathbf{G}} \mathbf{F} \ , \qquad \mathbf{G} = \mathrm{ISL}(2,\mathrm{C}) \ . \tag{4.1}$$

This construction gives rise to the generalized soldering maps (cf. (4.3.2) and (5.1.5))

$$\boldsymbol{\sigma}_x^{\mathbf{U}^{(\pm)}} : \boldsymbol{\Psi} \mapsto \Psi_r^{(\pm)}(\zeta) = \left\langle \boldsymbol{\Phi}_{x;\zeta,r}^{(\pm)} \middle| \boldsymbol{\Psi} \right\rangle \ , \qquad \boldsymbol{\Psi} \in \mathbf{F}_x \subset D\mathbf{M} \ , \tag{4.2}$$

which assign for each choice of Dirac quantum frame

$$\mathbf{U}^{(\pm)} = \left\{ \boldsymbol{\Phi}_{x;\zeta,r}^{(\pm)} \middle| \ \zeta = (q,v) \in \mathbf{R}^4 \times V^+, r = 1,2,3,4 \right\} \in \boldsymbol{\Pi}^{-1}(x) \subset DF\mathbf{M} \ , \tag{4.3}$$

coordinate wave function amplitudes belonging to an element $\Psi = \Psi^{(+)} \oplus \Psi^{(-)}$ in the standard fibre $\mathbf{F}$. In the transition from one choice of section s of $DF\mathbf{M}$ to another section s', these coordinate wave functions are related by a purely quantum counterpart of the relations (5.1.7), which involve the classical frames in (5.1.7b):

$$\boldsymbol{\sigma}_x^{\mathbf{U}'^{(\pm)}} \circ (\boldsymbol{\sigma}_x^{\mathbf{U}^{(\pm)}})^{-1} : \Psi \mapsto \Psi' = U_\eta(b(x),A(x))\Psi \ , \tag{4.4a}$$

$$\boldsymbol{\Phi}_{x;\zeta,r}^{(\pm)} = \boldsymbol{\Phi}_{x;\zeta,r}'^{(\pm)} \cdot (b(x),A(x)) = (\boldsymbol{\sigma}_x^{\mathbf{U}'^{(\pm)}})^{-1} U_\eta^{-1}(b(x),A(x)) \boldsymbol{\sigma}_x^{\mathbf{U}'^{(\pm)}} \boldsymbol{\Phi}_{x;\zeta,r}'^{(\pm)} \ . \tag{4.4b}$$

We note that in (4.4b) the action from the right ensures that the Dirac quantum bundle $D\mathbf{M}$ is a fibre bundle associated with the principal bundle $DF\mathbf{M}$ in the sense of the definition in Sec. 2.2. Indeed, for each choice of section s of the Dirac frame bundle $DF\mathbf{M}$, the soldering maps in (4.2) provide the local trivialization maps

$$\boldsymbol{\phi}^s : \boldsymbol{\Psi} \mapsto (x, \boldsymbol{\sigma}_x^{\mathbf{U}^{(\pm)}} \boldsymbol{\Psi}) \in \mathbf{M}^s \times \mathbf{F}^{(\pm)} \ , \quad \boldsymbol{\Psi} \in \mathbf{F}_x \subset \boldsymbol{\pi}^{-1}(\mathbf{M}^s) \ , \quad \mathbf{U}^{(\pm)} = s(x) \ , \tag{4.5}$$

generically defined in (2.2.4).

All the geometric considerations carried out in Chapter 5 in the context of Klein-Gordon quantum bundles can be transferred, generally with at most minor alterations, to the present case of Dirac quantum bundles. We shall, therefore, restrict the remaining considerations in this chapter to the definition of the parallel transport and of the related connection in a Dirac quantum bundle $D\mathbf{M}$.

The definitions of parallel transport and connection in the affine spin frame bundle $SF\mathbf{M}$, or equivalently, in the quantum Dirac frame bundle $DF\mathbf{M}$, can be carried out in accordance with the general procedures outlined in Secs. 2.4 and 2.5. Thus, let us consider the spin frame bundle $SF_0\mathbf{M}$ over $\mathbf{M}$ [W], which can be viewed as a subbundle of $SF\mathbf{M}$ that corresponds to the choice $\boldsymbol{a}(x) \equiv 0$ in (3.10). The adoption of any connection in the spin frame bundle $SF_0\mathbf{M}$, given as in (2.4.1), leads to a diffeomorphism of the kind exhibited in (2.4.7). On the other hand, by using (1.3a) we can perform the identification

$$SF_0\mathbf{M} \times_{\mathbf{G}} (W^* \otimes W) \quad \leftrightarrow \quad T\mathbf{M} \ , \qquad \mathbf{G} = \mathrm{SL}(2,\mathbf{C}) \ . \tag{4.6}$$

In case that we are dealing with a connection which gives rise, on account of this identification, to a connection compatible with the Lorentz metric on $\mathbf{M}$, this diffeomorphism can be extended, in accordance with (2.6.17), to a diffeomorphism

$$\tilde{\tau}_\gamma(x'',x') : \quad (\underset{\sim}{a}(x'), \xi'_A(x')) \ \mapsto \ (\underset{\sim}{a}(x''), \xi''_A(x'')) \ , \tag{4.7}$$

between the fibres of the affine spin frame bundle $SF\mathbf{M}$. In turn, by means of (3.12), this leads to a diffeomorphism between the respective fibres of the quantum Dirac frame bundle $DF\mathbf{M}$, which gives rise to the maps

$$\boldsymbol{\tau}_\gamma(x'',x') : \quad \boldsymbol{\Phi}^{(\pm)}_{x';\zeta,r} \ \mapsto \ \boldsymbol{\Phi}^{(\pm)}_{x'';\zeta,r} \ , \tag{4.8}$$

between corresponding elements of quantum Dirac frames. Upon taking into account that, according to (3.2) and (4.2),

$$\boldsymbol{\Psi}^{(\pm)} = \pm \sum_{r,s=1}^{4} \gamma^0_{rs} \int d\Sigma(\zeta)\, \Psi^{(\pm)}_s(\zeta)\, \boldsymbol{\Phi}^{(\pm)}_{x';\zeta,r} \in \mathbf{F}^{(\pm)}_{x'}, \qquad \boldsymbol{\Psi} \in \mathbf{F}_{x'} \ , \tag{4.9}$$

the map (4.8) can be extended into parallel transport within the Dirac quantum bundle $D\mathbf{M}$, given by the unitary map

$$\begin{aligned} &\sum_{r,s=1}^{4} \gamma^0_{rs} \int d\Sigma(\zeta) \Big(\Psi^{(+)}_s(\zeta) \boldsymbol{\Phi}^{(+)}_{x';\zeta,r} - \Psi^{(-)}_s(\zeta) \boldsymbol{\Phi}^{(-)}_{x';\zeta,r} \Big) \\ &\qquad \mapsto \sum_{r,s=1}^{4} \gamma^0_{rs} \int d\Sigma(\zeta) \Big(\Psi^{(+)}_s(\zeta) \boldsymbol{\Phi}^{(+)}_{x'';\zeta,r} - \Psi^{(-)}_s(\zeta) \boldsymbol{\Phi}^{(-)}_{x'';\zeta,r} \Big) \ , \end{aligned} \tag{4.10}$$

between the respective fibres of $D\mathbf{M}$.

The covariant derivatives corresponding to the above defined parallel transport in $D\mathbf{M}$, can be (densely) defined by the same type of strong limit as in (5.2.16). They can be then expressed in the form

$$\nabla_X \Psi_x^{(\pm)} = [\partial_X - i\tilde{\theta}^j(X) P_{j,u} + \tfrac{i}{2}\tilde{\omega}_{jk}(X) M_u^{jk}] \Psi_x^{(\pm)} , \tag{4.11a}$$

$$\partial_X \Psi_x^{(\pm)} = \pm \sum_{r,s=1}^{4} \gamma_{rs}^0 \int d\Sigma(\zeta) X(\Psi_{\zeta,s}^{(\pm)}) \Phi_{x;\zeta,r}^{(\pm)} , \qquad \Psi_{\zeta,s}^{(\pm)}(x) = \left\langle \Phi_{x;\zeta,s}^{(\pm)} \middle| \Psi_x \right\rangle , \tag{4.11b}$$

where the connection coefficients are the same ones as in (5.2.27), whereas the infinitesimal generators are defined as in (2.7) in terms of the coordinates with respect to the Poincaré frame $\boldsymbol{u}$ in $T_x\mathbf{M}$ that corresponds to the affine spin frame in the chosen section of $SF\mathbf{M}$.

With this definition of a connection within the quantum Dirac bundle $D\mathbf{M}$, the counterparts of (5.2.26) and (5.2.28) remain valid. The definition of a propagator for parallel transport can be carried out as in (5.4.2). Counterparts of (5.4.3)-(5.4.5) then immediately follow, in which the ζ-integration obviously has to be replaced by the combination of the ζ-integration and of the summation involving the matrix elements of $\pm\gamma^0$ that occur in (3.2a). The treatment of strictly causal GS propagation therefore remains basically the same as in the spin-zero case considered in Sec. 5.4. On the other hand, since the treatment of the weakly causal GS propagation was based in Chapter 5 on the use of diffusion processes, the usual path-integral subterfuge ([ST], p. 229) to a "squared" Dirac equation is required. Such an equation is also necessitated in SQM by the consistent treatment of interactions of single (stochastically extended) spin-1/2 quantum particles with external electromagnetic fields (Ali and Prugovečki, 1981). However, a treatment of the many-body problem for the spin-1/2 case which is mathematically elegant, as well as physically more satisfactory, is afforded by the field-theoretical techniques presented in Chapter 8.

Notes to Chapter 6

[1] Rigorous proofs of this and of the other statements concerning the Wigner-type momentum representation can be found in Chapter 6 of [BL].

[2] The proofs of this and of the remaining statements in this section can be found in Chapter 7 of [BL].

[3] All the proofs can be found in the original paper on this subject (Prugovečki, 1980). The Wigner-type representation based on (2.1)-(2.3) has been used (Ali and Prugovečki, 1981) in setting up models of stochastically extended spin-1/2 particles in external electromagnetic fields, which are consistent as single-particle external-field models in the sense that they do not give rise to spontaneous transition of particle states into antiparticle states – as is generically the case with models based on the Dirac equation in (1.13) or in (2.5).

[4] These formulae can be easily derived from the results reported in (Prugovečki, 1980) and in (Ali and Prugovečki, 1986c).

[5] Naturally, the definition of the Q-operators in (2.7b) requires the off-shell extrapolation carried out in (5.2.23), and discussed in Note 13 to Chapter 5.

[6] The Wigner-type GS spin-1/2 framework also possesses a probability current. That current is analogous to the one in (3.5.13), and is also covariant and conserved (Prugovečki, 1980).

[7] The proofs of the results stated in this section can be found in (Prugovečki, 1980), where these results are derived from the Wigner-type representation based on (2.1)-(2.3), by using the transformation (2.4). However, all the results in this section can be also verified by direct computation, using the well-known algebraic properties [IQ,SI] of the Dirac γ-matrices in a manner totally analogous to that of deriving the corresponding results in the momentum representation based on (1.10)-(1.11).

[8] Alternatively, one can associate, as in (Prugovečki and Warlow, 1989b), the negative energy solutions with values of v on the backward 4-velocity hyperboloid by setting for them $\Psi(q,-v) = \Psi(-q,v)$, so that the form (2.5) of the Dirac equation is preserved.

[9] However, the factor $1/2m$ was introduced, which is not present in the conventional approach. This factor is designed to compensate the $2m$ factor in (2.9) (which also occurs in (1.16)). In addition, there is also

the renormalization constant in (3.8), which is absolutely essential for securing the strict reproducibility of the integral kernel in (5.1.18) – from which (3.6) follows on account of the well-known algebraic relations for the Dirac γ-matrices [BL,IQ,SI].

10 Spin structures and their most basic properties are described in Chapter 13 of [W]. Detailed accounts of spinors and spin structures in curved (classical) spacetimes, and their use in CGR are provided by Carmeli (1982) at a level best suited to readers with a primary interest in physics and in the formulation of classical field theories, as well as by Penrose and Rindler (1986) in a form that might be more attractive to readers with a strong interest in the mathematics of the subject.

11 The concept of "null flag" (cf. [W]; [Penrose and Rindler, 1986]) associated with a spinor is introduced to deal with this problem, but it still leaves an irremovable ambiguity in sign in the two-to-one map between spinors and such null flags.

Chapter 7

Relativistic Quantum Geometries for Spin-0 Massive Fields

In the case of interacting quantum fields in Minkowski space the difficulties with the conventional schemes of quantization are very fundamental. In fact, they are intrinsically unsolvable in that context, on account of the physical and mathematical reasons mentioned in Sec. 1.2, and further discussed in Sec. 7.6. Hence, the last word on this subject has to go to the acknowledged founder of relativistic quantum field theory as well as of relativistic quantum mechanics, P.A.M. Dirac, whose insightful and uncompromisingly forthright assessments of these two disciplines have greatly inspired the present work.

In his very last paper, published posthumously[1] under the title "The Inadequacies of Quantum Field Theory", Dirac reiterated, in his very characteristically straightforward and clear-cut manner, his general assessment of the conventional renormalization program and the admonishments which he had publicly voiced on numerous occasions (cf., e.g., Dirac, 1951, 1962, 1965, 1973, 1977, 1978) throughout the last thirty years of his life[2]:

> Just because the results [of the conventional renormalization procedures in quantum field theory] happen to be in agreement with observation does not prove that one's theory is correct. After all, the Bohr theory was correct in simple cases. It gave very good answers, but still the Bohr theory had the wrong concepts. Correspondingly, the renormalized kind of quantum theory with which physicists are working nowadays is not justifiable by agreement with experiments.
>
> In spite of this, physicists have gone a long way in developing this theory; in fact, most of the physics of elementary particles over about forty years has been along these lines. People work with a Hamiltonian which, used in a direct way, would give the wrong results, and then they supplement it with these rules for subtracting infinities. I feel that, under those conditions, you do not have a correct mathematical theory at all. You have a set of working rules. So the quantum mechanics that most physicists are using nowadays is just a set of working rules, and not a complete dynamical theory. In spite of that, people have developed it in great detail.
>
> I want to emphasize that many of these modern quantum field theories are not reliable at all, even though many people are working on them and their work sometimes gets detailed results. (Dirac, 1987, p. 196)

It is, of course, generally understood that this mathematical "unreliability" of conventional quantum field theory in Minkowski space does not extend to the case of free quantum fields. Such fields are physically trivial, but their mathematical formulation is at least totally consistent when properly formulated in terms of operator-valued distributions [BL], which form the basis of the Wightman axiomatic approach[3]. In fact, free quantum fields, and so-called generalized free quantum fields [BL], still provide the *only* models that are known to satisfy those axioms in four-dimensional Minkowski space[4]. Hence, the earlier quoted negative assessment by Dirac obviously refers to the treatment of interacting quantum fields within the conventional approach to quantum field theory in general, and to conventional quantum electrodynamics in particular.

As is well-known, the foundations of this subject were laid by Dirac himself – and, as a matter of fact, "the first steps toward renormalization go back once again to Dirac" (Pais, 1987, p. 102). However, by 1936 "Dirac turned highly critical of quantum electrodynamics ... and adopted the strategy of attempting to modify the classical theory first, so as to rid it of *its* infinities. There are overwhelming reasons, however, why a return to classical theory is the wrong way to go." (*ibid.*, p. 105).

Dirac's many attempts to rid conventional quantum field theory of infinities in a mathematically consistent, and yet physically nontrivial manner, provide by themselves convincing evidence that his final verdict on this issue should be taken very seriously when approaching the construction of models for interacting quantum fields. On the other hand, it is not *a priori* clear whether his observations might also apply to the conventional formulation of quantum fields in a curved classical spacetime, as long as those fields do not interact amongst themselves, but rather interact only with an external gravitational field supplied by a Lorentzian metric **g** over a given differential manifold **M**.

Thus, it could be hoped that the same second quantization procedure that works rather well when constructing free quantum fields might work equally well when Minkowski space is replaced by a generic curved classical spacetime.

That turns out, however, not to be the case – and, this time, for rather evident *physical* reasons, which reveal and illustrate rather dramatically the epistemological inadequacies of the conventional mode of thinking about second quantization procedures.

To reveal and analyze the sources of these inadequacies, we shall review in Sec. 7.1 the salient features of the most fundamental quantization scheme for quantum fields in curved spacetime, namely the canonical method of second-quantization. Then in Sec. 7.2 we shall show that, even in classical *Minkowski spacetime* (i.e., in Minkowski space treated as a manifold, rather than as a linear space in which *global* Lorentz frames can be introduced as vector bases) they lead to a surprising conclusion: the conventional concept of quantum particle, based on the canonical method of quantization, leads to ambiguities. Such difficulties are usually interpreted [BD] as reflecting an observer-dependent quality in the concept of quantum particle. However, these ambiguities and difficulties are exacerbated in the context of curved spacetimes – as we shall show in Sec. 7.3, where we discuss the difficulties and inconsistencies resulting from the use of conventional schemes of second-quantization in a generic globally hyperbolic curved spacetime.

The GS method of quantization resolves these fundamental difficulties by replacing the concept of quantum particle by that of geometro-stochastic exciton, in accordance with the GS schemes of first quantization discussed in Secs. 3.4-3.6, and in Chapters 5-6. In turn, this leads to a GS second quantization method, which is presented in Sec. 7.4 for the case of neutral scalar fields. As in the single-exciton case discussed in the preceding three chapters, the propagation of such GS fields is formulated in geometro-stochastic terms, rather than being derived from action principles. In Sec. 7.5 it is shown, however, that action principles emerge naturally from the underlying definition of propagators for parallel transport. In Secs. 7.6–7.8 we compare the conventional approach with the GS approach to microcausality and quantum field locality, and to quantum field interactions, respectively.

*7.1. Canonical Second-Quantization in Curved Spacetime

The first attempts at formulating quantum fields in a curved spacetime background can be traced to the work of Schrödinger (1939). However, a canonical scheme for second-quantization, in which a curved spacetime background replaces Minkowski space, was first formulated and studied in a systematic manner by L.E. Parker (1966, 1968, 1971). The entire subject eventually came very much into vogue with the paper by S.W. Hawking (1975a) on thermal black hole radiation. Although the interest gradually waned in subsequent years[5], this subject has remained a *de rigueur* item in conference proceedings and memorial volumes on general relativity and cosmology (Parker, 1978, 1983; Gibbons, 1979; Davies, 1980, 1984; etc.). It has also become the sole topic of two monographs ([BD]; Fulling, 1989), and it has received prominent mention in recent textbooks on general relativity [W] and cosmology [N].

Let us first review canonical second-quantization in Minkowski space, where this procedure is straightforward and very well understood. In the spin-zero case, it relies on the possibility of *unambiguous* formulation of the Hilbert spaces

$$\hat{\mathcal{H}}^{(\pm)} = \left\{ \hat{\varphi}_t^{(\pm)} \middle| \hat{\varphi}_t^{(\pm)} = \exp(\mp i P_0 t)\hat{\varphi} \ , \ P_0 = \left(\mathbf{P}^2 + m^2\right)^{1/2} \ , \ \mathbf{P}^2 = -\Delta \right\} \ , \tag{1.1}$$

that consist of positive-energy and negative-energy solutions of the Klein-Gordon equation (cf. (3.3.5)),

$$\left(\eta^{\mu\nu}\partial_\mu\partial_\nu + m^2\right)\hat{\varphi}^{(\pm)}(x) = 0 \ , \qquad \hat{\varphi}^{(\pm)}(x^0,\mathbf{x}) = \hat{\varphi}_{x^0}^{(\pm)}(\mathbf{x}) \ . \tag{1.2}$$

These Hilbert spaces carry inner products which can be densely defined by the respective positive-definite sesquilinear forms (cf. (3.3.6a)),

$$\left\langle \hat{\varphi}_1^{(\pm)} \middle| \hat{\varphi}_2^{(\pm)} \right\rangle = \pm i \int_{\sigma_t} \hat{\varphi}_1^{(\pm)*}(x) \overleftrightarrow{\partial}_\mu \hat{\varphi}_2^{(\pm)}(x)\, d\sigma^\mu(x) \ , \qquad d\sigma^\mu(x) = n^\mu(x) d\sigma(x) \ , \tag{1.3}$$

in which the integrations can be performed along the maximal spacelike hypersurfaces σ_t, belonging to any foliation of the Minkowski space **M**, which is of the generic type (5.4.7). The conservation of the Klein-Gordon current in (3.3.9) ensures that the values assumed by these sequilinear forms are indeed independent of the choice of reference hypersurface σ_t. Hence, these forms define norms that can be used in constructing (cf. [PQ], Ch. I, Secs. 3.3 and 4.1) the completions that lead from the pre-Hilbert spaces consisting of the respective smooth solutions of (1.2), which are of faster than polynomial decrease at infinity, to the corresponding Hilbert spaces in (1.1) of much more general wave functions.

In the case of charged spin-zero particles and antiparticles, whose states are described by the elements of the respective completions of the pre-Hilbert spaces of positive-energy and negative-energy solutions of the Klein-Gordon equation in (1.2), the canonical method of second quantization is usually carried out [BD,BL,IQ,SI,N] by first constructing the *Fock space*

$$\hat{\mathcal{F}} = \bigoplus_{m,n=0}^{\infty} \hat{\mathcal{F}}_{m,n} \ , \qquad \hat{\mathcal{F}}_{m,n} = (\hat{\mathcal{H}}^{(+)} \underset{S}{\otimes} \cdots \underset{S}{\otimes} \hat{\mathcal{H}}^{(+)}) \otimes (\hat{\mathcal{H}}^{(-)} \underset{S}{\otimes} \cdots \underset{S}{\otimes} \hat{\mathcal{H}}^{(-)}) \ . \tag{1.4}$$

This space is a direct sum, whose (0,0)-component is a one-dimensional complex Hilbert space spanned by a normalized vector Ψ_0, called the *Fock vacuum*, and whose (m,n)-component is a tensor product, whose first factor consists of m symmetrized tensor products [PQ] of the single-particle Hilbert space in (1.1), and whose second factor consists of n symmetrized tensor products of the single-antiparticle Hilbert space in (1.1). Thus, the Fock space in (1.4) is itself a Hilbert space, whose elements generically represent superpositions of state vectors with arbitrarily large numbers of particles and antiparticles of the species described by the respective solutions of the Klein-Gordon equation in (1.2).

In the case of neutral spin-zero particles which possess no distinct antiparticles, the corresponding Fock space

$$\hat{\mathcal{F}} = \bigoplus_{n=0}^{\infty} \hat{\mathcal{F}}_n \ , \qquad \hat{\mathcal{F}}_n = \hat{\mathcal{H}}^{(+)} \underset{S}{\otimes} \cdots \underset{S}{\otimes} \hat{\mathcal{H}}^{(+)} \ , \tag{1.5}$$

is a direct sum whose generic term consists of n symmetrized tensor products of the single-particle Hilbert space in (1.1), for the obvious reason that there are no distinct antiparticles. For the sake of simplicity in exposition, and in view of the fact that the charged case presents no additional conceptual or mathematical difficulties, we shall restrict all the subsequent considerations in this chapter to the case of neutral scalar quantum fields acting in such Fock spaces for neutral spin-zero particles.

In that case the second quantization is achieved [BD,BL,IQ,SI] by first constructing the creation and annihilation operators for a quantum particle in the generic state f, whose momentum space representative is given by (3.3.10), by defining in the following manner,

$$\left(a(f)\tilde{\Psi}_n\right)_{n-1}(k_1,\ldots,k_{n-1}) = n^{1/2} \int_{k^0>0} \tilde{f}^*(k)\tilde{\Psi}_n(k,k_1,\ldots,k_{n-1})\, d\Omega_m(k) \ , \tag{1.6a}$$

$$\left(a^\dagger(f)\tilde{\Psi}_n\right)_{n+1}(k_1,\ldots,k_{n+1}) = (n+1)^{-1/2}\sum_{j=1}^{n+1}\tilde{f}(k_j)\,\tilde{\Psi}_n(k_1,\ldots,k_{j-1},k_{j+1},\ldots,k_{n+1}) \ , \tag{1.6b}$$

their actions on the momentum space representatives of the n-particle wave functions belonging to the state vectors of the Fock space in (1.5). These operators, whose definition can be extended by linearity to all finite linear combinations of various n-particle states, are unbounded but closable [PQ], so that they are well-defined on maximal dense domains of the Fock space in (1.5). On the other hand, the scalar quantum fields usually defined by[6]

$$\phi(x) = \phi^{(-)}(x) + \phi^{(+)}(x) \ , \qquad x \in \mathbf{M} \ , \tag{1.7a}$$

$$\phi^{(-)}(x) = \sum_{\alpha=1}^{\infty} f_\alpha(x)a(f_\alpha) \ , \qquad \phi^{(+)}(x) = \sum_{\alpha=1}^{\infty} f_\alpha^*(x)a^\dagger(f_\alpha) \ , \tag{1.7b}$$

for any choice of orthonormal basis $\{f_1, f_2, \ldots\}$in the single particle space in (1.1), are not actually well-defined by the second infinite series in (1.7b). In fact, it can be proved (cf. [BL], Sec. 10.4) that no "local" quantum field operators are well-defined at the points x of Minkowski space, so that some form of "smearing" with test functions is required to make the expression in (1.7) well-defined. Hence, Wightman (1956) proposed a "smearing" with test functions f from the Schwartz space $\mathcal{S}(\mathbf{R}^4)$ [BL]. Indeed, if such "smearings" are performed on each term in the series in (1.7b) prior to taking the strong limit that is implicit in computing the sum of such infinite series [PQ], then the ensuing weak limit [PQ]

$$\phi(f) = \int_{\mathbf{M}} f(x)\phi(x)d^4x := \underset{n\to\infty}{\text{w-lim}} \int_{\mathbf{M}} d^4x f(x) \sum_{\alpha=1}^{n} \left(f_\alpha(x)a(f_\alpha) + f_\alpha^*(x)a^\dagger(f_\alpha) \right), \quad (1.8)$$

is well-defined on a dense set in Fock space, and it determines on such a dense set an operator-valued distribution in $f \in \mathcal{S}(\mathbf{R}^4)$.

Despite the fact even free quantum fields are not well-defined at single points x in Minkowski space as bona fide operators in Fock space, some other basic expressions, such as the following one for the *total particle number operator*

$$\hat{\mathrm{N}} = \sum_{\alpha=1}^{\infty} a^\dagger(f_\alpha)a(f_\alpha) = i\int_{\sigma_t} \phi^{(+)}(x)\overleftrightarrow{\partial}_\mu \phi^{(-)}(x)\, d\sigma^\mu(x) \ , \quad (1.9)$$

can be defined in mathematically rigorous terms by means of integrals over hypersurfaces in Minkowski space. In the case of the integral in (1.9) this can be achieved by setting,

$$\lim_{n\to\infty} \left\langle \hat{\Psi}' \middle| \int_{\sigma_t} \sum_{\alpha=1}^{n} f_\alpha^*(x)a^\dagger(f_\alpha)\overleftrightarrow{\partial}_\mu \sum_{\beta=1}^{n} f_\beta(x)a(f_\beta)\, d\sigma^\mu(x)\, \hat{\Psi}'' \right\rangle$$
$$= \left\langle \hat{\Psi}' \middle| \int_{\sigma_t} \phi^{(+)}(x)\overleftrightarrow{\partial}_\mu \phi^{(-)}(x)\, d\sigma^\mu(x)\, \hat{\Psi}'' \right\rangle \ , \qquad \hat{\Psi}', \hat{\Psi}'' \in \hat{\mathcal{F}} \ , \quad (1.10)$$

so that, on account of (1.6) and of the orthonormality of the set $\{f_1, f_2, \ldots\}$, the so defined expectation values of the operator-valued integral in (1.9) coincide with those of the series in (1.9) – which in turn provides the mean value of the number of particles in a state represented by a normalized vector in Fock space. Furthermore, by using the following conventional definition of *normal ordering*,

$$:\phi^{(-)}(x)\phi^{(+)}(x): = :\phi^{(+)}(x)\phi^{(-)}(x): = \phi^{(+)}(x)\phi^{(-)}(x) \ , \quad (1.11)$$

as well as the easily verifiable fact that

$$\int_{\sigma_t} \hat{\varphi}_1^{(\pm)}(x)\overleftrightarrow{\partial}_\mu \hat{\varphi}_2^{(\pm)}(x)\, d\sigma^\mu(x) = 0 \ , \qquad \hat{\varphi}_1^{(\pm)}, \hat{\varphi}_2^{(\pm)} \in \hat{\mathcal{H}}^{(\pm)} \ , \quad (1.12)$$

we can rewrite (1.9) in the following form:

$$\hat{\mathrm{N}} = \tfrac{i}{2}\int_{\sigma_t} :\phi(x)\overleftrightarrow{\partial}_\mu \phi(x):\, d\sigma^\mu(x) \ . \quad (1.13)$$

Similarly, the total 4-momentum operators and the relativistic total angular momentum operators (whose components also represent the generators of spacetime translations of the unitary representation of the Poincaré group to which the single-particle representation in (3.3.7) gives rise in the Fock space defined in (1.5)) can be expressed as follows [SI]:

$$\hat{P}_\mu = \int_{\sigma_t} :T_{\mu\nu}[\phi(x)]:\, d\sigma^\nu(x) \ , \qquad \hat{M}_{\mu\nu} = \int_{\sigma_t} :x_\mu T_{\nu\lambda} - x_\nu T_{\mu\lambda}:\, d\sigma^\lambda(x) \ , \quad (1.14)$$

$$T_{\mu\nu} = \eta_{\mu\lambda} \frac{\partial \mathcal{L}}{\partial \phi_{,\lambda}} \phi_{,\nu} - \eta_{\mu\nu} \mathcal{L} = \phi_{,\mu}\phi_{,\nu} + \tfrac{1}{2}\eta_{\mu\nu}\left(m^2\phi^2 - \eta^{\kappa\lambda}\phi_{,\kappa}\phi_{,\lambda}\right) , \tag{1.15}$$

$$\mathcal{L} = \tfrac{1}{2}\left(\eta^{\mu\nu}\phi_{,\mu}\phi_{,\nu} - m^2\phi^2\right) , \qquad \phi_{,\lambda} := \partial\phi/\partial x^\lambda . \tag{1.16}$$

On the other hand, the fact that the quantum field operators at the individual points x of Minkowski space are not well-defined, so that at a mathematically rigorous level they have to be treated as operator-valued distributions, causes severe mathematical difficulties in the case of interacting quantum fields. Indeed, in that case the above simple-minded procedures, based on normal ordering, do not produce well-defined expressions, and additional divergencies have to be removed. However, the fundamentally distributional nature of quantum fields "at a point" is not the principal cause of difficulties when the above formalism for free quantum fields is extrapolated from Minkowski space to a curved spacetime manifold. Hence in the sequel, we shall not bother with this "smearing" procedure for quantum field operators. Rather, the reader interested in this *mathematically* [7] well-understood point should consult the papers by Isham (1978) and by Kay (1978, 1980) for a rigorous treatment of quantum fields as operator-valued distributions over spaces of test functions in a curved (classical) spacetime (**M**,g).

The original formulation by Parker (1966, 1968) of spin-zero quantum fields in curved spacetime (**M**,g) was mathematically very formal, and dealt only with the very special case of a spatially flat Robertson-Walker metric expressed in its best-known form [W], namely the one in which its components are in their outward appearance very similar to those for the Minkowski metric. However, it was soon noted by Fulling (1973) that such a construction was inherently nonunique. It is often stated that this intrinsic ambiguity of the canonical second-quantization procedure in curved spacetime originates from the fact that "*the notion of a vacuum or no-particle state in a curved spacetime is inherently ambiguous*" (Gibbons, 1979, p. 639). In fact, as we shall soon see, this ambiguity occurs already at the single-particle level, due to the lack of a *physically* unambiguous manner of splitting the space of solutions of the general relativistic Klein-Gordon equation with minimal coupling to an external gravitational field[8],

$$\left(g^{\mu\nu}\nabla_\mu\nabla_\nu + m^2\right)\varphi(x) = 0 , \tag{1.17}$$

into subspaces representing positive-energy and negative-energy solutions, that would represent counterparts of those in (1.1).

Indeed, to make the construction mathematically as rigorous and as unambiguous as possible, let us first consider only the family of all solutions of the Klein-Gordon equation in (1.17) that correspond to smooth initial conditions with compact support along an initial-data reference hypersurface σ_0. By analogy with (1.3), we can introduce in this family the sesquilinear form[9]

$$\left(\varphi_1 | \varphi_2\right) = i\int_{\sigma_t} \varphi_1^*(x)\overleftrightarrow{\nabla}_\mu \varphi_2(x)\, d\sigma^\mu(x) , \qquad d\sigma^\mu(x) = n^\mu(x) d\sigma(x) . \tag{1.18}$$

By using the general form of Gauss's law [W] it can be immediately deduced from (1.17) that the values assumed by the sesquilinear form in (1.18) are independent of the choice of hypersurface σ_t for an arbitrary foliation of the globally hyperbolic Lorentzian manifold $(\mathbf{M},g)$ in which the Klein-Gordon equation (1.17) is defined. However, this sesquilinear form is not positive definite, so that it cannot be used for introducing a norm in the considered family of solutions of (1.17), whose completion would yield a Hilbert space. Indeed, as can be seen from (1.3), a splitting of the considered family into positive-energy and negative-energy solutions of the Klein-Gordon equation in (1.17) is mandatory, in order to extrapolate the canonical second-quantization procedure from the Minkowski case to the present situation.

As presented in the original mathematically rigorous papers on this subject by Ashtekar and Magnon (1975) and by Kay (1978), as well as in the mathematically oriented text by Fulling (1989), the extrapolation of the canonical second-quantization procedure primarily deals with the case of stationary classical spacetimes $(\mathbf{M},g)$. Hence, we shall restrict the discussion in this section to the case where $(\mathbf{M},g)$ is stationary, and then discuss the generic case in Sec. 7.3, in the context of scrutinizing the physical implications of this extrapolation of the canonical second quantization procedure in Minkowski space.

By definition, a classical spacetime $(\mathbf{M},g)$ is *stationary* [M,W] if there is a globally defined timelike Killing vector field $\boldsymbol{t}$, i.e., if the diffeomorphisms $\phi_t : \mathbf{M} \to \mathbf{M}$ generated by the flow lines (namely integral curves) of $\boldsymbol{t}$ constitute a one-parameter group of isometries in $(\mathbf{M},g)$. Hence, in the presence of such a field, for a globally hyperbolic $(\mathbf{M},g)$ the choice of foliation in (5.4.7) can be related to the Killing vector field $\boldsymbol{t}$ by specifying any Cauchy surface in that foliation as the initial-data hypersurface σ_0, and then setting

$$\mathbf{M} = \bigcup_{t \in \mathrm{I}} \sigma_t \quad , \qquad \sigma_t = \left\{ x = \phi_t(\bar{x}) \middle| \, \bar{x} \in \sigma_0 \right\} \ , \qquad \boldsymbol{t}(\bar{x}) = \phi_t{}^* \boldsymbol{t}(x) \ . \tag{1.19}$$

We can then introduce along σ_0 a scalar field N and a vector field $\boldsymbol{N}$ (known in the ADM formulation of the initial-value problem in CGR as the lapse and shift fields, respectively – cf. Sec. 11.1), for which, in terms of the future-pointing normals $\boldsymbol{n}$ to σ_0,

$$\boldsymbol{t}(\bar{x}) = N(\bar{x})\boldsymbol{n}(\bar{x}) + \boldsymbol{N}(\bar{x}) \ , \qquad \forall \bar{x} \in \boldsymbol{\sigma}_0 \ . \tag{1.20}$$

In that case, for any choice of coordinate chart in σ_0, we can extend those coordinates along the flow lines of $\boldsymbol{t}$ to the corresponding points in (Cauchy) reference surfaces σ_t in (1.19). In such coordinate charts that are adapted to $\boldsymbol{t}$, the metric g in $\mathbf{M}$ can be expressed in the following form in terms of the Riemannian metric $\boldsymbol{h}$, which it induces along σ_0 (cf. [M], p. 507, or [W], p. 255),

$$g = g_{00} dt \otimes dt + g_{0a}\, dt \otimes d\bar{x}^a + g_{a0}\, d\bar{x}^a \otimes dt - h_{ab}\, d\bar{x}^a \otimes d\bar{x}^b \ , \tag{1.21a}$$

$$g_{00} = g(\boldsymbol{t},\boldsymbol{t}) = N^2 - h_{ab} N^a N^b \ , \qquad g_{0a} = g_{a0} = -h_{ab} N^b \ , \qquad a,b = 1,2,3 \ . \tag{1.21b}$$

Hence, in such specifically chosen coordinates all its components are t-independent[10].

Fundamentally, it is this t-independence of the metric components for these special types of foliation of $\mathbf{M}$ that enables the introduction (Ashtekar and Magnon, 1975; Kay, 1978) of a t-independent complex structure in a suitably restricted family of real solutions of the Klein-Gordon equation in (1.17), and therefore a t-independent mode of decomposition of the corresponding complex solutions of (1.17) into positive-frequency and negative-frequency solutions. Indeed, in any globally hyperbolic (**M**,g) we can consider the family of all real solutions of (1.17) that result in a unique manner (Choquet-Bruhat, 1968) from the set of all smooth and real-valued initial data that have compact support in σ_0, namely the following set of pairs of functions on σ_0:

$$\hat{C}(\sigma_0) = \left\{ (\hat{f},\hat{g}) \middle| \hat{f}(\bar{x}) = \hat{\varphi}(0,\bar{x}),\ \hat{g}(\bar{x}) = \partial_t \hat{\varphi}(0,\bar{x}),\ \hat{f},\hat{g} \in C_0^\infty(\sigma_0) \right\} . \tag{1.22}$$

This set constitutes a real vector space, in which we can introduce the symplectic form [C]

$$\left((\hat{f}_1,\hat{g}_1) \middle| (\hat{f}_2,\hat{g}_2) \right)_{\sigma_0} = \int_{\sigma_0} \left[\hat{f}_1(x)\hat{g}_2(x) - \hat{g}_1(x)\hat{f}_2(x) \right] d\sigma(x) . \tag{1.23}$$

In a stationary spacetime (**M**,g), the time-evolution in this vector space can be related to the Killing field $\boldsymbol{t}$ via the aforementioned coordinate charts as follows,

$$\mathcal{T}(t):\ (\hat{f},\hat{g}) \ \mapsto\ (\hat{f}_t,\hat{g}_t) \in \hat{C}(\sigma_0)\ ,\ \hat{f}_t(\bar{x}) = \hat{\varphi}(t,\bar{x}),\ \ \hat{g}_t(\bar{x}) = \partial_t \hat{\varphi}(t,\bar{x}) . \tag{1.24}$$

Hence, it constitutes a one-parameter group, that preserves the symplectic form in (1.23). Upon completing the real vector space in (1.22) with respect to an E-norm whose square

$$\left\| (\hat{f},\hat{g}) \right\|_E^2 = \int_{\sigma_0} \hat{T}_{\mu\nu}[\hat{\varphi}(x)]\, t^\mu(x)\, d\sigma^\nu(x)\ ,\ \ \hat{f}(\bar{x}) = \hat{\varphi}(0,\bar{x}),\ \ \hat{g}(\bar{x}) = \partial_t \hat{\varphi}(0,\bar{x}), \tag{1.25}$$

is determined by the energy density

$$\hat{T}_{\mu\nu}[\hat{\varphi}] = \nabla_\mu \hat{\varphi} \nabla_\nu \hat{\varphi} - \tfrac{1}{2} g_{\mu\nu} (\nabla^\lambda \hat{\varphi} \nabla_\lambda \hat{\varphi} - m^2 \hat{\varphi}^2)\ ,\qquad \nabla^\kappa \hat{\varphi} = g^{\kappa\lambda} \nabla_\lambda \hat{\varphi}\ , \tag{1.26}$$

of the real-valued classical field in (1.24) which obeys (1.17), we arrive at an auxiliary Hilbert space $\mathcal{A}(\boldsymbol{t};\sigma_0)$, whose inner product corresponds to that norm. In that completion $\mathcal{A}(\boldsymbol{t};\sigma_0)$ we can introduce a complex structure (cf. Note 35 to Chapter 3), compatible with the symplectic form in (1.23), by means of the following unitary operator J_t,

$$J_t = h_t |h_t|^{-1}\ ,\qquad J_t^2 = -\mathbf{1}\ ,\qquad h_t(\hat{f},\hat{g}) := -d(\mathcal{T}(t)(\hat{f},\hat{g}))/dt\big|_{t=0} . \tag{1.27}$$

It can be then easily proved (Ashtekar and Magnon, 1975; Kay, 1978) that this particular operator J_t provides the only complex structure on $\mathcal{A}(t;\sigma_0)$ that preserves the symplectic form in (1.23), and which also conserves the total classical energy E defined by the integral in (1.25).

We can now isometrically embed the real Hilbert space $\mathcal{A}(\boldsymbol{t};\sigma_0)$ into a complex Hilbert space $\mathcal{H}(\boldsymbol{t};\sigma_0)$ by means of an identification map $I_t : \mathcal{A}(t;\sigma_0) \to \mathcal{H}(t;\sigma_0)$ which is such that

$$(a+ib) I_t((\hat{f},\hat{g})) = I_t(a(\hat{f},\hat{g})) + I_t(bJ_t(\hat{f},\hat{g}))\ ,\qquad \forall a,b \in \mathbf{R}^1 . \tag{1.28}$$

This complex Hilbert space $\mathcal{H}(t;\sigma_0)$ can be then obtained by carrying out, in the customary manner [PQ], the completion of $I_t\mathcal{A}(t;\sigma_0)$ with respect to the norm defined by the inner product

$$\left\langle I_t(\hat{f}_1,\hat{g}_1)\middle|I_t(\hat{f}_2,\hat{g}_2)\right\rangle_{\sigma_0} = \tfrac{1}{2}\left((\hat{f}_1,\hat{g}_1)\middle|J_t(\hat{f}_2,\hat{g}_2)\right)_{\sigma_0} + \tfrac{i}{2}\left((\hat{f}_1,\hat{g}_1)\middle|(\hat{f}_2,\hat{g}_2)\right)_{\sigma_0} \quad . \tag{1.29}$$

We then define the quantum Hamiltonian H_t associated with the Killing field t by setting

$$H_t\hat{\varphi} := -I_t J_t h_t(\hat{f},\hat{g}) \quad , \qquad \hat{\varphi} = I_t(\hat{f},\hat{g}) \ . \tag{1.30}$$

In this manner we arrive at an essentially self-adjoint operator which governs the time-evolution in the complex Hilbert space $\mathcal{H}(t;\sigma_0)$, which is then adopted as the single-particle Hilbert space corresponding to the chosen Killing field. Finally, by identifying in a natural manner, namely in accordance with (1.24), the wave functions on various reference hypersurfaces σ_t, we can eliminate the dependence of the construction of $\mathcal{H}(t;\sigma_0)$ on the choice of the initial-data reference hypersurface σ_0. Hence, the resulting single particle Hilbert space $\mathcal{H}(t)$ can be deemed to be dependent only on the timelike Killing field t.

From this point onwards, canonical second-quantization for neutral spin-zero particles can proceed as in the Minkowski case. Thus, we can introduce in the Fock space

$$\hat{\mathcal{F}}(t) = \bigoplus_{n=0}^{\infty} \hat{\mathcal{F}}_n(t) \ , \qquad \hat{\mathcal{F}}_n(t) = \mathcal{H}(t) \underset{S}{\otimes} \cdots \underset{S}{\otimes} \mathcal{H}(t) \ , \tag{1.31}$$

creation operators for a particle in the generic state f that belongs to $\mathcal{H}(t)$:

$$\left(a^{\dagger}(f)\Psi_n\right)_{n+1}(x_1,\dots,x_{n+1}) = (n+1)^{-1/2}\sum_{j=1}^{n+1} f(x_j)\,\Psi_n(x_1,\dots,x_{j-1},x_{j+1},\dots,x_{n+1}) \ . \tag{1.32}$$

We can then define the corresponding annihilation operators $a(f)$ as their Hilbert space adjoints. Thus, for any choice of orthonormal basis $\{f_1, f_2, \dots\}$in the single particle space $\mathcal{H}(t)$ we obtain a standard representation of the canonical commutation relations:

$$[a_\alpha, a_\beta^*] = \delta_{\alpha\beta} \quad , \qquad [a_\alpha, a_\beta] = [a_\alpha^*, a_\beta^*] = 0 \quad , \qquad a_\alpha^* = a^{\dagger}(f_\alpha) \ . \tag{1.33}$$

The definition of the corresponding scalar quantum field in the classical stationary spacetime $(\mathbf{M},g)$ then proceeds as in (1.7), namely be setting

$$\phi(x) = \sum_{\alpha=1}^{\infty}\left(f_\alpha(x)\, a_\alpha + f_\alpha^*(x)\, a_\alpha^* \right) \ . \tag{1.34}$$

On the surface, it might appear that after arriving at this definition of scalar quantum field in a stationary spacetime $(\mathbf{M},g)$, the definitions of other basic quantities, such as that of the normally ordered stress-energy tensor that occurs in (1.14), can proceed along the same lines as in Minkowski space. This, however, is not the case due to a number of difficulties and ambiguities, to whose examination we turn next.

*7.2. Spontaneous Rindler Particle Creation in Minkowski Spacetime

There are numerous difficulties and *physical* inconsistencies to which the canonical scheme of second quantization described in the preceding section, as well as all the other conventional schemes of quantization of fields in curved spacetime [BD], give rise even in the absence of mutual interactions between quantum fields. In a section of the well-known review article "Quantum Field Theory in Curved Spacetime" by B. S. DeWitt (1975), which is suitably entitled "Failure of conventional procedures", some of these difficulties are aptly and succinctly summarized as follows (with italics retained as in the original):

"This [canonical quantization procedure of fields in curved spacetime] is just as in conventional particle physics. The trouble with it is: *it's wrong*. It is not wrong in a technical mathematical sense. It simply provides a grossly inadequate foundation for the theory. Here are just some of the situations where it fails:

1. There may be no Killing vector at all, timelike or spacelike. This is the generic situation. How to deal with it is unknown, except possibly when there is an approximate Killing vector that becomes exact asymptotically. ...

2. There may be a global Killing vector, but it may not be everywhere timelike. ...

3. Spacetime may be stationary only in limited regions. If each region possesses complete Cauchy hypersurfaces then a local timelike Killing vector field may be set up in each and vacuum defined for each [such region]. ... " (DeWitt, 1975, p. 302).

DeWitt then continues to illustrate how, as a consequence of the third of the above points, the existence of these distinct vacuum states gives rise to ambiguities and anomalies in the definition of the stress-energy tensor. Later work by Wald (1977, 1978) has suggested axioms[11] for the stress-energy tensor that are capable of removing most of these ambiguities[12], but certain anomalies still remain (cf. [BD], pp. 220-221). Furthermore, the dual role which the stress-energy tensor plays in quantum field theory gives rise to an even more serious type of *physical* ambiguity which, as will be discussed in the next section, brings into question the very meaning of the conventional concept of quantum particle.

These ambiguities manifest themselves already in Minkowski spacetime, i.e., in $\mathbf{R}^4$ treated as a flat Lorentzian manifold in which *local* holonomic or nonholonomic frames play a central physical role – rather than just as a vector space in which global Lorentz frames are of primary physical significance. We can then consider the question of the physical significance of the Fock space in (1.31), that is constructed for a Killing field other than the one corresponding to a coherent flow (cf. Sec. 5.4) of classical test particles moving by inertia along parallel worldlines, which obviously produces the Fock space in (1.5).

A simple example of such a noninertial flow of classical test particles (or classical "observers") in Minkowski spacetime can be easily provided in the Rindler coordinates (t,χ,y,z), which are related as follows to the Minkowski coordinates with respect to a global Lorentz frame $(e_i(O))$ with origin at O (cf. (2.3.12)):

$$x^0 = \pm\chi\sinh at\,,\quad x^1 = \pm\chi\cosh at\,,\quad x^2 = y\,,\; x^3 = z\,,\qquad \pm x^1 > |x^0|\,, \tag{2.1}$$

$$x^0 = \pm\chi\cosh at\,,\quad x^1 = \pm\chi\sinh at\,,\quad x^2 = y\,,\; x^3 = z\,,\qquad \pm x^0 > |x^1|\,, \tag{2.2}$$

In these coordinates the Minkowski spacetime metric assumes the form

$$\boldsymbol{g} = \eta_{\mu\nu} \boldsymbol{dx}^{\mu} \otimes \boldsymbol{dx}^{\nu} = \pm a^2 \chi^2 \boldsymbol{dt} \otimes \boldsymbol{dt} \mp \boldsymbol{d\chi} \otimes \boldsymbol{d\chi} - \boldsymbol{dy} \otimes \boldsymbol{dy} - \boldsymbol{dz} \otimes \boldsymbol{dz} \ , \tag{2.3}$$

so that its components in Rindler coordinates are static. Hence the field $\partial/\partial t$ tangential to the timelike coordinate lines obtained by keeping χ, y and z fixed, while varying t, is indeed a Killing field. It is therefore possible to apply the construction outlined in the preceding section using reference hypersurfaces to which this particular Killing field gives rise, i.e., hypersurfaces that correspond to constant values of t. As a matter of fact, it is possible to explicitly perform [BD,N] the decomposition of a positive-energy solution of the Klein-Gordon equation (1.2) in terms of positive-frequency and negative frequency solutions with respect to the Killing field $\partial/\partial t$. One can then not only demonstrate that $\boldsymbol{t} = \partial/\partial t$ gives rise to a Fock space in (1.31) which is distinct from the one in (1.5), but one can then actually compute the expectation value of the total particle number operator

$$\hat{\mathbf{N}}_{\text{Rindler}} = \sum_{\alpha=1}^{\infty} a_{\alpha}{}^{*} a_{\alpha} \ , \qquad a_{\alpha}{}^{*} = a^{\dagger}(f_{\alpha}) \ , \qquad f_{\alpha} \in \mathcal{H}(\partial/\partial t) \ , \tag{2.4}$$

for "Rindler particles" in the "Minkowski vacuum" state $\Psi_{0;\text{Mink}}$ of inertial observers, i.e., the vacuum state corresponding to the Fock space in (1.5).

It turns out (Unruh, 1976) that this distribution of Rindler particles in the Minkowski vacuum state displays over the entire available range of frequencies ν the same type of distribution that Hawking (1975a) has computed for his well-known prediction of thermal emission from black holes, namely that (cf. [BD], pp. 53-54; or [N], p. 282)

$$\left\langle \Psi_{0;\text{Mink}} \middle| \hat{\mathbf{N}}_{\text{Rindler}}(\nu,\mathbf{k}) \Psi_{0;\text{Mink}} \right\rangle = \left(\exp(2\pi\nu/a) - 1\right)^{-1} \ , \qquad \nu \in (0,\infty) \ , \tag{2.5}$$

$$\hat{\mathbf{N}}_{\text{Rindler}}(\nu,\mathbf{k}) = a^{\dagger}(f_{\nu,\mathbf{k}}) a(f_{\nu,\mathbf{k}}) \ , \qquad \mathbf{k} = k_2 \boldsymbol{e}_2(O) + k_3 \boldsymbol{e}_3(O) \ , \ k_2, k_3 \in \mathbf{R}^1 \ , \tag{2.6}$$

$$f_{\nu,\mathbf{k}}(t,\chi,y,z) = \left((\sinh \nu\pi)^{1/2}/2\pi^2\right) \exp(-i\nu t + ik_2 y + ik_3 z) K_{i\nu}(|\mathbf{k}| a\chi) \ . \tag{2.7}$$

Hence, the ν-distribution in (2.5) coincides with that of thermal radiation at temperature

$$T_{\text{Rindler}} = a/2\pi k_{\text{B}} = (\hbar a/2\pi c k_{\text{B}})^{\text{o}}\text{K} \approx \left(4 \times 10^{-23} a\right)^{\text{o}}\text{K} \ , \tag{2.8}$$

where k_{B} denotes Boltzmann's constant, and the first expression for that temperature is in Planck units, whereas the last one is in Kelvin degrees when a is expressed in cgs units.

A classical test particle or detector (i.e., "Rindler observer" [BD]) whose world line is one of the timelike flow lines of this Killing field within the Rindler wedge in (2.1), and passes through a point with Rindler coordinates χ, y and z, displays a uniform proper acceleration of magnitude a/χ (Rindler, 1969). Thus, we see that the transference of the conventional special relativistic concept of quantum propagation of spin-zero massive particles to a Minkowski spacetime background in which the process of quantization is carried out in relationship to uniformly accelerated detectors leads to the prediction that those detectors will register spontaneous pair creation *ex nihilo* .

As analyzed by Unruh and Wald (1984), this type of spontaneous pair creation *ex nihilo* is not part of the energy-conserving and well established phenomenon of pair creation for quantum fields in Minkowski space. In fact, the final conclusion of their "analysis suggests a rather surprising viewpoint of this [spontaneous Rindler] radiation: it seems as though the detector is excited by swallowing part of the vacuum fluctuation of the field in the region of spacetime containing the detector. This liberates the correlated fluctuations in a noncausally related region of [Minkowski] spacetime to become a real particle." (Unruh and Wald, 1984, p. 1055).

Thus, according to this type of analysis, in Minkowski spacetime "a uniformly accelerated observer will 'see' thermal radiation (Davies, 1975; Unruh, 1976) even though the field is in a vacuum state and, as far as inertial observers are concerned, no particles are detected whatever" ([BD]. p. 54). Such a strange phenomenon has not been observed, however, in high energy particle accelerators. Hence, it is argued that, in view of the numerical value of the factor in (2.8) for the temperature of thermal radiation in Kelvin degrees, "this [Rindler radiation] effect is much too small to be perceived by an ordinary laboratory detector", but that nevertheless "the effect of this thermal bath on the spin of accelerating electrons may be measurable (Bell and Leinaas, 1983)" (cf. [W], p. 415). On the other hand, *if* this Rindler radiation effect *were* observed[13], it would imply that "*both* the detector and the field [producing the particles] gain energy" ([BD], p. 55). Consequently, it is conceded that, possibly, "basing one's treatment of these [pair creation *ex nihilo*] concepts on the considerations of accelerated observers is a fraud, because inertial observers occupy a special status in most physical theory" ([BD]. p. 55).

Indeed, ever since the inception of classical mechanics by Newton, inertial frames have enjoyed a special status in physics. In particular, the laws set by Newton were presumed to hold true *only* in inertial frames, with *fictitious* forces (such as Coriolis forces, centrifugal forces, etc.) making their appearance if a transition to accelerated frames were performed. This special status of inertial frames and inertial observers has been retained in special relativity – and in fact it has been implicitly transferred by Einstein also to general relativity (cf. Chapter V in Friedmann, 1983). It might therefore appear prudent to interpret the fact that the conventional methods of quantization have brought into doubt that special status of inertial observers as a reflection on the physical validity of those schemes, rather than as proof the most basic principle of *all* forms of mechanics, which has received overwhelming experimental support for well over three centuries, is to be discarded. In the absence of *any* form of experimental evidence, such a drastic action does not seem at all justified, despite the fact that "there is now broad agreement on most of the technical results" ([BD], p. 8) of conventional quantum field theory in curved spacetime amongst the leading contemporary researchers in this area of quantum physics (cf. Secs. 12.2-12.3).

In addition to violating the special status of inertial observers by putting their frames of reference on an equal footing with those of any other observers, the point of view that "Rindler particles", as well as all the other "particles" allegedly observable *only* by select classes of noninertial observers in Minkowski spacetime, actually exist, has to contend also with a fundamental epistemological difficulty. This difficulty is actually shared by other present-day theories (such as the cosmological scenario by Vilenkin (1982, 1989) of *ex nihilo* creation of our universe) that are based on the idea that spontaneous creation *ex nihilo* is a *physically meaningful* concept: since *something* is presumed to be literally created out of what is *physically* nothing, but exists only in our minds as a mathematical construct, what is there to prevent the creation in this manner of *any* kind of object, i.e., in the present context, the creation of Rindler particles of any mass, or spin, or charge, etc.?

Despite its obvious nature, this question does not appear to have been raised and discussed in the literature. Perhaps that is due to the fact that the Fock vacuum in (1.5), which in that context represents a true state of *zero particles*, is mentally identified with the "dressed vacuum" of quantum field theories for interacting fields, which, on account of the customary renormalization procedures, is not a no-particle state, but rather it is inhabited by an infinity of "bare" particles. If that is the case, note should be taken of the fact that the arguments leading to (2.5) apply to *free* quantum fields, so that the "dressed vacuum" of quantum field theories for interacting fields has no bearing on the present subject.

An easy way out of the above epistemological puzzle might be to invoke some *ad hoc* principle – such as a "censorship hypothesis", whereby noninertial observers are forbidden to "see" any species of quantum particles which are not also observable to inertial observers. This *ad hoc* censorship hypothesis would be also useful in reinstating the special status of inertial observers – without which the argument leading to Rindler radiation effects could be reversed, so that a "bath" of Minkowski particles would inhabit the world of inertial observers who are surrounded by a Rindler vacuum. However, such a proposal would still not solve for Rindler particles the conceptual problem in a *physically* satisfactory manner, as it should be evident from a second glance at equation (2.5).

Indeed, the rate of production of Rindler particles is the same for all quantum fields, since, according to (2.5), it depends only on the acceleration of the noninertial observers, i.e., it is "independent of the means used to accelerate the detector, but depends only on the acceleration itself" (Unruh, 1976, p. 885) – and in fact it is the same also for charged particles of arbitrary spins and rest masses. Since neither special nor general relativity imposes any limits on achievable magnitudes a of acceleration, the fact that at the present time the Rindler radiation effect is "too small to be perceived by an ordinary laboratory detector" ([W], p. 415) is only a temporary anthropic limitation[14], rather than a fundamental one. Hence, there might exist in our universe "observers" who are so highly accelerated in relation to inertial frames that they are perpetually immersed in a very hot "bath" of Rindler particles, in which various masses, spins, charges, etc, occur in *equal* proportions. In particular, such observers would be surrounded by equal numbers of Rindler particles and antiparticles belonging to all species that are "in existence", and might or might not have been already discovered by inertial observers in terrestrial laboratories. On the other hand, "Minkowski" particles and antiparticles have a tendency to annihilate each other as soon as they come in close proximity. Assuming that the same is true of Rindler particles, such highly accelerated Rindler "observers" would literally inhabit an explosively unstable environment merely by virtue of the fact that they are accelerated in relation to inertial frames.

In view of such paradoxical situations to which the conventional methods of second-quantization have given rise, what is the alternative provided by the geometro-stochastic method of second-quantization?

The answer to this question lies already in the GS method of first quantization, elaborated in the preceding four chapters. Thus, in Secs. 3.5 we have seen that the origins of GS quantization are from a mathematical point of view group-theoretical, and from a physical point of view operational: a proper state vector, representing a quantum test body marking the origin of a quantum Lorentz frame, has been subjected to all possible spacetime translations and Lorentz boosts, in order to use the corresponding duplicates of those bodies as markers of all possible locations and inertial states of motion in relation to that frame. Hence, as opposed to the canonical method of quantization, which is heavily coordinate-dependent from the outset, already the initial step in GS first-quantization is coordinate-independent – and neither Killing fields, nor any other kind of fields, play a favored role in it. Furthermore, this initial step produces a family of quantum Lorentz frames that are related in a purely kinematical manner – to precisely the same extent to which relationships of classical Lorentz frames are of a purely kinematical nature.

As presented in a general relativistic context in Chapter 5, the next step in GS first quantization is also coordinate-independent and "observer"-independent: quantum bundles are constructed, and in these bundles frame-independence as well as observer-independence are secured by adopting the Poincaré group as a structure group.

It is only in the subsequent step that an implicit observer dependence makes its appearance. Indeed, the formulation of the strictly causal GS propagation in Sec. 5.4 requires a foliation of the base manifold **M** of the quantum bundle. Such a foliation was also required in formulating a weak notion of relativistic causality in Sec. 5.7. However, as explained in that same section, in accordance with the view that "the four-dimensional spacetime geometry simply does not exist, except in a classical approximation" ([M], p. 1183), the reference hypersurfaces σ_t should be viewed as being the constituents of a geometrodynamic evolution, rather than of a foliation of an *already* existing classical spacetime manifold. Their observer-dependence then merely reflects the kind of observer-dependence that is unavoidable in any quantum framework, on account of the active role that observers can play in quantum theory in general. As such, from the point of view of the full-fledged theory of quantum spacetime, which of necessity incorporates quantum gravity, such choices of geometrodynamic evolutions result in part from particular quantum gravity gauge choices (cf. Chapter 11). In part they also reflect the intervention of *real* observers by means of measurement acts which, on the macroscopic level, produce the kind of effects that are commonly described as "reductions of the wave packet".

In any event, as we have seen already in Sec. 5.4, and as it will be further discussed in Sec. 7.7, the stongly causal formulation of GS propagation for free quantum fields in Minkowski spacetime is observer-independent, since in the absence of gravity (i.e., of a dynamical interplay between matter and geometry) we can indeed envisage a world in which the base quantum spacetime manifold unfolds in a strictly deterministic manner. On the other hand, the weakly causal formulation of GS propagation does allow effects which to a classically conditioned observer might *appear* as creation *ex nihilo* – but in actuality represent quantum tunneling effects which take place along those stochastic paths which are not *classically* causal. As such, such tunneling effects still abide by strict energy conservation, and are as observable to inertial observers as they are to noninertial observers.

Consequently, even if some form of *apparently* spontaneous particle production would be observed in high-energy accelerators, that still would not represent a true test for the existence of Rindler particle radiation that is in accordance with (2.5). Indeed, its hallmark is not only a violation of *local* energy conservation[15], as it "liberates the correlated fluctuations in a noncausally related region of [Minkowski] spacetime to become a real particle" (Unruh and Wald, 1984, p. 1055), but also the *indiscriminate nature of Rindler particle production*, whereby each species of particle has an equal chance of being produced literally out of a vacuum, rather than as a result of a collision process. As a corollary of these manifestations, it might even appear that a Rindler perpetuum mobile could be created, whereby unlimited amounts of energy could be produced by the simple expedient of accelerating any material object, of however small rest mass, to sufficiently high acceleration in relation to any inertial frame – such as a (terrestrial) laboratory frame.

The only way out of this ultimate "paradox" is to suggest that in a particle accelerator the energy for producing the Rindler particles, which an accelerated micro-detector would register, might come from the accelerator itself, since as it accelerates the micro-detector, "the work done by [it] ... supplies the missing energy that feeds into the [quantum] field via the quanta emitted from the detector" ([BD], p. 55). But in that case the whole Rindler particle production phenomenon is due to the fact that an open system, capable of receiving energy from the outside, i.e., from the accelerator, was treated as a closed system – whereas the physically *correct* treatment should have incorporated from the outset all the quantum fields created by the accelerator during the acceleration process[16]. Indeed, on one hand it is claimed that "the energy for this [Rindler] emission, as far as the Minkowski observer is concerned, comes from the external field accelerating the detector" (Unruh, 1976, p. 885), whereas on the other hand the actual derivation of (2.5) requires absolutely no *physical* external fields that would be *coupled* to Rindler particles. In fact, *only* the *mathematical adoption* of a noninertial Killing field in Minkowski spacetime in applying the canonical second-quantization procedure described in the preceding section is required. Thus, it would appear that the mere *mathematical* presence of such Killing fields is viewed as a reflection of a *physical* reality that represents the presence of an external agent.

However, there is a more sensible alternative: the presence of Rindler radiation when a non-inertial Killing field in Minkowski spacetime is adopted in the course of the canonical second-quantization procedure represents a *fundamental inconsistency* of this very same procedure, due to the fact that the *localization problem for relativistic pointlike quantum particles has not been solved* prior to implementing it. Indeed, in the best-known studies of this "phenomenon", the mathematical analysis of various *models* of detectors of Rindler radiation are based on a key *tacit* assumption – namely that the Minkowski coordinate variables in (2.1) mark points in the spectrum of relativistic quantum position observables related to the *sharp* localization of the presumed detection process. For example, a model detector might be envisaged as a *nonrelativistic* free quantum particle within a rigid box with sharply delineated boundaries (cf. Unruh, 1976, Sec. III; Wald and Unruh, 1984, p. 1051); or the coupling of the model detector to the quantum field might be taken to incorporate "a smooth function [of $\mathbf{x}$, $x = (t,\mathbf{x})$] which vanishes outside the detector, [where] we use x to denote a [Minkowski] spacetime point and $\mathbf{x}$ to denote a point on [the static initial-data slice] Σ" (Wald and Unruh, 1984, p. 1051). Alternatively: "functions $x^m(\tau)$ define the world-line of the detector (idealized as a pointlike object) and the operator $\boldsymbol{m}(\tau)$ represents its monopole moment at proper time τ [and moreover] the detector has a discrete set of internal energy eigenstates" (DeWitt, 1979, p. 693), so that it is "a classical point-like object with internal quantum states" (Davies, 1984, p. 66). Hence, all such *assumptions* about the *physical* nature of the detector contravene the no-go results on sharp relativistic quantum localization discussed in Secs. 1.2, 3.3 and 3.5. On the other hand, Einstein causality is tacitly assumed to hold, contrary to Hegerfeldt's theorem. It is therefore of no surprise that, although such studies "have resolved some apparent paradoxes regarding energy conservation and causality" (Wald and Unruh, 1984, p. 1055), the central ones, pertaining to the indiscriminate nature of Rindler radiation and its spatio-temporal origin, still remain. In particular, as P.C.W. Davies points out, in comparing the calculations leading to (2.5) with the calculations based on such detectors, "caution is necessary, for [these two] calculations are really addressing quite different issues. One is the experiences of a detector along a given worldline, the other concerns particles that are not localized but are defined over the entire Rindler wedge [specified by (2.1)]. ... Thus the experiences of a specific detector are in general no guide at all to the 'particle content' as defined by the Bogoliubov transformations [such as those in (3.3) below]!" (Davies, 1984, p. 71).

Finally, let us ignore for a moment these last several points, and accept the explanation that, after all, the Rindler radiation phenomenon is being caused by the "external field accelerating the detector". In that case the same argument cannot be applied to a similar micro-detector in *free* fall, whose behavior will be examined in the next section – despite all the manifest formal analogies between Rindler particle production and the particle creation *ex nihilo*, which will emerge during that examination, and which are very much stressed in most of the literature on the subject. Indeed, if the equivalence principle of general relativity is correct, then *"observers" in free fall are truly inertial* – and as such they are not to be viewed as being accelerated by an external gravitational field, as it is the case in Newtonian mechanics.

*7.3. Ambiguities in the Concept of Quantum Particle in Curved Spacetime

The fundamental difficulties to which the conventional methods of second-quantization in curved spacetime give rise when they are specialized to Minkowski spacetime represent only a preamble to the even graver difficulties to which they give rise in the context of a curved spacetime. As a consequence, according to Narlikar and Padmanabhan (1986), p. 277: "*There does not exist a quantum field theory formalism in an arbitrary curved spacetime*. This problem is deep rooted and arises from the fact that standard formalisms of field theory require a preferential slicing in spacetime."

Basically, this is due to the fact that in quantum field theory in Minkowski space the components $T_{\mu\nu}(x)$ of the stress-energy tensor, such as those in (1.14), not only supply such physically measurable values as the expectation values $\langle P_\mu \rangle$ of total 4-momentum, but they also provide the generators of spacetime translations. In particular, for a choice of reference hypersurfaces σ_t which in some global Lorentz frame correspond to the hyperplanes consisting of all points with Minkowski coordinates $x^0 = t$, the operator P_0 plays on one hand the role of Hamiltonian H governing time-evolution, whereas on the other hand it plays the role of operator that represents the total energy of the system as measured by *global inertial observers* stationary with respect to that frame.

It therefore *appears* most natural to generalize the canonical second-quantization method, described in Sec. 7.1, to an arbitrary globally hyperbolic classical spacetime manifold $(\mathbf{M},g)$ by preserving as much as possible this dual role of the stress-energy tensor. Hence, upon introducing a foliation (5.4.7) of $\mathbf{M}$ into hypersurfaces σ_t with future-pointing unit normals $\boldsymbol{n}$, Ashtekar and Magnon (1975) have proposed casting

$$E(\boldsymbol{n},t) = \int_{\sigma_t} \widetilde{T}_{\mu\nu}[\widehat{\varphi}(x)]\, N(x) n^\mu(x) d\sigma^\nu(x) \ , \qquad N\boldsymbol{n} = \partial/\partial t \ , \tag{3.1}$$

into the role of "total classical energy" that would serve, as it did in (1.25), as a basic ingredient in the definition of a norm, which, however, might be t-dependent. If that is the case, then the construction of Sec. 7.1 will result in a t-dependent complex structure $\boldsymbol{J_n}(t)$, which in turn gives rise to a t-dependent quantum Hamiltonian $\boldsymbol{H_n}(t)$ such that

$$\langle\hat{\varphi}|H_{n}(t)\hat{\varphi}\rangle_{\sigma_t} = \int_{\sigma_t} \hat{T}_{\mu\nu}[\hat{\varphi}(x)]\, N(x) n^{\mu}(x) d\sigma^{\nu}(x) \;, \qquad \hat{\varphi} = I_{n}(t)(\hat{f}_t, \hat{g}_t) \;. \tag{3.2}$$

This last condition selects, as it did in Sec. 7.1, a unique complex structure $J_n(t)$. However, in case that Nn is not a Killing field, this complex structure is not only n-dependent but also t-dependent, so that it gives rise to "Bogoliubov transformations" in the corresponding t-dependent Fock spaces $\mathcal{F}(n;t)$. These types of transformations are then conventionally interpreted as representing spontaneous pair creation *ex nihilo*, which can be "seen" by "observers" who travel along the flow lines of the adopted timelike vector field n. Thus, a *mathematical* analogy with the case of noninertial observers in Minkowski spacetime does exist, and it is very much emphasized in the literature [BD,N]. However, there is also a notable difference: for special choices of $(\mathbf{M},g)$ and n, some of the flow lines of the vector field n might be geodesics, so that the observers travelling along them would actually be in free fall, i.e, they would be inertial observers, rather than accelerated observers.

On the other hand, from a purely mathematical point of view, the occurrence of Bogoliubov transformations is very easy to explain. Due to the dependence of the complex structure $J_n(t)$ on the global time-parameter t, the decomposition of the family of complex solutions of (1.17) into positive-frequency and negative-frequency solutions becomes t-dependent. Consequently, in general the complex Hilbert space $\mathcal{H}(n;\sigma_t)$ resulting from the use of $J_n(t)$ with $t \neq 0$ in the construction described in Sec. 7.1 does not coincide with $\mathcal{H}(n;\sigma_0)$ upon relating respective initial conditions on the basis of (1.24). Rather, the elements of an orthonormal basis $\{f_1(\sigma_t), f_2(\sigma_t), \ldots\}$ in the complex Hilbert space $\mathcal{H}(n;\sigma_t)$ contain both positive as well as negative frequency components when expressed as elements of the Hilbert spaces resulting from the use of $J_n(0)$, so that they are related to orthonormal bases in the latter by a *Bogoliubov transformation*:

$$f_\alpha(\sigma_t) = \sum_{\beta=1}^{\infty}\left(c_{\alpha\beta}(t) f_\beta(\sigma_0) + d_{\alpha\beta}(t) f_\beta^*(\sigma_0)\right) \;. \tag{3.3}$$

The orthonormality conditions for these sets of orthonormal bases lead to the following constraints on the Bogoliubov coefficients in (3.3):

$$\sum_{\gamma=1}^{\infty}\left(c_{\alpha\gamma}(t) c_{\beta\gamma}^*(t) - d_{\alpha\gamma}(t) d_{\beta\gamma}^*(t)\right) = \delta_{\alpha\beta} \;, \tag{3.4a}$$

$$\sum_{\gamma=1}^{\infty}\left(c_{\alpha\gamma}(t) d_{\beta\gamma}(t) - d_{\alpha\gamma}(t) c_{\beta\gamma}(t)\right) = 0 \;. \tag{3.4b}$$

The presence of the Bogoliubov transformations in (3.3) has profound physical implications. Indeed, by the construction in (1.32) of creation operators, as well as of the corresponding annihilation operators, defined in Sec. 7.1 as their Hilbert space adjoints, we obtain:

$$a_\alpha(\sigma_0) = \sum_{\beta=1}^{\infty}\left(c_{\alpha\beta}(t) a_\beta(\sigma_t) + d_{\alpha\beta}^*(t) a_\beta^\dagger(\sigma_t)\right) \;, \tag{3.5a}$$

$$a_\alpha^\dagger(\sigma_0) = \sum_{\beta=1}^{\infty}\left(c_{\alpha\beta}^*(t) a_\beta^\dagger(\sigma_t) + d_{\alpha\beta}(t) a_\beta(\sigma_t)\right) \;. \tag{3.5b}$$

Therefore, the expectation value of the total particle number operator along σ_t ,

$$\hat{N}(\sigma_t) = \sum_{\alpha=1}^{\infty} a^{\dagger}{}_{\alpha}(\sigma_t)\, a_{\alpha}(\sigma_t) \quad , \qquad a_{\alpha}(\sigma_t) := a(f_{\alpha}(\sigma_t)) \quad , \tag{3.6}$$

in the state represented by the state vector $\Psi_0(\sigma_0)$, which used to represent the vacuum state for observations performed along σ_0 , is equal to

$$\left\langle \Psi_0(\sigma_0) \middle| \hat{N}(\sigma_t) \Psi_0(\sigma_0) \right\rangle = \sum_{\alpha,\beta=1}^{\infty} \left| d_{\alpha\beta}(t) \right|^2 . \tag{3.7}$$

Since in the generic situation the right-hand side of (3.7) is not zero, this mathematical fact is routinely interpreted [BD,N] as evidence of spontaneous particle creation that is visible only to those observers who travel along the worldlines determined by the vector field $\boldsymbol{n}$. The analogy is then drawn with the phenomenon of spontaneous Rindler particle creation, that can be allegedly detected by accelerated detectors in Minkowski spacetime – such as those within a terrestrial particle accelerator.

As we discussed at the end of the preceding section, in a particle accelerator the energy required for such Rindler particle production *ex nihilo* might be *assumed* to come from the accelerator itself. However, let us consider any *curved* classical spacetime, and adopt in it a choice of vector field which is tangential at least in part to timelike geodesics. Those observers who travel along those geodesics would then be in *free* fall, so that the assumption that their apparatus might detect particles materializing out of vacuum would lead, via the equivalence principle, to a clear-cut violation of local energy conservation. Indeed, observers in free fall are *not* accelerated observers, but inertial observers. And, by the equivalence principle, inertial observers do occupy a very special status in the general relativistic theory (Friedmann, 1983, p. 195): to them, locally, the world should appear as it does to inertial observers in Minkowski space.

Thus, even if one subscribes to conventional quantization procedures for fields in a curved classical spacetime, and accepts the verdict that quantum particles have "an essential observer-dependent quality about them", and that "part of the reason for the nebulousness of the particle concept is its *global* nature" ([BD], p. 49), and even if one furthermore accepts the verdict that an *accelerated* observer in Minkowski spacetime will actually "see" spontaneous pair creation *ex nihilo*, that still does not explain why an *inertial* observer in curved spacetime is also liable to "see" spontaneous pair creation *ex nihilo*.

On the other hand, if it is assumed that a freely falling observer in curved spacetime is, contrary to Einstein's point of view, an accelerated observer, the question arises: in relation to what is he accelerated? The only *mathematical* answer to that question that can be found in the literature is that, in the case of curved spacetimes that are *asymptotically* flat, an observer in free fall is accelerated in relation to an observer in the *fictional* flat spacetime that asymptotically merges with the considered curved spacetime. However, since no acceptable cosmological models of our universe as a whole are asymptotically flat, even such an *ad hoc* "solution" does not provide a truly satisfactory answer.

These physical inconsistencies of the conventional framework for quantum fields in curved spacetime extend also to its formulation of the vacuum state. Indeed, it is again acknowledged that "as far as Minkowski space is concerned, the [Fock] vacuum is a strong candidate for the 'correct' or 'physical' vacuum – the experiences of the accelerated observers being 'distorted' by the effects of their non-uniform motion. The trouble is that when gravitational fields are present, inertial observers become free-falling observers, and in general no two free-falling detectors will agree on a choice of vacuum" ([BD], p. 55).

This "lack of agreement" between various free-falling detectors is due, however, to the fact that in CGR all inertial observers are *local* observers, whereas the only "vacuum" considered in [BD], as well as in all other literature on conventional quantum fields in curved spacetime, is a *global* vacuum. Indeed, in the earlier discussed generic situation of an arbitrary foliation of a curved classical spacetime manifold $(\mathbf{M},\boldsymbol{g})$, such a vacuum state $\Psi_0(\sigma_t)$ is associated with each one of the reference hypersurfaces σ_t as a whole, rather than being defined in a local manner, which would be in keeping with the formulation of all fundamental physical concepts in general relativity.

In the preceding two chapters we have seen, however, that an alternative to the global quantum concept of "particle" can be formulated which is of a *local* nature, and which is capable of occupying a *quantum spacetime exciton state* – and to which we therefore refer as a *GS exciton* (cf. Sec. 1.4 and Note 20 to Chapter 1). Such an exciton state is not described merely by the specification of its location in relation to a Lorentzian base manifold (that coincides with a classical spacetime), but also by its location within the fibres *above* that base manifold. In the next section we shall see that an epistemologically sound quantum field theoretical framework can be found which is based on such a physical concept, and which requires only the introduction of *local* vacuum states.

The justification for introducing such local vacuums is that, from the point of view of *local* observers, the concept of a global vacuum is epistemologically unsound at the cosmological level: its actual *physical* existence would require the realizability, at least in principle, of a state of our entire universe in which there is no matter whatsoever for the duration of *global* observations required for establishing its existence. But how could, even in principle, such a state $\Psi_0(\sigma_t)$ be detected by local observers, given the fact that it would require the placement of detecting apparatuses (which are themselves material objects) at all points along those supposedly "vacuous" *reference* hypersurfaces σ_t ? And how could our universe be, even in principle, "temporarily" brought into such a state $\Psi_0(\sigma_{t'})$ with absolutely no matter in it? And even if it could, how would matter be recreated[17] in it at some later $t'' > t'$? In other words, what is the *epistemological meaning* of the gratuitous *assumption* that there *can be* a global vacuum state $\Psi_0(\sigma_t)$ for our *entire* universe at any stage in its development?

It might be answered that such an assumption is acceptable, since it is routinely made in the Minkowski case – and in that context it gives rise to no epistemic inconsistencies. So, the following counter-question could be posed: why would the assumption of a global vacuum give rise to any epistemological inconsistencies in the general relativistic context?

One of the answers to that last question becomes obvious when it is recalled that there is no *global* 4-momentum in a closed universe, since "around a closed universe there is no place to put a test object or gyroscope into Keplerian orbit to determine either any so-called 'total mass' or 'rest frame' or '4-momentum' or 'angular momentum' of the system." ([M], p. 458). However, the existence of a global vacuum $\Psi_0(\sigma_t)$ implies, even in the case of a closed universe, the existence of expectation values of *global* 4-momentum operators, such as the curved spacetime counterparts of those in (1.14) (cf. also (3.2)).

In fact, the epistemological inadequacies of describing the states of any quantum objects (whether they are "particles", "strings", "excitons", or any other presumed physical entities) by wave functions which are supposedly defined everywhere along each reference hypersurface σ_t , on grounds that one can extrapolate from the Minkowski case, become apparent as soon as it is recalled that CGR models can serve as full-fledged cosmological models. On the other hand, if one subscribes to the basic tenets of CGR, *no* flat spacetime model can be used for that purpose. Indeed, within the mathematical framework of CGR, Minkowski space plays the role of typical fibre for the tangent bundle over a classical spacetime, but it can *never* also play the role of base manifold, since the presence of any amount of matter, however small, gives rise to curvature effects.

True, from a *pragmatic* point of view, Minkowski space can indeed provide a good *approximation* for the treatment, under *local terrestrial* conditions, of all physical processes in which gravitational effects play a negligible role. But what could be the possible epistemological interpretation of wave functions which are defined along slices of our entire universe, when in reality such slices have to include regions of very strong gravitational fields, where such an approximation would not be even pragmatically admissible? And what is one to make of such a formulation in the presence of particle and observer horizons? For example, according to the analysis on p. 667 of (Narlikar and Padmanabhan, 1988), "considering observers inside the event horizon makes problems of interpretation even more difficult" already in the classical context. However, at the same time, according to the basic tenets of quantum mechanics, a wave function which cannot be, at least in principle, observed in its entirety fails the test of being physically meaningful.

Indeed, in the presence of horizons, globally defined wave functions can become even in principle unobservable in their *entirety* by a family of local observers, since some of those observers cannot communicate with each other. For this reason, in one of his studies of the implications of his treatment of black hole evaporation, Hawking was forced to postulate "two sets of observables, observables at infinity which describe the outgoing particles and observables inside the black hole which describe what fell through the event horizon." (Hawking, 1984, p. 396). However, communication between the observers measuring those two sets of observables is absolutely necessary in order to execute a *complete* act of observation in the case of those wave functions which do not have compact supports. On the other hand, we have seen already in the case of positive-energy solutions of Klein-Gordon and Dirac wave equations in Minkowski space that no single-particle wave functions can have compact supports (Hegerfeldt, 1974, 1989; Thaller, 1984). In turn, the fact that the supports of such wave functions are not compact implies that it is impossible to prepare states that could be observable in their entirety by a proper subset of the family of local observers following the flow lines of an earlier considered global vector field $\boldsymbol{n}$.

In his article entitled "The Unpredictability of Quantum Gravity", S.W. Hawking offers as a resolution to this kind of problem a very radical departure from conventional quantum concepts, whereby the possibility of measuring pure quantum states, which is central to all quantum theories of measurement, is to be abandoned on the following grounds: "The system [in a classical spacetime region outside a black hole] would still be in a pure quantum state but an observer at infinity could measure only part of the state; he could not even in principle measure what fell into the black hole. Such an observer would have to describe his observation by a mixed state which was obtained by summing with equal probability over all black hole states. One could still claim that the system was in a pure state though this would be rather metaphysical because it could be measured only by an angel and not by a human observer." (Hawking, 1982, p. 396).

On the other hand, P.C.W. Davies concludes an article, which he provocatively but nevertheless appropriately entitled "Particles do not Exist", with the following words: "The study of DeWitt-style particle detectors has exposed the nebulousness of the particle concept and suggests that it should be abandoned completely." (Davies, 1984, p. 76).

Thus, together with many other researchers in the field, we have to conclude that, due to its global nature, the conventional notion of quantum particle is indeed "nebulous" – and in fact untenable in the general relativistic regime. However, as it will be shown in Sec. 7.6, due to foundational epistemological reasons the solution to this fundamental difficulty certainly does not lie, as claimed by some of these researchers, in "concentrating instead on observables built out of [a quantum] field itself" (Fulling, 1989, p. 217). Furthermore, quantum fields themselves require some kind of test bodies for their localization (Bohr and Rosenfeld, 1933, 1950). And where are such test bodies to be found on micro-scales whose units dip down to atomic sizes, and then further, down to the size of hadrons, and then still further to that of leptons and of quarks, and possibly further still to preons, pre-preons, etc., in a conceivably unending chain of ever more refined localizations?

One possible solution might be thought to be provided by string theory. However, as we pointed out in the discussion towards the end of Sec. 1.2, "the fundamental physical and geometric principles that lie at the foundation of superstring theory are still unknown" (Kaku, 1988, p. viii). Some of the most pertinent reasons behind that fact are described as follows by one of the discoverers of the crucial fact that superstring theory is anomaly-free, and possibly finite to all orders in perturbation theory (namely of the theoretical features that initiated the great surge of interest in superstring theory in the first place):

"The anomaly cancellation works for all [the various superstring theories]. ... But string theory is really much deeper than that, [since] it really ought to alter what we mean by space and time as well as altering what we mean by particles. Now that aspect of string theory, this really deep aspect whereby the spacetime in which the string is moving is itself altered by strings, is not really contained within the present formulation of string theory. ... One ought to formulate string theory in a much larger space – something like the space of all possible positions of a string. In fact, this is an *infinitely* larger space ... . The reason we use the language of ten dimensions or four dimensions is because we have so far been forced to talk about string theories in an approximate way, and it's only in this approximation that the whole notion of a small finite number of spacetime dimensions makes sense. ... Talking about four or ten dimensions at all is itself only an approximation to this much larger stringy space which really has an infinite number of dimensions." (Green, 1989, pp. 130-131).

The fundamental hypothesis outlined in Sec. 1.5, and more fully discussed in Chapter 12, is that the construction of a "much larger stringy space which really has an infinite number of dimensions" might lie in the concept of *quantum spacetime exciton states*: such exciton states might represent locations in quantum spacetime that can be actually occupied by material entities, to which we shall refer in the sequel as *GS excitons*, since the established similarities to those of conventional string theory are still rather superficial and remote[18]; or they might represent, as is the case with the points of classical spacetimes in the absence of actually pointlike classical objects, mere mathematical idealizations whose properties, in the aforementioned unending chain of refinements of localization procedures, approximate better and better (rather than faithfully mirror) those of actual quantum objects.

As pointed out in Sec. 1.5, the main goal of the considerations in the remainder of this monograph is not to attempt to supply a definitive answer to the difficult question as to which of these two alternatives might be actually realized in nature. Rather, it is to construct a mathematical framework for GS second quantization which would avoid the physical and conceptual pitfalls encountered by the more conventional second-quantization procedures, as they were outlined and discussed in the preceding two sections. And, even in the absence of a definitive answer to the above question, the central thesis underlying this construction is clear-cut: such a construction is to be based on quantization schemes which resolve the difficulties with the concept of localization in relativistic quantum mechanics by transcending the classical mode of thinking about geometry, and developing methods of second quantization within the context of suitable quantum geometries.

7.4. Fock Quantum Bundles for Spin-0 Neutral Quantum Fields

In the case of spin-0 neutral quantum fields, a Fock quantum bundle $\mathcal{E}$ can be constructed in a straightforward manner (Prugovečki, 1987b), by first building from the Klein-Gordon standard fibre $\mathbf{F}$ in (5.1.1) a corresponding standard Fock fibre $\mathcal{F}$, in accordance with the general procedure used in (1.5):

$$\mathcal{F} = \bigoplus_{n=0}^{\infty} \mathcal{F}_n \quad , \qquad \mathcal{F}_n = \mathbf{F} \underset{S}{\otimes} \cdots \underset{S}{\otimes} \mathbf{F} \ . \tag{4.1}$$

Then, for a given Lorentzian base manifold $(\mathbf{M},g)$, we can set $\mathcal{E}$ equal to its bundle product with the Poincaré principal frame bundle $P\mathbf{M}$:

$$\mathcal{E} = P\mathbf{M} \times_{\mathbf{G}} \mathcal{F} \quad , \qquad \mathbf{G} = \mathrm{ISO}_0(3,1) \ . \tag{4.2}$$

In turn, this implies that the Fock quantum bundle is a Whitney direct sum of symmetrized Whitney direct products[19] of duplicates of the Klein-Gordon quantum bundles $\mathbf{E}$,

$$\mathcal{E} = \widehat{\bigoplus}_{n=0}^{\infty} \mathcal{E}_n \quad , \qquad \mathcal{E}_n = \mathbf{E} \underset{S}{\widehat{\otimes}} \cdots \underset{S}{\widehat{\otimes}} \mathbf{E} \ , \tag{4.3}$$

in very much the same manner in which the Fock space in (4.1) is a (Hilbert) direct sum of symmetrized (Hilbert) tensor products [PQ] of the one-particle space in (5.1.1).

We note that the above construction implicitly assigns a *local* vacuum state vector $\Psi_{0;x}$ to each base location $x \in \mathbf{M}$. This vector, which we shall always deem to be normalized, spans the one-dimensional vacuum sector $\mathcal{F}_{0;x}$ of the Fock fibre $\mathcal{F}_x$ above that base location. Thus, one of the fundamental difficulties encountered by the conventional theory of quantum fields in curved spacetime is resolved: the vacuum state vectors of the Fock quantum bundle $\mathcal{E}$ are geometrically local objects, in the sense that they are associated with single points x in $\mathbf{M}$, rather than with all of $\mathbf{M}$, or reference hypersurfaces in $\mathbf{M}$. Hence, $\Psi_{0;x}$ unambiguously represents a quantum state with no excitons above x, as measurable by any *local* observer equipped with a local quantum detector attached to a quantum Lorentz frame at x – such as one of the frames $\Phi^{u(x)}$ in (5.1.21). Of course, the outcomes of the measurements of the relative base locations of the origins of such frames are subject to fundamental indeterminacies, which are embodied in the confidence functions associated with those frames (cf. Sec. 3.2, as well as Principle 1 in Sec. 1.3). Hence, the implicit notion of locality is not "sharp", and therefore it does not reflect a classical notion of locality[20]. On the other hand, these fundamental indeterminacies in measurement outcomes can be deduced from the adopted values of the quantum spacetime form factor f in (5.1.2), in accordance with the physical interpretation of the probability amplitudes in (5.5.3) proposed in Sec. 5.5. In fact, in a neighborhood $\mathcal{N}_x$ of some $x \in \mathbf{M}$ where the relative curvature effects are small (in the sense discussed in Secs. 2.7 and 5.5), these uncertainties are mirrored by the *local quantum fluctuation amplitudes* (cf. (5.1.18) and (6.3.8))

$$\Delta_x^{(+)}(\boldsymbol{\zeta}';\boldsymbol{\zeta}) = -i\, \boldsymbol{\Phi}_{f,m;\zeta}^{\boldsymbol{u}(x)}(\zeta') \,, \qquad \zeta = (q,v) \,,\quad \zeta' = (q',v') \,, \tag{4.4a}$$

$$\boldsymbol{\zeta} = (\boldsymbol{a} + q^i \boldsymbol{e}_i, v^i \boldsymbol{e}_i) \in T_x\mathbf{M} \times V_x^+ \,, \qquad \zeta = (q,v) \in \mathbf{R}^4 \times V^+ \,. \tag{4.4b}$$

Indeed, in that case, the above local quantum fluctuation amplitudes can be transferred, by means of the exponential map $\exp_x : T_x\mathbf{M} \to \mathbf{M}$, to the neighborhood $\mathcal{N}_x$ from the corresponding neighborhood $(\exp_x)^{-1}\, \mathcal{N}_x$ of the point of contact in the tangent space $T_x\mathbf{M}$.

In each Fock fibre $\mathcal{F}_x$, the n-exciton sector $\mathcal{F}_{n;x}$, as well as any n-exciton state vector $\boldsymbol{\Psi}_{n;x}$ in it, are also unambiguously defined. By extending in a natural manner the generalized soldering map in (5.1.5) to $\mathcal{F}_{n;x}$, we can assign to each $\boldsymbol{\Psi}_{n;x} \in \mathcal{F}_{n;x}$ a *coordinate wave function* $\Psi_{n;x} \in \mathcal{F}_n$ with respect to each quantum frame $\boldsymbol{\Phi}^{\boldsymbol{u}(x)}$, so that

$$\sigma_x^{\boldsymbol{u}} : \ \boldsymbol{\Psi}_{n;x}(\boldsymbol{\zeta}_1,\ldots,\boldsymbol{\zeta}_n) \mapsto \Psi_{n;x}(\zeta_1,\ldots,\zeta_n) \,, \qquad \boldsymbol{\zeta}_1,\ldots,\boldsymbol{\zeta}_n \in T_x\mathbf{M} \times V_x^+ \,. \tag{4.5}$$

Hence, for any given $\boldsymbol{\zeta}_1,\ldots,\boldsymbol{\zeta}_n$, the complex number $\boldsymbol{\Psi}_{n;x}(\boldsymbol{\zeta}_1,\ldots,\boldsymbol{\zeta}_n)$ in (4.5) can be regarded as the frame-independent value of $\boldsymbol{\Psi}_{n;x} \in \mathcal{F}_{n;x}$, to which the soldering map in (4.5) assigns the *coordinate wave function amplitude* $\Psi_{n;x}(\zeta)$ belonging to the element $\Psi_{n;x} \in \mathcal{F}_n$ in the standard Fock fibre $\mathcal{F}$. Note should be made of the fact that the coordinate-independent wave functions in (4.5) are perfectly well-defined functions of the coordinate-independent variables $\boldsymbol{\zeta}_1,\ldots,\boldsymbol{\zeta}_n$, on account of the Poincaré gauge invariance of the entire framework, as well as of the fact that these functions belong to reproducing kernel Hilbert spaces – so that they are continuous and everywhere well defined[21].

Using the extension principle for bounded linear transformations (cf. [PQ], p. 188), we can extend the generalized soldering map in (4.5) to the entire Fock fibre $\mathcal{F}_x$, so that

$$\sigma_x^{\boldsymbol{u}} : \ \boldsymbol{\Psi}_x \mapsto \Psi \in \mathcal{F} \,, \qquad \forall \boldsymbol{\Psi}_x \in \mathcal{F}_x, \qquad \boldsymbol{u} = (\boldsymbol{a},\boldsymbol{e}_i) \in \boldsymbol{\Pi}^{-1}(x) \,. \tag{4.6}$$

We can then define exciton annihilation operators by analogy with (1.6a), namely by

$$\left(\varphi^{(-)}(x;\boldsymbol{\zeta})\boldsymbol{\Psi}_{n;x}\right)_{n-1}(\boldsymbol{\zeta}_1,\ldots,\boldsymbol{\zeta}_{n-1}) = i n^{1/2} \int \Delta_x^{(+)}(\boldsymbol{\zeta};\boldsymbol{\zeta}_n)\boldsymbol{\Psi}_{n;x}(\boldsymbol{\zeta}_1,\ldots,\boldsymbol{\zeta}_n)\, d\Sigma(\boldsymbol{\zeta}_n) \,, \tag{4.7}$$

where the integration can be performed, as was the case in Sec. 5.1, over any hypersurface $\Sigma = \sigma \times V_x^+$ with σ equal to a maximal spacelike hypersurface in $T_x\mathbf{M}$. In view of the reproducibility properties of the integral kernels defined by (4.4a), we have

$$\left(\varphi^{(-)}(x;\boldsymbol{\zeta})\boldsymbol{\Psi}_{n;x}\right)_{n-1}(\boldsymbol{\zeta}_1,\ldots,\boldsymbol{\zeta}_{n-1}) = \sqrt{n}\, \boldsymbol{\Psi}_{n;x}(\boldsymbol{\zeta},\boldsymbol{\zeta}_1,\ldots,\boldsymbol{\zeta}_{n-1}) \,. \tag{4.8}$$

The domain of definition of these exciton annihilation operators can be extended, by using their linearity as well as standard procedures for constructing closures of operators (cf. [PQ], Ch. III, Sec. 3.5), to that of unique densely defined closed (but unbounded) operators in each Fock fibre $\mathcal{F}_x$ of the Klein-Gordon Fock bundle $\mathcal{E}$.

We can similarly define exciton creation operators by analogy with (1.6b):

$$\left(\varphi^{(+)}(x;\zeta)\Psi_{n;x}\right)_{n+1}(\zeta_1,\dots,\zeta_{n+1})$$
$$= i(n+1)^{-1/2}\sum_{j=1}^{n+1}\Delta_x^{(+)}(\zeta_j;\zeta)\,\Psi_{n;x}(\zeta_1,\dots,\zeta_{j-1},\zeta_{j+1},\dots,\zeta_{n+1}) \ . \quad (4.9)$$

The domain of definition of these creation operators can also be extended, by linearity and by taking operator-closures, to that of uniquely and densely defined operators in $\mathcal{F}_x$. In fact, they then become equal to the adjoints of the corresponding annihilation operators:

$$\varphi^{(+)}(x;\zeta) = \varphi^{(-)*}(x;\zeta) \ , \qquad \forall\zeta\in T_x\mathbf{M}\times V_x^+ \ , \qquad \forall x\in\mathbf{M} \ . \quad (4.10)$$

Furthermore, it is easily verified that[22]

$$\left[\varphi^{(-)}(x;\zeta),\varphi^{(+)}(x;\zeta')\right] = i\Delta_x^{(+)}(\zeta;\zeta') \ , \qquad \left[\varphi^{(\pm)}(x;\zeta),\varphi^{(\pm)}(x;\zeta')\right] = 0 \ . \quad (4.11)$$

In view of all this, the *scalar quantum frame fields*

$$\varphi(x;\zeta) = \varphi^{(+)}(x;\zeta) + \varphi^{(-)}(x;\zeta) \ , \qquad x\in\mathbf{M} \ , \qquad \zeta\in T_x\mathbf{M}\times V_x^+ \ , \quad (4.12)$$

are bona fide self-adjoint operators, densely defined within the Fock fibre $\mathcal{F}_x$. On account of (4.11), they satisfy the commutation relations

$$\left[\varphi(x;\zeta),\varphi(x;\zeta')\right] = i\Delta_x(\zeta;\zeta') \ , \qquad \Delta_x(\zeta;\zeta') = \Delta_x^{(+)}(\zeta;\zeta') + \Delta_x^{(-)}(\zeta;\zeta') \ , \quad (4.13)$$

whereas on account of (5.1.18) and (4.4) (cf. also (3.4.3)),

$$(\partial^i\partial_i + m^2)\varphi(x;\zeta) = 0 \ , \qquad \partial_i = \partial/\partial q^i \ , \qquad \partial^i = \eta^{ij}\partial_j \ . \quad (4.14)$$

However, note should be taken of the fact that these quantum frame field operators do not embody any dynamical properties. Rather, their main role will be to enable the construction of second-quantized frames in each Fock fibre $\mathcal{F}_x$ (which will be carried out in the remainder of this section), and to serve as basic ingredients for the definition of stress-energy and Lagrangian operators – which will be carried out in the next section.

Let us introduce the creation operator of an exciton in state $\mathbf{f}$, which can be (densely) defined in $\mathcal{F}_x$ by the following Bochner integrals[23],

$$\varphi^{(+)}(\mathbf{f}) = \int \varphi^{(+)}(x;\zeta)\mathbf{f}(\zeta)\,d\Sigma(\zeta) \ , \qquad \mathbf{f}\in\mathbf{F}_x \ , \quad (4.15)$$

where the integration can be carried out along any of the aforementioned hypersurfaces Σ, as was the case in (4.7) – and as it will be the case everywhere else in similar contexts in the sequel. We can define the *second-quantized frames* at x as vector-valued functionals of single-exciton state vectors $\mathbf{f}$, that assume the following values in a given Fock fibre $\mathcal{F}_x$:

$$\boldsymbol{\Phi}_{\mathbf{f}} = \exp\left[-\tfrac{1}{2}\int|\mathbf{f}(\zeta)|^2 d\Sigma(\zeta) + \varphi^{(+)}(\mathbf{f})\right] \boldsymbol{\Psi}_{0;x} , \qquad \mathbf{f} \in \mathbf{F}_x . \tag{4.16a}$$

The above exponential can be expanded into a strongly convergent power series:

$$\boldsymbol{\Phi}_{\mathbf{f}} = \exp\left[-\tfrac{1}{2}\|\mathbf{f}\|^2\right] \left(\sum_{n=0}^{\infty} \varphi^{(+)}(\mathbf{f})^n / n!\right) \boldsymbol{\Psi}_{0;x} . \tag{4.16b}$$

If we use (4.7) and (4.11), as well as the fact that, according to (5.1.16) and (4.4),

$$\mathbf{f}(\zeta) = i\int \Delta_x^{(+)}(\zeta;\zeta')\mathbf{f}(\zeta')d\Sigma(\zeta') , \qquad \mathbf{f} \in \mathbf{F}_x , \tag{4.17}$$

we can easily establish in the present context that these second-quantized frames have the following important property,

$$\varphi^{(-)}(\mathbf{f})\boldsymbol{\Phi}_{\mathbf{g}} = \langle \mathbf{f} | \mathbf{g} \rangle \boldsymbol{\Phi}_{\mathbf{g}} , \qquad \mathbf{f},\mathbf{g} \in \mathbf{F}_x , \tag{4.18}$$

in relation to the annihilation operators of excitons in state $\mathbf{f}$, densely defined in $\mathcal{F}_x$ by

$$\varphi^{(-)}(\mathbf{f}) = \int \varphi^{(-)}(x;\zeta)\mathbf{f}^*(\zeta)d\Sigma(\zeta) = \varphi^{(+)*}(\mathbf{f}) , \qquad \mathbf{f} \in \mathbf{F}_x . \tag{4.19}$$

In particular, upon noting that on account of (4.4) and of (4.7)-(4.9),

$$\varphi^{(\pm)}(\boldsymbol{\Phi}^{u(x)}_{f,m;\zeta}) = \varphi^{(\pm)}(x;\zeta) , \quad \zeta = (q,v) , \quad \boldsymbol{\zeta} = (\boldsymbol{a} + q^i \boldsymbol{e}_i, v^i \boldsymbol{e}_i) , \quad u(x) = (\boldsymbol{a},\boldsymbol{e}_i), \tag{4.20}$$

we immediately obtain from (5.1.16b) and (4.18) that

$$\varphi^{(-)}(x;\zeta)\boldsymbol{\Phi}_{\mathbf{f}} = \mathbf{f}(\zeta)\boldsymbol{\Phi}_{\mathbf{f}} , \qquad \mathbf{f} \in \mathbf{F}_x . \tag{4.21}$$

This last relationship demonstrates that the second-quantized frames in (4.16) are *bona fide*[24] eigenvectors of the exciton annihilation operators. Hence, as we shall see in the next section, they give rise to expectation values for the stress-energy tensor that are the quantum analogues of classical stress-energy tensor values. This observation will turn out to be of essential importance for the existence of an action-based formulation of GS propagation, which will be made possible by the fact that the family

$$\mathcal{Q}_x = \left\{\boldsymbol{\Phi}_{\mathbf{f}} \,\middle|\, \mathbf{f} \in \mathbf{F}_x\right\} , \qquad x \in \mathbf{M} , \tag{4.22}$$

of all second-quantized frames in a given Fock fibre $\mathcal{F}_x$ gives rise to a continuous resolution of the identity operator $\mathbf{1}_x$ within that fibre:

$$\int_{\mathbf{F}_x} |\boldsymbol{\Phi}_{\mathbf{f}}\rangle \, \mathbf{df}\, \mathbf{df}^* \langle \boldsymbol{\Phi}_{\mathbf{f}}| = \mathbf{1}_x \ . \tag{4.23}$$

A precise mathematical meaning can be assigned to the functional integral in (4.23) by using a method advocated by Berezin (1966). To apply this method, we introduce in the Klein-Gordon quantum fibre $\mathbf{F}_x$ an orthonormal basis $\{\mathbf{w}_1, \mathbf{w}_2, \dots\}$, so that

$$\mathbf{f} = \sum_{\alpha=1}^{\infty} z_\alpha \mathbf{w}_\alpha \ , \qquad z_\alpha = \langle \mathbf{w}_\alpha | \mathbf{f} \rangle \in \mathbf{C}^1 \ , \qquad \alpha = 1,2,\dots \ . \tag{4.24}$$

It is noteworthy that the above series converges to $\mathbf{f}$ not only in the Hilbert space norm of $\mathbf{F}_x$, but that it also converges pointwise to the continuous function $\mathbf{f}(\boldsymbol{\zeta})$, and that this pointwise convergence is actually uniform in $\boldsymbol{\zeta}$. Indeed, on one hand we have

$$\left|\sum_{\alpha=m}^{n} z_\alpha \mathbf{w}_\alpha(x;\zeta)\right| = \left|\left\langle \boldsymbol{\Phi}^{u(x)}_{f,m;\zeta} \middle| \sum_{\alpha=m}^{n} z_\alpha \mathbf{w}_\alpha \right\rangle\right| \le \left\|\boldsymbol{\Phi}^{u(x)}_{f,m;\zeta}\right\| \left\|\sum_{\alpha=m}^{n} z_\alpha \mathbf{w}_\alpha\right\| \ , \tag{4.25}$$

whereas on the other hand, by (5.1.16b) and (5.1.18),

$$\left\|\boldsymbol{\Phi}^{u(x)}_{f,m;\zeta}\right\|^2 = Z^{-2}_{f,m} \int_{u^0>0} \left|f(u^0)\right|^2 d\Omega(u) \ , \qquad \forall \zeta = (q,v) \in \mathbf{R}^4 \times V^+ \ , \tag{4.26}$$

so that the partial sums on the left-hand side of (4.25) are uniformly bounded by a $\boldsymbol{\zeta}$-independent expression on its right-hand side, which goes to zero as $m,n \to \infty$. This pointwise uniform convergence ensures that no ambiguities or inconsistencies occur in any of the subsequent expressions in which we will be dealing with the values of single exciton wave functions, as well as their expansions (4.24), at individual points $\boldsymbol{\zeta}$.

Let us now introduce the canonical creation and annihilation operators

$$a_\alpha(x) = \varphi^{(-)}(\mathbf{w}_\alpha) \ , \qquad a_\alpha^*(x) = \varphi^{(+)}(\mathbf{w}_\alpha) \ , \qquad \mathbf{w}_\alpha \in \mathbf{F}_x \ . \tag{4.27}$$

These operators satisfy the following set of canonical commutation relations

$$[a_\alpha(x), a_\beta^*(x)] = \delta_{\alpha\beta} \ , \qquad [a_\alpha(x), a_\beta(x)] = [a_\alpha^*(x), a_\beta^*(x)] = 0 \ . \tag{4.28}$$

Indeed, the first family of these relations can be derived as follows, with the help of (4.17),

$$\begin{aligned}[a_\alpha(x), a_\beta^*(x)] &= i\int \overline{\mathbf{w}}_\alpha(x;\zeta) \Delta_x^{(+)}(\zeta;\zeta') \mathbf{w}_\beta(x;\zeta') \, d\Sigma(\zeta) d\Sigma(\zeta') \\ &= \int \overline{\mathbf{w}}_\alpha(x;\zeta) \mathbf{w}_\beta(x;\zeta) \, d\Sigma(\zeta) = \delta_{\alpha\beta} \ , \end{aligned} \tag{4.29}$$

whereas the second family follows in an obvious manner from (4.7)-(4.9).

The existence of the canonical commutation relations in (4.28) enables us to introduce the following Glauber (1963) coherent states,

$$|\hat{z}\rangle = \exp\left[-\tfrac{1}{2}\sum_{\alpha=1}^{\infty}|z_\alpha|^2 + \sum_{\alpha=1}^{\infty} z_\alpha\, a_\alpha^*(x)\right]\Psi_{0;x} \ , \qquad \hat{z} = (z_1, z_2, \dots) \ . \tag{4.30}$$

As originally proved by Bargmann (1961), coherent states of this type provide continuous resolutions of the identity in any Hilbert space in which they can be realized. In the present context, that Hilbert space is equal to the Fock fibre $\mathcal{F}_x$.

Upon inserting (4.24) into (4.16), it is easy to establish that the coherent state in (4.30) coincides with the corresponding element of the second-quantized frame in (4.22). This means that the functional integral in (4.23) can be indeed interpreted as a "continual integral" in the sense described in [B]. More specifically, as an operator in $\mathcal{F}_x$, the functional integral in (4.23) is equal to the following Hilbert-space weak limit [PQ],

$$\int_{\mathbf{F}_x} |\boldsymbol{\Phi}_\mathbf{f}\rangle\, \mathbf{df}\, \mathbf{df}^* \langle \boldsymbol{\Phi}_\mathbf{f}| \ = \ \mathrm{w}\text{-}\lim_{n\to\infty} \pi^{-n} \int_{\mathbf{R}^{2n}} |\hat{z}_n\rangle \mathbf{d}\hat{z}_n \mathbf{d}\hat{z}_n^* \langle \hat{z}_n| \ , \tag{4.31a}$$

$$\hat{z}_n = (z_1, \dots, z_n, 0, 0, \dots) \ , \qquad \mathbf{d}\hat{z}_n \mathbf{d}\hat{z}_n^* = \prod_{i=1}^{n} d(\mathrm{Re}\, z_i) d(\mathrm{Im}\, z_i) \ , \tag{4.31b}$$

where the integrals on the right-hand side of (4.31a) are Bochner integrals. As shown in Sec. 1.2 of [B], this definition of the functional integral in (4.23) is independent of the choice of orthonormal basis $\{\mathbf{w}_1, \mathbf{w}_2, \dots\}$ in $\mathbf{F}_x$.

7.5. Parallel Transport and Action Principles in Fock Quantum Bundles

According to (4.3), the Fock bundle $\mathcal{E}$ can be obtained from the Klein-Gordon quantum bundle **E** by taking Whitney sums of symmetrized Whitney products of **E**. Hence, any operator for parallel transport in (5.2.14) gives rise to a unitary operator

$$\tau_\gamma(x'', x') : \quad \mathcal{F}_{x'} \ \to \ \mathcal{F}_{x''} \tag{5.1}$$

for parallel transport within $\mathcal{E}$. Covariant derivatives can be then defined as in (5.2.16), with the result that, for a given section $\boldsymbol{s}$ of PM, (5.2.24) is now replaced by

$$\boldsymbol{\nabla} = \boldsymbol{d} - i\tilde{\theta}^i \boldsymbol{P}_{i;\boldsymbol{u}} + \tfrac{i}{2}\tilde{\omega}_{jk} \boldsymbol{M}_{\boldsymbol{u}}^{jk} \ , \qquad \boldsymbol{d} = \theta^i \partial_i \ , \qquad \partial_i := \partial_{e_i} \ . \tag{5.2}$$

The connection coefficients in (5.2) are the same as in (5.2.19) or (5.2.24), whereas the infinitesimal generators are those of the representation

$$\boldsymbol{U}_{x;\boldsymbol{u}}(b, \Lambda) = \bigoplus_{n=0}^{\infty} U_{x;\boldsymbol{u}}(b, \Lambda)^{\otimes n} \ , \qquad \boldsymbol{u} = \boldsymbol{s}(x) \ , \tag{5.3}$$

induced in $\mathcal{F}_x$ by the representation in (5.2.20).

In order to establish a relationship between this purely geometric formulation of parallel transport in (5.1) and the action principles used in the customary Lagrangian formulations of quantum field theories [IQ,SI], it is essential that these generators be related to an operator-valued stress-energy tensor. To achieve that, we first note that

$$P_{j;u} = i\int \varphi^{(+)}(x;\zeta)\partial_j \varphi^{(-)}(x;\zeta)d\Sigma(\zeta)\ , \quad \zeta = (a + q^i e_i, v^i e_i)\ , \ \partial_j = \partial/\partial q^j\ . \quad (5.4)$$

The above relationship can be easily verified by taking matrix elements of the above operator in between any two n-exciton states, using (4.8) as well as (4.10), and then comparing the outcome with that obtained by using (5.2.22b). Upon expressing the inner product in **E** in the form appearing in (5.1.13), the Bochner integral in (5.4) can be recast in the form

$$P_{j;u} = -\hat{Z}_{f,m}\int \varphi^{(+)}(x;\zeta)\overleftrightarrow{\partial}_k \partial_j \varphi^{(-)}(x;\zeta)d\sigma^k(q)d\Omega(v)\ . \quad (5.5a)$$

Taking into account the easily verifiable fact that (cf. [P], p. 222)

$$\int \varphi^{(\pm)}(x;\zeta)\overleftrightarrow{\partial}_k \varphi^{(\pm)}(x;\zeta)d\sigma^k(q)d\Omega(v) = 0\ , \qquad q = a + q^i e_i\ , \ v = v^i e_i\ , \quad (5.5b)$$

and using the general definition of normal ordering, we can rewrite (5.5a) in the form:

$$P_{j;u} = -\tfrac{1}{2}\hat{Z}_{f,m}\int :\varphi(x;\zeta)\overleftrightarrow{\partial}_k \partial_j \varphi(x;\zeta):\, d\sigma^k(q)d\Omega(v)\ . \quad (5.6)$$

We can choose now as the surface σ for q-integration in (5.6) any of the hyperplanes corresponding to a constant value of q^0 in the local frame $s(x)$ – such as the hyperplane with $q^0 = 0$. In the case of $j = 1,2,3$, we can then integrate by parts along σ to arrive at

$$P_{a;u} = \hat{Z}_{f,m}\int_{q^0=0} :\varphi_{,a}(x;\zeta)\varphi_{,0}(x;\zeta):\, d^3\mathbf{q}\, d\Omega(v)\ , \qquad a = 1,2,3\ , \quad (5.7a)$$

whereas in the case of $j = 0$, we can use the equation (4.14) to write

$$P_{a;u} = \tfrac{1}{2}\hat{Z}_{f,m}\int_{q^0=0} :m^2\varphi^2(x;\zeta) + \sum_{j=0}^{3}\varphi_{,j}^2(x;\zeta):\, d^3\mathbf{q}\, d\Omega(v)\ . \quad (5.7b)$$

The net result of the above considerations is that we can express the operators in (5.4) in the following manifestly covariant form,

$$P_{j;u} = \int :T_{jk}[\varphi(x;\zeta)]:\, d\sigma^k(q)d\Omega(v)\ , \quad (5.8a)$$

$$T_{jk}[\varphi] = \hat{Z}_{f,m}\left(\varphi_{,j}\varphi_{,k} + \tfrac{1}{2}\eta_{jk}(m^2\varphi^2 - \eta^{il}\varphi_{,i}\varphi_{,l})\right)\ , \quad (5.8b)$$

which involves a renormalized version of the stress-energy tensor in (1.15).

In a completely analogous manner we can derive that (5.3), taken in conjunction with (5.2.20) and (5.2.22a), implies that

$$M_u^{ij} = \int :Q_u^i T^{jk}[\varphi(x;\zeta)] - Q_u^j T^{ik}[\varphi(x;\zeta)] : d\sigma_k(q) d\Omega(v) \quad , \tag{5.9}$$

where the above Q-operators act in accordance with (5.2.22b) on the variables ζ, whose components are equal to the components of ζ with respect to the Poincaré frame in (5.4).

These results are very analogous to similar ones in the conventional theory of the neutral scalar field [SI], with one crucial difference: in the latter case the energy-momentum tensor operator that occurs in (1.14) is a purely formal entity, since the quantum fields at a point x in Minkowski space are not well defined – but rather the smeared version (1.8) is well defined; on the other hand, the stress-energy operator in (5.8b) is perfectly well defined as it stands. Furthermore, conventional quantum field theory encounters severe difficulties, mentioned at the beginning of Sec. 7.2, when the definition of the stress-energy tensor operator in (1.15) is extended to the case where x belongs to a curved classical spacetime. On the other hand, in the present GS context, the stress-energy tensor representing the actual energy-momentum exciton density at any $x \in \mathbf{M}$ in relation to the quantum frame $\boldsymbol{\Phi}^u$ in (5.1.21) (associated with the Poincaré frame $\boldsymbol{u}$ used in (5.4)), can be obtained by setting $\boldsymbol{q} = -\boldsymbol{a}$ in (5.8b), and then integrating over all possible values of $\boldsymbol{v}$.

We can introduce, by analogy with (5.4.2), the Poincaré gauge invariant *second-quantized frame propagator*

$$K_\gamma(\mathbf{f}'';\mathbf{f}') = \langle \boldsymbol{\Phi}_{\mathbf{f}''} | \tau_\gamma(x'',x') \boldsymbol{\Phi}_{\mathbf{f}'} \rangle \quad , \qquad \mathbf{f}' \in \mathbf{F}_{x'} \quad , \qquad \mathbf{f}'' \in \mathbf{F}_{x''} \quad , \tag{5.10}$$

for parallel transport along any smooth curve $\gamma = \{x(t) | \, t' \le t \le t''\}$, joining two base points $x' = x(t') \in \mathbf{M}$ and $x'' = x(t'') \in \mathbf{M}$. Upon inserting into the right-hand side of the inner product in (5.10) the expression $\tau_\gamma(x'', x') = \tau_\gamma(x'', x)\, \tau_\gamma(x, x')$, and then using (4.23) at any given point $x \in \gamma$, we arrive at the following counterpart of (5.4.4):

$$K_\gamma(\mathbf{f}'';\mathbf{f}') \;=\; K_\gamma^*(\mathbf{f}';\mathbf{f}'') = \int_{\mathbf{F}_x} K_\gamma(\mathbf{f}'';\mathbf{f})\, K_\gamma(\mathbf{f};\mathbf{f}') \mathbf{df}\, \mathbf{df}^* \,. \tag{5.11}$$

Consequently, by iterating (5.11) for any subdivision $t' = t_0 < t_1 < \cdots < t_N = t''$, we conclude that even in the limit $\varepsilon = \max(t_n - t_{n-1}) \to +0$ we can write, as in (5.4.5),

$$K_\gamma(\mathbf{f}'';\mathbf{f}') = \lim_{\varepsilon \to +0} \int K_\gamma(\mathbf{f}_N;\mathbf{f}_{N-1}) \prod_{n=N-1}^{1} K_\gamma(\mathbf{f}_n;\mathbf{f}_{n-1})\, \mathbf{df}_n \mathbf{df}_n^* \quad , \quad \mathbf{f}' = \mathbf{f}_0 \;, \quad \mathbf{f}'' = \mathbf{f}_N \;. \tag{5.12}$$

We shall now demonstrate that, on account of (5.12) and of the basic properties of the second-quantized frames in (4.22), the second-quantized frame propagator, defined in a purely geometric manner in (5.10), can be expressed in terms of an action integral. We, therefore, first observe that, on account of the smoothness properties of the curve γ as well as of any appropriately chosen moving Poincaré reference frame $\boldsymbol{s}$ (namely of a section $\boldsymbol{s}$ of PM whose domain of definition $\mathbf{M}^s$ contains the curve γ), and of the fact that all elements of all second-quantized frames lie within the domains of definition of the infinitesimal generators of (5.3) that appear in (5.2), we have that

$$K_\gamma(\mathbf{f}_n;\mathbf{f}_{n-1}) = \left\langle \Phi_{\boldsymbol{\varphi}_n(x_n)} \middle| \left(\mathbf{1}_{x_n} - i\delta x_n^i\, \boldsymbol{P}_{i;\boldsymbol{u}} + \tfrac{i}{2}\tilde{\boldsymbol{\omega}}_{jk}(\delta x_n)\boldsymbol{M}_{\boldsymbol{u}}^{jk}\right)\Phi_{\boldsymbol{\varphi}_{n-1}(x_n)}\right\rangle$$

$$+\ O\left((\delta t_n)^2\right) \ , \qquad \delta t_n = t_n - t_{n-1} \ , \tag{5.13a}$$

$$\delta x_n^i = \theta_\mu^i(x_n)\dot{x}^\mu(t_n)\delta t_n \ , \qquad \delta x_n = e_i(x_n)\,\delta x_n^i \ , \qquad x_n = x(t_n) \ , \tag{5.13b}$$

$$\boldsymbol{\varphi}_n(x_n) = \mathbf{f}_n \ , \qquad \boldsymbol{\varphi}_{n-1}(x_n) = \left((\boldsymbol{\sigma}_{x_n}^{\boldsymbol{u}})^{-1} \circ \boldsymbol{\sigma}_{x_{n-1}}^{\boldsymbol{u}}\right)\mathbf{f}_{n-1} \ , \qquad \boldsymbol{u}(x_n) = \left(\mathbf{0}, e_i(x_n)\right) . \tag{5.13c}$$

For the sake of notational convenience, in (5.13) we have restricted our attention to sections $\boldsymbol{s}$ of the Lorentz frame subbundle of PM. We note that, in terms of the coordinate wave function amplitudes in (4.5), which represent elements of the typical fibre $\mathbf{F}$, we have

$$\varphi_n(x_n;\zeta) = f_n(\zeta) := \varphi(x_n;\zeta) \ , \qquad \varphi_{n-1}(x_n;\zeta) = f_{n-1}(\zeta) := \varphi(x_{n-1};\zeta) \ , \tag{5.14}$$

for all $\zeta \in \mathbf{R}^4 \times V^+$. Hence, the use of (5.8) and (5.9) in (5.13a) leads to,

$$K_\gamma(\mathbf{f}_n;\mathbf{f}_{n-1}) = \left[1 - i\delta x_n^i\, P_i(\varphi(x_n);\varphi(x_{n-1}))\right]\left\langle \Phi_{\boldsymbol{\varphi}_n(x_n)} \middle| \Phi_{\boldsymbol{\varphi}_{n-1}(x_n)}\right\rangle$$

$$+ i\tilde{\omega}_{jk}(\delta x_n) M^{jk}(\varphi(x_n);\varphi(x_{n-1}))\left\langle \Phi_{\boldsymbol{\varphi}_n(x_n)} \middle| \Phi_{\boldsymbol{\varphi}_{n-1}(x_n)}\right\rangle + O\left((\delta t_n)^2\right) , \tag{5.15}$$

where, on account of (4.10) and (4.21), as well as of the definition (1.11) of normal quantum field ordering, we have

$$P_i(\varphi(x_n);\varphi(x_{n-1})) = \int T_{ik}[\overline{\varphi}(x_n;\zeta) + \varphi(x_{n-1};\zeta)]\ d\sigma^k(q)d\Omega(v) \ , \tag{5.16a}$$

$$M^{jk}(\varphi(x_n);\varphi(x_{n-1})) = \int Q^{[j}T^{k]l}[\overline{\varphi}(x_n;\zeta) + \varphi(x_{n-1};\zeta)]d\sigma_l(q)d\Omega(v) \ , \tag{5.16b}$$

$$Q^{[j}T^{k]l} := \tfrac{1}{2}\left(Q^j T^{kl} - Q^k T^{jl}\right) \ , \tag{5.16c}$$

with the bar denoting complex conjugation.

On the other hand, it follows from (4.13), (4.16) and (4.17) that

$$\left\langle \Phi_{\mathbf{f}} \middle| \Phi_{\mathbf{g}}\right\rangle = \exp\left[-\tfrac{1}{2}\left(\|\mathbf{f}\|^2 + \|\mathbf{g}\|^2\right) + \langle \mathbf{f}|\mathbf{g}\rangle\right] \ , \qquad \forall \mathbf{f},\mathbf{g} \in \mathbf{F}_x \ , \qquad \forall x \in \mathbf{M} \ . \tag{5.17}$$

Consequently, (5.12) can be rewritten in the form

$$K_\gamma(\boldsymbol{\varphi}(x'');\boldsymbol{\varphi}(x')) = \lim_{\varepsilon \to +0} \int \prod_{n=N}^{1}{}' \mathcal{D}[\varphi(x_n)] \exp[i\delta t_n\, L_\gamma(\varphi(x_n);\varphi(x_{n-1}))] \ , \tag{5.18a}$$

$$\boldsymbol{\varphi}(x') = \mathbf{f}' \ , \qquad \boldsymbol{\varphi}(x'') = \mathbf{f}'' \ , \qquad \mathcal{D}[\varphi(x_n)] = \mathbf{d}f_n \mathbf{d}f_n^* \ , \qquad n = 1,\ldots,N-1 \ , \tag{5.18b}$$

where the prime indicates the absence of the functional integration over the N-th mode, and

$$L_\gamma(\varphi(x_n);\varphi(x_{n-1})) = \tfrac{i}{2}\big[\langle\varphi(x_n)|\,\pi(x_n)\rangle - \langle\pi(x_n)|\,\varphi(x_{n-1})\rangle\big]$$
$$- X_n{}^i P_i(\varphi(x_n));\varphi(x_{n-1})) + \tfrac{1}{2}\tilde{\boldsymbol{\omega}}_{jk}(\boldsymbol{X}_n)M^{jk}(\varphi(x_n);\varphi(x_{n-1}))\ , \qquad (5.19a)$$

$$\pi(x_n) = \big(\varphi(x_n) - \varphi(x_{n-1})\big)/(t_n - t_{n-1})\ , \quad \boldsymbol{X}_n = \dot{x}^\mu(t_n)\partial_\mu\ , \quad x_n = x(t_n)\ . \qquad (5.19b)$$

To recast (5.18a) in the form of an action integral, we note that, according to the form (5.1.13) of the inner product in the typical fibre **F** of the quantum Klein-Gordon bundle **E**,

$$\langle\varphi(x_n)|\,\pi(x_n)\rangle = i\hat{Z}_{f,m}\int\overline{\varphi}(x_n;\zeta)\overset{\leftrightarrow}{\partial}_k\pi(x_n;\zeta)\,d\sigma^k(q)\,d\Omega(v)\ . \qquad (5.20)$$

Consequently, we can express the second-quantized frame propagator for parallel transport along any piecewise smooth path γ, which was geometrically defined in (5.10), in the form of the functional integral

$$K_\gamma(\boldsymbol{\varphi}(x'');\boldsymbol{\varphi}(x')) = \int\mathcal{D}\varphi\,\exp\big(iS_\gamma[\varphi,\pi]\big)\ , \qquad \mathcal{D}\varphi = \prod\nolimits_{t''>t\geq t'}\mathcal{D}[\varphi(x(t))]\ , \quad (5.21)$$

which incorporates the action

$$S_\gamma[\varphi,\pi] = \int_{t'}^{t''}dt\int\mathcal{L}_k^{(\gamma)}[\varphi(x(t);\zeta),\pi(x(t);\zeta)]\,d\sigma^k(q)\,d\Omega(v)\ , \qquad (5.22a)$$

$$\mathcal{L}_k^{(\gamma)}[\varphi(x(t);\zeta),\pi(x(t);\zeta)] =$$
$$\tfrac{1}{2}\hat{Z}_{f,m}\big[\overline{\varphi}_{,k}(x(t);\zeta)\,\pi(x(t);\zeta) + \varphi_{,k}(x(t-0);\zeta)\overline{\pi}(x(t);\zeta)\big]$$
$$-\tfrac{1}{2}\hat{Z}_{f,m}\big[\overline{\varphi}(x(t);\zeta)\,\pi_{,k}(x(t);\zeta) + \varphi(x(t-0);\zeta)\,\overline{\pi}_{,k}(x(t);\zeta)\big]$$
$$-X^i(t)T_{ik}[\overline{\varphi}(x(t);\zeta) + \varphi(x(t-0);\zeta)]$$
$$+\,\eta_{kl}\tilde{\boldsymbol{\omega}}_{ij}(\boldsymbol{X}_n(t))Q^{[i}T^{j]l}[\overline{\varphi}(x(t);\zeta) + \varphi(x(t-0);\zeta)]\ , \quad k = 0,1,2,3. \quad (5.22b)$$

In writing down the expression (5.22b), which immediately follows from (5.19) in the passage to the limit $\varepsilon = \max(t_n - t_{n-1}) \to +0$, we have followed universally adopted traditions in the theory of path integrals, and denoted the field contributions at $x(t_n)$ from field values at the preceding point $x(t_{n-1})$ as if they behaved smoothly. However, the discontinuous nature of the field "history" has to be kept in mind, so that those contributions have to be treated as independent while we functionally integrate, for each t_n, over all possible modes in the typical fibre **F** prior to taking that limit. In addition to notational clarity and convenience, the adoption of this notation underscores the formal similarities with similar action integrals in conventional quantum field theory [IQ]. The appearance of not just one, but four "Lagrangian" terms in (5.22b), as well as the last angular momentum dependent term in those "Lagrangians", might be disconcerting to those already well familiar with the standard expressions for action integrals in conventional quantum field theory.

Their presence is due, however, to the great generality of (5.22), which reflects full Poincaré gauge invariance. In fact, a moment's reflection indicates that, on geometric grounds alone, even in conventional quantum field theory similar terms would make their appearance if global Lorentz frames would be replaced with general *local* moving Lorentz frames in an attempt to achieve a theory that is Poincaré *gauge* invariant, and not merely *special* relativistically invariant – i.e., invariant only under *global* Poincaré transformations.

In curved spacetime *global* Poincaré invariance is a physically and mathematically meaningless concept. Hence, when we shall extrapolate in the next section the functional integral in (5.21) for parallel transport into one for actual geometro-stochastic propagation, the form of the action in (5.22) will have to be employed. However, as long as we are dealing with a single path γ, we can always choose a section $\mathbf{s}$ of LM which is adapted to γ (cf. Sec. 2.7), so that all the local Lorentz frames $\mathbf{s}(x(t))$ along γ are parallel transports of each other. If we then also choose the normals to the surfaces $\sigma(x(t))$, $t \in \gamma$, for q-integration along the time axes of those frames, and use a parameter t which is identified along γ with the variable q^0 in the typical fibre of TM, then the expressions in (5.22) assume the following simpler and more easily recognizable form:

$$S_\gamma[\varphi,\pi] = \int_{t'<q^0<t''} \mathcal{L}_0^{(\gamma)}[\varphi(x(q^0);\zeta),\pi(x(q^0);\zeta)]\,d^4q\,d\Omega(v) \quad , \tag{5.23a}$$

$$\begin{aligned} &\mathcal{L}_0^{(\gamma)}[\varphi(x(q^0);\zeta),\pi(x(q^0);\zeta)] = \\ &\quad \tfrac{1}{2}\hat{Z}_{f,m}\Big[\dot{\overline{\varphi}}(x(q^0);\zeta)\,\pi(x(q^0);\zeta) + \dot{\varphi}(x(q^0-0);\zeta)\overline{\pi}(x(q^0);\zeta)\Big] \\ &\quad -\tfrac{1}{2}\hat{Z}_{f,m}\Big[\overline{\varphi}(x(q^0);\zeta)\,\dot{\pi}(x(q^0);\zeta) + \varphi(x(q^0-0);\zeta)\,\dot{\overline{\pi}}(x(q^0);\zeta)\Big] \\ &\quad -X^i(t)T_{i0}[\overline{\varphi}(x(q^0);\zeta) + \varphi(x(q^0-0);\zeta)] \quad . \end{aligned} \tag{5.23b}$$

The above expressions for Lagrangians in the GS quantum field theory of a neutral spin-0 field remain distinct from their counterparts [IQ,SI] in conventional quantum field theory in two essential respects: 1) their terms contain the Poincaré gauge variables ζ, and as such are *bona fide* continuous functions of those variables – and, in fact, for the exciton states of the Born quantum metric operator (cf. Sec. 3.6 and [P], Sec. 4.5), those functions are actually smooth; 2) finite renormalization factors that multiply the *geometric phase* component of these GS quantum field Lagrangians emerge naturally from the framework itself, on the basis of purely geometric considerations.

These geometric phases are given in (5.22b) and (5.23b) by the expressions between the first two sets of square brackets on their right-hand sides. Comparison with the general theory of the geometric phase recounted in Secs. 3.8 and 3.9, and in particular with (3.8.7) and (3.8.23), reveals the clear-cut geometric nature of those phase terms in the present field theoretical context.

The fact that action principles are consequences of geometric ones, rather than having to be postulated as primary – as has become customary in the path integral quantization methods developed during the last fews decades – should come as no surprise if it is recalled that the origin of action principles lies in the application by Maupertuis in the mid-seventeenth century of *teleological* arguments to classical mechanics, and that those arguments were based on the scientifically rather dubious doctrine of "final causes" (cf. Barrow and Tipler, 1986, p. 66): a classical particle supposedly *chooses* to follow, amongst all kinematically acceptable paths, the one which is, within the Newtonian framework, dynamically correct *because* its *purpose* is to minimize its own action. If one does not adopt, however, such a teleological attitude, then there are no compelling epistemological grounds for considering variational action principles as being fundamental. In fact, in CGR action principles historically emerge as secondary: the Hilbert action principle was independently discovered by Einstein and Hilbert in 1915 only *after* the geometric foundations of the theory had been already worked out by Einstein over a period of many years, and the Palatini variant of the CGR action came four years later (cf. Chapter 21 in [M], and the references cited therein). The recent formulation by Witten (1988) of topological quantum field theories with no classical analogues indicates that field Lagrangians are secondary entities, completely determined by symmetries (Montano and Sonnenschein, 1989). As a matter of fact, in three spacetime dimensions, the aforementioned Einstein-Hilbert action turns out (Witten, 1989) to be equal to the Chern-Simons action for the Poincaré group ISO(2,1).

On the other hand, as originally formulated by Dirac (1933), and later elaborated by Feynman (1948), the Lagrangian approach to quantum mechanics has removed the teleological undertones of action principles: all kinematically acceptable broken paths are allowed in the path integral formulation of quantum mechanics, but only those which are close to the classical ones contribute most on account of the superposition principle. However, as the formulation in Sec. 3.8 of the geometric aspects of the quantum formalism makes it clear, it is not so much the Lagrangian, but rather quantum geometry at its most basic level, namely the level of geometric phases, taken in conjunction with Hamiltonians which are geometrically viewed as generators of time translations, that permits this formulation – as the explicit or implicit presence of geometric phases in *all* types of path integrals makes it abundantly clear.

7.6. Relativistic Microcausality and Geometro-Stochastic Field Locality

Relativistic microcausality and the locality of quantum field theoretical models are treated by many authors as synonymous terms – cf., e.g., (Streater and Wightman, 1964), p. 100, or (Bogolubov *et al.*, 1990), p. 325; additional aspects of classical and quantum causality are discussed in (Blokhintsev, 1973). For example, in the case of a single neutral scalar field in Minkowski space, these two fundamental physical concepts are identified, at the mathematical level, with the following well-known *local commutativity* condition [BL]:

$$[\phi(x),\phi(x')]=0 \ , \qquad \forall\ (x-x')^2<0 \ . \tag{6.1}$$

Similar identifications of "microcausality" with "local" (anti)commutativity axioms are imposed on other boson and fermion quantum fields, respectively [BL].

The justification for this identification is believed to emerge from a conventional quantum field theoretical "principle of locality", which "states that it is meaningful to talk of observables which can be measured in a specific space-time region and that observables in causally disjoint regions are always compatible." (Haag and Kastler, 1964, p. 848).

At first glance, the above clear-cut formulation of such a "principle of locality" appears eminently sensible, and supported by all our seemingly intuitive notions of locality. In actuality, however, these "intuitive" notions represent concepts to which everybody who has mastered the basic aspects of conventional quantum theory becomes very much *accus-*

tomed in the process of coping with its mathematical formalism, and with any of the prevailing variants of the orthodox interpretation of quantum mechanics. Once this hidden bias is uncovered, closer scrutiny of the above formulation of "principle of locality" reveals that the physical interpretations of all its key elements rely on *assumptions* which, however sensible they might appear at first sight, give rise to fundamental epistemological questions.

The epistemic roots of those questions emerge, in part, from the difficulties with the conventional notions of relativistic quantum particle localizability discussed in Chapter 1. In the general relativistic regime, these difficulties are compounded by additional difficulties, which are due to the global nature of the concept of quantum particle, that was discussed in Secs. 7.2 and 7.3. In turn, these latter difficulties are greatly exacerbated in the quantum gravity context – as it will be demonstrated and discussed in Chapters 11 and 12.

In addition, other fundamental deficiencies of the conventional approach to quantum field theory emerge already in the special relativistic context. In the context of the, by general consensus, most successful conventional quantum field theory to date, namely quantum electrodynamics, the crux of the matter has been aptly and succinctly described by J. Schwinger (1958) as follows:

"The observational basis of quantum electrodynamics is self-contradictory. The fundamental dynamical variables of the electron-positron field, for example, have meaning only as symbols of the localized creation and annihilation of charged particles, to which are ascribed a definite mass without reference to the electromagnetic field. Accordingly it should be possible, in principle, to confirm these properties by measurements, which, if they are to be uninfluenced by the coupling of the particles to the electromagnetic field, must be performed instantaneously. But there appears to be nothing in the formalism to set a standard for arbitrarily short times and, indeed, the assumption that over sufficiently small intervals the two fields behave as though free from interaction is contradicted by evaluating the supposedly small effect of the coupling... . It seems that we have reached the limits of the quantum theory of measurement, which asserts the possibility of instantaneous observations, without reference to specific agencies. The localization of charge with indefinite precision requires for its realization a coupling with the electromagnetic field that can attain arbitrarily large magnitudes. The resulting appearance of divergences, and contradictions, serves to deny the basic measurement hypothesis. We conclude that a convergent theory cannot be formulated consistently within the framework of present space-time concepts." (Schwinger, 1958, pp. xv-xvi).

In view of the above deep-rooted causes of the difficulties experienced by quantum field theory, it should be clear that no new merely computational techniques, such as those initiated by Feynman (1948), can provide the required cure. In fact, it was pointed out by Feynman himself that: "Actually, the path integral did not then [namely in 1948] provide, nor has it since provided, a truly satisfactory method of avoiding the divergence difficulties of quantum electrodynamics" – cf. (Feynman and Hibbs, 1965), p. viii.

Clearly, the GS approach to quantum field theory, whose most basic aspects have been outlined in the preceding two sections, eliminates "the possibility of instantaneous observations" in the classical sense, by relating all observational acts to "specific agencies" which do not incorporate classical objects – namely to *quantum* test bodies and to *quantum* frames constructed from such test bodies, that are not predicated on any classical concept of "localization with indefinite precision". In so doing, this approach brings forth subtle epistemological distinctions that exist between the GS concepts of "locality", "separability" and "causality" – three fundamental physical concepts which, although interrelated, are not synonymous in the present GS framework.

To better understand the epistemological nature of these distinctions, let us first briefly reconsider the reasons for their identification within the conventional framework for quantum field theory in Minkowski space.

All conventional frameworks for relativistic quantum field theory are predicated on two fundamental *assumptions*: 1) Classical relativistic geometries provide suitable modes of description of operationally-based geometric relationships even at the microlevel – which, in the special relativistic context, are supplied by the geometry of Minkowski space. 2) Quantum measurement schemes can be associated with *any* bounded open region B in the adopted classical manifold – namely, in the present context, in Minkowski space – and those schemes are represented by an *algebra of local observables* $\mathcal{A}(B)$ over B [BL].

The epistemologically unwarranted nature of the first of these assumptions, and the inconsistencies to which it gives rise, were already discussed in Chapter 1. Consequently, let us now concentrate on the second assumption.

At closer scrutiny, the second assumption does not fare any better than the first. For example, in the Haag-Kastler (1964) quantum field theoretical scheme, the "algebras of observables" are C^*–algebras, whereas within the Wightman (1956) scheme, they are the W^*–algebras generated [BL] by all the quantum fields "smeared" with test functions f with supports within B. The fact these algebras have to incorporate an operation of involution, i.e., a $*$–operation, in order to allow for their realizations in terms of informationally complete algebras of operators in complex Hilbert spaces, leads to severe difficulties in justifying their adoption on measurement-theoretical grounds. For example, in attempting to deduce from first principles an "algebra of microscopic measurement", Schwinger (1959) had to adopt a multiplication law[25] between "elementary selective measurements". But that law incorporates *complex*-valued probability amplitudes which do not emerge from the adopted elementary measurement procedures, and cannot be justified on operational grounds. Alternatively, as first proposed by Dirac (1945), "complex probabilities" could be introduced in quantum theory, as a means of dealing with this type of problem. The physical meaning of such "complex probabilities" can be then operationally related to measurements of "incompatible observables" by a procedure described in (Prugovečki, 1967). Furthermore, consistent systems of axioms can be formulated for these "complex probabilities", which then in turn yield C^*–algebras of observables for nonrelativistic quantum mechanics, as well as for relativistic quantum field theory (Prugovečki, 1966, 1969a). However, one of the key axioms[26] again cannot be justified on measurement-theoretical grounds alone.

Thus, despite the deceptively "natural" mode of *a posteriori* emergence of involutive operator algebras *from* the conventional Hilbert space formalism for nonrelativistic quantum theory, their adoption in quantum field theory as "algebras of observables" cannot be justified on *a priori* foundational grounds, i.e., as an intrinsic feature of quantum frameworks that follows from the very nature of the quantum theory of measurement. On the contrary, in conventional quantum theory, *"only quantities which commute with all additive conserved quantities are precisely measurable"* (Wigner, 1981, p. 298), so that the *assumption*[27] of such an involutive algebraic structure for the *family* of all *actually* observable quantities "contradicts some of the basic principles of quantum theory" (*ibid.*, p. 298). Thus, their adoption in quantum physics reflects, on the part of their advocates a *belief* in their methodological advantages[28], rather than the outcome of a compelling epistemological analysis based on a careful analysis of the quantum theory of measurement.

This last fact can be easily illustrated in the case of the neutral scalar field defined in (1.7). The "algebra of local observables" $\mathcal{A}(B)$ over any bounded open region B in Minkowski space can be then easily constructed – at least at the abstract level (cf. [BL], Sec 24.1). Since, in *this* particular case, the quantum field is constructed out of particle creation and annihilation operators, it could be expected that, at the very least, each $\mathcal{A}(B)$ would contain observables pertaining to quantum particles localized within B. *If* that were the case, then the quantum particle localizability problem could be solved very simply, namely by identifying, for each bounded open region B in Minkowski space, the operators within $\mathcal{A}(B)$ that represent those particle localization observables. However, such an identification of specific elements in $\mathcal{A}(B)$ with these (in the laboratory most certainly) observable quantities (without which the entire area of elementary particle physics would be experimentally unapproachable) has *not* been carried out anywhere in the literature. In fact, if it *were* carried out, according to Hegerfeldt's (1974) theorem, a contradiction would result with the aforementioned conventional "principle of locality". Indeed, that principle claims that "observables in causally disjoint regions are always compatible" (Haag and Kastler, 1964, p. 848). However, Hegerfeldt's theorem establishes that, as long as the classical geometry of Minkowski space is accepted as fundamental in relativistic quantum physics, quantum particle observables in purportedly "causally disjoint regions" cannot be causally disjoint since, under *any* mode of relativistically covariant sharp localization, wave packet propagation proceeds acausally. At the other extreme, when quantum gravity is taken into account, it has been acknowledged by Haag *et al.* (1984) that "algebras of local observables" totally lose their motivation, since then all possible spacetime metrics can become "available".

In fact, it would be most tempting to interpret the Wightman two-point function [BL]

$$\langle \Psi_0 | \phi(x)\phi(x')\Psi_0 \rangle = \langle \Psi_0 | \phi^{(-)}(x)\phi^{(+)}(x')\Psi_0 \rangle = i\Delta^{(+)}(x-x') \quad , \tag{6.2}$$

from which the Fock space and the entire theory of the neutral scalar free field $\phi(x)$ can be reconstructed, as pertaining to the creation at x' of a particle described by that quantum field, and its subsequent annihilation at x. However, such an interpretation would run counter to the fact that the operators of multiplication by the components of x' and x with respect to any global Lorentz frame cannot be interpreted as spacetime observables for relativistic quantum particles. Moreover, even if such an interpretation were arbitrarily imposed, it would run counter to the relativistic causality that is allegedly embedded in the theory of the neutral scalar free field $\phi(x)$, since the Pauli-Jordan function in (5.7.3) and (6.2) does not vanish for spacelike separated x' and x. We are thus left in the paradoxical situation that *none* of the so-called "algebras of local observables" $\mathcal{A}(B)$ contains even the most basic quantum particle observables[29]. In particular, they contain *none* of the observables that are routinely measured in the laboratory whenever a particle detection is carried out within a macroscopic region B in Minkowski space!

It is therefore becoming clear that it is not just the "global nature of quantum particles", considered and analyzed in Secs. 7.2 and 7.3, that is to blame for the difficulties which the conventional mode of thinking encounters in quantum field theory. Rather, that blame is to be shared also by the "global nature of locality" (Bogolubov *et al.*, 1990, p. 373) that emerges from the adoption of the local (anti)commutativity axioms – such as the one incorporated into (6.1) in the case of neutral scalar fields. Indeed, as otherwise devoted advocates of that axiom, Bogolubov *et al.* (1990) have the following to say:

"The postulate of local commutativity is one of the most restrictive principles of quantum field theory. Misgivings could arise concerning the experimental justification of this postulate: we have no special reason for supposing that the measurement of a component of a Hermitian field at some point has no influence on the value of the components of this field at another point separated from the first point by an arbitrarily small spacelike interval. It turns out, however, that the property of local commutativity can be *proved* if we make the assumption, which at a first glance seems to be weaker, that the fields commute only at sufficiently large spacelike separations. It follows from this that if the remaining requirements of relativistic quantum theory hold in the non-local theory, then, roughly speaking, the commutator of the fields must be non-zero everywhere. It is therefore no surprise that the attempts to introduce 'non-locality in the small' at the same time require a rejection of some other requirements of the Wightman formalism, for example, 'renormalizability'." (Bogolubov *et al.*, 1990, p. 373).

Briefly stated, "local commutativity" and "renormalizability" – namely the two cornerstones of conventional quantum field theory – necessarily give rise to a "global type of locality", which has absolutely no experimental justification "in the small".

It is interesting to note that, about the same time when an axiomatic quantum field theoretical framework, which assigned to the local (anti)commutativity axiom center stage, was launched by Wightman (1956), Schwinger had the following to say about developments in conventional quantum electrodynamics – which is implicitly based on that axiom, and which to this day remains the crowning success of conventional quantum field theory:

"The post-war developments of quantum electrodynamics have been largely dominated by questions of formalism and technique, and do not contain any fundamental improvement in the physical foundations of the theory. Such a situation is not new in the history of physics; . . . But, we may ask, is there a fatal fault in the structure of field theory? Could it not be that the divergences – apparent symptoms of malignancy – are only spurious by-products of an invalid expansion in powers of the coupling constant and that renormalization, which can change no physical implication of the theory, simply rectifies this mathematical error? This hope disappears on recognizing that the observational basis of quantum electrodynamics is self-contradictory." (Schwinger, 1958, p. xv).

It was clear from the outset that the Wightman axiomatic framework represented a systematic effort to impart much-needed mathematical rigor to conventional quantum field theory, without, however, in any way enriching and improving the *physical* foundations of the theory. Therefore, in comparing Schwinger's well-set emphasis, in the late 1950s, on the need for new physical principles in quantum field theory, with the beliefs of those who since that time have put their entire faith into mathematical techniques, it is important to assess the situation that emerged three decades later.

After almost a full decade devoted to perfecting the axiomatic method (Streater and Wightman, 1964), and after an additional two decades of concerted effort to produce, within the confines of the constructive quantum field theoretical framework that emerged from the Wightman axioms, at least a single nontrivial and yet mathematically rigorous model in four-dimensional Minkowski space, the two leading initiators of the constructive quantum field theoretical programme that embodied that effort arrived at the following conclusion:

"At present the arguments favoring triviality [of the ϕ^4 quantum field theory] seem to be stronger, but a definitive answer seems to be out of reach of available methods. These arguments apply equally to the four-dimensional Yukawa and electrodynamic (QED) interactions. If these interactions are all trivial, it would mean that a short distance cutoff resulting from the quark interactions is essential to a theory of protons, photons, mesons and electrons as elementary particles. Since it is known experimentally that protons and mesons are not elementary particles, but composites formed from quarks and gauge fields ('gluons'), such a short distance cutoff, set at the proton radius, for example, would not violate [the] experimental facts." (Glimm and Jaffe, 1987, p. 120).

However, such a short distance cutoff *would* violate Wightman's key axioms, and most of all the local (anti)commutativity one – not to mention the fact that it could not be justified in the case of photons and electrons, in particular, and in the case of gauge particles and leptons, in general, on the same grounds that can be used for "protons and mesons", or for other hadrons.

Thus, although this is not yet openly acknowledged by many researchers in the field, historical developments over the past three decades have not only supported Schwinger's 1958 unfavorable evaluation of the prospects of the conventional approach to quantum field theory, which we cited on two earlier occasions in this section, but have also vindicated Dirac's many similar public assessments[30], made by him since 1951. These assessments culminated in his final verdict on the subject of "inadequacies" of the conventional framework for quantum field theory, which he himself had founded. This is the verdict which we cited in the introduction to this chapter, and which concludes with the following very significant statement, which we reproduce now in italics:

"I want to emphasize that many of these modern quantum field theories are not reliable at all, even though many people are working on them and their work sometimes gets detailed results." (Dirac, 1987, p. 196).

The GS approach to quantum field theory, whose initial stages we outlined in the preceding two sections – and which will be further elaborated in Chapters 8 to 11 – has emerged from a systematic effort at *improving not just the mathematical, but also the physical foundations of quantum field theory* – without, however, changing its structure to the extent that all connection with the conventional formalism would be lost, and that all its *numerical* successes would be thus forfeited. Basically, the proposed improvements emerge from the rejection of the *conventional* conceptualization of locality for quantum fields, that is grounded in the retaining of classical geometries in the quantum regime. The concurrent adoption in the GS approach of quantum geometries enables the introduction of concepts of locality and microcausality, which are not only distinguishable from each other in their epistemologically formative stages, but remain mathematically distinguishable even in the present more advanced stages, leading to the formulation of a GS framework for quantum field theory. Consequently, on a purely epistemological level, GS locality and GS microcausality are closer to the CGR concept of locality and of Einstein causality than any of the aforementioned conventional concepts of locality for quantum fields.

At first sight, this epistemic feature of GS locality and microcausality might appear paradoxical, in view of the aforementioned rejection of classical geometries as accurate reflections of physical reality, and therefore of the *classical* notion of locality. However, in fact there is no paradox, since CGR locality and Einstein causality have emerged from an *epistemologically* consistent framework of ideas, that preceded the development of modern quantum mechanics in general, and of the discovery of the uncertainty principle in particular. In formulating the concepts of GS locality, as well as of weak and strong GS microcausality, the present approach is trying to enrich rather than reject the epistemological basis of classical relativity. In so doing, it is striving to retain the criteria of epistemological consistency, that represented a guiding light in the development of both relativity and quantum theory, even if that entails a fundamental *modification* of the mathematical apparatus of conventional quantum field theory, and occasionally even the sacrifice of some of its most cherished *mathematical* techniques.

Admittedly, such a strategy does not abide by the canons of the conventional mode of thinking that prevailed in the post-World War II years, and still enjoys great popularity at present. Indeed, in the conventional approach the above priorities are reversed: formal mathematical techniques are deemed essential to progress, while foundational considerations are dismissed as sterile or irrelevant (cf. Secs. 12.2 and 12.3). Therefore, it is necessary to further elucidate the above points.

We first note that the classical relativistic concept of locality is based on the concept of pointlike *event* (Einstein, 1905, 1916), which in turn is predicated on the assumption of the physical existence of deterministically behaving test bodies which are pointlike – or at least can approximate to an arbitrary degree of accuracy the conventional mental image of a pointlike object viewed as a kind of limit of extended classical bodies. Thus, the classical relativistic concept of locality is, first of all, totally unrelated to one of the key *assumptions* in the earlier cited Haag-Kastler formulation of the "principle of locality", namely that "observables in causally disjoint regions are always compatible" (Haag and Kastler, 1964, p. 848). Indeed, first of all, such a statement is epistemically vacuous in the classical regime, since *all* classical observables are "compatible". Second, it is totally independent of another key *assumption* that is routinely used in justifying local commutativity, namely that "the measurement of a component of a Hermitian field at some point has no influence on the value of the components of this field at another point separated from the first point by an arbitrarily small spacelike interval" (Bogolubov *et al.*, 1990, p. 373): nonrelativistic theories can also be local, and the assumption that light has the same velocity in all inertial frames is independent of the assumption that no signal can propagate with superluminal velocities (cf. Friedman, 1983, Ch. IV, Sec. 6). In fact, as argued by Recami (1987), Fanchi (1988, 1990), and many others, the existence of superluminal particles, namely tachyons, whose behavior would violate Einstein causality, is not at odds with the other basic principles of relativistic physics. The main point is, however, that in *classical* special relativity, the assumption that such objects can interact with ordinary matter would enable the transmission of signals backwards in time (cf. Friedman, 1983, p. 162), and thus give rise to various physically unacceptable paradoxes.

Most telling, however, is the fact that the formulation of *Einstein causality*[31] in either special or general classical relativity theory does not have to be carried out in measurement-theoretical terms, and in fact it is usually formulated in purely dynamical terms [M,W]: relativistic *causal* signals are assumed to propagate only along causal worldlines, where, by definition, a *causal worldline* is a smooth path in some classical spacetime manifold whose tangent vector at each one of its points is future-directed, and it is either null or timelike. The relation of this formulation to the theory of measurement then leads to a categorical choice, subject to a yes-no type of experimental verification, in the case of any field or particle theory formulated in a classical spacetime manifold $(\mathbf{M},g)$ which is globally hyperbolic (cf. Theorem 8.3.14 in [W]): either that theory *does* obey Einstein causality, in which case the fields or particles whose behavior it purportedly governs cannot be used to transmit signals from a point $x \in \mathbf{M}$ to some other point $x' \in \mathbf{M}$ which does not lie within its causal future $J^+(x)$ [W]; or it *does not* obey Einstein causality, in which case it should be possible, at least in principle, to envisage experimental arrangements, based on that theory, in which a signal transmitted from some point $x \in \mathbf{M}$ would be *definitely* received at some point $x' \notin J^+(x)$. Hence, in the second instance, an experimental test can be then carried out to confirm or refute such a theoretical prediction, and thereby empirically falsify the theory, if the latter turns out to be the case. Due to the yes-no nature of such tests, if the latter would *not* turn out to be the case, that would either signify a serious challenge to classical relativity theory as a whole, or the possibility of transmission of signals backwards in time – both alternatives difficult to consider as serious possibilities.

On the other hand, this yes-no nature of the tests of CCR's fundamental principles is not necessarily shared by those physical theories which are basically statistical in nature – as it was already realized in the last century by Boltzmann (cf. Kac, 1964), in the course of interpreting the second law of classical thermodynamics from the point of view of classical statistical mechanics. However, albeit conventional quantum field theory is a statistical theory at the measurement theoretical level, it is nevertheless compatible only with yes-no formulations of relativistic microcausality due to the earlier mentioned *global nature of the concept of local (anti)commutativity*, which lies at its very foundation. In turn, the identification of the *mathematical* concept of local (anti)commutativity with the *physical* concept of relativistic microcausality is contingent upon the adoption of underlying classical (and therefore deterministic) geometries, as well as upon the adoption of a deterministic mode of quantum field propagation. And indeed, deterministic field equations remain the hallmark of the conventional conceptualization of quantum field propagation despite the, by now, routine use of Feynman path integral formulations for that propagation (Faddeev and Slavnov, 1980; Rivers, 1987; [IQ]). This is due to the fact that, in the conventional mode of conceptualizing quantum field propagation, path integral formulations are primarily viewed as sources of mathematical techniques for computing functional integrals, rather than as accurate reflections of the physical nature of that propagation.

If, on the other hand, quantum geometries based on the three principles stated in Sec. 1.3 are adopted, then the ensuing quantum field theoretical concept of locality is no longer of a global nature, since the measurement-theoretical *separability* of regions in **M** is no longer sharp: in accordance with the principle of irreducible indeterminacy (cf. Principle 1 in Sec. 1.3) the geometric structure of quantum fibres (cf. Secs. 3.7-3.9) modifies the *operational interpretation* of the geometry of regions in the "classical spacetime" manifold **M** by "blurring" their boundaries. Furthermore, in the context of quantum geometries, path integrals can be then viewed as accurate reflections of the physical nature of the *actual* propagation of GS excitons. Indeed, such quantum objects possess a *proper* state vector, and therefore are not pointlike, so that their propagation can be envisaged as taking place along various individual paths, without running into inconsistencies with the uncertainty principle[32]. On account of the superposition principle, these paths no longer have to be causal in a classical sense, since it is well-known that superimposed wave packets can lead not only to amplification effects, but also to reduction effects. One can therefore distinguish between strong and weak versions of relativistic microcausality – in the same manner as we did in Sec. 5.7 – without arriving at any conflict with the well established experimental data on Einstein causality at the macroscopic level. Consequently, in the GS framework, the ultimate choice between these two possibilities can be left to future theoretical research, as well as to experimental investigations.

Let us now make these observations more clear and specific, by explicitly defining the concepts of locality and microcausality of GS quantum fields.

In the present approach based on quantum geometries, all *GS quantum fields are local* by virtue of the fact that they assume well-defined values within fibres above various points x in a Lorentzian base manifold **M**. On the other hand, their microcausality features reflect only an aspect of the *propagation* of GS quantum fields, namely the type of path along which they propagate: *strong GS microcausality* results from GS quantum field propagation along strictly causal paths (i.e., paths resulting as limits of broken paths constructed from arcs of timelike or null geodesics), whereas *weak GS microcausality* results from GS quantum field propagation along broken paths which also incorporate arcs of certain spacelike geodesics, of the type described in Sec. 5.7, and further discussed later in this section. In turn, microcausality should be distinguished from the aforementioned measurement-theoretical concept of *separability*, which deals only with the possibility of exchanging *reliable* signals (i.e., signals clearly distinguishable from the "background noise" of geometro-stochastic fluctuations) between various *operationally distinguishable* base locations in a quantum spacetime.

The above described feature of GS locality is shared by the concept of local classical fields over a classical spacetime manifold **M**. Indeed, classical fields are local only because they assume values in fibres above various points $x \in \mathbf{M}$ – except that those "classical" fibres are finite-dimensional real vector spaces, rather than the infinite-dimensional spaces constituting operator algebras over Fock fibres in GS quantum field theories. By contrast, *conventional* quantum fields, such as those in (1.7), are "local" only by virtue of the fact that, when rigorously defined as operator-valued *distributions*, as is the case in (1.8), they are well-defined for test functions f of arbitrarily small support within **M** (Jaffe, 1967) – but can never be defined as *bona fide* operator-valued *functions* of $x \in \mathbf{M}$, except in the physically and mathematically trivial case that they are equal to constant multiples of the identity operator (cf. [BL], Sec. 10.4).

7.7. Strongly and Weakly Causal Geometro-Stochastic Field Propagation

The strong microcausality of GS relativistic quantum fields is reflected by the type of GS free-fall propagation that generalizes the one described in Sec. 5.4 in the case of single spin-0 exciton states. As a matter of fact, for GS fields that interact only with the prescribed external gravitational field, which supplies the metric as well as the Levi-Civita connection to the base Lorentzian manifold **M** of the quantum Fock bundle $\mathcal{F}$, all the features of that propagation follow from the GS propagator in (5.4.8). That single-exciton propagator obviously gives rise to the following type of *strongly causal GS quantum field propagator* (Prugovečki, 1988b),

$$\boldsymbol{K}(\boldsymbol{\varphi}(x''); \boldsymbol{\varphi}(x')) = \lim_{\varepsilon \to +0} \int \prod_{n=N}^{1}{}' \, d\sigma(x_n) \mathcal{D}[\varphi(x_n)] \, K(\varphi(x_n); \varphi(x_{n-1})) \, , \tag{7.1}$$

from the base location $x' \in \mathbf{M}$ to some future location $x'' \in I^+(x')$ in the base manifold **M**. Its definition is based on the following averaging procedure, that is carried out over all relativistically causal broken geodesic paths connecting x' to $x'' \in I^+(x')$: prior to taking the limit $\varepsilon = \max(t_n - t_{n-1}) \to +0$ in (7.1), the integration in x_n is carried out, for each $n = 1,\ldots,N$, over $\sigma_{t_n} \cap I^+(x_{n-1}) \cap I^-(x'')$ for integrands $K(\varphi(x_n); \varphi(x_{n-1}))$ that are equal to the propagator in (5.21) for parallel transport along the timelike geodesic $\gamma(x_{n-1}, x_n)$, which connects the respective points x_{n-1} and x_n on the two consecutive slices $\sigma_{t_{n-1}}$ and σ_{t_n}, renormalized by division with the area of $\sigma_{t_n} \cap I^+(x_{n-1}) \cap I^-(x'')$.

To express the strongly causal GS propagation of an arbitrary state within the Fock bundle (4.2)–(4.3) in terms of the propagator in (7.1), let us expand any such state as follows, in accordance with (4.6) and (4.23):

$$\boldsymbol{\Psi} = \int_{\mathbf{F}} \mathcal{D}[f]\, \Psi(f)\, \boldsymbol{\Phi}_{\mathbf{f}} \quad , \qquad f \in \mathbf{F} \, , \qquad \boldsymbol{\Psi}, \boldsymbol{\Phi}_{\mathbf{f}} \in \mathcal{F}_x \quad , \tag{7.2a}$$

$$\Psi(f) = \langle \Phi_f | \Psi \rangle = \langle \boldsymbol{\Phi}_{\mathbf{f}} | \boldsymbol{\Psi} \rangle \, , \qquad \mathbf{f} \in \mathbf{F}_x \, , \qquad \Psi, \Phi_f \in \mathcal{F} \, . \tag{7.2b}$$

We can then arrive at the following generalization of (5.5.2), describing the propagation of multi-exciton states in $\mathcal{E}$:

$$\Psi(\varphi(x(t))) = \int_{\mathbf{F}_{x(0)}} \boldsymbol{K}(\varphi(x(t)); \varphi(x(0)))\, \Psi(\varphi(x(0))) \mathcal{D}[\varphi(x(0))] \, . \tag{7.3}$$

The GS quantum field propagators in (7.1) are therefore completely described by the strongly causal free-fall propagation of any quantum field state corresponding to any initial exciton configuration within a Fock fibre $\mathcal{F}_{x(0)}$.

As discussed in Sec. 5.7, within the context of a given geometrodynamic evolution (5.4.6), weakly causal GS free-fall propagation results by the superposition principle from stochastic parallel transport along stochastic paths whose broken-path approximations might also contain arcs of spacelike geodesics connecting points on two *successive* hypersurfaces in that geometrodynamic evolution. Hence, weak causality emerges from the global time-ordering governed by that geometrodynamic evolution. The resulting *weakly causal GS quantum field propagator* is given by (Prugovečki, 1987b)

$$\mathcal{K}(\boldsymbol{\varphi}(x''); \boldsymbol{\varphi}(x')) = \lim_{\varepsilon \to +0} \int \prod_{n=N}^{1}{}' \, d\sigma(x_n) \mathcal{D}[\hat{\varphi}(x_n)] \, K_\gamma(\hat{\varphi}(x_n); \hat{\varphi}(x_{n-1})) \quad , \tag{7.4a}$$

$$\hat{\varphi}(x_n; v) = f_n(\hat{\zeta}(x_n)) \quad , \qquad v \in \mathbf{V}^+, \qquad \hat{\zeta}(x_n) = (-a^i(x_n), v^i) \quad , \tag{7.4b}$$

where the integration in x_n is carried out, for each $n = 1,...,N$, over the entire hypersurface σ_{t_n}, and K_γ denotes the propagator in (5.21) for the parallel transport along a given geodesic $\gamma(x_{n-1},x_n)$, connecting the points x_{n-1} and x_n on the two consecutive slices $\sigma_{t_{n-1}}$ and σ_{t_n}, of the field modes in (7.4b). We note that those field modes are *localized* at the respective points of contact between the base manifold and its tangent spaces. On account of (5.21)–(5.23), such a weakly causal propagator can be expressed in the form

$$\mathcal{K}(\boldsymbol{\varphi}(x''); \boldsymbol{\varphi}(x')) = \int_{\mathbf{M}(t',t'')} \mathcal{D}\hat{\varphi} \, \exp(iS[\hat{\varphi}, \hat{\pi}]) \quad , \qquad \mathcal{D}\hat{\varphi} = \prod_x \mathcal{D}[\hat{\varphi}(x)] \quad . \tag{7.5}$$

On the surface, this functional integral has the same appearance as the one in (5.21), but, in fact, there are several fundamental distinctions[33].

First of all, the functional integration in (7.5) is over those *external* field modes resulting from the weakly causal GS propagation described in Sec. 5.6, and localized at points of contact with base points x that lie within the segment between the reference hypersurfaces $\sigma_{t'}$ and $\sigma_{t''}$, containing x' and x'', respectively, rather than over all possible *internal* modes at points along a fixed curve γ in the base space. Second, the action term that makes its appearance in (7.5), i.e.,

$$S[\hat{\varphi},\hat{\pi}] = \int_{t'}^{t''} dt \int_{\Sigma_t} \mathcal{L}_k[\hat{\varphi}(x(t);v),\hat{\pi}(x(t);v)]\, d\sigma^k(x(t))\, d\Omega(v) \;, \tag{7.6}$$

incorporates the Lagrangian emerging from the parallel transport along special curves, namely geodesics connecting two points on successive reference hypersurfaces σ_t within that segment, thus reflecting free fall conditions. Moreover, for each fixed value of t, the integration is performed over the hypersurfaces $\Sigma_t = \sigma_t \times V^+$ in general relativistic phase space, rather than over hypersurfaces contained within the tangent and cotangent spaces of $\mathbf{M}$ at points along a fixed piecewise smooth curve γ. Consequently, (7.5) shares many common formal features with the path integrals formulated in conventional quantum field theory (Rivers, 1987). As mentioned at the end of Sec. 5.7, this is a feature that makes weakly causal GS propagation worth considering as a very serious alternative to the strongly causal one, leaving it to future theoretical and experimental investigations to decide on the correct choice between them.

In this context it should be noted that the weakly causal counterpart of (7.3), namely

$$\hat{\Psi}(\varphi(x(t))) = \int_{\mathbf{F}_{x(0)}} \mathcal{K}(\varphi(x(t)); \varphi(x(0)))\, \Psi(\varphi(x(0)))\mathcal{D}[\varphi(x(0))] \;, \tag{7.7}$$

supplies within $I^+(x(0))$ the same probability amplitudes as (7.3) does in the case where $(\mathbf{M},g)$ is the Minkowski spacetime. Indeed, in each of the Poincaré gauges displayed in (2.3.12), the weakly causal propagator then transcends, on account of (7.4), into a propagator which for each single exciton state coincides with the one in (3.6.21)-(3.6.25), with η equal to the proper state vector of that exciton. Due to the path-independence of parallel transport in the case of such a flat $(\mathbf{M},g)$, the same is true of the the strongly causal propagator in (7.3) – which for single exciton states always reduces to the one in (5.4.8).

Thus, in those regions where gravitational curvature effects are negligible, the above two forms of GS propagation are experimentally indistinguishable within the chronological future $I^+(x(0))$ of any base point $x(0)$ within those regions. Furthermore, in the case of the weakly causal propagation, the "spillovers" resulting from propagation outside $I^+(x(0))$ for excitons whose proper linear dimensions approach the Planck length are exceedingly small, and approach zero as the Minkowski time coordinate t tends to infinity (Greenwood and Prugovečki, 1984). Consequently, the design of actually feasible experimental tests meant to distinguish between these two modes of propagation on causality grounds alone might be very difficult to achieve in a contemporary terrestrial setting.

The empirically observable distinctions between the weak and strong modes of GS propagation become, however, more emphatic in those strong gravitational fields where cumulative effects due to violations of strict causality can clearly manifest themselves. The most clear-cut such instance is presented by the gravitational field of a black hole: strongly causal GS propagation obviously cannot give rise to black hole evaporation of the Hawking (1975a) type, since all excitons within the event horizon [M,W] cannot escape. However, weakly causal GS propagation can give rise to such evaporation due to the possibility of "tunnelling" through that event horizon, by propagation along stochastic paths that can be approximated by broken piecewise smooth curves incorporating spacelike geodesics.

This brings to mind the question as to whether weakly causal GS propagation might not give rise to paradoxes that are similar to those displayed by the conventional approach – as discussed in Sec. 7.3. Indeed, superficially it might appear that the above described violations of *strict* classical causality are of the same type as those due to spontaneous particle creation in the conventional approach. However, that is not actually the case, due to the fact that in conventional quantum field theory in curved spacetime field propagation is *globally* implemented, whereas in GS quantum field theory weakly causal propagation is based on local concepts, formulated in accordance with *local* energy-momentum conservation. Indeed, it is well-known that global conservation of energy and momentum cannot be meaningfully imposed in CGR (cf. [W], pp. 286-287). On the other hand, a *local* conservation holds for the classical energy-momentum, which is reflected by the condition in (2.7.9) on the stress-energy tensor (cf. [M], p. 386; [W], pp. 69-70). According to (4.12), (4.21) and (5.8), in GS quantum field theory we have

$$\langle \Phi_{\mathbf{f}} | T_{jk}[\varphi(x;\zeta)] \Phi_{\mathbf{g}} \rangle = \hat{T}_{jk}[\bar{\mathbf{f}}(\zeta) + \mathbf{g}(\zeta)] \langle \Phi_{\mathbf{f}} | \Phi_{\mathbf{g}} \rangle , \qquad \mathbf{f},\mathbf{g} \in \mathbf{F}_x , \tag{7.8a}$$

$$\hat{T}_{jk}[\mathbf{f}(\zeta)] = \hat{Z}_{f,m}\left(\mathbf{f}(\zeta)_{;j}\mathbf{f}(\zeta)_{;k} + \tfrac{1}{2} g_{jk}(m^2 \mathbf{f}(\zeta)^2 - g^{il}\mathbf{f}(\zeta)_{;i}\mathbf{f}(\zeta)_{;l})\right) , \tag{7.8b}$$

where the semicolons indicate covariant derivatives of the classical field $\mathbf{f}(\zeta)$ obtained by keeping in $\mathbf{f}(x;\zeta)$ the components ζ of ζ in (4.4b) fixed while varying x. We note that (7.8b) are the components of a classical stress-energy tensor (cf. Eq. (4.3.10) in [W]), for which we have in accordance with (2.7.9),

$$\hat{T}_{jk}{}^{;k}[\mathbf{f}(\zeta)] = \overline{\nabla}^k \hat{T}_{jk}[\mathbf{f}(\zeta)] \equiv 0 , \tag{7.9}$$

due to the fact that, on account of (3.3.5), (5.1.1) and (5.1.5),

$$\left(\overline{\nabla}^k \overline{\nabla}_k + m^2\right)\mathbf{f}(\zeta) = 0 . \tag{7.10}$$

Since weakly as well as strongly causal GS propagation is based on parallel transport, both these forms of propagation respect *local* energy-momentum conservation.

7.8. Interacting Quantum Fields in Extended Fock Bundles

The concept of locality of quantum fields has to be distinguished from the concept of local field interactions. In special relativity nonlocal interactions of local classical fields over Minkowski space can be defined by introducing nonlinear interaction terms which contain the values of those fields at distinct points in Minkowski space. Of course, usually there is no physical motivation for studying such nonlocal interaction between local fields in

Minkowski space. On the other hand, ever since the early efforts of Yukawa (1950), there have been advocates of nonlocal quantum field theories that employ nonlocal interactions in constructing nonlocal interacting field Lagrangians (Efimov, 1977; Namsrai, 1986; Evens *et al.*, 1991), while retaining local Lagrangians for the free fields. Hence, in such quantum field theories nonlocal interactions emerge from *ad hoc* "nonlocal regularization" rulings for the removal of ultraviolet divergencies from the corresponding "local" models.

On the other hand, the corresponding conventionally "local" quantum field theoretical models are themselves not mathematically well defined in their usual form [IQ,SI]. Hence, their rigorous *mathematical* interpretation within the constructive quantum field theory programme (Glimm and Jaffe, 1987) proceeds by first introducing various field "cutoffs", which violate relativistic invariance as well as locality. Their "removal" is then attempted by constructing "renormalized" Hamiltonians that act in a new Hilbert space, which carries a non-Fock representation of the canonical (anti)commutation relations. However, as mentioned in the preceding section, this procedure has failed to arrive at *any* physically nontrivial relativistic quantum field theoretical models in four-dimensional Minkowski space.

The question of interacting fields in curved spacetime has not been faced at a mathematically rigorous level in either of the above approaches. This is understandable in view of the fundamental obstacles which are encountered in that context not only at the mathematical but also at the physical level. For example, even in the case of those classical fields over a curved classical spacetime $(\mathbf{M},g)$ which are not scalar fields, the nonlocal interactions taken from the special relativistic regime are not well defined, since they involve field values at distinct points in $\mathbf{M}$. On a purely mathematical level, this difficulty could be dealt with by implementing parallel transport between those distinct points in $\mathbf{M}$. This, however, would raise the question as to what paths should be used in performing such parallel transport. Hence, the definition of nonlocal field interactions over a curved classical spacetime $(\mathbf{M},g)$ would have to involve physically *ad hoc* assumptions. This problem is shared by any quantum field theory with nonlocal interactions, so that there is no physically convincing way of transferring the nonlocal models of Efimov (1977) and others to the curved spacetime regime.

On the other hand, *local GS interactions* can be defined without any difficulty by nonlinear terms of GS quantum fields acting in Fock fibres above the *same* base location $x \in \mathbf{M}$, once those Fock fibres are appropriately extended. There is no physical motivation at present for considering such interactions between massive fields mediated by fields which are also massive, since it is by now generally believed that all fundamental interactions in nature are mediated by gauge fields. Hence, the most appropriate place for discussing local GS quantum field interactions would be in Chapters 10-11 – where that becomes mandatory due to the unavoidable presence of self-interactions exhibited by the non-Abelian gauge fields treated in those chapters. However, if we concentrate only on key mathematical points, then the basic ideas for treating local GS interactions can be readily explained already at this stage. For the sake of notational simplicity, we shall use for the purpose of illustration the well-known φ^4 model[34] – but all our remarks remain equally valid for any interaction Lagrangian which is a polynomial in the interacting fields.

As in conventional quantum field theory, we shall treat each quantum field interaction by means of a Lagrangian density supplying a source term, which in the case of the φ^4 model is equal to (cf. Nash, 1978)

$$\mathcal{L}_I[\varphi(x;\zeta)] = -(\lambda/4!) :\varphi(x;\zeta)^4: \quad . \tag{8.1}$$

We note that the normal ordering in (8.1) is well-defined due to the presence of a *local* vacuum in each Fock fibre $\mathcal{F}_x$. Moreover, that normally-ordered operator in (8.1) is densely defined in $\mathcal{F}_x$ for each value of ζ. In particular, all the elements of the second-quantized frames in (4.16) are within its domain of definition. On the other hand, a lengthy but straightforward computation, based on the repeated use of the commutation relations in (4.11), shows that (cf. (5.1.14), (5.1.18) and (4.4a)):

$$\left\| :\varphi(x;\zeta)^4: \boldsymbol{\Phi}_{\mathfrak{f}} \right\| \geq 4\left[\Delta_x^{(+)}(\zeta;\zeta)\right]^4 = 2^{-10}(m^2/\pi^3)^4 \hat{Z}_{f,m}^{-4} \quad . \tag{8.2}$$

Consequently, the source term that modifies the time-evolution governed by the free Hamiltonian $\boldsymbol{P}_{0;\boldsymbol{u}}$ in (5.8), namely the operator

$$\hat{H}_{I;\boldsymbol{u}} = -(\lambda/4!) \int :\varphi(x;\zeta)^4: \, d\sigma^0(\boldsymbol{q})d\Omega(\boldsymbol{v}) \quad , \tag{8.3}$$

is not well defined as a Bochner integral whose integrand acts on the elements $\boldsymbol{\Phi}_{\mathfrak{f}}$ of the second-quantized frames in $\mathcal{F}_x$, since in the present context that definition requires (cf. [PQ], p. 479) that the norm in (8.2) be integrable over the hypersurface Σ over which the integration in (8.3) is carried out. Moreover, if we try to deal with this problem by introducing a space cutoff in (8.3) and expanding the quantum frame field (4.12) in accordance with (4.24) and (4.27),

$$\varphi(x;\zeta) = \sum_{\alpha=1}^{\infty}\left(\mathrm{w}_\alpha(x;\zeta)a_\alpha^*(x) + \overline{\mathrm{w}}_\alpha(x;\zeta)a_\alpha(x)\right) \quad , \tag{8.4}$$

we observe that the removal of that space cutoff cannot be carried out within $\mathcal{F}_x$, since in that limit probability amplitudes for states with N excitons do not approach zero as N is allowed to go to infinity. In other words, the interaction Hamiltonian in (8.3) gives rise to states for which the probability of observing N excitons is bounded from below by a strictly positive number, so that it does not approach zero, as in the case of Fock space states.

This mathematical fact is a reflection of the underlying reason for the presence of the well-known Haag theorem [BL] in conventional quantum field theory, which prohibits the mathematically rigorous formulation of such theories for interacting fields within any single Fock space. The proof of Haag's theorem provided by Streit (1969), as further analyzed in (Emch, 1972), Ch. 3, Sec. 1d, reveals that this theorem is *not* a reflection either of local (anti)commutativity or of relativistic invariance of quantum theories for interacting fields. Rather, it merely represents a reflection of the required uniqueness of a *global* vacuum state and of the (explicit or implicit) imposition of the *canonical* formalism via the demand that the generator H of time translations satisfy, for some choice of real pre-Hilbert space $\mathfrak{R}$,

$$P(f) = i[H,Q(f)] \quad , \qquad f \in \mathfrak{R} \quad , \tag{8.5}$$

in relation to an irreducible representation of the canonical commutation relations (CCR's) associated with an infinite numbers of degrees of freedom (cf. Emch, 1972, p. 251). Such a representation is the following one (cf. Emch, 1972, p. 234),

$$[Q(f),P(g)] = i\langle f|g\rangle \ , \quad [Q(f),Q(g)] = [P(f),P(g)] = 0 \ , \quad f,g \in \Re \ , \tag{8.6}$$

where, in general, the above commutators are defined in a dense Gårding domain within the Hilbert space **H** in which they act.

The Fock representations are singled out by the requirement that in **H** there is a vector Ψ_0 such that

$$a(f)\Psi_0 = \mathbf{0} \ , \qquad a(f) = 2^{-1/2}(Q(f) + iP(f)) \ , \qquad \forall f \in \Re \ . \tag{8.7}$$

Of course, in physical terms this vector represents the Fock vacuum state – cf. (1.6). For an infinite numbers of degrees of freedom, the Fock representations are the only ones for which total particle number operators, such as those in (1.9), are well-defined (cf. Emch, 1972, p. 262). Thus, in all the other representations that are unitarily *non*-equivalent to Fock representations all quantum states display an infinite number of particles[35]. On the other hand, Haag's theorem essentially asserts that if, at any single instant t_0, a Hamiltonian H satisfies (8.5) for any representation (8.6) of the CCR's that is unitarily equivalent to a Fock representation, then H has to be unitarily equivalent to the free Hamiltonian H_0 of that Fock representation. Taken in conjunction with the (postulated) uniqueness of the (global) vacuum state as the only state left invariant by spacetime translations, this leads to the conclusion that H cannot describe interactions (cf. [BL], Sec. 21.2).

Clearly, for a finite number of degrees of freedom, the conditions in (8.5) reflect in the nonrelativistic context the standard canonical quantization procedure, whereby in the classical Hamilton equations for position and momentum the Poisson brackets are replaced with operator commutator brackets. However, the Hamiltonian formalism encounters in the relativistic regime severe difficulties already at the classical level, where no-go theorems demonstrate its incompatibility with the presence of interactions between classical particles (Currie *et al.*, 1963; Kerner, 1965; Hill, 1967). Consequently, there certainly are no compelling *physical* reasons to insist on the retention of (8.5) in the quantum relativistic regime. Rather, the contrary is the case: the entire problem with position operators for sharp localization in the relativistic regime, briefly recounted in Secs. 1.2 and 3.4, indicates that the relations (8.5), which have their roots in nonrelativistic classical mechanics, cannot be retained in the relativistic regime. Naturally, in conventional quantum field theory there are no plausible alternatives to (8.5), since in that context a *global* vacuum state is postulated, and a single Hilbert space is adopted for quantum fields, so that (8.5) emerges from the requirement of *global* Poincaré covariance of the conventional quantum field theoretical formalism. On the other hand, in GS quantum field theory these problems do not exist, since only *local* Poincaré gauge invariance is instead required, and only *local* vacuum states are envisaged. Consequently, the difficulties created by Haag's theorem can be circumvented – as it will transpire from the subsequent considerations in this section.

In the present GS context the CCR's in (8.6) are embedded within each Fock fibre $\mathcal{F}_x$ via the commutation relations in (4.28). Let us therefore adopt within the typical fibre **F** an orthonormal basis, such as the one in (3.7.17). We can then introduce the real Hilbert space in (3.8.4a), so that we can set within the standard Fock fibre $\mathcal{F}$,

$$Q(f) = 2^{-1/2}[a^*(f) + a(f)] \ , \quad P(f) = 2^{-1/2}i[a^*(f) - a(f)] \ , \quad f \in \Re(\{w_\alpha\}) \ , \tag{8.8a}$$

$$a(f) = q^\alpha a_\alpha \ , \qquad a_\alpha = \varphi^{(-)}(w_\alpha) \ , \qquad w_\alpha = \sigma_x^u \mathbf{w}_\alpha \in \mathbf{F} \ , \tag{8.8b}$$

$$[a_\alpha, a_\beta^*] = \delta_{\alpha\beta} \ , \qquad [a_\alpha, a_\beta] = [a_\alpha^*, a_\beta^*] = 0 \ , \qquad \alpha, \beta = 1, 2, \ldots \ . \tag{8.8c}$$

Corresponding relations are obtainable within each Fock fibre $\mathcal{F}_x$, by using the generalized soldering map in (4.6) and its inverse. However, as pointed out earlier, when the operator in (8.3) acts upon $\boldsymbol{\Phi}_\mathbf{f}$, the contribution from the pure creation terms do not belong to $\mathcal{F}_x$.

This problem can be handled by introducing a Naimark extension $\mathcal{F}^\uparrow$ of the typical Fock fibre $\mathcal{F}$, and then building a *Naimark bundle* $\mathcal{E}^\uparrow$ whose fibre $\mathcal{F}_x^\uparrow$ above each $x \in \mathbf{M}$ equals the corresponding extension of $\mathcal{F}_x$. Indeed, upon introducing, in accordance with (3.8.2), (3.8.4) and (4.31), the POV measures

$$E_{\mathcal{F}_x}(B_n) = (2\pi)^{-n} \int_{B_n} |\hat{z}_n\rangle d^n\hat{q}_n d^n\hat{p}_n \langle \hat{z}_n| \ , \qquad B_n \subset \mathbf{R}^{2n} \ , \tag{8.9}$$

we can construct[36] the Naimark extension $\mathcal{F}_x^\uparrow$. This extension is a nonseparable Hilbert space which carries non-Fock representations of the CCR's, in addition to the Fock representation in (4.28) that corresponds to the one given within the typical fibre $\mathcal{F}$ by (8.8). In each one of these Naimark fibres $\mathcal{F}_x^\uparrow$ the operators

$$\mathbf{P}_{\mathcal{F}_x} = \int_{\mathbf{F}_x} |\boldsymbol{\Phi}_\mathbf{f}\rangle \, d\mathbf{f}\, d\mathbf{f}^* \langle \boldsymbol{\Phi}_\mathbf{f}| = \text{w-}\lim_{n\to\infty} E_{\mathcal{F}_x}(\mathbf{R}^{2n}) \tag{8.10}$$

act as orthogonal projectors onto $\mathcal{F}_x$, so that the original Fock bundle $\mathcal{E}$ emerges as a sub-bundle of the Naimark bundle $\mathcal{E}^\uparrow$.

The application of the Naimark extension procedure leading from $\mathcal{F}$ to $\mathcal{F}^\uparrow$, and from $\mathcal{F}_x$ to $\mathcal{F}_x^\uparrow$, respectively, ensures that the Naimark bundle is associated to the principal Klein-Gordon quantum frame bundle in (5.1.22). Hence, the connection as well as the parallel transport described in Sec. 7.5 can be routinely extended to the Naimark bundle $\mathcal{E}^\uparrow$, in whose fibres the interaction Hamiltonian (8.3) acts in a mathematically meaningful manner. However, as we shall now demonstrate, the projection of (8.3),

$$H_{I;u} = \mathbf{P}_{\mathcal{F}_x} \hat{H}_{I;u} \mathbf{P}_{\mathcal{F}_x} = -(\lambda/4!) \int \mathbf{P}_{\mathcal{F}_x} : \varphi(x;\zeta)^4 : \mathbf{P}_{\mathcal{F}_x} d\sigma^0(q) d\Omega(v) \ , \tag{8.11}$$

back into the Fock fibres can represent the correct interaction Hamiltonian. This means that we can rigorously deal with the problem posed by Haag's theorem in conventional theory in a manner which is very close to the customary mathematically nonrigorous treatments [SI,IQ] of quantum field theoretical interactions, which habitually ignore[37] that theorem.

The justification for the adoption of (8.11) as the interaction Hamiltonian in the present GS context emerges by drawing suitable parallels with the treatment of interactions in nonrelativistic SQM (cf. [P], p. 44). Thus, for a stochastically extended particle with proper state vector ξ, whose states are in general represented by wave functions ψ from the Hilbert space $\mathcal{H} = \mathbf{P}_\xi L^2(\mathbf{R}^6)$ described in Sec. 3.2, the Naimark extension $\mathbf{H}$ equals $L^2(\mathbf{R}^6)$. Hence, the projector $\mathbf{P}_\xi$ represents a mathematical counterpart of the one in (8.10).

For an external field represented by a potential V we could now attempt to define an interaction Hamiltonian H_I in the manner customary in conventional nonrelativistic quantum mechanics, namely by multiplication of ψ with $V(\mathbf{q})$. However, in general such an operator will not leave $\mathcal{H} = \mathbf{P}_\xi L^2(\mathbf{R}^6)$ invariant. On the other hand, it would be perfectly well-defined on $\mathbf{H} = L^2(\mathbf{R}^6)$. The problem is that $\mathcal{H} = \mathbf{P}_\xi L^2(\mathbf{R}^6)$, rather than $\mathbf{H} = L^2(\mathbf{R}^6)$, carries an irreducible representation of the CCR's. Therefore, in conformity with basic quantum mechanical principles, it is $H_{I,\xi} = \mathbf{P}_\xi H_I \mathbf{P}_\xi$ that has to be adopted as the correct interaction Hamiltonian.

Of course, the spectral analysis in (3.2.9) shows that $\mathbf{H} = L^2(\mathbf{R}^6)$ carries an infinity of other representations of the Galilei group. In fact, most of them are, on account of (3.2.12), unitarily inequivalent to the one in the particular subspace $\mathcal{H} = \mathbf{P}_\xi L^2(\mathbf{R}^6)$. Hence, if we followed in the footsteps of the mathematical type of arguments used in constructive quantum field theory, we might argue that some of the irreducible subspaces for these other representations are more suitable. Indeed, although those other representations correspond to proper state vectors and/or spins which are different from the ones of the considered quantum particle, the rationale for such an *ad hoc* procedure might lie in the claim that the interaction gives rise to a "renormalization" of those quantities. Fortunately, in the present nonrelativistic context the sharp-point limits discussed in Sec. 3.5 exist, and the fallacy of such a rationalization can be explicitly demonstrated: the transition to the configuration representation reveals that $H_{I,\xi} = \mathbf{P}_\xi H_I \mathbf{P}_\xi$ is the only interaction Hamiltonian which corresponds to the "smearing" of the potential acting upon a pointlike particle with the confidence function in (3.2.17), and which therefore transcends in the sharp-point limit in (3.5.4) into the correct local interaction Hamiltonian for point quantum particles – cf. [P], Ch. 1, Eqs. (8.22) and (8.23).

Naturally, it could be argued that the above simple-minded argument is irrelevant to quantum field theory, since it does not involve self-interactions. However, such self-interactions could be introduced upon extending the above model into one of mutually interacting stochastically extended particles, in the same manner as it is done in the BCS model – cf. (Emch, 1972), Sec. 1.f. In fact, such considerations indicate that the *local* Fock vacuum plays in the GS quantum field theoretical regime a role similar to that played by the proper state vector ξ in nonrelativistic SQM, since the latter gives rise to various mean values around which *finite* fluctuations in spatial localization take place, whereas the former gives rise to various mean values around which *finite* fluctuations in exciton numbers occur.

The conventional image of *infinite* "clouds" of virtual quanta, which purportedly "dress" bare particles in conventional quantum field theory, reflects a *global* spacetime point of view, as it becomes evident as soon as the Feynman diagram technique that underlies such imagery is examined within Minkowski space (cf. [SI], Sec. 14b). Indeed, from an epistemic point of view, such an approach treats all virtual pair creations and annihilations that accompany the propagation of any single particle in spacetime *as if* they could be *collectively* incorporated into a single entity, which *concurrently* exists in *all* of spacetime – rather than as an *evolutionary* process, whereby finite numbers of pairs are constantly *being* created and annihilated, while the process of quantum propagation is unabatedly taking place. This is largely due to the adoption of the concept of the S-matrix, on which the Feynman diagram techniques rely, as epistemologically fundamental, rather than a mere idealization, which is, strictly speaking, physically unrealizable – albeit it can provide a convenient description of scattering processes in asymptotically *flat* spacetimes.

However, this concept becomes *physically* meaningless in generic curved spacetimes, including those which serve as realistic cosmological models [N]. As an extreme case, what could be the possible physical meaning of asymptotic states in a closed Robertson-Walker model, which starts with a "Big Bang", and ends in a "Big Crunch"? In general, what could be the physical meaning of free asymptotic states [PQ] and of an S-matrix in the global ever-presence of gravitation, from which there is no escape and no shielding?

Thus, the fundamental epistemological distinction between the two points of view – infinite clouds of virtual quanta dressing "bare" particles for infinite spans of time vs. dynamic processes involving perpetual but *localized* virtual pair creation and annihilation constituting an integral part of GS propagation – reflects fundamental disparities in modes of *physical* conceptualization: the conventional point of view is intrinsically global, and as such it can at best be in keeping with the conceptualizations of spacetime predating *general* relativity; whereas the GS point of view is intrinsically local, and as such it is in keeping with general relativity – albeit the underlying notion of locality is fundamentally quantum, rather than classical, in its physical as well as mathematical nature.

In accordance with (4.12), (4.21) and (7.8), we have,

$$\langle \Phi_{\mathbf{f}} | H_{I;u} \Phi_{\mathbf{f}'} \rangle = -(\lambda/4!) \langle \Phi_{\mathbf{f}} | \Phi_{\mathbf{f}'} \rangle \int \left(\bar{\mathbf{f}}(\zeta) + \mathbf{f}'(\zeta) \right)^4 d\sigma^0(\boldsymbol{q}) d\Omega(\boldsymbol{v}) \quad . \tag{8.12}$$

In view of the fact that the functions within the round bracket in (8.12) are continuous, bounded and square-integrable over any of the Σ-hypersurfaces of integration, the integral in (8.12) converges for all $\mathbf{f},\mathbf{f}' \in \mathbf{F}_x$. In accordance with (3.6.18) and (4.4.8), the interaction Hamiltonian in (8.11) contributes the additional term

$$\tfrac{1}{2}\hat{Z}_{f,m}\left(\bar{\varphi}_{,k}\,\Delta\varphi + \varphi'_{,k}\,\Delta\bar{\varphi} - \bar{\varphi}\,\Delta\varphi_{,k} - \varphi'\,\Delta\bar{\varphi}_{,k}\right) - (\lambda/4!)\eta_{0k}\left(\bar{\varphi} + \varphi'\right)^4 \quad , \tag{8.13a}$$

$$\varphi = \varphi(x(t);\zeta) \quad , \qquad \varphi' = \varphi'(x(t);\zeta) \quad , \qquad \Delta\varphi = \varphi - \varphi' \quad , \tag{8.13b}$$

to the functional integral in (7.1) (which is obtained from the action integral in (5.22) after all $\varphi(x(t);\zeta)$ are replaced by $\varphi'(x(t);\zeta)$), and a corresponding term to the functional integral in (7.6). Hence, the superposition of proper state vectors resulting from strong or weak GS propagation along various paths, from some $x \in \mathbf{M}$ to a given $x' \in \mathbf{M}$, will be modified by the process of pair creation and annihilation embedded in (8.11)-(8.13). This propagation takes place between distinct Fock fibres, so that the underlying assumptions of Haag's theorem are not applicable: in GS quantum field theory there are no infinite renormalizations, but rather an uncountable infinity of distinct *local* Fock fibres. These fibres are the seats of physically *feasible* states, describing (in principle) observable excitons above various base locations $x \in \mathbf{M}$ within a given quantum spacetime geometry.

Thus, by adopting a fundamentally local point of view, GS quantum field theory dispenses with the conceptualization of an infinity of virtual quanta that are postulated by conventional renormalization procedures. In this manner it abides by Dirac's many admonishments[38] against the *ad hoc* subtraction of infinities, cited in this chapter as well as in Chapter 12. Indeed, all renormalizations in GS quantum field theory are finite, as reflected by the finite renormalization factors in the Lagrangian in (5.22). It is only in the (relativistically unacceptable) sharp-point limit, discussed in the nonrelativistic as well as relativistic context in Sec. 3.5 and 3.6, that these renormalizations diverge to infinity.

Notes to Chapter 7

1 This article was originally intended for a special issue commemorating Dirac's 80th year: "A special Festshrift prepared for him under the authorship of some of his friends, which was to be published in 1984 with a contribution by him entitled 'The Inadequacies of quantum field theory', will now, most regrettably, appear as a memorial to him." (Kursunoglu, 1987, p. xiii).

2 "Paul Adrien Maurice Dirac, born in Bristol, England, on August 8, 1902, died on October 20, 1984, just two months after his 82th birthday. His loss ranks with the loss of Isaac Newton, James Clerk Maxwell, and Albert Einstein." (Kursunoglu, 1987, p. ix).

3 This *mathematically* very satisfactory axiomatic formulation of conventional quantum field theory stemmed from the thesis that, by the early 1960s, "the Main Problem of quantum field theory turned out to be to kill it or cure it: either to show that idealizations involved in the fundamental notions of the theory (relativistic invariance, quantum mechanics, local fields, etc.) are incompatible in some physical sense, or to recast the theory in such a form that it provides a practical language for the description of elementary particle dynamics." (Streater and Wightman, 1964, p. 1). For some time many mathematical physicists, including the present author (Prugovečki, 1964), were convinced the latter will turn out be the case. However, as discussed in Sec. 1.2, that hope proved illusory for many fundamental reasons – all centered around the concept of locality in relativistic quantum theory.

4 In *two*-dimensional Minkowski space the ϕ^4 and the Yukawa model appear to satisfy the modification of those axioms corresponding to the *unphysical* choice of dimension. There have also been some partial successes with these two models in *three*-dimensional Minkowski space (Glimm and Jaffe, 1987). However, in the physically realistic case of *four*-dimensional Minkowski space, the mathematically rigorous methods developed by Glimm and Jaffe (1987), in conjunction with their many followers, lead to a ϕ^4 model which is *physically* trivial (Fröhlich, 1991), and the same seems to be true even of QED – cf. (Glimm and Jaffe, 1987), p. 120.

5 In fact, even nowadays there still are no astronomical observations that can be *indubitably* interpreted as black holes. Hence, even though the prediction of thermal black hole radiation has obviously captured the imagination of many physicists, it is still not supported by hard observational data.

6 Note that throughout this monograph we follow a convention whereby fields that create particles carry a "plus" sign, whereas those which annihilate particles carry a "minus" sign. This convention is the opposite to the one adopted in most physics textbooks on conventional quantum field theory [IQ,SI].

7 Physically this point is not that well understood, since there is no physical justification for the choice of any one particular space of test functions over another. Sometimes the well-known results of Bohr and Rosenfeld (1950) are invoked, since they imply that quantum fields over classical spacetime regions with sharp boundaries are not physically well-defined. However, even assuming that the smearing with test functions which are *not generically positive definite* represents a legitimate way of taking care of this problem, that still does not single out a particular space of test functions as physically acceptable.

8 The ν-derivative in (1.17) is in fact an ordinary derivative, since it is applied to a scalar field, whereas the covariant μ-derivative in (1.17) is the one defined by (2.6.23)-(2.6.25) for the Levi-Civita connection in the Lorentzian manifold $(\mathbf{M},g)$. The case of conformal coupling [BD], which requires the insertion within the bracket in (1.17) of the additional term $R/6$, is also very popular on account of conformal invariance, and the fact that the Feynman path integrals require its presence (cf. [ST], Ch. 24). However, that term leads to no new fundamental insights. Since the formulation of GS propagation does not require it, we shall omit it in the subsequent considerations.

9 With the convention most common in physics literature, a *sesquilinear form* [PQ] is antilinear in its first argument, and linear in the second argument, but it is nevertheless also often called a bilinear form. Note that the covariant derivatives in (1.18) coincide, in any coordinate chart, with ordinary derivatives, since the solutions of (1.17) are scalar fields.

10 A *static* classical spacetime manifold $(\mathbf{M},g)$ is a stationary one in which in addition the Killing field t is orthogonal to the (Cauchy) reference surfaces σ_t in (1.18), so that in (1.21) we have $g_{a0} = 0$ for all a = 1,2,3 [M,W].

11 The most recent developments with regard to the renormalization of the stress-energy tensor based on improved versions of these axioms in conventional quantum field theory in curved spacetime were recently reviewed by Fulling (1989). Several years after the appearance of the review article by DeWitt

(1975), papers by Isham (1978) and Dimock (1980) also provided Wightman and Haag-Kastler axioms, respectively, for quantum field theory in curved spacetime. These mathematical formulations do not remove, however, any of the foundational difficulties in this area (cf. Sec. 7.6). For example, the mere *postulation* of an "algebra of observables" guarantees neither its feasibility, nor the uniqueness of its construction *if* it is actually implementable, in which case "the resulting *representation* of the algebra is not guaranteed to have any physical relevance (in the nonstatic case)." (Fulling, 1989, p. 127).

12 The present situation is aptly summarized by Wald (1984) as follows: "In curved spacetime, there is, in general, no meaningful notion of an 'instantaneous vacuum state' with respect to which one could 'normal order' the expression for the [stress-energy tensor $T_{\mu\nu}$]. ... [To remove its divergencies] one finds that one must introduce two new, nonclassical parameters into the expression for $T_{\mu\nu}$ corresponding to the freedom of adding in the identity operator times multiples of the two conserved local curvature terms of dimension (length)$^{-4}$. Thus, there is a two-parameter ambiguity in the expression for $T_{\mu\nu}$. However, $T_{\mu\nu}$ satisfies a list of physically reasonable properties which uniquely determine it up to this two-parameter ambiguity (Wald, 1977, 1978), so it appears that this is the only ambiguity present." ([W], p. 410).

13 According to Bell and Leinaas (1983), very large storage rings, such as SPEAR at Stanford, *could* produce thermal radiation temperatures in the 1200° K range. However, as we shall point out later in this section, the presence of such radiation in an accelerator or storage ring might be of no relevance to the Rindler radiation effect based on (2.5).

14 In fact, as mentioned in the preceding note, this limitation might have been actually already overcome in case of nonlinear accelerations.

15 In an article provocatively entitled "Particles do not exist", P.C.W. Davies states the following: "These 'particles' [in Minkowski spacetime] carry no energy or momentum, yet they still excite a *static* detector! ... The 'particles' that excite a Rindler (accelerating) detector do not carry energy or momentum either. ... So, in a sense, the energy is not where the particles are, and may have long departed from the space-time region of the detector." (Davies, 1984, pp. 72-73).

16 This is actually the case in the earlier cited considerations by Bell and Leinaas (1983), who point out that electrons in a linear accelerator have to be accelerated by an electric field of approximately 1400 MV/m in order to register a Rindler temperature of about 1°K, whereas the SLAC linear accelerator produces only 7 MV/m; on the other hand, ultra-relativistic electrons in very large storage rings, such as SPEAR at Stanford, achieve accelerations which, according to (2.8), correspond to Rindler radiation temperatures in the 1200°K range.

17 Of course, the proponents of "scenarios" of the creation of the universe *ex nihilo* might answer this last question by invoking "vacuum fluctuations". However, as it was pointed out in Sec. 1.5, and as it further discussed in Chapter 12, such creation *ex nihilo* "scenarios" warrant very close scrutiny.

18 They were first pointed out in (Prugovečki, 1988a), and then further discussed in (Prugovečki, 1989b), Appendix A. A summary of the conclusions reached thus far will be provided in Chapter 12.

19 The concepts of Whitney direct sum and of Whitney direct product are usually defined for real and complex vector bundles with finite-dimensional fibres (cf. [KN], p. 306; Osborn, 1982, pp. 110 and 170, respectively), but those definitions can be extended in a most obvious manner to all Hilbert bundles with infinite-dimensional fibres (cf. Fell and Doran, 1988, pp. 149-151).

20 As discussed in (Prugovečki, 1991a), as well as in Chapter 12, one should distinguish between the concepts of "locality" and "separability" in general relativistic theories of measurement. One way of underlining this distinction is to say that *locality* pertains to measurement procedures dealing with spatio-temporal relationships between physical systems and local frames of reference (which are the only kind of frames admissible in general relativistic theories), whereas *separability* pertains to measurement procedures dealing with spatio-temporal relationships between various local frames. In CGR the origins of such frames are taken to be sharply separable. For example, it is taken for granted that, for two given inertial frames, amongst the measurement procedures that can be performed by a third party, there are always some that can distinguish with absolute certainty between their origins. In the present quantum geometry framework, this kind of *assumption* is relinquished on account of the epistemological reasons explained in Sec. 1.3, and further elaborated in Chapter 12.

21 This is not the case in Hilbert spaces of functions which are square integrable with respect to Lebesgue measure, such as those in (3.1.1), or those in the Hilbert space in (1.1). The wave functions belonging

to them actually belong to equivalence classes of "almost everywhere" equal functions [PQ], so that their values at each point can be changed without changing the equivalence class. Hence, with such functions the subsequent construction in (4.7)-(4.9), and therefore also all the main considerations in this chapter, would not be feasible.

22 As is generally the custom even in mathematically rigorous treatments of quantum field theory [BL], throughout this monograph such commutation relations as those in (1.33), (4.11) and (4.13), for which the left-hand side is only densely defined, whereas a real or complex number appears on the right-hand side, are to be interpreted as holding with that number multiplying the restriction of the identity operator to the dense domain of definition of the left-hand side.

23 The concept of *Bochner integral* can be treated most generally in the context of Banach spaces (Hille and Phillips, 1968). However, in the case of Hilbert spaces, its basic theory simplifies considerably – as shown in [PQ], Ch. V, Sec. 3.7. Readers not interested in its mathematical aspects can interpret Bochner integrals by integrating matrix elements of the operator-valued integrand with respect to the specified measure, in order to arrive at the corresponding matrix element of the operator represented by that Bochner integral.

24 Strictly speaking, that is not the case when a similar construction is carried out with the creation and annihilation operators in (1.7b), since not only are those operators not well defined in Fock space, but the "eigenvalue" $f(x)$ is also not well defined for an arbitrary element of the Hilbert space in (1.1). Indeed, the elements of those Hilbert spaces are equivalence classes of functions, rather than single functions, well defined at each point x in Minkowski space. Mathematically, this is one of the fundamental reasons why many of the constructions carried out in the remainder of this chapter cannot be duplicated within the conventional formalism. Physically, the reasons run deep, namely to the very notion of locality in relativistic quantum mechanics – as our analysis of these problems in Chapters 3 and 5 has made it at least in part clear, and as later considerations will further illucidate this point.

25 Cf. Eq. (5) of Schwinger (1959), which contains "a number characterizing the statistical relation between [two] states". As such, this number might be expected to be real, but in fact it equals the complex probability amplitude in Eq. (12). Since Eq. (5) is fundamental for all that follows in that paper, that means that complex probability amplitudes, rather than mere probabilities, are introduced from the outset in an *implicit* manner. Hence, the ultimate promise that "further analysis of the measurement algebra leads to a geometry associated with the states of the system" (Schwinger, 1959, p. 1553) points in the direction of a quantum geometry, rather than in the direction of a classical geometry.

26 Cf. Axiom IV in the third paper in (Prugovečki, 1966) – which is analogous in both appearance as well as its *ad hoc* nature to similar axioms in "quantum logic" formulations (Mackey, 1963). The possibility of implementing far-reaching structural features in a quantum theoretical framework by imposing *seemingly* innocuous *assumptions*, labelled "axioms", attests to the intrinsic and fundamental weakness of the axiomatic approach in the formative stages of any physical framework of ideas.

27 This assumption is often justified by the *postulate* that the set of all observables "can be identified with the set of all self-adjoint elements of a real or complex, associative, and involutive algebra" (cf. the structure axiom in Emch, 1972, p. 71), so that, by the well-known GNS construction, it can be then associated with "the algebra of all self-adjoint operators in a uniformly closed involutive algebra of bounded operators acting in a real (or complex) Hilbert space" (*ibid.*, p. 75). In turn, the roots for this postulate can be traced to von Neumann's (1932) *assumption* that *all* self-adjoint operators in the Hilbert space of a nonrelativistic quantum system represent observables. However, as pointed out by Wigner (1976, 1981), not only do we *not know how* to measure even such simple "observables" as $Q^2P^2+P^2Q^2$ or QP^2Q , but even if we knew, we *could not measure them even in principle*, in the conventional sense of obtaining "sharp" (or, at least, arbitrarily sharp) measurement outcomes – cf. also (Busch, 1985b).

28 The roots of this *belief* can be traced to the fact that von Neumann's theorem on the uniqueness, up to unitary equivalence, of irreducible representations of the canonical commutation relations (cf., e.g., [PQ], Ch. IV, Sec. 6), does not remain valid in the case of infinitely many degrees of freedom (cf., e.g., (Emch, 1972), Ch. 3). As a matter of fact, in the latter case there are infinitely many representations which are not equivalent to Fock representations, and, according to Haag's theorem [BL], a Wightman theory of interacting fields does *not* correspond to any Fock representation of the canonical (anti)-commutation relations. This *mathematical* fact is, however, of no relevance in the present context, which relies on Fock bundles rather than Fock spaces, and in which the role played by the canonical commutation relations in conventional theory is taken over by geometro-stochastic methods.

29 In fact, in this context, even purely quantum field "local observables" (cf., e.g., Horuzhy, 1990) do not fare better in literature than particle "observables". Indeed, the considerations carried out more than two decades ago by the present author (Prugovečki, 1969a), seem to have remained over the intervening years the only published attempt at establishing that some of the elements in $\mathcal{A}$(B) might be indeed related to operationally meaningful concepts, rather than merely an *ad hoc* attachment of the label "observable" to an abstract mathematical construct. In hindsight, their lack of success at the physical level can be understood in view of the later published theorem of Hegerfeldt (1974).

30 One of the first of these assessments by Dirac that appeared in print stated that: "Recent work by Lamb, Schwinger and Feynman and others has been very successful . . . but the resulting theory is an ugly and incomplete one." (Dirac, 1951). Other unfavorable assessments continued to appear in print at regular intervals over the years (Dirac, 1962, 1965, 1973, 1977, 1978). Their essence became crystalized in his very last paper, whose title "The Inadequacies of Quantum Field Theory" reflects his life-long concern with the modern developments in this most important area of quantum theory, which he had founded in 1927 – cf. Note 36 to this chapter, as well as Chapter 12.

31 Unfortunately, the term "Einstein causality" is sometimes used as yet another synonym for the local commutativity of quantum fields in Minkowski space – cf. e.g., (Emch, 1972, p. 278). However, it is not only historically inaccurate to identify local commutativity with Einstein's name, but in view of Einstein's well-known attitude towards quantum theory in general, it represents a disservice to the memory of some of his deepest epistemological convictions (Einstein, 1949).

32 It has been argued by Aharanov and Vardi (1980) that they can "give an operational meaning to an individual 'Feynman path' . . . [by] a process of dense measurements, made in temporal sequence, which check whether the particle moves along any given trajectory in space-time." However, such a determination clearly involves the simultaneous measurement of position and momentum, so that it cannot be carried out with arbitrary precision once the effects of the disturbances created by the measurement act are taken into account. In fact, the subsequently discovered "new interpretation of the scalar product in Hilbert space" (Aharanov *et al.*, 1981) turned out to be "the one underlying the stochastic phase-space approach to nonrelativistic quantum mechanics" (Prugovečki, 1982b) – namely the one in (3.2.4). Similarly, the "surprising quantum effect" (Aharonov *et al.*, 1987), of apparently non-disturbing spin measurements, turned out "to be an instance of the generalized (stochastic, fuzzy, unsharp, imprecise, ...) measurements which were systematically introduced in the operational approach to quantum mechanics" (Busch, 1988).

33 On the other hand, the interpretation of the functional integral in (7.4a) can proceed along the same general lines as in the case of the one in (5.21): the space of all single-exciton wave functions in (7.4b) is introduced, which are square-integrable in the measure $d\Sigma$ over each relativistic phase space slice Σ_t, whose elements are obtained from those at the starting point x' by means of the weakly causal GS propagation described in Secs. 5.6-5.7; the functional integration is then performed over all such *external* modes along each slice Σ_t.

34 The treatment of the φ^4 model in conventional quantum field theory is discussed in detail by Nash (1978) at the mathematically formal level, and by Glimm and Jaffe (1987) within the context of constructive quantum field theory. As mentioned in the preceding section, in the latter case this model turns out to be physically trivial in four spacetime dimensions, despite the apparent nontriviality of its self-interaction term. Thus, according to Glimm and Jaffe (1987), p. 433: "Constructive field theory provides a standard of absolute truth for testing ideas and hypotheses concerning quantum fields. ... Consider $d \geq 5$ space-time dimensions. We show in this case that the continuum ϕ^4 field theory is trivial. The method of proof strongly suggests that the result may also extend to pure scalar ϕ^4 models in $d = 4$." The same triviality verdict is extended even to quantum electrodynamics, which is otherwise well-known for its numerically nontrivial and experimentally well-verified results when it is treated by the usual (i.e., mathematically nonrigorous) Dyson perturbation series methods, rather than by the Glimm-Jaffe methods. In their monograph Glimm and Jaffe support their astounding conclusions with the claim that in quantum field theory "truth is more easily accessible by a mathematical proof then it would be by a laboratory measurement" (*ibid.*, p. 433), and that "twenty years after its inception constructive quantum field theory is on the threshold of achieving its major goals" (*ibid.*, p. v). However, to those who approach the study of Nature with the same kind of humility and reverence that has been in past centuries exemplified by the statements and actions of all great scientists since Newton – and in this century most notably by Einstein and Dirac – the kind of almost "unprecedented level of success"

claimed in the Preface of (Glimm and Jaffe, 1987) seems to be more indicative of the inadequacy of the Wightman axioms underlying the constructive quantum field theory program, and of the mathematical techniques which its chief advocates employ, rather than of a physically intrinsic triviality of various quantum field theoretical interactions, reflecting the self-proclaimed discovery of a "standard of absolute truth" in quantum field theory.

35 In heuristic terms, these are the "clouds of virtual quanta" which, according to conventional quantum field theoretical imagery [SI, IQ], purportedly "dress" all "bare" particles, due to self-interactions. This raises the question as to what might be the *direct* physical meaning of a literal infinity of "virtual", and therefore unobservable particles – since, after all, the vacuum polarization effects to which they purportedly give rise could manifest themselves equally well if only a finite number of "virtual quanta" were present. On this score, it has to be recalled that no rigorous mathematical proof of the presence for such representations exists in conventional relativistic quantum field theory, since no mathematically rigorous model of interacting conventional quantum fields in four spacetime dimensions has *ever* been actually constructed (cf. Note 34).

36 This follows from a very general "principal theorem" by Sz.-Nagy (1960), proven in a separately published appendix to the well-known textbook "Functional Analysis" by F. Riesz and B. Sz.-Nagy. However, in that publication the Germanic spelling "Neumark" is used by Sz.-Nagy in referring to the Russian mathematician M. A. Naimark, who first established the following theorem: Let $\mathcal{B}$ be a family of subsets of a set Ω closed under taking finite numbers of intersections and unions, and incorporating the empty set $\emptyset$ as well as Ω itself, and let a nonnegative self-adjoint operator $F(B) \leq \mathbf{1}$ in a Hilbert space $\mathcal{H}$ be assigned to each $B \in \mathcal{B}$ in such a manner that $F(\emptyset) = \mathbf{0}$, $F(\Omega) = \mathbf{1}$, and $F(B_1 \cup B_2) = F(B_1) + F(B_2)$ whenever $B_1 \cap B_2 = \emptyset$; then there is a spectral measure $E(B)$ on the Boolean σ-algebra generated by $\mathcal{B}$ [PQ], consisting of projectors acting in an extension Hilbert space $\mathbf{H}$ of $\mathcal{H}$, and such that $F(B) = \mathbf{P}E(B)$ for all $B \in \mathcal{B}$, where $\mathbf{P}$ is the projector of $\mathbf{H}$ onto $\mathcal{H}$, and such that $\mathbf{H}$ is spanned by all vectors of the form $E(B)f$, with $B \in \mathcal{B}$ and $f \in \mathcal{H}$. We note that in the present context Ω is the family of all sequences $(q^1,p_1,q^2,p_2,\ldots)$ with $q^\alpha,p_\alpha \in \mathbf{R}^1$, and $\mathcal{B}$ can be obtained by taking finite unions and intersections of those sets in Ω whose elements are sequences whose first $2n$ elements belong, for some positive integer value of n, to one of the Borel sets B_n in (8.9).

37 Indeed, the treatment of interactions in conventional quantum field theory is still presented, even in such recent textbooks as [IQ], in the interaction picture in Fock space (cf., e.g., [IQ], Sec. 4-1). Hence, such treatments rely *exclusively* on Green's functions based on vacuum expectation values of time-ordered quantum field operators, such as the scalar quantum field $\phi(x)$ in (1.7), which (formally) acts within the Fock space in (1.5) (cf., e.g., [IQ], Sec. 6-1). This means that, formally speaking, the conventionally adopted interaction Lagrangian implicitly incorporates the counterpart of the projector in (8.11), which projects objects from outside the Fock space back into Fock space. As a consequence, it can be immediately seen that, when the base space $\mathbf{M}$ equals Minkowski space, and when global Lorentz frames are adopted (so that the identifications in (2.3.12) can be carried out), then, in the formal sharp-point limit $f \to 1$, the interaction Hamiltonian in (8.11) transcends into the conventional one [IQ] upon performing the infinite renormalization required for the removal of the v-integration in (8.11). On account of such formal manipulations, it is easily seen how, in the special-relativistic context, the S-matrix elements of the conventional quantum field theories can be recovered from their GS counterparts.

38 Dirac has described very explicitly his personal views on physics, that eventually led him to his uncompromising opposition to the conventional renormalization programme, in a 1968 lecture at the International Center for Theoretical Physics at Trieste – entitled "Methods in Theoretical Physics", and reprinted in (Salam, 1990), pp. 125-143. The lecture begins with the following statements: "I shall attempt to give you some idea how a theoretical physicist works ... One can distinguish between two main procedures for a theoretical physicist. One of them is to work from an experimental basis. ... The other procedure is to work from the mathematical basis. One examines and criticizes the existing theory. One tries to pin-point the faults in it and then tries to remove them. The difficulty here is to remove the faults without destroying the great successes of the existing theory." Dirac's lecture ends with the following statement: "However, the present quantum electrodynamics does not conform to the high standard of mathematical beauty that one would expect for a fundamental physical theory, and leads one to suspect that a drastic alteration of basic ideas is still needed."

Chapter 8

Relativistic Quantum Geometries for Spin-1/2 Massive Fields

The formulation of the canonical quantization procedure for spin-1/2 quantum fields proceeds in curved spacetime in very much the same manner as for the spin-0 case described in Sec. 7.1, as can be best seen from the comparative summary of the two respective procedures provided by Gibbons (1979). It is based on the curved spacetime counterpart of the Dirac equation, obtained by replacing in (6.1.13) the partial derivatives ∂_μ by the covariant derivatives ∇_μ, and then adopting as an inner product in the single fermion-antifermion space $\mathcal{H}$ the counterpart of (6.1.16) along the spacelike reference hypersurfaces σ_t in a foliation (5.4.7) of a globally hyperbolic spacetime $(\mathbf{M},g)$. As opposed to the situation in the spin-0 case, that inner product is positive definite in the spin-1/2 case. However, that does not help with the difficulties of the canonical approach described in Secs. 7.1-7.3, since an unambiguous decomposition $\mathcal{H} = \mathcal{H}^{(+)} \oplus \mathcal{H}^{(-)}$ into a subspace $\mathcal{H}^{(+)}$ of positive-frequency solutions representing fermion state vectors, and a subspace $\mathcal{H}^{(-)}$ of negative-energy solutions representing antifermion state vectors, does not exist in the spin-1/2 case either, due to the absence of global Poincaré invariance.

This fundamental problem is "solved" in the same *ad hoc* manner as in the spin zero case: it is first *assumed* that $(\mathbf{M},g)$ is stationary, so that it possesses a global timelike Killing vector field generating a one-parameter group of isometries ϕ_t, and then the choice of foliation is arbitrarily *restricted* to the very special case of Cauchy surfaces corresponding to the various constant values of t in synchronous (Gaussian normal) coordinates to which the global timelike Killing field gives rise. If one *further assumes* asymptotic flatness of $(\mathbf{M},g)$ in the "distant past", i.e., as $t \to -\infty$, one can single out an "incoming" fermion subspace $\mathcal{H}^{(+)}{}_{in}$ and an "incoming" antifermion subspace $\mathcal{H}^{(-)}{}_{in}$, constructed from the respective momentum space solutions in (6.1.9b) by retaining (6.1.12) in that limit for a suitable choice of synchronous coordinates for $x \in \sigma_t$. Not surprisingly, if one also assumes asymptotic flatness of $(\mathbf{M},g)$ in the "distant future", i.e., as $t \to +\infty$, and then constructs in a similar *ad hoc* manner "outgoing" subspaces $\mathcal{H}^{(+)}{}_{out}$ and $\mathcal{H}^{(-)}{}_{out}$ of $\mathcal{H}$, it turns out that in general $\mathcal{H}^{(+)}{}_{in} \neq \mathcal{H}^{(+)}{}_{out}$ and $\mathcal{H}^{(-)}{}_{in} \neq \mathcal{H}^{(-)}{}_{out}$, so that the corresponding asymptotic quantum fields are related by Bogoliubov transformations – just as in the spin-0 case discussed in Sec. 7.3. This *mathematical* feature can be then interpreted – exactly as in the spin-zero case – as an indication of *ex nihilo* creation of fermion-antifermion pairs within the *actual* universe in which we live. Thus, formal procedural techniques, with no deeper physical roots, can be again used to arrive at physically implausible "predictions".

The physical, mathematical and epistemological objections of this chain of *ad hoc* procedures remains the same as those presented in Secs. 7.2 and 7.3 in the spin-0 context. Consequently, we shall dispense with any reviews of the conventional quantization procedure for spin-1/2 fields in curved spacetime.

We begin, therefore, right away by describing in Sec. 8.1 the first step in the GS second quantization for spin-1/2 fermions, namely the construction of Fock-Dirac quantum bundles. In Sec. 8.2 we discuss the parallel transport, and the corresponding connection within such bundles, deriving in that context the expressions for the GS stress-energy tensor of fermion fields. In Sec. 8.3 we introduce Z_2-graded standard Berezin-Dirac superfibres over Grassmann algebras of supernumbers, which we then use in Sec. 8.4 in the formulation of path-integral formulae for propagators for the parallel transport of Berezin-Dirac quantum frames within the corresponding Berezin-Dirac superfibre bundles.

8.1. Fock-Dirac Bundles for Spin-1/2 Charged Quantum Fields

In the case of spin-1/2 charged quantum fields, a Fock-Dirac quantum bundle $\mathcal{E}$ can be constructed along the same general lines as in the spin-0 case (Prugovečki, 1987b; Prugovečki and Warlow, 1989), provided allowances are made for the presence of Fermi-Dirac instead of Bose-Einstein statistics. Thus, a standard Fock-Dirac fibre $\mathcal{F}$ is first constructed as follows from the single-fermion and single-antifermion subspaces of the standard Dirac fibre $\mathbf{F}$ in (6.3.1) by means of antisymmetric tensor products[1]:

$$\mathcal{F} = \bigoplus_{m,n=0}^{\infty} \mathcal{F}_{m,n} \quad , \qquad \mathcal{F}_{m,n} = (\mathbf{F}^{(+)} \underset{A}{\otimes} \cdots \underset{A}{\otimes} \mathbf{F}^{(+)}) \otimes (\mathbf{F}^{(-)} \underset{A}{\otimes} \cdots \underset{A}{\otimes} \mathbf{F}^{(-)}) \, . \tag{1.1}$$

Then, for a given Lorentzian base manifold $\mathbf{M}$, we can set $\mathcal{E}$ equal to its bundle product with the principal frame bundle $\boldsymbol{SFM}$ of affine spin frames in (6.3.10):

$$\mathcal{E} = \boldsymbol{SFM} \times_{\mathbf{G}} \mathcal{F} \quad , \qquad \mathbf{G} = \mathrm{ISL}(2,\mathbf{C}) \, . \tag{1.2}$$

Alternatively, we could have defined the Fock-Dirac bundle $\mathcal{E}$ as equal to the Whitney direct sum of appropriately antisymmetrized Whitney direct products of duplicates of the fermion and antifermion subbundles of the Dirac quantum bundle $\boldsymbol{DM}$ in (6.4.1):

$$\mathcal{E} = \bigoplus_{m,n=0}^{\infty} (\boldsymbol{DM}^{(+)} \underset{A}{\hat{\otimes}} \cdots \underset{A}{\hat{\otimes}} \boldsymbol{DM}^{(+)}) \hat{\otimes} (\boldsymbol{DM}^{(-)} \underset{A}{\hat{\otimes}} \cdots \underset{A}{\hat{\otimes}} \boldsymbol{DM}^{(-)}) \, . \tag{1.3}$$

As in the spin-0 case, the above construction implicitly assigns a normalized local vacuum state vector $\boldsymbol{\Psi}_{0;x}$ to each base location $x \in \mathbf{M}$. This vector spans the (complex) one-dimensional vacuum sector $\mathcal{F}_{0;x}$ of the Fock fibre $\mathcal{F}_x$ above that base location.

The generalized soldering map in (6.4.2) can be extended in a natural manner to the Fock-Dirac fibre $\mathcal{F}_x$ as follows: in the m-exciton and n-antiexciton sector $\mathcal{F}_{m,n;x}$ we assign to each state vector $\boldsymbol{\Psi}_{m,n;x}$ a coordinate wave function $\Psi_{m,n} \in \mathcal{F}_{m,n}$, with respect to each Dirac quantum frame defined in (6.3.2), in such a manner that

$$\sigma_x^u : \boldsymbol{\Psi}_{m,n;x}(\zeta_1,\ldots,\zeta_m;\zeta_1',\ldots,\zeta_n') \mapsto \Psi_{m,n}(\zeta_1,\ldots,\zeta_m;\zeta_1',\ldots,\zeta_n') \, . \tag{1.4}$$

By the extension principle for bounded linear operators (cf. [PQ], p. 188), we can then extend this generalized soldering map to the entire Fock-Dirac fibre $\mathcal{F}_x$:

$$\sigma_x^u : \boldsymbol{\Psi}_x \mapsto \Psi \in \mathcal{F} \ , \qquad \boldsymbol{\Psi}_x \in \mathcal{F}_x, \qquad \boldsymbol{u} = (\underset{\sim}{a}, \xi_A) \in \Pi^{-1}(x) \subset \boldsymbol{SFM} \ . \tag{1.5}$$

The local quantum fluctuation amplitudes for spin-1/2 (fermion)-excitons are the counterparts of the ones in (7.4.4) for spin-0 (boson)-excitons, and are given by (cf. (6.3.7))

$$S_x^{(+)}(\zeta;\zeta') = -(1/2m)(m + i\gamma^j \partial/\partial q^j)\Delta_x^{(+)}(\zeta;\zeta') \ . \tag{1.6}$$

In terms of them we can define spin-1/2 fermionic exciton creation operators by

$$\begin{aligned}\left(\tilde{\psi}^{(+)}(x;\zeta,r)\boldsymbol{\Psi}_{m,n;x}\right)_{m+1,n}&(\zeta_1,r_1,\ldots,\zeta_{m+1},r_{m+1};\ldots) = i(m+1)^{-1/2}\\ &\times\sum_{j=1}^{m+1}(-1)^j S_x^{(+)}(\zeta_j,r_j;\zeta,r)\,\boldsymbol{\Psi}_{m,n;x}(\zeta_1,r_1,\ldots,\hat{\zeta}_j,\hat{r}_j,\ldots,\zeta_{m+1},r_{m+1};\ldots) \ ,\end{aligned} \tag{1.7a}$$

where the hats signify that the variables under them are to be omitted, and the r-variables refer to matrix components. On the other hand, for spin-1/2 antifermion creation operators we have

$$\begin{aligned}\left(\tilde{\psi}^{(+)}(x;-\zeta,r)\boldsymbol{\Psi}_{m,n;x}\right)_{m,n+1}&(\ldots;\zeta_1,r_1,\ldots,\zeta_{n+1},r_{n+1}) = i(n+1)^{-1/2}\\ &\times\sum_{j=1}^{n+1}(-1)^{j+m} S_x^{(+)}(\zeta_j,r_j;\zeta,r)\,\boldsymbol{\Psi}_{m,n;x}(\ldots;\zeta_1,r_1,\ldots,\hat{\zeta}_j,\hat{r}_j,\ldots,\zeta_{n+1},r_{n+1}) \ .\end{aligned} \tag{1.7b}$$

In either case, the fermion annihilation operators can be defined by taking adjoints, so that in view of the reproducibility properties in (6.3.6) we have:

$$\begin{aligned}\left(\psi^{(-)}(x;\zeta,r)\boldsymbol{\Psi}_{m,n;x}\right)_{m-1,n}&(\zeta_1,r_1,\ldots,\zeta_{m-1},r_{m-1};\ldots)\\ &= m^{1/2}\,\boldsymbol{\Psi}_{m,n;x}(\zeta,r,\zeta_1,r_1,\ldots,\zeta_{m-1},r_{m-1};\ldots) \ ,\end{aligned} \tag{1.8a}$$

$$\begin{aligned}\left(\psi^{(-)}(x;-\zeta,r)\boldsymbol{\Psi}_{m,n;x}\right)_{m,n-1}&(\ldots;\zeta_1,r_1,\ldots,\zeta_{n-1},r_{n-1})\\ &= n^{1/2}(-1)^m\,\boldsymbol{\Psi}_{m,n;x}(\ldots;-\zeta,r,\zeta_1,r_1,\ldots,\zeta_{n-1},r_{n-1}) \ .\end{aligned} \tag{1.8b}$$

The following anticommutation relations are then easily derived:

$$\{\psi^{(-)}(x;\pm\zeta,r),\,\tilde{\psi}^{(+)}(x;\pm\zeta',r')\} = -iS_x^{(\pm)}(\zeta,r;\zeta',r') \ , \tag{1.9a}$$

$$\{\psi^{(-)}(x;\pm\zeta,r),\,\psi^{(+)}(x;\mp\zeta',r')\} = \{\psi^{(\pm)}(x;\zeta,r),\,\psi^{(\pm)}(x;\zeta',r')\} = 0 \ . \tag{1.9b}$$

The *Dirac quantum frame field* can be now defined as the one-column matrices

$$\psi(x;\boldsymbol{q},\boldsymbol{v}) = \psi^{(-)}(x;\boldsymbol{q},\boldsymbol{v}) + \psi^{(+)}(x;\boldsymbol{q},-\boldsymbol{v}) \ , \tag{1.10a}$$

$$\psi^{(+)}(x;\boldsymbol{\zeta}) = \tilde{\psi}^{(+)}(x;\boldsymbol{\zeta})\gamma^0 \ , \qquad \boldsymbol{\zeta} = (\boldsymbol{q},\boldsymbol{v}) \in T_x\mathbf{M} \times \boldsymbol{V}_x^+ \ , \tag{1.10b}$$

of superpositions of creation and annihilation operators. They are ISL(2,C)-gauge covariant objects, i.e., with the change of choice of section of the affine spin frame bundle in (6.3.10), such as the one in (6.4.4), they transform as follows

$$\psi(x;\zeta,r) \ \mapsto \ \psi'(x;\zeta,r') = \sum_{r=1}^{4} S_{r'r}(A(x))\psi(x;\zeta,r) \ , \tag{1.11}$$

where the matrix $S(A)$ is related to A as in (6.1.10b). Their components are bounded operators, which as such are defined in the Fock-Dirac fibre $\mathcal{F}_x$ – as opposed to their spin-0 counterparts in (7.4.12), which are only densely defined. On account of (1.6)-(1.8), it is immediately seen that the quantum frame field above each $x \in \mathbf{M}$ satisfies the Dirac equation

$$(i\gamma^j \partial/\partial q^j - m)\psi(x;\zeta) = 0 \ , \qquad \zeta = (\boldsymbol{a}+q^j\boldsymbol{e}_j, v^j\boldsymbol{e}_j) \ , \tag{1.12}$$

as it acts within the Fock-Dirac fibre $\mathcal{F}_x$.

8.2. Parallel Transport and Stress-Energy Tensors in Fock-Dirac Bundles

The parallel transport within any Fock-Dirac bundle $\mathcal{E}$ can be derived from the parallel transport operators in (6.4.8) that act in the Dirac quantum bundle $\boldsymbol{D}\mathbf{M}$. Thus, according to (1.3), $\mathcal{E}$ is constructed by taking Whitney sums of antisymmetrized Whitney products of $\boldsymbol{D}\mathbf{M}$, so that the operator for parallel transport in (6.4.8) gives rise to a unitary operator

$$\tau_\gamma(x'',x') : \quad \mathcal{F}_{x'} \ \to \ \mathcal{F}_{x''} \ , \tag{2.1}$$

acting between the Fock-Dirac fibres above the end points x' and x'' of a given piecewise smooth curve γ, which lies within the base manifold $\mathbf{M}$. As usual, covariant derivatives can be then defined in accordance with (5.2.16). For a given section $\boldsymbol{s}$ of $\boldsymbol{SF}\mathbf{M}$ in (1.2), they give rise to the covariant differentiation operator (cf. Sec. 2.4)

$$\boldsymbol{\nabla} = \boldsymbol{d} - i\tilde{\boldsymbol{\theta}}^i \boldsymbol{P}_{i;\boldsymbol{u}} + \tfrac{i}{2}\tilde{\boldsymbol{\omega}}_{jk}\boldsymbol{M}_{\boldsymbol{u}}^{jk} \ , \qquad \boldsymbol{d} = \boldsymbol{\theta}^i\partial_i \ , \qquad \partial_i := \partial_{\boldsymbol{e}_i} \ , \tag{2.2}$$

where the connection coefficients are the same as in (6.4.11a), whereas, the infinitesimal generators are those of the representation

$$\mathbf{U}_{x;\boldsymbol{u}}(b,A) = \bigoplus_{m,n=0}^{\infty} \left(\mathbf{U}_{x;\boldsymbol{u}}^{(+)}(b,A)^{\otimes m} \otimes \mathbf{U}_{x;\boldsymbol{u}}^{(-)}(b,A)^{\otimes n}\right) \ , \tag{2.3}$$

induced in $\mathcal{F}_x$ by the representations in (6.4.4b) – cf. also (7.5.3).

We shall now relate these generators to an operator-valued stress-energy tensor acting within the Fock-Dirac fibre $\mathcal{F}_x$. The procedure is in most respects quite similar to the one in Sec. 7.5 for the spin-0 case, despite the presence of Fermi-Dirac statistics. Thus, the generators of spacetime translations with respect to the Poincaré frame associated with the affine spin frame $\boldsymbol{u}$, which correspond to the representation in (2.3), are given by

$$\boldsymbol{P}_{j;\boldsymbol{u}} = \tfrac{i}{2}\int \tilde{\psi}^{(+)}(x;\zeta)\overleftrightarrow{\partial}_j \psi^{(-)}(x;\zeta)\,d\Sigma(\zeta)\ , \qquad \overleftrightarrow{\partial}_j = \overleftrightarrow{\partial}/\partial q^j . \tag{2.4}$$

This can be easily verified by taking matrix elements of the above operator in between any two states with m fermions and n antifermions, and using (1.6)-(1.8) in conjunction with the expression of the type (6.2.8) for the inner product in the Fock-Dirac fibre $\mathcal{F}_x$. In this context it should be observed that the operator in between round brackets in (6.2.8) has the value $\pm\mathbf{1}$ for fermion and antifermion states, respectively. Upon making the transition to the manifestly covariant form of the inner product corresponding to (6.2.9), we can recast (2.4) into the form

$$\boldsymbol{P}_{j;\boldsymbol{u}} = im\hat{Z}_{f,m}\int \tilde{\psi}^{(+)}(x;\zeta)\gamma_k\overleftrightarrow{\partial}_j\psi^{(-)}(x;\zeta)\,d\sigma^k(\boldsymbol{q})\,d\Omega(\boldsymbol{v})\ . \tag{2.5}$$

Taking into account that

$$\int \tilde{\psi}^{(\pm)}(x;\zeta)\overleftrightarrow{\partial}_k\,\psi^{(\pm)}(x;\zeta)\,d\sigma^k(\boldsymbol{q})\,d\Omega(\boldsymbol{v}) = 0\ , \tag{2.6}$$

and using normal ordering, we can rewrite (2.5) in the manifestly covariant form

$$\boldsymbol{P}_{j;\boldsymbol{u}} = \int :T_{jk}[\psi(x;\zeta)]:\, d\sigma^k(\boldsymbol{q})\,d\Omega(\boldsymbol{v})\ , \tag{2.7}$$

which involves a renormalized stress-energy tensor

$$T_{jk}[\psi] = im\hat{Z}_{f,m}\left(\tilde{\psi}\gamma_k\psi_{,j} - \tilde{\psi}_{,j}\gamma_k\psi\right)\ . \tag{2.8}$$

On account of (6.2.7) we can derive, in a manner analogous to that used in the derivation of (2.4), that the generators of infinitesimal Lorentz transformations are given by

$$\begin{aligned}\boldsymbol{M}_{\boldsymbol{u}}^{ij} &= i\int \tilde{\psi}^{(+)}(x;\zeta)\left(Q^i_{\boldsymbol{u}}\partial^j - Q^j_{\boldsymbol{u}}\partial^i\right)\psi^{(-)}(x;\zeta)\,d\Sigma(\zeta) + \boldsymbol{S}_{\boldsymbol{u}}^{ij}\\ &= -i\int\left[\left(Q^i_{\boldsymbol{u}}\partial^j - Q^j_{\boldsymbol{u}}\partial^i\right)\tilde{\psi}^{(+)}(x;\zeta)\right]\psi^{(-)}(x;\zeta)\,d\Sigma(\zeta) + \boldsymbol{S}_{\boldsymbol{u}}^{ij}\ ,\end{aligned} \tag{2.9a}$$

$$\boldsymbol{S}_{\boldsymbol{u}}^{ij} = \tfrac{i}{4}\int \tilde{\psi}^{(+)}(x;\zeta)\left[\gamma^i,\gamma^j\right]\psi^{(-)}(x;\zeta)\,d\Sigma(\zeta)\ . \tag{2.9b}$$

After carrying out deductive steps which are totally analogous to those leading to (2.5) and (2.6), we arrive at the following manifestly covariant form for these generators,

$$M_u^{ij} = \int :M_u^{ijk}[\psi(x;\zeta)] : d\sigma_k(q)d\Omega(v) \quad , \tag{2.10}$$

which contains *bona fide* relativistic angular momentum tensor operators:

$$M_u^{kij}[\psi] = im\hat{Z}_{f,m}\left(\tfrac{1}{2}[\tilde{\psi}(Q_u^i \overleftrightarrow{\partial^j} - Q_u^j \overleftrightarrow{\partial^i})\gamma^k\psi] + \tfrac{1}{8}\tilde{\psi}\{[\gamma^i,\gamma^j],\gamma^k\}\psi\right) \quad . \tag{2.11}$$

As it was the case for the boson fields treated in Sec. 7.5, the operator-valued stress-energy tensor in (2.8), as well as the operator-valued angular momentum tensor in (2.11), are perfectly well defined as they stand – in contrast to their counterparts in conventional quantum field theory [SI]. This will enable us to obtain in Sec. 8.4 well-defined action integrals for the propagator for the parallel transport based on (2.1). However, before such a derivation is possible, we require second quantized fermionic frames similar to those in (7.4.16) for the bosonic case. It is at this stage that one of the fundamental mathematical differences between Bose-Einstein and Fermi-Dirac statistics emerges at the technical level.

8.3. Second-Quantized Frames in Berezin-Dirac Superfibre Bundles

The fact that the excitons and antiexcitons produced by the creation operators in (1.7) obey Fermi-Dirac statistics, giving rise to the anticommutation relations in (1.9), implies that immediate counterparts of the states in (7.4.16) do not provide the continuous resolutions of the identity in the Fock-Dirac fibre $\mathcal{F}_x$, which are required in order to implement the path integral methods introduced in Chapter 7. There are, however, various methods for defining coherent states for fermions, of which the one developed by Berezin [B,BI] is the best known, as well as the best suited to our purposes[2].

The adaptation of these techniques to the present situation can be achieved by first extending the typical Fock-Dirac fibre $\mathcal{F}$ into a typical *Berezin-Dirac superfibre* $\mathcal{B}$, and then constructing a Berezin-Dirac superfibre bundle which represents an extension[3] of the Fock-Dirac fibre bundle $\mathcal{E}$. The typical superfibre $\mathcal{B}$ of this Berezin-Dirac superfibre bundle can be defined as a Hilbert superspace over an infinite-dimensional Grassmann algebra Λ with involution[4]. Generically, such a Grassmann algebra is a topological vector space over the field $\mathbf{C}^1$ of complex numbers, within which an operation of algebraic multiplication is defined in such a manner that its general element, to which we shall refer as a *supernumber*[5], can be written in the form (cf. [B], p. 60; or [D], p. 1; or [BI], p. 36)

$$\xi = c_0 + \sum_{n=1}^{\infty} \frac{1}{n!} c_{\alpha_1 \dots \alpha_n} \theta^{\alpha_1} \cdots \theta^{\alpha_n} \quad , \qquad c_0, \dots, c_{\alpha_1 \dots \alpha_n}, \dots \in \mathbf{C}^1 \quad , \tag{3.1}$$

where the above complex coefficients are antisymmetric under the permutation of any pair of their indices, and $\{\theta_\alpha \mid \alpha = 1,2,\dots\}$ represents a family of generators of Λ. As a consequence of the basic definition of the involution operation within a Grassmann algebra (cf. [B], pp. 66-67), under any given involution all the above coefficients will be mapped into their complex conjugates. Furthermore, if a bar above a supernumber is used to denote its conjugate, i.e., its image under that involution, then we shall have

$$\{\theta^\alpha,\theta^\beta\} = \{\overline{\theta}^\alpha,\overline{\theta}^\beta\} = \{\theta^\alpha,\overline{\theta}^\beta\} \ , \qquad \alpha,\beta = 1,2,3,\dots \ , \tag{3.2}$$

where the brace brackets in (3.2) stand for anticommutators.

Any supernumber can be obviously decomposed into its *even* or *grade-0* component, obtained by replacing with zero all the coefficients in (3.1) corresponding to odd values of n, and an *odd* or *grade-1* component, obtained by replacing with zero all the coefficients in (3.1) corresponding to even values of n, including the one with $n = 0$. In turn, this gives rise to the decomposition of Λ into a commutative subalgebra Λ^0 consisting of all even supernumbers, and the vector subspace Λ^1 of all odd supernumbers, which does not constitute a subalgebra, since the product of two odd supernumbers is always an even supernumber. Thus, a Grassmann algebra is a Z_2-graded vector space.

Superanalysis [BI] and functional superanalysis (Khrennikov, 1988) became well-developed in the eighties. However, for our needs we shall require only the formal concept of Berezin integration [B,BI], based on the following well-known basic definitions [B,IQ]:

$$\int d\theta^\alpha = \int d\overline{\theta}^\alpha = 0 \ , \qquad \int \theta^\alpha d\theta^\alpha = \int \overline{\theta}^\alpha d\overline{\theta}^\alpha = 1 \ , \qquad \alpha = 1,2,3,\dots \ . \tag{3.3}$$

These rules can be extended to multiple integrals of arbitrary supernumber-valued functions by linearity and by iteration, in accordance with the rules (cf. [B], p. 52)

$$\{d\theta^\alpha,d\theta^\beta\} = \{\theta^\alpha,d\theta^\beta\} = \{\overline{\theta}^\alpha,d\theta^\beta\} = 0 \ , \qquad \alpha,\beta = 1,2,3,\dots \ , \tag{3.4}$$

as well as the rules obtained by taking the involutes of the above relations.

The typical Berezin-Dirac superfibre $\mathcal{B}$ can be now constructed by first extending the standard Dirac fibre into a Hilbert superspace[6] $\mathbf{B}$, with inner product that has the property

$$\langle \xi_1 f \,|\, \xi_2 g \xi_3 \rangle = \overline{\xi}_1 \langle f \overline{\xi}_2 | g \rangle \xi_3 \ , \qquad f,g \in \mathbf{B} \ , \qquad \xi_1,\xi_2,\xi_3 \in \Lambda \ . \tag{3.5}$$

Then, upon constructing $\mathcal{B}$ from $\mathbf{B}$ in accordance with (1.1), we easily deduce that

$$[\xi, \psi^{(\pm)}(x;\zeta)] = 0 \ , \qquad \xi \in \Lambda^0 \ , \qquad \zeta \in \mathbf{R}^4 \times \boldsymbol{V} \ , \tag{3.6a}$$

$$\{\xi, \psi^{(\pm)}(x;\zeta)\} = 0 \ , \qquad \xi \in \Lambda^1 \ , \qquad \zeta \in \mathbf{R}^4 \times \boldsymbol{V} \ , \tag{3.6b}$$

holds for the extensions to $\mathcal{B}$ of the fermion and antifermion creation or annihilation operators in the typical Fock-Dirac fibre $\mathcal{F}$, if we allow all supernumbers to commute with the Fock-Dirac vacuum state vector $\Psi_{0,0}$. We obtain the desired Berezin-Dirac superfibre bundle by extending in a similar manner each fibre $\mathbf{F}_x$ into a corresponding superfibre $\mathbf{B}_x$, and then constructing in the above manner a Berezin-Dirac superfibre $\mathcal{B}_x$ out of $\mathbf{B}_x$.

Let us now choose within the standard Dirac fibres $\mathbf{F}^{(\pm)}$ orthonormal bases, so that by combining them we can decompose the identity operator $\mathbf{1_F}$ in $\mathbf{F}=\mathbf{F}^{(+)}\oplus\mathbf{F}^{(-)}$ as follows:

$$\sum_{\alpha=1}^{\infty} |w_\alpha\rangle\langle w_\alpha| \ = \ \mathbf{1_F} \ , \qquad \langle w_\alpha | w_\beta \rangle = \delta_{\alpha\beta} \ , \qquad \alpha,\beta = 1,2,\dots \ . \tag{3.7}$$

The inverse of the generalized soldering map in (1.4)-(1.5) maps this basis into a corresponding orthonormal basis within $\mathbf{F}_x = \mathcal{F}_{1,1;x} = \mathcal{F}_{1,0;x} \oplus \mathcal{F}_{0,1;x}$:

$$(\sigma_x^{\boldsymbol{u}})^{-1}: \ w_\alpha \mapsto \boldsymbol{w}_\alpha \in \mathbf{F}_x \ , \qquad \alpha = 1,2,\ldots \ , \qquad \boldsymbol{u} \in \boldsymbol{\Pi}^{-1}(x) \subset \boldsymbol{SFM} \ . \tag{3.8}$$

If we now set, in accordance with (6.2.9) and (6.2.10),

$$\psi^{(+)}(x;\boldsymbol{f}) = \int \tilde{\psi}^{(+)}(x;\boldsymbol{\zeta})\gamma_k \boldsymbol{f}(\boldsymbol{\zeta})d\sigma^k(\boldsymbol{q})d\hat{\Omega}(\boldsymbol{v}) \ , \qquad \boldsymbol{f} \in \mathbf{F}_x \ , \tag{3.9a}$$

$$\psi^{(-)}(x;\boldsymbol{f}) = \int \bar{\boldsymbol{f}}(\boldsymbol{\zeta})\gamma_k \psi^{(-)}(x;\boldsymbol{\zeta})d\sigma^k(\boldsymbol{q})d\hat{\Omega}(\boldsymbol{v}) \ , \qquad \bar{\boldsymbol{f}} = \boldsymbol{f}\gamma^0 \ , \tag{3.9b}$$

then we can deduce by using (1.9) that

$$\begin{aligned}\{\psi^{(-)}(x;\boldsymbol{w}_\alpha), \psi^{(+)}(x;\boldsymbol{w}_\beta)\} = -i\int \overline{\boldsymbol{w}}_\alpha(\boldsymbol{\zeta})\gamma_j[S_x^{(+)}(\boldsymbol{\zeta};\boldsymbol{\zeta}') + S_x^{(-)}(\boldsymbol{\zeta};\boldsymbol{\zeta}')] \\ \times \gamma_k \boldsymbol{w}_\beta(\boldsymbol{\zeta}')d\sigma^j(\boldsymbol{q})d\hat{\Omega}(\boldsymbol{v})d\sigma^k(\boldsymbol{q}')d\hat{\Omega}(\boldsymbol{v}') \ .\end{aligned} \tag{3.10}$$

Hence, in view of (6.3.5) and (6.3.6), we conclude that

$$\{\psi^{(-)}(x;\boldsymbol{w}_\alpha), \psi^{(+)}(x;\boldsymbol{w}_\beta)\} = \langle \boldsymbol{w}_\alpha | \boldsymbol{w}_\beta \rangle = \delta_{\alpha\beta} \ , \qquad \alpha,\beta = 1,2,\ldots \ . \tag{3.11a}$$

Furthermore, (1.9b) also implies that

$$\{\psi^{(+)}(x;\boldsymbol{w}_\alpha), \psi^{(+)}(x;\boldsymbol{w}_\beta)\} = \{\psi^{(-)}(x;\boldsymbol{w}_\alpha), \psi^{(-)}(x;\boldsymbol{w}_\beta)\} = 0 \ , \tag{3.11b}$$

so that a representation of the canonical anticommutation relations is thus obtained in each one of the Fock-Dirac fibres $\mathcal{F}_x$.

This means that we can now introduce in the corresponding Berezin-Dirac fibres $\mathcal{B}_x$ the following Berezin coherent states [B],

$$\boldsymbol{\Phi}_{\boldsymbol{f}} = \exp\left(-\tfrac{1}{2}\sum_{\alpha=1}^{\infty}\bar{\theta}_\alpha\theta_\alpha\right)\exp\left(\sum_{\alpha=1}^{\infty}\langle \boldsymbol{w}_\alpha | \boldsymbol{f}\rangle \psi^{(+)}(x;\boldsymbol{w}_\alpha)\theta_\alpha\right)\boldsymbol{\Psi}_{0,0;x} \ , \tag{3.12}$$

associated with the Grassmann algebra generators adopted in (3.2). In formal appearance these coherent states resemble the Glauber coherent states in (7.4.30). Hence, let us associate with each $\boldsymbol{f} \in \mathbf{F}_x$ the following supernumber-valued functions that represent elements within the superfibre $\mathbf{B}_x$, and its dual, respectively:

$$\xi(\boldsymbol{\zeta},\boldsymbol{\theta}) = \sum_{\alpha=1}^{\infty}\theta_\alpha\langle \boldsymbol{w}_\alpha | \boldsymbol{f}\rangle \boldsymbol{w}_\alpha(\boldsymbol{\zeta}) \ , \qquad \boldsymbol{\theta} = (\theta_1,\theta_2,\ldots) \ , \tag{3.13a}$$

$$\bar{\xi}(\boldsymbol{\zeta},\bar{\boldsymbol{\theta}}) = \sum_{\alpha=1}^{\infty}\bar{\theta}_\alpha\langle \boldsymbol{f} | \boldsymbol{w}_\alpha\rangle \overline{\boldsymbol{w}}_\alpha(\boldsymbol{\zeta}) \ , \qquad \bar{\boldsymbol{\theta}} = (\bar{\theta}_1,\bar{\theta}_2,\ldots) \ . \tag{3.13b}$$

We can then introduce the following creation operators acting in the Berezin-Dirac superfibre $\mathcal{B}_x$, and its dual, respectively,

$$\boldsymbol{\psi}^{(+)}(x;\xi) = \int \tilde{\psi}^{(+)}(x;\zeta)\gamma_k \xi(\zeta,\boldsymbol{\theta}) d\sigma^k(\boldsymbol{q}) d\hat{\Omega}(\boldsymbol{v}) , \qquad \xi \in \mathbf{B}_x , \tag{3.14a}$$

$$\boldsymbol{\psi}^{(-)}(x;\bar{\xi}) = \int \bar{\xi}(\zeta,\bar{\boldsymbol{\theta}})\gamma_k \psi^{(-)}(x;\zeta) d\sigma^k(\boldsymbol{q}) d\hat{\Omega}(\boldsymbol{v}) , \qquad \bar{\xi} \in \mathbf{B}_x^\dagger . \tag{3.14b}$$

In turn, we can use these operators to define supernumber-valued *Berezin-Dirac quantum superframes*, and their duals, in a manner formally analogous to the Klein-Gordon second quantized frames in (7.4.16a):

$$\left|\boldsymbol{\Phi}_\xi\right\rangle = \exp\left(-\tfrac{1}{2}\int \bar{\xi}(\zeta,\bar{\boldsymbol{\theta}})\gamma_k \xi(\zeta,\boldsymbol{\theta}) d\sigma^k(\boldsymbol{q}) d\hat{\Omega}(\boldsymbol{v})\right)\exp\left(\boldsymbol{\psi}^{(+)}(x;\xi)\right)\left|\Psi_{0,0;x}\right\rangle , \tag{3.15a}$$

$$\left\langle\boldsymbol{\Phi}_\xi\right| = \left\langle\Psi_{0,0;x}\right|\exp\left(\boldsymbol{\psi}^{(-)}(x;\xi)\right)\exp\left(-\tfrac{1}{2}\int \bar{\xi}(\zeta,\bar{\boldsymbol{\theta}})\gamma_k \xi(\zeta,\boldsymbol{\theta}) d\sigma^k(\boldsymbol{q}) d\hat{\Omega}(\boldsymbol{v})\right) . \tag{3.15b}$$

We observe that the elements of the Berezin-Dirac quantum superframe specified in (3.15a) can be set in one-to-one correspondence with the Berezin coherent states in (3.12),

$$\boldsymbol{\Phi}_f \leftrightarrow \left|\boldsymbol{\Phi}_\xi\right\rangle , \qquad f \in \mathbf{F}_x \cong \hat{\mathbf{F}}_x , \quad f \leftrightarrow \xi \in \hat{\mathbf{F}}_x \subset \mathbf{B}_x , \tag{3.16}$$

on account of the one-to-one map between the single fermion or antifermion state vectors $\boldsymbol{f} \in \mathbf{F}_x$ and those $\boldsymbol{\xi} \in \mathbf{B}_x$ represented by the supernumber-valued functions in (3.13a). Hence, we can use the Berezin method of integration to set, in formal analogy with (7.4.31),

$$\int_{\hat{\mathbf{F}}_x} \left|\boldsymbol{\Phi}_\xi\right\rangle d\bar{\xi}\, d\xi \left\langle\boldsymbol{\Phi}_\xi\right| \;=\; \text{w-}\lim_{n\to\infty} \int \left|\boldsymbol{\Phi}_{\xi_n}\right\rangle d\hat{\bar{\boldsymbol{\theta}}}_n\, d\hat{\boldsymbol{\theta}}_n \left\langle\boldsymbol{\Phi}_{\xi_n}\right| , \tag{3.17a}$$

$$\xi_n(\zeta,\boldsymbol{\theta}) = \sum_{\alpha=1}^{n} \theta_\alpha \langle \boldsymbol{w}_\alpha | \boldsymbol{f}\rangle \boldsymbol{w}_\alpha(\zeta) , \qquad d\hat{\bar{\boldsymbol{\theta}}}_n\, d\hat{\boldsymbol{\theta}}_n = \prod_{\alpha=n}^{1} d\bar{\theta}_\alpha\, d\theta_\alpha , \tag{3.17b}$$

where, due to the one-to-one map in (3.16), the integration can be viewed as taking place over the Grasmannian image of $\mathbf{F}_x$. According to a basic theorem of Berezin (cf. [B], p. 83), the Berezin functional integral in (3.17) then supplies a continuous resolution of the identity operator in the Fock-Dirac fibre $\mathcal{F}_x$:

$$\int_{\hat{\mathbf{F}}_x} \left|\boldsymbol{\Phi}_\xi\right\rangle d\bar{\xi}\, d\xi \left\langle\boldsymbol{\Phi}_\xi\right| \;=\; \mathbf{1}_{\mathcal{F}_x} . \tag{3.18}$$

Furthermore, by using (3.2), it is easily verified that the constituents of the Berezin-Dirac quantum superframes in (3.15a) are eigenvectors of the natural extension to $\mathcal{B}_x$ of the annihilation operators in (1.8):

$$\psi^{(-)}(x;\zeta)\left|\boldsymbol{\Phi}_\xi\right\rangle = \xi(\zeta,\boldsymbol{\theta})\left|\boldsymbol{\Phi}_\xi\right\rangle , \qquad \xi \in \hat{\mathbf{F}}_x \subset \mathbf{B}_x . \tag{3.19}$$

We can also extend in a natural manner the generalized soldering map in (1.4)-(1.5) to the fibres of the Berezin-Dirac superfibre bundle[7]. This is most easily achieved by using the map in (3.8) to introduce the counterparts

$$\xi(\zeta,\boldsymbol{\theta}) = \sum_{\alpha=1}^{\infty} \theta_\alpha \langle w_\alpha | f \rangle w_\alpha(\zeta) \ , \qquad \bar{\xi}(\zeta,\bar{\boldsymbol{\theta}}) = \sum_{\alpha=1}^{\infty} \bar{\theta}_\alpha \langle f | w_\alpha \rangle \bar{w}_\alpha(\zeta) \ , \tag{3.20}$$

of the supernumber-valued functions in (3.13), and then proceeding as in (3.14)-(3.18), so as to arrive at the following relation and definitions, respectively:

$$\int_{\hat{\mathbf{F}}} |\boldsymbol{\Phi}_\xi\rangle \, d\bar{\xi} \, d\xi \langle \boldsymbol{\Phi}_\xi| = \mathbf{1}_{\mathcal{F}} \ , \tag{3.21a}$$

$$|\boldsymbol{\Phi}_\xi\rangle = \exp\left(-\tfrac{1}{2}\int \bar{\xi}(\zeta,\bar{\boldsymbol{\theta}})\gamma_k \xi(\zeta,\boldsymbol{\theta}) d\sigma^k(q) d\hat{\Omega}(v)\right) \exp\left(\boldsymbol{\psi}^{(+)}(x;\xi)\right) |\Psi_{0,0}\rangle \ , \tag{3.21b}$$

$$\boldsymbol{\psi}^{(+)}(x;\xi) = \int \bar{\psi}^{(+)}(x;\zeta)\gamma_k \xi(\zeta,\boldsymbol{\theta}) d\sigma^k(q) d\hat{\Omega}(v) \ , \qquad \xi \in \mathbf{B} \ . \tag{3.21c}$$

We can subsequently incorporate the Berezin-Dirac quantum superframes into the domain of the generalized soldering map in (1.5) by setting

$$\boldsymbol{\sigma}_x^{\boldsymbol{u}} : \ \boldsymbol{\Phi}_\xi \mapsto \Phi_\xi \in \mathcal{B} \ , \qquad \boldsymbol{\Phi}_\xi \in \mathcal{B}_x \ . \tag{3.22a}$$

On account of (3.18) and (3.21b), we then arrive by linearity at the desired extension of the generalized soldering map to all of $\mathcal{B}_x$:

$$\boldsymbol{\sigma}_x^{\boldsymbol{u}} : \ \boldsymbol{\Psi} \mapsto \int_{\hat{\mathbf{F}}} d\bar{\xi} \, d\xi \, \Phi_\xi \langle \boldsymbol{\Phi}_\xi | \boldsymbol{\Psi} \rangle \in \mathcal{B} \ , \qquad \boldsymbol{\Psi} \in \mathcal{B}_x \ . \tag{3.22b}$$

This construction results in a Berezin-Fock bundle which is associated to the Dirac quantum frame bundle ***DFM*** in (6.3.12), or its classical counterpart, namely the affine spin frame bundle ***SFM*** in (6.3.10), so that, from either point of view, ISL(2,C) can be regarded as its structure group. Indeed, under a change of section of ***SFM***, giving rise to the change of coordinate wave functions in (6.4.4a), we shall have a corresponding change of supernumber-valued functions (i.e., "classical" supernumber fields) in (3.20),

$$\boldsymbol{\sigma}_x^{\boldsymbol{u}'} \circ (\boldsymbol{\sigma}_x^{\boldsymbol{u}})^{-1} : \ \xi(\zeta,\boldsymbol{\theta}) \mapsto \xi'(\zeta,\boldsymbol{\theta}) = \sum_{\alpha=1}^{\infty} \theta_\alpha \langle w_\alpha | f \rangle w'_\alpha(\zeta) \ , \tag{3.23a}$$

$$\boldsymbol{\sigma}_x^{\boldsymbol{u}'} \circ (\boldsymbol{\sigma}_x^{\boldsymbol{u}})^{-1} : \ \bar{\xi}(\zeta,\bar{\boldsymbol{\theta}}) \mapsto \bar{\xi}'(\zeta,\bar{\boldsymbol{\theta}}) = \sum_{\alpha=1}^{\infty} \bar{\theta}_\alpha \langle f | w_\alpha \rangle \bar{w}'_\alpha(\zeta) \ , \tag{3.23b}$$

where the primed basis elements in (3.23) are obtained from their unprimed counterparts in accordance with (6.4.4). The above extended transition maps then act on the elements of the standard Berezin-Dirac superfibre $\mathcal{B}$ in accordance with (3.22) and (3.23). Consequently, the formulation of connection and parallel transport introduced in the preceding section can be immediately extended from the Fock-Dirac bundle to the present Berezin-Fock bundle.

8.4. Geometro-Stochastic Propagation in Fock-Dirac Bundles

The natural extension of the operator for parallel transport in (2.1) to the Berezin-Dirac superfibre bundle defined in the preceding section enables the introduction of the Poincaré (in the ISL(2,C)-sense) gauge invariant *Berezin-Dirac superframe propagator*,

$$K_\gamma(\xi'';\xi') = \langle \Phi_{\xi''} | \tau_\gamma(x'',x') \Phi_{\xi'} \rangle \ , \qquad \xi' \in \hat{\mathbf{F}}_{x'} \subset \mathbf{B}_{x'} \ , \quad \xi'' \in \hat{\mathbf{F}}_{x''} \subset \mathbf{B}_{x''} \ , \tag{4.1}$$

for parallel transport along any smooth curve $\gamma = \{x(t) | \, t' \leq t \leq t''\}$, joining any two points $x' = x(t') \in \mathbf{M}$ and $x'' = x(t'') \in \mathbf{M}$. The values assumed by such a propagator are in general supernumbers with nonvanishing odd components. However, on account of (3.18), the parallel transports of Fock-Dirac state vectors, which are themselves Fock-Dirac state vectors, can be expressed directly in terms of Berezin functional integrals:

$$\tau_\gamma(x'',x')\Psi = \int_{\hat{\mathbf{F}}_{x''}} d\bar{\xi}'' d\xi'' \Phi_{\xi''} \int_{\hat{\mathbf{F}}_{x'}} d\bar{\xi}' d\xi' \, K_\gamma(\xi'';\xi') \langle \Phi_{\xi'} | \Psi \rangle \ . \tag{4.2}$$

Indeed, the above integration results in supernumbers with vanishing soul component (cf. Note 5) on account of the fact that (3.18) provides a resolution of the identity in the Fock-Dirac fibre proper. On the other hand, upon inserting $\tau_\gamma(x'', x') = \tau_\gamma(x'', x)\, \tau_\gamma(x\ , x')$ into the right-hand side of the inner product in (4.1), and then using (3.18) at any given point $x \in \gamma$ between its two endpoints, we arrive at the following reproducibility relation,

$$K_\gamma(\xi'';\xi') \ = \ \int_{\hat{\mathbf{F}}_x} K_\gamma(\xi'';\xi)\, d\bar{\xi}\, d\xi\, K_\gamma(\xi;\xi') \ , \tag{4.3}$$

in which, on account of (3.15), the left-hand side assumes supernumber values that have nonzero odd parts. Hence, by using iteration in the same manner as in the bosonic case treated in Sec.7.5, we conclude that in the limit of sequences of subdivisions $t' = t_0 < t_1 < \cdots < t_N = t''$ of the interval $[t', t'']$ for which $\varepsilon = \max(t_n - t_{n-1}) \to +0$, we have that

$$K_\gamma(\xi'';\xi') = \lim_{\varepsilon \to +0} \int K_\gamma(\xi_N;\xi_{N-1}) \prod_{n=N-1}^{1} d\bar{\xi}_n\, d\xi_n\, K_\gamma(\xi_n;\xi_{n-1}) \ . \tag{4.4}$$

Let us choose now some Poincaré gauge which, on account of Geroch's theorem mentioned in Sec. 6.3, can be given in the present context by specifying a global section $\boldsymbol{s}$ of the affine spin frame bundle $SF\mathbf{M}$. The elements of the Berezin-Dirac superframes in (3.15) lie in the domains of definition of the extensions to the fibres of the Berezin-Dirac bundle of the infinitesimal generators of (2.3) which enter (2.2), so that we can write

$$\begin{aligned} K_\gamma(\xi_n;\xi_{n-1}) = \Big\langle \Phi_{\Psi_n(x_n)} \Big| \Big(\mathbf{1}_{x_n} - i\,\delta x_n^i\, P_{i;u} + \tfrac{i}{2} \tilde{\omega}_{jk}(\delta x_n) M_u^{jk} \Big) \Phi_{\Psi_{n-1}(x_n)} \Big\rangle \\ + \ O\big((\delta t_n)^2\big) \ , \qquad \delta t_n = t_n - t_{n-1} \ , \end{aligned} \tag{4.5a}$$

$$\boldsymbol{\psi}_n(x_n)=\xi_n\ ,\quad \boldsymbol{\psi}_{n-1}(x_n)=\left((\boldsymbol{\sigma}^{\boldsymbol{u}}_{x_n})^{-1}\circ\boldsymbol{\sigma}^{\boldsymbol{u}}_{x_{n-1}}\right)\xi_{n-1}\ ,\quad \boldsymbol{u}(x_n)=\left(\mathbf{0},\xi_A(x_n)\right)\ ,\tag{4.5b}$$

where, for the sake of simplicity, we have restricted our attention to Poincaré gauges corresponding to sections of the Lorentz frame bundle $\boldsymbol{LM}$, so that $\boldsymbol{a}(x)\equiv\mathbf{0}$ (cf. (6.3.10).

On the formal level, the procedure leading to an action integral for the Berezin-Dirac superframe propagator in (4.1) is the same as in the case of the second quantized Klein-Gordon frame propagator in (7.5.10). Nevertheless, basic differences exist, since in the present context we are dealing with supernumber-valued functions and functionals.

To begin with, the coordinate wave function amplitudes of the superfibre elements in (4.5) represent elements of the typical superfibre $\mathbf{B}$ and its adjoint, so that they assume values specified for all $\zeta\in\mathbf{R}^4\times\boldsymbol{V}^+$ in terms of one-column or one-row matrices,

$$\psi(x_{n-1};\zeta,\boldsymbol{\theta})=\sum_{\alpha=1}^{\infty}\theta_\alpha\left\langle w_\alpha\middle|\boldsymbol{\sigma}^{\boldsymbol{u}}_{x_n}\boldsymbol{\psi}_{n-1}(x_n)\right\rangle w_\alpha(\zeta)\ ,\tag{4.6a}$$

$$\overline{\psi}(x_n;\zeta,\overline{\boldsymbol{\theta}})=\sum_{\alpha=1}^{\infty}\overline{\theta}_\alpha\left\langle\boldsymbol{\sigma}^{\boldsymbol{u}}_{x_n}\boldsymbol{\psi}_n(x_n)\middle|w_\alpha\right\rangle\overline{w}_\alpha(\zeta)\ ,\tag{4.6b}$$

with supernumber entries. Consequently, it follows from (2.7)-(2.9) and (4.5) that

$$\begin{aligned}\boldsymbol{K}_\gamma(\xi_n;\xi_{n-1})&=\left[1-i\delta x^i_n\,P_i(\overline{\psi}(x_n);\psi(x_{n-1}))\right]\left\langle\boldsymbol{\Phi}_{\boldsymbol{\psi}_n(x_n)}\middle|\boldsymbol{\Phi}_{\boldsymbol{\psi}_{n-1}(x_n)}\right\rangle\\&\quad+i\tilde{\boldsymbol{\omega}}_{jk}(\boldsymbol{\delta}x_n)M^{jk}(\overline{\psi}(x_n);\psi(x_{n-1}))\left\langle\boldsymbol{\Phi}_{\boldsymbol{\psi}_n(x_n)}\middle|\boldsymbol{\Phi}_{\boldsymbol{\psi}_{n-1}(x_n)}\right\rangle+O\left((\delta t_n)^2\right)\ ,\end{aligned}\tag{4.7}$$

where, on account of (3.19) and its adjoint relation, the functionals of the supernumber-valued fields in (4.6) are themselves supernumber-valued:

$$P_i(\psi(x_n);\psi(x_{n-1}))=\int T_{ik}[\psi(x_n;q,-v,\boldsymbol{\theta});\psi(x_{n-1};q,v,\boldsymbol{\theta})]\,d\sigma^k(q)d\Omega(v)\ ,\tag{4.8}$$

$$M^{jk}(\psi(x_n);\psi(x_{n-1}))=\int M^{jkl}[\psi(x_n;q,-v,\boldsymbol{\theta});\psi(x_{n-1};q,v,\boldsymbol{\theta})]d\sigma_l(q)d\Omega(v)\,.\tag{4.9}$$

As a matter of fact, in view of (2.8) and (2.11), we have

$$\begin{aligned}&T_{jk}[\psi(x_n;q,-v,\boldsymbol{\theta});\psi(x_{n-1};q,v,\boldsymbol{\theta})]\\&=im\hat{Z}_{f,m}[(\overline{\psi}^{\uparrow}_n+\overline{\psi}^{\downarrow}_{n-1})\gamma_k(\psi^{\downarrow}_n+\psi^{\uparrow}_{n-1})_{,j}-(\overline{\psi}^{\downarrow}_n+\overline{\psi}^{\uparrow}_{n-1})_{,j}\gamma_k(\psi^{\downarrow}_n+\psi^{\uparrow}_{n-1})],\end{aligned}\tag{4.10a}$$

$$\begin{aligned}&M^{jkl}[\psi(x_n;q,-v,\boldsymbol{\theta});\psi(x_{n-1};q,v,\boldsymbol{\theta})]\\&\qquad=im\hat{Z}_{f,m}\big(\tfrac{1}{2}[(\overline{\psi}^{\uparrow}_n+\overline{\psi}^{\downarrow}_{n-1})(\overleftrightarrow{Q^j\partial^k}-\overleftrightarrow{Q^k\partial^j})\gamma^l(\psi^{\downarrow}_n+\psi^{\uparrow}_{n-1})]\\&\qquad\qquad+\tfrac{1}{8}(\overline{\psi}^{\uparrow}_n+\overline{\psi}^{\downarrow}_{n-1})\{[\gamma^j,\gamma^k],\gamma^l\}(\psi^{\downarrow}_n+\psi^{\uparrow}_{n-1})\big)\ ,\end{aligned}\tag{4.10b}$$

$$\psi^{\uparrow\downarrow}_{n-1}=\psi(x_{n-1};q,\pm v,\boldsymbol{\theta})\ ,\qquad \psi^{\downarrow\uparrow}_n=\psi(x_n;q,\mp v,\boldsymbol{\theta})\ ,\qquad v\in\boldsymbol{V}^+\ ,\tag{4.10c}$$

so that the values they assume are even supernumbers with nonzero soul components.

The rules for Berezin integration in (3.3) and (3.4), taken in conjunction with the definitions in (3.15) for the elements of Berezin-Dirac quantum superframes and their duals, yield in a very straightforward manner a formal analogue of (7.5.17):

$$\langle \Phi_{\xi} | \Phi_{\xi'} \rangle = \exp[\int \bar{\xi}(\zeta,\bar{\theta})\gamma_k \xi'(\zeta,\theta) d\sigma^k(q) d\hat{\Omega}(v)$$
$$-\tfrac{1}{2}\int (\bar{\xi}(\zeta,\bar{\theta})\gamma_k \xi(\zeta,\theta) + \bar{\xi}'(\zeta,\bar{\theta})\gamma_k \xi'(\zeta,\theta)) d\sigma^k(q) d\hat{\Omega}(v)] \,. \quad (4.11)$$

When we use this result in (4.7), we end up with a formal counterpart of (7.5.18),

$$K_\gamma(\psi(x''); \psi(x')) = \lim_{\varepsilon\to+0} \int \prod_{n=N}^{1}{}' \mathcal{D}[\psi(x_n)] \exp[i\delta t_n\, L_\gamma(\psi(x_n); \psi(x_{n-1}))] \,, \quad (4.12a)$$

$$\psi(x') = \xi', \quad \psi(x'') = \xi'', \quad \mathcal{D}[\psi(x_n)] = d\bar{\xi}_n\, d\xi_n \,, \quad n = 1,\ldots,N-1 \,, \quad (4.12b)$$

where the prime indicates the absence of functional integration over the N-th mode, and[8]

$$L_\gamma(\psi(x_n); \psi(x_{n-1}))$$
$$= \tfrac{i}{2}\int (\bar{\psi}(x_n;\zeta,\bar{\theta})\gamma_k \dot{\psi}(x_n;\zeta,\theta) + \dot{\bar{\psi}}(x_{n-1};\zeta,\bar{\theta})\gamma_k \psi(x_{n-1};\zeta,\theta)) d\sigma^k(q) d\hat{\Omega}(v)$$
$$-X_n^{\,i} P_i(\psi(x_n)); \psi(x_{n-1})) + \tfrac{1}{2}\tilde{\omega}_{jk}(X_n) M^{jk}(\psi(x_n); \psi(x_{n-1})) \,, \quad (4.13a)$$

$$\dot{\psi}(x_n) = (\psi(x_n) - \psi(x_{n-1}))/\delta t_n \,, \qquad \dot{\bar{\psi}}(x_{n-1}) = (\bar{\psi}(x_{n-1}) - \bar{\psi}(x_n))/\delta t_n \,. \quad (4.13b)$$

Consequently, upon using (4.8)-(4.10), we arrive at the following functional integral

$$K_\gamma(\psi(x''); \psi(x')) = \int \mathcal{D}\psi \exp\big(iS_\gamma[\bar{\psi},\psi]\big), \quad \mathcal{D}\psi = \prod_{t''>t\geq t'} \mathcal{D}[\psi(x(t))], \quad (4.14)$$

for the Berezin-Dirac superframe propagator for parallel transport, which was defined in purely geometric terms in (4.1), but which in its above form incorporates the action

$$S_\gamma[\bar{\psi},\psi] = \int_{t'}^{t''} dt \int \mathcal{L}_k^{(\gamma)}[\bar{\psi}(x(t);\zeta,\bar{\theta}), \psi(x(t);\zeta,\theta)]\, d\sigma^k(q) d\Omega(v) \,, \quad (4.15a)$$

$$\mathcal{L}_k^{(\gamma)}[\bar{\psi}(x(t);\zeta,\bar{\theta}), \psi(x(t);\zeta,\theta)] =$$
$$m\hat{Z}_{f,m}\Big[\bar{\psi}(x(t);\zeta,\bar{\theta})\gamma_k \dot{\psi}(x(t);\zeta,\theta) + \dot{\bar{\psi}}(x(t-0);\zeta,\bar{\theta})\gamma_k \psi(x(t-0);\zeta,\theta)\Big]$$
$$-X^i(t) T_{ik}[\psi(x(t);\zeta,\theta), \psi(x(t-0);\zeta,\theta)]$$
$$+\eta_{kl}\tilde{\omega}_{ij}(X_n(t)) M^{ijl}[\psi(x(t);\zeta,\theta), \psi(x(t-0);\zeta,\theta)] \,, \qquad k = 0,1,2,3 \,. \quad (4.15b)$$

We observe that, as it was the case in Sec. 7.5, the action in (4.15a) contains four "Lagrangian" terms. The physical reasons for that are the same as in the bosonic case: the enforcement in the present context of Poincaré gauge invariance, based on *local* moving spin frames, necessitates the appearance of terms that are absent under *global* Poincaré transformations of *global* frames. Indeed if we proceed as in the derivation of (7.5.23a), i.e., if we choose a section $\boldsymbol{s}$ which is adapted to γ, normals to the surfaces $\sigma(x(t))$ for q-integration that point along the time axes of the corresponding Lorentz frames along γ, and if we use a parameter t which is identified along γ with the variable q^0 in the typical fibre of $T\mathbf{M}$, then three of these "Lagrangian" terms disappear from (4.15a):

$$S_\gamma[\overline{\psi},\psi] = \int_{t'<q^0<t''} \mathcal{L}_0^{(\gamma)}[\overline{\psi}(x(q^0);\zeta,\overline{\boldsymbol{\theta}}),\psi(x(q^0);\zeta,\boldsymbol{\theta})]\,d^4q\,d\Omega(v) \ , \tag{4.16a}$$

$$\begin{aligned} &\mathcal{L}_0^{(\gamma)}[\overline{\psi}(x(q^0);\zeta,\overline{\boldsymbol{\theta}}),\psi(x(q^0);\zeta,\boldsymbol{\theta})] = \\ &m\hat{Z}_{f,m}\Big[\psi^\dagger(x(q^0);\zeta,\overline{\boldsymbol{\theta}})\,\dot{\psi}(x(q^0);\zeta,\boldsymbol{\theta}) + \dot{\psi}^\dagger(x(q^0{-}0);\zeta,\overline{\boldsymbol{\theta}})\,\psi(x(q^0{-}0);\zeta,\boldsymbol{\theta})\Big] \\ &-X^i(t)T_{i0}[\psi(x(q^0);\zeta,\boldsymbol{\theta}),\psi(x(q^0-0);\zeta,\boldsymbol{\theta})] \ , \qquad \psi^\dagger = \overline{\psi}\gamma^0 \ . \end{aligned} \tag{4.16b}$$

Naturally, the geometric phase (cf. Secs. 3.8 and 3.9) represented by the term between the first set of square brackets on the right-hand side of (4.16b) remains. In fact, similar geometric phases appear also in conventional quantum field theory when coherent states are used (cf. [IQ], p. 438).

With the action functional integral for the propagator of parallel transport of Berezin-Dirac quantum superframes thus established, the formulation of strongly and weakly causal GS propagation proceeds in exactly the same manner as in the bosonic case treated in Sec. 7.7: the expression in (4.14) is inserted in the fermionic counterparts of the definitions based on (7.7.1) and (7.7.4), respectively. We thus arrive at the fermionic GS quantum field propagators for the Berezin-Dirac quantum superframes in (3.15), which in the weakly causal case can be expressed in a form analogous to (7.7.4)-(7.7.5):

$$\mathcal{K}(\boldsymbol{\psi}(x'');\boldsymbol{\psi}(x')) = \int_{\mathbf{M}(t',t'')} \mathcal{D}\hat{\psi}\,\exp\!\big(iS[\hat{\overline{\psi}},\hat{\psi}]\big) \ , \qquad \mathcal{D}\hat{\psi} = \prod\nolimits_x \mathcal{D}[\hat{\psi}(x)] \ , \tag{4.17a}$$

$$\hat{\psi}(x;v,\boldsymbol{\theta}) = \xi(\hat{\zeta}(x),\boldsymbol{\theta}) \ , \qquad \hat{\overline{\psi}}(x;v,\overline{\boldsymbol{\theta}}) = \overline{\xi}(\hat{\zeta}(x),\overline{\boldsymbol{\theta}}) \ , \qquad \hat{\zeta}(x) = (-a^i(x),v^i) \ . \tag{4.17b}$$

Furthermore, the discussion of interaction carried out in the bosonic context in Sec. 7.8 can be carried out equally well in the present fermionic context, the only emerging difference being that the commutation relations in (7.8.6) have to be replaced by the corresponding anticommutation relations realized in (3.11).

Notes to Chapter 8

[1] Cf. [PQ], Ch. IV, Sec. 4.5. The construction in (1.1) follows [SI]. In some textbooks the definition of the Fock space for fermionic fields is carried out by taking direct sums of antisymmetric tensor products of the Hilbert space for both single fermions and single antifermions (cf. [BL],Sec. 8.3), which in the present GS context equals $\mathcal{F}$. The two definitions lead to the same fermion creation and annihilation operators, due to the presence of the factor $(-1)^m$ in (1.7b) and (1.8b), and to the mutual orthogonality of the single-fermion and single-antifermion subspaces of $\mathcal{F}$.

[2] For example, as opposed to the Berezin method adopted in this chapter, the method of Blaizot and Orland (1980, 1981) does not make use of "anticommuting numbers", i.e., of Hilbert superspaces over

Grassmann algebras. Rather, it uses duplicates of the fermionic Fock space to construct a bosonic Fock space, which carries the required coherent states, and into which the original fermionic Fock space can be embedded in a natural manner. However, a Blaizot-Orland coherent state $|f\rangle$ cannot be an eigenvector of the annihilation operator $\psi^{(-)}(x;\zeta)$, as it is the case in (3.19) for Berezin coherent states. Indeed, an application of $\psi^{(-)}(x;\zeta)$ to both sides of (3.19) immediately yields $f(\zeta)^2 = 0$. This implies that, if $f(\zeta)$ were an ordinary complex number, then we would necessarily have that $f(\zeta) = 0$.

3 The methods for constructing such extensions are very similar to those for constructing the Naimark extensions in Sec. 7.8. The fundamental difference is, however, that in Sec. 7.8 we were dealing with extensions of Fock fibres that were Hilbert spaces over the field $\mathbf{C}^1$ of complex numbers, whereas at present we are dealing with extensions which are Hilbert superspaces over a Grassmann algebra Λ, which contains anticommuting elements.

4 Most of the basic definitions and results on Grassmann algebras and on superanalysis that are required in the sequel can be found in [B]. A mathematically more rigorous exposition, which also deals with supermanifolds, is presented in [BI]. A presentation of superanalysis and of supermanifold theory that is more accessible to physicists can be found in [D]. This last presentation can be suitably supplemented with the treatment of Berezin-type coherent states given by Ohnuki and Kashiwa (1978).

5 It should be noted, however, that a Grassmann algebra does not constitute an (algebraic) field, since not every nonzero supernumber has a unique inverse. Indeed, the supernumber in (3.1) has an inverse if and only if its *body* (i.e., the term c_0 [D]) is non-zero, so that a supernumber which has only a non-zero *soul* (i.e., the term represented by the sum in (3.1) [D]) does not possess an inverse.

6 Cf. Sec. 5.2 in [D], where the term *super-Hilbert space* is used instead. A definition of Hilbert superspaces which completely parallels the conventional one [PQ], by not involving duals from the outset, can be found in (Khrennikov, 1988). In fact, readers concerned with making use of mathematically sound methods in quantum physics should take note of the fact that the dual $\mathcal{H}^*$ of a Hilbert space $\mathcal{H}$ is a *topological* rather than an algebraic dual, so that it consists of only those linear functionals which are continuous. Indeed, by Riesz's theorem (cf. [PQ], p. 184), it is due to this continuity that each element $\langle f|$ of that dual $\mathcal{H}^*$ can be identified with a vector $|f\rangle$ in the original Hilbert space $\mathcal{H}$. Thus, the presence of an inner product is required *prior* to defining duals of Hilbert spaces. The tradition of reversing the order in which these two concepts are introduced, exemplified in many physics textbooks on quantum mechanics (cf. Sec. 12.3), reflects a lack of understanding of the fundamental mathematical differences between the theory of finite-dimensional Hilbert spaces (where that is indeed possible, since topological and algebraic duals are identical), and the theory of infinite-dimensional Hilbert spaces (where that is *not* possible, since the algebraic dual is much larger than the topological dual).

7 We make a distinction between *superfibre bundles*, which, by definition, are bundles whose base spaces are ordinary manifolds, but whose fibres are superspaces, and *superbundles*, which are *bona fide* supermanifolds for which, therefore, the fibres as well as the base spaces are themselves supermanifolds. The structure groups of the former can be ordinary Lie groups, as it is the case with the present Berezin-Dirac superfibre bundles, whereas the structure groups of the latter are generically Lie supergroups [BI,D].

8 The conventional Lagrangian density (cf. [SI], p. 218; or [IQ], p. 143) for Dirac fields is of first order in $\partial_0\psi$, so that the canonically conjugate field momenta are not independent of the (non-quantized) Dirac field ψ. As a matter of fact, that Lagrangian density vanishes when ψ satisfies the Dirac equation. At the quantum level, however, the normal ordering of the energy-momentum tensor and the presence of Fermi-Dirac statistics, imposed via canonical anticommutation relations, prohibits the same conclusion. However, in deference to the conventional point of view, we denote the objects in (4.13b) as if they were "time derivatives", despite the fact that the very notion of derivative along a typical stochastic path, which as a rule is not smooth, has no well-defined mathematical meaning. On the other hand, for the sake of notational simplicity, we denote the measure for the functional integration in (4.14) and (4.17a) by the $\mathcal{D}\psi$ defined in (4.12b) and (4.14), rather than, as customary (Rivers, 1987), by $\mathcal{D}\bar{\psi}\mathcal{D}\psi$.

Chapter 9

Quantum Geometries for Electromagnetic Fields

The geometro-stochastic quantization of the electromagnetic field gives rise to a kind of problem not treated thus far in this monograph, namely the formulation of quantum geometries for quanta of zero mass. Even at a classical level the localization of particles of zero mass poses special problems due to the absence of rest frames for such objects. This means the notion of proper time is meaningless for zero-mass particles, and that such particles can be localized only in relation to frames constructed out of massive particles.

The problems encountered by the conventional approaches with the formulation of localizability of quantum particles of non-zero mass are exacerbated in the zero-mass case. Hence, although there is a significant amount of literature attempting to deal with the problem of photon localization, and of mass-0 quantum particles in general, those treatments have made no impact on the conventional methods of quantization of the electromagnetic field in Minkowski space.

The Dirac-Schwinger and the Gupta-Bleuler methods of quantization of the electromagnetic field provide the two best known frameworks for the formulation of quantum electrodynamics (QED) in Minkowski space. The Dirac-Schwinger framework employs a positive-definite inner product in the space of state vectors of the quantum electromagnetic field, and can be in fact mathematically related to the Gupta-Bleuler framework (Dürr and Rudolph, 1969; Ferrari *et al.*, 1974). However, the latter has by now totally prevailed in physics literature due to its manifest gauge covariance and to the formal "locality" displayed in it by all electromagnetic quantum field modes. These features have been purchased, however, at the expense of a positive-definite inner product, so that in the Gupta-Bleuler framework the space of single-photon states carries an indefinite inner product. On the other hand, the fundamental classification [BR] of the irreducible unitary representations of the Poincaré group, first carried out by Wigner (1939), supplies for the mass-0 case only representations in infinite-dimensional Hilbert spaces with positive-definite inner product.

We shall therefore begin by demonstrating in Sec. 9.1 the mathematical equivalence of the depiction of the quantum states of a single photon by equivalence classes of elements of a pseudo-Hilbert space with indefinite metric (introduced in physics by Dirac (1942), and called a Krein space in mathematics literature) with their depiction by elements of an ordinary Hilbert space, in which a mass-0 and spin-1 unitary and irreducible representation of the (orthochronous) Poincaré group acts in the manner originally formulated by Wigner. In Sec. 9.2 we construct the standard fibre for the bundle of local single-photon states. The Gupta-Bleuler bundle of multi-photon local states is then constructed in Sec. 9.3, with special attention being devoted to key mathematical aspects that have thus far remained unnoticed in the mathematically rigorous literature on the Gupta-Bleuler formalism. In Sec. 9.4 we study parallel transport within such bundles, and in Sec. 9.5 we formulate action integrals for that parallel transport and the corresponding GS propagation. We then discuss in Sec. 9.6 the relationship of the GS formulation of quantum electrodynamics, based on the framework developed in the present and in the preceding chapter, to conventional QED, comparing them on the basic issues of locality, microcausality and occurrence of infinities.

9.1. Krein Spaces for Momentum Space Representations of Photon States

The Gupta-Bleuler (1950) formalism for the quantization of a free classical electromagnetic field in Minkowski space requires the introduction of a global Lorentz frame $\{e_\mu|\mu = 0,1,2,3\}$, and the replacement of that field with 4-potentials, so that the Maxwell equations in vacuo assume in relation to that frame the form [SI, IQ] :

$$\partial^\mu \hat{F}_{\mu\nu} = \partial^\mu \partial_\mu \hat{A}_\nu - \partial^\nu \partial_\mu \hat{A}^\mu = 0 \ , \qquad \hat{F}_{\mu\nu} = \partial_\mu \hat{A}_\nu - \partial_\nu \hat{A}_\mu \ . \tag{1.1}$$

This formulation remains invariant under the general Maxwell gauge transformations

$$\hat{A}_\mu(x) \mapsto \hat{A}'_\mu(x) = \hat{A}_\mu(x) + \partial_\mu \hat{\lambda}(x) \ , \qquad \partial_\mu = \partial/\partial x^\mu \ . \tag{1.2}$$

An especially convenient choice is the family of Lorenz[1] gauges, in which the four equations for the 4-potential in (1.1) assume the decoupled form

$$\partial^\mu \partial_\mu \hat{A}_\nu = 0 \ , \qquad \partial^\mu \hat{A}_\mu = 0 \ . \tag{1.3}$$

A mathematically rigorous formulation of Gupta-Bleuler quantization[2] can be arrived at by considering the linear space of all solutions of the wave equations in (1.3), and disregarding the Lorenz gauge condition (Fermi, 1932), but requiring that the sesquilinear form

$$\left(\hat{f}\middle|\hat{f}'\right)_{\hat{J}} = i\sum_{\mu=0}^{3} \int_{x^0=0} \hat{f}^*_\mu(x)\, \overleftrightarrow{\partial}_0\, \hat{f}'_\mu(x)\, d^3\mathbf{x} \tag{1.4}$$

be well-defined and non-negative definite, so that it gives rise to a so-called *J-inner product* (Bognar, 1974). The completion [PQ] of the resulting pre-Hilbert space with respect to the norm defined by this J-inner product leads to a Krein space[3], i.e., to a pseudo-Hilbert space in which the following additional *indefinite* inner product is defined:

$$\left\langle\hat{f}\middle|\hat{f}'\right\rangle = -i\int_{x^0=0} \eta^{\mu\nu} \hat{f}^*_\mu(x)\, \overleftrightarrow{\partial}_0\, \hat{f}'_\nu(x)\, d^3\mathbf{x} \ . \tag{1.5}$$

We can decompose this Krein space of single-photon states into the direct sum

$$\hat{\mathbf{K}} = \hat{\mathbf{K}}^+_{\hat{J}} \oplus \hat{\mathbf{K}}^-_{\hat{J}}, \quad \hat{\mathbf{K}}^+_{\hat{J}} = \left\{\hat{f} \in \hat{\mathbf{K}}\middle|\hat{f}_0 \equiv 0\right\}, \quad \hat{\mathbf{K}}^-_{\hat{J}} = \left\{\hat{f} \in \hat{\mathbf{K}}\middle|\hat{f}_a \equiv 0,\, a = 1,2,3\right\}, \tag{1.6}$$

so that the indefinite inner product in (1.5) is positive definite on the first of the subspaces in this direct sum, and negative definite on the second of these subspaces. Furthermore, by using the fundamental projections onto these subspaces, obtained by setting equal to zero, respectively, the first component or the last three of the components of any given 4-vector-valued wave function that satisfies (1.1), we can write

$$\left(\hat{f}\middle|\hat{f}'\right)_{\hat{J}} = \left\langle\hat{f}\middle|\hat{J}\hat{f}'\right\rangle \ , \qquad \hat{J} := \mathbf{P}_{\hat{J}}^{+} - \mathbf{P}_{\hat{J}}^{-} \ . \tag{1.7}$$

On account of (2.3.7), the representation of the orthochronous[4] Poincaré group, defined on the above constructed Krein configuration-representation space for photons by

$$\hat{U}(a,\Lambda) \ : \ \hat{f}_\mu(x) \mapsto \hat{f}'_\mu(x) = \Lambda_\mu{}^\nu \hat{f}_\nu(\Lambda^{-1}(x-a)) \ , \qquad (a,\Lambda) \in \mathrm{ISO}^\uparrow(3,1) \ , \tag{1.8}$$

is pseudo-unitary, i.e., it preserves the indefinite inner product in (1.5) – but it is not unitary with respect to the J-inner product in (1.4). In other words, the J-inner product is not left invariant by all the transformations in (1.8), since obviously the adoption in Minkowski space of global Lorentz frames which are not stationary with respect to each other gives rise to distinct J-inner products. This means that, in physical terms, a Lorentz boost gives rise to a new Krein configuration-representation space for photons. On the other hand, the norm topology defined by these J-inner products is left invariant, since the operators in (1.8) are J-bounded, i.e., they are bounded with respect to a given J-inner product in (1.4). Hence, although J-inner products can play no direct physical role, similar to that played by inner products in the Hilbert spaces of massive quantum particles, they do play the very important mathematical role of determining the *majorant topology* (Bognar, 1974) required in defining continuity properties of representations of the Poincaré group – as well as other mathematical features which cannot be defined by the indefinite inner product in (1.5).

Although we followed conventional terminology by referring to the Krein space in (1.6) as the photon "configuration representation" space, there is no formulation of photon localization in Minkowski space that can be associated with the wave functions of this Krein space. Indeed, all the problems with such localization for massive particles, described in Chapter 3, not only remain unresolved, but in the mass-0 case they are further exacerbated by the lack of existence of a positive-definite relativistically invariant inner product in the spaces of such wave functions. The problem of localization in momentum space can be nevertheless solved, on account of the possibility of introducing polarization tetrads in a corresponding momentum representation for single-photon states.

The photon wave functions in this momentum-representation space are defined by

$$\tilde{f}_\mu(k) = 2k_0(2\pi)^{-3/2}\int_{x^0=0} \exp(ik\cdot x)\hat{f}_\mu(x)\,d^3\mathbf{x} \ , \tag{1.9a}$$

$$\hat{f}_\mu(x) = (2\pi)^{-3/2}\int_{V_0^+} \exp(-ix\cdot k)\tilde{f}_\mu(k)d\Omega_0(k) \ , \tag{1.9b}$$

$$\mathbf{V}_0^+ = \left\{k = (k_0,\mathbf{k})\middle| k_0 = |\mathbf{k}|\right\} \ , \qquad d\Omega_0(k) = \delta(k^2)\,d^4k \ , \quad k^2 = k\cdot k \ . \tag{1.9c}$$

The transformation defined by (1.9a) leads to the Krein space

$$\tilde{\mathbf{K}} = \tilde{\mathbf{K}}_{\tilde{J}}^{+} \hat{\oplus} \tilde{\mathbf{K}}_{\tilde{J}}^{-}, \quad \tilde{\mathbf{K}}_{\tilde{J}}^{+} = \left\{ \tilde{f} \in \tilde{\mathbf{K}} \middle| \tilde{f}_0 \equiv 0 \right\}, \; \tilde{\mathbf{K}}_{\tilde{J}}^{-} = \left\{ \tilde{f} \in \tilde{\mathbf{K}} \middle| \tilde{f}_a \equiv 0,\, a = 1,2,3 \right\}, \tag{1.10}$$

which carries the following indefinite inner product and J-inner product, respectively,

$$\langle \tilde{f} | \tilde{f}' \rangle = -\int_{\mathbf{V}_0^+} \eta^{\mu\nu} \tilde{f}_\mu^*(k)\, \tilde{f}'_\nu(k)\, d\Omega_0(k) \ , \tag{1.11}$$

$$(\tilde{f} | \tilde{f}')_{\tilde{J}} = \sum_{\mu=0}^{3} \int_{\mathbf{V}_0^+} \tilde{f}_\mu^*(k)\, \tilde{f}'_\mu(k)\, d\Omega_0(k) = \langle \tilde{f} | \tilde{J} \tilde{f}' \rangle \ , \qquad \tilde{J} = \mathbf{P}_{\tilde{J}}^{+} - \mathbf{P}_{\tilde{J}}^{-} \ . \tag{1.12}$$

The transform of the representation in (1.8) is easily computed to be given by:

$$\tilde{U}(a,\Lambda) : \ \tilde{f}_\mu(k) \mapsto \tilde{f}'_\mu(k) = \exp(ia\cdot k)\Lambda_\mu{}^\nu \tilde{f}_\nu(\Lambda^{-1}k) \ . \tag{1.13}$$

To elucidate the physical interpretation of this pseudo-unitary momentum space representation of the orthochronous Poincaré group, and to show how it can lead to a representation[5] which is equivalent to the unitary and irreducible representation of Wigner-type for mass-0 and spin-1 quantum particles [BR], let us restrict ourselves to the *Lorenz space*

$$\tilde{\mathbf{K}}^L = \left\{ \tilde{f} \in \tilde{\mathbf{K}} \middle| k_\mu \tilde{f}^\mu(k) = 0 \right\} \subset \tilde{\mathbf{K}} \ , \qquad \tilde{f}^\mu := \eta^{\mu\nu} \tilde{f}_\nu \ , \tag{1.14}$$

consisting of all momentum space photon wave functions whose configuration space counterparts satisfy (1.3). This subspace is obviously left invariant by the pseudo-unitary representation in (1.13). The same is true of the subspace

$$\tilde{\mathbf{K}}^0 = \left\{ \tilde{f} \in \tilde{\mathbf{K}} \middle| k_\mu \tilde{f}_\nu(k) - k_\nu \tilde{f}_\mu(k) = 0 \ , \ \mu,\nu = 0,1,2,3 \right\} \subset \tilde{\mathbf{K}}^L \ , \tag{1.15}$$

which consists of all momentum space photon wave functions whose configuration space counterparts correspond to a vanishing electromagnetic field $F_{\mu\nu}$. It is easily checked that this subspace of the Krein space in (1.10) is indeed also a subspace of the Lorenz space defined in (1.14).

For a given global Lorentz frame $\boldsymbol{u} = \{\boldsymbol{e}_\mu | \mu = 0,1,2,3\}$, both these subspaces can be conveniently characterized by introducing a family of corresponding *polarization frames* or *tetrads* $\{\boldsymbol{\varepsilon}_{(\alpha)}(k) | \ \alpha = 0,1,2,3\}$, attached to all forward light-cone points, and such that

$$\boldsymbol{\varepsilon}_{(0)}(k) = \boldsymbol{e}_0 \ , \ \ \boldsymbol{\varepsilon}_{(3)}(k) = -|\mathbf{k}|^{-1} \sum_{a=1}^{3} k_a \boldsymbol{e}_a \ , \qquad k_\mu \varepsilon_{(1)}^\mu(k) = k_\mu \varepsilon_{(2)}^\mu(k) = 0 \ . \tag{1.16}$$

Since according to the above construction $k = k_0(\boldsymbol{\varepsilon}_{(0)}(k) + \boldsymbol{\varepsilon}_{(3)}(k))$, it is easily seen that

$$\tilde{\mathbf{K}}^L = \left\{ \tilde{f} = \tilde{f}^{(\alpha)}(k) \boldsymbol{\varepsilon}_{(\alpha)}(k) \in \tilde{\mathbf{K}} \middle| \tilde{f}^{(0)}(k) = \tilde{f}^{(3)}(k) \right\} \ , \tag{1.17}$$

$$\tilde{\mathbf{K}}^0 = \left\{ \tilde{f} \in \tilde{\mathbf{K}}^L \middle| \tilde{f}^{(1)}(k) = \tilde{f}^{(2)}(k) = 0 \right\} = \left\{ \tilde{f} \in \tilde{\mathbf{K}}^L \middle| \langle \tilde{f} | \tilde{f} \rangle = 0 \right\} . \tag{1.18}$$

Furthermore, by means of such polarization tetrads the indefinite inner product and J-inner product can be expressed in the Lorenz space in the following respective forms,

$$\langle \tilde{f} | \tilde{f}' \rangle = -\int_{\mathbf{V}_0^+} \eta_{\alpha\beta} \tilde{f}^{(\alpha)*}(k)\, \tilde{f}^{(\beta)\prime}(k)\, d\Omega_0(k) \quad , \qquad \tilde{f}, \tilde{f}' \in \tilde{\mathbf{K}}^L \ , \tag{1.19}$$

$$\left(\tilde{f} | \tilde{f}' \right)_{\tilde{J}} = \sum_{\alpha=0}^{3} \int_{\mathbf{V}_0^+} \tilde{f}^{(\alpha)*}(k) \tilde{f}^{(\alpha)\prime}(k) d\Omega_0(k) \ , \qquad \tilde{f}, \tilde{f}' \in \tilde{\mathbf{K}}^L \ , \tag{1.20}$$

provided that the remaining freedom in choosing the transversal elements of the polarization frames in (1.16) is carried out in accordance with the later imposed relation (1.29). As a consequence, the Lorenz space can be decomposed into the direct sum

$$\tilde{\mathbf{K}}^L = \tilde{\mathbf{K}}^0 \underset{J}{\oplus} \tilde{\mathbf{K}}_J^L \ , \quad \tilde{\mathbf{K}}_J^L = \left\{ \tilde{f} = \tilde{f}^{(\alpha)}(k)\, \boldsymbol{\varepsilon}_{(\alpha)}(k) \in \tilde{\mathbf{K}}^L \middle| \tilde{f}^{(0)}(k) = \tilde{f}^{(3)}(k) = 0 \right\}, \tag{1.21}$$

with respect to the J-inner product in the chosen Lorentz frame $\boldsymbol{u}$. Its subspace $\tilde{\mathbf{K}}_J^L$ is not left invariant by the representative (1.13) of a general Poincaré transformation, but it is left invariant by all those transformations for which $\Lambda\, \boldsymbol{e}_0 = \boldsymbol{e}_0$, so that it is only J-dependent.

On the other hand, the quotient space

$$\tilde{\mathbf{K}}^L / \tilde{\mathbf{K}}^0 = \left\{ [\tilde{f}] \middle| \tilde{f} \in \tilde{\mathbf{K}}^L ,\ [\tilde{f}] = [\tilde{f}'] \Leftrightarrow \langle \tilde{f} - \tilde{f}' | \tilde{f} - \tilde{f}' \rangle = 0 \right\} \tag{1.22}$$

is a *bona fide* Hilbert space with respect to the inner product induced on the equivalence classes constituting its elements by the indefinite inner product on the Lorenz space in (1.17). This Hilbert space is obviously left invariant by the representation of the Poincaré group induced by the pseudo-unitary representation in (1.13), since both spaces in that quotient are left invariant by it. Furthermore, the representation thus induced on the Hilbert space in (1.22) is unitary since it can be expressed in the form (Warlow, 1992)

$$\tilde{U}_{\boldsymbol{u}}(a,\Lambda) = V_{\boldsymbol{u}} [\tilde{U}(a,\Lambda)] V_{\boldsymbol{u}}^{-1} \ , \quad V_{\boldsymbol{u}} : [\tilde{f}] \ \mapsto \ \mathbf{P}_{\boldsymbol{u}} \tilde{f} \in \tilde{\mathbf{K}}_J^L \ , \quad [\tilde{f}] \in \tilde{\mathbf{K}}^L / \tilde{\mathbf{K}}^0 \ , \tag{1.23}$$

where $V_{\boldsymbol{u}}$ is a unitary transform, and $\mathbf{P}_{\boldsymbol{u}}$ is the J-orthogonal projection onto the second subspace in the J-orthogonal decomposition in (1.21). An explicit computation shows that

$$\tilde{U}_{\boldsymbol{u}}(a,\Lambda) : \tilde{f} \ \mapsto \ \exp(ia \cdot k) \left(\Lambda\, \tilde{f}(\Lambda^{-1}k) - \tfrac{1}{2} |\mathbf{k}|^{-2} \sum_{\mu=0}^{3} k_\mu \Lambda_\mu{}^\nu \tilde{f}_\nu(\Lambda^{-1}k) k \right) . \tag{1.24}$$

The fact that (1.24) is an irreducible representation of the orthochronous Poincaré group can be established by proving its unitary equivalence to the Wigner representation

$$\hat{U}_{\boldsymbol{u}}(a,\Lambda) \ : \hat{f}(k) \ \mapsto \ \exp(ia \cdot k) D_{\hat{\boldsymbol{k}}}(\hat{\Lambda}_{\boldsymbol{k}}^{-1} \Lambda \hat{\Lambda}_{\Lambda^{-1}\boldsymbol{k}}) \hat{f}(\Lambda^{-1}k) \ , \tag{1.25a}$$

$$\widehat{\Lambda}_k = R(\hat{\mathbf{k}})\Lambda_{\gamma(k)e_3} \ , \quad k = (k_0,\mathbf{k}) \ , \quad \hat{\mathbf{k}} = \mathbf{k}/|\mathbf{k}| \ , \quad \widehat{k} = \tfrac{1}{2}(e_0 + e_3) \in V_0^+ \ , \tag{1.25b}$$

where $R(\hat{\mathbf{k}})$ rotates $\mathbf{e}_3$ into $\hat{\mathbf{k}}$ around an axis which is orthogonal to these two 3-vectors, and $\Lambda_{\gamma e_3}$ is the boost along $e_3 = (0,\mathbf{e}_3)$ which takes $\widehat{k}$ into $(k_0, k_0\mathbf{e}_3)$. This Wigner representation is realized on the Hilbert space of two-component wave functions with inner product

$$\langle \widehat{f} | \widehat{f}' \rangle = \sum_{r=\pm 1} \int_{V_0^+} \widehat{f}_r^{\,*}(k)\, \widehat{f}_r'(k)\, d\Omega_0(k) \ , \tag{1.26}$$

and for $\Lambda \in \mathrm{SO}_0(3,1)$ it incorporates the well-known representation [BR]

$$D_{\widehat{k}} : \ \Lambda(A(\theta,z)) \ \mapsto \ A(2\theta,0) \in E_{\widehat{k}}(2) \ , \quad A(\theta,z) = \begin{pmatrix} e^{i\theta/2} & z e^{-i\theta/2} \\ 0 & e^{-i\theta/2} \end{pmatrix} \ , \tag{1.27}$$

of the little group[6] for photons, which is defined in terms of the map in (6.1.3a) as follows:

$$E_{\widehat{k}}(2) = \left\{ A \in \mathrm{SL}(2,\mathbf{C}) \middle| A \underset{\sim}{\widehat{k}} A^* = \underset{\sim}{\widehat{k}} \right\} = \left\{ A(\theta,z) \middle| \ \theta \in [0,4\pi),\ z \in \mathbf{C}^1 \right\} \ . \tag{1.28}$$

The proof of the unitary equivalence of the representations in (1.24) and (1.25) can be then obtained by introducing the circular polarization 4-vectors

$$\varepsilon_{[\pm 1]}(k) = \widehat{\Lambda}_k\, \varepsilon_{[\pm 1]}(\widehat{k}) \ , \qquad \varepsilon_{[\pm 1]}(\widehat{k}) = \mp \tfrac{1}{\sqrt{2}}(e_1 \pm i e_2) \ , \tag{1.29}$$

and verifying (cf. Sec. 11.8) that the linear map

$$W_u : [\tilde{f}] \mapsto \widehat{f} = \begin{pmatrix} \widehat{f}^{(+1)} \\ \widehat{f}^{(-1)} \end{pmatrix}, \quad \left(V_u[\tilde{f}]\right)(k) = \widehat{f}^{(+1)}(k)\, \varepsilon_{[+1]}(k) + \widehat{f}^{(-1)}(k)\, \varepsilon_{[-1]}(k), \tag{1.30}$$

is unitary, and that it is such that

$$\widehat{U}_u(a,\Lambda) = W_u\, \tilde{U}_u(a,\Lambda) W_u^{-1} \ . \tag{1.31}$$

The two-component wave functions of the Wigner representation provide a physical interpretation of the above formalism, since those components can be interpreted as being the momentum space probability amplitudes for the single-photon states represented by those wave functions. Thus, neither the elements of the Krein spaces in (1.6) and (1.10), nor those of the Lorenz space in (1.17), possess a direct physical meaning as probability amplitudes, but they can be grouped into equivalence classes, which can be then set in one-to-one correspondence with wave functions which do have a direct physical meaning. We observe that this correspondence intrinsically necessitates a choice of global Lorentz frame in Minkowski space, and a corresponding choice of polarization tetrads in (1.16), whose complete specification is fixed by (1.29).

9.2. The Typical Krein-Maxwell Fibre for Single-Photon States

In the context of classical special relativity theory, a mass-0 particle, such as the photon, possesses no rest frame. As a consequence, the concept of 4-velocity has no physical meaning for a photon, albeit the concept of its 4-momentum is physically meaningful – as confirmed experimentally by the well-known Compton effect [SI], which establishes its conservation in collision processes. Its localization can be carried out therefore only at the instant of its collision with a massive particle, by observing the recoil of the latter.

This suggests that the localization of photons should be mathematically treated by a limiting procedure in the momentum representation, whereby one begins with a system of covariance for the localization of a spin-1 particle of rest mass $\mu_0 > 0$ with respect to a quantum frame of spin-0 test particles of fixed rest mass – which we choose to be equal to one – and then one considers the limit $\mu_0 \to +0$. In order to recover in such a limit the Lorenz spaces introduced and studied in Sec. 9.1, we have to begin by considering the solutions of the Proca equations for mass $\mu_0 > 0$:

$$\partial^\mu \hat{F}^{(\mu_0)}_{\mu\nu} + \mu_0^2 \hat{A}^{(\mu_0)}_\nu = 0 \ , \qquad \hat{F}^{(\mu_0)}_{\mu\nu} = \partial_\mu \hat{A}^{(\mu_0)}_\nu - \partial_\nu \hat{A}^{(\mu_0)}_\mu \ . \tag{2.1}$$

These equations are easily seen to imply that (cf. [IQ], Sec. 3-2-3)

$$\left(\partial^\mu \partial_\mu + \mu_0^2\right)\hat{A}^{(\mu_0)}_\nu = 0 \ , \qquad \partial^\mu \hat{A}^{(\mu_0)}_\mu = 0 \ , \tag{2.2}$$

so that, as $\mu_0 \to +0$, any photon wave functions that are obtained by taking such a limit are automatically in the Lorenz gauge.

In the momentum representation, the Proca space of solutions of (2.2), namely

$$\tilde{\mathbf{K}}^{(\mu_0)} = \left\{ \tilde{f}^{(\mu_0)} \middle| \sum_{\mu=0}^{3} \int_{V^+_{\mu_0}} \left|\tilde{f}^{(\mu_0)}_\mu(k)\right|^2 d\Omega_{\mu_0}(k) < +\infty \ , \ k^\mu \tilde{f}^{(\mu_0)}_\mu(k) = 0 \right\} \ , \tag{2.3}$$

carries the following Proca inner product and J-inner product, respectively,

$$\left\langle \tilde{f}^{(\mu_0)} \middle| \tilde{f}^{(\mu_0)\prime} \right\rangle = -\int_{V^+_{\mu_0}} \eta^{\mu\nu} \tilde{f}^{(\mu_0)*}_\mu(k) \ \tilde{f}^{(\mu_0)\prime}_\nu(k)\, d\Omega_{\mu_0}(k) \ , \tag{2.4}$$

$$\left(\tilde{f}^{(\mu_0)} \middle| \tilde{f}^{(\mu_0)\prime}\right)_{\hat{J}_{\mu_0}} = \sum_{\mu=0}^{3} \int_{V^+_{\mu_0}} \tilde{f}^{(\mu_0)*}_\mu(k) \ \tilde{f}^{(\mu_0)\prime}_\mu(k)\, d\Omega_{\mu_0}(k) = \left\langle \tilde{f}^{(\mu_0)} \middle| \tilde{J}_{\mu_0} \tilde{f}^{(\mu_0)\prime} \right\rangle \ . \tag{2.5}$$

The counterpart of the representation in (1.13) for the case of massive Proca particles of spin one is

$$\tilde{U}^{(\mu_0)}(a,\Lambda) \ : \ \tilde{f}^{(\mu_0)}_\mu(k) \ \mapsto \ \tilde{f}^{(\mu_0)\prime}_\mu(k) = \exp(ia\cdot k)\Lambda_\mu{}^\nu \tilde{f}^{(\mu_0)}_\nu(\Lambda^{-1}k) \ . \tag{2.6}$$

By using the same method as in establishing the physical equivalence of the representations in (1.13) and (1.25), we can demonstrate the physical equivalence of the indefinite-metric representation in (2.6) to the positive-definite metric Wigner-type representation

$$\widehat{U}_{u}^{(\mu_0)}(a,\Lambda)\ :\widehat{f}^{(\mu_0)}(k)\ \mapsto\ \exp(ia\cdot k)D^1(R(\Lambda,k))\widehat{f}^{(\mu_0)}(\Lambda^{-1}k)\ , \tag{2.7a}$$

$$R(\Lambda,k) = \Lambda_k^{-1}\Lambda\Lambda_{\Lambda^{-1}k}^{-1} \in \mathrm{SO}(3) \subset \mathrm{SO}(3,1)\ , \qquad k = (k_0,\mathbf{k}) \in V_{\mu_0}^{+}\ , \tag{2.7b}$$

which acts on three-component wave functions from the direct sum of three Hilbert spaces, each one of which carries the inner product in (3.3.13). The matrix $R(\Lambda,k)$ in (2.7) represents a Wigner rotation, considered as an element of the Lorentz group[7], and Λ_k is the pure Lorentz transformation which boosts a particle of mass m , that is at rest, into one that has 4-momentum k .

The little group for particles of rest mass $\mu_0 > 0$ contracts[8], in the appropriately taken limit $\mu_0 \to +0$, into that of the corresponding massless particles. The corresponding contraction of SO(3) into SE(2) can be heuristically envisaged by embedding a sphere of large radius R into a Euclidean 3-space, and then considering around its "north pole" an area which is so small that it can be deemed to be almost flat, so that those rotations which move the locations of the axis of a Cartesian frame passing through the pole to a location within that area can be approximated with translations in the tangent plane at the "north pole". In a strict sense, the contractions of SO(3) into SE(2) ensue as we take the limit $R \to +\infty$. The detailed study of the zero-mass and infinite-momentum contractions of the little groups of massive particles shows (Kim and Wigner, 1987) that the contraction procedure requires a choice of global Lorentz frame, and that it is dependent on that choice to the extent that the procedure is not left invariant by boosts of the originally chosen frame.

Indeed, geometrically this procedure depends in an intrinsic manner on the choice of point on the forward hyperboloid for rest mass $\mu_0 > 0$, which in the limit $\mu_0 \to +0$ merges into the vertex of the forward light cone – i.e., in physical terms, on the choice of massive particles at rest which in that limit "merge" into mass-0 particles. Thus, the mathematical nature of the contraction reflects the fact that, from the physical point of view, the limit involved in it is not operationally realizable. Indeed, a mass-0 particle has no rest frame, since from the classical point of view it travels with the same speed in relation to any global Lorentz frame. Thus, although such a limit is mathematically well-defined, and it is routinely invoked in quantum field theory [IQ] when dealing with mass-0 particles and their propagators, it is of importance for later considerations to keep in mind that it represents a mathematically asymptotic procedure, which as such has no *exact* counterpart in physical reality.

The frame-dependence of this zero-mass limit is reflected quantum mechanically in the manner in which the elements of the Lorenz space in (1.14) are set in one-to-one correspondence with the elements of a J_{μ_0}-dense set in the Proca space in (2.3), namely by the fact that the map

$$\tilde{f}^{(\mu_0)}(\sqrt{\mathbf{k}^2+\mu_0^2},\mathbf{k})\ \leftrightarrow\ \tilde{f}(|\mathbf{k}|,\mathbf{k})\ , \qquad \mathbf{k} = k^a \boldsymbol{e}_a = -\sum_{a=1}^{3} k_a \boldsymbol{e}_a\ , \tag{2.8}$$

is not left invariant by Lorentz boosts. Consequently, distinct J_{μ_0}-dense sets are obtained in global Lorentz frames which are not at rest with respect to each other.

All this is also reflected in the concept of photon GS localization which emerges[9] when, for a given quantum spacetime form factor f, we construct a counterpart of (5.1.2), namely the phase-space transform

$$f^{(\mu_0)}(q,v) = Z_f^{(\mu_0)-1} \int_{\mathbf{V}_{\mu_0}^+} \exp[-iq\cdot k]\, f(k\cdot v)\tilde{f}^{(\mu_0)}(k)\; d\Omega_{\mu_0}(k) \; . \tag{2.9}$$

This construction determines a transformation of the Proca momentum representation space in (2.3) into a space of wave functions with inner product

$$\left\langle f^{(\mu_0)} \middle| f^{(\mu_0)\prime} \right\rangle = -Z_f^{(\mu_0)-2} \int_\Sigma \eta^{\mu\nu} f_\mu^{(\mu_0)*}(q,v) f_\nu^{(\mu_0)\prime}(q,v)\, d\Sigma(q,v) \; . \tag{2.10}$$

As it was shown to be the case in (5.1.9), the integration in (2.10) can be carried out over any hypersurface $\Sigma = \sigma\times V^+$ in relativistic phase space whose spacetime component σ is maximally spacelike.The value of the renormalization constant in (2.9) is determined by the requirement that (2.4) and (2.10) should be equal, so that the resulting transformation would be pseudo-unitary, namely that

$$W_f^{(\mu_0)}:\;\; \tilde{f}^{(\mu_0)} \mapsto f^{(\mu_0)} \;, \qquad \left\langle f^{(\mu_0)} \middle| f^{(\mu_0)\prime} \right\rangle = \left\langle \tilde{f}^{(\mu_0)} \middle| \tilde{f}^{(\mu_0)\prime} \right\rangle \; . \tag{2.11}$$

Let us now take the mass-0 limit in a given global Lorentz frame $\boldsymbol{u}$:

$$\begin{aligned} f(q,v) = \lim_{\mu_0\to+0} Z_f^{(\mu_0)} f^{(\mu_0)}(q,v) &= \lim_{\mu_0\to+0} \int_{\mathbf{R}^3} \exp\left[i\left(\mathbf{q}\cdot\mathbf{k} - q^0\sqrt{\mathbf{k}^2+\mu_0^2}\right)\right] \\ &\times f(v^0\sqrt{\mathbf{k}^2+\mu_0^2} - \mathbf{v}\cdot\mathbf{k})\tilde{f}^{(\mu_0)}(\sqrt{\mathbf{k}^2+\mu_0^2},\mathbf{k}) \frac{d^3\mathbf{k}}{2\sqrt{\mathbf{k}^2+\mu_0^2}} \; . \end{aligned} \tag{2.12}$$

Then, in accordance with (2.8), we obtain that

$$f(q,v) = \int_{\mathbf{V}_0^+} \exp(-iq\cdot k)\, f(k\cdot v)\tilde{f}(k)\; d\Omega_0(k) \;, \tag{2.13}$$

since the normalization constant in (2.9) was removed by a wave function amplitude renormalization, because it diverges in the mass-0 limit.

To explicitly see that such is the case, let us specialize the above considerations to the case of the *fundamental* quantum spacetime form factor in (5.5.5). Upon inserting

$$f_{\ell\mu_0}(v\cdot k) = \exp(-\ell\; v\cdot k) \;, \qquad \ell > 0 \;, \qquad k\in\mathbf{V}_{\mu_0}^+ \;, \tag{2.14}$$

into (2.9) (cf. Sec. 12.5 for foundational arguments supporting this choice), we obtain

$$f^{(\mu_0)}(\zeta) = Z_\ell^{(\mu_0)-1} \int_{\mathbf{V}_{\mu_0}^+} \exp\left(-i\bar{\zeta}\cdot k\right)\tilde{f}^{(\mu_0)}(k)\; d\Omega_{\mu_0}(k) \;, \qquad \zeta = q + i\ell v \; . \tag{2.15}$$

The adaptation to the case of (2.15) of the computations leading to (3.6.26) or to (5.5.6), under the requirement that the equality in (2.11) be satisfied, leads to the conclusion that

$$Z_\ell^{(\mu_0)} = (8\pi^4 \mathrm{K}_2(2\ell\mu_0)/\ell\mu_0^2)^{1/2} = 2\pi^2/\ell^{3/2}\mu_0^2 + O(1/\mu_0) \xrightarrow[\mu_0 \to +0]{} +\infty \ . \quad (2.16)$$

As a feature characteristic of zero mass, the renormalization of the amplitude of the wave function in (2.15) belongs to the category of infrared renormalizations carried out in conventional QED (cf. [IQ], Sec. 5-1-7), as well as in all non-Abelian quantum gauge field theories (Curci and Ferrari, 1976). However, as opposed to the latter cases, the present renormalization is the outcome of a well-defined limit that requires no *ad hoc* subtractions[10]. As such, it is in no way dependent on formal regularization procedures, but it is a renormalization in the strict mathematical sense of the word, i.e., it represents a change of norm resulting from the multiplication of an inner product by a positive number, coupled with the subsequent taking of the well-defined limit in (2.16).

For reasons discussed in Sec. 12.5, the quantum spacetime form factor in (2.14), with ℓ equal to the Planck length, is the most natural choice of fundamental quantum spacetime form factor. As such, it will enter into the later derived propagator for massless GS excitons. Consequently, we shall henceforth focus all our attention on it, albeit most of the results reached in the present chapter, as well as in the subsequent two chapters, could be equally well derived for generic quantum spacetime form factors. On the other hand, although $\ell = 1$ in the Planck natural units, which we have adopted throughout this monograph, we shall retain in all of the subsequent formulae the general value of ℓ . This will enable us to later illustrate with greater ease the effects of taking in them the sharp-point limit $\ell \to +0$.

We now adopt as *standard single-photon fibre* the Krein-Maxwell space $\mathbf{K}$ of all the photon wave functions

$$f(\zeta) = \int_{V_0^+} \exp\left(-i\bar{\zeta}\cdot k\right)\tilde{f}(k)\, d\Omega_0(k) \ , \qquad \zeta = q + i\ell v \ , \quad q \in \mathbf{R}^4 \ , \quad v \in \mathbf{V}^+ , \quad (2.17)$$

that are obtained as $\tilde{f}$ varies over all elements of the Krein space (1.10). These functions are analytic in a neighborhood of the domain specified in (2.17) for the complex variable ζ, but their indefinite as well as J-inner products,

$$\langle f | f' \rangle = -\int_\Sigma \eta^{\mu\nu} \bar{f}_\mu(\zeta) f_\nu'(\zeta) d\tilde{\Sigma}(\zeta) \ , \qquad d\tilde{\Sigma}(\zeta) = Z_\ell^{(0)-2} d\Sigma(\zeta) \ , \quad (2.18a)$$

$$(f|f')_J = \sum_{\mu=0}^{3} \int_\Sigma \bar{f}_\mu(\zeta) f_\mu'(\zeta) d\tilde{\Sigma}(\zeta) = \langle f | J f' \rangle \ , \quad J = \mathbf{P}_J^+ - \mathbf{P}_J^- \ , \quad (2.18b)$$

require an infinite renormalization, formally absorbed into the measure of integration, in order to retain their equality to those in (1.11) and (1.12), respectively.

The Klein-Gordon form of the inner product in (2.10), which represents a counterpart of the alternative form (5.1.13) of the inner product in (5.1.9), is given by

$$\left\langle f^{(\mu_0)} \middle| f^{(\mu_0)\prime} \right\rangle = -i\hat{Z}_\ell^{(\mu_0)} \int_\Sigma \eta^{\mu\nu} f_\mu^{(\mu_0)*}(q,v) \overleftrightarrow{\partial}_\lambda f_\nu^{(\mu_0)\prime}(q,v) d\sigma^\lambda(q)\, d\Omega(v) \ . \quad (2.19)$$

It is defined on the direct sum of four of the Hilbert spaces in (5.1.1), so that the family of wave functions which in addition satisfies (with respect to q) the Lorenz-type condition in (2.2) constitutes a proper subspace. That subspace carries a pseudo-unitary irreducible representation of the orthochronous Poincaré group which is equivalent to the one in (2.6). The inner product in (2.19) contains the renormalization constant

$$\hat{Z}_\ell^{(\mu_0)} = \mathrm{K}_2(2\ell\mu_0)/\mu_0\mathrm{K}_1(2\ell\mu_0) = 1/2\ell\mu_0^2 + O(1/\mu_0) \xrightarrow[\mu_0\to+0]{} +\infty \ , \tag{2.20}$$

so that in the zero-mass limit it leads to the following expression for the inner product in the standard single-photon fibre $\mathbf{K}$,

$$\langle f|f'\rangle = -i\int_\Sigma \eta^{\mu\nu}\bar{f}_\mu(\zeta)\overleftrightarrow{\partial}_\lambda f_\nu{}'(\zeta)\,d\sigma^\lambda(q)\,d\tilde{\Omega}(v) \ , \qquad d\tilde{\Omega}(v) = \hat{Z}_\ell^{(0)}d\Omega(v) \ . \tag{2.21}$$

As it was the case in (2.18), the infinite renormalization constant resulting from the limiting procedure that leads from (2.19) to (2.21) was again incorporated into the measure of integration – a feature symbolized by a tilde above Ω, so as to distinguish it from the hat that symbolizes the finite renormalization carried out in (6.2.10) for the case of massive fermions.

By using such relationships as (3.6.24), it is easily verified that Proca phase space representation spaces with the inner product (2.10) are reproducing kernel spaces. In the zero-mass limit the corresponding reproducing kernels in the Krein-Maxwell space $\mathbf{K}$ are given by the 4×4 matrix-valued functions whose matrix elements have the form

$$iD^{(+)}(\zeta'';\zeta')\eta_{\mu\nu} = \eta_{\mu\nu}\int_{V_0^+}\exp\left[-i(\bar{\zeta}''-\zeta')\cdot k\right]\,d\Omega_0(k) \ . \tag{2.22a}$$

The value of the integral in (2.22a) can be explicitly computed by the procedure used in [P] for computing (3.6.27), with the following result:

$$D^{(+)}(q'',v'';q',v') = -2\pi/(q''-q'-i\ell(v''+v'))^2 \ , \qquad q',q'',\in \mathbf{R}^4, \ v',v''\in \mathbf{V}^+ . \tag{2.22b}$$

The function in (2.22b) obeys the reproducibility relations

$$\begin{aligned} iD^{(+)}(\zeta'';\zeta') &= -\int_\Sigma D^{(+)}(\zeta'';\zeta)D^{(+)}(\zeta;\zeta')\,d\tilde{\Sigma}(\zeta) \\ &= -i\int_\Sigma D^{(+)}(\zeta'';\zeta)\overleftrightarrow{\partial}_\lambda D^{(+)}(\zeta;\zeta')\,d\sigma^\lambda(q)\,d\tilde{\Omega}(v) \ , \end{aligned} \tag{2.23}$$

from which those for the phase space photon propagator in (2.22a) immediately follow.

The transform of the representation in (1.13) under the mapping of the Krein space (1.10) onto $\mathbf{K}$, determined by (2.17), yields the following pseudo-unitary representation of the orthochronous Poincaré group:

$$U(a,\Lambda) \ : \ f_\mu(q,v) \mapsto f_\mu'(q,v) = \Lambda_\mu{}^\nu f_\nu(\Lambda^{-1}(q-a),\Lambda^{-1}v) \ . \tag{2.24}$$

Its existence enables us to construct the single-photon fibre bundle

$$SP\mathbf{M} = P^{\uparrow}\mathbf{M} \times_{\mathbf{G}} \mathbf{K} \ , \qquad P^{\uparrow}\mathbf{M} \cong \mathbf{P}(\mathbf{M},\mathbf{G}) \ , \qquad \mathbf{G} = \mathrm{ISO}^{\uparrow}(3,1) \ , \tag{2.25}$$

associated to the principal Poincaré frame bundle $P^{\uparrow}\mathbf{M}$, whose typical fibre is diffeomorphic to the orthochronous Poincaré group $\mathrm{ISO}^{\uparrow}(3,1)$. The theory of propagators for the parallel transport of single photon states within $\mathbf{K}$ can be easily developed (Prugovečki, 1988b). However, we shall not reproduce those results here, since it is the propagation of multi-photon Gupta-Bleuler quantum frames, rather than of single-photon states, that is of primary interest in quantum electrodynamics. Such quantum frames will be treated as elements of an enveloping Gupta-Bleuler bundle $\mathcal{G}_e$, which will be constructed in the next section.

9.3. Gupta-Bleuler Quantum Bundles and Frames

At a purely formal level, the construction of a Gupta-Bleuler bundle, in which a GS quantum electromagnetic field could act, might appear to be able to proceed along lines very similar to the construction of the Fock quantum bundle in Sec. 7.4. Indeed, we could use the single-photon standard fibre $\mathbf{K}$ to construct the Gupta-Bleuler space

$$\breve{\mathcal{K}} = \breve{\bigoplus}_{n=0}^{\infty} \mathcal{K}_n \ , \qquad \mathcal{K}_n = \mathbf{K} \underset{S}{\otimes} \cdots \underset{S}{\otimes} \mathbf{K} \ , \tag{3.1}$$

in which the representation in (2.24) then gives rise to the pseudo-unitary representation

$$\mathbf{U}_J(a,\Lambda) = \bigoplus_{J}{}_{n=0}^{\infty} U(a,\Lambda)^{\otimes n} \ , \qquad (a,\Lambda) \in \mathrm{ISO}^{\uparrow}(3,1) \ , \tag{3.2}$$

of the orthochronous Poincaré group. It might therefore appear that, as it was the case in (7.4.2), we could define the Gupta-Bleuler bundle as being equal to the $\mathbf{G}$-product of the extended Poincaré principal frame bundle $P^{\uparrow}\mathbf{M}$ with the Gupta-Bleuler space in (3.1).

At the mathematically rigorous level, special care is, however, required on account of the fact that the representation in (3.2) is unbounded in the J-norm topology of the Krein space in (3.1). In fact, the conventional counterparts of the operators in (3.2), acting in the Gupta-Bleuler spaces constructed out of the Krein spaces in (1.6) or (1.10), are also unbounded[11], so that these Gupta-Bleuler spaces cannot be treated as ordinary Krein spaces. Rather, their mathematically rigorous formulation requires a generalization of the concept of Krein space[12], whereby all possible J-inner products have to be considered simultaneously, each one giving rise to a separate branch of that Gupta-Bleuler space. Each one of these branches is by itself a Krein space, and they all have in common the algebraic direct sum of n-photon subspaces for $n = 0,1,2,\dots$, but they all have distinct J-closures.

In the case of the Gupta-Bleuler space in (3.1), these branches are Krein spaces that equal the J-direct sums

$$\mathcal{K}_J = \bigoplus_{J}{}_{n=0}^{\infty} \mathcal{K}_n \ , \qquad J \in \mathcal{L}_{\mathrm{B}} \leftrightarrow \mathcal{L}^{(3)} := \left\{ \Lambda_{\vartheta}^{(3)} \middle| \vartheta \in \mathbf{R}^1 \right\} \ , \tag{3.3}$$

and have in common the algebraic[13] direct sum $\mathcal{K}_\infty$ of the n-photon spaces $\mathcal{K}_n$, which the representation in (3.2) leaves invariant:

$$\breve{\mathcal{K}} = \bigcup_{J\in\mathcal{L}_B} \mathcal{K}_J, \qquad \mathbf{U}_J(a,\Lambda) : \mathcal{K}_\infty \to \mathcal{K}_\infty = \bigoplus_{a\ n=0}^{\infty} \mathcal{K}_n \subset \bigcap_{J\in\mathcal{L}_B} \mathcal{K}_J. \tag{3.4}$$

The family $\mathcal{L}_B$ of all distinct J-operators in (3.3) can be set in one-to-one correspondence with the family of J-operators in (1.7) or (1.12). In turn, this latter family can be set in one-to-one correspondence with the family of all Lorentz boosts of a given frame (such as the canonical basis in $\mathbf{R}^4$, giving rise to the fundamental symmetry J_0) in the direction of its third spatial axis – cf. (5.2.21). Thus, the pseudo-unitary operators in (3.2) corresponding to such boosts provide isomorphisms that relate pairs of Krein-Gupta-Bleuler spaces:

$$\mathbf{U}_J(a,\Lambda_\vartheta^{(3)}) : \mathcal{K}_J \to \mathcal{K}_{J'}, \qquad \Lambda_\vartheta^{(3)} : \boldsymbol{u} \mapsto \boldsymbol{u}' . \tag{3.5}$$

All this suggests[14] the introduction of a base manifold $\mathbf{L}$ identifiable, for any given global section $\boldsymbol{s}_0$ of $P^\uparrow\mathbf{M}$, with the principal bundle of all Lorentz frames obtained from each other by boost operations along their $\boldsymbol{e}_3$ axes:

$$\mathbf{L} \cong (\mathbf{P}_0, \Pi, \boldsymbol{s}_0, \mathcal{L}^{(3)}) \subset P\mathbf{M}, \qquad \mathbf{M} \leftrightarrow \boldsymbol{s}_0, \quad \mathrm{SO}_0(1,1) \leftrightarrow \mathcal{L}^{(3)} \cong \mathcal{L}_B . \tag{3.6}$$

Using the well-known fact that any proper Lorentz transformation can be decomposed as follows (cf. Scharf, 1989, Theorem 1.1, p. 4),

$$\Lambda = R_l^{-1}\Lambda_\vartheta^{(3)}R_r, \qquad \Lambda_\vartheta^{(3)} \in \mathcal{L}^{(3)}, \qquad R_l, R_r \in \mathrm{SO}(3) \subset \mathrm{SO}_0(3,1), \tag{3.7}$$

we can carry out in a natural manner the identification

$$P^\uparrow\mathbf{M} \leftrightarrow (P\mathbf{L}, \Pi_0, \mathbf{L}, \mathbf{G}_0), \qquad \mathbf{G}_0 = \mathrm{O}(3)_l \times (\mathcal{T}_4 \wedge \mathrm{SO}(3))_r . \tag{3.8}$$

An *enveloping Gupta-Bleuler bundle* $\mathcal{G}_e$ can be then constructed, which has $\mathbf{L}$ as its base manifold, $\mathbf{G}_0$ as its structure group, and the Krein-Gupta-Bleuler space $\mathcal{K}$ corresponding to the canonical basis in $\mathbf{R}^4$ as its typical fibre:

$$\mathcal{G}_e = P\mathbf{L} \times_{\mathbf{G}_0} \mathcal{K}, \qquad P\mathbf{L} = (P^\uparrow\mathbf{M}, \Pi_0, \mathbf{L}, \mathbf{G}_0), \qquad \mathcal{K} = \bigoplus_{J_0\ n=0}^{\infty} \mathcal{K}_n . \tag{3.9a}$$

On account of (3.8), the *core Gupta-Bleuler bundle* $\mathcal{G}_c$, which has $\mathbf{M}$ as its base manifold, the orthochronous Poincaré group $\mathrm{ISO}^\uparrow(3,1)$ as its structure group, and the restricted Gupta-Bleuler space $\mathcal{K}_\infty$ in (3.4) as its typical fibre, can be then imbedded into it:

$$\mathcal{G}_c = P^\uparrow\mathbf{M} \times_{\mathbf{G}} \mathcal{K}_\infty \xrightarrow[I]{} \mathcal{G}_e, \qquad \mathbf{G} = \mathrm{ISO}^\uparrow(3,1) . \tag{3.9b}$$

The considerations in Sec. 7.4 can be now combined with those of the preceding section, and then extended to this nested type of Gupta-Bleuler bundle. Such a task can be most easily accomplished by first extending them, in an obvious manner, to the auxiliary second-quantized Proca bundles

$$\mathcal{G}^{(\mu_0)} = P\mathbf{L} \times_{\mathbf{G}_0} \mathcal{K}^{(\mu_0)}, \quad \mathcal{K}^{(\mu_0)} = \bigoplus\nolimits_{n=0}^{\infty}{}_{J_{\mu_0}} \mathcal{K}_n^{(\mu_0)}, \quad \mathcal{K}_n^{(\mu_0)} = \mathbf{K}^{(\mu_0)} \underset{S}{\otimes} \cdots \underset{S}{\otimes} \mathbf{K}^{(\mu_0)}, \tag{3.10}$$

constructed from the wave functions in (2.9), and then taking the zero-mass limits in accordance with (2.12). By this method we can arrive at the generalized soldering maps of the Gupta-Bleuler fibres,

$$\sigma_x^u : \boldsymbol{\Psi} \mapsto \Psi \in \mathcal{K}\ , \qquad \boldsymbol{\Psi} \in \mathcal{K}_{x,J}\ , \qquad \boldsymbol{u} = (\boldsymbol{a}, \boldsymbol{e}_i) \in \Pi^{-1}(x) \subset P\mathbf{L}\ , \tag{3.11}$$

by assigning to each n-photon local state $\boldsymbol{\Psi}_{n;x} \in \mathcal{K}_{n;x}$ a *coordinate wave function* $\Psi_{n;x} \in \mathcal{K}_n$,

$$\sigma_x^u : \boldsymbol{\Psi}_{n;x}(\zeta_1, i_1, \ldots, \zeta_n, i_n) \mapsto \Psi_{n;x}(\zeta_1, i_1, \ldots, \zeta_n, i_n)\ , \qquad \boldsymbol{\Psi}_{n;x} \in \mathcal{K}_{n;x}\ , \tag{3.12a}$$

$$\boldsymbol{\zeta}_r = (\boldsymbol{a} + q_r^i \boldsymbol{e}_i, v_r^i \boldsymbol{e}_i) \in T_x\mathbf{M} \times \mathbf{V}_x^+\ , \quad \zeta_r = q_r + iv_r, \quad r = 1, \ldots, n\ , \tag{3.12b}$$

and then extending the above map to the entire Krein-Gupta-Bleuler fibre in (3.11). Note should be made of the fact that n-photon subspaces in (3.10) are J-independent, so that the Gupta-Bleuler bundle displays ordinary Poincaré gauge invariance for states with finite numbers of photons, whereas novel fibre-theoretical features emerge for (countably) infinite superpositions of such states. As we shall see in the next chapter, these novel features are closely related to supergauge degrees of freedom.

The above construction also provides the *local photon fluctuation amplitudes*

$$D_x^{(+)}(\boldsymbol{\zeta}''; \boldsymbol{\zeta}') = -i\int_{\mathbf{V}_0^+} \exp\left[-i(\bar{\zeta}'' - \zeta') \cdot k\right] d\Omega_0(k) = 2\pi i \big/ \left(\bar{\zeta}'' - \zeta'\right)^2\ , \tag{3.13}$$

by means of which we can construct photon creation operators:

$$\begin{aligned}&\left(A_i^{(+)}(x;\boldsymbol{\zeta})\boldsymbol{\Psi}_{n;x}\right)_{n+1}(\boldsymbol{\zeta}_1, i_1, \ldots, \boldsymbol{\zeta}_{n+1}, i_{n+1}) \\ &= -i(n+1)^{-1/2} \sum\nolimits_{r=1}^{n+1} \eta_{ii_r} D_x^{(+)}(\boldsymbol{\zeta}_r; \boldsymbol{\zeta})\, \boldsymbol{\Psi}_{n;x}(\boldsymbol{\zeta}_1, i_1, \ldots, \boldsymbol{\zeta}_{r-1}, i_{r-1}, \boldsymbol{\zeta}_{r+1}, i_{r+1}, \ldots, \boldsymbol{\zeta}_{n+1}, i_{n+1}).\end{aligned} \tag{3.14}$$

The corresponding annihilation operators satisfy the relations

$$\left(A_i^{(-)}(x;\boldsymbol{\zeta})\boldsymbol{\Psi}_{n;x}\right)_{n-1}(\boldsymbol{\zeta}_1, i_1, \ldots, \boldsymbol{\zeta}_{n-1}, i_{n-1}) = \sqrt{n}\, \boldsymbol{\Psi}_{n;x}(\boldsymbol{\zeta}, i, \boldsymbol{\zeta}_1, i_1, \ldots, \boldsymbol{\zeta}_{n-1}, i_{n-1})\ , \tag{3.15}$$

and are equal to the pseudo-adjoints of the appropriate J-closures of the creation operators in each Krein-Gupta-Bleuler fibre. These families of creation and annihilation operators satisfy the following commutation relations:

$$\left[A_i^{(-)}(x;\zeta), A_{i'}^{(+)}(x;\zeta')\right] = -i\eta_{ii'} D_x^{(+)}(\zeta;\zeta') , \quad \left[A_i^{(\pm)}(x;\zeta), A_{i'}^{(\pm)}(x;\zeta')\right] = 0. \tag{3.16}$$

They also give rise to corresponding *Gupta-Bleuler quantum frame fields*

$$A_i(x;\zeta) = A_i^{(+)}(x;\zeta) + A_i^{(-)}(x;\zeta) \ . \tag{3.17}$$

The well-known Gupta-Bleuler type of subsidiary conditions

$$\partial^i A_i^{(-)}(x;\zeta)\boldsymbol{\Psi} = \mathbf{0} \ , \qquad \boldsymbol{\Psi} \in \mathcal{K}_{x,J} \ , \qquad \partial^i = \eta^{ij}\partial_j \ , \quad \partial_i = \partial/\partial q^i \ , \tag{3.18}$$

give rise, upon taking J-closures of the sets of local state vectors satisfying them, to a subbundle of the enveloping Gupta-Bleuler bundle $\mathcal{G}_e$. This subbundle coincides with the *Gupta-Bleuler-Lorenz bundle*

$$\mathcal{G}_e^L = P\mathbf{L} \times_{\mathbf{G}_0} \mathcal{K}^L \ , \qquad \mathcal{K}^L = \bigoplus_{J_0}{}_{n=0}^{\infty} \mathcal{K}_n^L \ , \qquad \mathcal{K}_n^L = \mathbf{K}^L \underset{S}{\otimes} \cdots \underset{S}{\otimes} \mathbf{K}^L \ , \tag{3.19}$$

which is constructed from the *Lorenz typical single-photon fibre*

$$\mathbf{K}^L = \left\{ f \in \mathbf{K} \,\middle|\, \partial^i f_i(q,v) \equiv 0 \right\} \tag{3.20}$$

in the same manner in which the Gupta-Bleuler bundle was constructed from the typical Krein-Gupta-Bleuler fibre.

Following a method of Gomatam (1971), used for constructing coherent states in Fock-type spaces with indefinite metric, we introduce the *Gupta-Bleuler quantum frames*

$$\boldsymbol{\Phi}_f = \exp\left[-\tfrac{1}{2}\langle f|f\rangle - \int f^i(\zeta) A_i^{(+)}(x;\zeta) d\tilde{\Sigma}(\zeta)\right] \boldsymbol{\Psi}_{0;x} \ , \qquad f \in \mathbf{K}_{x,J} \ , \tag{3.21}$$

in each of the Krein-Gupta-Bleuler fibres $\mathcal{K}_{x,J}$. The Berezin method of functional integration, adapted in Sec. 7.4 to the Fock quantum bundle for spin-0 massive fields, can be now further adapted to the present zero-mass situation by introducing an orthonormal basis in each one of the four components of the J-direct sum into which the typical Krein-Maxwell single-photon fibre decomposes with respect to the canonical frame,

$$f = \sum_{i=0}^{3} \sum_{\alpha=1}^{\infty} z_{i,\alpha} w_{i,\alpha} \ , \qquad z_{i,\alpha} = \left(w_{i,\alpha}|f\right)_{J_0} \ , \qquad f, w_{i,\alpha} \in \mathbf{K} \ . \tag{3.22}$$

It can be then derived from (2.23) and (3.16), by essentially the same procedure as the one used in (7.4.29), that the operators

$$a_{i,\alpha} = -\int \overline{w}_{i,\alpha}^j(\zeta) A_j^{(-)}(x;\zeta) d\tilde{\Sigma}(\zeta) \ , \qquad a_{i,\alpha}^\dagger = -\int w_{i,\alpha}^j(\zeta) A_j^{(+)}(x;\zeta) d\tilde{\Sigma}(\zeta) \ , \tag{3.23}$$

obey the following commutation relations:

$$\left[a_{i,\alpha}, a^{\dagger}_{j,\beta}\right] = -\eta_{ij}\delta_{\alpha\beta} \ , \qquad \left[a_{i,\alpha}, a_{j,\beta}\right] = \left[a^{\dagger}_{i,\alpha}, a^{\dagger}_{j,\beta}\right] = 0 \ . \tag{3.24}$$

Upon inserting (3.22) into (3.21), and then using the J-orthonormality properties of the basis in (3.22), as well as the algebraic relation between the indefinite and the J-inner products in (2.18), it can be easily verified that

$$\boldsymbol{\Phi}_f = \exp\left[\sum_{\alpha=1}^{\infty} \eta^{ij}\left(\tfrac{1}{2}\bar{z}_{i,\alpha} z_{j,\alpha} - z_{i,\alpha} a^{\dagger}_{j,\alpha}\right)\right] \boldsymbol{\Psi}_{0;x} \ \in \mathcal{K}_{x,J} \ , \tag{3.25a}$$

$$\boldsymbol{f} = \sum_{i=0}^{3} \sum_{\alpha=1}^{\infty} z_{i,\alpha}\, \boldsymbol{w}_{i,\alpha} \in \mathbf{K}_{x,J} \ , \qquad \boldsymbol{w}_{i,\alpha} = \left(\boldsymbol{\sigma}^{u}_{x}\right)^{-1} w_{i,\alpha} \ . \tag{3.25b}$$

From the above expression it can deduced in a routine manner that the elements of the Gupta-Bleuler quantum frames are eigenvectors of the photon annihilation operators:

$$A_i^{(-)}(x;\zeta)\boldsymbol{\Phi}_f = f_i(\zeta)\boldsymbol{\Phi}_f \ , \qquad \boldsymbol{f} = f_i\,\boldsymbol{\theta}^i \in \mathbf{K}_{x,J} \ . \tag{3.26}$$

On the other hand, when trying to derive a counterpart of (7.4.31), the fact that for each one of the timelike basis elements in (3.22) the commutator of the corresponding pair of annihilation and creation operators in (3.24) displays a negative rather than a positive value gives rise to the divergent factor

$$\exp\left[\tfrac{1}{2}\sum_{\alpha=1}^{\infty}\left|z_{0,\alpha}\right|^2\right] = \exp\left[\tfrac{1}{2}\sum_{\alpha=1}^{\infty}\int\left|f_0(\zeta)\right|^2 d\tilde{\Sigma}(\zeta)\right] \ , \tag{3.27}$$

if one proceeds formally, as it is customary in conventional quantum field theory [SI,IQ]. This factor can be, however, eliminated by introducing a compensating factor in the measure of integration, with the following final outcome:

$$\int_{\mathbf{K}_{x,J}} \left|\boldsymbol{\Phi}_f\right) \mathbf{d}\boldsymbol{f}\,\mathbf{d}\bar{\boldsymbol{f}} \left(\boldsymbol{\Phi}_f\right| \ = \ \text{w-}\lim_{n\to\infty} \pi^{-4n} \int_{\mathbf{R}^{8n}} \left|\hat{z}_n\right) \mathbf{d}\hat{z}_n \mathbf{d}\hat{z}_n^{\,*} \left(\hat{z}_n\right| \ = \ \mathbf{1}_{x,J} \ , \tag{3.28a}$$

$$\left|\hat{z}_n\right) = \left|z_{0,1},\dots,z_{3,n},0,0,\dots\right) = \exp\left[\sum_{\alpha=1}^{n} \eta^{ij}\left(\tfrac{1}{2}\bar{z}_{i,\alpha} z_{j,\alpha} - z_{i,\alpha} a^{\dagger}_{j,\alpha}\right)\right] \boldsymbol{\Psi}_{0;x} \ , \tag{3.28b}$$

$$\mathbf{d}\hat{z}_n \mathbf{d}\hat{z}_n^{*} = \exp\left[-\sum_{\alpha=1}^{n}\left|z_{0,\alpha}\right|^2\right] \prod_{\alpha=1}^{n}\prod_{i=0}^{3} d(\mathrm{Re}\, z_{i,\alpha})\, d(\mathrm{Im}\, z_{i,\alpha}) \ . \tag{3.28c}$$

The limit in (3.28a) can be taken in the weak topology induced by the indefinite metric in $\mathcal{K}_{x,J}$. However, the presence of the exponential factor in (3.28c) destroys any possibility of manifest covariance of the formalism. The same situation is encountered in the treatment of coherent states within the conventional Gupta-Bleuler framework (Gomatam, 1971). This is to be expected, in view of the general observations about frame-dependence of Gupta-Bleuler spaces presented at the beginning of this section.

Formal covariance can be restored to the above types of continuous resolutions of the identity if the functional integration is restricted to the *transversal-mode Gupta-Bleuler-Lorenz subbundle*

$$\mathcal{G}_e^{\mathrm{T}} = P\mathbf{L} \times_{\mathbf{G}_0} \mathcal{K}^{\mathrm{T}} , \qquad \mathcal{K}^{\mathrm{T}} = \bigoplus_{J_0\ n=0}^{\infty} \mathcal{K}_n^{\mathrm{T}} , \qquad \mathcal{K}_n^{\mathrm{T}} = \mathbf{K}_{J_0}^{\mathrm{T}} \underset{S}{\otimes} \cdots \underset{S}{\otimes} \mathbf{K}_{J_0}^{\mathrm{T}} , \tag{3.29}$$

of the Gupta-Bleuler-Lorenz bundle in (3.19), obtained by restricting the single-photon wave functions to the *transversal-polarization Lorenz subfibre*

$$\mathbf{K}_{J_0}^{\mathrm{T}} = \left\{ f \in \mathbf{K} \middle| \partial^i f_i(\zeta) \equiv 0 ,\, f^{(0)}(\zeta) \equiv f^{(3)}(\zeta) \equiv 0 \right\} \subset \mathbf{K}^L . \tag{3.30}$$

The elements of this subfibre can be obtained from those of the subspace $\tilde{\mathbf{K}}_{J_0}^L$ in the decomposition in (1.21), to which the fundamental symmetry J_0 corresponding to the canonical basis $\boldsymbol{u}_0$ in $\mathbf{R}^4$ gives rise, upon introducing within the typical single-photon fibre in (3.20) the *photon polarization tetrads*

$$\Phi_{(\mu)\zeta}^{\boldsymbol{u}_0}(\zeta') = \int_{\mathbf{V}_0^+} \exp\left[i(\zeta' - \bar{\zeta}) \cdot k \right] \boldsymbol{\varepsilon}_{(\mu)}(\boldsymbol{k})\, d\Omega_0(\boldsymbol{k}) , \qquad \mu = 0,1,2,3 , \tag{3.31}$$

and the corresponding *photon polarization coordinate wave functions*:

$$f^{(\mu)}(\zeta) = \eta^{\mu\nu} \left\langle \Phi_{(\nu)\zeta}^{\boldsymbol{u}_0} \middle| f \right\rangle , \qquad f \in \mathbf{K}^L . \tag{3.32}$$

Indeed, we can then expand any local single-photon state vector with respect to the above quantum polarization tetrads in the following manner :

$$\boldsymbol{f} = -\int d\hat{\Sigma}(\zeta) f^{(\mu)}(\zeta) \boldsymbol{\Phi}_{(\mu)\zeta}^{\boldsymbol{u}(x)} , \qquad \boldsymbol{f} \in \mathbf{K}_{x,J}^L , \tag{3.33a}$$

$$f^{(\mu)}(\zeta) = \eta^{\mu\nu} \left\langle \boldsymbol{\Phi}_{(\nu)\zeta}^{\boldsymbol{u}(x)} \middle| \boldsymbol{f} \right\rangle , \qquad \boldsymbol{\Phi}_{(\mu)\zeta}^{\boldsymbol{u}(x)} = \left(\boldsymbol{\sigma}_x^{\boldsymbol{u}} \right)^{-1} \Phi_{(\mu)\zeta}^{\boldsymbol{u}_0} . \tag{3.33b}$$

According to (1.17), (1.19) and (1.20), we can write

$$\langle \boldsymbol{f} | \boldsymbol{f}' \rangle = -\int \eta_{\mu\nu} \bar{f}^{(\mu)}(\zeta) f^{(\nu)\prime}(\zeta) d\tilde{\Sigma}(\zeta) = \sum_{\rho=1}^{2} \int \bar{f}^{(\rho)}(\zeta) f^{(\rho)\prime}(\zeta) d\tilde{\Sigma}(\zeta) , \tag{3.34a}$$

$$\left(\boldsymbol{f} | \boldsymbol{f}' \right)_J = \sum_{\mu=0}^{3} \int \bar{f}^{(\mu)}(\zeta) f^{(\mu)\prime}(\zeta) d\hat{\Sigma}(\zeta) , \qquad \boldsymbol{f}, \boldsymbol{f}' \in \mathbf{K}_{x,J}^L . \tag{3.34b}$$

Consequently, we can carry out a unique decomposition, corresponding to the J-direct sum in (1.21), of any single-photon local state vector in the Lorenz gauge:

$$\boldsymbol{f} = \boldsymbol{f}^{\mathrm{T}} + \boldsymbol{f}^{\perp} \in \mathbf{K}_{x,J}^L , \qquad \boldsymbol{f}^{\perp} \in \mathbf{K}_{x,J}^0 = \left\{ \boldsymbol{h} \in \mathbf{K}_{x,J}^L \mid \langle \boldsymbol{h} | \boldsymbol{h} \rangle = 0 \right\} , \tag{3.35a}$$

$$f^{\mathrm{T}} = -\int d\tilde{\Sigma}(\zeta) f^{(\rho)}(\zeta) \boldsymbol{\Phi}^{\boldsymbol{u}(x)}_{(\rho)\zeta} \in \mathbf{K}^{\mathrm{T}}_{x,J} \ , \ f^{\perp} = -\int d\tilde{\Sigma}(\zeta) f^{(0)}(\zeta) \left(\boldsymbol{\Phi}^{\boldsymbol{u}(x)}_{(0)\zeta} + \boldsymbol{\Phi}^{\boldsymbol{u}(x)}_{(3)\zeta} \right). \quad (3.35b)$$

When we insert it into (3.21), we find that there is a similar decomposition of the Gupta-Bleuler quantum frame for Lorenz gauge modes,

$$\boldsymbol{\Phi}_f = \boldsymbol{\Phi}_{f^{\mathrm{T}}} + \boldsymbol{\Phi}_f^{\perp} \ , \qquad f \in \mathbf{K}^{L}_{x,J} \ , \tag{3.36}$$

into a component that depends only on the two physical (i.e., transversal) polarization modes, and which, on account of (3.34), can be expressed also in the following alternative form,

$$\boldsymbol{\Phi}_{f^{\mathrm{T}}} = \exp\left[-\tfrac{1}{2}\langle f^{\mathrm{T}} | f^{\mathrm{T}} \rangle + \int \left[f^{(1)}(\zeta) + f^{(2)}(\zeta) \right] \cdot \mathbf{A}^{(+)}(x;\zeta) d\tilde{\Sigma}(\zeta) \right] \boldsymbol{\Psi}_{0;x} \ , \tag{3.37a}$$

$$f^{(\mu)}(\zeta) = f_{(\mu)}(\zeta) \boldsymbol{\Phi}^{\boldsymbol{u}(x)}_{(\mu)\zeta} \ , \qquad f_{(\mu)} = \eta_{\mu\nu} f^{(\nu)} \ , \qquad \mu = 0,1,2,3 \ , \tag{3.37b}$$

and a remainder obtained upon expanding the resulting exponential into a series:

$$\boldsymbol{\Phi}_f^{\perp} = \sum_{n=1}^{\infty} (n!)^{-1} \left(\int \left[f^{(0)}(\zeta) + f^{(3)}(\zeta) \right] \cdot \mathbf{A}^{(+)}(x;\zeta) d\tilde{\Sigma}(\zeta) \right)^n \boldsymbol{\Phi}_{f^{\mathrm{T}}} \ . \tag{3.38}$$

It is then easily seen that

$$\langle \boldsymbol{\Phi}_{f^{\mathrm{T}}} | \boldsymbol{\Phi}_{f^{\mathrm{T}}} \rangle = \langle \boldsymbol{\Phi}_f | \boldsymbol{\Phi}_f \rangle = 1 \ , \qquad \langle \boldsymbol{\Phi}_f^{\perp} | \boldsymbol{\Phi}_{f^{\mathrm{T}}} \rangle = \langle \boldsymbol{\Phi}_f^{\perp} | \boldsymbol{\Phi}_f^{\perp} \rangle = 0 \ . \tag{3.39}$$

We can now adopt, instead of (3.22), an expansion into a basis respecting the four polarization modes, so that

$$f = \sum_{\mu=0}^{3} \sum_{\alpha=1}^{\infty} z_{\alpha}^{(\mu)} w_{\alpha}^{(\mu)} \ , \qquad z_{\alpha}^{(\mu)} = \left(w_{\alpha}^{(\mu)} | f \right)_J \ , \qquad f, w_{\alpha}^{(\mu)} \in \mathbf{K} \ , \tag{3.40a}$$

$$\boldsymbol{f} = \sum_{\mu=0}^{3} \sum_{\alpha=1}^{\infty} z_{\alpha}^{(\mu)} \boldsymbol{w}_{\alpha}^{(\mu)} \in \mathbf{K}_{x,J} \ , \qquad \boldsymbol{w}_{\alpha}^{(\mu)} = \left(\boldsymbol{\sigma}_x^{\boldsymbol{u}} \right)^{-1} w_{\alpha}^{(\mu)} \ . \tag{3.40b}$$

We can then proceed as in (3.23)-(3.25), by first introducing the annihilation and creation operators

$$a_{\alpha}^{(\mu)} = -\int \overline{\boldsymbol{w}}_{\alpha}^{(\mu)}(\zeta) \cdot \mathbf{A}^{(-)}(x;\zeta) d\hat{\Sigma}(\zeta) \ , \ a_{\alpha}^{(\mu)\dagger} = -\int \boldsymbol{w}_{\alpha}^{(\mu)}(\zeta) \cdot \mathbf{A}^{(+)}(x;\zeta) d\hat{\Sigma}(\zeta) \ , \tag{3.41}$$

and then, upon verifying the counterparts of (3.24), deriving a counterpart of (3.25a). The essential new observation is that

$$\boldsymbol{\Phi}_{f^{\mathrm{T}}} = \exp\left[\sum_{\alpha=1}^{\infty} \sum_{\rho=1}^{2} \left(-\tfrac{1}{2} \left| z_{\alpha}^{(\rho)} \right|^2 + z_{\alpha}^{(\rho)} a_{\alpha}^{(\rho)\dagger} \right) \right] \boldsymbol{\Psi}_{0;x} \in \mathcal{K}^{\mathrm{T}}_{x,J} \ . \tag{3.42}$$

Hence, we obtain the following counterparts of (3.28),

$$\int_{\mathbf{K}^{\mathrm{T}}_{x,J}} |\Phi_f\rangle\, df\, d\bar{f} \langle\Phi_f| = \mathrm{w}\text{-}\lim_{n\to\infty} \pi^{-2n} \int_{\mathbf{R}^{4n}} |\hat{z}_n\rangle \mathbf{d}\hat{z}_n \mathbf{d}\hat{z}_n{}^* \langle\hat{z}_n| = \mathbf{1}^{\mathrm{T}}_{x,J} , \tag{3.43a}$$

$$|\hat{z}_n\rangle = |z_{1,1},\ldots,z_{2,n},0,0,\ldots\rangle = \exp\left[\sum_{\alpha=1}^{n}\sum_{\rho=1}^{2}\left(-\tfrac{1}{2}\left|z_\alpha^{(\rho)}\right|^2 + z_\alpha^{(\rho)} a_\alpha^{(\rho)\dagger}\right)\right]\Psi_{0;x} , \tag{3.43b}$$

$$d\hat{z}_n\, d\hat{z}_n^* = \prod_{\alpha=1}^{n}\prod_{\rho=1}^{2} d(\mathrm{Re}\, z_\alpha^{(\rho)})\, d(\mathrm{Im}\, z_\alpha^{(\rho)}) , \tag{3.43c}$$

in which the measure for integration contains no extraneous exponential factors.

Thus, we have arrived at continuous resolutions of the identity operators in the transversal-mode subfibres of the Gupta-Bleuler-Lorenz bundle which possess all the formal covariance features associated with those modes, namely manifest covariance under spatial rotations and spacetime translations. Moreover, since on account of (3.36) and (3.39),

$$\mathbf{K}^{L}_{x,J} = \mathbf{K}^{\mathrm{T}}_{x,J} \underset{J}{\oplus} \mathbf{K}^{0}_{x,J} , \qquad \mathcal{K}^{L}_{x,J} = \mathcal{K}^{\mathrm{T}}_{x,J} \underset{J}{\oplus} \mathcal{K}^{0}_{x,J} , \tag{3.44}$$

where the second terms in the above J-direct sums are null subfibres (i.e., subspaces of a Gupta-Bleuler-Lorenz fibre consisting of vectors of zero indefinite-metric norm), we can express the indefinite inner product in each Gupta-Bleuler-Lorenz fibre in terms of the functional integral in (3.43):

$$\int_{\mathbf{K}^{\mathrm{T}}_{x,J}} \langle\Psi|\Phi_f\rangle\, df\, d\bar{f} \langle\Phi_f|\Psi'\rangle = \langle\Psi|\Psi'\rangle , \qquad \forall\Psi,\Psi' \in \mathbf{K}^{L}_{x,J} . \tag{3.45}$$

Hence, although manifest Poincaré covariance is lacking, a physical type of Poincaré covariance prevails in Gupta-Bleuler-Lorenz bundles: the transversal-polarization Lorenz subfibres carry representative photon state vectors which fall into equivalence classes, and the equivalence relations defining them are based on representations of the Poincaré group for every state with a finite number of photons.

*9.4. Parallel Transport in Gupta-Bleuler Quantum Bundles

Parallel transport within the core of the Gupta-Bleuler quantum bundle defined in (3.9b), can be formulated as in the massive case, since that bundle is associated directly to the Poincaré frame bundle PM. Thus, the operator for parallel transport

$$\tau_\gamma(x'',x') : \Psi' \mapsto \Psi'' , \qquad \Psi' \in \mathcal{K}_{\infty;x'} , \qquad \Psi'' \in \mathcal{K}_{\infty;x''} , \tag{4.1}$$

is determined by the operator in (5.2.11) in terms of the following map:

$$\tau_\gamma(x'',x') = \left(\sigma_{x''}^{u''}\right)^{-1} \circ \sigma_{x'}^{u'} , \qquad u'' = \tilde{\tau}_\gamma(x'',x')u' . \tag{4.2}$$

The above linear map leaves unchanged the respective n-photon coordinate wave functions defined by (3.12), so that, in particular,

$$f'' = \tau_\gamma(x'',x')f' \in \mathbf{K}_{x'',J''} , \quad f' \in \mathbf{K}_{x',J'} \quad \Leftrightarrow \quad \sigma_{x'}^{u'} f' = \sigma_{x''}^{u''} f'' \in \mathbf{K} . \tag{4.3}$$

Consequently, it is immediately seen that, in each of the n-photon subfibres, the operator for parallel transport in (4.2) is pseudo-unitary, as well as J-unitary with respect to the J-inner products induced by the initial and final Poincaré frames.

One of the mathematical features of the present mass-0 case, which is absent in the massive case considered in the preceding two chapters, emerges from the need[15] to use as a base manifold of the enveloping Gupta-Bleuler bundle the principal bundle $\mathbf{L}$ in (3.6), rather than the Lorentz manifold $(\mathbf{M}, g)$ itself. This foreshadows the need of using appropriate extensions of the base manifold $\mathbf{M}$ in the construction, in the next chapter, of quantum bundles for non-Abelian gauge fields. Indeed, the extension of the base manifold $\mathbf{M}$ into either a principal bundle or superbundle necessitates the imposition of constraints, which will turn out to assume the form of a generalization of the well-known BRST constraints in the conventional theory of non-Abelian gauge fields [IQ].

In the GS framework the need for such constraints can be justified geometrically, by distinguishing between physical and unphysical smooth paths in the thus extended base manifold. Indeed, a path in a principal frame bundle with a total space $\mathbf{L}$, which lies entirely within a given fibre merely corresponds to a change of frame at a given base location $x \in \mathbf{M}$, and it is therefore unphysical from the point of view of quantum propagation. On the other hand, any physical process based on the application of the equivalence principle to the parallel transport of frames, should allow amongst the paths $\gamma^\uparrow$ which are lifts of paths γ in $\mathbf{M}$, namely which correspond to an assignment of a frame $u(x) \in \mathbf{L}$ to each $x \in \gamma$, only those $\gamma^\uparrow$ for which all the frames $u(x)$ along $\gamma^\uparrow$ are parallel transports of each other. In other words, a smooth path $\gamma^\uparrow$ in $\mathbf{L}$ is *physical* if and only if it equals the horizontal lift γ^* (cf. Sec. 2.4) of a smooth path γ in $\mathbf{M}$.

For such a smooth physical path γ^* in $\mathbf{L}$, which connects (x',J') to (x'',J''), parallel transport within the enveloping Gupta-Bleuler bundle in (3.9a) can be defined in a natural manner, and directly from (4.2) and (4.3), by setting

$$\tau_{\gamma^*}(x'',x') : \Phi_{f'} \mapsto \Phi_{f''} , \qquad f' = f'' \in \mathbf{K} , \quad f' = \sigma_{x'}^{u'} f' , \quad f'' = \sigma_{x''}^{u''} f'' . \tag{4.4}$$

This definition of parallel transport of Gupta-Bleuler quantum frames along physical paths can be then extended to all vectors within each fibre of $\mathcal{G}_c$ by using (3.28),

$$\tau_{\gamma^*}(x'',x') : \int_{\mathbf{K}_{x',J'}} \mathbf{d}f'\mathbf{d}\bar{f}' \, \Psi(f') \, \Phi_{f'} \mapsto \int_{\mathbf{K}_{x'',J''}} \mathbf{d}f''\mathbf{d}\bar{f}'' \, \Psi(f'') \, \Phi_{f''} . \tag{4.5}$$

It then gives rise to operators for parallel transport that are pseudo-unitary, as well as J-unitary with respect to the J-inner products induced by the initial and final Poincaré frames:

$$\left\langle \tau_{\gamma^*}(x'',x')\Psi \middle| \tau_{\gamma^*}(x'',x')\Psi' \right\rangle = \langle \Psi | \Psi' \rangle , \qquad \Psi, \Psi' \in \mathcal{K}_{x',J'} , \tag{4.6a}$$

$$\left(\tau_{\gamma *}(x'',x')\Psi \middle| \tau_{\gamma *}(x'',x')\Psi'\right)_{J''} = \left(\Psi \middle| \Psi'\right)_{J'} \ . \tag{4.6b}$$

Thereby, all functional-analytic ambiguities[16] due to the unbounded nature of the operators in (3.2) are avoided, as long as we restrict ourselves to smooth physical paths within $\mathbf{L}$.

It is of geometric interest, however, to define the notion of parallel transport within the enveloping Gupta-Bleuler bundle for arbitrary smooth, or piecewise smooth, paths within $\mathbf{L}$. This obviously necessitates the extension of the Levi-Civita connection for bundles over $\mathbf{M}$ to bundles over $\mathbf{L}$. Such considerations will throw light on the physical justification for replacing in the next chapter base manifolds with frame bundles, and then imposing BRST types of constraints on the connections in those frame bundles, as well as on the parallel transport of the local state vectors in the corresponding superfibre bundles.

In order to extend any connection form ω_1 on a principal bundle $(\mathbf{P}_1, \Pi_1, \mathbf{M}, \mathbf{G}_1)$ (cf. Sec. 2.2) to a new principal bundle $(\mathbf{P}_2, \Pi_2, \mathbf{P}_1, \mathbf{G}_2)$ over its total space $\mathbf{P}_1$ in a manner which reflects correctly the physical role played by parallel transport in $\mathbf{M}$, let us consider the principal bundle with total space given by the product manifold $\mathbf{P}_2 = \mathbf{P}_1 \times \mathbf{G}_2$, whose structure group $\mathbf{G}_2$ is in general a Lie subgroup of $\mathbf{G}_1$ (so that, in particular, we might have $\mathbf{G}_2 = \mathbf{G}_1$), which acts in $\mathbf{P}_2$ in a natural manner, namely by right translation upon the second element of each pair from the direct product $\mathbf{P}_1 \times \mathbf{G}_2$. In the case of exclusive importance to us, namely when $(\mathbf{P}_1, \Pi_1, \mathbf{M}, \mathbf{G}_1)$ is a principal frame bundle, the new bundle $(\mathbf{P}_2, \Pi_2, \mathbf{P}_1, \mathbf{G}_2)$ can be identified with a principal frame bundle,

$$\mathbf{P}_1 \times \mathbf{G}_2 \xrightarrow[I_2]{} \mathbf{P}_2 = \left\{ \, \boldsymbol{u}_h = \boldsymbol{u} \cdot h \;\middle|\; \boldsymbol{u} \in \mathbf{P}_1, \quad h \in \mathbf{G}_2 \, \right\} , \tag{4.7}$$

in which the action of $\mathbf{G}_2$ upon the frames $\boldsymbol{u}$ in the original frame bundle is viewed as giving rise to new degrees of freedom.

We observe that the manifold can be subjected to two consecutive fibrations

$$\mathbf{P}_2 \xrightarrow[\Pi_2]{} \mathbf{P}_1 \xrightarrow[\Pi_1]{} \mathbf{M} \ , \tag{4.8}$$

thus giving rise to a *doubly-fibrated principal bundle*, in which $\mathbf{P}_1$ can be identified with the global section $\mathbf{P}_1 \times \{\boldsymbol{e}\}$ of $\mathbf{P}_2$. We can, therefore, extend now the connection form ω_1 on $\mathbf{P}_1$ into a connection form ω_2 on $\mathbf{P}_2$ as follows (cf. [I], Eq. (3.5.10), p. 158),

$$\left(I_2^* \omega_2\right)_{\boldsymbol{u}_h}(\mathbf{X}_1, \mathbf{X}_2) = \mathrm{Ad}_{h^{-1}}\left(\omega_{1;\boldsymbol{u}}(\mathbf{X}_1)\right) + \boldsymbol{\Theta}_h(\mathbf{X}_2) \ , \tag{4.9a}$$

$$\boldsymbol{\Theta}_h(\mathbf{X}_2) = L_h{}^* \mathbf{X}_2 \quad , \qquad (\mathbf{X}_1, \mathbf{X}_2) \in T_{\boldsymbol{u}_h}\mathbf{P}_2 \cong T_{\boldsymbol{u}}\mathbf{P}_1 \oplus T_h \mathbf{G}_2 \ , \tag{4.9b}$$

where the adjoint representation $h \mapsto \mathrm{Ad}_{h^{-1}}$ is defined in accordance with (2.5.7b). The $L(\mathbf{G}_2)$-valued one-form is the well-known left-invariant *Maurer-Cartan form* on the Lie group $\mathbf{G}_2$ (cf. [C], p. 168, or [I], p. 85, as well as Chapter 10 – especially Eqs. (10.2.12) and (10.3.10)), and it obeys the *Maurer-Cartan structural equation* [C,I,KN]:

$$\mathbf{d}\boldsymbol{\Theta}_h(\mathbf{X}, \mathbf{Y}) + [\boldsymbol{\Theta}_h(\mathbf{X}), \boldsymbol{\Theta}_h(\mathbf{Y})] = 0 \ , \qquad \mathbf{X}, \mathbf{Y} \in T_h \mathbf{G}_2 \ . \tag{4.10}$$

Naturally, the coefficients in (4.10) are the structure constants of the Lie algebra $L(\mathbf{G}_2)$, so that comparison with (2.5.16) shows that the validity of these structural equations is equivalent to the statement that the extension of the original connection $\boldsymbol{\omega}_1$ in the direction of the fibres of the new bundle $(\mathbf{P}_2,\boldsymbol{\Pi}_2, \mathbf{P}_1,\mathbf{G}_2)$ is flat.

Another way of looking at this construction is achieved upon noting that the horizontal subspaces of the old connection are preserved within each $T_{\boldsymbol{u}_h}\mathbf{P}_1$ regarded as a subspace of $T_{\boldsymbol{u}_h}\mathbf{P}_2$. Consequently, parallel transport within the $\mathbf{G}_2$-enlargements of the fibres of $\mathbf{P}_1$ is path-independent, so that the only curvature effects in $\mathbf{P}_2$ are those inherited from $\mathbf{P}_1$. We can express this fact most clearly by choosing, as in (2.5.8), a basis in the Lie algebra $L(\mathbf{G}_1)$, whose elements within $L(\mathbf{G}_2)$ provide a basis for $L(\mathbf{G}_2)$, and then setting

$$\boldsymbol{\omega}_{\boldsymbol{u}\cdot h} = \mathrm{Ad}_{h^{-1}}\left(\boldsymbol{\omega}_{1;\boldsymbol{u}}\right) = \boldsymbol{\omega}^{a}_{\boldsymbol{u}\cdot h}\tilde{\mathbf{Y}}_a \ , \qquad a = 1,\ldots,dim\mathbf{G}_1 \ , \tag{4.11a}$$

$$\boldsymbol{\Theta}_h = \boldsymbol{\Theta}_h^{\alpha}\tilde{\mathbf{Y}}_\alpha \ , \qquad \alpha = 1,\ldots,dim\mathbf{G}_2 \le dim\mathbf{G}_1 \ , \qquad L(\mathbf{G}_2)\subset L(\mathbf{G}_1) \ . \tag{4.11b}$$

The Maurer-Cartan structural equation in (4.10) then becomes equivalent to the system of equations whose formal appearance resembles[17] that of the key BRST constraints in the theory of Yang-Mills fields (cf. Sec. 10.4):

$$\mathbf{d}\boldsymbol{\Theta}^\alpha + \tfrac{1}{2}C_{\beta\gamma}{}^{\alpha}\boldsymbol{\Theta}^\beta \wedge \boldsymbol{\Theta}^\gamma = 0 \ . \tag{4.12}$$

Naturally, in (4.12) the operator $\mathbf{d}$ stands for exterior differentiation on the manifold determined by the Lie group $\mathbf{G}_2$.

We can now extend the considerations carried out in (2.5.17)-(2.5.26) to a vector bundle $(\mathbf{E}_2,\boldsymbol{\pi}_2,\mathbf{P}_1)$ associated with $(\mathbf{P}_2,\boldsymbol{\Pi}_2, \mathbf{P}_1,\mathbf{G}_2)$, which is such that its restriction to any global section $\boldsymbol{s}_1$ of the principal bundle $\mathbf{P}_1$ gives rise to a bundle $(\mathbf{E}_1,\boldsymbol{\pi}_2,\boldsymbol{s}_1)$ associated with $(\mathbf{P}_1,\boldsymbol{\Pi}_1,\mathbf{M},\mathbf{G}_1)$. It is natural to say that such a doubly-fibrated vector bundle,

$$\mathbf{E}_2 \xrightarrow[\boldsymbol{\pi}_2]{} \mathbf{E}_1 \xrightarrow[\boldsymbol{\pi}_1(\boldsymbol{s}_1)]{} \mathbf{M} \ , \tag{4.13}$$

is a bundle associated with the doubly-fibrated principal bundle in (4.8). We can, therefore, introduce a connection which, to any two respective sections $\boldsymbol{s}_1$ and $\boldsymbol{s}_2$ of $\mathbf{P}_1$ and $\mathbf{P}_2$,

$$\boldsymbol{s}_1 : x \mapsto \boldsymbol{u} \in \mathbf{P}_1 \ , \quad \boldsymbol{s}_2 : \boldsymbol{u} \mapsto \boldsymbol{u}\cdot h \in \mathbf{P}_2 \ , \quad \boldsymbol{s} : x \mapsto \boldsymbol{u}\cdot h \in \mathbf{P}_1 \ , \quad x \in \mathbf{M} \ , \tag{4.14}$$

and to any section $\boldsymbol{\Psi}$ of the above doubly-fibrated vector bundle,

$$\boldsymbol{\Psi} : x \mapsto \boldsymbol{u} \mapsto \boldsymbol{\Psi}_{x,h} \in \boldsymbol{\pi}_2^{-1}(\boldsymbol{u}) \subset \mathbf{E}_2 \ , \tag{4.15}$$

assigns the following covariant derivative:

$$\nabla_{\boldsymbol{X}}\boldsymbol{\Psi}_{x,h} = [\partial_{\boldsymbol{X}} + \boldsymbol{\omega}^{a}_{\boldsymbol{u}\cdot h}(\boldsymbol{s}_*\boldsymbol{X}_1)\hat{\mathbf{A}}_{a;\boldsymbol{u}\cdot h} + \boldsymbol{\Theta}_h^{\alpha}(\boldsymbol{s}_{2*}\boldsymbol{X}_2)\hat{\mathbf{A}}_{\alpha;\boldsymbol{u}\cdot h}]\boldsymbol{\Psi}_{x,h} \ , \tag{4.16a}$$

$$\boldsymbol{X} \in T_{\boldsymbol{u}\cdot h}\mathbf{P}_1 \cong T_x\mathbf{M} \oplus T_h\mathbf{G}_2 \ , \qquad \boldsymbol{X} \cong (\boldsymbol{X}_1,\boldsymbol{X}_2) \in T_x\mathbf{M} \oplus T_h\mathbf{G}_2 \ , \tag{4.16b}$$

$$\partial_{\boldsymbol{X}}\boldsymbol{\Psi}_{x,h} = \left[\left(X_1^{\mu}\partial_{\mu} + X_2^{\alpha}\partial_{\alpha}\right)\Psi_{x,h}^{\rho}\right]\boldsymbol{\Phi}_{\rho}(\boldsymbol{u}\cdot h), \quad \partial_{\mu} = \partial/\partial x^{\mu}, \quad \partial_{\alpha} = \partial/\partial h^{\alpha} \ . \quad (4.16c)$$

The corresponding Koszul connection (cf. p. 47) in the doubly-fibrated vector bundle in (4.13) can be therefore expressed in a form analogous to that of (2.5.23):

$$\nabla\boldsymbol{\Psi} = [\mathbf{d} + \boldsymbol{A}^{\boldsymbol{s}} + \boldsymbol{C}^{\boldsymbol{s}_2}]\boldsymbol{\Psi} \ , \qquad \mathbf{d} = dx^{\mu}\partial_{\mu} + dh^{\alpha}\partial_{\alpha} \ , \quad (4.17a)$$

$$\boldsymbol{A}^{\boldsymbol{s}} = \boldsymbol{\omega}_{\boldsymbol{s}}^{\alpha}\hat{\mathbf{A}}_{\alpha;\boldsymbol{u}\cdot h} \ , \qquad \boldsymbol{\omega}_{\boldsymbol{s}}^{\alpha}(\boldsymbol{X}_1) = \boldsymbol{\omega}_{\boldsymbol{u}\cdot h}^{a}(\boldsymbol{s}_*\boldsymbol{X}_1) \ , \quad (4.17b)$$

$$\boldsymbol{C}^{\boldsymbol{s}_2} = \boldsymbol{\Theta}_{\boldsymbol{s}_2}^{\alpha}\hat{\mathbf{A}}_{\alpha;\boldsymbol{u}\cdot h} \ , \qquad \boldsymbol{\Theta}_{\boldsymbol{s}_2}^{\alpha}(\boldsymbol{X}_2) = \boldsymbol{\Theta}_{h}^{\alpha}(\boldsymbol{s}_{2*}\boldsymbol{X}_2) \ . \quad (4.17c)$$

The above construction can be used to define within the enveloping Gupta-Bleuler bundle a notion of parallel transport along arbitrary piecewise smooth curves in **L**, which for those paths that are physical coincides with the one defined by (4.1)-(4.5). However, in carrying out this task we have to extend the construction from the doubly-fibrated principal bundle in (4.8) to a *triply-fibrated principal bundle*,

$$\mathbf{P}_3 \xrightarrow[\Pi_3]{} \mathbf{P}_2 \xrightarrow[\Pi_2]{} \mathbf{P}_1 \xrightarrow[\Pi_1]{} \mathbf{M} \ , \quad (4.18)$$

on account of the fact that the group $\mathbf{G}_0$ in (3.8) is not a subgroup of the Poincaré group. As we shall see in the next chapter, the need to introduce a triply-fibrated bundle is very much an intrinsic feature of the geometric interpretation[18] of BRST transformations, with the second and third fibration corresponding to the Faddeev-Popov so-called "ghost" and "antighost" degrees of freedom.

Let us start the required construction by first setting

$$\mathbf{P}_1 = P^{\uparrow}\mathbf{M} \cong \mathbf{P}(\mathbf{M},\mathbf{G}_1) \, , \ \mathbf{G}_1 = \mathrm{ISO}^{\uparrow}(3,1) \ , \quad \mathbf{P}_2 \cong \mathbf{P}_1 \times \mathbf{G}_2 \ , \ \mathbf{G}_2 = \mathrm{SO}(3), \quad (4.19)$$

to arrive at a doubly-fibrated principal bundle (4.8). We can then construct the total space

$$\mathbf{P}_2 \times \mathbf{G}_3 \xrightarrow[I_3]{} \mathbf{P}_3 = \left\{ \boldsymbol{u}_{h,\bar{h}} = \bar{h}^{-1}\cdot\boldsymbol{u}_h \ \middle| \ \boldsymbol{u}_h \in \mathbf{P}_2, \ \bar{h} \in \mathbf{G}_3 = \mathrm{O}(3) \right\} , \quad (4.20)$$

of a triply-fibrated principal bundle (4.18). We note that the action of $\mathbf{G}_3$ upon the frames in $\mathbf{P}_2$ represents algebraically action-from-the-right (cf. Note 15 to Chapter 2), so that a principal bundle with SO(3) as structure group is indeed the outcome of the construction.

We can now introduce triply-fibrated vector bundles

$$\mathbf{E}_3 \xrightarrow[\boldsymbol{\pi}_3]{} \mathbf{E}_2 \xrightarrow[\boldsymbol{\pi}_2(\boldsymbol{s}_2)]{} \mathbf{E}_1 \xrightarrow[\boldsymbol{\pi}_1(\boldsymbol{s}_1)]{} \mathbf{M} \ , \quad (4.21)$$

associated to the triply-fibrated principal bundle given by (4.19) and (4.20) by setting

$$\mathbf{E}_3 = \mathbf{P}_3 \times_{\mathbf{G}_3} \mathcal{K}_{\infty} \ , \qquad \mathbf{E}_2 \cong \mathbf{P}_2 \times_{\mathbf{G}_2} \mathcal{K}_{\infty} \ , \qquad \mathbf{E}_1 \cong \mathcal{G}_c = P\mathbf{M} \times_{\mathbf{G}_1} \mathcal{K}_{\infty} \ . \quad (4.22)$$

We note that the fibres of all these bundles are pre-Hilbert spaces in the respective majorant topologies of those fibres, but that upon taking the completions of the fibres of the bundles $\mathbf{E}_2$ and $\mathbf{E}_3$ in the J-norms determined by the Poincaré frames above which those fibres are located, we obtain the *bona fide* Krein bundles

$$\overline{\mathbf{E}}_2 \cong \mathcal{G}_2 := \mathbf{P}_2 \times_{\mathbf{G}_2} \mathcal{K} \ , \qquad \overline{\mathbf{E}}_3 \cong \mathcal{G}_3 := \mathbf{P}_3 \times_{\mathbf{G}_3} \mathcal{K} \ , \tag{4.23}$$

associated with $\mathbf{P}_2$ and $\mathbf{P}_3$, respectively.

The technical problem to which the unboundedness of the operators in (3.2) had given rise, and which prevented us from constructing a fibre bundle with typical fibre $\mathcal{K}$ that would be associated directly to the Poincaré frame bundle, is thus by-passed: we have constructed the doubly-fibrated bundle in (4.23), which has $\mathcal{K}$ as typical fibre, and which incorporates the triply-fibrated subbundle in (4.22), that displays full Poincaré gauge covariance. Thus, any Koszul connection on the latter bundle (4.22), given by definitions that extrapolate the ones in (4.17) in an obvious manner, namely by

$$\nabla \Psi = [\mathbf{d} + A^{s} + C^{s_2} + \overline{C}^{s_3}]\Psi \ , \qquad \mathbf{d} = dx^{\mu}\partial_{\mu} + dh^{\alpha}\partial_{\alpha} + d\overline{h}^{\beta}\overline{\partial}_{\beta} \ , \tag{4.24a}$$

$$A^{s} = \omega_{s}^{\alpha}\hat{\mathbf{A}}_{\alpha;\overline{h}^{-1}\cdot u\cdot h} \ , \qquad C^{s_2} = \Theta_{s_2}^{\alpha}\hat{\mathbf{A}}_{\alpha;\overline{h}^{-1}\cdot u\cdot h} \ , \qquad C^{s_3} = \Theta_{s_3}^{\beta}\hat{\mathbf{A}}_{\beta;\overline{h}^{-1}\cdot u\cdot h} \ , \tag{4.24b}$$

$$s : x \mapsto \overline{h}^{-1}\cdot u\cdot h \in \mathbf{P}_1 \ , \quad s_2 : u \mapsto u\cdot h \in \mathbf{P}_2 \ , \quad s_3 : u \mapsto \overline{h}^{-1}\cdot u \in \mathbf{P}_3 \ , \tag{4.24c}$$

gives rise to a parallel transport on the former bundle (4.23). This parallel transport reduces to Poincaré-covariant parallel transport along piecewise smooth physical paths γ^{**} in the sense that it coincides with parallel transport along the projection γ of those paths in $\mathbf{M}$. Clearly, in the present context a physical path results from the double lift $\gamma \to \gamma^{*} \to \gamma^{**}$ of a piecewise smooth path in $\mathbf{M}$.

The construction of the triply-fibrated bundle $\mathcal{G}_3$ in (4.23) removes the apparent dependence of the enveloping Gupta-Bleuler bundle $\mathcal{G}_e$ in (3.9a) on the choice of section s_0 in (3.6). Indeed, in the present context any global section s_1 of $P\mathbf{M}$ can be chosen, so that the framework displays full Poincaré covariance. For $s_1 = s_0$ the identification of $\mathcal{G}_e$ with a subbundle of $\mathcal{G}_3$ can be carried out in an obvious manner on the basis of the following identification:

$$\mathbf{P}_1 = P\mathbf{M} \leftrightarrow (s_0 \times \mathbf{G}_2) \times \mathbf{G}_3 \ \to \ \mathbf{P}_3 \cong (\mathbf{P}_1 \times \mathbf{G}_2) \times \mathbf{G}_3 \ . \tag{4.25}$$

In this sense the present doubly-fibrated Gupta-Bleuler bundle in (4.23) can be said to impart to the core Gupta-Bleuler bundle in (3.9) manifest extended Poincaré covariance.

This extended Poincaré covariance can be formulated in a mathematically most elegant manner by extending the base Lorentz manifold $\mathbf{M}$ into a larger base manifold that coincides with the extended Poincaré frame bundle $P^{\uparrow}\mathbf{M}$, whose typical fibre can be identified with the orthochronous Poincaré group $\mathrm{ISO}^{\uparrow}(3,1)$. Basically, that is the approach adopted in the next chapter in the context of (non-Abelian) Yang-Mills fields, for which the Gupta-Bleuler formalism proves to be inadequate. The results of Chapter 10 can be, of course, specialized to the present (Abelian) electromagnetic case.

9.5. Stress-Energy Tensors and GS Propagation in Gupta-Bleuler Bundles

The concept of parallel transport, which served in the preceding two chapters as the cornerstone in the formulation of GS propagation for massive fields, retains that fundamental role in the case of gauge fields. This means that, since this concept is of a purely geometric nature, GS propagation can be formulated for all types of fields without any reference to the action principles of Lagrangian field theory. Epistemologically, this is very desirable, as the conventional use of Noether's theorem for classical fields (Bogolubov and Shirkov, 1959) is at best a heuristic device[19] in the quantum regime, and as such it is incapable of justifying the normal ordering of the stress-energy and the angular momentum tensors that provide the generators for infinitesimal parallel transport. By deriving these quantities from fundamental geometric principles, not only are all ordering ambiguities removed in the GS approach, but full mathematical rigor can be retained throughout the derivation.

Using the formulae (3.14) and (3.15) defining the action of the photon creation and annihilation operators on n-photon state vectors, it is easy to verify that within the core Gupta-Bleuler bundle $\mathcal{G}_c$ in (3.9b), all the generators of spacetime translations can be expressed in a form analogous to that in (7.5.4) and (8.2.4):

$$\boldsymbol{P}_{j;\boldsymbol{u}} = i\int \eta^{ik} A_i^{(+)}(x;\boldsymbol{\zeta})\partial_j A_k^{(-)}(x;\boldsymbol{\zeta})d\tilde{\Sigma}(\boldsymbol{\zeta}) \ , \qquad \partial_j = \partial/\partial q^j \ . \tag{5.1}$$

By employing the form (2.21) of the indefinite inner product in the Gupta-Bleuler fibres, and by choosing a hypersurface Σ corresponding to a constant q^0 value, such as $q^0 = 0$, with respect to the Poincaré frame $\boldsymbol{u}$, we can recast (5.1) in the form

$$\boldsymbol{P}_{j;\boldsymbol{u}} = -\int_\Sigma d^3\mathbf{q}\, d\tilde{\Omega}(v)\, \eta^{ik}\left[A_i^{(+)}(x;\boldsymbol{\zeta})\partial_0\partial_j A_k^{(-)}(x;\boldsymbol{\zeta}) - \partial_0 A_i^{(+)}(x;\boldsymbol{\zeta})\partial_j A_k^{(-)}(x;\boldsymbol{\zeta})\right]. \tag{5.2}$$

Let us now restrict ourselves to the Gupta-Bleuler-Lorenz bundle in (3.19), whose vectors satisfy (3.18), so that

$$\partial^i\partial_i A_k^{(-)}(x;\boldsymbol{\zeta})\,\boldsymbol{\Psi} = \mathbf{0} \ , \qquad \boldsymbol{\Psi} \in \mathcal{K}_{x,J}^L \ , \qquad k = 0,1,2,3 \ . \tag{5.3}$$

In that case, for $j = 0$ we can replace within the square bracket in (5.2) the expression on the left-hand side of the following equality with that on its right-hand side,

$$\partial_0\partial_0 A_k^{(-)}(x;\boldsymbol{\zeta})\,\boldsymbol{\Psi} = \sum_{b=1}^3 \partial_b\partial_b A_k^{(-)}(x;\boldsymbol{\zeta})\,\boldsymbol{\Psi} \ , \qquad \boldsymbol{\Psi} \in \mathcal{K}_{x,J}^L \ . \tag{5.4}$$

Upon carrying out an integration by parts with respect to the $\mathbf{q}$-variable, we obtain[20]:

$$\boldsymbol{P}_{0;\boldsymbol{u}}\,\boldsymbol{\Psi} = -\int_{q^0=0} d^3\mathbf{q}\, d\tilde{\Omega}(v) \sum_{j=0}^3 \partial_j A_i^{(+)}(x;\boldsymbol{\zeta})\,\eta^{ik}\partial_j A_k^{(-)}(x;\boldsymbol{\zeta})\,\boldsymbol{\Psi} \ . \tag{5.5a}$$

On the other hand, by using the same type of arguments which establish the validity of (7.5.5b), we can establish that

$$\int_{q^0=0} d^3\mathbf{q}\, d\tilde{\Omega}(v) \sum_{j=0}^{3} \partial_j A_i^{(\pm)}(x;\zeta)\, \eta^{ik} \partial_j A_k^{(\pm)}(x;\zeta)\, \boldsymbol{\Psi} = \mathbf{0} \ . \tag{5.5b}$$

Consequently, we arrive at the conclusion that

$$\boldsymbol{P}_{0;\boldsymbol{u}}\, \boldsymbol{\Psi} = -\int_{q^0=0} d^3\mathbf{q}\, d\tilde{\Omega}(v) : \tfrac{1}{2} \sum_{j=0}^{3} \partial_j A_k(x;\zeta)\, \partial_j A^k(x;\zeta) : \boldsymbol{\Psi} \, , \qquad \boldsymbol{\Psi} \in \mathcal{K}_{x,J}^{L} \ . \tag{5.6}$$

In the case of $j = 1,2,3$ we can apply (5.2) to state vectors in the Lorenz gauge, and then integrate by parts to obtain

$$\begin{aligned} \boldsymbol{P}_{b;\boldsymbol{u}}\, \boldsymbol{\Psi} = \int_{q^0=0} d^3\mathbf{q}\, d\tilde{\Omega}(v)\, \eta^{ik} [\partial_b A_i^{(+)}(x;\zeta)\, \partial_0 A_k^{(-)}(x;\zeta) \\ + \partial_0 A_i^{(+)}(x;\zeta)\, \partial_b A_k^{(-)}(x;\zeta)] \boldsymbol{\Psi} \, , \qquad b = 1,2,3 \ . \end{aligned} \tag{5.7}$$

Upon noting again that

$$\int_{q^0=0} d^3\mathbf{q}\, d\tilde{\Omega}(v)\, \eta^{ik} [\partial_b A_i^{(\pm)}(x;\zeta)\, \partial_0 A_k^{(\pm)}(x;\zeta) + \partial_0 A_i^{(\pm)}(x;\zeta)\, \partial_b A_k^{(\pm)}(x;\zeta)] = 0 \ , \tag{5.8}$$

and then taking advantage of the Gupta-Bleuler subsidiary condition (3.18), we arrive at the following result:

$$\boldsymbol{P}_{b;\boldsymbol{u}}\, \boldsymbol{\Psi} = \int_{q^0=0} d^3\mathbf{q}\, d\tilde{\Omega}(v) : \partial_0 A^k(x;\zeta)\, \partial_b A_k(x;\zeta) : \boldsymbol{\Psi} \, , \quad \boldsymbol{\Psi} \in \mathcal{K}_{x,J}^{L} \, , \ b = 1,2,3 \ . \tag{5.9}$$

The expressions in (5.6) and (5.9) are, at the formal level, totally analogous to those for the 4-momentum operator in the conventional Gupta-Bleuler formalism [SI,IQ]. Upon introducing the stress-energy tensor[21]

$$T_{ij}[\mathbf{A}] = \tfrac{1}{2} \eta_{ij} A_{k,l} A^{k,l} - A_{k,i} A^{k}{}_{,j} \ , \qquad A_{i,j} = \partial A_i / \partial q^j \ , \qquad \mathbf{A} = A_i \boldsymbol{\theta}^i \ , \tag{5.10}$$

they can be written in the following manifestly covariant form

$$\boldsymbol{P}_{j;\boldsymbol{u}} = \int : T_{jk}[\mathbf{A}(x;\zeta)] : d\sigma^k(q)\, d\tilde{\Omega}(v) \ , \quad j = 0,1,2,3 \ . \tag{5.11}$$

The derivation of the infinitesimal generators for spatial rotations and Lorentz boosts proceeds in very much the same manner. The end result can be expressed in the type of notation used in (7.5.9) and (8.2.10)-(8.2.11):

$$\boldsymbol{M}_{\boldsymbol{u}}^{ij} = \int : M_{\boldsymbol{u}}^{ijk}[\mathbf{A}(x;\zeta)] : d\sigma_k(q)\, d\tilde{\Omega}(v) \ , \tag{5.12a}$$

$$M_{\boldsymbol{u}}^{ijk}[\mathbf{A}] = Q_{\boldsymbol{u}}^{i} T^{jk}[\mathbf{A}] - Q_{\boldsymbol{u}}^{j} T^{ik}[\mathbf{A}] + \mathbf{A}^{\dagger} S_{\boldsymbol{u}}^{ijk} \mathbf{A} \ . \tag{5.12b}$$

We observe that the last term in (5.12b) incorporates the spin-1 contribution, and that it can be therefore derived from the Lie algebra of $SO_0(3,1)$.

The derivation of an action-based formula for the propagator for parallel transport of Gupta-Bleuler frames along physical paths,

$$K_\gamma(\boldsymbol{f}'';\boldsymbol{f}') = \left\langle \Phi_{f''} \middle| \tau_{\gamma*}(x'',x')\Phi_{f'} \right\rangle \quad , \qquad \Phi_{f'} \in \mathcal{K}^L_{x',J'} \ , \qquad \Phi_{f''} \in \mathcal{K}^L_{x'',J''} \ , \qquad (5.13)$$

proceeds along lines which are very similar to those of the massive spin-0 case treated in Sec. 7.5. Indeed, we can use (3.43) to write

$$K_\gamma(\boldsymbol{f}'';\boldsymbol{f}') \;=\; K^*_\gamma(\boldsymbol{f}';\boldsymbol{f}'') = \int_{\mathbf{K}^{\mathrm{T}}_{x,J}} K_\gamma(\boldsymbol{f}'';\boldsymbol{f})\, K_\gamma(\boldsymbol{f};\boldsymbol{f}')\boldsymbol{df}\, \boldsymbol{d\bar{f}} \ , \qquad (5.14)$$

for transversal modes $\boldsymbol{f}'$ and $\boldsymbol{f}''$, but on account of (3.39) the above relation remains true for arbitrary Lorenz-gauge modes $\boldsymbol{f}'$ and $\boldsymbol{f}''$. Consequently, we can, in general, write

$$K_\gamma(A(x'');A(x')) = \lim_{\varepsilon\to+0} \int K_\gamma(A(x_N);A(x_{N-1})) \prod_{n=N-1}^{1} K_\gamma(A(x_n);A(x_{n-1}))\, \mathcal{D}[A(x_n)] \ . \ (5.15)$$

In the above expression we have made the transition to the coordinate wave functions of the single-photon modes, so that

$$A(x') = \sigma^{u'}_{x'} \boldsymbol{f}',\ A(x'') = \sigma^{u''}_{x''} \boldsymbol{f}'' \in \mathbf{K} \ , \qquad \boldsymbol{f}' \in \mathbf{K}^L_{x'} \ , \qquad \boldsymbol{f}'' \in \mathbf{K}^L_{x''} \ , \qquad (5.16a)$$

$$A(x_n) = \sigma^{u_n}_{x_n} \boldsymbol{f}_n \in \mathbf{K}^{\mathrm{T}}_{J_0} \ , \qquad \boldsymbol{f}_n \in \mathbf{K}^{\mathrm{T}}_{x_n,J_n} \ , \qquad \boldsymbol{u}_n = \boldsymbol{s}(x_n) \in \gamma^* \ , \qquad (5.16b)$$

$$\boldsymbol{df}_n\, \boldsymbol{d\bar{f}}_n \quad \mapsto \quad \mathcal{D}[A(x_n)] \ , \quad x_n \in \gamma \ , \qquad n = 1,2,\ldots,N-1 \ . \qquad (5.16c)$$

When we apply (5.15) to a broken physical path, in which the parallel transport is carried out in $\mathbf{L}$ along horizontal lifts of the smooth segments of the path γ in $\mathbf{M}$, then we can work within the bundles (4.23) in the extended Poincaré gauges in which $\boldsymbol{s}_1 = \boldsymbol{s}_0$, whereas $\boldsymbol{s}_2$ and $\boldsymbol{s}_3$ are null sections, so that (4.24) reduces to the form

$$\nabla\Psi = [\boldsymbol{d} - i\tilde{\theta}^i\, \boldsymbol{P}_{i;u} + \tfrac{i}{2}\tilde{\omega}_{jk}\, \boldsymbol{M}^{jk}_u]\Psi \ , \qquad \boldsymbol{d} = \boldsymbol{dx}^\mu \partial_\mu \ . \qquad (5.17)$$

Hence, using (3.26) , and proceeding as in (7.5.13)-(7.5.16), we obtain that

$$K_\gamma(A(x_n);A(x_{n-1})) = \left[1 - i\delta x^i_n\, P_i(A(x_n);A(x_{n-1}))\right]\left\langle \Phi_{A_n(x_n)} \middle| \Phi_{A_{n-1}(x_n)} \right\rangle$$
$$+\, i\tilde{\omega}_{jk}(\delta x_n) M^{jk}(A(x_n);A(x_{n-1}))\left\langle \Phi_{A_n(x_n)} \middle| \Phi_{A_{n-1}(x_n)} \right\rangle + O\left((\delta t_n)^2\right) , \quad (5.18a)$$

$$P_i(A(x_n);A(x_{n-1})) = \int T_{ik}[\overline{A}(x_n;\zeta) + A(x_{n-1};\zeta)]\ d\sigma^k(q)d\tilde{\Omega}(v) \ , \qquad (5.18b)$$

$$M^{jk}(A(x_n);A(x_{n-1})) = \int M^{jkl}_u[\overline{A}(x_n;\zeta) + A(x_{n-1};\zeta)]\ d\sigma_l(q)d\tilde{\Omega}(v) \ . \qquad (5.18c)$$

On the other hand, it follows from (3.37)-(3.39) that

$$\left\langle \Phi_{A_n(x_n)} \middle| \Phi_{A_{n-1}(x_n)} \right\rangle = \exp\left\{-\tfrac{1}{2}\left[\langle A(x_n)|A(x_n)\rangle + \langle A(x_{n-1})|A(x_{n-1})\rangle\right]\right\}$$
$$\times \exp\left[\langle A(x_n)|A(x_{n-1})\rangle\right] , \qquad n = 1,2,\dots,N , \qquad (5.19)$$

and that, on account of (3.34) and (3.35), these relations remain true even at the two end-points of γ, where the chosen single-photon modes are not necessarily transversal – albeit in the Gupta-Bleuler formalism they have to be in the Lorenz gauge. Upon inserting this result into (5.18), we obtain

$$K_\gamma(A(x''); A(x')) = \lim_{\varepsilon \to +0} \int \prod_{n=N}^{1}{}' \mathcal{D}[A(x_n)] \exp[i\delta t_n L_\gamma(A(x_n); A(x_{n-1}))] , \qquad (5.20)$$

where, as in the analogous expression in (7.5.18a), the prime indicates the absence of the functional integration over the N-th mode, and

$$L_\gamma(A(x_n); A(x_{n-1})) = \tfrac{i}{2}\left[\langle \Pi(x_n)|A(x_{n-1})\rangle - \langle A(x_n)|\Pi(x_n)\rangle\right]$$
$$- X_n{}^i P_i(A(x_n)); A(x_{n-1})) + \tfrac{1}{2}\tilde{\omega}_{jk}(\mathbf{X}_n) M^{jk}(A(x_n); A(x_{n-1})) , \qquad (5.21a)$$

$$\langle A(x_n)|\Pi(x_n)\rangle = i\int \left[\overline{A}(x_n;\zeta)\cdot \Pi_{,k}(x_n;\zeta) - \overline{A}_{,k}(x_n;\zeta)\cdot \Pi(x_n;\zeta)\right] d\sigma^k(q)\, d\tilde{\Omega}(v), \ (5.21b)$$

$$A_{,k}(x_n;\zeta) := \partial_k A(x_n;\zeta) , \qquad \Pi(x_n;\zeta) := -\left(A(x_n;\zeta) - A(x_{n-1};\zeta)\right)/\left(t_n - t_{n-1}\right). \ (5.21c)$$

Consequently, we arrive at the following functional integral for the propagator describing the parallel transport of Gupta-Bleuler frames along any piecewise smooth physical path γ^* in **L**, with projection γ in **M**,

$$K_\gamma(A(x''); A(x')) = \int \mathcal{D}A \ \exp\left(iS_\gamma[A,\Pi]\right), \quad \mathcal{D}A = \prod\nolimits_{t''>t\geq t'} \mathcal{D}[A(x(t))] , \ (5.22)$$

with the mathematical interpretation of the notation being the same as for (7.7.5) in Sec. 7.7. On account of (5.20) and (5.21), this path integral is based on the following action,

$$S_\gamma[A,\Pi] = \int_{t'}^{t''} dt \int \mathcal{L}_k^{(\gamma)}[A(x(t);\zeta), \Pi(x(t);\zeta)]\, d\sigma^k(q)\, d\Omega(v) , \qquad (5.23a)$$

$$\mathcal{L}_k^{(\gamma)}[A(x(t);\zeta), \Pi(x(t);\zeta)] =$$
$$\tfrac{1}{2}\hat{Z}_\ell^{(0)}\left[\overline{A}_{,k}(x(t);\zeta)\cdot \Pi(x(t);\zeta) + A_{,k}(x(t-0);\zeta)\cdot \Pi(x(t);\zeta)\right]$$
$$-\tfrac{1}{2}\hat{Z}_\ell^{(0)}\left[\overline{A}(x(t);\zeta)\cdot \Pi_{,k}(x(t);\zeta) + A(x(t-0);\zeta)\cdot \Pi_{,k}(x(t);\zeta)\right]$$
$$- X^i(t)\hat{Z}_\ell^{(0)} T_{ik}[\overline{A}(x(t);\zeta) + A(x(t-0);\zeta)]$$
$$+ \eta_{kl}\tilde{\omega}_{ij}(\mathbf{X}_n(t))\hat{Z}_\ell^{(0)} M^{ijl}[\overline{A}(x(t);\zeta) + A(x(t-0);\zeta)] , \quad k = 0,1,2,3, \ (5.23b)$$

in which, in order to consider QED interactions in the next section, we have transferred the infinite renormalization constant from the measure in (2.21) to the "Lagrangian" terms.

Naturally, the above action is in outward appearance very analogous to the one in (7.5.22) – especially since the notation was so chosen as to underline the similarities. However, at the deeper level there are some fundamental differences, in addition to the more obvious ones, namely the presence in (5.23) of an infinite renormalization formally incorporated into the measure of integration, and of vector-valued fields corresponding to spin one, whose indefinite (Minkowski) inner products are denoted by a dot. Indeed, in the present case there is an intrinsic choice of extended Poincaré gauge, that has been made already in (5.17), and without which additional terms would appear in (5.23b). In a BRST formulation, these terms would incorporate counterparts of Faddeev-Popov ghost and antighost fields. We shall encounter them in the next chapter, when deriving similar action integrals for the propagators for the parallel transport of Yang-Mills fields, since those expressions can be always specialized to the Abelian gauge field with which we are dealing at the present time.

A simplification of (5.23), which is totally analogous to the one in (7.5.23), can be carried out in exactly the same manner as in the case of the massive spin-0 fields treated in Sec. 7.5. The comparison with the conventional formalism can be then easily implemented. As will be further discussed in the next section, this suggests the following *weakly causal GS quantum field propagator for Gupta-Bleuler frames*,

$$\mathcal{K}(A(x'');A(x')) = \int_{\mathbf{M}(t',t'')} \mathcal{D}\hat{A}\,\exp\left(iS[\hat{A},\hat{\Pi}]\right)\,, \qquad \mathcal{D}\hat{A} = \prod\nolimits_x \mathcal{D}[\hat{A}(x)]\,, \tag{5.24a}$$

$$\hat{A}(x;v) = f(\hat{\zeta}(x))\,, \qquad v \in \mathbf{V}^+, \qquad \hat{\zeta}(x) = (-a^i(x),v^i)\,, \tag{5.24b}$$

where the interpretation of the above notation is the same as in (7.7.4) and (7.7.5).

The corresponding *strongly causal GS quantum field propagator for Gupta-Bleuler frames* , from some base location $x' \in \mathbf{M}$ to a location $x'' \in J^+(x')$ within its causal future [M,W] in the base manifold $\mathbf{M}$, can be defined by analogy with (7.7.1) as follows:

$$\boldsymbol{K}(A(x'');A(x')) = \lim_{\varepsilon \to +0} \int \prod_{n=N}^{1}{}' ds(x_n)\mathcal{D}[A(x_n)]\,K(A(x_n);A(x_{n-1}))\;. \tag{5.25}$$

However, in view of the fact that photons have zero mass, the averaging procedure is in some of its basic respects different from the one used in the massive case treated in Sec. 7.7: in the present context the averaging has to be carried out over all broken null-geodesic paths connecting x' to $x'' \in J^+(x')$, rather than over all broken timelike geodesic paths. Consequently, prior to taking the limit $\varepsilon = \max(t_n - t_{n-1}) \to +0$ in (7.1), the integration in x_n has to be carried out, for each $n = 1,\dots,N$, over the curve $\sigma_{t_n} \cap C^+(x_{n-1}) \cap J^-(x'')$, where $C^+(x_{n-1}) = J^+(x_{n-1}) - I^+(x_{n-1})$ is the boundary of $I^+(x_{n-1})$. Thus, the integrand $K(A(x_n);A(x_{n-1}))$ in (5.25) is equal to the propagator in (5.22) for parallel transport along the null-geodesic $\gamma(x_{n-1},x_n)$, which connects the respective points x_{n-1} and x_n on the two consecutive slices $\sigma_{t_{n-1}}$ and σ_{t_n}, renormalized by division with the length of the spacelike curve $\sigma_{t_n} \cap C^+(x_{n-1}) \cap J^-(x'')$.

*9.6. Geometro-Stochastic vs. Conventional Quantum Electrodynamics

The quantization of the electromagnetic field was first considered by Born and Jordan (1925). However, the generally acknowledged founder of quantum electrodynamics (QED) is Dirac, and with very good reason: his was the first attempt at developing a method of quantization of the electromagnetic field which would be truly consistent with the tenets of special relativity. In the introduction to his pioneering 1927 paper, which *de facto* founded QED, Dirac himself correctly summarized the problems he encountered, and his degree of success at coping with them, as follows:

"Hardly anything has been done up to the present on quantum electrodynamics. The question of the correct treatment of a system in which forces are propagated with the velocity of light instead of instantaneously, of the production of an electromagnetic field by a moving electron, and of the reaction of this field on the electron have not yet been touched. In addition, there is a serious difficulty in making the theory satisfy all the requirements of the restricted principle of relativity, since a Hamiltonian function can no longer be used. This relativity question is, of course, connected with the previous one, and it will be impossible to answer any one question completely without at the same time answering them all. However, it appears possible to build a fairly satisfactory theory of the emission of radiation and of the reaction of the radiation field on the emitting system on the basis of kinematics and dynamics which are not strictly relativistic. This is the main object of the present paper. The theory is non-relativistic on account of the time being counted throughout as a c-number, instead of being treated symmetrically with the space co-ordinates. . . . The mathematical development of the theory has been made possible by the author's general transformation theory of the quantum matrices. Owing to the fact that we count time as a c-number, we are allowed to use the notion of the value of the dynamical variable at any instant of time. This value is a q-number, capable of being represented by a generalized 'matrix' according to many different matrix schemes, some of which may have continuous ranges of rows and columns, and may require the matrix elements to involve certain kind of infinities (of the type given by the δ functions)." (Dirac, 1927, pp. 244-246).

Half a century later, after having made many attempts[22] to deal with this very fundamental problem of infinities, which was left unresolved by his original attempt, and after having to witness, since the late 1940s, the constantly repeated claims as to the purported success of conventional renormalization theory in coping with those infinities (Schweber, 1990), Dirac assessed the situation as follows:

"[Quantum field] theory has to be based on sound mathematics, in which one neglects only quantities which are small. One is not allowed to neglect infinitely large quantities. The renormalization idea would be sensible only if it is applied with finite renormalization factors, not infinite ones.

"For these reasons I find the present quantum electrodynamics quite unsatisfactory. One ought not to be complacent about its faults. The agreement with observation is presumably a coincidence, just like the original calculation of the hydrogen spectrum with Bohr orbits. Such coincidences are no reason for turning a blind eye to the faults of a theory.

"Quantum electrodynamics is rather like the Klein-Gordon equation. It was built up from physical ideas that were not correctly incorporated into the theory and it has no sound mathematical foundation. One must seek a new relativistic quantum mechanics and one's prime concern must be to base it on sound mathematics." (Dirac, 1978b, p. 5).

Unfortunately, for historical reasons discussed in Chapter 12, Dirac's deen concern with sound mathematics was not shared by most physicists of the post-World War II generation[23]. In a 1975 lecture, devoted exclusively to QED, Dirac unwittingly revelas a most anomalous situation, whereby, motivated by a kind of scientific integrity[24] with perhaps no parallel in the history of science, one of the greatest physicists of this century publicly and repeatedly criticizes the very foundations of the most important theory which he had founded; whereas, also with no parallel in the history of science, the masses of professionals whom he addresses, ignore that criticism with utterly disinterested complacency:

"These calculations [of the Lamb shift, magnetic moment of the electron, etc.] do give results in agreement with observation. Hence most physicists are very satisfied with this situation. They say: 'Quantum electrodynamics is a good theory, and we do not have to worry about it any more.' I must say that I am very dissatisfied with the situation, because this so-called 'good theory' does involve neglecting infinities which appear in its equations, neglecting them in an arbitrary way. This is just not sensible mathematics. Sensible mathematics involves neglecting a quantity when it turns out to be small – not neglecting it just because it is infinitely great and you do not want it!" (Dirac, 1978a, p. 36).

These words were spoken more than a decade prior to the final admission by Feynman that "it's also possible that [conventional quantum] electrodynamics is not a consistent theory" (Davies and Brown, 1989, p. 199). They were also spoken a good many years before the constructive quantum field theory attempt at basing QED on "sensible mathematics" finally led to the conclusion that "arguments favoring triviality ... apply equally to the four-dimensional Yukawa and electrodynamic (QED) interactions" (Glimm and Jaffe, 1987, p. 120).

However, the fact that conventional QED (to which we shall refer in the remainder of this section as CQED) is fundamentally an inconsistent theory is still not generally acknowledged in physics literature. In fact, even some of the most recent textbooks dealing with this subject tend to leave their readers with the impression that conventional renormalization theory is based on acceptable and legitimate mathematics. The following type of balanced depiction of the present status of CQED is still the exception, rather than the rule:

"There is as yet no logically consistent and complete relativistic quantum theory. ... The [present] theory is largely constructed on the pattern of ordinary quantum mechanics. This structure of the theory has yielded good results in quantum electrodynamics. The lack of complete logical consistency in this theory is shown by the occurrence of divergent expressions when the mathematical formalism is directly applied, although there are quite well-defined ways of eliminating those divergencies. Nevertheless, such methods remain, to a considerable extent, semiempirical rules, and our confidence in the correctness of the results is ultimately based only on their excellent agreement with experiment, not on the internal consistency or logical ordering of the fundamental principles of the theory." (Berestetskii *et al.*, 1982, p. 4).

However, even this "excellent agreement with experiment" can represent a somewhat contentious issue if one is not predisposed to an unquestioning conformity to the type of conventional wisdom that has become entrenched in quantum physics during the post-World War II era (cf. Chapter 12). As we saw in an earlier quotation, Dirac has stated that the "agreement with observation [of renormalized perturbative series computations in CQED] is presumably a coincidence". The following account of the status of this agreement, cited from the second edition of one of the best known textbooks on CQED, and published at around the same time as the comment by Dirac (who, however, made many similar remarks on other occasions), explains how this might be possible, despite the commonly held belief as to the utterly unquestionable experimental verification of CQED:

"The most famous test of quantum electrodynamics is of course the Lamb-Rutherford shift in hydrogen ..., [and it represents] an example of an often encountered state of temporary uncertainty in the comparison between theory and experiment. The reader should realize that this comparison is often not clear cut either because different measurements disagree with one another or because different theorists in their independent computations obtain different results for these very complex calculations. Sometimes it takes years until such discrepancies are resolved. ... While it does not seem likely that there is a discrepancy [in the case of the Lamb shift], the present experimental and theoretical results do not permit the conclusion that there is complete agreement." (Jauch and Rohrlich, 1976, pp. 533-535). If one adds to these remarks the fact that the Lamb shift can be derived *without*[25] the aid of CQED, then it becomes evident that Dirac's criticism of CQED actually was not only justified, but that it warranted careful reconsideration of such CQED "successes".

In general, in CQED – not to mention the other areas of modern elementary particle physics to which the conventional renormalization schemes are applied – closer scrutiny reveals that the "excellent agreement" between theory and experiment involves a considerable amount of *subjective faith* in the theory, so that Dirac's remarks on this score were not unwarranted. Indeed, as we shall illustrate with examples in Chapter 12, it is not totally uncommon that what appears to be, during a given historical period, incontrovertible experimental evidence in favor of theories that were fashionable during that period, turns out later to have been based on extraneous factors, which primarily reflected the conceptual biases prevailing during that period. In fact, one does not have to go as far back into the history of science as the Ptolomaic system to find such examples. Rather, they are plentiful in this century – its second half included[26]. Consequently, to base one's "confidence in the correctness of the results" of CQED exclusively on "their excellent agreement with experiment" does not appear prudent even by the prevailing instrumentalist standards, that characterize the post-World War II approach to quantum physics. At the very least, the question should be asked as to how well do the computational schemes advanced by the renormalization program stand up to closer scrutiny in the case of CQED – which, according to Feynman (1985), represents "the jewel of physics – our proudest possession". The following is an assessment of the fundamental problems left open by the renormalization scheme in CQED, formulated from an instrumentalistic point of view:

"What has not been achieved so far [in CQED] is a rigorous derivation of [the Feynman rules for the S-matrix] which would include a consistent treatment of the renormalization terms, a general proof of the unitarity of the new S-matrix and a non-perturbative proof of the absence of infrared divergencies. It is obvious that the present treatment of the infrared problem comes as an afterthought. The coherent state space is not introduced into the theory from the beginning. Nor is the necessary infrared renormalization carried through and shown to be consistent with ultraviolet renormalization." (Jauch and Rohrlich, 1976, pp. 528-529).

Thus, from a foundational point of view it is not only justified, but even necessary[27] to search for a new formulation of QED, which is nonperturbational in its origins, so that it does not exhibit the fundamental physical and mathematical shortcomings of CQED, but which, in some kind of limit can be seen to reproduce its perturbative results. Clearly, if any such new formulation is mathematically consistent, the limit in which the CQED perturbation series is recovered would have to be merely formal, in the sense that, after all the appropriate "renormalizations" are performed, each term in the formal perturbation expansion of the S-matrix in the new theory approaches its counterpart in CQED. Indeed, in view of all the inconsistencies of CQED, the best that one can hope to achieve is to explain the numerical successes of CQED as being due to an *underlying* consistent theory, in which all original sources of inconsistencies that manifest themselves in the form of divergencies have been removed from the outset, rather than by an *ad hoc* renormalization procedure that merely discards them, without resolving any of the foundational problems of either a mathematical or a physical nature.

As a matter of fact, as long as the mere removal of divergencies is the main concern, there is no unique way in which this problem can be solved. On the other hand, as we have seen in the last of the above quotations, coherent states are essential to any proposal for a consistent solution. The GS approach to quantum field theory introduces such states, in the form of local quantum frames, from the outset and on fundamental measurement-theoretical grounds, rather than as a mere afterthought. In this, as well as other key aspects, it distinguishes itself from the nonlocal quantum field theories proposed in recent times (Efimov, 1977; Namsrai, 1986; Evens *et al.*, 1991), which try to deal with the problem of divergencies by the *ad hoc* introduction of cut-offs in the interaction term.

The geometro-stochastic version of QED (to which we shall refer in this section as GSQED) can be formulated over any Lorentz base manifold $(\mathbf{M},\boldsymbol{g})$ by introducing the Whitney product of the Fock-Dirac bundle in (8.1.3) that corresponds to the fundamental quantum spacetime form factor in (5.5.5), with the core Gupta-Bleuler bundle in (3.9b). For the purpose of formulating parallel transport, and therefore GS propagation in $\mathbf{M}$, the first one of these bundles has to be extended into the kind of Berezin-Dirac bundle described in Sec. 8.3, whereas the second one has to be nested into the enveloping Gupta-Bleuler bundle in (3.9a). The GS propagation for GSQED can be then formulated upon adopting a total action in which to the sum of the terms in (8.4.15b) and (5.23b) are added the following terms,

$$m\hat{Z}_{\ell,m}(\overline{\psi}\gamma_k\Delta\psi-\Delta\overline{\psi}\gamma_k\psi')+\tfrac{1}{2}\hat{Z}_\ell^{(0)}\left(\overline{A}_{,k}\cdot\Delta A+A'_{,k}\cdot\Delta\overline{A}-\overline{A}\cdot\Delta A_{,k}-A'\cdot\Delta\overline{A}_{,k}\right)$$
$$-2em\hat{Z}_{\ell,m}\eta_{k0}\left(\overline{\psi}^{\uparrow}+\overline{\psi}'^{\downarrow}\right)\gamma^i\left(\overline{A}_i+A'_i\right)\left(\psi^{\downarrow}+\psi'^{\uparrow}\right) \quad , \tag{6.1a}$$

$$\overline{\psi}=\overline{\psi}(x(t);\zeta,\overline{\boldsymbol{\theta}}) \quad , \qquad \psi'=\psi'(x(t);\zeta,\boldsymbol{\theta}) \quad , \qquad \Delta\psi=\psi-\psi' \quad , \tag{6.1b}$$

$$\overline{A}=\overline{A}(x(t);\zeta) \quad , \qquad A'=A'(x(t);\zeta) \quad , \qquad \Delta A=A-A' \quad , \tag{6.1c}$$

where the last term in (6.1a) stems from the standard QED interaction based on the charge current that emerges from (6.2.10),

$$J^i_{\ell,m}(x;\zeta)=2em\hat{Z}_{\ell,m}\,\overline{\psi}(x;\zeta)\gamma^i\psi(x;\zeta) \quad , \tag{6.2}$$

whereas the first two terms are the accompanying geometric phases (cf. Sec. 3.9) to which that interaction term gives rise as it acts on quantum frames that are coherent states – as was the case also in (7.8.13).

The distinction between the strongly and weakly causal GS propagation becomes now crucial: in the strongly causal situation the paths of propagation between any two consecutive slices of **M** is fundamentally different for fermions, as opposed to photons – being timelike geodesics in the massive case, as opposed to null geodesics in the massless case. In the weakly causal situation no such distinction exists, since the microcausality features emerge exclusively from the properties of all the modes that propagate to a given base location x : those that have propagated from the causal past of that point are the ones with significant amplitudes, and their superposition leads to mutual reinforcements rather than to amplitude cancellations.

The weakly causal propagator that emerges from the adoption of (6.1) represents the formal counterpart of that in the path integral formulation of CQED – cf. (Rivers, 1987), Secs. 4.7, 9.7 and 10.1, as well as (Popov, 1983), Sec. 8. To see that, we have to specialize the above considerations to the case where $(\mathbf{M}, g)$ is Minkowski spacetime, and work with special Poincaré gauges that correspond to the adoption of a global Lorentz frame $\mathcal{L}$. We can then carry out the identification of that frame with the moving frame in (2.3.12). In turn, this identification mediates the identification of the corresponding fibres above all points in **M** (and, of course, also above those in **L**) with the typical fibre of each one of the bundles involved. For such gauges the last terms in (8.4.15b) and (5.23b) vanish, so that the fermionic part of the action assumes the form in (8.4.16), which combined with the corresponding form for the bosonic part yields an action which is based on a total Lagrangian density that can be expressed in the following schematic form[28]:

$$\mathcal{L}_{\text{QED}} = -\tfrac{1}{4}\hat{Z}_{\ell}^{(\mu_0)}\left(\partial^{\mu}A^{\nu} - \partial^{\nu}A^{\mu}\right)\left(\partial_{\mu}A_{\nu} - \partial_{\nu}A_{\mu}\right) + \tfrac{1}{2}\mu_0^2\,\hat{Z}_{\ell}^{(\mu_0)}A^{\nu}A_{\nu}$$
$$-\tfrac{1}{2}\hat{Z}_{\ell}^{(\mu_0)}\left(\partial^{\nu}A_{\nu}\right)^2 + 2m\hat{Z}_{\ell,m}\left(\tfrac{i}{2}\overline{\psi}\gamma_{\nu}\overleftrightarrow{\partial}^{\nu}\psi - m\overline{\psi}\,\psi\right) - 2em\hat{Z}_{\ell,m}\overline{\psi}\gamma^{\nu}A_{\nu}\psi \,. \tag{6.3}$$

It should be noted that, in accordance with the considerations in (2.8)-(2.16), as well as with the renormalization procedures in Chapters 7 and 8 of [IQ], a photon mass μ_0 is provisionally introduced, pending the taking of the limit $\mu_0 \to +0$. Furthermore, in view of the aforementioned identification, the quantum fluctuation amplitudes become interrelated at all $x \in \mathbf{M}$. Consequently, they can be replaced by x-independent expressions, in which the previous internal gauge variables q^{μ} take over the role of Minkowski coordinates, so that, as it was the case in Chapter 3, the partial derivatives in (6.2) pertain to the latter. Thus, the local quantum fluctuation amplitudes in (8.1.6) can be identified with the SQM propagators (cf. Secs. 3.6 and 6.3),

$$iS^{(+)}(q,v;q',v') = -(i/2m)(m + i\gamma^{\mu}\partial/\partial q^{\mu})\Delta^{(+)}(q,v;q',v') \;, \tag{6.4a}$$

where, in view of (5.5.5)–(5.5.10) and (6.3.7)–(6.3.8),

$$\Delta^{(+)}(q,v;q',v') = -iZ_{\ell,m}^{-2}\int_{u^0>0}\exp[im(q'-q)\cdot u - \ell(v'+v)\cdot u]\,d\Omega(u) \;, \tag{6.4b}$$

$$Z_{\ell,m} = (8\pi^4 \mathrm{K}_2(2\ell)/\ell m^3)^{1/2} = (\ell^3 m^3/4\pi^4)^{-1/2} + O(\ell^{-1/2}) \xrightarrow[\ell\to+0]{} +\infty \;. \tag{6.4c}$$

Similarly, on account of (2.8)-(2.22), the SQM propagator for photons of mass μ_0, that obeys the Proca equation in the q-variables, is given by

$$iD^{(+)}_{\mu_0}(q,v;q',v')\eta_{\mu\nu} = \hat{Z}^{(\mu_0)^{-2}}_{\ell}\eta_{\mu\nu}\int_{V^{(+)}_{\mu_0}} \exp[i(q'-q)\cdot k - \ell(v'+v)\cdot k]\,d\Omega_{\mu_0}(k) \ , \quad (6.5a)$$

$$Z^{(\mu_0)}_{\ell} = (8\pi^4 K_2(2\ell\mu_0)/\ell\mu_0^2)^{1/2} = 2\pi^2/\ell^{3/2}\mu_0^2 + O(\ell^{-1/2}) \xrightarrow[\ell\to+0]{} +\infty \ . \quad (6.5b)$$

Clearly, the above quantum fields should assume in the formal sharp-point limit $\ell \to +0$ the role of their "renormalized" counterparts in CQED. To obtain GS fields which formally merge into "bare" quantum fields that give rise to particles in quantum states represented by the wave functions in (6.1.12) and in (2.2), corresponding to "bare" masses $\hat{m}_0$ and $\hat{\mu}_0$, respectively, we first have to perform a "mass renormalization",

$$\psi(q,v) \to \psi^{(m_\ell)}(q,v) \ , \qquad 2\hat{Z}_{\ell,m_\ell} = (2\pi)^{-6} m_\ell^3 Z^4_{\ell,m_\ell} \ , \quad (6.6a)$$

$$A(q,v) \to A^{(\mu_\ell)}(q,v) \ , \qquad 2\hat{Z}^{(\mu_\ell)}_{\ell} = (2\pi)^{-6}\mu_\ell^3 Z^{(\mu_\ell)4}_{\ell} \ , \quad (6.6b)$$

whereby we make the transition to the counterpart $\mathcal{L}_\ell$ of the Lagrangian in (6.3) that corresponds to the masses m_ℓ and μ_ℓ instead of m_0 and μ_0, respectively. Upon performing the wave function renormalization

$$\psi_\ell(q,v) = Z_2^{1/2}(\ell)\,\psi^{(m_\ell)}(q,v) \ , \qquad Z_2(\ell) = (2\pi)^{-6} m_\ell^4 Z^4_{\ell,m_\ell} \ , \quad (6.7a)$$

$$A_\ell(q,v) = Z_3^{1/2}(\ell) A^{(\mu_\ell)}(q,v) \ , \qquad Z_3(\ell) = (2\pi)^{-6}\mu_\ell^4 Z^{(\mu_\ell)4}_{\ell} \ , \quad (6.7b)$$

this Lagrangian for "bare" GS fields assumes the form (cf. Eq. (8-97) in [IQ])

$$\begin{aligned}\mathcal{L}_\ell = &-\tfrac{1}{4}\left(\partial^\mu A^\nu_\ell - \partial^\nu A^\mu_\ell\right)\left(\partial_\mu A_{\ell\nu} - \partial_\nu A_{\ell\mu}\right) + \tfrac{1}{2}\mu_\ell^2 A^\nu_\ell A_{\ell\nu} \\ &-\tfrac{1}{2}\left(\partial_\nu A^\nu_\ell\right)^2 + \left(\tfrac{i}{2}\overline{\psi}_\ell\gamma_\nu\overleftrightarrow{\partial}^\nu\psi_\ell - m_\ell\overline{\psi}_\ell\,\psi_\ell\right) - e_\ell\,\overline{\psi}_\ell\gamma_\nu A^\nu_\ell\,\psi_\ell \ , \end{aligned} \quad (6.8a)$$

$$e_\ell = Z_1(\ell)Z_3^{-1}(\ell)Z_3^{-1/2}(\ell)e = Z_3^{-1/2}(\ell)e \ , \qquad Z_1(\ell) = 2m_\ell\hat{Z}_{\ell,m_\ell} \ . \quad (6.8b)$$

Furthermore, on account of the equalities imposed in (6.6) we have $Z_1(\ell) = Z_2(\ell)$ (cf. Eq. (8-99) in [IQ]), and for "small" values of ℓ we can write[29]

$$\begin{aligned}Z_2^{1/2}(\ell)\Delta^{(+)}_{m_\ell}(q,v;q',v') &= -i(2\pi)^{-3}\int_{k^0>0}\exp[i(q'-q)\cdot k - \ell(v'+v)\cdot k]\,d\Omega_{m_\ell}(k) \\ &\approx -i(2\pi)^{-3}\int_{k^0>0}\exp[i(q'-q)\cdot k]\,\delta(k^2-m_\ell^2)d^4k = \Delta^{(+)}_{m_\ell}(q-q') \ , \end{aligned} \quad (6.9)$$

$$2m_\ell Z_2^{1/2}(\ell)S^{(+)}_{m_\ell}(q,v;q',v') \approx -(m_\ell + i\gamma^\mu\partial/\partial q^\mu)\Delta^{(+)}_{m_\ell}(q-q') = S^{(+)}_{m_\ell}(q-q') \ , \quad (6.10)$$

$$\begin{aligned}Z_3^{1/2}(\ell)D^{(+)}_{\mu_\ell}(q,v;q',v') &= -i(2\pi)^{-3}\int_{k^0>0}\exp[i(q'-q)\cdot k - \ell(v'+v)\cdot k]\,d\Omega_{\mu_\ell}(k) \\ &\approx -i(2\pi)^{-3}\int_{k^0>0}\exp[i(q'-q)\cdot k]\,\delta(k^2-\mu_\ell^2)d^4k = D^{(+)}_{\mu_\ell}(q-q') \ . \end{aligned} \quad (6.11)$$

On the other hand, in view of (6.4b) and (cf. Sec. 5.5)

$$\hat{Z}_{\ell,m} = \mathrm{K}_2(2\ell)/m\mathrm{K}_1(2\ell) = (\ell m)^{-1} + O(1) \xrightarrow[\ell \to +0]{} +\infty \ , \tag{6.12}$$

the "bare" mass m_ℓ diverges in the sharp-point limit,

$$m_\ell^2 = \pi^2/8\ell^5 \xrightarrow[\ell \to +0]{} +\infty \ , \tag{6.13}$$

so that, in a strict mathematical sense, we cannot claim that the "bare" GS propagators converge to their conventional counterpart in the limit $\ell \to +0$. On the other hand, this feature of m_ℓ, together with the fact that by (6.4b), (6.7), (6.8) and (6.12),

$$Z_1(\ell) = Z_2(\ell) \xrightarrow[\ell \to +0]{} +\infty \quad , \qquad Z_3(\ell) \xrightarrow[\ell \to +0]{} +\infty \ , \tag{6.14}$$

agrees with the formalistic treatment of "renormalization constants" in CQED. However, it does not, and it cannot, impart mathematical or physical legitimacy to that formalistic treatment, so that Dirac's earlier cited verdict remains in force: "One is not allowed to neglect infinitely large quantities"! The present treatment is therefore merely meant to illustrate the root causes for the appearance of those infinities, as well as to demonstrate why GSQED is capable of coping with them.

The main value of CQED resides exclusively in its numerical successes, obtained from its well-known perturbation series. The formal arguments leading to that series (cf. [SI], Secs. 11f and 13a) are equally well applicable to the interaction picture of GSQED on Minkowski space, and provide the following formula for the S-matrix[30]:

$$S = \sum_{n=0}^{\infty} \frac{(-i)^n}{n!} \int d\mu(\zeta_1) \cdots \int d\mu(\zeta_n) T\left[\mathcal{H}_I^{(\ell)}(\zeta_1) \cdots \mathcal{H}_I^{(\ell)}(\zeta_n) \right] \ , \tag{6.15a}$$

$$d\mu(\zeta) = d^4 q \, d\Omega(v) \quad , \qquad \zeta = (q,v) \in \mathbf{R}^4 \times \mathbf{V}^+ \ , \tag{6.15b}$$

$$\mathcal{H}_I^{(\ell)}(\zeta) = -e_\ell : \tilde{\psi}_\ell(\zeta) \gamma_\nu A_\ell^\nu(\zeta) \, \psi_\ell(\zeta) : \ . \tag{6.15c}$$

Furthermore, Wick's theorem [SI,IQ] is applicable to the time-ordered products in (6.15a), so that those products can be expressed in terms of normal products and of the following "causal"[31] two-point functions (cf. [SI], pp. 437–446; [IQ], pp. 92, 150, 133–134):

$$\left\langle \Psi_0 \middle| T\left[\psi_\ell(\zeta,r) \, \tilde{\psi}_\ell(\zeta',r') \right] \Psi_0 \right\rangle = i S_F^{(\ell)}(q,v,r;q',v',r') \ , \tag{6.16a}$$

$$\begin{aligned} T\left[\psi_\ell(\zeta,r) \, \tilde{\psi}_\ell(\zeta',r') \right] &= \theta(q^0 - q'^0) \, \psi_\ell(q,v,r) \, \tilde{\psi}_\ell(q',v',r') \\ &\quad - \theta(q'^0 - q^0) \, \tilde{\psi}_\ell(q',v',r') \, \psi_\ell(q,v,r) \ , \qquad r,r' = 1,2,3,4 \ , \end{aligned} \tag{6.16b}$$

$$-i S_F^{(\ell)}(q,v;q',v') = \theta(q^0 - q'^0) S_{m_\ell}^{(+)}(q,v;q',v') - \theta(q'^0 - q^0) S_{m_\ell}^{(-)}(q,v;q',v') \ , \tag{6.16c}$$

$$\left\langle \Psi_0 \middle| T\left[A_{\ell\mu}(\zeta) A_{\ell\nu}(\zeta') \right] \Psi_0 \right\rangle = i D_F^{(\ell)}(q,v;q',v') \eta_{\mu\nu} \ , \tag{6.17a}$$

$$T\left[A_{\ell\mu}(\zeta)A_{\ell\nu}(\zeta')\right] = \theta(q^0 - q'^0)A_{\ell\mu}(q,v)A_{\ell\nu}(q',v')$$
$$+\theta(q'^0 - q^0)A_{\ell\nu}(q',v')A_{\ell\mu}(q,v) \ , \qquad \mu,\nu = 0,1,2,3 \ , \quad (6.17b)$$

$$-iD_F^{(\ell)}(q,v;q',v') = \theta(q^0 - q'^0)D_{\mu_\ell}^{(+)}(q,v;q',v') - \theta(q'^0 - q^0)D_{\mu_\ell}^{(-)}(q,v;q',v') \ . \quad (6.17c)$$

Expressions for these "causal" functions, that are in closed analytic form based on modified Bessel functions, follow directly from (5.5.8). Consequently, to all the standard CQED Feynman rules in configuration space (cf. [SI], p. 471) we can assign faithful GSQED counterparts, whose numerical values can be computed from those functions.

The GSQED counterparts of the CQED rules in momentum space (cf. [SI, p. 478, and [IQ], p. 275) – which enjoy much greater popularity – can be also derived in such a manner that the formal GSQED perturbation expansion can be then compared, diagram by diagram for each S-matrix element, with the corresponding CQED perturbation expansion. Indeed, in the momentum representation, each incoming GSQED line representing a "bare" GS fermion, having spin component s along the third spatial axis of the chosen global Lorentz frame, and impacting at q with stochastic 4-velocity v, gives rise (in accordance with the transformations to the momentum representation discussed in Sec. 3.4, with sharply defined 4-momentum p) to the following matrix-valued function (with $w^s(\mathbf{p})$ defined as in [SI], pp. 87 and 478):

$$w^s(\mathbf{p})\exp(-ip\cdot\bar{\zeta}) \ , \qquad \bar{\zeta} = q - i\ell v \ , \qquad p = (p^0,\mathbf{p}) \in V_{m_\ell}^+ \ . \quad (6.18)$$

As such it contributes to the integrals for a GSQED S-matrix element, computed on the basis of (6.15a) and of the inner product in the form (6.2.9). Similarly, an outgoing photon in a given transversal linear polarization mode (cf. (1.16) and (1.29)), can be deemed to give rise, in the momentum representation, to a wave function in the sharp 4-momentum variable k :

$$\varepsilon_{(\rho)}(k)\exp(ik\cdot\zeta) \ , \qquad \zeta = q + i\ell v \ , \qquad k = (k^0,\mathbf{k}) \in V_{\mu_\ell}^+ \ . \quad (6.19)$$

Analogous representatives can be written down for all other states of incoming or outgoing fermions, antifermions, and photons.

To deal with the internal lines in the momentum representation, we first observe that, according to (6.3.7) and (6.3.8), the momentum space representatives of the GSQED causal fermion propagator in (6.16c) can be also computed from

$$S_{m_\ell}^{(\pm)}(q,v;q',v') = -(m_\ell + i\gamma^\mu\partial/\partial q^\mu)\Delta_{m_\ell}^{(\pm)}(q,v;q',v') \ , \qquad q^0 \gtrless q'^0 \ , \quad (6.20a)$$

$$\Delta_{m_\ell}^{(\pm)}(q,v;q',v') = -\frac{i}{(2\pi)^3}\int_{\mathbf{R}^3}\exp[i(q'-q)\cdot p - \ell\varepsilon(\operatorname{Re}p^0)(v'+v)\cdot p]\frac{d^3\mathbf{p}}{2p^0} \ , \quad (6.20b)$$

$$\varepsilon(\operatorname{Re}p^0) = \theta(\operatorname{Re}p^0) - \theta(-\operatorname{Re}p^0) \ , \quad (6.20c)$$

where p^0 is on the m_ℓ-mass shell. The GSQED causal photon propagator in (6.17a) can be expressed in a corresponding form. Hence, at each ζ-vertex, the integration with respect to the measure μ, when first carried out in the υ-variable, produces the Lorentz-invariant vertex factor

$$V_\ell(p,p',k) = \int_{\mathbf{V}^+} \exp[-\ell\,\upsilon\cdot(\varepsilon(p^0)p+\varepsilon(p'^0)p'+\varepsilon(k^0)k)]\,d\Omega(\upsilon)\ . \tag{6.21}$$

This vertex factor can be computed by using the same method as in the computation leading to (3.6.28) (cf. [P], Sec. 2.9), with the following outcome:

$$V_\ell(p,p',k) = \frac{2\pi}{\ell\sqrt{\mathsf{p}^2}}\,\mathrm{K}_1(\ell\sqrt{\mathsf{p}^2}) = \frac{\sqrt{2\pi^3}}{\ell^3}e^{-\ell\sqrt{\mathsf{p}^2}}\left(\frac{1}{(\mathsf{p}^2)^{3/2}} + O\left(\frac{1}{(\mathsf{p}^2)^{5/2}}\right)\right)\ , \tag{6.22a}$$

$$\mathsf{p}^2 = \mathsf{p}\cdot\mathsf{p} > 0\ , \quad \mathsf{p} = \varepsilon(p^0)p+\varepsilon(p'^0)p'+\varepsilon(k^0)k,\quad p,p' \in \mathbf{V}_{m_\ell},\ k \in \mathbf{V}_{\mu_\ell}. \tag{6.22b}$$

The exponential decrease which this vertex factor exhibits as $\mathsf{p}^2 \to +\infty$ provides an overall momentum cutoff. Hence, the situation in GSQED is rather different from that in any of the "nonlocal" versions of QED, where "local" propagators are implicitly retained in the free Hamiltonian, whereas a momentum cutoff is imposed on each Feynman propagator in the interaction Hamiltonian – cf., e.g., (Namsrai, 1986).

In the sharp-point limit $\ell \to +0$ the above vertex factor behaves as $(\ell^2 \mathsf{p}^2)^{-1}$ for any fixed value of p^2, so that singularities reemerge. If the resulting divergencies are absorbed into the renormalization constants, then the well-known representations by means of contour integrals (cf. [SI], Sec. 13d) of the various two-point functions of conventional theory can be applied to the present situation. This can be achieved by expressing the remaining q-integrations in terms of such auxiliary "functions" (i.e., actual distributions) as

$$\Delta^{(\pm)}_{m_\ell}(q;q') = -(2\pi)^{-4}\lim_{\delta\to+0}\int_{\mathbf{R}^3} d^3\mathbf{k}\int_{C^{(\pm)}_\delta} dk^0\left(k^2 - m_\ell^2\right)^{-1}\exp[i(q'-q)\cdot k]\ . \tag{6.23}$$

In (6.23) the p^0-integration is performed along contours for which $\mp C^{(\mp)}_\delta$ result from the combination of the straight lines from $(-R-i)\delta$ to $-i\delta$, from $-i\delta$ to $+i\delta$, and from $+i\delta$ to $(+i+R)\delta$, with semicircles of radius R in the upper and lower half-planes of the complex variable p^0, respectively – with the limit $R \to +\infty$ being taken first, and the limit $\delta \to +0$ being taken afterwards. Indeed, for fixed $\delta > 0$, the integrand in (6.23) is analytic in the complex variable k^0 everywhere in the complex plane, and the two integrals rapidly approach zero as $R \to +\infty$, when they are confined to the respective semicircles of radius R, in the corresponding contexts with $q^0 > q'^0$ and $q^0 < q'^0$. Hence, by a standard application of the calculus of residues, we find that

$$\Delta^{(\ell)}_F(q;q') = 2i[\theta(q^0 - q'^0)\Delta^{(+)}_{m_\ell}(q;q') - \theta(q'^0 - q^0)\Delta^{(-)}_{m_\ell}(q;q')] \tag{6.24}$$

can be expressed in the alternative form

$$\Delta_F^{(\ell)}(q;q') = -\frac{1}{(2\pi)^4} \lim_{\varepsilon\to+0} \int_{\mathbf{R}^4} \frac{d^4k}{k^2 - m_\ell^2 + i\varepsilon} \exp[i(q'-q)\cdot k] \,, \tag{6.25}$$

where, according to (6.13), $m_\ell \to +\infty$ in the limit $\ell \to +0$, so that (6.25) formally plays the role of the Feynman propagator $\Delta_F(q-q')$.

Carried out in this manner, the q-integration with respect to the measure μ produces the familiar singular distribution $\delta^4(p - p' + k)$ for fermion lines carrying 4-momenta p and p', that meet at that vertex a photon line carrying the 4-momentum k (cf. [SI], Sec. 14c). The auxiliary "causal" two-point "function"

$$S_F^{(\ell)}(q;q') = (2\pi)^{-4} \int_{\mathbf{R}^4} S_F^{(\ell)}(p) \exp[i(q'-q)\cdot p]\, d^4p \quad , \tag{6.26a}$$

$$S_F^{(\ell)}(p) = (2\pi)^{-4}\left(\gamma^\mu p_\mu - m_\ell + i0\right)^{-1} \,, \qquad p \in \mathbf{R}^4 \,, \tag{6.26b}$$

supplies the Feynman "propagator" for fermions in the momentum representation, whereas

$$D_F^{(\ell)}(q;q') = -(2\pi)^{-4} \int_{\mathbf{R}^4} D_F^{(\ell)}(k) \exp[i(q'-q)\cdot k]\, d^4k \quad , \tag{6.27a}$$

$$D_F^{(\ell)}(k) = \left(k^2 - \mu_\ell^2 + i0\right)^{-1} \,, \qquad k \in \mathbf{R}^4 \,, \tag{6.27b}$$

essentially gives rise to the Feynman "propagator" for photons in momentum space, and in the so-called Feynman "gauge" (cf., e.g., [IQ], p. 134).

The above considerations demonstrate that the formal GSQED perturbation series can reproduce all the numerical results of that in CQED for sufficiently small values of the fundamental length ℓ. Of course, detailed quantitative studies are desirable, but the main issues are foundational: how do the two approaches to QED compare on the fundamental issues of locality, microcausality, covariance and the removal of mathematical infinities, namely on all the issues that raised problems since the inception of QED by Dirac in 1927, and which were discussed in the first part of this section?

The CQED idea of locality can be traced to "a certain oscillation between *de facto* operationalism and declared but undeveloped realism" (d'Espagnat, 1989, p. 156), which, in turn, has its roots in a "naive" realism based on "very simple idealizations of everyday experience" (*ibid.*, p 12). This epistemological approach postulates that the *sharp* separability of spacetime events, which has proven very useful in classical physics, can be extended to the microworld. However, such underpinnings of this naive idea of locality have been proven to be without support by the experiments of Aspect *et al.* (1981, 1982) and many others. Furthermore, the various theorems on the inconsistencies of the concept of sharp localization of relativistic quantum systems, discussed in Secs. 3.3, 7.2, 7.3 and 7.6, indicate that such a naive extrapolation of the macroscopic notion of locality is totally untenable. As a matter of fact, such indications emerge even from past attempts (Strocchi, 1967, 1970, 1978) at formulating CQED in mathematically rigorous terms, which have eventually resulted in proofs of the fact that conventional "*locality is necessarily lost at the level of asymptotic fields*" (Fröhlich *et al.*, 1979, p. 245) – emphasis as in the original.

The GSQED concept of locality relies on a series of mathematical results, reviewed in Chapter 3, proving the feasibility of a concept of *unsharp* separability associated with the introduction of a fundamental length ℓ, which avoids the aforementioned inconsistencies and is fully consistent with relativistic covariance. In the limit $\ell \to +0$, the mathematical underpinnings of this SQM locality merge in the nonrelativistic regime into those of conventional quantum mechanics, whereas in the relativistic context the appearance of divergencies prevents the rigorous and unconditional taking of such a sharp-point limit. Rather, a general relativistic concept of GS locality can be then introduced, in which global Poincaré covariance transcends into local Poincaré gauge covariance that is based on operational procedures with local *quantum* frames. These frames incorporate the physical meaning of this concept of GS locality in basically the same manner as their classical counterparts do in CGR.

The CQED idea of relativistic microcausality is conventionally identified with "local" (anti)commutativity, which not only lacks direct experimental support at the microlevel (cf. Sec. 7.6), but becomes *operationally* meaningless in the absence of a viable special relativistic notion of sharp locality and separability. Furthermore, it gets to be also *mathematically* meaningless at the general relativistic level. Indeed, in general relativity non-scalar fields *at* different points in space-time cannot be *directly* compared with respect to any of their algebraic properties, since their mathematical representatives exist in distinct fibres *above* those points, and any direct algebraic relations between elements of such distinct fibres are totally undefined. The brute force attempt at preempting this basic fact, which has been carried out in conventional quantum field theory in curved spacetime [BD] by introducing fields associated with Cauchy hypersurfaces in that curved spacetime, has merely resulted in a host of inconsistencies (cf. Secs. 7.2 and 7.3). Such inconsistencies are then compiled on top of the already very severe problems encountered in quantum field theories with the conventional concept of locality, borrowed from the classical regime.

The GSQED concept of relativistic microcausality is based exclusively on the features of the adopted mode of propagation – exactly as it is the case with the concept of Einstein causality in *classical* special or general relativistic physics. To achieve direct concordance with the numerical successes of CQED, a weakly causal form of GS propagation has been adopted in the present version of GSQED, whereby propagation is allowed along *all* paths that obey the temporal ordering imposed by the geometrodynamic evolution entailed by the GS form of quantum gravity studied in Chapter 11. An attenuation of GS wave functions above base points lying outside causal futures, which is *macroscopically* in accordance with Einstein causality, results then from the superposition of all the proper state vectors that have propagated from various locations along past reference hypersurfaces to the fibres above such points.

It might be thought that, in the special relativistic limit where a flat base manifold **M**, which is identifiable with Minkowski space, is chosen, the adoption of a geometrodynamically induced time ordering might lead to inconsistencies with the time ordering adopted in CQED. However, that is not the case at a fundamental level. Indeed, to the geometrodynamically inherited time ordering adopted in (6.16) and (6.17) corresponds in CQED a *natural* time ordering, which in the case of charge current operators leads to the following expression (cf. Eq. (5-92) in [IQ]):

$$\tilde{T}\left[j_\mu(x)\, j_\nu(x')\right] = \theta(x^0 - x'^0)\, j_\mu(x)\, j_\nu(x') + \theta(x'^0 - x^0)\, j_\nu(x')\, j_\mu(x) \quad . \tag{6.28}$$

However, after the discovery[32] by Schwinger (1959b) of terms in the vacuum expectation of the right-hand side of (6.27) which are *not* special relativistically covariant (since they are not covariant under Lorentz boosts), this natural time ordering was *conventionally* dismissed as "naive", and it was thereby replaced by a "covariant time-ordering" operation T, so that in present-day treatments of CQED we have (cf. Eq. (11-85) in [IQ]):

$$\langle 0|T[j_\mu(x)\,j_\nu(x')]|0\rangle = \langle 0|\tilde{T}[j_\mu(x)\,j_\nu(x')]|0\rangle - i(\eta_{\mu\nu} - \eta_{\mu 0}\eta_{\nu 0})\delta^4(x-x')\xi \ . \tag{6.29}$$

It is by such *ad hoc* artifices that *global* Poincaré covariance is rescued in CQED. In contradistinction, the version of GSQED presented in this section enjoys full *local* Poincaré (gauge) covariance. Hence, its *natural* time-ordering is of a geometrodynamic origin, that can be traced to the semiclassical approximation of the GS formulation of quantum gravity in Chapter 11. As such, it requires no *ad hoc* discarding of any kind of undesired terms. This fact is in general agreement with the observation that the argument demonstrating the seemingly unavoidable presence of Schwinger terms in (6.28) "is no longer valid if quantum gravity is taken into account" (Nakanishi, 1990, p. 59).

Finally, on the issue of infinities, the basic facts also favor GSQED. Indeed, the conventional renormalization procedure in CQED is based on *ad hoc* rules, rather than on an attempt to assign a legitimate mathematical meaning to quantum fields at a point in Minkowski space, that would endow the CQED interaction Lagrangian density with a mathematically meaningful and consistent interpretation. The constructive quantum field theory attempt in this latter direction has merely resulted in the speculation cited in Sec. 1.2 as well as earlier in this section, to the effect that the CQED S-matrix is trivial. Since this is very hard to believe in view of the highly nontrivial numerical successes of CQED, the only *physically* sensible conclusion is that CQED is indeed internally inconsistent – due to the singular nature of its interaction Hamiltonian density – but that there is an underlying consistent theory, for which CQED supplies suitable *numerical* approximations in a sharp-point limit of only *partial* validity. The present underlying thesis is that GSQED is that consistent theory, and that, in its formal perturbative expansions, the formal sharp-point limit is valid only on a term-by-term basis when combined with suitable renormalization procedures, which diverge in that limit. On the other hand, the GSQED interaction term in (6.2) has a precise mathematical interpretation in the context of the Naimark extensions discussed in Sec. 7.8, so that the GSQED propagators can be introduced nonperturbatively, by functional integrals whose integrands are mathematically well-defined.

The conclusion is that, by basing GSQED on foundations that combine a mathematically with an epistemically sound approach, it is possible to retain the numerical successes of CQED, and yet arrive at an internally consistent theory. Naturally, detailed studies of the mathematical existence and properties of GSQED nonperturbative solutions are required before a full understanding of all the mathematical and physical implications of the present approach to QED can be claimed.

Notes to Chapter 9

[1] This is not a misspelling of the name of Lorentz, but rather a rectification of a wide-spread historical misconception, whereby in practically all contemporary physics literature the name of H. A. Lorentz is confused with that of "Ludwig Lorenz (1829-1891) of Copenhagen, who independently developed an

electromagnetic theory of light a few years after the publication of Maxwell's memoirs." (Whittaker, 1951, pp. 267-268). It is to L. Lorenz that this gauge condition has to be rightfully attributed, since he first introduced it in 1867 – cf. also (Penrose and Rindler, 1984), p. 321.

2 Nakanishi (1972) provides a detailed review of this entire subject. The earliest attempt at a mathematically rigorous formulation of the Gupta-Bleuler framework was made by Wightman and Gårding (1964). However, this as well as subsequent studies by Strocchi (1967, 1970), Strocchi and Wightman (1974), Yngvason (1977), Rideau (1978), Mintchev (1980), Morchio and Strocchi (1980), Araki (1985), Cassinelli *et al.* (1989, 1991), Pierotti (1990), Strube (1990), and others, have left open some key mathematical points, which have been recently dealt with by Warlow (1992) – cf. Notes 5, 11 and 12.

3 A *Krein space* is defined [A] as a topological vector space $\mathcal{K}$ (over **R** or C) which is complete with respect to a positive-definite inner product, called its J-inner product, and on which there is specified a non-degenerate Hermitian inner product (which, as such, is in general given by an indefinite sesquilinear form), as well as two closed subspaces $\mathcal{K}^{\pm}$ on which that form is positive and negative definite, respectively, and which are mutually orthogonal so that their direct sum equals $\mathcal{K}$ – as it is the case in (1.6). A decomposition of the form (1.6) is a *fundamental decomposition* of $\mathcal{K}$. The orthogonal projectors $\mathbf{P}^{\pm}$ onto $\mathcal{K}^{\pm}$, respectively, are called *fundamental projections*, and the operator $J = \mathbf{P}^{+} - \mathbf{P}^{-}$ is the called the *fundamental symmetry* corresponding to the fundamental decomposition in (1.6). The norm corresponding to the J-inner product is called the J-norm. It is obvious that the present construction, based on (1.4)–(1.5), satisfies all the above stipulations. A Minkowski space provides an example of a 4-dimensional Krein space if its J-inner product is chosen to be the Euclidean inner product corresponding to a given choice of global Lorentz frame. Note should be taken of the fact that, as opposed to the terminology adopted in (Bognar, 1974), as well as in the present monograph, Azizov and Iokhvidov (1989) refer to an indefinite metric as a J-metric. Despite the fact that this latter reference is more recent, its terminology appears to be less suitable in applications to quantum field theory, so that we rarely refer to it.

4 By definition, space reflections (but no time reversals) are included in the orthochronous Poincaré group $\mathrm{ISO}^{\uparrow}(3,1)$. The use of the orthochronous Poincaré group is necessary in the present context, because the operators in (1.8) give rise to a reducible representation of $\mathrm{ISO}_0(3,1)$, whereas its extension to $\mathrm{ISO}^{\uparrow}(3,1)$ is irreducible. This can be most easily seen by considering the helicity eigenstates – where *helicity* is defined as being equal to the spin in the direction of the momentum, so that if spin is parallel or antiparallel to the momentum the helicity is said to be positive or negative, respectively. For all *restricted* Poincaré transformations, i.e., for all $U(a,\Lambda)$ with $(a,\Lambda) \in \mathrm{ISO}_0(3,1)$, the helicity sign is preserved, so that $U(a,\Lambda) = U^{(+)}(a,\Lambda) \oplus U^{(-)}(a,\Lambda)$, where $U^{(\pm)}(a,\Lambda)$ are irreducible when considered as representations of the (restricted) Poincaré group $\mathrm{ISO}_0(3,1)$. However, helicity is a pseudo-scalar, so that the inversions in the orientation of the spatial axes of a Lorentz frame interchange those eigenstates – cf., e.g., (Kim and Noz, 1986), Chapter VII, Sec. 3. It should be noted that pseudo-unitary representations of the form (1.8) have been studied on previous occasions and at a mathematically rigorous level by many authors – such as Shaw (1964, 1965), Bertrand (1971), Barut and Raczka (1972), Parvizi (1975), Rideau (1978), and others. Furthermore, Wigner's well-known theorem on ray representations [BL,BR] has been extended to the case of indefinite metric (i.e., to Krein spaces) by Bracci *et al.* (1975), so that the physical significance of such representations is completely assured.

5 A detailed and mathematically rigorous proof of the equivalence of the pseudo-unitary representation in (1.13) to the mass-0 and spin-1 unitary and irreducible Wigner representation of the orthochronous Poincaré group, acting in the Hilbert space with positive-definite inner product (1.26) in accordance with (1.25), was given by Bertrand (1971) in the radiation gauge. A more general proof, valid for any combination of the Lorenz with an axial gauge, has been recently provided by Warlow (1992). It is the principal lines of argument in this latter proof that are followed in Sec. 9.1. Note should be made of the fact that the elements of the Krein space in (1.10) are equivalence classes of functions on the forward light cone which are equal almost everywhere with respect to the invariant measure in (1.9c), so that equalities between some of these functions, such as the ones between the brace brackets in (1.14) or (1.15), in general hold only almost everywhere along the forward light cone.

6 Cf. [BR], Ch. 11 and (Kim and Noz, 1986), Ch. VII. The little group of mass-0 particles of integer spin, such as the photon and the graviton, is isomorphic to the Euclidean group E(2) in two dimensions, or to its subgroup SE(2), in case that space reflections are not incorporated. We shall therefore refer also to SE(2) as the Euclidean group in two dimensions.

7 Cf. (6.1.5b), which shows how the corresponding Wigner SU(2)-element for the case of spin-1/2 is embedded into SL(2,C). Since SU(2) and SL(2,C) are the covering groups of SO(3) and $SO_0(3,1)$, respectively, the map of SL(2,C) onto $SO_0(3,1)$ supplied by (6.1.3c) provides the corresponding embedding of Wigner rotations into $SO_0(3,1)$.

8 As shown and discussed by Inönü and Wigner (1953), and more recently by Kim and Wigner (1987), the little group of massless particles of integer spin, which is isomorphic to the Euclidean group SE(2) in two dimensions, can be obtained by a contraction of the little group of massive particles of integer spin, which is isomorphic to the rotation group SO(3) in three dimensions. A review of the basic results on such contractions can be found in Chapter VIII of (Kim and Noz, 1986), whereas more recent studies of these little groups, that are relevant to the considerations in this chapter, can be found in (Vassiliadis, 1989) and in (Kim and Wigner, 1990).

9 The stochastic localization framework for photons originally developed in (Prugovečki, 1978c) was based on the concept of frontal localizability, suggested by Acharya and Sudarshan (1960). Frontal localizability is, however, a physically unsatisfactory concept, since it presupposes a perfectly sharp measurement of the direction of motion of a photon, and it provides no information about its location in directions spatially orthogonal to that. The approach to photon localization described in the present section was therefore adopted instead in the GS quantization of the electromagnetic field (Prugovečki, 1988b). It involves an infinite wave function renormalization in (2.12), but note should be taken of the fact that when the inner product is written in the momentum representation no infinities occur, and that the entire procedure is mathematically well defined. Hence, this type of infinite renormalization is of a totally different nature from the one in CQED (cf. Sec. 9.6), where "infinities" are subtracted in an *ad hoc* manner from such basic physical parameters of the theory as rest masses and charges.

10 One such *ad hoc* subtraction consists of discarding Schwinger terms, containing δ-functions, from Feynman propagators, on the grounds that they are not manifestly covariant, so that "we can therefore define a covariant propagator by simply omitting these terms without affecting its properties" ([IQ], p. 136). However, this "simple omission" disguises the unavoidable mathematical and physical *fact* that the map in (2.8) is not left invariant by boosts. It therefore actually represents an unwarranted violation of mathematical consistency – i.e., an instance of what Dirac describes as "just not sensible mathematics" (Dirac, 1978a, p. 36). As such, it contributes to the multitude of all the other inconsistencies in conventional QED, to which *ad hoc* "renormalization" procedures give rise, and which were repeatedly criticized by Dirac ever since the inception of the renormalization programme.

11 The fact that the customary [SI,IQ] Fock-type representation $U_u(a,\Lambda)$, constructed from the momentum space representation in (1.13) by taking a counterpart of the direct sum in (3.2), is unbounded, has remained basically unnoticed in the mathematically rigorous literature on the Gupta-Bleuler formalism, probably due to the fact that the restrictions of the operators $U_u(a,\Lambda)$, to each one of the n-photon subspaces $\mathcal{K}_n$, constructed out of the single-photon space in (1.10), are J-bounded operators in the customary combination of the Lorenz with timelike axial gauges, which is used in the present chapter. [However, it turns out (Warlow, 1992) that even that feature is not generally true, since in fact the topological structure of a single-photon Krein space depends on the chosen axial gauge in the null and the spacelike case]. Only Strocchi (1978) allows for the possibility of J-unboundedness in his extension of the Wightman axioms to quantum gauge fields, without providing, however, further analysis, examples, or proofs. It was, however, noticed by Warlow (1992) that when Λ is a Lorentz boost, such as the one in (5.2.21), then we have the following inequalities, $\exp(-n\vartheta) \leq \|U_u(a,\Lambda)f_n\|_J/\|f_n\|_J \leq \exp(n\vartheta)$, $\vartheta \geq 0$, where the lower and upper bounds are actually infima and suprema, respectively, when f_n varies over all of $\mathcal{K}_n$ – as can be easily established by considering in (1.18) all J-normalized wave functions of increasingly small support around a given value of $k \neq 0$ on the forward light-cone. This means, however, that whereas the restriction of $U_u(a,\Lambda)$, to the momentum-space counterpart of each subspace $\mathcal{K}_n$ is bounded, that is not the case for the Gupta-Bleuler space constructed by taking for $n =$ 0,1,2,... the J-direct sum of all $\mathcal{K}_n$. Hence, such claims as the one that "Lorentz covariance is manifest in the Gupta-Bleuler formalism" (Kaku, 1991, p. 13) are incorrect not only in string theory, but even in conventional QED. [An argument to this effect was first put forward by Sunakawa (1958), but it was not based on topological arguments, and it contained a basic error, so that it was later easily refuted by Gupta (1959)]. Indeed, for such a manifest covariance to exist, we need within a Gupta-Bleuler space a one-to-one correspondence between *all* the multi-photon state vector representatives given by

momentum-space wave functions in relation to two Lorentz frames in a state of relative motion. Such a one-to-one correspondence exists between the state vectors within the *core* Gupta-Bleuler space, which consists of *finite* linear superpositions of n-photon state vectors for various values of n, and that correspondence is indeed given by (3.2). However, that one-to-one correspondence cannot be extended in a *unique* manner compatible with linearity, and therefore with the superposition principle. Such an extension would amount to taking, in a given J-norm, the graph-closures [PQ] of the restrictions of the unbounded operators $U_u(a, \Lambda)$, to the linear space spanned by the aforementioned finite linear superpositions, and that would contradict the closed-graph theorem [PQ]. Indeed, that most fundamental theorem of functional analysis stipulates that such graph-closures can be taken only for *bounded* operators. On the other hand, the indefinite inner product does not provide a separable topological vector space (i.e., one containing a countable dense set, and therefore enabling expansions in countable bases), so that it cannot be used to give meaning even to the infinite series defining coherent states for photons – such as those in (3.21). Thus we are left with the various J-topologies, which are, however, inequivalent, so that a J-specification is required each time an infinite superposition of state vectors from the core Gupta-Bleuler space is considered.

12 The mathematical construction required for dealing with the unboundedness problem described in Note 11, involves the use of a type of generalized inner product space (Prugovečki, 1969c) which displays the structure of the partial inner product spaces formulated by Antoine and Grossmann (1976, 1978), and further studied by Antoine (1980). The detailed construction has been carried out by Warlow (1992).

13 In an algebraic direct sum, as opposed to a Hilbert direct sum [PQ], the sequences of vectors from the constituent spaces, which represent elements of that algebraic direct sum, contain only a finite number of non-zero entries. Thus, an *algebraic* direct sum of an infinite number of Hilbert spaces is only a pre-Hilbert space, whereas the corresponding *Hilbert* direct sum is equal to its completion. It might appear that, after constructing the core Gupta-Bleuler bundle in (3.9b), all we have to do is take the closure of its fibres, patterned after the algebraic sum $\mathcal{K}_\infty$, in order to obtain a Gupta-Bleuler bundle associated directly to PM, whose fibres incorporate the Gupta-Bleuler quantum frames constructed in (3.21). However, for the convergence of the operator-valued power series in (3.21) we need a separable topology, which in Krein spaces is provided by the J-norms. However, in the present case, the question becomes, which J-norm to choose, since the corresponding J-topologies are not equivalent (cf. Note 11). In (3.9a) that ambiguity is resolved by adopting a base manifold which entails the specification of $\mathcal{J}$.

14 Further motivation for this construction can be found in Sec. 9.4, as well as in the next chapter. The construction could be carried out even in the absence of a global section, by using a covering with local trivialization maps, as in (2.2.4)-(2.2.7). However, since any form of QED over a Lorentzian manifold $(\mathbf{M},g)$ requires introduction of the spin-frames in (6.3.10), namely the existence of a spin structure, in view of Geroch's theorem (cf. Sec. 6.3) the Poincaré frame bundle PM has to be trivial, so that the stipulated global section $\mathbf{s}_0$ must exist. The subsequent construction of $\mathbf{L}$ might appear to be dependent on the choice of $\mathbf{s}_0$, but as seen from the considerations in the latter half of Sec. 9.4, this is not actually the case.

15 Naturally, a corresponding need would be revealed in conventional QED if proper attention were paid to the underlying mathematics (cf. Note 11). The use of heuristic Lagrangian procedures has served to obscure that need.

16 An unbounded operator does not necessarily map a Cauchy sequence into another Cauchy sequence. Hence, distinct choices of Cauchy sequences from a dense set, such as the core of a Gupta-Bleuler fibre, converging to the same vector outside it, but within the Gupta-Bleuler fibre, might lead to distinct definitions of its extension outside the core. It should be noted that a counterpart of Stone's theorem [PQ] was proven by Naimark (1966) in the context of Krein spaces. This guarantees the existence of the (densely defined) self-adjoint infinitesimal generators that enter in all expressions for covariant derivatives in this section – namely those in (4.16), as well as those that emerge from (4.17) and (4.24). Naturally, the same type of mathematical existence is also assured in the case of the infinitesimal generators in (5.1) and (5.12), as well as in the case of all their counterparts in the next two chapters.

17 Cf., e.g., (Baulieu, 1985), p. 9. In the next chapter considerations similar to the present ones will enable us to impart a purely geometric interpretation to BRST types of constraints.

18 The first such interpretation is due to Thierry-Mieg (1980), who considered only a single Faddeev-Popov ghost field giving rise to BRST operators. It was later realized (Baulieu and Thierry-Mieg, 1982) that in addition to BRST operators, anti-BRST operators corresponding to an antighost field (Curci and Ferrari,

1976) had to be also introduced. The use of triply-fibrated principal bundles in the mathematically rigorous treatment of BRST and anti-BRST transformations was first advocated by Quirós *et al.* (1981) – who, however, did not use the present terminology. An alternative, based on a gauge group approach, will be presented in the next chapter.

19 The following is a fair presentation of its epistemic assets and liabilities, as well as of its basic features: "The logical sequence of steps [in Lagrangian quantum field theory] is as follows: given a *classical* Lagrangian of a field or a system of interacting fields one begins with an action principle and derives the field equations and conservation laws by means of the calculus of variations (Noether's theorems are here the key). The resulting equations and conservation laws are then regarded as operator equations and form the starting point of a quantum field theory. One must keep in mind that this 'derivation' takes place with classical (c-number) fields, since the calculus of variations is not defined for a Lagrangian constructed from operators in a Hilbert space. Consequently, the derived field equations and conservation laws can in general not be regarded as equations for operators without ambiguity of ordering and, usually, additional lack of precise mathematical meaning. Nevertheless, Lagrangian field theory is used extensively as a heuristic tool: it permits intuitive input, it allows an easy way to include invariance properties, and it gives a certain assurance that the theory will have a classical counterpart." (Jauch and Rohrlich, 1976, p. 480).

20 This, as well as all remaining formal steps involving integrations in the present mass-0 case can be carried out in a mathematically rigorous manner, by first writing out the corresponding integrals for the case based on (2.9), carrying out the depicted procedures, and in the end taking the mass-0 limit. In the present case this procedure involves applying the resulting Bochner integral to a massive counterpart of an n-photon state vector $\boldsymbol{\Psi}$, and then taking the inner products with another arbitrary massive state vector, which is automatically in the Lorenz gauge. The rapid decrease at infinity of the coordinate wave functions in the $\mathbf{q}$-variables ensures that the surface terms resulting from the integration by parts contribute zero.

21 This stress-energy tensor is not $U(1)$-gauge invariant, as it is the case with some of the momentum-energy tensors which can be introduced in classical electromagnetism – cf., e.g., [IQ], Sec. 1-2-2. This is, however, a general feature of the second-quantization of all gauge field theories, which involves a breaking of the gauge invariance of their classical counterparts – cf. [SI], p. 241 and [IQ], Eq. (3-120). This gauge breaking was implicit in the construction of the single-photon space in (1.6), whose elements are wave functions that satisfy the first, but not the necessarily second of the equations in (1.3).

22 Cf. (Pais, 1987) as well as (Kragh, 1990). Dirac himself had the following to say towards the end of his life: "I really spent my life mainly trying to find better equations for quantum electrodynamics, and so far without success, but I continue to work on it." (Dirac, 1979, p. 653).

23 The opposite was true in the pre-World War II years. As described in a recent biography of Dirac, not only Dirac, but also Bohr, Born, Fock, Heisenberg, Pauli, Peierls, and many other leading theoretical physicists of the generation that founded quantum mechanics came to believe that "the failure of quantum electrodynamics at high energies would require a revolutionary break with current theory", so that Dirac's "critical attitude toward quantum electrodynamics during the period from 1935 to 1947 was neither unique nor particularly remarkable" (Kragh, 1990, p. 166). That was in sharp contrast to the attitude of those in the younger generation who "adapted themselves to the new situation without caring too much about the theory's lack of formal consistency and conceptual clarity. The pragmatic attitude of the 'quantum engineers', including Fermi, Bethe, Heitler, and a growing number of young American physicists, proved to be of significant value, but did not eliminate the fundamental problems that continued to worry Dirac, Pauli, and others." (*ibid.*, p. 166). On the other hand, the attitude of the "quantum engineers" prepared the ground for the reception of what Schweber (1991) describes as a realization of "the pragmatic ideal of American physics that ... [became] hegemonic worldwide." (Schweber, 1989, p. 673), namely the launching in 1947 of renormalization theory. However, whereas the majority in the new crop of theoretical physicists readily "agreed that everything was fine and that the long-awaited revolution was unnecessary", Dirac and other founders of quantum theory realized that the true sources of difficulties have been left untouched. Thus: "Although acknowledging the empirical success and social appeal of [renormalization] theory, Dirac found it completely unacceptable. ... He never fully accepted that the problems of quantum electrodynamics had been solved by a new generation of physicists whose approach was essentially conservative and instrumentalistic" (Kragh, 1990, p. 183). On the other hand, amongst this "new generation of physicists", it appears that at least its most outstanding member agreed

with Dirac, as he stated the following: "The observational basis of quantum electrodynamics is self-contradictory... The localization of charge with indefinite precision requires for its realization a coupling with the electromagnetic field that can attain arbitrarily large magnitudes. The resulting appearance of divergences, and contradictions, serves to deny the basic measurement hypothesis. We conclude that a convergent theory cannot be formulated consistently within the framework of present space-time concepts." (Schwinger, 1958, pp. xv-xvi).

24 Dirac's most recent biographer assesses this truly extraordinary situation as follows: "[Dirac's] proposal to discard the relativistic quantum theory was drastic and essentially negative. It was only a state of intellectual despair that led him to the proposal, which, if taken seriously, would ruin the very achievements on which so much of his reputation rested. The relativistic version of quantum mechanics was Dirac's Nobel Prize-rewarded brainchild. Now he felt forced to renounce it without being able to offer an alternative. No wonder most of his colleagues considered his proposal a retrograde step." (Kragh, 1990, pp. 170-171).

25 Indeed, for the Lamb shift, as well as some other "predictions" ascribed exclusively to CQED, it is not only the case that "different theorists in their independent computations obtain different results", but also that different theorists in their independent computations obtain the *same result* from *different theories*. For example, the Lamb shift was derived by Ali and Prugovečki (1981) in the context of a quantum model in which the electron interacts *exclusively* with an external (i.e., non-quantized) Coulomb field. Even more revealing of the theoretical status of the Lamb shift is its recent derivation by Jaynes (1990), who uses a purely classical model, so that he demonstrates dramatically how the "successes" of the CQED "subtraction physics" can be matched by those of a purely "classical subtraction physics". This is especially significant in view of the fact that CQED does not treat the hydrogen bound-state problem in a consistent quantum-field theoretical fashion, whereby the proton constituting its nucleus and its interaction with the electron would correspond to quanta of a quantum field theory; rather, the conventional treatment also uses a semiclassical method [SI,IQ], whereby the Coulomb field of the nucleus is treated as an external field, with CQED being used only for the computation of "radiative corrections". It is therefore of interest to mention that, after listing some of the "weight of authority" that had gone since the late 1940s into convincing the entire physics community that the "explanation" of the Lamb shift is a most remarkable CQED "success", Jaynes points out the following: "The problem has been that these calculations have been done heretofore only in the quantum theory context. Because of this, people jumped to the conclusion that they were quantum effects (i.e., effects of field quantization), without taking the trouble to check whether they were present also in classical theory. As a result, two generations of physicists have regarded the Lamb shift as a deep, mysterious effect that ordinary people cannot hope to understand. So we were facing not so much a weight of authority and facts as a mass of accumulated folklore." (Jaynes, 1990, p. 394).

26 For example, two recent historical studies (Franklin, 1986, 1990) dedicated to demonstrating that in contemporary physics "experiment plays a legitimate role in theory choice and confirmation" nevertheless reveal "not only that experimental results can be wrong, but also that theoretical calculations and the comparison between experiment and theory can also be incorrect" (Franklin, 1990, pp. 4-5). The two principal case histories to which these observations pertain are: "(1) the interaction of theory and experiment in the development of the theory of weak interactions from Fermi's theory in 1934 to the *V-A* theory in 1957 and (2) atomic parity violating experiments in the 1970s and 1980s and their interaction with the Weinberg-Salam unified theory of electroweak interactions" (*ibid.*, p. 4). Another relevant quotation is the following: "Schweber (1989) has made a case that can be interpreted as showing that the very nature of scientific practice changed significantly with the advent of 'big science' after the Second World War. This becomes especially apparent in some of the recent large collaborative efforts in high-energy physics in which experiments may come to be performed just *once*. Some of the danger in unrepeated experiments is evident in the discovery of the weak neutral currents by the groups in Europe and in America (Galison, 1987). If the American group had been the *only* one doing the experiment and analyzing the data, the final verdict could have been very different from what is accepted today." (Cushing, 1990, p. 248).

27 Indeed, in a later article, Rohrlich (1980) has pointed out that "our present formulation [of QED] is *physically* incorrect", and expanded the above list of deficiencies to include other glaring failures, such as the lack of a "clean proof" that CQED merges into classical electrodynamics in some suitably formulated classical limit. A "Critical Review of the Theory of Quantum Electrodynamics" was recently pub-

lished by Nakanishi (1990), which discusses CQED "difficulties" and other topics not treated by Rohrlich. In the same volume Yokorama and Kubo (1990) provide an extensive and up-to-date bibliography, citing publications dealing with all the various theoretical and/or experimental aspects of CQED.

28 Cf. Eqs. (7-70) and (8-97) in [IQ]. Throughout this monograph we have strived to use a notation that underlines all the essential details. Since, as a rule, stochastic paths are not smooth, the ordering of terms in action integrals is essential, so that we have indicated in all the preceding formulae for "Lagrangians" whether a particular term belongs to the initial or the the final end of a smooth segment in a broken path approximating a stochastic path. However, for the sake of easy comparison with the mainstream literature on CQED, which tends to ignore such "details", in the remainder of this chapter we shall employ the same type of schematic notation that is prevalent in the textbooks and monographs cited in this section.

29 Cf. Eq. (75) on p. 168 and Eq. (55) on p 226 of [SI]. The factor $2m$ can be eliminated from the left-hand side of (6.10) by renormalizing the inner product in the typical Dirac fibre, and thus eliminating it also from (6.1.16) – cf. Note 9 to Chapter 6.

30 We abide by conventional terminology, although the expression occurring in (6.15) is the one for a scattering *operator* [PQ], rather than for a "matrix" in a mathematically acceptable sense. In fact, even what is conventionally called in physics literature "S-matrix elements" are not "matrix" elements in a mathematically meaningful way. A mathematically rigorous treatment of these and related topics in scattering theory is provided in Chapter V of [PQ] in a nonrelativistic context, but the principal theorems of that treatment can be easily extended to the present single Hilbert space case, in which $\mathbf{M}$ is a Minkowski space.

31 The term "causal" Green's function is used for the CQED counterparts of these functions, namely the Feynman propagators, because (cf. [SI], p. 438) they reflect the correct time ordering in relation to a *given* global Lorentz frame, and not because the supports of these Green's functions lie in either one of their spacetime variables within the causal future and/or past of the other one: the Feynman propagators are *not* causal in the latter sense (cf. [IQ], Eq. 1-180), so that vis-à-vis Einstein causality their status is exactly the same as those in (6.16) in (6.17). Of course, the fermion field satisfies "local" anticommutativity, but, as argued in Sec. 7.6, it is not at all evident even for conventional quantum field theories in Minkowski space that such anticommutativity is in any deeper *physical* sense related to actual Einstein causality. In fact, it was recently remarked by an S-matrix specialist that: "The *exact* meaning of this commutator [or anticommutator] relation is not clear. As we mentioned earlier, this is underscored in the 1957 LSZ paper in which an operator $A(x)$ satisfying Eq. [(7.6.1) of the present monograph] for spacelike separations is *defined* as a *causal* operator – without discussing here (or anywhere in that paper) the physical interpretation (Lehmann, Symanzik, and Zimmermann, 1957, p. 323). It seems *plausible* that it may have something to do with first-signal-principle [i.e., Einstein causality], but the equivalence of these two definitions of causality is not established. The main justification of this new criterion for a causal theory is that it produces a desired result (dispersion relations), while not obviously conflicting with a more direct notion of causality." (Cushing, 1990, p. 214). Thus, as it will be discussed in Chapter 12, it appears likely that the adoption of such a criterion of "causality" is merely a reflection of an instrumentalist attitude, which has become "a hallmark of much of theoretical physics after the Second World War (as contrasted with the period before the War)" (*ibid.*, p. 214).

32 Strictly speaking, "historically the existence of [this] inconsistency was found first by Goto and Imamura (1955) in the discussion of criticizing [sic] Källen's (1954) work. Later it was rediscovered by Schwinger (1959b) by using a somewhat different reasoning, and he proposed to introduce a non-canonical term in the equal time current-current commutator. Therefore, it is not appropriate to call this term [the] 'Goto-Imamura-Schwinger term', [so that] we call the inconsistency [the] 'Goto-Imamura difficulty' and the non-canonical term [we call the] 'Schwinger term'." (Nakanishi, 1990, p. 55). Unfortunately, this method of resolving inconsistencies merely supplies a prime example of the actual effects in everyday practice of the adoption of the instrumentalist criteria for judging the validity of scientific theories (cf. Sec. 12.3), whereby the concept of "truth", including mathematical truth, is identified with "consensus" based on professionally accepted *conventions.* When faced with it, one cannot help but be reminded of Dirac's many public statements, made over a thirty-year period following the inception of the renormalization program, in which he urged that "sound and sensible" mathematics, as well as the laws of "regular" logic, not be sacrificed for the sake of "working rules", designed to merely produce numerical results in CQED.

Chapter 10

Classical and Quantum Geometries for Yang-Mills Fields

The physical and mathematical origins of Yang-Mills fields can be traced to the work of H. Weyl, who in an unsuccessful but otherwise well-known[1] attempt to unify CGR with classical electromagnetism introduced the idea of a "gauge" field in 1918, and a decade later pinpointed U(1) as a "gauge group"[2] in the quantum regime (Weyl, 1929). Subsequently, O. Klein (1939) considered a non-Abelian gauge theory for the first time. However, Yang and Mills (1954) were apparently unaware of all this when they made their proposal to replace global SU(2) isospin invariance with local gauge invariance. Although the geometric origins of their proposal had eluded them at the time, the purely geometric nature of the entire idea of "gauge symmetry" was becoming increasingly clear as "their" idea became popular amongst elementary particle physicists in the late 1960s and in the 1970s – and especially when it led to the Weinberg-Salam model of electroweak interactions, and ultimately to QCD and the well-known "standard model" based on SU(3)×SU(2)×U(1) gauge symmetry, which at present dominates the theory of strong interactions. Eventually Wu and Yang (1975) established the full equivalence between the concepts used in elementary particle physics and their mathematical equivalents in the theory of connections on principal fibre bundles. It has, therefore, become customary to refer by the name of *Yang-Mills theory* to any field theory that is based on a "gauge group" $\mathbf{G}_0$ which is a compact Lie group (Utiyama, 1956), i.e., which in the classical context gives rise to fields that are sections of vector bundles associated with the principal bundle over Minkowski space $\mathbf{M}$ with structure group equal to $\mathbf{G}_0$. Thus, in the quantum field theories used at present in elementary particle physics the elements of the Poincaré group are deemed to represent global symmetries, whereas such groups as SU(3)×SU(2)×U(1), as well as other even larger compact groups, based on grand-unified models, are viewed as being the gauge symmetry groups for interactions between "elementary particles" (Nachtman, 1990).

It was noticed, however, already by Feynman (1963) that despite many common features that such Yang-Mills theories share with electromagnetic theory in Minkowski space, where U(1) plays the exclusive role of structure group, the Gupta-Bleuler formalism is not suitable in the more general case where the gauge groups are non-Abelian, since it leads to violations of the "unitarity" of the "physical" S-matrix (Kugo and Ojima, 1979). The method of Faddeev and Popov (1967), based on the introduction of additional "ghost" fields assuming values in Grassmann algebras, and suggested by a method of Dirac (1958) of dealing with Hamiltonian systems under constraints, eventually paved the way for the formulation of special relativistic Yang-Mills quantum field theories which are renormalizable in the conventional sense ([IQ], Faddeev and Slavnov, 1980; Gitman and Tyutin, 1990; Nakanishi and Ojima, 1990). The subsequent independent discoveries by Becchi, Rouet and Stora (1976), and by Tyutin (1975), of global symmetries associated with what became known as "BRST charges" in such theories, led to a "BRST quantization method", that eventually acquired great popularity during the subsequent developments in superstring theory in the 1980s (Green *et al.*, 1987; Brink and Henneaux, 1988; Kaku, 1989, 1991).

A geometric interpretation of the Faddeev-Popov "ghost" fields as left-invariant forms on a group of gauge transformations was apparently first proposed by R. Stora (1976). However, it was first noticed by Thierry-Mieg (1980) that the constraint equations satisfied by the BRST operators bear a strong *formal* resemblance to the Maurer-Cartan equations in the theory of Lie algebras of Lie groups [C,I]. Hence, he speculated that such charges can be interpreted as the generators for parallel transport along the fibres of the principal bundle with which the given Yang-Mills fields are associated. This interpretation was subsequently criticized by Leinaas and Olaussen (1982) on the grounds that if the Faddeev-Popov ghost fields were identified, as suggested by Thierry-Mieg, with differential forms on such principal fibre bundles, then they would have to span finite-dimensional spaces, since the exterior algebra generated by such forms is finite-dimensional – whereas that was not actually the case. However, subsequent work by Bonora *et al.* (1981–1983), Hoyos *et al.* (1982), Mayer *et al.* (1981, 1983), and others, brought considerable mathematical clarification to these issues, so that the geometric interpretation of BRST symmetries has become by now quite standard in the perturbative treatment of Yang-Mills gauge fields at the classical as well as at the quantum level (Baulieu *et al.*, 1985-1991).

We shall outline in Sec. 10.4 the most essential aspects of this work in the context of classical Yang-Mills fields, after preparing the ground by reviewing in Sec. 10.2 the generic features of such theories, and describing in Secs. 10.2 and 10.3 the required mathematical techniques that deal with groups of gauge transformations on principal fibre bundles, and with external covariant differentiation on graded Lie algebras constructed from families of connections on such bundles. We shall then present in the remaining two sections of this chapter the GS adaptation (Prugovečki, 1988b, 1989b) of the ensuing geometric treatment of BRST symmetries to the formulation of Yang-Mills *quantum* geometries over Lorentzian manifolds.

10.1. Basic Geometric Aspects of Classical Yang-Mills Fields

In purely geometric terms, classical *Yang-Mills gauge fields* in the Minkowski space $\mathbf{M}$ are gauge potentials that are assigned, in accordance with the general procedure described in Sec. 2.5, to sections $\boldsymbol{s}$ of a principal bundle $\mathbf{P}(\mathbf{M},\mathbf{G}_0)$ over $\mathbf{M}$, whose structure group $\mathbf{G}_0$ is a compact Lie group. In physics literature it is customary to introduce a basis in the Lie algebra $L(\mathbf{G}_0)$ of $\mathbf{G}_0$, as well as a global Lorentz frame $\{e_\mu \mid \mu = 0,1,2,3\}$ in $\mathbf{M}$, and then expand these Yang-Mills gauge fields in a basis related to the direct product of those two bases (Nakahara, 1990).

To see how that is done, let us consider the generic situation where a Yang-Mills gauge field acts on some "matter field" Ψ (such as the wave function of a quantum particle), given by a section of some associated vector bundle $(\mathbf{E}, \boldsymbol{\pi}, \mathbf{M}, \mathbf{F})$, via a faithful representation U of $\mathbf{G}_0$. Then, in accordance with (2.5.8), (2.5.9) and (2.5.17-(2.5.20), we can express as follows the pull-back of the connection form $\boldsymbol{\omega}$, defined as in (2.5.7), to the corresponding principal bundle whose typical fibre is $U(\mathbf{G}_0)$ (i.e., which coincides with the group to which that representation gives rise):

$$\hat{\mathcal{A}} = \hat{A}^a_\mu T_a \otimes dx^\mu , \qquad \hat{A}^a_\mu = A^a(\partial_\mu) \in \mathbf{R}^1 , \qquad \hat{A}^a := \hat{A}^a_\mu dx^\mu , \tag{1.1}$$

$$[T_a, T_b] = C_{ab}{}^c T_c , \qquad a,b,c = 1,\ldots,n , \quad n = dim\, \mathbf{G}_0 , \tag{1.2a}$$

$$\tilde{\mathbf{Y}}_a \leftrightarrow T_a = \partial U(g)/\partial g^a \big|_{g=e} , \qquad g \mapsto (g^1,\ldots,g^n) \in \mathbf{R}^n . \tag{1.2b}$$

Consequently, by (2.5.22), the components of the covariant derivative which this pull-back determines for the vector field $\boldsymbol{\Psi}$ over the Minkowski space $\mathbf{M}$, given by a section of an associated vector bundle, are given by

$$D_\mu \boldsymbol{\Psi}_x = \partial_\mu \boldsymbol{\Psi}_x + \hat{A}^a_\mu \hat{\mathbf{T}}_a \boldsymbol{\Psi}_x , \qquad T_a = \tilde{\mathbf{Y}}_a \mapsto \hat{\mathbf{T}}_a = \hat{\mathbf{A}}_{a;\boldsymbol{u}} , \tag{1.3}$$

where the map in (1.3) is defined at each $\boldsymbol{u} = \mathbf{s}(x) \in \mathbf{P}(\mathbf{M},\mathbf{G}_0)$ in accordance with (2.5.21). Under Poincaré transformations of the adopted global Lorentz frames, the Yang-Mills gauge fields transform as 4-vector fields,

$$\hat{A}^b_\mu(x) \mapsto \hat{A}'^b_\mu(x) = \Lambda_\mu{}^\nu \hat{A}^b_\nu(\Lambda^{-1}(x-a)) , \qquad b = 1,\ldots,n , \tag{1.4}$$

whereas under a change of section $\mathbf{s}$ (i.e., under a change of Yang-Mills gauge) they transform in accordance with (2.5.7b) and (2.5.26):

$$\hat{A}_\mu \mapsto \hat{A}'_\mu = \mathrm{Ad}_{U^{-1}(g)} \hat{A}_\mu + U^{-1}(g)\partial_\mu U(g) , \qquad \hat{A}_\mu := \hat{A}^a_\mu \hat{\mathbf{T}}_a \in T_{\boldsymbol{u}}\mathbf{P} . \tag{1.5}$$

The *Yang-Mills field tensor* (also called a *Yang-Mills field strength*) is identified with the curvature tensor corresponding to the gauge potential in (1.1), and can be expressed as follows [C,I,NT]:

$$\hat{F} = \tfrac{1}{2}\hat{F}_{\mu\nu} dx^\mu \wedge dx^\nu , \qquad \hat{F}_{\mu\nu} := \hat{F}^a_{\mu\nu} \hat{\mathbf{T}}_a , \tag{1.6a}$$

$$\hat{F}_{\mu\nu} = \partial_\mu \hat{A}_\nu - \partial_\nu \hat{A}_\mu + \left[\hat{A}_\mu, \hat{A}_\nu\right] = \partial_\mu \hat{A}_\nu - \partial_\nu \hat{A}_\mu + \hat{A}^a_\mu \hat{A}^b_\nu C_{ab}{}^c \hat{\mathbf{T}}_c . \tag{1.6b}$$

Under Poincaré transformations its components in (1.6a) transform as a *(0,2)*-tensor field in Minkowski space,

$$\hat{F}^b_{\mu\nu}(x) \mapsto \hat{F}'^b_{\mu\nu}(x) = \Lambda_\mu{}^\kappa \Lambda_\nu{}^\lambda \hat{F}^b_{\kappa\lambda}(\Lambda^{-1}(x-a)) , \tag{1.7}$$

whereas, under a change of Yang-Mills gauge, they transform as a Yang-Mills tensor

$$\hat{F}_{\mu\nu} \mapsto \hat{F}'_{\mu\nu} = \mathrm{Ad}_{U^{-1}(g)} \hat{F}_{\mu\nu} = U(g^{-1})\hat{F}_{\mu\nu} U(g) . \tag{1.8}$$

The curvature 2-form in (1.6a), generally representing a non-Abelian version of the electromagnetic field (which obviously emerges as a very special case, corresponding to $\mathbf{G}_0 = \mathrm{U}(1)$), is taken to satisfy a straightforward generalization of the Maxwell equations in (9.1.1) [NT] :

$$D^{\mu}\hat{F}_{\mu\nu} = \partial^{\mu}\hat{F}_{\mu\nu} + \left[\hat{A}^{\mu},\hat{F}_{\mu\nu}\right] = 0 \quad , \qquad \hat{A}^{\mu} = \eta^{\mu\nu}\hat{A}_{\nu} \ . \tag{1.9}$$

In case that $\mathbf{G}_0$ is equal to SU(N), or some other Lie group consisting of $N \times N$ (real or complex) matrices $g = U(g)$ that can act directly upon a "matter field", so that for any two other $N \times N$ matrices A and B,

$$\mathrm{Tr}[\mathrm{AB}] = \mathrm{Tr}[(g\mathrm{A}g^{-1})(g\mathrm{B}g^{-1})] \quad , \qquad g \in \mathbf{G}_0 \ , \tag{1.10}$$

the above equation can be derived from the following gauge invariant Yang-Mills (classical) action integral[3],

$$\hat{S}_{\rm cl}(\hat{A}) = -\tfrac{1}{4}\int_{\mathbf{M}} \eta^{\kappa\mu}\eta^{\lambda\nu}\hat{F}_{\kappa\lambda} \bullet \hat{F}_{\mu\nu}\, d^4x \quad , \qquad \hat{F}_{\kappa\lambda} \bullet \hat{F}_{\mu\nu} := \sum_{a=1}^{n} \hat{F}^{a}_{\kappa\lambda}\hat{F}^{a}_{\mu\nu} \ , \tag{1.11a}$$

by its variation with respect to $\hat{A}_{\mu}$. This action is often expressed in the alternative form

$$\hat{S}_{\rm cl}(\hat{A}) = -\tfrac{1}{4}\int_{\mathbf{M}} \mathrm{Tr}\left(\hat{F}_{\mu\nu}\hat{F}^{\mu\nu}\right) d^4x = \tfrac{1}{2}\int_{\mathbf{M}} \mathrm{Tr}\left(\hat{F} \wedge {}^*\hat{F}\right) d^4x \ , \tag{1.11b}$$

$$\hat{F}^{\kappa\lambda} = \eta^{\kappa\mu}\eta^{\lambda\nu}\hat{F}_{\mu\nu} \ , \qquad {}^*\hat{F}_{\kappa\lambda} = \tfrac{1}{2}\varepsilon_{\kappa\lambda\mu\nu}\hat{F}^{\mu\nu}, \qquad \mathrm{Tr}[T_a T_b] = -\tfrac{1}{2}\delta_{ab} \ , \tag{1.11c}$$

where ${}^*\hat{F}$ is the Hodge-dual of $\hat{F}$, in which the Levi-Civita antisymmetric symbol $\varepsilon_{\kappa\lambda\mu\nu}$ is zero whenever any of its two indices are equal, and it assumes the values +1 or –1 when κ-λ-μ-ν represents, respectively, an even or an odd permutation of 0-1-2-3.

By analogy with the case of electromagnetic fields, the above formulation of a classical Yang-Mills theory remains invariant under the "infinitesimal" gauge transformations

$$\hat{A}^{a}_{\mu}(x) \mapsto \hat{A}'^{a}_{\mu}(x) = \hat{A}^{a}_{\mu}(x) + D_{\mu}\hat{\lambda}^{a}(x) \quad , \qquad \hat{\lambda}^{a}(x) \in \mathbf{R}^1 \ . \tag{1.12}$$

In fact, such "infinitesimal" gauge transformations can be derived from those in (1.5), once those transformations are transcribed in component form with respect to the basis in (1.1). This kind of approach to gauge transformations can be used to arrive at convenient choices of gauge families, such as the *generalized Lorenz*[4] *gauges*, obtained by using an auxiliary "gauge-fixing" field (Nakanishi, 1966; Lautrup, 1967; Kugo and Ojima, 1979) to impose the following type of subsidiary conditions:

$$\partial^{\mu}\hat{A}_{\mu} = \hat{B} \quad , \qquad \hat{B} = \hat{B}^{a}\mathbf{T}_{a} \ . \tag{1.13}$$

The *Lorenz gauge* is a special case, that corresponds to a "gauge-fixing" field which is zero everywhere. In the electromagnetic case, it produced the four decoupled equations in (9.1.3). In the present non-Abelian case, with which we are exclusively concerned in this chapter, the components of the equations resulting from (1.9) are no longer decoupled:

$$\partial^\mu \partial_\mu \hat{A}_\nu = -\partial^\mu[\hat{A}_\mu, \hat{A}_\nu] + [\hat{A}^\mu, (\partial_\nu \hat{A}_\mu - \partial_\mu \hat{A}_\nu - [\hat{A}_\mu, \hat{A}_\nu])] , \quad \partial^\mu \hat{A}_\mu = 0 . \quad (1.14)$$

It is this fact that makes the quantization of Yang-Mills fields in many respects radically different than that of classical electromagnetic fields. In particular, the Gupta-Bleuler method of quantization, which makes essential use of the Lorenz gauge in the form (of the conventional counterpart of the) Gupta-Bleuler subsidiary condition in (9.3.18), becomes physically ineffectual, since it cannot cope with the nonlinearity of the term on the right-hand side of the first of the equations in (1.14). Indeed, if its application is attempted, then, formally speaking, the physical space corresponding to transversal polarization modes is not left invariant by the nonlinear self-interaction term in the Yang-Mills action in (1.11) – a fact which Feynman (1963) *interpreted*[5] as leading to a violation of the "unitarity" of the *S*-matrix for Yang-Mills theories.

The method that has become standard, ever since a well-known proposal by Faddeev and Popov (1967) based on Dirac's (1958) ideas, is to introduce "ghost" fields for fictitious "particles" in order to (formally) cope with this situation. However, as mentioned in the introduction to this chapter, this idea turned out to possess an underlying geometric interpretation. After developing in the next two sections the general mathematical tools required by this interpretation, we shall outline it in Sec. 10.4 in the classical context, before turning to the quantum situation in the last two sections of this chapter.

10.2. Gauge Groups of Global Gauge Transformations in Principal Bundles

The geometric definition and construction of classical "ghost" fields, which correspond to the Faddeev-Popov fields in the quantum theory of Yang-Mills fields, and which will be studied in Sec. 10.4, requires the introduction of the concept of "gauge group" in a sense[6] that is distinct from that of the structure group $\mathbf{G}$ of a principal bundle (cf. Secs. 2.2 and 2.5). Consequently, in this and the next section we will study this and related concepts in a general fibre theoretical setting, which is also readily applicable in the quantum regime.

Let $\mathbf{P}(\mathbf{M},\mathbf{G})$ denote any given principal bundle $(\mathbf{P}, \Pi, \mathbf{M}, \mathbf{G})$ over a differential manifold $\mathbf{M}$. With each one of its sections $\boldsymbol{s}$, defined as in (2.5.18), we can associate the local trivialization map defined, in accordance with (2.2.4), by

$$\boldsymbol{\phi}^{s} : \boldsymbol{u} \mapsto (x,g) \in \mathbf{M}^{s} \times \mathbf{G} , \qquad \boldsymbol{u} \in \Pi^{-1}(\mathbf{M}^{s}) = \mathbf{U}^{s} , \qquad (2.1a)$$

$$(\boldsymbol{\phi}^{s} \circ \boldsymbol{s})(x) = \boldsymbol{\phi}^{s}(\boldsymbol{s}(x)) = (x,e) , \qquad x \in \mathbf{M}^{s} , \qquad (2.1b)$$

$$(\boldsymbol{\phi}^{s})^{-1} : (x,g) \mapsto \boldsymbol{s}(x) \cdot g \in \mathbf{U}^{s} \subset \mathbf{P}(\mathbf{M},\mathbf{G}) , \quad x \in \mathbf{M}^{s} , \; g \in \mathbf{G} , \qquad (2.1c)$$

where e denotes the unit element of $\mathbf{G}$. Any two such local trivializations $\boldsymbol{\phi}^{s}$ and $\boldsymbol{\phi}^{s'}$ are then related to the local gauge transformation in (2.5.25) by means of the map

$$g(\boldsymbol{\phi}^{s};\boldsymbol{\phi}^{s'}) : \boldsymbol{u} \mapsto g(\boldsymbol{u}) \in \mathbf{G} , \quad \boldsymbol{\phi}^{s'}(\boldsymbol{u}) = \boldsymbol{\phi}^{s}(\boldsymbol{u}) \cdot g(\boldsymbol{u}) , \quad \boldsymbol{u} \in \mathbf{U}^{s} \cap \mathbf{U}^{s'} . \quad (2.2)$$

Such a map can be viewed (cf. Göckeler and Schücker, 1987, p. 160) to be a restriction of some smooth map from the total space **P** into the structure group **G**, defined globally on **P**:

$$g : \boldsymbol{u} \mapsto g(\boldsymbol{u}) \in \mathbf{G} \ , \qquad g(\boldsymbol{u} \cdot h) = h^{-1} g(\boldsymbol{u})\, h \ , \qquad \boldsymbol{u} \in \mathbf{P}, \qquad h \in \mathbf{G}. \tag{2.3}$$

Within the family $C^\infty(\mathbf{P}, \mathbf{G})$ of all such smooth maps we can define in the following natural manner the operations of taking the inverse and of group multiplication,

$$g^{-1} := \{\, g^{-1}(\boldsymbol{u}) \in \mathbf{G} \,|\, \boldsymbol{u} \in \mathbf{P} \,\}, \qquad g_1 g_2 := \{\, g_1(\boldsymbol{u}) g_2(\boldsymbol{u}) \in \mathbf{G} \,|\, \boldsymbol{u} \in \mathbf{P} \,\}, \tag{2.4}$$

so that it becomes a group $\mathcal{G}(\mathbf{P})$, which we shall call the *gauge group of the principal bundle* **P(M,G)**. As pointed out in (Atiyah *et al.*, 1978), such a gauge group is in general an infinite-dimensional Lie group[7] – as opposed to the specific instances of structure groups **G** encountered in classical or quantum field theories.

An *automorphism of the principal bundle* **P(M,G)** is defined to be a pair (ϕ,ψ) of diffeomorphisms $\phi : \mathbf{P} \to \mathbf{P}$ and $\psi : \mathbf{M} \to \mathbf{M}$ which are such that $\phi(\boldsymbol{u} \cdot h) = \phi(\boldsymbol{u}) \cdot h$ for all $\boldsymbol{u} \in \mathbf{P}$ and all $h \in \mathbf{G}$, and that $\Pi \circ \phi = \psi \circ \Pi$. The family **AutP** of all such automorphisms constitutes a group under the group multiplication operations $(\phi_1,\phi_2) \mapsto \phi_1 \circ \phi_2$ and $(\psi_1,\psi_2) \mapsto \psi_1 \circ \psi_2$, provided by the composition of diffeomorphisms. A subgroup $\mathbf{Aut_M P}$ of **AutP** is supplied by all the *vertical automorphisms* defined as those automorphisms which induce in **M** the identity transformation, i.e., for which $\Pi(\boldsymbol{u}) = \Pi(\phi(\boldsymbol{u}))$, so that ψ is the identity map on **M**. The gauge group $\mathcal{G}(\mathbf{P})$ is isomorphic to the group $\mathbf{Aut_M P}$, since to each $\phi \in \mathbf{Aut_M P}$ we can obviously associate a $g_\phi \in \mathcal{G}(\mathbf{P})$ which is such that

$$\phi(\boldsymbol{u}) = \boldsymbol{u} \cdot g_\phi(\boldsymbol{u}) \ , \qquad g_\phi(\boldsymbol{u} \cdot h) = h^{-1} g_\phi(\boldsymbol{u})\, h \ , \qquad \boldsymbol{u} \in \mathbf{P}, \qquad h \in \mathbf{G}, \tag{2.5}$$

and vice versa – cf. [BG], Thm. 3.2.2.

Let us construct now a bundle, associated to the principal bundle **P(M,G)** by means of the adjoint action $\mathrm{ad}_h : g \mapsto hgh^{-1}$ of **G** onto itself, as follows:

$$ad\mathbf{P} = \mathbf{P} \times_{\mathrm{ad}} \mathbf{G} = \{\, \lceil \boldsymbol{u}, g \rceil \,|\, \boldsymbol{u} \in \mathbf{P}, g \in \mathbf{G} \} \ , \tag{2.6a}$$

$$\lceil \boldsymbol{u}, g \rceil := \{\, (\boldsymbol{u} \cdot h, \mathrm{ad}_{h^{-1}} g) \,|\, h \in \mathbf{G} \} \ , \qquad \mathrm{ad}_{h^{-1}} g = h^{-1} g h \ . \tag{2.6b}$$

It is then immediately seen that $\mathcal{G}(\mathbf{P})$, and therefore also $\mathbf{Aut_M P}$, can be identified with the set $\Gamma(ad\,\mathbf{P})$ of all global sections of this latter bundle,

$$\mathcal{G}(\mathbf{P}) \ \leftrightarrow \ \mathbf{Aut_M P} \ \leftrightarrow \ \Gamma(ad\,\mathbf{P}) \ , \tag{2.7a}$$

by associating with each gauge group element $g_\phi \in \mathcal{G}(\mathbf{P})$ a corresponding cross-section $s_\phi \in \Gamma(ad\,\mathbf{P})$ of $ad\,\mathbf{P}$:

$$s_\phi : \boldsymbol{u} \mapsto \lceil \boldsymbol{u}, g_\phi(\boldsymbol{u}) \rceil \in ad\,\mathbf{P} = \mathbf{P} \times_{\mathrm{ad}} \mathbf{G} \ , \qquad \boldsymbol{u} \in \mathbf{P}. \tag{2.7b}$$

Consequently, we could use the same symbol g to refer to the map $\mathbf{P} \to \mathbf{G}$, defined in (2.3), to the vertical automorphism g_Φ , defined in (2.5), as well as to the global section s_Φ of the bundle $ad\,\mathbf{P}$ in (2.6) – and we could call any of them a *global gauge transformation*[8] *of the principal bundle* **P(M,G)**. According to (2.1)-(2.3), a local gauge transformation, defined as in (2.5.18), is related to such a global gauge transformation of the corresponding principal bundle by

$$\boldsymbol{s}'(x) = \boldsymbol{s}(x) \cdot g(\boldsymbol{s}(x)) \quad , \qquad g(\boldsymbol{s}(x)) \in \mathbf{G} \quad , \; x \in \mathbf{M}^{s} \cap \mathbf{M}^{s'} \quad . \tag{2.8}$$

To obtain a computationally convenient as well as an explicit representation of the Lie algebra $L(\mathcal{G}(\mathbf{P}))$ of the gauge group $\mathcal{G}(\mathbf{P})$, which shall be called the *gauge algebra of the principal bundle* **P(M,G)**, let us consider the associated bundle

$$Ad\,\mathbf{P} = \mathbf{P} \times_{\mathrm{Ad}} L(\mathbf{G}) = \{ \lceil \boldsymbol{u}, f \rceil \mid \boldsymbol{u} \in \mathbf{P}, f \in L(\mathbf{G}) \} \quad , \tag{2.9a}$$

$$\lceil \boldsymbol{u}, f \rceil := \{ (\boldsymbol{u} \cdot g, \mathrm{Ad}_{g^{-1}} f) \mid g \in \mathbf{G} \} , \qquad \mathrm{Ad}_g : f \mapsto (\mathrm{ad}_g)_* f \, , \tag{2.9b}$$

determined, in accordance with (2.5.7), by the adjoint representation of the group **G** on its Lie algebra $L(\mathbf{G})$. It is very easy to establish that this associated bundle becomes a Lie algebra if its Lie bracket is defined, at each one of its points, by the corresponding Lie brackets in $L(\mathbf{G})$,

$$[\lceil \boldsymbol{u}, f \rceil, \lceil \boldsymbol{u}, f' \rceil] := \lceil \boldsymbol{u}, [f, f'] \rceil \quad , \tag{2.10a}$$

and that the following family

$$\mathrm{Exp}(t \lceil \boldsymbol{u}, f \rceil) := \lceil \boldsymbol{u} \, , \exp(tf) \rceil \, , \qquad t \in \mathbf{R}^1 \, , \tag{2.10b}$$

determines an additive one-parameter subgroup of $\mathcal{G}(\mathbf{P})$ viewed as a family of cross-sections of the bundle in (2.6). Hence, the Lie algebra $L(\mathcal{G}(\mathbf{P}))$ can be identified with the infinite-dimensional vector space of cross-sections $\Gamma(Ad\,\mathbf{P})$ of the associated bundle in (2.9) – cf. [BG], Thms. 3.2.10 and 3.2.11:

$$\mathcal{L}(\mathbf{P}) := L(\mathcal{G}(\mathbf{P})) \quad \leftrightarrow \quad \Gamma(Ad\,\mathbf{P}) \ . \tag{2.11}$$

The introduction of the gauge group $\mathcal{G}(\mathbf{P})$ affords an "active" approach (cf. [I], p. 159) to the theory of connections on any given principal bundle **P(M,G)**, which is mathematically equivalent to the "passive" point of view adopted in Chapter 2, but opens new perspectives, which are of importance in the theory of Yang-Mills fields as well as in general relativity. Indeed, in the case of manifolds the *passive* point of view is the one we adopted in Sec. 2.1, whereby the map in (2.1.4) is viewed as a change of coordinates for a given point $x \in \mathbf{M}$; the corresponding *active* point of view consists of considering the same map within the context of a given chart, as providing a transformation taking the point $x \in \mathbf{M}$ into the point $x' \in \mathbf{M}$ having the primed coordinates in (2.1.4) in that chart. Similarly, for tensor fields the passive point of view is the one adopted in Sec. 2.2, whereby the map in (2.2.18) is viewed as corresponding to the change of components above $x \in \mathbf{M}$, such as that in (2.1.17); the corresponding active point of view is to consider the same map within the context of a given section $\boldsymbol{s}$ of $GL\mathbf{M}$, namely as supplying a transformation that takes the tensor field T into a new tensor field T' having the primed coordinates defined in (2.1.17) with respect to the frames provided by that section $\boldsymbol{s}$.

To be able to adopt the same "active" attitude in the case of connections, let us consider the family $\mathcal{C}(\mathbf{P})$ of all connection forms on a given principal bundle $\mathbf{P}(\mathbf{M},\mathbf{G})$. This is not a vector space, but it is a convex (affine) space, since $(1-t)\boldsymbol{\omega}'+t\boldsymbol{\omega}'' \in \mathcal{C}(\mathbf{P})$ for any two connection forms $\boldsymbol{\omega}',\boldsymbol{\omega}'' \in \mathcal{C}(\mathbf{P})$ and $0 \leq t \leq 1$. It can be also easily checked, using the same method as in proving (2.5.26a) [BG,C,I], that for any $\boldsymbol{\omega} \in \mathcal{C}(\mathbf{P})$ and any $\phi \in \mathrm{Aut}_{\mathbf{M}}\mathbf{P}$, we have

$$\phi^*\boldsymbol{\omega} = \mathrm{Ad}_{g^{-1}}\,\boldsymbol{\omega} + g_\phi^{\;*}\,\boldsymbol{\Theta}^{\mathbf{G}} \ , \qquad g = g_\phi \in \mathcal{G}(\mathbf{P}) \, , \tag{2.12a}$$

where the $L(\mathbf{G})$-valued one-form $\boldsymbol{\Theta}^{\mathbf{G}}$is the Maurer-Cartan form [C,I] of the structure group $\mathbf{G}$, which, by definition, maps the values of any left-invariant vector field in $\mathbf{G}$ into the value it assumes in $T_e\mathbf{G}$. In the case where $\mathbf{G}$ is a group of matrices or operators, as it is generically the case in physics, the above relation can be written in the form (cf. [I], p. 161, and [BG], p. 31)

$$^{g}\boldsymbol{\omega} \ := \phi^*\boldsymbol{\omega} = \mathrm{Ad}_{g^{-1}}\,\boldsymbol{\omega} + g^{-1}\boldsymbol{d}g \ , \tag{2.12b}$$

where $g^{-1}\boldsymbol{d}g$ is the Maurer-Cartan form of $\mathbf{G}$ expressed, in accordance with (2.5.26), in some given chart in $\mathbf{G}$ that contains the unit element e of $\mathbf{G}$, whereas $g^{-1}(\boldsymbol{u}) := g(\boldsymbol{u})^{-1}$ and

$$\boldsymbol{\Theta}^{\mathcal{G}} = g^{-1}\boldsymbol{d}g \ , \qquad g \in \mathcal{G}(\mathbf{P}) \, , \qquad \boldsymbol{d}g \in \mathcal{L}(\mathbf{P}) = \Gamma(Ad\,\mathbf{P}) \, . \tag{2.12c}$$

In general, the pull-back $^{g}\boldsymbol{\omega}$ of $\boldsymbol{\omega}$ satisfies (2.5.7a), and it maps the elements of the fundamental field in (2.5.5) into the corresponding element of $T_e\mathbf{G}$. This means that $^{g}\boldsymbol{\omega}$ represents an Ehresmann connection on $\mathbf{P}(\mathbf{M},\mathbf{G})$ (cf. [SC], p. 359; or [I], p. 156), so that $^{g}\boldsymbol{\omega} \in \mathcal{C}(\mathbf{P})$. The curvature form $^{g}\boldsymbol{\Omega}$ of $^{g}\boldsymbol{\omega}$, defined in accordance with (2.5.14) by means of the horizontal components of vector fields in $\mathbf{P}$ that are determined by $^{g}\boldsymbol{\omega}$, is easily seen to be related to the curvature form $\boldsymbol{\Omega}$ of $\boldsymbol{\omega}$ as follows:

$$^{g}\boldsymbol{\Omega} \ := {}^{g}\mathbf{D}\,{}^{g}\boldsymbol{\omega} = \mathrm{Ad}_{g^{-1}}\,\boldsymbol{\Omega} \ , \qquad g = g_\phi \in \mathcal{G}(\mathbf{P}) \ . \tag{2.13}$$

The relationship of the "active" point of view, presented thus far in this section, to the "passive" point of view adopted in Chapter 2, becomes very obvious once the counterparts on $\mathbf{P}(\mathbf{M},\mathbf{G})$,

$$\mathbf{g}_{\beta\alpha}(x) = \boldsymbol{\varphi}_\beta(\boldsymbol{u}) \circ \boldsymbol{\varphi}_\alpha^{-1}(\boldsymbol{u}) \ : \ \mathbf{G} \to \mathbf{G} \ , \qquad \boldsymbol{u} \in \Pi^{-1}(x) \ , \quad x \in \mathbf{M}_\alpha \cap \mathbf{M}_\beta \ , \tag{2.14a}$$

$$\boldsymbol{\phi}_\alpha \colon \ \boldsymbol{u} \mapsto (x, \boldsymbol{\varphi}_\alpha(\boldsymbol{u})) \in \mathbf{M}_\alpha \times \mathbf{G} \ , \qquad \boldsymbol{u} \in \Pi^{-1}(\mathbf{M}_\alpha) \subset \mathbf{P} \ , \tag{2.14b}$$

of the transition functions in (2.2.6) are introduced and, for the corresponding covering of **M**, the following preferred set of local sections of **P(M,G)**,

$$\boldsymbol{s}_\alpha : \; x \mapsto \boldsymbol{u}\cdot \boldsymbol{\varphi}_\alpha^{-1}(\boldsymbol{u}) \in \Pi^{-1}(x) , \qquad x \in \mathbf{M}_\alpha \subset \mathbf{M} \;, \tag{2.15}$$

is used to define the following local $L(\mathbf{G})$-valued pull-backs:

$$\boldsymbol{\omega}_\alpha = \boldsymbol{s}_\alpha{}^* \boldsymbol{\omega} \;, \qquad \boldsymbol{\omega}_\beta = \boldsymbol{s}_\beta{}^* \boldsymbol{\omega} \;. \tag{2.16}$$

Indeed, the maps defined by (2.15) are smooth and, on account of the fact that $\boldsymbol{\varphi}_\alpha(\boldsymbol{u}\cdot h) = \boldsymbol{\varphi}_\alpha(\boldsymbol{u})\cdot h$ for all $h \in \mathbf{G}$, they are $\boldsymbol{u}$-independent

$$\boldsymbol{u}\cdot \boldsymbol{\varphi}_\alpha^{-1}(\boldsymbol{u}) = \boldsymbol{u}'\cdot \boldsymbol{\varphi}_\alpha^{-1}(\boldsymbol{u}') \;, \qquad \forall\, \boldsymbol{u},\boldsymbol{u}' \in \Pi^{-1}(x) \;, \tag{2.17}$$

so that they indeed provide sections of **P(M,G)**. Let us again assume that **G** is a Lie group of matrices. It is then very easy to establish (cf. Daniel and Viallet, 1980, p. 183) that the following relations are satisfied,

$$\boldsymbol{\omega}_\beta = \mathrm{Ad}_{\mathbf{g}_{\alpha\beta}^{-1}}\, \boldsymbol{\omega}_\alpha + \mathbf{g}_{\alpha\beta}^{-1}\, \boldsymbol{d}\mathbf{g}_{\alpha\beta} \;, \tag{2.18}$$

and that, conversely, if any given collection of local $L(\mathbf{G})$-valued one-forms associated with some covering of **M** satisfies the compatibility relations in (2.18), then there is a unique connection form $\boldsymbol{\omega}$ of **M** related to them by (2.16).

These considerations show that "passive" counterparts of (2.12b) can be set up, so that each (global) gauge transformation, defined on the principal bundle by (2.3), can be identified with a family of (in general local) maps from the base space **M** to the structure group **G**,

$$\mathcal{g} : \mathbf{P} \to \mathbf{G} \qquad \leftrightarrow \qquad \left\{ \hat{\mathcal{g}}_\alpha \middle| \; \hat{\mathcal{g}}_\alpha : \mathbf{M}_\alpha \to \mathbf{G} \;, \; \bigcup\nolimits_\alpha \mathbf{M}_\alpha = \mathbf{M} \right\} \;, \tag{2.19a}$$

that satisfy the following compatibility relations:

$$\hat{\mathcal{g}}_\beta(x) = \mathbf{g}_{\beta\alpha}(x)\hat{\mathcal{g}}_\alpha(x)\mathbf{g}_{\beta\alpha}^{-1}(x) \;, \qquad \forall x \in \mathbf{M}_\alpha \cap \mathbf{M}_\beta \;. \tag{2.19b}$$

For any given covering of **P(M,G)** with a family of local trivialization maps, such as those in (2.14b), these local mappings are obtained by constructing the corresponding preferred family of sections in (2.15), and then casting $\mathcal{g} \circ \boldsymbol{s}_\alpha$ in the role of the local maps in (2.19a) – cf. (Göckeler and Schücker, 1987), p. 160. Conversely, for any covering of **P(M,G)**, and any family of local maps from **M** to **G** that satisfy the compatibility relations in (2.19b), the gauge transformation in (2.3) can be constructed by piecing together the corresponding compositions of $\boldsymbol{s}_\alpha$ and the maps in (2.19b). Thus, the "active" and the "passive" treatments of connections are indeed equivalent from a mathematical point of view.

*10.3. Graded Lie Algebras Generated by Connection Forms

The fact that each connection form ω on a principal bundle $\mathbf{P}(\mathbf{M},\mathbf{G})$ is an $L(\mathbf{G})$-valued 1-form, whereas its corresponding curvature form is an $L(\mathbf{G})$-valued 2-form, suggests that we should study the collections of all such mathematical entities as subsets of the set of all $L(\mathbf{G})$-valued one-forms and two-forms, respectively, over the manifold $\mathbf{P}$ equal to the total space of $\mathbf{P}(\mathbf{M},\mathbf{G})$.

Let L be any Lie algebra, and M any manifold. By definition, an *L-valued k-form* a on M is given, for any fixed value $k = 0,1,\ldots,\dim M$, by a map from M into L, which is such that in some basis $\{\tilde{\mathbf{Y}}_a \mid a = 1,\ldots,n\}$ of L the coefficients in the following expansion (in which the Einstein summation convention is used for the index a),

$$a : x \mapsto a^a_x \otimes \tilde{\mathbf{Y}}_a \in L \ , \qquad x \in M \ , \tag{3.1}$$

represent real-valued k-forms a^a over M – in the sense defined in Sec. 2.5. Clearly, if that is the case in one basis of L, then the same remains true in all the other bases, so that such a definition is actually basis independent.

The *exterior product* of any L-valued k-form a with any L-valued l-form b can be defined by the expression

$$[a,b] = \left(a^a \wedge b^b\right) \otimes [\tilde{\mathbf{Y}}_a, \tilde{\mathbf{Y}}_b] = C_{ab}{}^c \left(a^a \wedge b^b\right) \otimes \tilde{\mathbf{Y}}_c \ , \qquad [\tilde{\mathbf{Y}}_a, \tilde{\mathbf{Y}}_b] = C_{ab}{}^c \tilde{\mathbf{Y}}_c \ , \tag{3.2}$$

so that it represents an L-valued $(k+l)$-form. Furthermore, it is easily checked that, on account of (2.5.12), we have

$$[a,b](\boldsymbol{X}_1,\ldots,\boldsymbol{X}_{k+l}) = \frac{1}{k!l!} \sum_\pi (\mathrm{sign}\pi) \left[a(\boldsymbol{X}_{\mu_1},\ldots,\boldsymbol{X}_{\mu_k}), b(\boldsymbol{X}_{\nu_1},\ldots,\boldsymbol{X}_{\nu_l})\right] \ , \tag{3.3}$$

for any vectors $\boldsymbol{X}_1,\ldots,\boldsymbol{X}_{k+l}$ tangent to M at any of its points x. Hence, this definition is basis-independent. It then follows immediately that if L is the Lie algebra of a Lie group of matrices (or operators), so that its elements are also matrices (or operators), and the Lie bracket of any two of its elements is equal to the commutator bracket of the matrices (or operators) representing them, then

$$[a,b] = a \wedge b - (-1)^{kl} b \wedge a \ , \tag{3.4}$$

where the wedge product of two matrix-valued (or operator-valued) forms is defined by the same algebraic expression as in (2.5.12), but with matrix (or operator) multiplication replacing ordinary multiplication.

The real vector space of all L-valued k-forms on M is usually denoted by $\Lambda^k(M,L)$. We shall therefore denote by $\Lambda(M,L)$ the collection of all such spaces $\Lambda^k(M,L)$, obtained for $k = 0,1,\ldots,\dim M$. It is easily checked that for any $a \in \Lambda^k(M,L)$, $b \in \Lambda^l(M,L)$ and $c \in \Lambda^m(M,L)$ we have (cf. [BG], Thm. 2.1.3),

$$[a,b]+(-1)^{kl}[b,a]=0 \ , \tag{3.5a}$$

$$(-1)^{km}[[a,b],c]+(-1)^{ml}[[c,a],b]+(-1)^{lk}[[b,c],a]=0 \ , \tag{3.5b}$$

so that $\Lambda(M,L)$ has the structure of a graded Lie algebra[9].

In case that a metric g is supplied on a manifold M of dimension N, then to each real-valued k-form $\boldsymbol{\omega}$, expressed as in (2.5.11), we can assign a Hodge-dual $(N-k)$-form $*\boldsymbol{\omega}$ by setting it equal to (cf., e.g., Benn and Tucker, 1987, Sec. 1.4)

$$*\boldsymbol{\omega} = |g|^{1/2}\frac{1}{k!}\omega^{\mu_1\cdots\mu_k}\varepsilon_{\mu_1\cdots\mu_k\mu_{k+1}\cdots\mu_N}\, \mathbf{d}x^{\mu_{k+1}}\otimes\cdots\otimes\mathbf{d}x^{\mu_N} \ , \tag{3.6a}$$

$$\boldsymbol{\omega} = \omega_{\mu_1\cdots\mu_k}\mathbf{d}x^{\mu_1}\otimes\cdots\otimes\mathbf{d}x^{\mu_k} \ , \qquad |g| = \det\|g_{\mu\nu}\| \ , \tag{3.6b}$$

where the completely antisymmetric ε-tensor is defined by generalizing the definition of the one in (1.11c) from four to N dimensions. This definition of the dual, as well as that of the exterior derivative in (2.5.13), can be immediately extended to any $a\in\Lambda^k(M,L)$ by setting

$$*a := *a^a\otimes\tilde{\mathbf{Y}}_a \ , \qquad \mathbf{d}a := \mathbf{d}a^a\otimes\tilde{\mathbf{Y}}_a \ . \tag{3.7}$$

They both provide maps from $\Lambda(M,L)$ into $\Lambda(M,L)$, although in general they do not leave the modules $\Lambda^k(M,L)$ invariant. It is easily seen, however, that on account of (3.2), for any $a\in\Lambda^k(M,L)$ and $b\in\Lambda^l(M,L)$,

$$\mathbf{d}[a,b]=[\mathbf{d}a,b]+(-1)^k[a,\mathbf{d}b] \ . \tag{3.8}$$

Let us now specialize these considerations to the case where M is equal to the total space of a principal bundle $\mathbf{P}(\mathbf{M},\mathbf{G})$, and L is equal to the Lie algebra $L(\mathbf{G})$ of its structure group. Then each connection form $\boldsymbol{\omega}$ on a principal bundle $\mathbf{P}(\mathbf{M},\mathbf{G})$ gives rise to an *exterior covariant derivative* $\mathbf{D}^{\omega}$ on the graded Lie algebra $\Lambda(\mathbf{P},L(\mathbf{G}))$. This external covariant derivative is defined for any $a\in\Lambda^k(\mathbf{P},L(\mathbf{G}))$ in accordance with (2.5.14) and (2.5.15), namely it is such that

$$\mathbf{D}^{\omega}a(\mathbf{X}_1,\ldots,\mathbf{X}_{k+1}) = \mathbf{d}a(\overline{\mathbf{X}}_1^{\omega},\ldots,\overline{\mathbf{X}}_{k+1}^{\omega}) \ , \qquad \mathbf{X}_1,\ldots,\mathbf{X}_{k+1}\in T_u\mathbf{P} \ , \tag{3.9}$$

for any set $\{\mathbf{X}_1,\ldots,\mathbf{X}_{k+1}\}$ of vector fields in $\mathbf{P}$, whose horizontal components appearing in (3.9) are those determined by the connection form $\boldsymbol{\omega}$ in accordance with (2.5.6).

In view of (2.5.15), (2.5.16a) and (3.3), we can express the *Cartan structural equations* for the corresponding curvature form as follows (cf. also [BG], Thm. 2.2.4)

$$\boldsymbol{\Omega}^{\omega} := \mathbf{D}^{\omega}\boldsymbol{\omega} = \mathbf{d}\boldsymbol{\omega} + (1/2)[\boldsymbol{\omega},\boldsymbol{\omega}] \ , \tag{3.10a}$$

or, by employing (3.3), in terms of the $L(\mathbf{G})$-values which the above forms assume:

$$\Omega^{\omega}(\mathbf{X},\mathbf{Y}) = \mathbf{d}\omega(\mathbf{X},\mathbf{Y}) + [\omega(\mathbf{X}), \omega(\mathbf{Y})] , \qquad \mathbf{X},\mathbf{Y} \in T_u\mathbf{P} . \tag{3.10b}$$

The corresponding Bianchi identities in (2.5.16c) can be then written in the form (cf. also [BG], Thm. 2.2.8)

$$\mathbf{D}^{\omega}\Omega^{\omega} = \mathbf{d}\Omega^{\omega} + [\omega , \Omega^{\omega}] = \mathbf{0} . \tag{3.11}$$

In case that $\mathbf{G}$ is a Lie group of matrices or operators, then on account of (3.4) the Cartan structural equation in (3.10) assumes the form

$$\Omega^{\omega} = \mathbf{d}\omega + \omega \wedge \omega . \tag{3.12}$$

Let us now consider a family U of operators $U(g): \mathcal{V} \to \mathcal{V}$, $g \in \mathbf{G}$, which provide a faithful representation of the structure group $\mathbf{G}$ on some vector space $\mathcal{V}$ – such as the space in which the "matter fields" of classical Yang-Mills theories assume their values (cf. Sec. 10.1), or such as the typical fibres of quantum geometries in the preceding six chapters – in which case that vector space is actually an infinite-dimensional (pseudo-)Hilbert space, and the representation is (pseudo-)unitary. The fact that such a representation is faithful means that it provides an isomorphism between the structure group $\mathbf{G}$ and the group U of all such operators. In turn, this implies that there is a Lie algebra isomorphism $L(\mathbf{G}) \to L(U)$ between the Lie algebras of these two groups. Consequently, we can introduce in $L(U)$ a basis $\{T_a \mid a = 1,\dots,n\}$, and then consider the following special case of (3.1),

$$\mathcal{A} : \boldsymbol{u} \mapsto \mathcal{A}^a_{\boldsymbol{u}} T_a \in L(U) , \qquad \boldsymbol{u} \in \mathbf{P} , \tag{3.13}$$

as a generalization of the $L(\mathbf{G}_0)$-valued one-form (1.1) to arbitrary structure groups $\mathbf{G}$ and arbitrary base manifolds $\mathbf{M}$, as well as an extension of such $L(\mathbf{G})$-valued k-forms from the base manifold $\mathbf{M}$ to the total space of $\mathbf{P}(\mathbf{M},\mathbf{G})$. Naturally, all the definitions and results in (3.2)-(3.5) immediately apply to the present case of the graded Lie algebra $\Lambda(\mathbf{P},L(U))$ – except that in the case where the vector space under consideration is the infinite-dimensional typical fibre of a quantum geometry, care has to exercised by defining in (3.4) the products of the operator-valued elements of $L(U)$ on suitable common cores [PQ] of the chosen basis $\{T_a \mid a = 1,\dots,n\}$. The definitions of exterior differentiation as well as exterior covariant differentiation can be then immediately extended to $\Lambda(\mathbf{P},L(U))$.

To extend these latter concepts, for all orders k, also to associated vector bundles

$$\mathbf{P}\times_{\mathbf{G}} \mathcal{V} = \{ \lceil \boldsymbol{u},\mathbf{f} \rceil \mid \boldsymbol{u} \in \mathbf{P} , \mathbf{f} \in \mathcal{V} \} , \qquad \lceil \boldsymbol{u},\mathbf{f} \rceil := \{ (\boldsymbol{u} \cdot g, U(g^{-1})\mathbf{f}) \mid g \in \mathbf{G} \} , \tag{3.14}$$

whose sections play in physics the role of "matter fields", let us also consider the families $\Lambda^k(U)$ of all k-forms on $L(U)$, for all $k = 0,1,\dots,\dim\mathbf{P}$. We can then generalize (3.14) by defining the associated bundles

$$\mathbf{P}\times_{\mathbf{G}}(\mathcal{V}\otimes\Lambda^k(U)) = \{ \lceil \boldsymbol{u};\mathbf{f},\phi \rceil \mid \boldsymbol{u} \in \mathbf{P}, \ (\mathbf{f},\phi) \in \mathcal{V} \otimes \Lambda^k(\mathbf{R}^N) \} , \tag{3.15a}$$

$$\varphi = \lceil \boldsymbol{u};\mathbf{f},\phi \rceil := \{ (\boldsymbol{u} \cdot g, U(g^{-1})\mathbf{f}, U(g^{-1})_* \phi) \mid g \in \mathbf{G} \} , \tag{3.15b}$$

of $\mathcal{V}$-valued k-forms on $\mathbf{P}$, amongst which the one for $k = 0$ consists of 0-forms, so that it can be identified with the one in (3.14). We shall denote by

$$\Lambda^k(\mathbf{P},\mathcal{V}) = \Gamma(\,\mathbf{P}\times_{\mathbf{G}}(\mathcal{V}\otimes\Lambda^k(U))) \ , \tag{3.15c}$$

the family of all sections of the respective bundles in (3.15a), so that $\Lambda^0(\mathbf{P},\mathcal{V})$ coincides with the family of all $\mathcal{V}$-valued global vector fields on $\mathbf{M}$. For the remaining values of $k = 1,\dots,\dim\mathbf{P}$, we can consider those $\boldsymbol{\varphi}\in \mathbf{P}\times_{\mathbf{G}}(\mathcal{V}\otimes\Lambda^k(U))$ for which $\boldsymbol{\varphi}(\mathbf{X}_1,\dots,\mathbf{X}_k) = 0$ if any one of the vectors $\mathbf{X}_1,\dots,\mathbf{X}_k$ is vertical. We shall denote by $\mathit{\Lambda}^k(\mathbf{P},\mathcal{V})$ the family of sections of these subbundles

$$\left\{\boldsymbol{\varphi}\middle|\ \boldsymbol{\varphi}(\mathbf{X}_1,\dots,\mathbf{X}_k) = \mathbf{0},\ \mathbf{X}_1\in V_{\boldsymbol{u}}(\mathbf{P}),\ \mathbf{X}_2,\dots,\mathbf{X}_k\in T_{\boldsymbol{u}}(\mathbf{P})\right\} \ , \tag{3.16}$$

so that $\mathit{\Lambda}^k(\mathbf{P},\mathcal{V})$ are subfamilies of $\Lambda^k(\mathbf{P},\mathcal{V})$. Upon introducing in each $\Lambda^k(\mathbf{P},\mathcal{V})$ exterior derivatives defined by

$$\boldsymbol{d\varphi} = \lceil \boldsymbol{u};\mathbf{f},\mathbf{d}\boldsymbol{\phi}\rceil \in \Lambda^{k+1}(\mathbf{P},\mathcal{V}) \ , \qquad \boldsymbol{\varphi} = \lceil \boldsymbol{u};\mathbf{f},\boldsymbol{\phi}\rceil \in \Lambda^{k}(\mathbf{P},\mathcal{V}) \ , \tag{3.17}$$

we can assign to any connection form $\boldsymbol{\omega} \in \mathcal{C}(\mathbf{P})$ an *external covariant derivative* $\boldsymbol{D}^{\boldsymbol{\omega}}$ on the collection $\mathit{\Lambda}(\mathbf{P},\mathcal{V})$ of all these families $\mathit{\Lambda}^k(\mathbf{P},\mathcal{V})$ by means of the maps

$$\boldsymbol{D}^{\boldsymbol{\omega}} : \mathit{\Lambda}^k((\mathbf{P},\mathcal{V}) \to \mathit{\Lambda}^{k+1}(\mathbf{P},\mathcal{V}) \ , \qquad k = 0,1,\dots,\dim\mathbf{P} \ , \tag{3.18a}$$

which are such that, as it is the case in (3.9),

$$\boldsymbol{D}^{\boldsymbol{\omega}}\,\boldsymbol{\varphi}(\mathbf{X}_1,\dots,\mathbf{X}_{k+1}) = \boldsymbol{d\varphi}(\overline{\mathbf{X}}{}_1^{\boldsymbol{\omega}},\dots,\overline{\mathbf{X}}{}_{k+1}^{\boldsymbol{\omega}}) \ , \qquad \mathbf{X}_1,\dots,\mathbf{X}_{k+1} \in T_{\boldsymbol{u}}\mathbf{P} \ . \tag{3.18b}$$

It can then be shown (cf. [BG], Thm. 3.1.5) that

$$\boldsymbol{D}^{\boldsymbol{\omega}}\boldsymbol{\varphi} = \boldsymbol{d\varphi} + \mathcal{A}^{\boldsymbol{\omega}}\wedge\boldsymbol{\varphi} \ , \qquad \mathcal{A}^{\boldsymbol{\omega}} = \boldsymbol{\omega}^a T_a \ , \qquad \boldsymbol{\varphi} \in \mathit{\Lambda}^k(\mathbf{P},\mathcal{V}) \ , \tag{3.19}$$

provided that we extend the definition of the wedge product in (2.5.12) to all $\boldsymbol{\varphi}\in\mathit{\Lambda}^k(\mathbf{P},\mathcal{V})$ and all $\mathcal{A}\in\Lambda^k(\mathbf{P},L(U))$ as follows :

$$(\mathcal{A}\wedge\boldsymbol{\varphi})(\boldsymbol{X}_1,\dots,\boldsymbol{X}_{k+l}) = \frac{1}{k!l!}\sum_{\pi}(\mathrm{sign}\,\pi)\ \mathcal{A}(\boldsymbol{X}_{\mu_1},\dots,\boldsymbol{X}_{\mu_k})\boldsymbol{\varphi}(\boldsymbol{X}_{\nu_1},\dots,\boldsymbol{X}_{\nu_l}) \ . \tag{3.20}$$

Let us now specialize the above considerations to the case where $\mathcal{V}$ is equal to the Lie algebra $L(\mathbf{G})$ treated as a vector space, and U is the representation of $\mathbf{G}$ that acts upon $L(\mathbf{G})$ from the left as the adjoint representation, so that $U(g) = \mathrm{Ad}_{g^{-1}}$ as it acts on the elements of $L(\mathbf{G})$. In that case, it follows from (3.19) and (3.20) that

$$\boldsymbol{D}^{\boldsymbol{\omega}}\boldsymbol{\tau} = \boldsymbol{d\tau} + [\boldsymbol{\omega}\,,\,\boldsymbol{\tau}] \ , \qquad \boldsymbol{\tau} \in \mathit{\Lambda}^k(\mathbf{P},L(\mathbf{G})) \ , \tag{3.21}$$

so that an identification of $\Lambda^1(\mathbf{P},L(\mathbf{G}))$ with the family $C(\mathbf{P})$ of all connection forms on $\mathbf{P}$ can be carried out: for any given $\omega_0 \in C(\mathbf{P})$, there is a one-to-one map

$$\tau \mapsto \omega = \omega_0 + \tau \in C(\mathbf{P}), \qquad \tau \in \Lambda^1(\mathbf{P},L(\mathbf{G})), \tag{3.22}$$

of $\Lambda^k(\mathbf{P},L(U))$ onto $C(\mathbf{P})$ (cf. [BG], Thm. 3.2.6). Since $\Lambda^k(\mathbf{P},L(U))$ is a linear space, this shows that $C(\mathbf{P})$ is an affine space [C].

In terms of the operator-valued counterpart $\mathcal{A}^\omega$ in (3.19) of the connection form ω, that acts directly on $\varphi \in \Lambda^l(\mathbf{P},\mathcal{V})$, we can rewrite (3.22) as

$$\mathcal{T} \mapsto \mathcal{A}^\omega = \mathcal{A}^{\omega_0} + \mathcal{T} \in C(\mathbf{P},\mathcal{V}), \qquad \mathcal{T} \in \Lambda^1(\mathbf{P},L(U)), \tag{3.23}$$

where $C(\mathbf{P},\mathcal{V})$ can be identified with the set of all Koszul connections (i.e., covariant differentiation operators) acting on the $\mathcal{V}$-valued "matter fields" belonging to a given vector bundle $(\mathbf{E},\pi,\mathbf{M},\mathbf{F})$ associated to $\mathbf{P}(\mathbf{M},\mathbf{G})$. Under the earlier made stipulations on the representation U, it constitutes a manifold diffeomorphic to the family $C(\mathbf{P})$ of all connections on $\mathbf{P}(\mathbf{M},\mathbf{G})$.

10.4. BRST Transforms and Ghost Fields in Classical Yang-Mills Theories

The considerations in the second half of Sec. 2.5 have shown that if we expand any field Ψ, given by a section of a vector bundle $(\mathbf{E},\pi,\mathbf{M},\mathbf{F})$ associated to a principal frame bundle $\mathbf{P}(\mathbf{M},\mathbf{G}_0)$, in terms of local frames in general associated to those in $\mathbf{P}(\mathbf{M},\mathbf{G})$, as it was done in (2.5.17), then its covariant derivatives are given by the expressions in (2.5.22) for any given moving frame – i.e., section $\boldsymbol{s}$ of $\mathbf{P}(\mathbf{M},\mathbf{G}_0)$. Those covariant derivatives contain the pull-backs by $\boldsymbol{s}$, given in (2.5.23), of the Cartan connection forms in (2.5.10).

In the theory of Yang-Mills fields it is convenient to make the transition from the above "passive" point of view to an "active" one, and to consider any "matter field" Ψ on which the Yang-Mills field acts as being defined on the total space $\mathbf{P}$ of the principal bundle $\mathbf{P}(\mathbf{M},\mathbf{G}_0)$, rather than just on the Minkowski base space $\mathbf{M}$. In that case that matter field can be deemed to represent a *generalized Higgs field*[10] Ψ, defined as being a field on $\mathbf{P}$ which is equivariant under the action of the structure group $\mathbf{G}_0$, i.e., which satisfies

$$\Psi(\boldsymbol{u}\cdot g) = U(g^{-1})\Psi(\boldsymbol{u}), \qquad \boldsymbol{u} \in \mathbf{P}, \quad g \in \mathbf{G}_0 . \tag{4.1}$$

For any given covering of $\mathbf{P}(\mathbf{M},\mathbf{G}_0)$ with a family of local trivialization maps, the transition from the passive to this active point of view can be effected by using the local gauges provided by the preferred set of local sections of $\mathbf{P}(\mathbf{M},\mathbf{G})$ in (2.15). Thus, we can define, for each local section $\{\Psi_x \mid x \in \mathbf{M}^s\}$ of the vector bundle $(\mathbf{E},\pi,\mathbf{M},\mathbf{F})$, the push-forwards

$$\Psi(\boldsymbol{u}) = \Psi_{x,0} = \boldsymbol{s}_\alpha{}^* \Psi_x, \qquad \boldsymbol{u} = \boldsymbol{s}_\alpha(x), \quad x \in \mathbf{M}^s \cap \mathbf{M}_\alpha, \tag{4.2}$$

and then take advantage of the stipulation of equivariance in (4.1) to extend their domains of definition to $\Pi^{-1}(\mathbf{M}^s \cap \mathbf{M}_\alpha)$.

For a given connection $\omega \in \mathcal{C}(\mathbf{P})$, the action of the operators of covariant differentiation given in (1.3), which in the passive approach were acting on matter fields that were $\mathcal{V}$-valued sections of the vector bundle $(\mathbf{E},\boldsymbol{\pi},\mathbf{M},\mathbf{F})$, have to be extended in the present active framework everywhere along the fibres of $\mathbf{P}(\mathbf{M},\mathbf{G}_0)$; moreover, such an extension should enable them to mediate parallel transport in arbitrary directions within the total space $\mathbf{P}$. In principle, such an extension can be deduced from (2.12), since according to (2.5) we can reach all locations in $\mathbf{P}$ by means of various gauge transformations $\mathcal{g}$ of $\mathbf{P}(\mathbf{M},\mathbf{G}_0)$,

$$^{\mathcal{g}}\!\mathcal{A}^{\omega} = (\mathrm{Ad}_{\mathcal{g}^{-1}}\,\omega + \mathcal{g}^{-1}\boldsymbol{d}\mathcal{g})^a\, T_a \quad , \qquad \mathcal{g} \in \mathcal{G}(\mathbf{P}) \quad . \tag{4.3}$$

As originally conjectured by Thierry-Mieg (1980), as a by-product of such an extension the Maurer-Cartan form of the structure group $\mathbf{G}_0$, which is implicitly contained in (4.3), can be formally related to the Faddeev-Popov "ghost" fields, which emerge naturally when Dirac's (1950, 1958, 1964) formalism of constrained systems is applied to the Yang-Mills classical action in (1.11) – cf. (Babelon and Viallet, 1981), Sec. 2. However, as can be seen from (2.12c), such an identification has to be carried out from the point of view of the infinite-dimensional gauge group $\mathcal{G}(\mathbf{P})$, rather than just that of the structure group $\mathbf{G}_0$ (Leinaas and Olaussen, 1982). Furthermore, in order to underline the formal analogy with the situation in conventional quantum field theory, where additional "antighost" fields make their appearance, some authors (Quirós *et al.*, 1981; Bonora and Cotta-Ramusino, 1983) have introduced such fields also in the classical context although, as pointed out by Mayer (1983), that is not absolutely necessary in that case. Moreover, Bonora *et al.* (1981b, 1982), Hoyos *et al.* (1982), and others, have also extended these formal considerations to a superfield framework – although that again is not absolutely necessary. We shall, therefore, not follow a superfield approach, but on account of the requirements of the GS quantization of Yang-Mills fields, which is treated in the next section, we shall present a formulation that incorporates "antighost" as well as "ghost" fields.

We begin by carrying out the identification

$$c := (\mathcal{g}^{-1}\boldsymbol{d}\mathcal{g})^a\, T_a \quad \leftrightarrow \quad \boldsymbol{\Theta}^{\mathcal{G}} = \mathcal{g}^{-1}\boldsymbol{d}\mathcal{g} \ . \tag{4.4}$$

We shall refer to c, regarded as a field defined on the total space $\mathbf{P}$ of the principal bundle $\mathbf{P}(\mathbf{M},\mathbf{G}_0)$, as a *ghost field*. Gauge transformations which are "infinitesimal", namely

$$\mathcal{A}^{\omega} \mapsto \mathcal{A}^{\omega} + s\mathcal{A}^{\omega} \quad , \qquad s = \boldsymbol{d}\mathcal{g}^a\, \partial_a \quad , \quad \partial_a = \partial U(\mathcal{g})\,/\,\partial \mathcal{g}^a \quad , \tag{4.5a}$$

can be then expressed by means of the mapping

$$\mathcal{A}^{\omega} \mapsto \mathcal{A}^{\omega} + \boldsymbol{D}^{\omega}c \quad , \qquad \boldsymbol{D}^{\omega}c \ \leftrightarrow \ \mathbf{D}^{\omega}\boldsymbol{\Theta}^{\mathcal{G}} \quad , \tag{4.5b}$$

computed at $\mathcal{g}(\boldsymbol{u}) = e$. The exterior differential form s in (4.5a) is called in the present context of Yang-Mills field theories a *BRST operator*. We note that $\boldsymbol{D}^{\omega}c$ assumes values equal to those of $s\mathcal{A}^{\omega}$ for vectors that are tangent to $\mathbf{P}$ and point in ω-horizontal directions, whereas in all vertical directions the Maurer-Cartan structural equations are satisfied. This means that the external covariant derivative of the Maurer-Cartan form has to vanish, so that, in accordance with (3.10), we have

$$sc + \tfrac{1}{2}[c,c] = 0 \quad , \qquad s^2 = 0 \ , \tag{4.6}$$

where the second equation merely indicates the fact that repeated external differentiation of any forms always produces a null result, so that the BRST operator s is nilpotent.

As we mentioned previously, the family $\mathcal{C}(\mathbf{P},\mathcal{V})$ of all Koszul connections (i.e., the set of covariant differentiation operators acting on the $\mathcal{V}$-valued matter fields) is an infinite-dimensional manifold diffeomorphic to the family $\mathcal{C}(\mathbf{P})$ of all Ehresmann connections (i.e., the set of all connection forms) on $\mathbf{P}(\mathbf{M},\mathbf{G}_0)$. Since the elements of the gauge group $\mathcal{G}(\mathbf{P})$ map the latter set onto itself, in accordance with (2.12) the elements of $\mathcal{C}(\mathbf{P},\mathcal{V})$ fall naturally into equivalence classes – which we shall call *gauge orbits* within $\mathcal{C}(\mathbf{P},\mathcal{V})$. Thus, these gauge orbits are the elements of the family

$$\mathcal{C}_{\mathcal{G}}(\mathcal{V}) = \mathcal{C}(\mathbf{P},\mathcal{V})/\mathcal{G}(\mathbf{P}) \ , \tag{4.7}$$

of left cosets of $\mathcal{C}(\mathbf{P},\mathcal{V})$ modulo $\mathcal{G}(\mathbf{P})$.

The introduction of "antighost" fields occurs naturally in Yang-Mills theories when gauge fixing is considered. A *gauge-fixing* procedure [IQ] amounts to a choice of a unique representative from each such equivalence class. One way of achieving that in the theory of Yang-Mills fields is to introduce an auxiliary scalar field b that indexes the gauge orbits, and to specify the values of additional parameters for choosing a representative $\mathcal{A}^{\omega}$ in each gauge orbit, so that we obtain a map ϕ_C such that

$$\phi_C : \mathcal{C}_{\mathcal{G}}(\mathcal{V}) \rightarrow \mathcal{C}(\mathbf{P},\mathcal{V}) \ , \qquad \pi_C \circ \phi_C = I_C \ , \tag{4.8}$$

with π_C denoting the natural projection of $\mathcal{C}(\mathbf{P},\mathcal{V})$ onto $\mathcal{C}_{\mathcal{G}}(\mathcal{V})$, and I_C the identity map on $\mathcal{C}_{\mathcal{G}}(\mathcal{V})$. If this procedure is meant to produce a continuous map ϕ_C for the family of Yang-Mills gauge potentials which give rise to finite action integrals, then the existence of such a gauge fixing is precluded by a "no-go" theorem of Singer (1978). This theorem confirmed the existence of Gribov (1978) "ambiguities" in SU(N) Yang-Mills theories[11], and its basic conclusion is that if for such theories $\mathcal{C}(\mathbf{P},\mathcal{V})$ is considered to supply the total space of a principal bundle having $\mathcal{C}_{\mathcal{G}}(\mathcal{V})$ as base space and $\mathcal{G}(\mathbf{P})$ as structure group, or if a suitable modification[12] is considered, then such principal bundles do not possess global sections, i.e., they are not trivial.

Such problems are dealt with in the conventional theory of Yang-Mills fields by restricting attention to suitable subfamilies of $\mathcal{C}(\mathbf{P})$, that give rise to gauge potentials which satisfy the Lorenz gauge condition, and to positive Faddeev-Popov operators (Zwanziger, 1989). However, within a sufficiently small neighborhood of the unit element of the gauge group $\mathcal{G}(\mathbf{P})$ the Gribov ambiguity does not occur. Hence, we can add to the "infinitesimal" gauge transformation in 4.5a) another term, to obtain the gauge transformation

$$\mathcal{A}^{\omega} \mapsto \mathcal{A}^{\omega} + s\mathcal{A}^{\omega} + \bar{s}\mathcal{A}^{\omega} , \qquad \bar{s} = \boldsymbol{d}\bar{g}^{\alpha}\bar{\partial}_{\alpha} , \qquad \bar{\partial}_{\alpha} = \partial U(\bar{g})/\partial \bar{g}^{\alpha} , \tag{4.9a}$$

which incorporates an "infinitesimal" gauge fixing field b , so that it can be expressed by means of the mapping

$$\mathcal{A}^{\omega} \mapsto \mathcal{A}^{\omega} + \boldsymbol{D}^{\omega}c + \boldsymbol{D}^{\omega}\bar{c} , \qquad s\bar{c} = b . \tag{4.9b}$$

We shall refer to $\bar{c}$, regarded as a field on a second copy $\overline{\mathbf{P}}$ of the total space of the principal bundle $\mathbf{P}(\mathbf{M},\mathbf{G}_0)$, as an *antighost field*, and to $\bar{s}$ in (4.9a) as an *anti-BRST operator*.

Let us use a section $\boldsymbol{s}$ of $\mathbf{P}(\mathbf{M},U)$ to pull $\mathcal{A}^{\omega}$ back to $\mathbf{M}$, so as to obtain a Yang-Mills gauge field $\boldsymbol{s}_*\mathcal{A}^{\omega}$, as in (1.1). Then, after some series of "infinitesimal" gauge transformations, given as in (4.9a), are integrated over a finite "stretch" in the gauge group, the outcome is in general the following Yang-Mills gauge potential

$$A(x,g,\bar{g}) = \hat{A}_{\mu}(x,g,\bar{g})\,\boldsymbol{dx}^{\mu} + c(x,g,\bar{g}) + \bar{c}(x,g,\bar{g}) , \qquad x \in \mathbf{M}^{s} \subset \mathbf{M} , \tag{4.10a}$$

$$c(x,g,\bar{g}) = c_{\alpha}(x,g,\bar{g})\,\boldsymbol{d}g^{\alpha}, \quad \bar{c}(x,g,\bar{g}) = \bar{c}_{\alpha}(x,g,\bar{g})\,\boldsymbol{d}\bar{g}^{\alpha}, \quad g,\bar{g} \in \mathcal{G}(\mathbf{P}) . \tag{4.10b}$$

On the other hand, each element g of the gauge group, given by the map in (2.3), can be identified with a section of the trivial principal bundle $\mathbf{P}\times\mathbf{G}_0$. Hence, the gauge potential in (4.10a) can be viewed as giving rise to a covariant differentiation operator

$$\nabla = \boldsymbol{d} + s + \bar{s} + A , \qquad \boldsymbol{d} = \boldsymbol{dx}^{\mu}\partial_{\mu} , \tag{4.11}$$

on a triply fibrated vector bundle with typical fibre $\mathcal{V}$, which is associated to a triply-fibrated principal bundle with $\mathbf{P}_1 = \mathbf{P}$, $\mathbf{P}_2 = \mathbf{P}\times\mathbf{G}_0$ and $\mathbf{P}_2 = (\mathbf{P}\times\mathbf{G}_0)\times\mathbf{G}_0$ in (9.4.18).

On account of (4.6), and of their counterparts for the antighost fields, we have

$$sc = -\tfrac{1}{2}[c,c] , \qquad \bar{s}\bar{c} = -\tfrac{1}{2}[\bar{c},\bar{c}] , \tag{4.12a}$$

$$s^2 = s\bar{s} + \bar{s}s = \bar{s}^2 = 0 , \tag{4.12b}$$

where the second equality in (4.12b) follows from the fact that $s + \bar{s}$ has to be also nilpotent, since it is a one-form. Furthermore, the Maurer-Cartan equations on $\mathbf{G}_0\times\mathbf{G}_0$ imply that in addition to (4.12b) we must also have

$$s\bar{c} + \bar{s}c = -[\bar{c},c] . \tag{4.12c}$$

The fact that $\boldsymbol{D}^{\omega}c$ assumes values equal to those of $s\mathcal{A}^{\omega}$ for vectors that are tangent to $\mathbf{P}$ and point in ω-horizontal directions, implies that

$$sA_{\mu} = D_{\mu}c , \qquad \bar{s}A_{\mu} = D_{\mu}\bar{c} , \tag{4.12d}$$

where the second set of equations in (4.12d) follows from the "antighost" counterpart of this observation. Finally, the second equation in (4.9b), taken in conjunction with the equation in (4.12), implies that

$$s\bar{c} = b \ , \qquad \bar{s}c = -[\bar{c},c] - b \ , \tag{4.13a}$$

$$sb = 0 \ , \qquad \bar{s}b = -[\bar{c},b] \ . \tag{4.13b}$$

The set of equations in (4.12) and (4.13) constitutes the basis of the geometric treatment of Faddeev-Popov fields and BRST transformations for classical Yang-Mills fields in some of the physics literature on this subject (Baulieu and Thierry-Mieg, 1982; Baulieu, 1985; Baulieu and Singer, 1988, 1991). When such fields act as external fields on matter fields $\boldsymbol{\Psi}$ representing wave functions of quantum particles, then the following subsidiary conditions are imposed in order for the matter fields to be "physical" (Baulieu, 1985):

$$(s+c)\boldsymbol{\Psi} = \mathbf{0} \ , \qquad (\bar{s}+\bar{c})\boldsymbol{\Psi} = \mathbf{0} \ . \tag{4.14}$$

In view of (4.11), we see that in geometric terms these conditions amount to requiring that matter fields represented by generalized Higgs fields should have vanishing covariant derivatives in vertical directions.

*10.5. Lorenz and Transverse Gauges in Typical Weyl-Klein Fibres

The GS method of quantization exhibits additional new features when contrasted with the conventional methods of quantizing Yang-Mills theories – cf. [IQ], (Kugo and Ojima, 1979), (Faddeev and Slavnov, 1980), (Popov, 1983), (Gitman and Tyutin, 1990), (Nakanishi and Ojima, 1990). Indeed, in addition to being readily applicable to the generic situation of a curved spacetime **M**, and involving the fundamental spacetime form factor in (9.2.14), which contains the fundamental length ℓ, so that no quantum field singularities are bound to occur, the GS method also affords the possibility[13] of working *within* each quantum fibre with Abelian *internal* gauges, which can be then consistently integrated into the external non-Abelian gauges that characterize Yang-Mills fields.

This can be achieved by adapting to the quantum regime a procedure of Deser (1970), whereby in the classical regime one starts by replacing (1.11) with the first-order action

$$\hat{S}_0(\hat{A},\hat{F}) = -\tfrac{1}{2}\int_{\mathbf{M}}\Big[\big(\partial_\mu \hat{A}_\nu - \partial_\nu \hat{A}_\mu\big)\bullet \hat{F}^{\mu\nu} - \tfrac{1}{2}\hat{F}_{\mu\nu}\bullet \hat{F}^{\mu\nu}\Big]d^4x \ , \tag{5.1}$$

in the Minkowski space **M**. By the independent variation of the gauge potential and field strength components one then arrives at the field equations

$$\partial^\mu \hat{F}^a_{\mu\nu} = 0 \ , \qquad \hat{F}^a_{\mu\nu} = \partial_\mu \hat{A}^a_\nu - \partial_\nu \hat{A}^a_\mu \ , \qquad a = 1,\ldots,n \ , \tag{5.2}$$

which are identical with the ones in (9.1.1) for each value assumed by the index a. This set of equations is invariant under the Maxwell gauge transformations

$$\hat{A}^a_\mu(x) \mapsto \hat{A}'^a_\mu(x) = \hat{A}^a_\mu(x) + \partial_\mu \hat{\lambda}^a(x) \ , \qquad a = 1,\dots,n \ . \tag{5.3}$$

To produce SU(N)-gauge invariance, the action in (5.1) is augmented by the addition of a self-coupling term

$$\hat{S}(\hat{A},\hat{F}) = \hat{S}_0(\hat{A},\hat{F}) + \int_{\mathbf{M}} \hat{j}_\mu \bullet \hat{A}^\mu \, d^4x \ , \qquad \hat{j}_\mu \bullet \hat{A}^\mu = \sum_{a=1}^{n} \eta^{\mu\nu} \hat{j}^a_\mu \hat{A}^a_\nu \ , \tag{5.4a}$$

based on the conserved current

$$\hat{j}_\nu(x) = \left[\hat{F}_{\mu\nu}(x) \, , \, \hat{A}^\mu(x)\right] \ . \tag{5.4b}$$

This leads to the field equations

$$\partial^\mu \hat{F}_{\mu\nu} = \hat{j}_\nu \ , \qquad \hat{F}_{\mu\nu} = \partial_\mu \hat{A}_\nu - \partial_\nu \hat{A}_\mu + \left[\hat{A}_\mu, \hat{A}_\nu\right] \ , \tag{5.5}$$

which are identical to the Yang-Mills equations in (1.9).

We shall reinterpret this procedure in the GS framework, by treating the Maxwell transformations in (5.3) as *internal* gauge transformations, which involve only the gauge variables q and v, and those based on the remaining degrees of freedom in the structure group $\mathbf{G}_0$ as *external* gauge transformations, involving the base variable x, and pertaining to a GS propagation that gives rise to self-interactions.

We start by considering, as in Sec. 9.1, the space of solutions of the equations[14]

$$\partial^\mu \partial_\mu \hat{f}^a_\nu(x) = 0 \ , \qquad \nu = 0,1,2,3 \ , \qquad a = 1,\dots,n \ , \tag{5.6}$$

for which the following sesquilinear form,

$$\left(\hat{f}\middle|\hat{f}'\right)_{\hat{J}} = i\sum_{\mu=0}^{3} \int_{x^0=0} \hat{f}^*_\mu(x) \bullet \overleftrightarrow{\partial}_0 \hat{f}'_\mu(x) d^3\mathbf{x} \ , \tag{5.7}$$

is well-defined and non-negative definite, so that it gives rise to a J-inner product. Hence, the completion of the resulting pre-Hilbert space with respect to the norm defined by this J-inner product gives rise to a Krein space, whose *indefinite* inner product is provided by the following related integral:

$$\left\langle\hat{f}\middle|\hat{f}'\right\rangle = -i\int_{x^0=0} \eta^{\mu\nu} \hat{f}^*_\mu(x) \bullet \overleftrightarrow{\partial}_0 \hat{f}'_\nu(x) \, d^3\mathbf{x} \ . \tag{5.8}$$

The transition to the momentum representation is effected as in Sec. 9.1 by setting

$$\tilde{f}_\mu(k) = 2k_0(2\pi)^{-3/2} \int_{x^0=0} \exp(ik\cdot x) \hat{f}_\mu(x) \, d^3\mathbf{x} \ , \tag{5.9}$$

so that we obtain the new Krein space

$$\tilde{\mathbf{W}} = \tilde{\mathbf{W}}_{\tilde{J}}^{+} \hat{\oplus} \tilde{\mathbf{W}}_{\tilde{J}}^{-}, \quad \tilde{\mathbf{W}}_{\tilde{J}}^{+} = \left\{\tilde{f} \in \tilde{\mathbf{W}} \middle| \tilde{f}_0 \equiv 0\right\}, \ \tilde{\mathbf{W}}_{\tilde{J}}^{-} = \left\{\tilde{f} \in \tilde{\mathbf{W}} \middle| \tilde{f}_A \equiv 0, A = 1,2,3\right\}, \tag{5.10}$$

which carries the following indefinite inner product and J-inner product, respectively,

$$\langle \tilde{f} | \tilde{f}' \rangle = -\int_{V_0^+} \eta^{\mu\nu} \tilde{f}_\mu^*(k) \bullet \tilde{f}'_\nu(k) \, d\Omega_0(k) \ , \tag{5.11}$$

$$\left(\tilde{f} | \tilde{f}' \right)_{\tilde{J}} = \sum_{\mu=0}^{3} \int_{V_0^+} \tilde{f}_\mu^*(k) \bullet \tilde{f}'_\mu(k) \, d\Omega_0(k) \ , \qquad \tilde{J} = \mathbf{P}_{\tilde{J}}^{+} - \mathbf{P}_{\tilde{J}}^{-} \ . \tag{5.12}$$

In this space the transformations

$$\tilde{U}(a,\Lambda) \ : \ \tilde{f}_\mu(k) \mapsto \tilde{f}'_\mu(k) = \exp(ia \cdot k) \Lambda_\mu{}^\nu \tilde{f}_\nu(\Lambda^{-1}k) \ , \qquad \tilde{f} \in \tilde{\mathbf{W}} \ . \tag{5.13}$$

provide a pseudo-unitary (reducible) representation of the orthochronous Poincaré group.

Let us now introduce a *gauge-fixing field* $\tilde{b}$, whose components $\tilde{b}^a$ behave as scalar fields under Poincaré transformations, and let us consider the (in general affine) subspaces

$$\tilde{\mathbf{W}}(\tilde{b}) = \left\{ \tilde{f} \in \tilde{\mathbf{W}} \middle| \ k_\mu \tilde{f}^\mu(k) = \tilde{b}(k) \right\} \subset \tilde{\mathbf{W}} \ , \qquad \tilde{b}(k) := \tilde{b}^a(k) \hat{\mathbf{T}}_a \ , \tag{5.14}$$

corresponding to the above *generalized Lorenz gauges*. The Lorenz gauge itself is obviously obtained when this gauge-fixing field is required to vanish (almost) everywhere in Minkowski space. Upon introducing the polarization tetrads in (9.1.16), it can be again established that the *Lorenz space* obtained by imposing the Lorenz gauge condition, can be decomposed as the direct sum,

$$\tilde{\mathbf{W}}(0) = \tilde{\mathbf{W}}^0 \underset{J}{\oplus} \tilde{\mathbf{W}}_J(0) \ , \qquad \tilde{\mathbf{W}}^0 = \left\{ \tilde{f} \in \tilde{\mathbf{W}}(0) \middle| \langle \tilde{f} | \tilde{f} \rangle = 0 \right\} \ , \tag{5.15a}$$

$$\tilde{\mathbf{W}}_J(0) = \left\{ \tilde{f} = \tilde{f}^{(\alpha)}(k) \, \varepsilon_{(\alpha)}(k) \in \tilde{\mathbf{W}}(0) \middle| \tilde{f}^{(0)}(k) = \tilde{f}^{(3)}(k) = 0 \right\} \ , \tag{5.15b}$$

on account of being able to express the inner products in (5.11) and (5.12) in the following respective forms:

$$\langle \tilde{f} | \tilde{f}' \rangle = -\int_{V_0^+} \eta_{\alpha\beta} \tilde{f}^{(\alpha)*}(k) \bullet \tilde{f}^{(\beta)\prime}(k) \, d\Omega_0(k) \ , \qquad \tilde{f}, \tilde{f}' \in \tilde{\mathbf{W}}(0) \ , \tag{5.16}$$

$$\left(\tilde{f} | \tilde{f}' \right)_{\tilde{J}} = \sum_{\alpha=0}^{3} \int_{V_0^+} \tilde{f}^{(\alpha)*}(k) \bullet \tilde{f}^{(\alpha)\prime}(k) d\Omega_0(k) \ , \qquad \tilde{f}, \tilde{f}' \in \tilde{\mathbf{W}}(0) \ . \tag{5.17}$$

We shall refer to the additional two gauge-fixing conditions in (5.15b), which together with the Lorenz gauge conditions totally "fix the gauge" in the sense defined in the preceding section, as *transverse gauge* conditions. Thus, in the resulting *transverse Lorenz gauge* the state vector of Yang-Mills quanta is uniquely determined, modulo the usual multiplicative factor. However, note should be taken of the fact that the transverse Lorenz gauge conditions are not left invariant by Lorentz boosts. On the other hand, it can be established in the same manner as in Sec. 9.1 that the restriction of the representation in (5.13) to the Lorenz space obtained by imposing the Lorenz gauge condition is, modulo "pure" gauge modes as in (9.1.15), the direct sum over $a = 1,..,n$ of unitary and irreducible Wigner-type representations of the orthochronous Poincaré group for mass-0 and spin-1.

The transition to the stochastic phase space representation based on the fundamental quantum spacetime form factor in (5.5.5) and in (9.2.14) can be effected as in Sec. 9.2, namely by setting

$$f(\zeta) = \int_{V_0^+} \exp\left(-i\bar{\zeta}\cdot k\right)\tilde{f}(k)\, d\Omega_0(k)\ , \qquad \zeta = q + i\ell v\ , \quad q \in \mathbf{R}^4\ , \quad v \in \mathbf{V}^+, \tag{5.18}$$

and using the set of Proca equations which are the counterparts of those in (5.6), to define the following renormalized indefinite inner products and J-inner products, respectively:

$$\begin{aligned}\langle f \,|\, f'\rangle &= -\int_\Sigma \eta^{\mu\nu}\bar{f}_\mu(\zeta)\bullet f_\nu'(\zeta)d\tilde{\Sigma}(\zeta)\\ &= -i\int_\Sigma \eta^{\mu\nu}\bar{f}_\mu(\zeta)\bullet\overset{\leftrightarrow}{\partial}_\lambda f_\nu'(\zeta)d\sigma^\lambda(q)\, d\tilde{\Omega}(v)\ ,\end{aligned} \tag{5.19}$$

$$(f|f')_J = \sum_{\mu=0}^{3}\int_\Sigma \bar{f}_\mu(\zeta)\bullet f_\mu'(\zeta)d\tilde{\Sigma}(\zeta) = \langle f \,|\, Jf'\rangle\ . \tag{5.20}$$

We shall denote with $\mathbf{W}$ the Krein space based on these two inner products, and we shall refer to it as the *Weyl-Klein*[15] *typical fibre* for Yang-Mills single exciton states.

The transformation to which (5.18) gives rise, namely

$$\mathrm{T}_\mathbf{W} : \tilde{f} \mapsto f \in \mathbf{W}\ , \qquad \tilde{f} \in \tilde{\mathbf{W}}\ , \tag{5.21}$$

maps momentum-space wave functions into Yang-Mills exciton wave functions in a manner that is pseudo-unitary with respect to the indefinite metrics and unitary with respect to the J-metrics. The same pseudo-unitarity and J-unitarity features are displayed by the respective restrictions of this transformation to the subspaces in (5.14), which therefore give rise to maps onto the typical subfibres

$$\mathbf{W}(b) = \left\{ f \in \mathbf{W} \,\middle|\, \partial_\mu f^\mu(q,v) = b(q,v)\right\} \subset \mathbf{W}\ , \qquad \partial_\mu = \partial/\partial q^\mu\ , \tag{5.22}$$

corresponding to all the various generalized Lorenz gauges. In particular, the restriction to the Lorenz space in (5.15a) gives rise to the *Weyl-Lorenz typical fibre* $\mathbf{W}(0)$, which can be viewed as a subfibre of the Weyl-Klein fibre $\mathbf{W}$.

Let us introduce the *polarization coframes* $\{\varepsilon^{(\alpha)}(k) | \alpha = 0,1,2,3\}$, defined in relation to a global Lorentz coframe $\{\theta^\mu | \mu = 0,1,2,3\}$ by

$$\varepsilon^{(0)}(k) = \theta^0\ , \quad \varepsilon^{(3)}(k) = |\mathbf{k}|^{-1}\sum_{A=1}^{3} k_A\,\theta^A\ , \quad k\cdot\varepsilon^{(1)}(k) = k\cdot\varepsilon^{(2)}(k) = 0\ , \tag{5.23a}$$

so that they are the duals of the polarization tetrads attached in Sec. 9.1 to a global Lorentz frame $\boldsymbol{u}$,

$$\boldsymbol{\varepsilon}^{(\alpha)}{}_{\mu}(k)\,\boldsymbol{\varepsilon}_{(\beta)}{}^{\mu}(k) = \delta^{\alpha}{}_{\beta} \quad , \qquad k \in \mathbf{V}^{+} \; , \qquad \alpha,\beta = 0,\dots,3 \; . \tag{5.23b}$$

Then the *Weyl-Klein quantum frames*

$$\Phi_{\boldsymbol{u}}^{\alpha\zeta}(\zeta') = \int_{V_0^+} \exp\left[i(\zeta' - \bar{\zeta})\cdot k\right] \boldsymbol{\varepsilon}^{(\alpha)}(k)\, d\Omega_0(k) \; , \qquad \zeta,\zeta' \in \mathbf{R}^4 \times \mathbf{V}^{+} \; , \tag{5.24}$$

supply continuous resolutions of the identity in the Weyl-Klein typical fibre $\mathbf{W}$,

$$f = \sum_{\alpha=0}^{3} \int_{\Sigma} d\tilde{\Sigma}(\zeta) f^{(\alpha)}(\zeta) \Phi_{\boldsymbol{u}}^{\alpha\zeta} , \quad f^{(\alpha)a}(\zeta) = \left(\Phi_{\boldsymbol{u}}^{\alpha\zeta} \middle| f^{a}\right)_{J}^{a} , \quad f \in \mathbf{W} \; , \tag{5.25}$$

provided that for each $a = 1,..,n$ we use in (5.25) the J-inner product in (9.2.18b).

Of course, the above continuous resolution is not Poincaré covariant, since it depends intrinsically on the adopted J-inner product, so that it is not left invariant by Lorentz boosts. Furthermore, in view of the fact that

$$\langle f | f' \rangle = -\int_{\Sigma} \eta_{\alpha\beta} \bar{f}^{(\alpha)}(\zeta) \bullet f'^{(\beta)}(\zeta) d\tilde{\Sigma}(\zeta) \; , \qquad f,f' \in \mathbf{W} \; , \tag{5.26}$$

$$\left(f \middle| f'\right)_{J} = \sum_{\alpha=0}^{3} \int_{\Sigma} \bar{f}^{(\alpha)}(\zeta) \bullet f'^{(\beta)}(\zeta) d\tilde{\Sigma}(\zeta) \; , \qquad f,f' \in \mathbf{W} \; , \tag{5.27}$$

it is immediately seen that, if we restrict ourselves to the Weyl-Lorenz fibre $\mathbf{W}(0)$, the indefinite-metric counterpart of (5.25) has the form

$$\mathbf{P}_J(0) f = -\int_{\Sigma} d\tilde{\Sigma}(\zeta) f_{(\alpha)}(\zeta) \Phi_{\boldsymbol{u}}^{\alpha\zeta} , \quad f_{(\alpha)}^{a}(\zeta) = \eta_{\alpha\beta} \left\langle \Phi_{\boldsymbol{u}}^{\beta\zeta} \middle| f^{a} \right\rangle^{a} , \quad f \in \mathbf{W}(0) \; , \tag{5.28}$$

where $\mathbf{P}_J(0)$ denotes the J-orthogonal projector onto the subspace

$$\mathbf{W}_J(0) = \left\{ f \in \mathbf{W}(0) \middle| f^{(0)}(\zeta) = f^{(3)}(\zeta) = 0 \right\} \; , \tag{5.29}$$

which obviously corresponds to the transverse Lorenz gauge.

For a given fundamental symmetry J in the Weyl-Klein typical fibre $\mathbf{W}$ we decompose the Weyl-Lorenz typical fibre $\mathbf{W}(0)$ into the J-direct sum of the subfibre $\mathbf{W}_J(0)$ in (5.29) consisting of all state vectors displaying only transverse polarization modes, and the space $\mathbf{W}^0$ of null state vectors. Hence, upon introducing in the latter space a J-orthonormal basis $\{g_1, g_2, \dots\}$, we can express any element of $\mathbf{W}(0)$ as follows:

$$f \oplus \sum_{\gamma=1}^{\infty} y_\gamma g_\gamma \in \mathbf{W}_J(0) \underset{J}{\oplus} \mathbf{W}^0 \; , \qquad y_1, y_2, \dots \in \mathbf{C}^1 \; , \quad \sum_{\gamma=1}^{\infty} \left| y_\gamma \right|^2 < \infty \; . \tag{5.30}$$

On the other hand, an arbitrary element of the Weyl-Klein typical fibre $\mathbf{W}$ can be expressed as a sum whose first component lies within $\mathbf{W}(0)$, whereas the second is a representative element h of $\mathbf{W}$ that obeys the generalized Lorenz subsidiary condition in (5.22). For example, a unique choice of such a representative h can be arrived at by requiring that all its three spacelike polarization components vanish. Upon introducing a J-orthonormal basis $\{h_1,h_2,\ldots\}$ in the Hilbert space of all such representatives, we reach the conclusion that all the elements of the Weyl-Klein typical fibre $\mathbf{W}$ are represented by (cf. Sec. 11.8)

$$f \oplus g \oplus h \ , \qquad f \in \mathbf{W}_J(0) \ , \qquad g = \sum_{\gamma=1}^{\infty} y_\gamma g_\gamma \in \mathbf{W}^0 \ , \tag{5.31a}$$

$$h = \sum_{\gamma=1}^{\infty} z_\gamma h_\gamma \in \overline{\mathbf{W}}_J \ , \qquad z_1, z_2, \ldots \in \mathbf{C}^1 \ , \quad \sum_{\gamma=1}^{\infty} |z_\gamma|^2 < \infty \ , \tag{5.31b}$$

as f varies over all those elements of the Weyl-Klein typical fibre $\mathbf{W}$ that obey the transversal Lorenz condition in a global Lorentz frame $\boldsymbol{u}$ compatible with J, whereas the coefficients in the above expansions vary over all possible complex values for which the sums of their squared absolute values remain finite.

To understand the physical significance of the Weyl-Klein typical fibre and its elements from the point of view of Dirac's (1958, 1964) Hamiltonian framework for systems with constraints, let us consider a set of canonical constraint equations, symbolically represented by $\varphi^a(F,A) = 0$, $a = 1,\ldots,n$ – cf., e.g., [IQ] or (Faddeev and Slavnov, 1980). In accordance with (1.9), their explicit form is given in the present context by:

$$\varphi(F,A) := \partial^s F_{s0}(q,v) + \left[A^s(q,v), F_{s0}(q,v)\right] = 0 \ , \qquad s = 1,2,3 \ . \tag{5.32}$$

Let us consider now a set of internal gauge conditions $\chi^a(A) = 0$, such as the following,

$$\chi^a(A) = \partial^s A_s^a(q,v) = 0 \ , \qquad a = 1,\ldots,n \ , \tag{5.33}$$

for the Coulomb gauge. From the point of view of the Hamiltonian formalism (cf., e.g., Faddeev and Slavnov, 1980, Sec. 3.2), it is the set

$$\left\{(F_{s0}^a, A_s^a) \mid s = 1,2,3, \ a = 1,\ldots,n \right\} \tag{5.34}$$

that constitutes the canonical variables of the system. Upon setting

$$\partial^0 A_0^a(q,v) := \chi^a(A(q,v)) \ , \qquad a = 1,\ldots,n \ , \tag{5.35}$$

we have completely fixed a gauge in the sense of fixing at all $q^0 \in \mathbf{R}^1$ the values of Lagrange multipliers, collectively represented by A, once their values at $q^0 = 0$ are given. This is due to the fact that, in accordance with (5.6), (5.9) and (5.18),

$$\partial^\mu \partial_\mu A_\nu^a(q,v) = 0 \ , \qquad \nu = 0,1,2,3 \ , \quad a = 1,\ldots,n \ . \tag{5.36}$$

Since the constraints in (5.32) determine the corresponding canonical F-modes, the possibility of transition to a "true" Hamiltonian system (i.e., one in which only independent modes occur, such as the linear polarization modes in (5.15b)) is then dependent on a nonzero value for the Dirac (1964) determinant $\det M_{\mathrm{D}}$, which regulates the transition to these gauges, such as those in (5.31). In terms of Poisson brackets, this determinant is given by:

$$M_{\mathrm{D}}^{ab}(F,A) = \{\varphi^a(F,A),\chi^b(A)\} \quad . \tag{5.37}$$

As is well-known [IQ], it is the values of the above Dirac determinant that the Faddeev-Popov (1967) ghost fields are meant to reproduce by means of the Berezin (1964) method of integration over Grassmannian "ghost" degrees of freedom.

*10.6. Geometro-Stochastic Quantization of Yang-Mills Fields

For a Weyl-Klein typical fibre $\mathbf{W}$ of non-Abelian gauge exciton states, constructed as in the preceding section, the GS quantization of a Yang-Mills theory, corresponding to a principal bundle $\mathbf{P}(\mathbf{M},\mathbf{G}_0)$ over a Lorentzian manifold $\mathbf{M}$, is arrived at by combining the geometric techniques for the gauge group $\mathcal{G}(\mathbf{P})$, described earlier in this chapter, with the various techniques introduced in the preceding three chapters. Consequently, we shall present in this section only the most salient points in the construction of a GS Yang-Mills quantum field theory, reserving a fuller presentation of some of the technical details for the case of quantum gravity, with which such theories have many points in common[16].

As in Chapter 9, we shall first single out the fundamental symmetry J_0 corresponding to the canonical basis $\boldsymbol{u}_0$ in $\mathbf{R}^4$, and then rewrite the decomposition in (5.31) in the form

$$f \oplus g \oplus h \in \mathbf{W}^{\mathrm{P}} \oplus \mathbf{W}^0 \oplus \overline{\mathbf{W}} \,, \qquad \mathbf{W}^{\mathrm{P}} := \mathbf{W}_{J_0}(0) \,, \qquad \overline{\mathbf{W}} := \overline{\mathbf{W}}_{J_0} \,. \tag{6.1}$$

We can then introduce the *typical Yang-Mills fibre*

$$\mathcal{W} = \mathcal{W}^{\mathrm{P}} \otimes \left(\mathcal{Y} \underset{A}{\otimes} \overline{\mathcal{Y}}\right), \quad \mathcal{W}^{\mathrm{P}} = \underset{J_0}{\bigoplus}{}_{n=0}^{\infty} \mathcal{W}_n^{\mathrm{P}} \,, \quad \mathcal{Y} = \underset{J_0}{\bigoplus}{}_{n=0}^{\infty} \mathcal{Y}_n \,, \quad \overline{\mathcal{Y}} = \underset{J_0}{\bigoplus}{}_{n=0}^{\infty} \overline{\mathcal{Y}}_n \,, \tag{6.2a}$$

$$\mathcal{W}_n^{\mathrm{P}} = \mathbf{W}^{\mathrm{P}} \underset{S}{\otimes} \cdots \underset{S}{\otimes} \mathbf{W}^{\mathrm{P}} \,, \qquad \mathcal{Y}_n = \mathbf{W}^0 \underset{A}{\otimes} \cdots \underset{A}{\otimes} \mathbf{W}^0 \,, \qquad \overline{\mathcal{Y}}_n = \overline{\mathbf{W}} \underset{A}{\otimes} \cdots \underset{A}{\otimes} \overline{\mathbf{W}} \,, \tag{6.2b}$$

by means of which we can construct a triply-fibrated *quantum Yang-Mills bundle*

$$\mathbf{E}_2 \xrightarrow[\pi_2]{} \mathbf{E}_1 \xrightarrow[\pi_1(s_1)]{} \mathbf{E}_0 \xrightarrow[\pi_0(s_0)]{} \mathbf{M} \,, \tag{6.3a}$$

$$\overline{\mathbf{E}}_2 = \mathbf{P}_2 \times_{\mathbf{G}_2} \mathcal{W} \,, \qquad \overline{\mathbf{E}}_1 = \mathbf{P}_1 \times_{\mathbf{G}_1} \mathcal{W} \,, \qquad \mathbf{E}_0 = \mathbf{P}(\mathbf{M},\mathbf{G}^{\uparrow}) \times_{\mathbf{G}^{\uparrow}} \mathcal{W}_\infty \,, \tag{6.3b}$$

$$\mathcal{W}_\infty = \mathcal{W}_\infty^{\mathrm{P}} \otimes \left(\mathcal{Y}_\infty \underset{A}{\otimes} \overline{\mathcal{Y}}_\infty\right), \qquad \mathcal{W}_\infty^{\mathrm{P}} = \underset{a}{\bigoplus}{}_{n=0}^{\infty} \mathcal{W}_n^{\mathrm{P}}, \qquad \mathcal{Y}_\infty = \underset{a}{\bigoplus}{}_{n=0}^{\infty} \mathcal{Y}_n \,, \tag{6.3c}$$

associated to the triply-fibrated principal bundle (cf. Sec. 9.4)

$$\mathbf{P}_2 \xrightarrow[\Pi_2]{} \mathbf{P}_1 \xrightarrow[\Pi_1]{} \mathbf{P}_0 \xrightarrow[\Pi_0]{} \mathbf{M} \ , \tag{6.4a}$$

$$\mathbf{P}_0 = P^{\uparrow}\mathbf{M} \cong \mathbf{P}(\mathbf{M},\mathbf{G}^{\uparrow}) \ , \qquad \mathbf{G}^{\uparrow} = \mathrm{ISO}^{\uparrow}(3,1) \ , \tag{6.4b}$$

$$\mathbf{P}_1 = \mathbf{P}_0 \times \mathbf{G}_1 \ , \ \ \mathbf{G}_1 = \mathbf{G}_0 \ , \qquad \mathbf{P}_2 = \mathbf{P}_1 \times \mathbf{G}_2 \ , \ \ \mathbf{G}_2 = \mathbf{G}_0 \ . \tag{6.4c}$$

We shall refer to $\mathcal{W}^{\mathrm{P}}$ as the *standard physical Yang-Mills subfibre*, and we shall call $\bar{\mathcal{Y}}$ and $\mathcal{Y}$ the *standard Faddeev-Popov antighost* and *ghost subfibres*, respectively. The use of the algebraic direct sum in (6.3c) for the construction of the core Yang-Mills bundles that occur in (6.3a) (as opposed to the use of Hilbert direct sum in (6.2a) for building the enveloping Yang-Mills bundles in (6.3b)), is required in order to deal with the problem of non-invariance of the typical fibres of the latter bundles under the representation

$$\mathbf{U}_0(a,\Lambda) = \bigoplus_{J_0\ n=0}^{\infty} U(a,\Lambda)^{\otimes n} \ , \qquad (a,\Lambda) \in \mathrm{ISO}^{\uparrow}(3,1) \ , \tag{6.5a}$$

$$U(a,\Lambda) \ : f_\mu(q,v) \ \mapsto \ f_\mu'(q,v) = \Lambda_\mu{}^\nu f_\nu(\Lambda^{-1}(q-a),\Lambda^{-1}v) \ , \qquad f \in \mathbf{W} \ , \tag{6.5b}$$

of the orthochronous Poincaré group – whose single-exciton component U is derived from (5.13), (5.18) and (5.21). On the other hand, all the typical fibres of the enveloping as well as of the core bundles are left invariant by the representation

$$\mathbf{U}_{\mathrm{Y}}(g) = \bigoplus_{J_0\ n=0}^{\infty} U_{\mathrm{Y}}(g)^{\otimes n} \ , \qquad g \in \mathbf{G}_0 \ . \tag{6.6}$$

of the Yang-Mills structure group $\mathbf{G}_0$ – where the action of the operators $U_{\mathrm{Y}}(g)$ on the elements of the Weyl-Klein typical fibre is that inherited from the corresponding action of the operators $U(g)$, $g \in \mathbf{G}_0$, introduced in (1.5).

Let us consider now a principal frame bundle $\mathbf{P}(\mathbf{M},\mathbf{G})$ whose elements consist of Poincaré frames $(\boldsymbol{a},\boldsymbol{e}_i)$ that belong to $P^{\uparrow}\mathbf{M}$, as well as of Yang-Mills frames $\boldsymbol{u}$ that belong to a principal frame bundle which is isomophic to $\mathbf{P}(\mathbf{M},\mathbf{G}_0)$, so that

$$\mathbf{P}(\mathbf{M},\mathbf{G}) \supset P^{\uparrow}\mathbf{M} \ , \qquad \mathbf{P}(\mathbf{M},\mathbf{G}) \supset \mathbf{P}(\mathbf{M},\mathbf{G}_0) \ , \qquad \mathbf{G} = \mathbf{G}_0 \times \mathrm{ISO}^{\uparrow}(3,1) \ . \tag{6.7}$$

For a given choice of cross-section $\mathbf{s}_0$ of $P^{\uparrow}\mathbf{M}$, and for a preferred set of local sections $\boldsymbol{s}_\alpha$ of $\mathbf{P}(\mathbf{M},\mathbf{G}_0)$, we can identify, in accordance with (2.19), $\mathbf{P}$ in (6.7) with $\mathbf{P}_1$ in (6.4). In turn, $\mathbf{P}_1$ can be identified with the section $\mathbf{P}_1 \times \{\boldsymbol{e}\}$ of $\mathbf{P}_2$. Thus, we can regard the core bundle $\mathbf{E}_1$ in (6.3) as being associated to $\mathbf{P}(\mathbf{M},\mathbf{G})$.

The generalized soldering maps for the Yang-Mills fibres of this core quantum bundle are given by

$$\sigma_x^{\boldsymbol{u}} : \ \boldsymbol{\Psi} \mapsto \Psi \in \mathcal{W}_\infty \ , \qquad \boldsymbol{\Psi} \in \mathcal{W}_{\infty,x} \ , \qquad \boldsymbol{u} = (u,\boldsymbol{a},\boldsymbol{e}_i) \in \Pi^{-1}(x) \subset \mathbf{P} \ . \tag{6.8}$$

On the other hand, the generalized soldering maps of the core bundle $\mathbf{E}_2$ can be immediately extended into those for its enveloping bundle, which are then given by

$$\sigma_{\boldsymbol{u}}^{g}:\ \boldsymbol{\Psi}\mapsto\ \Psi\in\mathcal{W}\,,\quad \boldsymbol{\Psi}\in\mathcal{W}_{\boldsymbol{u}}\subset\overline{\mathbf{E}}_2\,,\quad \boldsymbol{u}\in\mathbf{P}\,,\quad g\in\mathcal{G}(\mathbf{P}(\mathbf{M},\mathbf{G}_0))\,, \tag{6.9}$$

so that they assign to local states $\boldsymbol{\Psi}_{k;m,n;\boldsymbol{u}}\in\mathcal{W}_{k;m,n;\boldsymbol{u}}$ of k Yang-Mills excitons, m Faddeev-Popov "ghosts" and n "antighosts" the *coordinate wave functions* $\Psi_{k;m,n;\boldsymbol{u}}\in\mathcal{W}_{k;m,n}$:

$$\sigma_{\boldsymbol{u}}^{g}\ :\ \boldsymbol{\Psi}_{k;m,n;\boldsymbol{u}}(\zeta_1,i_1,\dots,\zeta_k,i_k;\zeta_{k+1},\dots,\zeta_{k+m};\zeta_{k+m+1}\dots,\zeta_{k+m+n})$$
$$\mapsto\ \Psi_{k;m,n}(\zeta_1,i_1,\dots,\zeta_k,i_k;\zeta_{k+1},\dots,\zeta_{k+m};\zeta_{k+m+1}\dots,\zeta_{k+m+n})\,, \tag{6.10a}$$

$$\zeta_r=(\boldsymbol{a}+q_r^i\boldsymbol{e}_i,v_r^i\boldsymbol{e}_i)\in T_x\mathbf{M}\times\mathbf{V}_x^+\ ,\quad r=1,\dots,k+m+n\,,\quad x=\Pi(\boldsymbol{u})\ . \tag{6.10b}$$

We can now define the *Yang-Mills quantum frame fields* by

$$A_i(x;\zeta)=A_i^{(+)}(x;\zeta)+A_i^{(-)}(x;\zeta)\quad,\qquad A_i^{(+)}(x;\zeta)=A_i^{(-)\dagger}(x;\zeta)\ , \tag{6.11}$$

where the action of the Yang-Mills exciton annihilation operator is given by

$$\left(A_i^{(-)}(x;\zeta)\boldsymbol{\Psi}_{k;x}\right)_{k-1}(\zeta_1,i_1,\dots,\zeta_{k-1},i_{k-1})=\sqrt{k}\ \boldsymbol{\Psi}_{k;x}(\zeta,i,\zeta_1,i_1,\dots,\zeta_{k-1},i_{k-1})\,, \tag{6.12}$$

for states with k Yang-Mills excitons, regardless of the number of "ghosts" or "antighosts" in them. Similarly, the *Faddeev-Popov quantum frame* and *coframe fields* are given by[17]

$$C(x;\zeta)=C^{(+)}(x;\zeta)+C^{(-)}(x;\zeta)\quad,\qquad C^{(+)}(x;\zeta)=C^{(-)\dagger}(x;\zeta)\ , \tag{6.13a}$$

$$\overline{C}(x;\zeta)=\overline{C}^{(+)}(x;\zeta)+\overline{C}^{(-)}(x;\zeta)\quad,\qquad \overline{C}^{(+)}(x;\zeta)=-\overline{C}^{(-)\dagger}(x;\zeta)\ , \tag{6.13b}$$

where, for example, the action of the "ghost" annihilation operator is as follows:

$$\left(C^{(-)}(x;\zeta)\boldsymbol{\Psi}_{k;m;n;\boldsymbol{u}}\right)_{m-1}(\zeta_1,i_1,\dots,\zeta_k,i_k;\zeta_{k+1},\dots,\zeta_{k+m-1};\dots,\zeta_{k+m+n})$$
$$=\sqrt{m}\ \boldsymbol{\Psi}_{k;\boldsymbol{u}}(\zeta_1,i_1,\dots,\zeta_k,i_k;\zeta,\zeta_{k+1},\dots,\zeta_{k+m-1};\dots,\zeta_{k+m+n})\ . \tag{6.14}$$

The Yang-Mills quantum frame operators give rise to Yang-Mills quantum frames

$$\boldsymbol{\Phi}_f=\exp\left[-\tfrac{1}{2}\langle\boldsymbol{f}|\boldsymbol{f}\rangle-\int\boldsymbol{f}(\zeta)*\boldsymbol{A}^{(+)}(\boldsymbol{u};\zeta)d\tilde{\Sigma}(\zeta)\right]\,\boldsymbol{\Psi}_{0;\boldsymbol{u}}\quad,\qquad \boldsymbol{f}\in W_{\boldsymbol{u}}\ , \tag{6.15a}$$

$$\boldsymbol{f}*\boldsymbol{g}:=f^i\bullet g_i=\sum_{a=1}^{n}\eta^{ij}f_i^a g_j^a\ , \tag{6.15b}$$

in the physical Yang-Mills subfibre of each $\mathcal{W}_{\boldsymbol{u}}$. All the considerations carried out in Sec. 9.3 for the Gupta-Bleuler frames apply equally well to the present Yang-Mills frames, which therefore give rise to continuous resolutions of the identity in those physical fibres.

In order to use the Faddeev-Popov quantum frame and coframe fields in a similar manner, we have to extend the Faddeev-Popov ghost and antighost subfibres into Berezin-Faddeev-Popov superfibres, by using the method of Sec. 8.3, and then set, as in (8.3.15),

$$\left|\Phi_\xi\right\rangle = \exp\left(-\tfrac{1}{2}\int \bar{\xi}(\zeta,\bar{\theta}) * \overleftrightarrow{\partial}_k \xi(\zeta,\theta) d\sigma^k(q) d\tilde{\Omega}(v)\right) \exp\left(C^{(+)}(u;\xi)\right)\left|\Psi_{0;u}\right\rangle , \quad (6.16a)$$

$$\left\langle\Phi_\xi\right| = \left\langle\Psi_{0;u}\right| \exp\left(\bar{C}^{(-)}(u;\xi)\right) \exp\left(-\tfrac{1}{2}\int \bar{\xi}(\zeta,\bar{\theta}) * \overleftrightarrow{\partial}_k \xi(\zeta,\theta) d\sigma^k(q) d\tilde{\Omega}(v)\right) . \quad (6.16b)$$

In fact, the presence of a graded algebra structure, of the type studied in Sec. 10.3, makes the formulation of such an extension very natural. The resulting Berezin integrals then supply the required continuous resolutions of the identity in the Faddeev-Popov subfibres in a manner totally analogous to that in (8.3.17).

Let $\boldsymbol{\omega} \in C(\mathbf{P})$ be a connection form on $\mathbf{P}(\mathbf{M},\mathbf{G})$, in which the extension of the Levi-Civita connection to the Poincaré frame bundle $\mathbf{P}_0$ is combined with a classical Yang-Mills connection. We can then define, for any section $\boldsymbol{s}$ of $\mathbf{P}(\mathbf{M},\mathbf{G})$, the following operator for covariant differentiation on the triply-fibrated associated bundle in (6.3),

$$\boldsymbol{\nabla} = \boldsymbol{d} + \boldsymbol{\delta} + \bar{\boldsymbol{\delta}} + \boldsymbol{A} , \quad \boldsymbol{d} = \boldsymbol{d}x^\mu \partial_\mu, \quad \boldsymbol{\delta} = \boldsymbol{d}g^b\, \partial/\partial g^b, \quad \bar{\boldsymbol{\delta}} = \boldsymbol{d}\bar{g}^b\, \partial/\partial \bar{g}^b . \quad (6.17)$$

Its gauge potential $\boldsymbol{A}$ incorporates the infinitesimal generators of the representation $\boldsymbol{U}$, that corresponds to the direct product of the representations in (6.5a) and (6.6), so that

$$\boldsymbol{A}(x,g,\bar{g}) = A_\mu(x,g,\bar{g})\, \boldsymbol{d}x^\mu + \boldsymbol{C}(x,g,\bar{g}) + \bar{\boldsymbol{C}}(x,g,\bar{g}) , \quad x \in \mathbf{M}^s \subset \mathbf{M} , \quad (6.18a)$$

$$\boldsymbol{C}(x,g,\bar{g}) = C_b(x,g,\bar{g})\boldsymbol{d}g^b , \quad \bar{\boldsymbol{C}}(x,g,\bar{g}) = \bar{C}_b(x,g,\bar{g})\boldsymbol{d}\bar{g}^b , \quad g,\bar{g} \in \mathcal{G}(\mathbf{P}), \quad (6.18b)$$

satisfy the following counterparts[18] of (4.12) and (4.13):

$$\boldsymbol{\delta} A_\mu = \boldsymbol{\nabla}_\mu \boldsymbol{C} , \qquad \bar{\boldsymbol{\delta}} A_\mu = \boldsymbol{\nabla}_\mu \bar{\boldsymbol{C}} , \quad (6.19a)$$

$$\boldsymbol{\delta C} = -\tfrac{1}{2}[\boldsymbol{C},\boldsymbol{C}] , \qquad \bar{\boldsymbol{\delta}}\bar{\boldsymbol{C}} = -\tfrac{1}{2}\left[\bar{\boldsymbol{C}},\bar{\boldsymbol{C}}\right] , \quad (6.19b)$$

$$\boldsymbol{\delta}^2 = \boldsymbol{\delta}\bar{\boldsymbol{\delta}} + \bar{\boldsymbol{\delta}}\boldsymbol{\delta} = \bar{\boldsymbol{\delta}}^2 = \boldsymbol{\delta}\bar{\boldsymbol{C}} + \bar{\boldsymbol{\delta}}\boldsymbol{C} + \left[\bar{\boldsymbol{C}},\boldsymbol{C}\right] = \mathbf{0} , \quad (6.19c)$$

$$\boldsymbol{\delta}\bar{\boldsymbol{C}} = \boldsymbol{B} , \quad \boldsymbol{\delta B} = \bar{\boldsymbol{\delta}}\boldsymbol{B} + \left[\bar{\boldsymbol{C}},\boldsymbol{B}\right] = \boldsymbol{\delta}\bar{\boldsymbol{C}} + \left[\bar{\boldsymbol{C}},\boldsymbol{C}\right] + \boldsymbol{B} = \mathbf{0} . \quad (6.19d)$$

We note that in (6.17) and (6.18) the index b incorporates the Poincaré as well as the Yang-Mills gauge degrees of freedom, so that the parallel transport determined by (6.17) can be carried out in all directions within the bundle $\mathbf{P}(\mathbf{M},\mathbf{G})$, which serves as base manifold to $\mathbf{E}_2$ and to its envelope.

On the other hand, the parallel transport that would give rise to GS propagation, in accordance with the general principles expounded in the preceding six chapters, has to be carried out under subsidiary conditions incorporating those of the Kugo-Ojima (1979) type – which, in the non-Abelian case, replace the Gupta-Bleuler type of subsidiary conditions in (9.3.18). These subsidiary conditions[19] are usually formulated (Kugo and Ojima, 1979) in terms of BRST generators and Faddeev-Popov "ghost charge" operators, and require that physical states should lie in the kernels of these operators. The geometric meaning of these conditions is, however, quite straightforward. Thus, in the present context, the subsidiary condition that under *physical* parallel transport the local state vectors should lie within the kernels of the BRST generators and of the anti-BRST generators (i.e., that the operators representing those generators should map them into zero vectors), simply amounts to the counterparts of (4.14), namely to the stipulation that[20]

$$(\delta + C)\Psi = 0 \ , \qquad \left(\bar{\delta} + \bar{C}\right)\Psi = 0 \ . \tag{6.20}$$

In other word, physical parallel transport should *not* be carried out along vertical directions within the principal bundle **P(M,G)**, which serves as base manifold to the quantum bundle $\mathbf{E}_1$ and its envelope. Rather, GS propagation should proceed only along *physical paths* within **P(M,G)**, which, as in Sec. 9.4, are the horizontal lifts of the physical paths in the base Lorentzian manifold **M** of $\mathbf{E}_0$.

The additional subsidiary condition that physical states should also lie in the kernel of the Faddeev-Popov "ghost charge" operator is optional, since it is meant to merely reduce the "size" of the physical sector (Kugo and Ojima, 1979, p. 20). In the present GS context it amounts to requiring that the number of "ghosts" and "antighosts" be kept constant under the physical parallel transport that gives rise to GS propagation.

Given all these basic facts, the detailed formulation of GS propagation within Yang-Mills quantum bundles can be carried out as in the preceding chapters, by simply taking into account the above additional features. Consequently, we shall not spell out the details, especially since these features are also present in the quantum gravity case, treated in detail in the next chapter (cf. especially Secs. 11.7 to 11.10).

It is, however, of interest to note that the present approach to the theory of Yang-Mills fields is nonperturbative. Hence, like all the other GS reformulations of conventional relativistic quantum field theories treated in this monograph, it does not require any of the tools of conventional renormalization theory. This means that it does not necessitate the *ad hoc* introduction of Higgs bosons, which give rise to some serious epistemological as well as observational difficulties – cf. Note 10, as well as the discussion of the cosmological constant problem in Secs. 11.12 and 12.3.

Notes to Chapter 10

1 The original 1918 article by Weyl was reprinted in the original 1923 edition of *The Principle of Relativity* (cf. Einstein, 1916), which was reissued by Dover Publications in 1952. This pioneering formulation of a gauge theory is reported also in various editions of Weyl (1923).

2 As pointed out in Note 30 to Chapter 2, the term *gauge group* is used in the literature in two different senses: as synonymous to the term "structure group" **G** of a principal bundle **P(M,G)** (in which case in this monograph we either set it between quotation marks, or talk of a gauge group of the first kind), and in the sense defined in Sec. 10.2, namely as the group $\mathcal{G}(\mathbf{P})$ of all vertical automorphisms of **P(M,G)** which induce the identity on **M** . As a rule, in physical models the former group is a finite-dimensional Lie group, whereas the latter is infinite-dimensional, so that despite their close relationship, the two notions of gauge group are mathematically very distinct. Consequently, when they treat Yang-Mills theories, some authors (cf., e.g., Daniel and Viallet, 1980) call the structure group **G** a "gauge group", and refer to $\mathcal{G}(\mathbf{P})$ as "the group of gauge transformations". Others, such as Trautman (1980, 1981), talk about gauge transformations of the first kind and of the second kind, respectively (cf. Sec. 11.3). The reader is also cautioned that due to the prevalence of Lagrangian methods in conventional physics literature on quantum Yang-Mills fields the term "gauge" is also related to the specification of "gauge

fixing" in Lagrangians, so that a profusion of "noncovariant gauges" has emerged during the last couple of decades. A survey of the ensuing "gauge zoo" can be found in (Leibbrandt, 1987).

3 In the generic situation of an arbitrary Lie group $\mathbf{G}_0$, the scalar product in (1.11a), constructed by using the trace in its Lie algebra $L(\mathbf{G}_0)$, has to be replaced by some other form of non-degenerate scalar product left invariant by the adjoint map. In the case when $\mathbf{G}_0$ is semi-simple, such a scalar product can be obtained by expanding the elements of $L(\mathbf{G}_0)$ with respect to a fixed basis, such as the one in (1.2), and then using the Killing form of $\mathbf{G}_0$ – cf. [NT], p. 256.

4 We remind the reader that this is not a misspelling of the name of Lorentz, but represents a rectification of a widespread historical misconception – cf. Note 1 to Chapter 9.

5 Actually, this interpretation is a mere conjecture, since "for a relativistic quantized field theory it has *never* been demonstrated that any realistic interacting system of fields has a unitary solution. While formal power series solutions have been generated that satisfy unitarity to each order of the expansion, the series itself cannot be shown to converge." (Cushing, 1990, p. 177). Indeed, as we have stressed on several previous occasions in this monograph (cf.,e.g., Sec. 1.2), in conventional quantum field theory the *existence* of the S-matrix has not been *proved* for *any* realistic theory of interacting quantum fields – least of all for conventional QED, after which all the other quantum gauge field theories are patterned.

6 Cf. Note 2. The most basic features of this idea of a gauge group are derived in [I], Sec. 3.5, whereas a systematic treatment can be found in Chapters 2 and 3 of [BG], which contains the detailed proofs of most of the results presented without proof in this and the next section. The most basic of those results can be found, however, in [KN], Ch. 2, Sec. 11. They served in such mathematically rigorous treatments of this idea as those by Atiyah *et al.* (1978) and by Trautman (1979, 1980). An elaboration, carried out in the context of Yang-Mills theories, can be found in the review article by Daniel and Viallet (1980). Further relevant results are presented in (Babelon and Viallet, 1981) and (Socolovsky, 1991).

7 A detailed and mathematically rigorous study of the topological and functional analytic aspects of $\mathcal{G}(\mathbf{P})$ and its gauge orbit space, treated as infinite-dimensional manifolds modeled on Hilbert spaces (cf. Note 32 to Chapter 3), can be found in (Mitter and Viallet, 1981). In the sequel we shall concentrate, however, only on those geometric features of these mathematical objects which are most essential for applications.

8 This is the terminology adopted by most other authors [BG,I], whereby any gauge transformation of a principal bundle $\mathbf{P}(\mathbf{M},\mathbf{G})$ is called a vertical automorphism. However, in the approach of Trautman (1976, 1980, 1981), a gauge transformation of a principal bundle $\mathbf{P}(\mathbf{M},\mathbf{G})$ can also be given by an automorphism of $\mathbf{P}(\mathbf{M},\mathbf{G})$ which is not vertical, so that such "vertical" gauge transformations are called "pure" (cf. also Socolovsky, 1991). Trautman's terminology can be very useful when CGR is discussed from the point of view of the present approach to gauge groups (cf. Sec. 11.3).

9 A *graded algebra* Λ is a collection $\{\Lambda^k\}$ of modules, indexed by the integers k, for which there is an associative bilinear mapping $\Lambda \times \Lambda \to \Lambda$ (cf. [C], p. 197). A *module* over a ring R is an Abelian group together with an (external) scalar multiplication operation $R \times \Lambda \to \Lambda$ which is associative and distributive (cf. [C], p. 9) – so that in the case where R is a field (such as the field of real or complex numbers) then a module is actually a vector space. Indeed, a *ring* is defined as a set on which some addition and multiplication operations $R \times R \to R$ are defined which are associative and distributive, and in which addition is Abelian, but not necessarily also multiplication. Hence, a ring is a field only when multiplication is also Abelian, when there is in addition a unit element, and when each non-zero element in R has an inverse (cf. [C], p. 8). In the case of a *graded Lie algebra* the associativity property is replaced by the Jacobi identity, which in the present instance of $\Lambda(M,L)$ is given by (3.5b).

10 We follow the type of terminology used by Trautman (1980, 1981). The original conclusions of Higgs (1964), reached also by Englert and Brout (1964), pertained to massless gauge fields becoming massive vector fields as a result of "spontaneous symmetry breaking" – cf. [IQ], Sec. 12-5-3, [BG], Sec. 10.3 – and do not actually pertain to the present mathematical developments. Thus, in high-energy physics the "Higgs mechanism" is used as an *ad hoc* device to make the Salam-Weinberg electroweak model as well as QCD "renormalizable" and "in agreement with observations" (Pickering, 1984). This makes these well-known models highly dependent on this "mechanism" (Wali, 1986) but, as one of the main contributors to the quantum theory of Yang-Mills fields has pointed out, "the only legitimate reason for introducing the Higgs boson is to make the standard model mathematically consistent ... [in the sense that it] makes the theory renormalizable" (Veltman, 1986, pp. 76 and 81). On the other hand: "The

biggest drawback of the Higgs boson is that so far no evidence of its existence has been found. Instead, a fair amount of indirect evidence already suggests that this elusive particle does not exist." (*ibid.*, p. 76).

11 A lucid summary of these difficulties is presented in Sec. VI of (Daniel and Viallet, 1980).

12 The family $C(\mathbf{P},\mathcal{V})$ cannot supply the total space of a principal bundle with structure group $\mathcal{G}(\mathbf{P})$ if the center $Z(\mathbf{G})$ of $\mathbf{G}$ is non-trivial since, by definition, $Z(\mathbf{G})$ contains any element of $\mathbf{G}$ which commutes with all elements of $\mathbf{G}$, so that the adjoint action is not free (cf. Note 27 to Chapter 3). In that case the subfamily $C_0(\mathbf{P})$ of all *irreducible* connections (i.e., those connections whose holonomy group – cf. Sec. 3.8 for the definition – is equal to $\mathbf{G}$) is used instead of $C(\mathbf{P})$, and the quotient $\mathcal{G}_0(\mathbf{P})$ of $\mathcal{G}(\mathbf{P})$ by the family of $Z(\mathbf{G})$-valued sections of the adjoint representation is used instead of $\mathcal{G}(\mathbf{P})$ – cf. Sec. VI of (Daniel and Viallet, 1980). The "no-go" theorem of Singer (1978), as well as the version by Narasimhan and Ramadas (1979) independently derived for the case of $\mathbf{G} = SU(2)$ and $\mathbf{M} = S^3$, basically use a principal bundle with $C_0(\mathbf{P})$ as total space, and having $\mathcal{G}_0(\mathbf{P})$ as structure group.

13 The first-order formalism contained in Eqs. (5.1)-(5.5) is implicit also in the conventional quantization of Yang-Mills fields (cf., e.g., Faddeev and Slavnov, 1980, Sec. 3.2), but due to the absence of internal gauge degrees of freedom, the first order approximation is deemed to be related to the asymptotic fields (Kugo and Ojima, 1978, 1979). However, no proof of the existence of such fields is known (cf., e.g., Nakanishi and Ojima, 1990, p. 285), and the inconsistencies which the conventional quantum field theory encounters with other nonlinear terms makes it likely that the same inconsistencies are also hidden in conventional "perturbative" treatments of Yang-Mills quantum fields, but are simply ignored – as it has become the custom since the advent of conventional renormalization theory (cf. Secs. 9.6 and 12.3).

14 These equations are also implicit in the conventional quantization of Yang-Mills fields, in the *presumed* asymptotic form of the quantized Yang-Mills fields – cf., e.g., (Faddeev and Slavnov, 1980), p. 78.

15 As pointed out in the introduction to this chapter, it was not Yang and Mills (1954) who first introduced the concept of gauge transformation and gauge invariance in physics. Rather, H. Weyl (1918, 1929) was the actual pioneer of these ideas, and O. Klein (1939) was the first one to consider a non-Abelian gauge theory outside the realm of general relativity. Hence, although we have strived to follow conventional terminology as far as possible, this is one instance where we are able to accurately reflect the historical record, without giving rise to too much confusion.

16 Cf., e.g., (Zardecki, 1988), and the references cited therein. However, as emphasized by Trautman (1980, 1981), there are also many key distinctions between a classical Yang-Mills theory and a classical general relativistic theory viewed as a gauge theory. They originate primarily from the fact that the bundles for the latter case are *soldered*, thus admitting the notion of torsion, whereas that is not the case for Yang-Mills theories. Naturally, these differences are inherited by their quantum counterparts, but those points whose detailed description is omitted in the present section reflect aspects which they have in common in the context of the GS formulation of quantum Yang-Mills fields in this chapter, and of quantum gravity in the next chapter. The similarites between these two cases could have been further brought out by the use in the present Yang-Mills context of the techniques used in Secs. 11.8-11.9, instead of triply-fibrated bundles. However, the use of the latter brings out in stronger relief the common points shared with the treatment of quantum electromagnetic fields in the preceding chapter.

17 The anti-hermiticity assignment for the coframe fields is in accordance with the suggestion first made by Kugo and Ojima (1978), pp. 1871 and 1873, which is imposed on account of the requirement of having a (formally) pseudo-unitary S-matrix in the special relativistic regime.

18 Cf. Sec. 11.10, where the superfield approach to BRST symmetries developed by Bonora *et al.* (1981b, 1982a), Delbourgo and Jarvis (1982), Hoyos *et al.* (1982), Baulieu (1984, 1985), and others, is applied to GS quantum gravity. Note 63 to Chapter 11 explains the relationship of these two approaches.

19 The basic properties of BRST generators Q_B and Faddeev-Popov "ghost charge" operators Q_c can be deduced on general mathematical grounds (Henneaux, 1985, 1986; Horuzhy and Voronin, 1989; Azizov and Khoruzhii, 1990), but in the case of Yang-Mills theories explicit formulae have been also derived – cf. (Kugo and Ojima, 1979), Eq. (2.20) for Q_B, Eq. (2.23b) for Q_c, and Eqs. (2.25) for their BRST algebra (cf. also Nakanishi and Ojima, 1990). In the present GS context, the formulae for these operators can be derived in the same manner as in the quantum gravitational case studied in Secs. 11.8-11.10.

20 Cf. also Sec. 11.10. The more customary form of the first subsidiary condition in (6.20) would be $Q_B \Psi = \mathbf{0}$, as originally stated by Curci and Ferrari (1976c), as well as by Kugo and Ojima (1978a, 1979).

Chapter 11

Geometro-Stochastic Quantum Gravity

The decision as to what is observable in a physical theory affects in a most fundamental manner its interpretation, and even the future course of its structural development. Hence, some recent studies (Howard and Stachel, 1989) of the development by Einstein of classical general relativity (CGR) during the 1907-1915 period have devoted particular attention to the question as to what are the quantities that are "observable" in CGR. These studies point to a fundamental feature of CGR, that underlies its principle of equivalence (Norton, 1989), and which eventually led Einstein to adopting in CGR the principle of general covariance, despite some temporary reservations that stemmed from his well-known "hole" argument[1]. This fundamental general covariance feature of CGR reflects the fact that the *only* fundamental observable entities in CGR are spacetime coincidences (Norton, 1987), which are represented by the points of a Lorentzian manifold. In Einstein's own words: "*All our space-time verifications invariably amount to a determination of space-time coincidences. If, for example, events consisted merely in the motion of material points, then ultimately nothing would be observable but the meetings of two or more of these points.*" (Einstein, 1916, 1952, p. 117) – emphasis added.

The individuation of the spacetime points in CGR is, however, guaranteed not by the manifold structure itself, but rather by the presence of a metric tensor field, reflecting the presence of gravity. The presence of matter is represented in CGR by various tensor fields, which are ultimately incorporated into the stress-energy tensor field. In developing a framework for quantum general relativity (QGR), the GS strategy is not to alter the above interpretational basis of CGR in any basic manner, but rather to find its *extrapolation* to the quantum regime, by *combining* basic quantum principles with the original "positivistic" epistemology of Einstein (cf. Sec. 12.1). It is on account of such a basic strategy that the questions concerning the adaptation of the equivalence principle to the quantum regime, reformulated in the light of previous studies and results on quantum localizability (cf. Chapter 3), have supplied the central theme of this monograph.

In the most recent reviews of quantum gravity (Alvarez, 1989; Ashtekar, 1990) it is acknowledged that there is as yet no successful and generally accepted framework for this discipline, despite the multitude of schemes for quantizing gravity that have been proposed since the 1960s – as reviewed, for example, by Isham (1975, 1987). These schemes exhibit a remarkable variety of *formally* distinctive features[2], but epistemologically they all share one crucial feature: the problem of quantizing gravity is approached on purely formal grounds, as if it were a problem of mere mathematical (and at that largely algorithmic) technique which, once resolved, would automatically provide the solution to the plethora of foundational problems encountered by QGR at the measurement-theoretical as well as at the mathematically rigorous level. Thus, according to some of the prevailing conventional wisdom, "we all know that quantum gravity has to be formulated in the Euclidean domain. There, it is no problem: it is just a question of plumbing." (Hawking, 1988, p. 904).

From a foundational perspective, however, the basic issues are not quite that simple. As can be seen from Chapters 1, 3 and 12, the origins of some of the foundational problems encountered in QGR run deep, as they can be traced to the special relativistic quantum regime, and even beyond that, to the quantum theory of measurement in general. Consequently, the GS approach considers the formulation of a quantum theory of gravity as being as much a problem of formulating a consistent quantum theory of measurement, as of developing a suitable mathematical formalism that faithfully incorporates such a theory. In view of that, the geometro-stochastic quantum gravity (GSQG) presented in this chapter is unrelated to currently fashionable studies of "wormholes" and "baby universes", and at best marginally related to the "loop representations" (Rovelli and Smolin, 1988; Rovelli, 1991) which have emerged from the recent discoveries of new variables in canonical gravity (Ashtekar, 1987, 1988, 1991)[3]. Indeed, instead of searching for mathematically exotic formulations of long-standing physical problems in quantum gravity[4], the GS approach to quantum gravity attempts to reflect *quantum* reality in a physically cogent and operationally meaningful manner, which draws its inspiration directly from Einstein's epistemology, as presented in his key writings. In this spirit, the basic GS epistemology requires that, in the development of a framework which is best suited to the quantization of gravity, foundational problems should be resolved *concurrently* with the creation and the forging of the mathematical techniques most appropriate for this task. In other words, the thesis underlying GSQG is that sound epistemology has to go hand-in-hand with sound mathematics, if a *consistent* unification of general relativity and quantum theory is to be achieved.

The epistemological weaknesses of the conventional mode of thinking about these issues are well illustrated by the various treatments of the oldest scheme for quantizing gravity, namely of the canonical scheme, that have culminated in the current debate on the meaning of "time" in quantum gravity[5]. Since the basic ideas of this scheme are most essential to an understanding of many fundamental issues in the quantization of gravity, they will be the first to be reviewed in Sec. 11.1. We shall then discuss in Sec. 11.2 the more recent developments in quantum gravity, presenting the points of view of the most fashionable contemporary schools of thought on this subject, as well as of the much less fashionable point of view that gravity need not be quantized in the first place. After concluding that, on primarily foundational grounds, gravity has to be quantized, we extend in Sec. 11.3 the presentation of the basic principles of quantum geometry, begun in Sec. 1.3, to the case of quantum gravity. In Sec. 11.4 we discuss the question of what are the "observables" of general relativity. After finding that many of the recent treatments of this important issue are based on formal rather than on foundational criteria, so that they represent an unwarranted departure from Einstein's epistemology on these issues in CGR, we present a GS conceptualization of what is observable in QGR, that is meant to represent an adaptation to the quantum regime of the essence of Einstein's epistemology. In order to mathematically implement that conceptual adaptation, we reformulate in Sec. 5 the *internal* quantum geometry of would-be graviton fibres in a self-contained manner – namely without any appeal to a metric or affine structure of a base spacetime manifold – as it was the case in Chapters 5 to 10. The outcome is a *pre*geometry for quantum spacetime, in which quantum gravitational fibres with a Poincaré internal structure are associated to points in the principle bundle $GA\mathbf{S}$ of general affine frames over a manifold $\mathbf{S}$, which supplies a typical *segment* (cf. Sec. 11.1) in a would-be base Lorentzian manifold. Metric structures that give rise to vierbein fields are introduced in Sec. 11.6, and the corresponding quantum pregeometry fibres are soldered to **S** by means of soldering forms provided by those fields. Internal gauge transformations, related to infinitesimal coordinate changes in small neighborhoods of each point in **S**, are introduced and analyzed in Sec. 11.7. Graviton polarization frames related to the TT gauge and to null polarization tetrads are presented in Sec. 11.8. In Sec 11.9 they serve in the formulation of a quantum gravitational gauge group which incorporates Poincaré degrees of freedom as well as segmental diffeomorphisms, and gives rise to geometrized versions of Faddeev-Popov fields. Thus, a corresponding quantum gravitational bundle a GSQG connection can be formulated, which is covariant with respect to both the external Poincaré gauge group, as well as to diffeomorphism-related gauges. The parallel transport and propagation of quantum gravitational states is then formulated in Sec. 11.10 by means of such frames. In Sec. 11.11 we discuss the geometro-stochastic dynamics of the mutual interaction between the Lorentzian geometries of the geometro-stochastically *emerging* base manifold and the quantum states of the gravitational field. Finally, in Sec. 11.12 we compare the resulting geometro-stochastic concept of quantum spacetime with some of the other recently advocated approaches to quantum gravity and cosmology. In particular, we discuss the issue of a GS counterpart of the "wave function of the universe", introduced by Hartle and Hawking (1983).

11.1. Canonical Gravity and the Initial-Value Problem in CGR

The oldest approach to the quantization of gravity is based on the canonical formulation of CGR. Like so many other schemes and ideas that are currently in vogue, this one was initiated by Dirac (1948, 1949). However, after subsequently pioneering related research into Hamiltonian dynamics with constraints, Dirac did not further pursue the canonical quantization of gravity, since he eventually reached the following conclusion: "There does not exist any general method for handling [in canonical quantum gravity] quadratic quantities in the δ-function, free from inconsistencies. ... The problem of the quantization of the gravitational field is thus left in a rather uncertain state. If one accepts Schwinger's plausible [but nonrigorous] methods, the problem is solved. But one cannot be happy with such methods without having a reliable procedure for handling quadratic expressions in the δ-function." (Dirac, 1968, p. 543). And indeed, recent work by Goroff and Sagnotti (1986), that deals with two-loop formal perturbative expansions in pure quantum gravity treated as a gauge theory of the Lorentz group, confirmed the impossibility of arriving at even formally renormalizable expressions in quantum gravity.

On the other hand, Dirac's (1950, 1958, 1959) pioneering work on Hamiltonian dynamics with constraints has led to the Arnowitt-Deser-Misner (1962) approach to the initial-value problem in CGR – often referred to also as the CGR Cauchy problem. From the point of view of an already specified globally hyperbolic[6] Lorentzian manifold $(\mathbf{M}, g)$, which represents a spacetime in CGR, this ADM method deals with the problem of recovering $(\mathbf{M}, g)$ from data prescribed on an initial-data Cauchy surface σ_0. The points of such an initial-data hypersurface can be always[7] labeled by means of an atlas of coordinate charts, which are given by maps of the general form $(0, x^1, x^2, x^3) \leftrightarrow x \in \sigma_0$. A foliation of the globally hyperbolic Lorentzian manifold $(\mathbf{M}, g)$ into diffeomorphic maximal spacelike hypersurfaces σ_{x^0} can be then envisaged, so that $\mathbf{M}$ becomes equal to a union of such disjoint reference surfaces, which is of the form (5.4.7). It then proves convenient to express the metric in the form of the following block-matrix,

$$\|g_{\mu\nu}(x^0,\mathbf{x})\| = \begin{pmatrix} N^2 - N_a N_b \gamma^{ab} & N_b \\ N_a & -\gamma_{ab} \end{pmatrix} , \quad \mu,\nu = 0,1,2,3 \ , \quad a,b = 1,2,3 \ , \tag{1.1}$$

where N and N_a are called the ADM *lapse* and *shift* functions, respectively, whereas

$$\gamma(x^0,\mathbf{x}) = \gamma_{ab}(x^0,\mathbf{x})dx^a \otimes dx^b \ , \qquad \mathbf{x} = (x^1,x^2,x^3) \ , \qquad a,b = 1,2,3 \ , \tag{1.2}$$

represents a Riemannian 3-metric $\gamma^{(x^0)}$ along each of the (three-dimensional) reference hypersurfaces σ_{x^0}.

To understand the reasons for the above terminology, let us consider the differentials

$$L_0 = N(x^0,\mathbf{x})dx^0 \ , \qquad L_a = N_a(x^0,\mathbf{x})dx^0 \ , \qquad a = 1,2,3 \ , \tag{1.3}$$

and view them as being the components of an "infinitesimal vector" in $\mathbf{M}$. Then, for two "infinitesimally" close reference hypersurfaces σ_{x^0} and $\sigma_{x^0+dx^0}$, L_0 represents, at a given x^0, the "lapse" in the proper time of a test particle in free fall, whose worldline is a geodesic orthogonal to σ_{x^0}, as it reaches $\sigma_{x^0+dx^0}$; whereas, L_a represents the a-th component of the "shift" 3-vector, which joins the point where that worldline meets $\sigma_{x^0+dx^0}$ with the point where the coordinate line obtained by keeping the coordinates of $\mathbf{x}$ fixed, and varying x^0, impinges upon $\sigma_{x^0+dx^0}$. Hence, we indeed have that, in accordance with (1.1),

$$ds^2 := g_{\mu\nu}dx^\mu dx^\nu = \left(N^2 - N_a N_b \gamma^{ab}\right)(dx^0)^2 + 2N_a dx^0 dx^a - \gamma_{ab}dx^a dx^b \ . \tag{1.4}$$

It is also easily verified that the definition of the inverse metric tensor in (2.6.24b) yields in the present case

$$\|g^{\mu\nu}\| = \begin{pmatrix} -N^{-2} & N^{-2}N^b \\ N^{-2}N^a & \gamma^{ab} - N^{-2}N^a N^b \end{pmatrix} , \quad \gamma_{ac}\gamma^{cb} = \delta_a{}^b, \quad N^a = \gamma^{ab}N_b \ . \tag{1.5}$$

As described in Sec. 5.4, synchronous (Gaussian normal) coordinates,

$$\left(x^0 = \tau, x^1, x^2, x^3\right) , \qquad N \equiv 1 \ , \qquad N_a \equiv 0 \ , \tag{1.6}$$

can be always introduced (Isham and Kuchař, 1985). They correspond to the special case where the lapse and shift functions are those for *coherent* flows of classical test particles adapted to the chosen foliation. Consequently, in them the metric tensor assumes the form

$$\mathbf{g} = d\tau \otimes d\tau - \gamma_{ab}\, dx^a \otimes dx^b \ , \tag{1.7}$$

so that τ represents the proper time of those test particles in free fall. Let $\boldsymbol{n}$ denote the field of future-oriented tangents to the worldlines of these classical test particles. Then, along each of the reference hypersurfaces σ_τ, the field $\boldsymbol{n}$ over $\mathbf{M}$ determines a field of unit timelike normals $\boldsymbol{n}^{(\tau)}$ to that hypersurface, in terms of which the *extrinsic curvature* of σ_τ is given by the *(0,2)*-tensor field [M,W]

$$\boldsymbol{K}^{(\tau)} = K^{(\tau)}_{ab}\, \boldsymbol{dx}^a \otimes \boldsymbol{dx}^b = -\tfrac{1}{2}\pounds_{\boldsymbol{n}}\boldsymbol{\gamma}^{(\tau)} \;, \qquad K^{(\tau)}_{ab} := -D_a n^{(\tau)}_b \;, \quad D_a = \overline{\nabla}_a \;. \tag{1.8}$$

All the covariant derivatives occurring in (1.8) correspond, in accordance with (2.6.25), to the Levi-Civita connection in $\mathbf{M}$; the second of the above expressions for $\boldsymbol{K}^{(\tau)}$ involves the Lie derivative of the 3-metric $\boldsymbol{\gamma}^{(\tau)}$ in the direction of $\boldsymbol{n}^{(\tau)}$. Hence, the covariant derivative of a vector field $\boldsymbol{X}$ in $\mathbf{M}$ can be expressed as follows (cf. [M], p. 512),

$$\overline{\nabla}_a \boldsymbol{X} = X^b{}_{|a}\, \partial_b - K^{(\tau)}_{ab} X^b\, \boldsymbol{n}^{(\tau)} \;, \qquad X^b{}_{|a} := \overline{\nabla}^{(\tau)}_a X^b \;, \qquad \partial_b := \partial / \partial x^b \;, \tag{1.9}$$

in terms of the extrinsic curvature of the reference hypersurfaces σ_τ, and of the Levi-Civita connections along those hypersurfaces that are compatible with Riemannian metrics $\boldsymbol{\gamma}^{(\tau)}$ – and whose covariant derivatives are conventionally denoted by using a vertical bar.

In regions free of nongravitational sources the stress-energy tensor is zero, so that the Einstein equations in (2.7.3) assume their "vacuum form"

$$G_{ij}(\boldsymbol{g}) := R_{ij} - \tfrac{1}{2} g_{ij} R = 0 \;. \tag{1.10}$$

These vacuum Einstein equations can be obtained from variations of the Hilbert-Palatini action [M,W]

$$S_{\mathrm{HP}} = (1/16\pi)\int R\,\sqrt{-g}\, d^4x \;, \qquad g := \det\left\| g_{\mu\nu} \right\| \;, \tag{1.11a}$$

which is based on the Riemann curvature scalar R in (2.7.2b).

The starting point of the ADM geometrodynamic formulation of CGR consists of the reformulation, in units with $16\pi G = 2\kappa = 1$, of this action in the (3+1)-form, so that

$$16\pi S_{\mathrm{HP}} = S_{\mathrm{ADM}} - 32\pi \int dx^0 \int d^3\mathbf{x} \left(\pi^{ab} N_b - \tfrac{1}{2}\pi^b{}_b N^a + N^{|a} \right)_{,a} \;. \tag{1.11b}$$

By the use of Stokes' theorem (cf. [C], p. 216), the last term in the above equality can be expressed as a surface integral, so that it is usually[8] dropped, and one is left with

$$S_{\mathrm{ADM}} = \int dx^0 \int d^3\mathbf{x} \left(-\gamma^{(x^0)}_{ab}\dot{\pi}^{ab} - N\mathcal{H}^0 - N_a \mathcal{H}^a \right) \,, \tag{1.12a}$$

$$\pi^{ab} := \delta S_{\mathrm{ADM}} \big/ \partial\dot{\gamma}^{(x^0)}_{ab} = \sqrt{\gamma^{(x^0)}} \left(\gamma^{(x^0)ab} K^{(x^0)c}_c - K^{(x^0)ab} \right) \,, \tag{1.12b}$$

$$\dot{\pi}^{ab}(x^0,\mathbf{x}):=\partial\pi^{ab}/\partial x^0,\quad \gamma^{(x^0)}(\mathbf{x}):=\det\left\|\gamma_{ab}^{(x^0)}\right\|,\quad N^a:=\gamma^{(x^0)ab}N_b\ ,\tag{1.12c}$$

in which the lapse and shift functions play the role of Lagrange multipliers, and

$$\mathcal{H}^a(\boldsymbol{\gamma}^{(x^0)},\boldsymbol{K}^{(x^0)})=-2\pi^{ab}{}_{|b}\ ,\tag{1.13a}$$

$$\mathcal{H}^0(\boldsymbol{\gamma}^{(x^0)},\boldsymbol{K}^{(x^0)})=(\gamma^{(x^0)})^{-1/2}\left[\tfrac{1}{2}(\pi^b{}_b)^2-\pi^{ab}\pi_{ab}\right]-(\gamma^{(x^0)})^{1/2}R^{(x^0)}\ ,\tag{1.13b}$$

play the role of the associated *supermomentum* and *super-Hamiltonian* [M], respectively. The curvature scalar in (1.13b) corresponds, of course, to the Riemannian 3-metric along the reference surface σ_{x^0}.

The variation of the ADM action with respect to the lapse and shift functions in (1.12) obviously yields the four constraint equations

$$\mathcal{H}^a(\boldsymbol{\gamma}^{(x^0)},\boldsymbol{K}^{(x^0)})=0\quad\Leftrightarrow\quad G_{a0}(\boldsymbol{g})=0\ ,\tag{1.14a}$$

$$\mathcal{H}^0(\boldsymbol{\gamma}^{(x^0)},\boldsymbol{K}^{(x^0)})=0\quad\Leftrightarrow\quad G_{00}(\boldsymbol{g})=0\ ,\tag{1.14b}$$

equivalent to four of the Einstein equations in (1.10). The remaining six Einstein equations reflect the actual CGR geometrodynamics. They can be regarded as a system of nonlinear equations for the 3-metric and the extrinsic curvature of the reference hypersurfaces σ_{x^0}. Therefore, we shall henceforth refer to them collectively as the *Einstein geometrodynamic equations*, whereas those in (1.14) we shall call the *Einstein constraint equations*[9]. From the point of view of Einstein's second-order formalism, the equations obtained in the ADM framework from variations of the action with respect to π^{ab}, namely the equations

$$\partial\gamma_{ab}^{(x^0)}/\partial x^0=2N(\gamma^{(x^0)})^{-1/2}\left[\pi_{ab}-\tfrac{1}{2}\gamma_{ab}^{(x^0)}\pi^c{}_c\right]+N_{a|b}+N_{b|a}\ ,\tag{1.15}$$

can be viewed as relationships that define π^{ab}. The remaining equations for $\partial\pi^{ab}/\partial x^0$ turn out to be linear combinations of Einstein's geometrodynamic and constraint equations – cf. Eqs. (7-3.15*b*) in (ADM, 1962), or Eq. (21.115) in [M]. Upon introducing the following "Hamiltonian" for CGR "in vacuum" (i.e., in the absence of nongravitational fields),

$$H_0^{(t)}=\int_{x^0=t}\mathcal{H}\,d^3\mathbf{x}\ ,\qquad \mathcal{H}=N\mathcal{H}^0+N_a\mathcal{H}^a\ ,\tag{1.16}$$

and henceforth dropping (as it is customary) the indices explicitly indicating dependence upon x^0, the dynamical equations of the ADM formalism can be formulated in terms of Poisson brackets, obtained by replacing partial derivatives with functional derivatives in the Hamiltonian classical mechanics definition (Fischer and Marsden, 1979, pp. 151-157):

$$\dot{\gamma}_{ab}=\{\gamma_{ab},\mathcal{H}\}=\delta\mathcal{H}/\delta\pi^{ab}\ ,\qquad \dot{\pi}^{ab}=\{\pi^{ab},\mathcal{H}\}=-\delta\mathcal{H}/\delta\gamma_{ab}\ .\tag{1.17}$$

It should be noted that, although the constraint equations (1.14) impose subsidiary conditions involving the supermomentum and the super-Hamiltonian, which sets those quantities equal to zero in the course of the geometrodynamic development, that does not imply that the Poisson brackets in (1.17), which govern that development, themselves vanish. On the other hand, as we shall see in the course of the discussions in the next three sections, this feature of the constraint equations gives rise to difficulties with the concept of "time" in canonical gravity, as well as with the decision as to what are the "observables" in the canonical quantization of gravity.

The basic idea as to how to formulate the initial-value problem in CGR can be arrived at by viewing in the foliation (5.4.7) the diffeomorphisms obtained from all the coordinate curves corresponding to points in the initial-data hypersurface σ_0 , during the flow of the "time" $t = x^0$, as a means of identification of the reference hypersurfaces σ_{x^0} with a single differential manifold Σ_0. On the other hand, in an **M**-independent formulation of the initial-value problem, the Lorentzian manifold $(\mathbf{M}, \boldsymbol{g})$ is no longer known *a priori*, but rather has to be computed from a complete Cauchy set of initial data prescribed on some initially given 3-manifold Σ_0. These CGR initial data are comprised of the initial-data manifold Σ_0 itself, of a given initial Riemannian metric γ_0 and of a given initial extrinsic curvature $\boldsymbol{K}_0$ on Σ_0, as well as of suitable initial data for all the nongravitational fields (which we shall denote collectively by $\boldsymbol{\phi}$), that enable the computation of the initial value $\boldsymbol{T}_0$ of the stress-energy tensor $\boldsymbol{T}$. For the vacuum case, the counterparts of the equations (1.17) can be expressed in the form of the following equations for the geometrodynamic evolution of the 3-metric $\boldsymbol{\gamma}$ and of the extrinsic curvature $\boldsymbol{K}$ (cf., e.g., Isenberg and Nester, 1980, Eq. (4.18))

$$\dot{\gamma}_{ab} = -2NK_{ab} + \pounds_{\vec{N}}\, \gamma_{ab} \ , \qquad \pounds_{\vec{N}}\, \gamma_{ab} = N_{a|b} + N_{b|a} \ , \tag{1.18a}$$

$$\dot{K}^a{}_b = N\left(R^a{}_b + K^c{}_c K^a{}_b\right) - D^a D_b N + \pounds_{\vec{N}} K^a{}_b \ . \tag{1.18b}$$

The covariant derivatives in (1.18) are exclusively those of the Levi-Civita connection compatible with the 3-metric at the considered global "time" t , provided by the chosen parametrization. In the case of nonderivative couplings[10] to nongravitational sources, these equations acquire additional terms – cf. (ADM, 1962), (Isenberg and Nester, 1980), or (Choquet-Bruhat and York, 1980). Naturally, they have to be then supplemented by the field equations expressing the classical dynamics of all the nongravitational fields $\boldsymbol{\phi}$.

The constraint equations in (1.13) are then also modified by the appearance of the proper energy density ρ and of the proper momentum components j^a of the nongravitational sources. The resulting constraint equations can be therefore written in the following form (cf. Choquet-Bruhat and York, 1980, Eqs. (3.1) and (3.2)):

$$R(\gamma) - K^{ab}K_{ab} + (K_a{}^a)^2 = 16\pi\rho \ , \qquad \rho = T_{\mu\nu}n^\mu n^\nu \ , \tag{1.19a}$$

$$D_a K^b{}_b - D_b K^b{}_a = 8\pi j_a \ , \qquad j_a = T^0{}_a \ . \tag{1.19b}$$

In case that would-be Gaussian normal coordinates are adopted, obtained by choosing, as in (1.6), $N = 1$ and $N_a = 0$, then in accordance with (1.7),

$$g_{00} = 1 \ , \quad g_{a0} = g_{0a} = 0 \ , \quad g_{ab} = -\gamma_{ab} \ , \quad K_{ab} = -\tfrac{1}{2}\dot{g}_{ab} \ . \tag{1.20}$$

Algorithmic schemes for the numerical solution of the above formulated CGR intial-value problem have been devised[11], whereby the would-be classical spacetime manifold $\mathbf{M}$ is envisaged as being sliced into "thin" *segments* $\mathbf{S}_n$, $n = 0, \pm 1, \pm 2,...$, corresponding to $t \in [t_n, t_{n+1})$. In that case the initial data are presented by replacing all the differentials with differences within the initial-data segment $\mathbf{S}_0$ corresponding to $t \in [t_0, t_1)$ with $t_0 = 0$. Initial data $\boldsymbol{\gamma}_0$, $\boldsymbol{K}_0$ and $\boldsymbol{\phi}_0$, obeying the constraints in (1.19), are then provided as the starting point of an algorithm to compute $\boldsymbol{\gamma}_1$, $\boldsymbol{K}_1$ and $\boldsymbol{\phi}_1$ at t_1, in which the dynamic equations in (1.18) are used in the vacuum case, or their counterparts incorporating nongravitational fields are used in the presence of sources. Since, in principle, the constraints remain satisfied for all values of t (with violations occurring only from computational large margin of errors), the process can be repeated, so that $\boldsymbol{\gamma}_n$, $\boldsymbol{K}_n$ and $\boldsymbol{\phi}_n$ can be computed in all the other segments $\mathbf{S}_n$ by an iteration of this entire algorithm.

Naturally, although such algorithms might be computationally satisfactory, the standard type of Cauchy problem requires proofs of existence and uniqueness of solutions. For the vacuum case proofs of great generality are already well-known in CGR. They confirm the existence of a unique "maximal Cauchy development"[12] $(\boldsymbol{\Sigma}_0, \boldsymbol{\gamma}, \boldsymbol{K})$, giving rise, for initial data $(\boldsymbol{\Sigma}_0, \boldsymbol{\gamma}_0, \boldsymbol{K}_0)$ that satisfy the constraints in (1.14), to a globally hyperbolic Lorentzian manifold $(\mathbf{M}, \boldsymbol{g})$, in which $\boldsymbol{\Sigma}_0$ emerges as a Cauchy surface. As could be expected on account of Einstein's "hole" argument (cf. Sec. 11.3), the uniqueness of such Cauchy developments can be established only modulo diffeomorphisms between the resulting Lorentzian manifold $(\mathbf{M}, \boldsymbol{g})$ and other Lorentzian manifolds $(\mathbf{M}', \boldsymbol{g}')$ isometric to $(\mathbf{M}, \boldsymbol{g})$, which in general result from diffeomorphic initial conditions. Similar theorems are much harder to prove in the presence of nongravitational sources, since the results depend critically on the dynamical equation of those nongravitational fields. In fact, the strong cosmic censorship hypothesis (cf. Note 6) is tantamount to the requirement that, modulo diffeomorphisms, the initial data *uniquely* determine the structure of a classical spacetime. Hence, only particular classes of solutions have been treated thus far [W], and the entire area is still very much under active investigation.

The most straightforward type of quantization of gravity consists of formally extending the canonical quantization procedure used in nonrelativistic quantum mechanics for n degrees of freedom, whereby to each "classical observable" represented by a polynomial, or more generally by an analytic function $F(q,p)$ of the canonical position and momentum variables q and p , one attempts to assign a self-adjoint operator $F(Q,P)$ in such a manner that Poisson brackets are transposed into commutator brackets, and, in particular, so that the canonical Poisson bracket relations

$$\{q^j, q^k\} = \{p_j, p_k\} = 0 \ , \quad \{q^j, p_k\} = \delta^j_{\ k} \ , \quad j,k = 1,...,n \ , \tag{1.21a}$$

are transposed into the well-known quantum canonical commutation relations for position and momentum operators:

$$[Q^j, Q^k] = [P_j, P_k] = 0 \ , \quad [Q^j, P_k] = i\delta^j_{\ k} \ , \quad j,k = 1,...,n \ . \tag{1.21b}$$

As is well-known, these canonical quantization rules give rise to ordering problems (Dirac, 1949), as well as to possible inconsistencies when classical canonical variables in some general curvilinear coordinates are used instead of those in Cartesian coordinates. However, if one ignores for the time being all these difficulties, and follows the canonical quantization procedure as usually applied in conventional quantum field theory [IQ,SI], one can use the formal appearance of Poisson brackets in (1.17) to rationalize the introduction of the following "equal-time" canonical commutation relations[13]:

$$\left[\hat{\gamma}_{ab}(t,\mathbf{x}),\hat{\gamma}_{cd}(t,\mathbf{y})\right]=\left[\hat{\pi}^{ab}(t,\mathbf{x}),\hat{\pi}^{cd}(t,\mathbf{y})\right]=0\ ,\qquad a,b,c,d=1,2,3\ , \tag{1.22a}$$

$$\left[\hat{\gamma}_{ab}(t,\mathbf{x}),\hat{\pi}^{cd}(t,\mathbf{y})\right]=\tfrac{1}{2}i\left(\delta_a{}^c\delta_b{}^d+\delta_a{}^d\delta_b{}^c\right)\delta^3(\mathbf{x},\mathbf{y})\ ,\qquad \mathbf{x},\mathbf{y}\in\Sigma_0\ . \tag{1.22b}$$

One can then *imagine* that these commutation relations are realized as linear "operators"[14] on some abstract space of "ket" vectors, which carries a yet unspecified inner product. Upon inserting these "operators" into (1.16), one obtains what can be expected to be quantum generators of infinitesimal hypersurface deformations

$$\hat{H}_0(N,N_a)=\int\left(N\hat{\mathcal{H}}^0+N_a\hat{\mathcal{H}}^a\right)d^3\mathbf{x}\ . \tag{1.23}$$

Indeed, since they are functionals in the lapse and the shift functions, they represent the quantum counterparts of the classical generators of infinitesimal spacetime diffeomorphisms. Consequently, one can attempt to enforce a quantum counterpart of the constraints in (1.14) by demanding that when "physical" states are represented by "ket" vectors, they should satisfy the following Dirac-type of subsidiary conditions,

$$\hat{\mathcal{H}}^a(\hat{\gamma}(t,\mathbf{x}),\hat{\pi}(t,\mathbf{x}))|\psi\rangle=0\ ,\qquad \hat{\mathcal{H}}^0(\hat{\gamma}(t,\mathbf{x}),\hat{\pi}(t,\mathbf{x}))|\psi\rangle=0\ , \tag{1.24}$$

which are based on quantum supermomentum and super-Hamiltonian "operators" obtained by the simple expedient of replacing in (1.13) the classical fields with their quantum counterparts in (1.22).

There are, of course, the usual difficulties which such formal procedures always encounter with operator orderings. However, even if one ignores them, but assumes that the "operators" γ_{ab} transform collectively as the components of a tensor field under spatial diffeomorphisms generated by "Hermitian" supermomenta (which are obtained by setting $N=0$ in (1.23), and varying N_a at will), then it immediately follows from (1.22) that, for all "physical" state vectors ψ and ψ' (Bergmann and Komar, 1980),

$$\langle\psi'|\gamma_{ab}(t,\mathbf{x})|\psi\rangle=\langle\psi'|\pi^{ab}(t,\mathbf{x})|\psi\rangle=0\ ,\qquad a,b=1,2,3\ . \tag{1.25}$$

This indicates "a disaster for any attempt to interpret this theory by means of a correspondence principle" (*ibid.*, p. 245), namely for any attempt to relate the expectation values of γ_{ab} to any measurable metric relationships emerging from quantum spacetime fluctuations.

The various "resolutions to this dilemma" that were suggested by Bergmann and Komar (1980) are closely related to the problem of identifying the "observables" or "quasi-observables" of a canonical approach to QGR – an important issue that will be discussed in Sec. 11.4. Their suggestions have ultimately resulted, however, only in the admission by these two authors that "we are not yet in the position to identify a satisfactory space of state vectors on which the algebra of operators appropriate to the quantization of general relativity may be realized" (*ibid.*, p. 251).

As reviewed in Sec. 5 of (Isham, 1975), one of the earliest proposed (Kuchař, 1971, 1973) resolutions for such problems consisted of solving (by using perturbation methods) the classical constraints given by (1.13) and (1.14), in order to eliminate four out of the twelve canonical variables (γ_{ab}, π^{ab}). An additional four of these canonical CGR variables were eliminated by a suitable choice of coordinate charts. Canonical commutation relations of the type (1.22) were then imposed on the remaining four canonical variables (γ_A, π^A), $A = 1,2$, and supplemented by appropriate Heisenberg equations of motion. Of course, all the mathematical difficulties, that usually plague conventional quantum field theory in the presence of nonlinear terms resulting from quantum field interactions and self-interactions, are very much in evidence in such an approach. That, however, would not have been a serious deterrent to those subscribing to the prevailing forms of conventionalistic instrumentalism in quantum field theory (cf. Secs. 12.2 and 12.3). The drawback which was deemed to be more serious, and which led to the abandonment of this approach, lies in its obvious dependence of the resulting quantum gravity on the manner in which the coordinates were chosen, with different choices leading to physically inequivalent theories. Consequently, other approaches to the canonical quantization of gravity, which will be discussed in the next two sections, eventually proved more popular: they did not suffer from the fatal drawback that in them "the difficulties appear most honestly!" (Isham, 1975, p. 51).

11.2. Contemporary Approaches to the Quantization of Gravity

As we shall see in the course of our discussion of the latest developments in the most fashionable approaches to quantum gravity and quantum cosmology, these are areas in which some of the, at present, most popular trends are based on hypotheses and interpretations of quantum gravity and cosmology that sometimes totally blur the boundary between science and science-fiction[15]. Consequently, albeit unpopular, sober and well-thought-out arguments to the effect that the gravitational field should not be quantized at all, such as those originally advanced by Møller (1952) and by Rosenfeld (1957), and more recently by von Borzeszkowski and Treder (1988), certainly deserve serious consideration.

The most direct argument in favor of quantizing gravity was given by Dirac: "There is no experimental evidence for the quantization of the gravitational field, but we believe quantization should apply to all fields of physics. They all interact with one another, and it is difficult to see how some could be quantized and others not." (Dirac, 1968, p. 539).

In opposing in their recent monograph the quantization gravity, von Borzeszkowski and Treder counter this type of argument in the following oblique manner: "In order to avoid physical and mathematical inconsistencies resulting otherwise from Einstein's equations, one has to consider [the] quantization of gravitational fields. The quantum procedure should unify or at least harmonize classical and quantum theory. On the other hand, GRT [i.e., general relativity theory] is not genuine field theory. This is due to (i) the identification of gravitational field and spacetime metric (statement of the weak principle of equivalence) and (ii) the universal gravitational coupling making gravity itself a source of the gravitational field (this is, together with (i), a formulation of the strong equivalence principle). As a consequence of this strong principle, *Einstein's equations show a typical nonlin-*

earity producing back reaction effects. It makes all quantization rules problematical which transform a usual classical field theory into a quantum field theory." (von Borzeszkowski and Treder, 1988, p. 56) – emphasis added. That contention is then taken as basic motivation for advancing and defending the thesis that "if the strong equivalence principle is satisfied, there is no difference between quantized and non-quantized gravitational theory" (*ibid.*, p. 3).

The historical origins of the method used in attempting to prove this thesis are outlined in the following passage:

"Landau and Peierls (1931) pretended to show that, contrary to Einstein's photon hypothesis, electromagnetic fields themselves need not be quantized because quantization does not result in measurable effects. Moreover, quantization generates a lot of technical difficulties, e.g., divergencies, so that it is an unhappy procedure. Today one knows of course that *these arguments must be wrong somewhere, because there exists a physically meaningful quantum electrodynamics*. Nevertheless, ... in accepting Bohr's and Rosenfeld's (1933) objections to the Landau-Peierls paper, the conclusion which Landau and Peierls wanted to draw for electromagnetism can really be drawn for gravity." (*ibid.*, p. 26) – emphasis added.

One of the fundamental physical differences between the theory of measurement in quantum electrodynamics and quantum gravity is that the latter contains, in addition to Planck's constant h and the speed of light in vacuum c, also the gravitational constant G. Therefore, in advancing their arguments about the purported indistinguishability "between quantized and non-quantized gravitational theory", von Borzeszkowski and Treder in fact recover results by DeWitt (1962, 1964), Mead (1964), and many others, to the effect that in quantum gravity "fundamental limitations on measurement occur, which are associated with Planck's length" (*ibid.*, p. 36). On the other hand, they claim that: "In DeWitt (1964) the conclusion is drawn ... that 'a single observable can always, in principle, be measured with arbitrary accuracy'. It is added that, if this assumption is invalid, the foundations of quantum theory itself must be altered. However, the point which we stress ... is that for gravity ... there also exist limitations on the measurability of a single observable." (*ibid.*, p. 38). Consequently, their final conclusion, stated in the very last section of their monograph, reads as follows:

"The limitations on measurement of gravitational fields are, due to the geometrical character of gravity, necessarily limitations on the size of spacetime regions over which the Bohr-Rosenfeld quantum measurement procedure is, in principle, feasible. One way of expressing this fact is to say that supplementary to the well-known nonlocalizability of classical, i.e., non-quantized gravitational fields (making the $L_0 \to 0$ limit for lengths L_0 [of test bodies] physically senseless), there exists a finite limit ℓ_P [i.e., the Planck length] on the localizability of quantized gravitational fields. This is owed to the fact that *localizability is inconsistent with Heisenberg's uncertainty principle taken in conjunction with the basic tenets of general relativity.*" (*ibid.*, p. 100) – emphasis added.

Thus, given the implicit fact that for von Borzeszkowski and Treder "localizability" means exclusively localizability in an idealized *classical* sense (namely a sense which entails an arbitrarily high degree of sharpness, which in theory transcends into perfect accuracy), the main thesis which they *actually* establish is that the *orthodox* quantum theory of measurement is incompatible with the quantization of gravity, and that "the foundations of quantum theory itself must be altered" in order to cope with this fact. But that is exactly one of the principal theses on which the present approach to quantum geometry is founded. In fact, it has already been pointed out in Sec. 7.6 that, on account of additional measurement theoretical results, the thesis that "a single observable can always, in principle, be measured with arbitrary accuracy" has to be discarded[16]. Hence, the measurement-theoretical arguments presented by von Borzeszkowski and Treder (1988) actually only underline the fact that the taking into account of gravitational effects definitely requires a *revision* of conventional relativistic quantum theory.

Indeed, in addition to the reason presented by Dirac for the quantization of gravity – which remains standing in the face of all arguments presented in the past against quantizing gravity – there is also the basic fact that the general relativistic gravitational field possesses "radiative degrees of freedom of its own, [which] must obey quantum laws" (Bergmann and Komar, 1980). Tied in with this fact there are deeper ontological reasons for quantizing gravity, which we shall discuss in the next section, and which pertain to the very nature of the theory of quantum measurement. Indeed, since some measurement procedures can result in a "reduction of the wave packet", which in turn effects the geometry of any theory in which matter and geometry are in *mutual interaction*, no deterministic geometric ingredients can be acceptable in QGR. Rather, any QGR framework in which *quantum* matter and *quantum* geometry are in *mutual* interaction has to exhibit, even at the semiclassical level, all the various *possible* and intrinsically distinct quantum states of the geometry, which can emerge from all the physically *possible* distinct outcomes in the "reduction" of the states of matter and radiation that supply the sources of gravitation. In fact, despite their basic reservations about quantizing gravity, von Borzeszkowski and Treder have arrived at a similar conclusion:

"Whether quantum effects of gravity are measurable or not, quantization of gravitational fields is, as stressed repeatedly above, unavoidable for a consistent description of the interaction of gravitational and quantized matter fields. There arise, however, problems in the application of the usual schemes to gravity, because such procedures have to evade modifications which destroy the classical [general relativistic theory which] one wants to quantize." (von Borzeszkowski and Treder, 1988, p. 49).

Thus, in the end the objections that can be raised against the quantization of gravity actually hold only as objections against the application of the "usual schemes" – namely all the schemes usually classified as "canonical" or "covariant".

The basic ideas of the most straightforward application of the canonical scheme were explained at the end of the preceding section, but were found wanting in all basic respects. A very popular modification of that method is based on the idea of a "superspace" (Wheeler, 1962, 1968) of all Riemannian 3-geometries $(\Sigma_0,\boldsymbol{\gamma})$ over a given 3-manifold Σ_0. At the formal level, it appears natural to associate a "wave function" $\Psi(\boldsymbol{\gamma},\boldsymbol{\phi})$ with each pair consisting of such a 3-metric $\boldsymbol{\gamma}$ and of a nongravitational field configuration $\boldsymbol{\phi}$ on Σ_0.

In this superspace approach to canonical gravity, the canonical commutation relations in (1.22) are realized by *assuming* that functional differentiation can be treated in the same manner as ordinary differentiation is used in arriving at realizations of the canonical commutation relations in (1.21b), so that on the purely formal level one would have:

$$\hat{\gamma}_{ab}(\mathbf{x})\Psi(\boldsymbol{\gamma},\boldsymbol{\phi}) = \gamma_{ab}(\mathbf{x})\Psi(\boldsymbol{\gamma},\boldsymbol{\phi}) \ , \tag{2.1a}$$

$$\hat{\pi}^{ab}(\mathbf{x})\Psi(\boldsymbol{\gamma},\boldsymbol{\phi}) = -i\delta\Psi(\boldsymbol{\gamma},\boldsymbol{\phi})/\delta\gamma_{ab}(\mathbf{x}) \ . \tag{2.1b}$$

The superspace versions of the constraints in (1.24) would then assume the form

$$\hat{\mathcal{H}}^a(\gamma_{bc}(\mathbf{x}),-i\delta/\delta\gamma_{bc}(\mathbf{x}))\Psi(\gamma,\phi)=0\ , \qquad a=1,2,3\ , \tag{2.2}$$

$$\hat{\mathcal{H}}(\gamma_{ab}(\mathbf{x}),-i\delta/\delta\gamma_{ab}(\mathbf{x}))\Psi(\gamma,\phi)=0\ . \tag{2.3}$$

The omission of a superscript in equation (2.3), which has become known as the *Wheeler-DeWitt equation*, is meant to indicate that the generic case with nongravitational sources is taken under consideration. A *formal* inner product[17] on the "Hilbert space" of this framework is given in the absence of nongravitational fields by the following functional integral:

$$\langle\Psi|\Psi'\rangle=\int\prod\nolimits_{\mathbf{x}\in\Sigma_0}\Psi^*[\gamma(\mathbf{x})]\,\Psi'[\gamma(\mathbf{x})]\prod\nolimits_{a\le b}d\gamma_{ab}(\mathbf{x})/\gamma^2(\mathbf{x})\ . \tag{2.4}$$

A Euclidean path-integral variant of the superspace approach to quantum gravity has been vigorously promoted for quite a long while by Hawking (1979-88). It is based on the *assumption* that the "transition amplitude" from a state with metric γ_1 and matter fields ϕ_1 on a surface Σ_1 to a state with metric γ_2 and matter fields ϕ_2 on a surface Σ_2 can be represented in the form of the "Euclidean path integral"

$$\langle\gamma_2,\phi_2,\Sigma_2|\gamma_1,\phi_1,\Sigma_1\rangle=\int\mathcal{D}[\gamma,\phi]\exp(-S_{\mathrm{E}}[\gamma,\phi])\ . \tag{2.5}$$

The above functional integral is supposed to be taken over all intermediate fields which have the above prescribed values on Σ_1 and Σ_2. In a later much-publicized and extensively cited proposal, Hartle and Hawking (1983) advanced the idea that the boundary manifolds in (2.5) should be identified with the spatial geometries in the Wheeler-DeWitt equation (2.3), and that the solution of that equation should be presented in the form (2.5), with the integration formally carried out *only* over all *compact* metrics in the Euclidean regime, and over all matter fields which are regular over the resulting compact manifolds. However, even at such a purely formal level, the indefiniteness of the Euclidean action S_{E} makes the functional integral in (2.5) ill-defined outside the semiclassical approximation.

In addition to difficult and unresolved mathematical problems, a central issue of these superspace-based approaches to quantum gravity is that of physical interpretation. In a paper reviewing related recent attempts at reconciling the observed very small upper bound on the cosmological constant with the enormous value to which the presence of Higgs bosons in the "vacuum" of conventional quantum gauge theories gives rise (cf. Secs. 11.12 and 12.3), Weinberg describes these difficulties as follows (with the notation adjusted to that in (2.5)): "Hawking has proposed (1984) that $\exp(-S_{\mathrm{E}}[\gamma,\phi])$ should be regarded as proportional to the probability of a particular metric and matter field history. It is not immediately clear what is meant by this – even supposing that we had a godlike ability to measure the gravitational and matter fields throughout space-time, it would be in a space-time of Lorentzian rather than Euclidean signature. ... [Other] questions still arise concerning the probabilistic interpretation of Ψ, particularly with regard to normalization ... because in [the] functional integral [in (2.5)] we are summing up possibilities which are not exclusive; if the universe has some γ and ϕ on some 3-surface, then it may also have some other γ' and ϕ' on some other 3-surface." (Weinberg, 1989, p. 15).

DeWitt (1967) and Wheeler (1968) were strong advocates of Everett's Many-World-Interpretation of quantum mechanics at the time they advanced their schemes for quantum gravity[18]. Subsequently Tipler (1986) forcefully advocated a Many-World-Interpretation (MWI) of the Hartle-Hawking "wave function of the universe", which is largely based on such schemes. In fact, in a recent monograph, MWI is promoted as being uncontestably *the* basic interpretation for quantum cosmology: "No additional laws [for wave function reduction] need be invoked if we adopt the MWI, for here all the points in the initial data space – classical universes – actually exist. The question of why does this universe rather than that universe exist is answered by saying that *all* possible universes do exist. What else could there possibly be?" (Barrow and Tipler, 1986, pp. 495-496).

However, in addition to violations of Ockham's principle (which the above cited authors were actually unsuccessful in refuting, but simply decided to dismiss), such an interpretation raises totally unresolved (and unresolvable) ontological problems: the idea of parallel universes might make for amusing science fiction, but what is the actual empirical *meaning* of assuming the *existence* of an uncountable[19] number of other universes? Thus, the "unsatisfactory status of the wave function of the Universe in canonical quantum gravity" remains very much of an open problem, as clearly pointed out in the following quotation: "The 'naive interpretation' obtained by straightforwardly applying the standard interpretive rules to the canonical quantization of general relativity is manifestly unacceptable; the 'WKB interpretation' has only a limited domain of applicability; and the 'conditional probability interpretation' requires one to pick out a 'preferred time variable' (or preferred class of such variables) from among the dynamical variables. Evidence against the possibility of using a dynamical variable to play the role of 'time' is provided by the fact (proven here) that in ordinary Schrödinger quantum mechanics for a system with a Hamiltonian bounded from below, no dynamical variable can correlate monotonically with the Schrödinger time parameter t, and thus the role of t in the interpretation of Schrödinger quantum mechanics cannot be replaced by that of a dynamical variable." (Unruh and Wald, 1989, p. 2598). The authors of this quotation consider the possibility of introducing a "coordinate time" (Unruh, 1989) in canonical quantum gravity. However, such a concept has no empirical support, and runs counter to the spirit of general relativity – so that, in fact, it is admitted that "our proposal fails to yield a quantum theory which corresponds classically to ordinary general relativity." (*ibid.*, p. 2599).

The fact that the Wheeler-DeWitt equation does not entail a concept of time is an issue which is at present hotly debated (Hartle, 1990). The fundamental questions asked are "about the nature of the 'intrinsic' time that is supposedly encoded into the γ in the $\Psi(\gamma,\phi)$" (Isham, 1987, p. 116). Another very fundamental epistemic question that can be also asked, however, is the following: even if it is granted that a $\aleph_1^{\aleph_1+}$ "number" (cf. Note 19) of universes *are* "coexisting", so that there is an actual MWI ensemble required for the probabilistic interpretation of $|\Psi(\gamma,\phi)|^2$, and that furthermore communication between them is somehow realizable for the sake of actually observing and recording the probabilities predicted by the Wheeler-DeWitt equation, at which stage of their *individual* geometrodynamic evolution are such "observations" to be performed? In other words, since the concept of *simultaneous* (!) access is meaningless in the absence of a "super-time" variable that manifests itself "simultaneously" as a global "intrinsic" time variable in each one of the infinity of $\aleph_1^{\aleph_1+}$ branch-universes, at which "instant" of the "intrinsic" time in his own universe does the individual "observer" in a given branch compare notes with his "doppelgängers" in all the other branches, for the purpose of empirically verifying his theoretically predicted values of $|\Psi(\gamma,\phi)|^2$ with the "actually observed" ones? In fact, any probability *density*, such as $|\Psi(\gamma,\phi)|^2$ is presumed to be, is by itself useless for computing probabilities: the *measure* with respect to which it is to be integrated over superspace also has to be supplied (cf. Note 16). Furthermore, an empirical method for determining those probabilities at a chosen value of a superspace parameter T should be supplied to all "observers" in a given superspace, so that they would know which of their measurement data refers to the *same* superspace configuration. But *what* does T stand for? Perhaps it stands for an "instant" – but what is the meaning of an "instant" in superspace? Perhaps it stands for a superspace configuration supplied by a specific (γ,ϕ) in a specific universe U – but how do the countless armies of observers in other branch-universes U' *know* which ones of *their* respective configurations (γ',ϕ') are to be compared with that specific (γ,ϕ) in U?

Other equally fundamental epistemological questions can also be asked: How does one measure the space metric over sub-Planckian "spatial" separations? How does one actually *measure*[20] topologies? And how can one *operationally* distinguish between them? For example, if one cannot empirically distinguish between those topologies which possess "wormholes" at the sub-Planckian length scale from those which do not, what about Born's maxims (cited in Sec. 1.1), which represent a modern version of Ockham's razor? Are we to overload our theories with *uncountable* infinities of mathematically distinct possibilities, between which no observational distinctions exist even in principle?

The purely mathematical problems are not any less severe. First of all, since not only the metric but, according to Wheeler's (1957) ideas about "wormholes" (cf. Note 3), the topology itself can fluctuate, how does one prescribe *all* possible topologies, and how does one parametrize them? Even if we assume that the topology is fixed, how does one define a Hilbert space of functionals in an uncountable number of degrees of freedom related to the problem of specifying the measure over which one integrates? And how does one actually define, in a mathematically meaningful manner, functional derivatives in such cases?

Instead of trying to tackle the apparently hopeless task of supplying answers to all these very fundamental epistemological and mathematical questions, researchers in this field have limited themselves to "toy models" (Hawking, 1984, 1985; Alvarez, 1989), in which the unwieldy *infinite* number of degrees of freedom is restricted on purely pragmatic grounds to a very small *finite* number of degrees of freedom[21]. This makes the resulting mathematics quite manageable, but conclusions reached by such a gross over-simplification are obviously even less reliable than those that would be reached by similar types of models in classical general relativity, whereby the *uncountable* infinity of points of a manifold would be replaced by a small finite number, with no hints provided as to how to even approach *realistic* models in which finiteness transcends into uncountable infinities.

In addition to the very problem of existence of a Hilbert space of functionals in an uncountable number of degrees of freedom, another open fundamental question that is the focus of attention of various schools of thought on that subject (Hartle and Hawking, 1983; Banks, 1985; etc.) lies in the very formulation of those degrees of freedom – a problem connected to the *ad hoc* manner in which the choice of boundary conditions is usually being made (Hartle, 1990). Therefore, in reviewing this state of affairs, the following has been recently pointed out by Alvarez (1989, p. 592):

"[I]n spite of very ingenious efforts of many people (including Hartle and Hawking as leaders of one of the most active groups), it is fair to say that there is not a single cosmological model valid in the quantum regime. This should not be surprising [since] we do not understand quantum gravity even perturbatively, and quantum cosmology is still more difficult, being essentially nonperturbative. One could even argue (see, for example, Barbour and Smolin, 1988) that we do not have an acceptable measurement theory for solutions of the Wheeler-DeWitt equation, and we therefore do not know whether or not quantum mechanics can be applied to the universe as a whole (see also Vilenkin, 1988)."

An entirely different type of approach to the quantization of gravity (DeWitt, 1967) has become known under the name of *covariant quantum gravity*. This approach assumes an already given Lorentzian manifold, so that the components of the quantum metric field can be then viewed as equal to the following sum (cf., e.g., Adler, 1982, Sec. VI),

$$\hat{g}_{\mu\nu}(x) = g^{\mathrm{B}}_{\mu\nu}(x) + \hat{h}_{\mu\nu}(x) \quad , \qquad x \in \mathbf{M} \quad , \qquad \mu, \nu = 0,1,2,3 \quad , \tag{2.6}$$

whose first term is the "classical" Lorentzian metric of that manifold, called the "background metric", and the second term is deemed to be a quantum "correction" – formally viewed as "operator"-valued function[22] on the given manifold **M**. For a while the covariant approach with a choice of background metric equal to the Minkowski metric enjoyed greater popularity than the canonical approach, since general coordinate invariance could be then formally embedded into the framework via a non-Abelian gauge group, so that the formal perturbative techniques based on the Feynman rules developed for Yang-Mills theories could be applied to it (cf., e.g., Duff, 1975, 1981; van Nieuwenhuizen, 1977). A host of questions concerning both the physical justifiability as well as the mathematical validity of this approach can be raised, but from the conventional point of view its fatal fault lies in the nonrenormalizability of the terms in its formal perturbation expansion (cf., e.g., Deser, 1975, Sec. 3). Hence, the following conclusion was ultimately reached: "The failure to combine the particle physics version of quantum theory and general relativity poses a fundamental problem, since gravitation undeniably exists as a force in nature. Either our quantum theory must be modified, or other gravitational models should be considered, or we must leave gravitation unquantized, which might be inconsistent according to a Bohr argument." (van Nieuwenhuizen, 1977, p. 24). In view of all that, for a while hopes were entertained that supergravity (cf., e.g., Wess and Bagger, 1983; West, 1986) might be able to supply models constructed in the same vein, and which would be renormalizable in the conventional sense – but those expectations were not fulfilled (cf. Ashtekar, 1991, p. 7).

For all these reasons, until the recent revival[23] of interest in a canonical formulation of quantum gravity based on "new variables" (discussed in suitable contexts in the next two sections), all the hopes for supplying a better framework for quantum gravity were pinned on superstring theory. At the surface[24], this was due to the fact that, at the level of formal perturbation theory, the vertices of diagrams in string theory contain exponential nonlocal factors that cause loops to converge in the Euclidean regime. However, the hope that string theory would provide the cure to the divergencies of quantum gravity has not been realized thus far – although it is claimed that the hopes are still high, despite the "remarkable drop of interest in [superstrings] in the past year or so" (Horwitz, 1990, p. 419).

The fact remains that superstring perturbation theory encounters a breakdown of uniqueness after the compactification to four spacetime dimensions is carried out (Narain, 1986). Moreover, it has been proven that its perturbation series not only is *not* convergent, but it is not even Borel summable (Gross and Periwal, 1988). On the other hand, the most recent review of quantum gravity acknowledges that "perturbative computations are the only ones we know how to do in superstrings, and even those are not fully understood for arbitrary genus" (Alvarez, 1989, p. 600). This review, otherwise very favorably predisposed towards string theory, concludes with the following admission: "It is true that we are not (yet) able to address the physically more interesting questions in quantum gravity. But this is mainly due to lack of technique (and probably also lack of some deep understanding of the fundamental physical principles)." (*ibid.*, p. 602).

Indeed, as far as the future prospects of superstring theory are concerned, even more telling is the huge methodological gap that lies between the manner in which classical general relativistic gravity theory was created and developed by Einstein, and the manner in which string theory originally emerged from the Veneziano model in the late 1960s, and was subsequently developed in the course of the 1970s and 1980s. For example, the author of one of the main textbooks on the subject points out the following:

"For a theory that makes the claim of providing a unifying framework for all physical laws, it is the supreme irony that [superstring] theory itself appears so disunited, ... [with] the fundamental physical and geometric principles that lie at [its] foundation... still unknown. ... By contrast, when Einstein first discovered general relativity, he started with physical principles, such as the equivalence principle, and formulated them in the language of general covariance." (Kaku, 1988, pp. viii, 5-6).

11.3. Basic Epistemic Tenets of Geometro-Stochastic Quantum Gravity

In the next chapter, we shall survey the historical developments which have led in the post-World War II years to a drastic shift in the methodology of developing relativistic quantum theories in general, and elementary particle models in particular – as compared to the methodologies followed by the founders of relativity and quantum theory in the first half of this century. From such a historical perspective, the present quantum geometry approach represents an effort to apply to relativistic quantum theories in general, and quantum gravity in particular, the same old, but nevertheless venerable epistemological methods, that were routinely employed by all the well-known founders of relativity and quantum theory, led by Einstein and Bohr, respectively. Their basic methodology consisted of building a new framework of ideas on a firm foundation of clearly stated fundamental premises, which reflected from the outset a cohesive philosophy – a *Weltanschauung*. Hence, such an approach to the foundational problems with which a theoretical framework was intended to deal most certainly did not leave, even in the initial stages of its development (not to mention a quarter-century later), all the "fundamental physical and geometric principles that lie at [its] foundation ... still unknown".

In keeping with that traditional but perpetually alive scientific spirit, the basic principles of quantum geometry were enunciated in Sec. 1.3. Subsequently, their mathematical implementation was carried out in the preceding eight chapters for the case where the gravitation is treated as an external classical field. The quantization of gravity *per se* presents new ontological and epistemological problems, to whose enunciation we now turn.

The general discussion in Sec. 1.3, as well as the one in the next section, suggests as a first tenet of GS quantum gravity that *any quantum spacetime should be formulated as a fibre bundle over a base manifold which is an affine frame bundle, and which incorporates a superspace of Lorentzian base metrics*. Such a superspace represents the *various* "classical metrics" compatible with the various distinct *quantum* initial conditions. Thus, as opposed, for example, to the superspace approach to canonical quantum gravity, the GS approach to quantum gravity incorporates from the outset Einstein's conclusions about the basic physical nature of general relativistic spacetime, as those conclusions were laid out by him in 1952, in a beautifully limpid form, in Appendix V of the fifteenth expanded (and final) edition of his popular exposition of CGR. There he states (Einstein, 1961, p. 155):

"On the basis of general relativity ... space as opposed to 'what fills space' ... has no separate existence ... If we imagine the gravitational field, *i.e.* the functions g_{ik}, to be removed, there does not remain a space of the type [of Minkowski space] but absolutely *nothing*, not even a 'topological space'. For the functions g_{ik} describe not only the field, but at the same time also the topological and metrical structural properties of the manifold ... There is no such thing as empty space, *i.e.* space without a field. Space-time does not claim existence on its own, but only as a structural quality of the field."

The first tenet of GS quantum gravity therefore incorporates the above epistemic idea that the concept of metric is fundamental to *any* concept of spacetime – including a quantum one. However, in keeping with the principle of irreducible indeterminacy (Principle 1 in Sec. 1.3), in a GS quantum spacetime the fluctuations of that metric manifest themselves even in regions where no matter fields might be actually present. That actually represents one of the foundational reasons for quantizing gravity – rather than settling for an extrapolation of the semiclassical treatment of the preceding six chapters, whereby a classical metric field would be coupled to the mean stress-energy tensor of nongravitational sources. According to an extrapolation of the GS quantum general relativity principle (Principle 3 in Sec. 1.3), which will be discussed in greater detail later in this chapter, those fluctuations around mean metric values will be provided by exciton states consisting of gravitons of mass-0 and spin-2. Such states will originate from GS gravitational fields in quantum fibres *above* base locations, which are coupled to nongravitational GS fields. Hence, such metric fluctuations can be deemed to originate from corresponding fluctuations in the nongravitational quantum stress-energy tensor in those regions where matter and nongravitational radiation are actually present and in mutual interaction.

The contemporary debates on the meaning of "time" in CGR, spurred by the lack of an intrinsic concept of "time" in the canonical quantum gravity, often do not seem to take into account Einstein's point of view, acquired over the span of the many years that it took him to arrive in 1916 at the final formulation of CGR. In fact, the implications of the above cited remarks seem to have been mostly ignored by the participants in these debates. Thus, whereas it might be justified to categorically state that "in quantum gravity there is no background spacetime" (as there is, for example, in the covariant approach – cf. Sec. 11.2), and that "spacetime itself is a quantum variable"[25], no operationally defined substitute for a 4-metric is proposed instead. On the other hand, as it will be shortly discussed, albeit a *globally* defined "background" metric is indeed at odds with certain basic features of quantum behavior, if there is no *four*-metric of *some* kind present in a general relativistic framework, then according to Einstein's point of view there is no spacetime in a general relativistic sense – be that classical or quantum.

In fact, the above quotation from Einstein's writings incorporates one of Einstein's deepest epistemological conclusions, which cannot be ignored by anybody who claims to construct some kind of quantum general relativity – as opposed to some type of theoretical framework divorced from Einstein's ideas. The protracted intellectual struggle, required to arrive at the above quoted fundamental conclusion, is best illustrated by the "hole" argument, which caused Einstein to temporarily abandon the principle of general covariance, and which delayed his completion of CGR by several years.

Indeed, Einstein and Grossmann argued in their well-known 1913 *Entwurf* paper that if the nongravitational stress-energy tensor vanishes identically within some region L

("Loch" in the German original – i.e., "hole" in English) of a classical spacetime manifold **M**, but it was fixed and nonzero outside L, then a coordinate transformation *inside* L could change the components of the metric tensor inside L despite the fact that the components of the stress-energy tensor would remain unchanged in the entire spacetime manifold **M**. They took that to imply that "if therefore... one maintains that the [metric tensor] should be completely determined by the [nongravitational stress-energy tensor], then one is forced to restrict the choice of the coordinate system" (Einstein and Grossmann, 1913, p. 260) – namely to give up the principle of general covariance. However, contrary to the impression, gained by some contemporary relativists, that this conclusion showed a lack of understanding on the part of Einstein of what "covariance" under coordinate transformation actually meant, the analyses by Norton (1987, 1989) and Stachel (1989) demonstrate that the "hole" argument actually established a most fundamental conclusion about the structural nature of CGR theories, which can be expressed in contemporary mathematical language by saying that all CGR theories have to be invariant under global diffeomorphisms of **M**.

This fact becomes very obvious as soon as one considers the customary manner in which the diffeomorphism-invariance of CGR is depicted (cf., e.g., Rovelli, 1991b): one introduces the space $\mathrm{Riem}_L\mathbf{M}$ of all Lorentzian metrics g^{L} over a 4-dimensional manifold **M**, and then one identifies[26] a classical general relativistic spacetime with one of the equivalence classes in the (homogeneous) quotient space $\mathrm{Riem}_L\mathbf{M}/\mathrm{Diff}\mathbf{M}$ – where $\mathrm{Diff}\mathbf{M}$ denotes the infinite-dimensional group of all diffeomorphisms of a given differential manifold **M**. In fact, at the mathematical level, such a quotient space arises naturally as a base space of the principle bundle with total space $\mathrm{Riem}_L\mathbf{M}$,

$$\mathrm{Diff}\mathbf{M} \rightarrow \mathrm{Riem}_L\mathbf{M} \rightarrow \mathrm{Riem}_L\mathbf{M}/\mathrm{Diff}\mathbf{M} \quad , \tag{3.1}$$

in which $\mathrm{Diff}\mathbf{M}$ plays the role of structure group (cf. [I], pp. 127 and 132). However, the natural physical justification for its introduction can be explained as follows: If $\psi\colon \mathbf{M} \rightarrow \mathbf{M}$ is a diffeomorphism, if $\{\boldsymbol{T}_\alpha\}$ represents the family of all (matter and radiation) tensor fields in a given CGR model, and if $\{x_\beta(t)\}$ represents the family of all test-particle worldlines in the same model, then one can impose the equivalence relation

$$(\mathbf{M}, g^{\mathrm{L}}, \{\boldsymbol{T}_\alpha\}, \{x_\beta(t)\}) \sim (\mathbf{M}, \psi_* g^{\mathrm{L}}, \{\psi_* \boldsymbol{T}_\alpha\}, \{\psi \circ x_\beta(t)\}) \quad , \tag{3.2}$$

since on the basis of Einstein's "hole" argument one can regard the above two *mathematically* distinct models as physically indistinguishable. Furthermore, as we saw in Sec. 11.1, the formulation of the initial-value problem in CGR is also incapable of distinguishing between the two solutions in (3.2). As a matter of fact, it cannot distinguish even between those solutions where $\mathbf{M}'$ is a manifold distinct from **M**, but diffeomorphic to it (cf. [W], Sec. 10.2). Consequently, (3.2) can be generalized into the equivalence relation

$$(\mathbf{M}, g^{\mathrm{L}}, \{\boldsymbol{T}_\alpha\}, \{x_\beta(t)\}) \sim (\mathbf{M}', \psi'_* g^{\mathrm{L}}, \{\psi'_* \boldsymbol{T}_\alpha\}, \{\psi' \circ x_\beta(t)\}) \, , \tag{3.3}$$

where $\psi' : \mathbf{M} \rightarrow \mathbf{M}'$ is any given diffeomorphism between two manifolds **M** and $\mathbf{M}'$.

All this demonstrates that "although the points of a [classical spacetime] manifold are *mathematically* individuated by a coordinate system, they are not *physically* individuated before one has a metric tensor field on that manifold." (Stachel, 1991, p. 35). Hence, in the fibre bundle approach to CGR (cf. Chapter 2), "the points of the base manifold do *not* represent physical events, which are instead represented by mappings from a point on the cross section of the bundle into a point of the base manifold" (*ibid.*, p.36).

This is a very important and basic point, which is misunderstood not only by those general relativists who claim that it was Einstein who showed in his "hole" argument a lack of understanding of general covariance, but also by those who unconditionally claim that the diffeomorphism group is *the* gauge group of general relativity – without drawing a clear distinction between a gauge group of the first kind, and a gauge group of the second kind. The need for making such a distinction clearly emerges from fibre-bundle formulations of gauge theories, such as those presented by Drechsler (1977b, 1984), Atiyah *et al.* (1978), and Trautman (1979). As emphasized especially by Trautman (1980-1982), this distinction is of crucial importance in understanding the key differences between classical gauge theories of the Yang-Mills type, and CGR treated as a gauge theory. Hence, let us first recapitulate the terminology which is of key relevance to this point – comparing it in the process with Trautman's terminology – and then explain the basic structural distinction between Yang-Mills and general-relativistic theories.

When considering a gauge theory based on a given principal bundle **P(M,G)**, Trautman calls the structure group **G** a "group of gauge transformations of the first kind". Hence, we shall henceforth refer to **G** simply as a *gauge group of the first kind.* Indeed, when some authors, such as Drechsler and Mayer (1977), Carmelli (1982), or Choquet *et al.* (1987), speak of a "gauge group", they have in mind exclusively a gauge group of the first kind. On the other hand, Trautman talks of a "group $\mathcal{G}$ of gauge transformations of the second kind", and describes it as being in general a subgroup of the group of automorphisms Aut**P** of the principal bundle **P(M,G)** (cf. Sec. 10.2) "preserving the absolute elements of a gauge theory" (Trautman, 1981, p. 422). Hence, we shall henceforth call $\mathcal{G}$ simply a *gauge group of the second kind.* Trautman does not give a general and precise criterion for those "absolute elements of a gauge theory" that decide what gauge transformations actually are – obviously so as to make allowances for the very diverse usages of this term in physics. Indeed, when some other authors, such as Göckeler and Schücker (1987), speak of a "gauge group", they have in mind exclusively the group $\mathrm{Aut}_{\mathbf{M}}\mathbf{P}$ of *vertical* automorphisms (which Trautman calls *pure* gauge transformations), or, equivalently the gauge group $\mathcal{G}(\mathbf{P})$ of the principal bundle **P(M,G)** defined in Sec. 10.2, which is isomorphic to $\mathrm{Aut}_{\mathbf{M}}\mathbf{P}$. When such a restriction is (implicitly or explicitly) imposed, then it indeed becomes true that "the diffeomorphism group [**DiffM**] is not a gauge group [in CGR] and [that the Levi-Civita connection] Γ is not its connection" (Göckeler and Schücker, p. 76). On the other hand, other authors unconditionally claim that "diffeomorphism invariance is a local gauge symmetry of general relativity" (Smolin, 1991, p. 443), and consequently insist, without qualifications, on "the replacement of the Poincaré group by the diffeomorphism group as the governing symmetry group of [any nonperturbative quantization of gravity] theory" (*ibid.*, p. 442).

Part of the explanation for this divergence of opinions can be ascribed to the fact that "there is no unique 'gauge theory of gravitation'" (Trautman, 1980, p. 305), if no guiding principles based on clearly stated operational principles are used, and if only the purely formal mathematical content of CGR is examined for the purpose of its comparison with gauge theories of the Yang-Mills type. Indeed, in that case "the most important difference between gravitation and other gauge theories is due to the soldering of the bundle ***GLM*** of [linear] frames to the base manifold **M**" (*ibid.*, p. 306). This "soldering" is present in CGR on account of the canonical form **θ** in (2.6.4) – called by Trautman and others the "soldering form" of the bundle. In accordance with (2.6.5), this form assigns to each vector **X** tangent to *GL***M** at one of its points $\boldsymbol{u}$, represented by a linear frame in $T_x\mathbf{M}$, $x = \Pi(\boldsymbol{u})$, the components with respect to that frame $\boldsymbol{u}$ of its projection $\Pi_*\mathbf{X}$. Hence: "The soldering form **θ** leads to torsion which has no analog in nongravitational theories. Moreover, it affects the [CGR gauge group of the second kind] $\mathcal{G}$, which now consists of the automorphisms of *GL***M** preserving **θ**. This group contains no vertical automophisms other than the identity; it is isomorphic to the group **DiffM** of all diffeomorphisms of **M**." (*ibid.*, p. 306). And indeed, according to the general definition in Sec. 10.2, any automorphism of a principal bundle **P(M,G)** is given by a pair (ϕ,ψ) of diffeomorphisms ϕ: $\mathbf{P} \to \mathbf{P}$ and ψ: $\mathbf{M} \to \mathbf{M}$, so that if ϕ equals the identity transformation, then the automorphism can be identified with the diffeomorphism ψ of **M** onto itself.

The above mathematical analysis of the term "gauge group" is totally concordant with the observation that, in CGR, "a space-time (manifold with a metric) corresponds to a gravitational field; but a gravitational field corresponds to an *equivalence class* of space-times. ... This is the true significance of the concept of general covariance... . Unfortunately, this profound insight of Einstein was originally expressed in the language of coordinate systems, rather than in the coordinate-free language of modern formulations of differential geometry. This was responsible for a great deal of misunderstanding of the subject." (Stachel, 1987, p. 204).

As we shall see in the next section, this "misunderstanding" persists to the present day, and is the cause of much of the contemporary debate over what is "observable" in CGR. However, Trautman's differentiation between gauge transformations of the first and of the second kind certainly clarifies the confusion over the true significance of Einstein's "hole" argument, as well as the meaning and implications of the principle of general covariance in CGR. In fact, it also implicitly reveals that in CGR the "*points of space-time may be represented not by points of the base manifold as is usually done, but by a mapping of points of a cross-section of the bundle into points of the base manifold* [emphasis added]. Events are then defined as sets of such mappings in the equivalence class of cross-sections corresponding to a unique gravitational field. Thus, if there is no gravitational field, there are no events." (*ibid.*, p. 207). In other words, it is *not* the classical spacetime manifold **M** that is of direct physical significance; rather there are certain principal bundles **P(M,G)** over **M** (or any diffeomorphic images of **M**), which together with some of their associated bundles embody the physics of general relativity.

Upon asking the question "What is the structure group **G** of the gravitational principal bundle **P**?", Trautman presents the following reasons for viewing either the Lorentz group or the Poincaré group as gauge groups of the first kind in CGR: "Since space-time **M** is four dimensional, if **P** = *GL***M** then **G** = GL(4,**R**). But one can equally well take for **P** the bundle *GA***M** of affine frames; in this case **G** is the affine group [namely GA(4,**R**) in (2.3.9)]. There is a simple correspondence between affine and linear connections which makes it really immaterial whether one works with *GL***M** or *GA***M** ([K], Trautman, 1973b). If one assumes – as usually one does – that [the connection form] $\boldsymbol{\omega}$ and [the metric] $\boldsymbol{g}$ are compatible, then the structure group of *GL***M** or *GA***M** can be restricted to the Lorentz or the Poincaré group, respectively." (Trautman, 1980, p. 306).

All this shows that opinions, such as the earlier cited one, to the effect that the role of the Poincaré group should be taken over in quantum gravity by the diffeomorphism group, have no valid mathematical foundation in general relativity. It can be also readily seen that they have no valid physical foundation either, if one takes into consideration the following key points: (i) on one hand, in CGR Lorentzian gauge invariance is *required* by the equivalence principle, whereas, on the other hand, according to the above cited explanatory remarks of Trautman, such Lorentzian gauge invariance can be always extended into Poincaré gauge invariance – as explicitly shown in Sec. 2.6; (ii) as we ordinarily observe them, the geometric relationships in the universe around us comply to a very high degree of accuracy with the equivalence principle, as well as with the existence[27] of a *mean* classical metric, so that quantum fluctuations around those mean values must indeed be in some sense "very small", rather than being so large as to totally invalidate semiclassical approximations; (iii) if those quantum fluctuations are embodied (as is commonly believed) in the behavior of gravitons of spin-2 and mass-0, then the presence of Poincaré gauge invariance is absolutely necessary, since it *locally* supplies the required irreducible representations of the $ISL(2,\mathbf{C})$ group that reflect a most fundamental principle of elementary particle physics, according to which all elementary quantum objects are associated with such representations.

Thus, any downgrading of the role of the Poincaré group in quantum gravity cannot be justified either on mathematical or on physical grounds, and it represents merely a rationalization of the impossibility of embedding Poincaré gauge invariance into some particular method of quantizing gravity. The GS framework does not have this problem, since it has been conceived on firmly grounded physical principles, which cast the equivalence principle in a central role. Its basic philosophy is based on a search for the *right* mathematics to serve those principles, rather than on the modification of any of the most basic physical principles carried out in order to serve some *ad hoc* mathematical scheme. Hence, the second tenet of QGR is that *the GS quantum gravity gauge group of the first kind has to incorporate phase space representations of the Poincaré group that describe spin-2 and mass-0 gravitons, as well as all additional internal gauge group elements, which can be interpreted as infinitesimally local four-dimensional diffeomorphisms* (cf. Secs. 11.7-11.9).

To explain why only "infinitesimally local" forms of diffeomorphisms (cf. (7.7)-(7.10) in Sec. 11.7) can be deemed of relevance in QGR, it has to be first pointed out that a *global* base-geometry of a quantum spacetime, in which matter and geometry are in *mutual* interaction, *cannot* be provided a priori[28]. This is an inescapable epistemic conclusion if, on one hand, quantum gravity is meant to be applicable to the universe in its entirety, and, on the other hand, the *freedom* to locally perform "quantum state preparations" and "detections-cum-registrations" (which are integral parts of *active* observational processes constituting an "experiment") is to be allowed within a QGR framework – thus reflecting physical reality as it is ordinarily conceived in quantum theory. Indeed, by their very nature, such experimental procedures necessarily *change* the existing states of the "system" and of the "apparatus", since they give rise to "reductions of wave packets" – and therefore necessarily affect the geometry, and possibly even the topology, of a quantum spacetime. The fact that such changes might be merely local, and extremely minuscule from a cosmic perspective, is not decisive from the point of view of principle: they *do* occur, and therefore they *do* locally affect the geometry. Furthermore, they subsequently propagate quantum mechanically, and taken in conjunction with each other, they can be cumulative, so that they might even eventually give rise to large-scale changes[29].

Consequently, the GS quantum gravity framework presented in (Prugovečki, 1989b) incorporated from the outset the basic tenet that *neither the geometry, nor the topology of a quantum spacetime can be given a priori, but that it emerges gradually from the prescribed initial conditions within a base-segment* (cf. Secs. 11.1 and 11.5) during a process of geometro-stochastic evolution of a quantum gravitational superfibre bundle. When the universe is considered in its entirety, such GS evolution gives rise to a *universal wave function,* which constitutes the GS counterpart of the "wave function of the universe" introduced by Hartle and Hawking (1983). In the last section of this chapter we shall argue that "reduction-of-the-wave-packet" changes might be interpreted as *local informational* and *organizational* changes, which do not affect globally the *universal* wave function, but only our *perception and treatment* of its *local* features. Nevertheless, even if that thesis proves correct and fruitful, the GS quantum gravity framework has to be sufficiently large and flexible to be able to accommodate them at the local level. In other words, if we borrow evocative terminology from Prigogine (1980), we can say that *a quantum spacetime should never be envisaged in a state of "being", but only in a perpetual state of "becoming".*

The underpinnings of the third epistemic tenet of GS quantum gravity can be actually found already in the way the initial-value problem is treated in CGR (cf. Sec. 11.1), as well as in the way CGR models are actually developed – as opposed to the way such developments are usually presented in textbooks. In fact, this latter distinction was recently emphasized by J. Stachel, as he presented an analysis of Einstein's views, in order to draw "some lessons from general relativity". Towards the end of that analysis, he points out the following: "Even relativists have not fully adopted [Einstein's] point of view, 'no metric, no nothing'. ... They start out by introducing a global manifold ..., and then put such structures on this manifold as the metric tensor field. Is that the way any one of us actually goes about solving the field equations of general relativity? Of course not. One first solves them on a generic patch, and *then* one tries to maximally extend the local solution ... from that patch to a global manifold which is not known ahead of time."[30]

If we do *not* adopt the deterministic ontology which is implicit in CGR, according to which theoretically the entire spacetime geometry "already exists", in its entirety, in the form of an equivalence class of Lorentzian manifolds $(\mathbf{M}, \boldsymbol{g}^{\mathrm{L}})$, then there is neither a physical nor an epistemological justification for assigning to such a *global* concept as the group $\mathrm{Diff}\mathbf{M}$ any *physical* significance, since no $\mathbf{M}$ *exists* either singly, or as a member of an equivalence class. In other words, the *real* spacetime is constantly being *generated,* but ontologically there *is* no spacetime in a global sense – although, of course, there might be theoretical *extrapolations* of what $(\mathbf{M}, \boldsymbol{g}^{\mathrm{L}})$ "might eventually look like" when the Omega Point is reached (Barrow and Tipler, 1986), provided that such a conjecture can be at all *reliably* inferred from some future pertinent observational data.

The fact that the diffeomorphism-invariance of CGR is totally dependent on a deterministic outlook becomes very obvious as soon as the earlier described customary manner of its implementation is reconsidered: in formulating the equivalence relation in (3.2), a manifold $\mathbf{M}$, a Lorentzian metric $\boldsymbol{g}^{\mathrm{L}}$, a set of tensor fields $\{\boldsymbol{T}_\alpha\}$, and a family $\{x_\beta(t)\}$ of test-particle worldlines are supposed to have been already *determined.* Without that neither (3.2), nor its generalization (3.3), would make physical or mathematical sense. Hence, *global* diffeomorphism invariance makes sense only if strict determinism is accepted.

Global invariance under "spatial" 3-diffeomorphisms might be deemed to provide an alternative, at the recommendations of the advocates of the canonical approach to quantum gravity. However, such invariance presupposes the existence of "the marvelous moment *now*" (Hartle, 1991, p. 178), namely of a global "instant" at which such diffeomorphisms are to be implemented. Mathematically, there is no problem with such a hypothesis. On the other hand, if the *physical* existence of a global "time" is assumed, that has to be reflected in the possibility of *operationally* specifying how each such "marvelous moment *now*" can be observed. Not only are such schemes not in existence, but a search for them would be foreign to the entire spirit of general relativity, and would lead to the type of paradoxes already discussed in Secs. 7.2 and 7.3. Of course, it can be argued that the *classical* cosmological models that are at present most popular do display a preferential slicing of spacetime based on their symmetry features that receive observational support at very large supragalactic scales. However, such arguments most certainly do not have any observational justification at the subgalactic level – not to mention at the microscopic level.

The existence, in the GS approach to QGR, of the fundamental length ℓ equal to the Planck length suggests that the foliation procedure underlying the canonical approach should be replaced with a segmentation procedure (cf. Sec. 11.1), since no operational differentiation can be then made between segments which can be traversed in less than a Planck *proper* time span. The formulation of GS quantum gravity in terms of bundles over (affine) frame bundles which have such segments as base manifolds, rather than directly over the base-segments themselves, automatically insures diffeomorphism-invariance within each one of those segments. At the semiclassical level, this can be extended into diffeomorphism-invariance between segments corresponding to distinct foliations of a classical spacetime manifold – cf. (5.31). Hence, the diffeomorphism-invariance will not turn out to be the central issue as long as it is limited to physical events that have taken place in a given "era" of a GS evolution, rather than in a fictional *global* spacetime manifold.

Thus, if an operational outlook that is in the spirit of Einstein's seminal articles on relativity theory is adopted, it becomes clear that in any formulation of quantum gravity, which abides by the basic tenets of the quantum theory of measurement, a concept of general covariance based either on *global* diffeomorphisms in 4-manifolds $\mathbf{M}$, or in 3-manifolds Σ, cannot play any significant *physical* role. In fact, once the same choice of "observables" is adopted in CGR as the one originally made by Einstein (1905, 1916), then the same observations can be applied to the role played by $\mathrm{Diff}\,\mathbf{M}$ viewed as a group of *active* gauge transformations of the second kind, as were in the past applied to coordinate transformations viewed as their local *passive* counterparts. Indeed: "As Kretschmann pointed out in 1917, the principle of general covariance has no physical content whatever: it specifies no particular physical theory; rather, it merely expresses our commitment to a certain style of formulating physical theories." (Friedman, 1983, p. 55).

Thus, the allocation to the diffeomorphism group of the role of "gauge group" in CGR represents a choice in its "style"; whereas its physical substance, already present in the formulation of CGR originally conceived by Einstein (1916), is primarily reflected in the *underlying* choice of the Lorentz group or of the Poincaré group as gauge groups of the first kind. It is these gauge groups of the first kind that embody the equivalence principle – and which, as such, have direct *observational* consequences. Of course, this issue is closely related to the question of what are the "observables" of CGR – as well as of QGR.

11.4. Observables and Their Physical Interpretation in CGR and QGR

A great amount of attention is being devoted at the present time to the question of "physical observables" in the new formulations of canonical quantum gravity that have emerged most recently[31]. However, in many of these studies Einstein's own point of view as to what are the basic *measurable* quantities in general relativity seems to be ignored, and a conventional protocol with regard to the meaning of "physical observable" is substituted instead, whose roots do not actually lie in CGR. Indeed, as we shall shortly demonstrate in this section, recent suggestions to the effect that "*the* observables" of CGR should be identified with functions on a so-called general relativistic "phase space" represents a departure from the basic measurement-theoretical principles on which general relativity was built by Einstein. Moreover, this contention does not appear to have any deeper epistemological or empirical justification that might have escaped the attention of Einstein, or that might be essential to quantum gravity *per se*[32]. Rather, the advocacy of such an identification represents a transference of the basic ideas of the canonical approach to *non*relativistic quantum mechanics, discussed at the end of Sec. 11.1, to the general relativistic regime, in order to foundationally justify the adoption of recent technical innovations in CGR (Ashtekar, 1987, 1989), relying on the diffeomorphism group as the only "gauge group" of importance in the formulation of new mathematical techniques in the quantization of gravity.

On the other hand, Einstein had made it perfectly clear in a well-known review paper, in which CGR was presented for the first time in a cogent and finished form, that "*the results of our measurings are nothing but verifications of [space-time coincidences which are] meetings of the material points of our instruments with other material points*, [such as] coincidences between the hands of a clock and points on the clock dial, and observed point-events happening at the same place at the same time" (Einstein, 1916, 1952, p. 117) – emphasis added. Upon using this very basic observation as a justification of the introduction of his principle of general covariance, in §4 of the same paper Einstein takes advantage of the equivalence principle to advance the thesis that "the quantities $g_{\tau\sigma}$ are to be regarded from the physical standpoint as the quantities which describe the gravitational field in relation to the chosen system of reference". Einstein then concludes that section on the CGR theory of measurement with the key observation that "according to the general theory of relativity, gravitation occupies an exceptional position with regard to other forces ..., since the ten functions representing the gravitational field at the same time define the metrical properties of the space measured." There is absolutely no mention in that entire paper of some other basic "physical observables" for the gravitational field, despite its otherwise indubitably operationalistic attitude – as exemplified by the statement that, on account of the reasons given in preceding quotation, "the requirement of general co-variance, takes away from space and time the last remnant of physical objectivity" (*ibid.*, p. 117).

As is well-known, and as it is further discussed and analyzed in Sec. 12.1, in his later years Einstein became a "physical realist", who repudiated some of the philosophical implications of the statement that spacetime has no "physical objectivity". Nevertheless, even in his autobiographical notes, written more than three decades later, Einstein (1949) never retracted the contention that spacetime coincidences are, in effect, the only ultimate *manifestation* of the physical reality which underlies the principle of equivalence, and which constitutes the basis of the principle of general covariance in its modern form, discussed in the preceding section. Indeed, functional operationalism and physical realism do not have to be at odds with each other, if advocated wisely and practiced judiciously.

Extensive studies[33] in the 1960s and 1970s of the operational aspects of CGR have succeeded in clarifying and axiomatizing Einstein's rather informal approach, as well as in developing new measurement-theoretical ideas (such as those about "geodesic clocks" – cf. Marzke and Wheeler, 1964; Harvey, 1976). However, they did not alter in any fundamental manner Einstein's original conceptualization of what is measurable in CGR. In particular, they did not uncover any fundamentally new observable aspects of the classical gravitational field in CGR. Hence, the basic inference drawn from Einstein's epistemology, that underlies CGR, was left untouched: the fundamental *gravitational observables* of general relativity have to be operationally based; as such, they should be defined exclusively in terms of the spacetime coincidences mentioned in the quotation from Einstein's key 1916 paper, that was italicized in the introduction to this chapter. Hence, for pure gravity, they are given exclusively in terms of the ten components of the metric in relation to suitable local frames of reference, that locally describe the spatio-temporal relationships between those coincidences. It was this most fundamental realization which constituted the underlying reason for Einstein's earlier cited observation that if in CGR "we imagine the gravitational field, i.e., the functions g_{ik} to be removed, there [remains] ... absolutely *nothing*, not even a 'topological space'." It then necessarily follows that all the fundamental *nongravitational observables* of CGR are *local* entities (such as 4-velocities, 4-momenta, the components of the nongravitational stress-energy tensor, etc.), whose mathematical representatives belong to the fibres of various bundles associated to the general linear frame bundle ***GLM***, or its affine counterpart ***GAM*** (cf. Secs. 2.2- 2.3).

The question can be therefore asked whether some new kind of "observables", in the sense of "smooth functions on the phase space of the theory [which, if physical] have vanishing Poisson brackets with the constraints" (Rovelli, 1991b, p. 301), might exist in CGR, despite the fact that in the case of "the vacuum Einstein equations in the spatially compact case, *not a single physical observable is known explicitly as a function of phase space variables.*" (Smolin, 1991, p. 447) – italics as in the original. This question is especially pertinent in light of the fact that the italicized part of this statement represents a reiteration of the observation that "one obvious difficulty with this [kind of] approach is that thus far no one has been able to discover a single classical observable" (Bergmann and Komar, 1980, p. 246) which is represented by a first class variable on the phase space of a CGR theory. Since the latter statement was actually made more than a decade prior to the previous one, this lack of explicit examples is obviously not due to the lack of serious attempts at trying to find such "physical observables". Rather, it reflects some deeper features of general relativistic theories, related to their underlying physical nature.

The origin of the above cited "difficulties" can be in part traced to the presence of constraints in the canonical formulation of gravity. For example, in trying to adapt the *non*-relativistic canonical quantization approach to the general relativistic regime, it has been postulated that, on one hand, "in the quantum version of a general-relativistic theory only observables should play a role", whereas, on the other hand, in CGR "only first-class variables are observable" (Bergmann and Komar, 1980, p. 243) – where, in accordance with Dirac's (1959, 1964) classification, a first-class canonical variable is defined as one whose Poisson brackets with the Hamiltonian constraints given by (1.13) vanish identically. However, as pointed out in Sec. 5 of (Bergmann and Komar, 1980), such a definition removes the status of "observable" from the *most basic* of Einstein's *truly* observable quantities in CGR, namely from the metric tensor. In fact, even in the most elementary but very basic case of the nonrelativistic mechanics of a single particle, if, as done by Bergmann and Komar (1980, p. 247), "constraints" are imposed in the classical regime in the form

$$\mathcal{H}_0(p_0,\mathbf{p}) := p_0 - (\mathbf{p}^2/2m) = 0 \quad , \tag{4.1}$$

and in the quantum regime by means of the Schrödinger equation expressed in the form

$$\mathcal{H}_0(P_0,\mathbf{P})|\mathbf{q},t\rangle = 0 \ , \qquad \mathcal{H}_0(P_0,\mathbf{P}) = i\,\partial/\partial t + (\nabla^2_{\mathbf{q}}/2m) \ , \tag{4.2}$$

then the position operators with the spectral measure in (3.1.6), commonly defined by multiplication with the components of $\mathbf{q}$, no longer qualify for their usual status of observables, since we have

$$\{q^j,\mathcal{H}_0\} = p_j/m \neq 0 \, , \qquad \{p_j,\mathcal{H}_0\} = 0 \ , \qquad j = 1,2,3 \ , \tag{4.3}$$

$$[Q^j,\mathcal{H}_0] = iP_j/m \neq 0 \ , \qquad [P_j,\mathcal{H}_0] = 0 \ , \qquad j = 1,2,3 \ , \tag{4.4}$$

in the classical and in the quantum case, respectively. To rectify this glaring inconsistency with orthodox theory, the artificial concept of "quasiobservable" has to be introduced, so that the nonrelativistic position operators can be then reinstated at least to the partial status of "quasiobservables", and so that the conjecture can be then advanced that in canonical gravity the twelve canonical variables (γ_{ab} ,π^{ab}), discussed in Sec. 11.1, are all at least "quasiobservables".

Clearly, the motivation for the recent reconsideration of CGR "observables" which are "smooth functions on the phase space of a theory" and are "gauge-invariant" under the diffeomorphism group **DiffM**, despite their mathematically elusive and physically unwarranted status, does not reside in CGR itself – which has not required them for its *successful* physical interpretation during the past eight decades, since its inception. Indeed, the *true* observables in CGR, which have served from the beginning (Einstein, 1916) in the prediction and empirical verification of *observable* effects in CGR, are closely related to *some* of the entities that occur prominently in the equivalence relations in (3.2) and (3.3). Thus, they are *not* **M**, *nor* its elements, *nor* its differential-manifold structure, and *generically not* even the holonomic metric components $g_{\mu\nu}$– all of which are indispensable *mathematical* objects, reflecting the manner (i.e., "style") in which CGR is formulated, but of no direct physical significance. Rather, the basic observables in CGR are the *non*holonomic metric components g_{ij} with respect to certain *local* classical frames (cf. Sec. 11.6), the components of various tensors $\boldsymbol{T}_\alpha$ (stress-energy, angular momentum, matter and nongravitational radiation fields, etc.) with respect to the same frames, and the *relative* positions of the test particles which are in the immediate neighborhoods of the locations of such frames – with those in free-fall playing an especially important role in providing the best-known directly observable predictions in CGR (cf., e.g., [W], Secs. 5.3 and 6.3). As correctly pointed out by Rovelli (1991b, p. 304), the fact that all the frames and all the test particles are *material* objects implies that in such an interpretation we have to "neglect" their energy-momentum tensor in the Einstein equations, as well as their contributions to the dynamical equations for matter fields. However, this problem is common to all field theories. It has always been resolved by envisaging a limiting procedure in which the rest masses, charges, etc., of "test bodies" tends to zero. If the geodesic postulate is not incorporated in the CGR framework, then this can create internal consistency problems[34]. On the other hand, if the geodesic postulate *is* incorporated, as in the original formulation of CGR by Einstein (1916), then these entities provide all the "observables" in CGR. In that case the question of diffeomorphism-invariance becomes, just as it was the case with Einstein's (1916) original formulation of general covariance from the "passive" point of view of coordinate transformations, a purely mathematical question of choice amongst *mathematically* equivalent ways of formulating a given CGR model. Such an equivalence cannot be, however, subjected to empirical verification, and therefore is devoid of true *physical* meaning.

All this shows that the motivation behind the renewed search for some "physical observables" invariant under **DiffM** cannot possibly reside in the interpretation of CGR. Rather, it resides exclusively in their conceived usefulness to some new approaches to the canonical quantization procedure, such as those based on "loop representations" (Rovelli and Smolin, 1988, 1990). These types of approaches are based on purported "quantum observables" represented by "diffeomorphism-invariant operators", for which, however, "the problem is that we do not know how to make a correspondence between any of them and the classical diffeomorphism-invariant observables." (Smolin, 1991, p. 468).

In assessing the justifiability of applying the term "observable" to any abstract mathematical object, for which neither measurement schemes nor correspondence with classical counterparts has been established, one has to bear in mind the fact that in quantum theory the term "observable" has undergone very drastic changes during its historical evolution in the post-World-War II era. Since it is stated in this context that[35] "following Dirac, we call these observables gauge-invariant observables, or physical observables, [or] 'Dirac observables' ", it is useful to first examine some of the original sources that might provide the justification for such terminology.

The formulation of the term "observable" originally provided by Dirac (1930), and retained by him even in the later editions of his well-known monograph on quantum mechanics, stipulates the following criteria: "We call a real dynamical variable whose eigenstates form a complete set an *observable.* Thus any quantity that can be measured is an observable. The question now presents itself – Can every observable be measured? The answer theoretically is yes." (Dirac, 1947, p. 37). Thus, in modern mathematical language, any quantum observable *has* to be represented by a self-adjoint (and not just symmetric) operator, since by the spectral theorem [PQ] for self-adjoint operators, to any such operator can be assigned a unique spectral measure, which formally (cf. Sec. 12.3) provides the "eigenstates" forming "a complete set". On the other hand, in order to qualify as an observable, such an operator has to be *measurable*. Probably for this reason, in his path-breaking papers on Hamiltonian dynamics with constraints, and on CGR in Hamiltonian form, Dirac avoids[36] the use of the term "observable" for the "Hamiltonian variables" which are functions of the "basic variables q's, p's and v's" (Dirac, 1950). Moreover, in his well-known lectures based on this work, he talks only of "dynamical variables" (Dirac, 1964, p. 8), of "Hamiltonian variables" (*ibid.*, p. 14), and of "dynamical coordinates" (*ibid.*, p. 53).

There is no mention in the seminal work by Bohr, Born, Dirac, or any of the other founders of quantum mechanics of any alternative definitions of what is a quantum observable. Of course, in theories with clear-cut gauge degrees of freedom, it can be expected that all the basic observables would be gauge-invariant in the absence of some gauge-breaking mechanism. However, that leaves open the epistemic question of which comes first in the developmental stages of a new physical theory: are all the *basic* observables of the theory to be prescribed first, so that what constitutes "gauge freedom" is then decided afterwards, or should the developmental process of a theory proceed the other way around?

A general answer is certainly not possible, since the history of physics contains examples of both kinds: in Maxwell's theory, which was founded in an era epistemologically dominated by classical realism (cf. Sec. 12.1), there is no doubt that it was the electromagnetic field which supplied the basic "observables" from the outset, and the idea of

U(1)-gauge freedom came much later. On the other hand, in Yang-Mills theories, which were developed in an era dominated by a conventionalistic form of instrumentalism (cf. Sec. 12.2), there is no doubt that the idea of SU(2)-gauge freedom came first. Thus, philosophical outlooks can strongly condition such developments – albeit in contemporary physics these influences are more often covert rather than overt (cf. Sec. 12.7). The only general conclusion that can be drawn is that the conventionalistic instrumentalism of some contemporary physicists tends to favor the second approach, whereby gauge groups are chosen first, and the question of observables is settled afterwards – if at all; whereas the "positivism" of Einstein (1905-1916), which he clearly displayed while he was developing relativity theory (cf. Sec. 12.1), as well as that of Bohr, which was clearly exemplified in his treatment of quantum electrodynamics (Bohr and Rosenfeld, 1933, 1950), certainly favored the first approach. In that traditional approach, what are the *basic* observables of a physical theory is decided[37] in its initial developmental stages, so that the pinpointing of all the residual gauge degrees of freedom in such cases comes afterwards.

Indeed, as we have seen from the brief recounting of Einstein's "hole" argument at the beginning of the preceding section, what were the basic "observables" in CGR was certainly decided during a stage when Einstein could be classified as a "positivist", as he was under the strong influence of Mach. That *fundamental* decision was of such paramount importance that for several years Einstein was willing to sacrifice the principle of general covariance for its sake. In fact, that decision was the source of most of the physical intuition which Einstein achieved while developing CGR by means of his well-known *gedanken* measurement procedures and experiments (Einstein, 1905, 1916, 1949; Einstein and Infeld, 1938). The principle of general covariance, formulated in its original "passive" form, *followed* that conceptual development, whereas its explicit formulation in its modern "active" form, based on the diffeomorphism group, came much later. Hence, attempts which try to reverse that very natural epistemological development represent an, explicit or implicit, philosophical outlook that is totally different from the one that underlays Einstein's epistemology, as he developed CGR over the span of almost an entire decade.

The philosophical origins and basic tenets of this outlook are described in Sec. 12.2, in their proper historical setting. For the sake of the present analysis it suffices to note that already during its early post-World War II stages it led to the establishment and eventual total dominance of a purely conventionalistic type of instrumentalism in relativistic quantum theory. One of its telltale manifestations lay in modifications of basic quantum terminology, which began to evolve in new directions, whereby form and convention completely won over the deep concern of many of the founders of relativity and quantum mechanics with epistemological considerations and foundational analyses.

The concept of "observable" provides an example pertinent to the present discussion: in relativistic quantum field theory it "progressed" to the point of *postulating* in the 1960s the existence of "algebras of local observables", in terms of which quantum field theories can be purportedly formulated without any appeal to the type of quantum fields first introduced by Dirac (cf. Secs. 7.6 and 9.6; as well as [BL], Part 6, and Horuzhy, 1990). Thus, during this evolutionary "progress" towards a purely instrumentalist point of view, quantum "observables" became merely ingredients of abstract formalisms[38], which were totally divorced from the question of their physical interpretation, and of the unambiguously specified operational procedures for their measurement, on which Bohr placed so much value. This fact is recognized in some of the more perceptive contemporary analyses of how the quantization of gravity should be approached – of which the one containing the following quotation is representative: "Bohr's insistence on the primary role of classical ideas in quantum theory was not without a point. For example, it is not trivial to formulate a well-defined 'measurement theory' that does not employ some kind of fixed reference frame for its rods and clocks. Traditional 'Copenhagen' instrumentalism seems especially reliant on such a classical background, but attempts to replace it with a more realistic interpretation produce difficulties of their own. This problem is particularly acute in quantum cosmology, and involves the whole question of how quantum theory is to be interpreted in such situations – an issue that is still much debated (Gell-Mann and Hartle, 1990, Hartle, 1991, Penrose, 1987)." (Isham, 1991, p. 360).

In view of all this, a less arbitrary formulation of the term "observable", which has demonstrated so much apparent "flexibility" after the dawn of the conventionalistically instrumentalist era in quantum physics (cf. Chapter 12), seems not only warranted, but absolutely necessary if physically sensible conclusions are to be eventually reached.

In nonrelativistic quantum mechanics, it was von Neumann (1932) who first explicitly *postulated* that (i) every observable of a quantum system can be related to a self-adjoint operator in a Hilbert space $\mathcal{H}$, and (ii) any self-adjoint operator in that Hilbert space of the system represents an observable. The existence of various types of superselection rules (Wick *et al.*, 1952; Hegerfeldt *et al.*, 1968; Zurek, 1982; Omnès, 1990) certainly proves that (ii) is generally false. The *belief*[39] nevertheless persists that, at least in the absence of superselection rules, postulate (ii) could be still maintained, whereas in their presence, (ii) remains "true" in the superselection sectors (i.e., coherent subspaces) compatible with those existing superselection rules – which customarily are assumed to be discrete.

As discussed in Sec. 7.6 (cf. especially Note 27 to Chapter 7), the fallacy of this claim becomes evident once a physically sensible operational restriction is imposed on the term "observable", by requiring that *at least one* type of apparatus for its *actual*[40] measurement should be *proven* to exist. Indeed, as pointed out by Wigner (1963), "for some observables, in fact for the majority of them (such as xyp_z), nobody seriously believes that a measuring apparatus exists" (cf. [WQ], p. 338). He then goes on to state, on the same page of the same article: "On the other hand, most quantities which we believe to be able to measure, and surely all the important quantities such as position, momentum, fail to commute with all the conserved quantities [such as total angular momentum], so that their measurement cannot be possible with a microscopic apparatus."

Naturally, when Wigner made the above statement, he had in mind[41] the possibility of measurement of *sharp* values of position, momentum, etc. – namely, if not literally perfectly "sharp" outcomes, at least the possibility of *indefinitely* improving the precision of their measurement by "perfecting" the apparatus. Indeed, as long as gravitational effects are ignored, the increase of measurement accuracy allowed by the Wigner-Araki-Yanase theorem can be improved indefinitely *if* we *assume* that the "size" of the apparatus can be increased indefinitely. When, however, gravitational effects are taken into account, then, as pointed out already in Sec. 1.2, Planck's length limits the accuracy of position measurements regardless of the "size" of the apparatus. The accuracy of momentum measurements is then also limited, since in the presence of gravity the time span between two consecutive position measurements in the time-of-flight – or any other measurements of velocity and momentum – cannot be increased indefinitely for the sake of improving momentum measurement accuracy at a given position measurement accuracy. Indeed, the equivalence principle has to be applied to such measurements. However, classically that principle is valid only locally, in sufficiently small neighborhoods of the worldline of the "apparatus", whose size is contingent on the desired measurement accuracy (cf. Sec. 2.7). Hence, as the aforementioned time span increases, the quantum "particle" representing the "system" is bound to cross the boundary of a thus prescribed neighborhood.

For all these reasons, any theory of quantum measurement based on the concept of *sharp* measurement outcomes is bound to fail in the general relativistic regime even more extensively than it has already failed in the special relativistic regime (cf. Sec. 3.3). There it has proved faulty in the case of spacetime position measurements (which, however, are of no real concern to any of the prevailing S-matrix formulations, on which the comparison of conventional relativistic quantum theories with experiments are based), and has retained its viability only for momentum measurements (which are instrumental to all S-matrix approaches). For that reason, the GS notion of *quantum* spacetime is based on a *fundamental quantum spacetime form factor*, which incorporates the principle of irreducible indeterminacy (cf. Sec. 1.3) in a form that sets the *optimal* accuracy ℓ of measured spacetime separations as equal to the Planck length – i.e., to $\ell = 1$ in Planck natural units. As discussed in Sec. 12.5, foundational considerations lead to the conclusion that the quantum spacetime form factor f_ℓ in (5.5.5) should be adopted as fundamental in any quantum spacetime.

That is not to say, of course, that as technologically achievable spatio-temporal measurement accuracies eventually begin to approach Planckian orders of magnitude, this choice cannot be subjected to experimental tests, based on the actual measurements of quantum metric fluctuations, which manifest themselves as statistical fluctuations in the Fubini-Study distances defined and discussed in Secs. 3.7 and 3.9. On the other hand, one of the basic criteria that restricts the field of choices from a measurement-theoretical point of view, is that the fundamental quantum spacetime form factor has to provide the answer to the following question: how can the measurement of Fubini-Study distances of a local quantum state from a local quantum frame, which constitute the basis of the GS quantum theory of measurement, completely determine (at least in principle) that quantum state, regardless of how strong the gravitational fields are in the region of spacetime where the measurements are performed?

It turns out that the aforementioned choice f_ℓ resolves this issue of essential foundational importance, thus providing further support in its favor.

Indeed, we have seen in Sec. 3.7 that, in the SQM context, the measurement of the Fubini-Study distance in (3.7.15) of a quantum state to a given quantum frame uniquely determines that state if the frame is informationally complete, since operationally such a measurement is then tantamount to determining the probability density in (3.7.8). However, the natural arena for SQM is special relativistic physics, where *global* quantum frames can be envisaged. On the other hand, in the presence of gravitational fields, the former SQM q-variables introduced in Sec. 3.4 become the *internal* gauge variables first introduced in Chapter 5, where they assume only values within the (abstract) tangent space $T_x\mathbf{M}$. Consequently, an operational meaning can be assigned to them only on the basis of the exponential map at x (cf. Sec. 2.7), by letting that map act in a sufficiently small neighborhood of $x \in \mathbf{M}$, where curvature effects are negligible for the desired level of accuracy. In can be, however, easily seen that in the case where the elements of the quantum frame are the generalized coherent states constructed from proper state vectors belonging to the fundamental quantum spacetime form factor f_ℓ in (5.5.5), the Fubini-Study distances measured in fibres above points within such (arbitrarily small) neighborhoods are totally sufficient for the complete determination of any local quantum state above x – although that would not be at all necessarily true for other choices of quantum spacetime form factors.

As an example illustrating this important fact, let us consider a local quantum state vector $\boldsymbol{\Psi}$ that belongs to a fibre $\mathbf{F}_x$ of the Klein-Gordon quantum bundle introduced in Sec. 5.1. If the elements of the Klein-Gordon quantum frames are given by (5.5.7), then according to (5.1.2), (5.1.9) and (5.1.12)

$$\left\langle \boldsymbol{\Psi} \middle| \boldsymbol{\Phi}^{u(x)}_{\ell;\zeta'} \right\rangle \left\langle \boldsymbol{\Phi}^{u(x)}_{\ell;\zeta} \middle| \boldsymbol{\Psi} \right\rangle = Z_\ell^{-2} \int \exp[i(\zeta^* \cdot u - \zeta' \cdot u')]\, \tilde{\Psi}(u)\, \tilde{\Psi}^*(u')\, d\Omega(u)\, d\Omega(u'). \quad (4.5)$$

This implies that the expression on the left-hand side of (4.5) is analytic in the complex variables[42] $\zeta^* = q - iv$ and $\zeta' = q' + iv'$ corresponding to values of v and v' from some neighborhood of the 4-velocity hyperboloid in $\mathbf{R}^4$ supplied with a Minkowski metric. Consequently, the values of the left-hand side of (4.5) are completely determined by the values it assumes for $\zeta^* = \zeta'$ from an arbitrarily small neighborhood of the point of contact between $T_x\mathbf{M}$ and $\mathbf{M}$. But these latter values are exactly those that can be, at least in principle, measured with arbitrary accuracy if that neighborhood is so small that the use of the exponential map at x can be related to an operationally viable procedure for the given strength of the gravitational field in the corresponding neighborhood of $x \in \mathbf{M}$. Indeed, the values which (4.5) assumes when $\zeta^* = \zeta'$ corresponds to points from that small neighborhood are equal to the measurable probability densities which, according to (3.7.8), assume the values:

$$\rho_{u(x)}(\Psi;\zeta) = \left\langle \boldsymbol{\Psi} \middle| \boldsymbol{\Phi}^{u(x)}_{\ell;\zeta} \right\rangle \left\langle \boldsymbol{\Phi}^{u(x)}_{\ell;\zeta} \middle| \boldsymbol{\Psi} \right\rangle . \quad (4.6)$$

Thus, the local quantum state represented by $\boldsymbol{\Psi}$ can be, at least in principle, *uniquely* determined by such measurements over "almost flat" neighborhoods. Clearly, the congruence relations in (5.1.20) make this conclusion valid for quantum frames constructed from constituents of arbitrary rest mass m, and not just for $m = 1$.

We are now finally in position to formulate the basic operational principles of a QGR framework which is, at an *epistemological* level, albeit neither at a physical nor at a mathematical level, totally analogous to the CGR framework.

First of all, note should be taken of the essential fact that even in nonrelativistic quantum mechanics the GS approach dispenses with the *need* for providing the status of "observable" to any functions $F(q,p)$, $q,p \in \mathbf{R}^n$, of the classical position and momentum variables. Consequently, GS quantization encounters none of the ordering problems which can make the canonical quantization procedure ill-defined in complex situations, such as those prevailing in a general relativistic regime. As a matter of fact, the following was pointed out by Dirac in his paper on the Hamiltonian dynamics with constraints: "The [canonical quantization procedure] of passing from the classical to the quantum theory is not well-defined, because whenever a classical quantity involves a product of two factors whose [Poisson bracket] does not vanish, there is an ambiguity in the order in which the two factors should appear in the corresponding quantum expression. In practice with simple examples one finds no difficulty in deciding what the order should be. With complicated examples it may be impossible to choose the order in each case so as to make all the quantum equations consistent, and then one would not know how to quantize the theory. The present-day methods of quantization are all of the nature of practical rules, whose application depends on considerations of simplicity." (Dirac, 1950, p. 145).

Second, the GS method of quantization also does not encounter any of the measurement-theoretical problems, pointed out by Wigner (1963, 1976, 1981), with finding an apparatus for the measurement of $F(Q,P)$ in the case of those "classical observables" in which both the position variables q and momentum variables p actually occur. Indeed, the quantum mechanical operators $F_1(Q)$ and $F_2(P)$, corresponding to the "classical observables" $F_1(q)$ and $F_2(p)$ in only the q or only the p variables, respectively, can be regarded as being functions of the respective joint spectral measures for position and momentum – and as such redundant from a measurement-theoretical point of view. On the other hand, the role of the PV measures for position and momentum, and of the associated Galilei systems of imprimitivity described in Sec. 3.1, is taken over in the GS approach by *informationally complete* POV measures (such as those described in Sec. 3.2), and by the associated Galilei systems of covariance. In turn, such systems of covariance give rise to the concept of *quantum* frame (described mathematically in Secs. 3.7 -3.9, and physically in the next section). Measurements with respect to such frames – whose specific instances are implicitly or explicitly discussed by Yamamoto and Haus (1986), Busch *et al.* (1989-91), Schroeck (1991), and others – suffice for the complete determination of a quantum state.

All this implies that in the GS approach one can dispense with the canonical concept of "observable", which creates so many difficulties when its extrapolation to the quantum general relativistic regime is attempted. Instead, *in QGR the concept of quantum frame takes over the role played by canonical "observables"*. However, the concept of quantum frame not only resolves a theoretical dilemma, but it also provides a most *natural* counterpart of the classical concept of frame of reference, which in the form of *local* frame plays a *natural and essential* role in the general relativistic regime. Indeed, local frames provide the basis of the equivalence principle at the measurement-theoretical level, and embody the essence of the general covariance principle: the spatio-temporal distance between two points in a classical spacetime manifold **M** has a direct metric significance only when they are "sufficiently close"; hence, upon using the inverse of the exponential map at one of those points, their spatio-temporal separation can be approximated by that of their images in the tangent space. However, what "sufficiently close" means cannot be defined in purely topological terms, i.e., in the sense of a "sufficiently small" neighborhood in some coordinate chart, since under some diffeomorphisms "small" can become "large", and vice versa. It is the metric prevailing in such a neighborhood that mediates the operational definition (cf. Sec. 2.7) of what is "small" up to the n-th order. But the metric itself is manifested as a property of local frames, by singling out those linear frames which are orthonormal. Thus, as Stachel (1989, 1991) has noted on the basis of Einstein's (1913-1954) remarks, the general covariance principle tells us in effect that it is not the Lorentzian manifold **M** that possesses in CGR any direct physical significance; rather, it is the maps into **M** from certain bundles over **M** that embody that physical significance. And, of all those bundles of direct physical significance, the most important is the Lorentz frame bundle L**M**, or, more generally, the Poincaré frame bundle P**M** – since those two principal bundles take over in general relativity the metrical role played by Lorentz and Poincaré global frames in special relativity. Upon using in CGR these bundles as the fundamental carriers of spatio-temporal relationships we can indeed state, paraphrasing Born's maxim in Sec. 1.1, that "all distinctions in spacetime localization that cannot even in principle be observed are meaningless, and have been therefore eliminated".

As discussed in the preceding section, the geometro-stochastic approach to the formulation of quantum geometries for the description of spacetime in the presence of a quantized gravitational field is predicated on the thesis that the mathematical aspects of CGR retain considerable relevance even at the microscopic level, at which quantum phenomena prevail. Technically, this thesis is reflected in the assumption that the infinite-dimensional supermanifold describing a quantum geometry can be fibrated into infinite-dimensional fibres containing purely quantum degrees of freedom, but situated above the general affine frame bundles $GA\mathbf{S}$ over four-dimensional manifolds $\mathbf{S}$ which can be viewed as embodiments of "classical" degrees of freedom. Indeed, each such base manifold $\mathbf{S}$ represents a *segment* in the solution of a Cauchy problem (cf. Sec. 11.1) that provides a family of *mean* Lorentzian metrics $\boldsymbol{g}^{\mathrm{M}}$ which are compatible with some given quantum gravitational initial values. As the GS evolution proceeds, it gives rise to a *conglomeration* of Lorentzian *developments* $(\mathbf{S}, \boldsymbol{g}^{\mathrm{M}})$, obtained by smoothly fitting together the various base-segments $\mathbf{S}$ which emerge during that evolution (cf. Sec. 11.11).

In those situations where the gravitational field is treated semiclassically, as it was the case in preceding chapters – namely as an external field that influences the behavior of quantum states of matter and radiation, but is not in turn influenced by it – the aforementioned conglomeration contains a single family of isometric Lorentzian manifolds $(\mathbf{M}, \boldsymbol{g}^{\mathrm{M}})$. Each one of those manifolds can be then deemed to represent the same classical general relativistic spacetime, albeit its elements x are not interpreted as the loci of classical point-like events, but rather as the *base* locations of quantum events. Thus, in that case the basic tenets of the GS theory of measurement are as outlined in Sec. 5.5.

In any truly quantum model of gravity the influence of quantum phenomena on the geometry of $(\mathbf{S}, \boldsymbol{g}^{\mathrm{M}})$ has to be taken into account. Consequently, as it will be discussed in Sec. 11.11, the metric structure within each base-segment $\mathbf{S}$ of a Lorentzian development $(\mathbf{S}, \boldsymbol{g}^{\mathrm{M}})$ is of necessity dependent on the actually *inflowing* quantum states of matter and radiation. Hence, in that case the conglomeration of Lorentzian developments must consist of entire equivalence classes of Lorentzian geometries assigned to various four-dimensional manifolds, which have only a differential structure in common. However, as opposed to the application of the Many-Worlds-Interpretation (MWI) of quantum mechanics to superspace formulations of quantum gravity (cf. Sec. 11.2), which ascribe a probability amplitude to each one of the 3-metrics $\boldsymbol{\gamma}$ in some chosen superspace of 3-geometries (Tipler, 1986), the GS approach treats each Lorentzian development $(\mathbf{S}, \boldsymbol{g}^{\mathrm{M}})$ as a *distinct possibility*, which is totally conditional on the *actually* realized quantum states of matter and radiation. In other words, each quantum state that is completely known along some Cauchy surface of initial data gives rise to a unique equivalence class of *mean* Lorentzian geometries $(\mathbf{S}, \boldsymbol{g}^{\mathrm{M}})$, so that no empirically unverifiable postulate of "parallel worlds" is required. This means that the actually observed *mean* spacetime geometry is determined by the *mean metric tensor* $\boldsymbol{g}^{\mathrm{M}}$, which equals the expectation value of a quantum metric tensor $\boldsymbol{g}$ with respect to the quantum gravitational states in the quantum spacetime fibre above each base spacetime location x . This mean metric then supplies a metric structure, and therefore, by the fundamental lemma of Riemannian geometry, also a unique mean torsion-free connection. Thus, at each x all the quantum fluctuations around that mean metric will be the outcome of quantum field theoretical couplings which occur within the quantum stress-energy tensor $\boldsymbol{T}$ for matter as well as for radiation. It is the components $\langle g_{ij} \rangle$ of the mean metric tensor $\boldsymbol{g}^{\mathrm{M}}$, the quantum fluctuations around those values, the corresponding components of the quantum stress-energy and angular-momentum tensors, as well as their quantum fluctuations around those values, that are the *basic* measurable quantities of GS quantum gravity. It is, therefore, the quantum metric and quantum stress-energy tensor fields that are the most fundamental observables of GS quantum gravity. The gauge-invariance of these quantities under the diffeomorphism group for each base-segment **S** is insured from the outset by the fact that **S** does not play any *direct* physical role, since it only provides the base manifold of relevant principal frame bundles that are the *true* carriers of physical information.

Thus, Einstein's earlier cited declaration that "space-time does not claim existence on its own, but only as a structural quality of [a metric] field" is thereby realized in the form of the earlier cited formulation advanced by Stachel (1989, 1991), which we now paraphrase as follows: *The points of a base-segment* **S** *do not represent physical events; those events are instead represented by mappings from points in the cross-sections of certain quantum frame bundles into points of a base-segment* **S**.

11.5. Quantum Pregeometries for GS Graviton States

As outlined in the last part of the preceding section, the quantum gravitational bundle is constructed out of *quantum-pregeometry bundles* $QP\mathbf{S}$, which provide a mathematical description of the *potential* physical situation prevailing *before* a process of metrization *solders* some of the single-graviton fibres $\mathbf{Z}_{\boldsymbol{u}}$ to the *base-segment* **S**. The topology of such a quantum-pregeometry bundle coincides with that of the topological product $\mathbf{A}(\mathbf{S})\times\mathbf{Z}$, where $\mathbf{A}(\mathbf{S})$ is a *bundle-segment* that equals the general affine bundle of coframes over the base-segment **S**, whereas **Z** will eventually emerge as the standard fibre of single-graviton states, which as such carries a Krein-space topology (cf. Sec. 9.1). In view of the terminology we used when describing in Sec. 11.1 the algorithmic segmentation scheme for solving the initial-value problem in CGR, a base-segment **S** will eventually emerge in a similar role in the quantum context. However, at the present quantum-pregeometry level, there are no metrics as yet to single out from amongst the elements of the principal bundle $GA\mathbf{S}$ of affine frames over **S** those which are Poincaré frames (cf. Sec. 2.3) – or, equivalently, to single out the Poincaré coframes in the bundle-segment $\mathbf{A}(\mathbf{S})$. It will be *physical* input, in the form of actual initial data, or of geometric features resulting from GS propagation that is already under way, that will *create* a metric structure in a base-segment **S**. Therefore, by definition, we shall assume that any base-segment **S** is merely a 4-dimensional differential manifold, whose boundary $\partial\mathbf{S}$ consists of two disjoint connected 3-manifolds $\boldsymbol{\Sigma}'$ and $\boldsymbol{\Sigma}''$, to which we shall henceforth refer as the *inflow* and the *outflow surfaces* of that base-segment.

The fact that the topology of a quantum-pregeometry bundle is given by a direct topological product $\mathbf{A}(\mathbf{S})\times\mathbf{Z}$, rather than by a **G**-product (cf. Sec. 4.3), reflects the fact that fibres of such a bundle are not yet soldered in a nontrivial manner to the base manifold **S**. Thus, heuristically speaking, a quantum pregeometry represents a state of "preparedness" to *become* a quantum geometry, once there is an inflow of quantum information into **S** through its inflow surface $\boldsymbol{\Sigma}_0'$. In that case, and provided that only gravitational sources are present above the points of a base-segment **S**, the quantum gravitational fibres $Z_{\boldsymbol{s}(x)}$ that will emerge above each base location $x\in\mathbf{S}$ for a chosen section $\boldsymbol{s}$ of the bundle segment $\mathbf{A}(\mathbf{S})$ will be constructed exclusively out of single-graviton fibres $\mathbf{Z}_{\boldsymbol{s}(x)}$ which are isomorphic to the single-graviton typical fibre **Z**. Otherwise, fibres of the type studied in Chapters 7-10 have to be included in order to incorporate nongravitational fields.

The elements $\boldsymbol{u}=\{(\boldsymbol{a}(x),\boldsymbol{\theta}^i(x))|\ i=0,1,2,3\}$ of the bundle-segment $\mathbf{A}(\mathbf{S})$ incorporate all the covector tetrads $\{\boldsymbol{\theta}^i(x)\}$ dual to the tetrads $\{\boldsymbol{e}_i(x)\}$ of vectors in the tangent space $T_x\mathbf{S}$ (cf. Sec. 2.3). Each Lorentzian metric $\boldsymbol{g}^{\mathrm{L}}$ on $\mathbf{S}$ singles out from amongst such tetrads those which are orthonormal in that metric:

$$\boldsymbol{g}^{\mathrm{L}}(\boldsymbol{e}_i(x),\boldsymbol{e}_j(x)) = \boldsymbol{e}_i(x)\cdot\boldsymbol{e}_j(x) := \eta_{ij} \quad , \qquad i,j = 0,1,2,3 \ . \tag{5.1}$$

Equivalently, it can be said that a Lorentzian metric singles out from amongst the elements of the bundle-segment $\mathbf{A}(\mathbf{S})$ of coframes, dual to those in GAS, those ones which represent Poincaré coframes, so that a *Poincaré bundle-segment* $\mathbf{A}(\mathbf{S},\boldsymbol{g}^{\mathrm{L}})$ is thus determined:

$$\boldsymbol{g}^{\mathrm{L}}_x = \eta_{ij}\boldsymbol{\theta}^i(x)\otimes\boldsymbol{\theta}^j(x) \ , \qquad \boldsymbol{u} = (\boldsymbol{a}(x),\boldsymbol{\theta}^i(x)) \in \mathbf{A}(\mathbf{S},\boldsymbol{g}^{\mathrm{L}}) \subset \mathbf{A}(\mathbf{S}) \ . \tag{5.2}$$

Conversely, if we assume[43] that the bundle-segment QPS is trivial, i.e., that it possesses a global section $\boldsymbol{s}_0$, then the assignment of the *(0,2)*-tensors defined by (5.2) to the elements of the corresponding global section of the Whitney product $QPS \otimes QPS$ will produce a Lorentzian metric on $\mathbf{S}$. Furthermore, analogous mappings

$$\boldsymbol{s} : \boldsymbol{u} \mapsto \eta_{ij}\boldsymbol{\theta}^i(x)\otimes\boldsymbol{\theta}^j(x) \quad , \qquad \boldsymbol{u} = \boldsymbol{s}_0(x)\cdot(b(x),A(x)) \in \mathbf{A}(\mathbf{S}) \ , \tag{5.3a}$$

$$(b(x),A(x)) \in \mathbf{P}(\mathbf{S},\mathrm{GA}(4,\mathbf{R})) \quad , \qquad \forall x \in \mathbf{S} \ , \tag{5.3b}$$

can be defined for all other global sections of QPS, that correspond to all cross-sections (5.3b) of the principal bundle $\mathbf{P}(\mathbf{S},\mathrm{GA}(4,\mathbf{R}))$ whose structure group is the general affine group $\mathrm{GA}(4,\mathbf{R})$ defined in Sec. 2.3. The following *Poincaré-equivalence relation*,

$$\boldsymbol{s}' \sim \boldsymbol{s}'' \quad \Leftrightarrow \quad \boldsymbol{s}''(x) = \boldsymbol{s}'(x)\cdot(b(x),\Lambda(x)) \ , \qquad \forall x \in \mathbf{S} \ , \tag{5.4a}$$

$$(b(x),\Lambda(x)) \in \mathbf{P}(\mathbf{S},\mathrm{ISO}^{\uparrow}(3,1)) \quad , \qquad \forall x \in \mathbf{S} \ , \tag{5.4b}$$

can be therefore defined between pairs of such cross-sections, which subsequently fall naturally into the equivalence classes of moving frames (cf. Sec. 2.2) over $\mathbf{S}$:

$$\mathbf{A}^{\boldsymbol{s}}(\mathbf{S},\boldsymbol{g}^{\mathrm{L}}) = \{\boldsymbol{s}' | \boldsymbol{s}' \sim \boldsymbol{s}\} \quad \leftrightarrow \quad \mathbf{A}(\mathbf{S},\boldsymbol{g}^{\mathrm{L}}) \ . \tag{5.5}$$

These equivalence classes can be set in one-to-one correspondence with the Poincaré bundle-segments $\mathbf{A}(\mathbf{S},\boldsymbol{g}^{\mathrm{L}})$, defined by (5.1)-(5.2). In other words, the family of Lorentz metrics $\boldsymbol{g}^{\mathrm{L}}$ over the base-segment $\mathbf{S}$ can be identified with the family of subbundles $\mathbf{A}(\mathbf{S},\boldsymbol{g}^{\mathrm{L}})$ of the bundle-segment $\mathbf{A}(\mathbf{S})$, which in turn can be identified with the family $\mathbf{A}^{\boldsymbol{s}}(\mathbf{S},\boldsymbol{g}^{\mathrm{L}})$ of all equivalence classes of *moving* frames over $\mathbf{S}$, whose elements are Poincaré-equivalent to some fixed global moving frame $\boldsymbol{s}$, and therefore also mutually Poincaré-equivalent.

By a *metrization* of the base-segment $\mathbf{S}$ we shall mean a selection of one of these equivalence classes. Thus, a metrization of $\mathbf{S}$ is tantamount to the choice of a Lorentz metric over $\mathbf{S}$. However, such a mathematical metrization procedure is conceived in a manner that is readily subjected to an operational interpretation, namely to an empirical verification as to which linear frames over the base-segment $\mathbf{S}$ turn into Lorentz frames, once a GS inflow of matter and radiation has taken place within $\mathbf{S}$.

This kind of verification can be operationally carried out at the macroscopic level as in CGR, namely by working with *linear macro-frames* consisting of "rods" and "clocks". Its outcome is a *macro-metrization* that can be realized in the operational sense stipulated by Einstein (1905, 1916): light signals are used to check which ones of the various ordered sets of "unit rods", mathematically labelled by triples $\{\boldsymbol{e}_1(x),\boldsymbol{e}_2(x),\boldsymbol{e}_3(x)\}$, actually consist of "rigid rods" that are of unit length, and which amongst these latter sets consist of unit rods which are at right angles to each other. Such "classical" set-ups and measurement procedures are, however, obviously not adequate at the microscopic level, and can serve only as a basis for an extrapolation of measurement results performed in *macro*-neighborhoods of preselected base-locations, that are at macroscopic distances from each other.

Within *micro*-neighborhoods of preselected base-locations which are at microscopic distances from each other, a *micro-metrization* can be carried out as an operational procedure consisting of spatio-temporal verifications, meant to establish which *quantum* frames are Lorentzian in the presence of the prevailing quantum states of matter and radiation. The general mathematical definition of a quantum frame was provided in Sec. 3.7. Consequently, we now turn to the formulation of its operational definition[44].

A *gedanken* operational definition of an *inertial quantum frame* can be provided in terms of an observational procedure consisting of observations of a neutral quantum test body O, that marks the origin of a quantum frame, as well as those of six other identical test bodies A_a , $a = \pm1,\pm2,\pm3$, in its immediate vicinity, that mark the positive and negative axes at a chosen unit distance from O, and which are all in free fall. According to de Broglie's (1923) original conceptualization, if any quantum particle is "regarded as containing the rest energy $M_0c^2 = h\nu_0$, it [is] natural to compare it with a small clock of proper frequency ν_0" (de Broglie, 1979, p. 7). Hence, such a quantum frame already has a concept of *proper* time embedded into all its elements. This feature can serve to operationally define the repetitions of measurements of mutual spatio-temporal relationship, that take place under *identical local* conditions, as those carried out at equal proper time intervals. The operational verification of whether any given frame is a *Lorentz quantum frame* consists then of the repeated verifications, under identical *local* conditions, of whether or not the test bodies A_a, $a = \pm1,\pm2,\pm3$, are indeed at the chosen unit distance from O, and whether the axes joining O to each A_a for $a = +1,+2,+3$, as well as for $a = -1,-2,-3$, are indeed at right angles to each other. Such an operational procedure can be in principle carried out by observing the recoil of photons in between the elements of a quantum frame, and using in the process the standard compensatory mechanisms described by Bohr (1949, 1961) to maintain their original geometric configuration within the optimally minimal bounds of geometro-stochastic fluctuations.

Despite very likely present-day technological difficulties of implementing such measurement schemes in practice, at the *gedanken*-experiment level this manner of operationally defining spatial and temporal separations is actually closer to contemporary experimental procedures than the ones originally described by Einstein (1905, 1916) are even at the macro-level. Indeed, the role of macroscopic "unit rods", such as the standard meter in Paris, has been supplanted in experimental praxis by microscopic standards that have emerged from the realization that "the best physical definition of length and time is provided by a particular light or radio source which acts as standard for both length and time" (Arzeliès, 1966, p. 21). The measurement of spatial separation between two nearby "material points" by means of electromagnetic radiation can be accomplished by taking advantage of interference phenomena, which enable the counting of the number of wave-lengths needed to cover the distance between those two "points". In principle, the same procedure can be carried out with beams of massive "test bodies", such as electrons, in case that electromagnetic phenomena can be ignored in a given measurement situation (if not, any of the stable neutral elementary particles discovered thus far might be used instead – cf., e.g., Kelly *et al.*, 1980). In fact, in an electron microscope the replacement of a source of electromagnetic radiation (which from the microscopic point of view emit photons, that can be used in the measurement of spatio-temporal separations) with a source of massive elementary particles is a *fait accompli*, and can be used in the construction of quantum gravimeters[44].

The measurement of time-intervals can be reduced in such set-ups to that of spatial separations. Indeed, the following was emphasized already by Einstein, while discussing the basic measurement procedures for classical relativity theory: "The presupposition of the existence (in principle) of (ideal, viz., perfect) measuring rods and clocks is not independent of each other; since a lightsignal, which is reflected back and forth between the ends of a rigid rod, constitutes an ideal clock, provided that the constancy of the light-velocity in vacuum does not lead to contradictions." (Einstein, 1949, p. 55). In principle, in order to adapt these observations to the quantum regime, all one has to do is replace in this quotation the term "light signal" with "photon", and the stipulation of the reflection taking place "between the ends of a rigid rod", with "between two markers, such as O and A_{+1}, at stochastic relative rest in a quantum frame". Naturally, compensating procedures (Bohr, 1949) have to be used in order to maintain such a kinematical state of stochastic relative rest.

In general, operational procedures for the measurement of spatio-temporal separations at the micro-level can be realized in a variety of ways. The essential point is, however, that an *intrinsic* standard for such measurements is embedded into all quantum frames, and therefore also in all GS propagators, from the outset (i.e., in the context of the present monograph, beginning with the quantum frame in (5.1.18), or the standard quantum frame in (5.1.19), and then proceeding through all their counterparts in Chapters 5 to 8). Clearly, it is the plane waves $\exp(iq{\cdot}k)$, with $k = mu$, that supply the characteristic wave-lengths which serve as *natural* standards, independent of any *conventions* as to the choice of units for length of time durations, or to any particular choice of actual measurement procedures. Moreover, no *ad hoc* choices of "global time", "coordinate time", or any other kind of framework-dependent "time" has to be made, given the undisputed fact that *all* quantum matter displays a characteristic "wave nature" in its behavior.

Thus, a specific *mean metric* $\mathbf{g}^{\mathrm{M}}$ can be, in principle, always measured in the sense of implementing a micro-metrization process as a selection procedure of inertial quantum frames, that are Lorentz quantum frames with respect to the actually *existing* mean metric. From a GS perspective, such measurements are determinations of Fubini-Study distances, obtained by measuring the probability densities in (4.6). In the course of many repetitions of such measurements, under *locally* identical conditions, there will be stochastic fluctuations in the measured relative spatio-temporal distances, which are due to the stochastically extended nature of the quantum test bodies constituting those frames. In addition, there will be fluctuations of a quantum field theoretical origin, that will manifest themselves in the form of pair annihilations and creations of the GS matter fields related to second-quantized frames coupled to the locally prevailing states of quantum gravitational radiation.

In the present GS approach it will be assumed that such gravitational radiation manifests itself exclusively as states of mass-0 gravitons with spin-2, represented by elements of the single-graviton fibres $\mathbf{Z}_u$. On account of this assumption, the general element $\boldsymbol{f}$ of such a *graviton pregeometry fibre* $\mathbf{Z}_u$ can be expressed in the following form:

$$f(u;\zeta) = \int_{V_0^+} \exp\left(-i\bar{\zeta}\cdot k\right)\tilde{f}(u;k)\, d\Omega_0(k) \ , \qquad \bar{\zeta} = q - iv \ , \tag{5.6a}$$

$$\tilde{f}(u;k) = \tilde{f}_{ij}(k)\, \theta^i(x) \otimes \theta^j(x) \ , \qquad \tilde{f}_{ij} = \tilde{f}_{ji} \ , \qquad \mathrm{Tr}\tilde{f} := \eta^{ij}\tilde{f}_{ij} = 0 \ , \tag{5.6b}$$

$$V_0^+ = \left\{(k_0, \mathbf{k}) \middle|\ k_0 = |\mathbf{k}|\right\}, \quad d\Omega_0(k) = \delta(k^2)\, d^4k, \quad u = (a(x), \theta^i(x)) \in \mathbf{A} \ . \tag{5.6c}$$

In view of the generally acknowledged nonrenormalizability of all conventional approaches to quantum gravity, the sharp-point limit of GS quantum gravity is of no interest even at the formal perturbative level. Consequently, we shall henceforth set $\ell = 1$ in Planck natural units. This implies that the 4-tuple $\zeta = (\zeta^0, \ldots, \zeta^3)$ can be deemed to constitute a 4-vector of the form

$$\zeta = q + iv \ , \quad q \in \mathbf{R}^4 \ , \quad v \in V^+ = \left\{v' \middle|\ \eta_{ij} v'^i v'^j = 1 \ , \ v'^0 > 0\right\} \tag{5.7}$$

in a complex 4-dimensional affine space.

The general momentum-space wave function in (5.6b) will eventually have to correspond to mass-0 and spin-2. Therefore, its components constitute a symmetric matrix of zero trace[45], which will behave as a tensor under the later introduced Poincaré gauge transformations. As in the mass-0 and spin-1 case treated in Sec. 9.1, the components of this internal momentum-space wave function are assumed to be square-integrable with respect to the invariant measure on the forward light-cone defined in (5.6c). In accordance with (1.11) and (1.12), the graviton pregeometry fibre $\mathbf{Z}_u$ carries the indefinite inner product

$$\langle f | f' \rangle = \int_{V_0^+} \eta^{ii'}\eta^{jj'} \tilde{f}^*_{ij}(k)\, \tilde{f}'_{i'j'}(k)\, d\Omega_0(k) \ , \qquad f, f' \in \mathbf{Z}_u \ , \tag{5.8}$$

as well as the J-inner product

$$(f | f')_J = \sum_{i,j=0}^{3} \int_{V_0^+} \tilde{f}^*_{ij}(k)\, \tilde{f}'_{ij}(k)\, d\Omega_0(k) \ , \qquad f, f' \in \mathbf{Z}_u \ . \tag{5.9}$$

The formulation of the basic Krein-space properties of the graviton pregeometry fibres $\mathbf{Z}_u$, which is preparatory to the formulation in the next section of a mass-0 and spin-2 representation of the orthochronous Poincaré group $\mathrm{ISO}^{\uparrow}(3,1)$, proceeds along very much the same lines as in the mass-0 and spin-1 case studied in Secs. 9.1 and 9.2. Hence, we shall mention only its most salient points.

One of its most important aspects is reflected by the existence of radiation gauges for gravitation states in various polarization modes. Therefore, let $k = k_i\theta^i(x) \in T_x{}^*\mathbf{S}$ be any null covector with $k_0 > 0$ in the Lorentzian metric defined in (5.1) and (5.2). Then we can introduce $\{\theta^i(x)\}$-dependent *linear polarization tetrads* of vectors, which are totally analogous to the ones in (9.1.16):

$$\varepsilon_{(0)}(u;k) = e_0(x) \ , \qquad \varepsilon_{(3)}(u;k) = -|\mathbf{k}|^{-1} \sum_{a=1}^{3} k_a\, e_a(x) \ , \tag{5.10a}$$

$$k_i \varepsilon^i_{(1)}(x;k) = k_i \varepsilon^i_{(2)}(x;k) = 0 \ , \qquad k = k_i\theta^i(x) = k_0\theta^0(x) + \mathbf{k} \ . \tag{5.10b}$$

Consequently, they obey the following algebraic equations:

$$\varepsilon_{(\mu)}(\boldsymbol{u};\boldsymbol{k})\cdot\varepsilon_{(\nu)}(\boldsymbol{u};\boldsymbol{k}) = \eta_{\mu\nu} \quad , \qquad \mu,\nu = 0,1,2,3 \ . \tag{5.11}$$

On the other hand, the elements of the dual of such a polarization tetrad satisfy the relations:

$$\varepsilon^{i}_{(\mu)}(\boldsymbol{u};\boldsymbol{k})\,\varepsilon^{(\nu)}_{i}(\boldsymbol{u};\boldsymbol{k}) = \delta_{\mu}{}^{\nu} \quad , \qquad \varepsilon^{i}_{(\mu)}(\boldsymbol{u};\boldsymbol{k})\,\varepsilon^{(\mu)}_{j}(\boldsymbol{u};\boldsymbol{k}) = \delta^{i}{}_{j} \ . \tag{5.12}$$

If we insert them in (5.6a), we find that the generic element of $\boldsymbol{f}$ of a graviton pregeometry fibre $\mathbf{Z}_{\boldsymbol{u}}$ can be expressed in the following alternative form:

$$\boldsymbol{f}(\boldsymbol{u};\zeta) = \int_{V_0^+} \exp\left(-i\bar{\zeta}\cdot k\right)\boldsymbol{\varepsilon}^{(\mu)}(\boldsymbol{u};\boldsymbol{k})\otimes\boldsymbol{\varepsilon}^{(\nu)}(\boldsymbol{u};\boldsymbol{k})\tilde{f}_{(\mu\nu)}(\boldsymbol{k})\,d\Omega_0(\boldsymbol{k}) \ . \tag{5.13}$$

Hence, each graviton pregeometry fibre $\mathbf{Z}_{\boldsymbol{u}}$ incorporates elements of the form

$$\begin{aligned}\boldsymbol{\theta}^{ij\zeta}(\boldsymbol{u};\zeta') &= \int_{V_0^+} \exp[i(\zeta'-\bar{\zeta})\cdot k]\ \boldsymbol{\theta}^{i}(x)\otimes\boldsymbol{\theta}^{j}(x)\,d\Omega_0(\boldsymbol{k}) \\ &= \int_{V_0^+} \exp[i(\zeta'-\bar{\zeta})\cdot k]\ \varepsilon^{i}_{(\mu)}(\boldsymbol{u};\boldsymbol{k})\,\varepsilon^{j}_{(\nu)}(\boldsymbol{u};\boldsymbol{k})\ \boldsymbol{\varepsilon}^{(\mu)}(\boldsymbol{u};\boldsymbol{k})\otimes\boldsymbol{\varepsilon}^{(\nu)}(\boldsymbol{u};\boldsymbol{k})\,d\Omega_0(\boldsymbol{k}) \ ,\end{aligned} \tag{5.14}$$

which provide the following *graviton pregeometry frame* associated with $\boldsymbol{u}$:

$$\mathrm{Grf}(\boldsymbol{u}) = \left\{\boldsymbol{\theta}^{ij\zeta}\,\middle|\, i,j = 0,1,2,3,\ \ \zeta = q + iv \in \mathbf{R}^4\times\boldsymbol{V}^+\right\} \ . \tag{5.15}$$

Indeed, the arguments in Sec. 9.2, based on taking the mass-0 limit in (9.2.12), can be easily adapted to the present situation, and lead to the conclusion that the inner products in (5.8) can be rewritten in the following form (cf. the notation introduced in (5.2.3)),

$$\langle \boldsymbol{f}|\boldsymbol{f}'\rangle = \bar{f}^{ij\zeta} f'_{ij\zeta} := \sum_{i,j=0}^{3}\int \bar{f}^{ij\zeta} f'_{ij\zeta}\,d\tilde{\Sigma}(\zeta) \ , \qquad \bar{f}^{ij\zeta} = \left\langle \boldsymbol{f}\,\middle|\,\boldsymbol{\theta}^{ij\zeta}\right\rangle \ ; \tag{5.16}$$

and that, similarly, (5.9) can be expressed as

$$\left(\boldsymbol{f}|\boldsymbol{f}'\right)_J = \sum_{i,j=0}^{3}\int \bar{f}_{ij\zeta} f'_{ij\zeta}\,d\tilde{\Sigma}(\zeta) \ , \qquad f_{ij\zeta} := \eta_{ik}\eta_{jl}f^{kl\zeta} \ . \tag{5.17}$$

Hence, any element $\boldsymbol{f}$ of the graviton pregeometry fibres $\mathbf{Z}_{\boldsymbol{u}}$ can be expressed as follows:

$$\boldsymbol{f} = f_{ij\zeta}\boldsymbol{\theta}^{ij\zeta} := \int d\tilde{\Sigma}(\zeta) f_{ij\zeta}\,\boldsymbol{\theta}^{ij\zeta} \ , \qquad f^{ij\zeta} = \left\langle \boldsymbol{\theta}^{ij\zeta}\,\middle|\,\boldsymbol{f}\right\rangle \ . \tag{5.18}$$

Naturally, as in all previous instances, the integration in (5.16)-(5.18) is carried out over any hypersurface $\sigma\times\boldsymbol{V}^+$, where σ is a spacelike hyperplane with respect to the internal metric introduced in (2.3), and it is executed with respect to the covariant and formally renormalized measure of integration

$$d\tilde{\Sigma}(\zeta) = 2Z_1^{(0)-2} v_i\, \delta(v^2 - 1)\, d\sigma^i(q) d^4 v = 2Z_1^{(0)-2} v_i\, d\sigma^i(q) d\Omega(v) \ , \qquad (5.19)$$

which is based on taking the limit in (9.2.16).

An alternative form of the inner product in (5.16) for local graviton states is given by

$$\langle f | f' \rangle = i \int \bar{f}^{ij\zeta} \overleftrightarrow{\partial}_k f'_{ij\zeta}\, d\sigma^k(q)\, d\tilde{\Omega}(v) \ , \qquad d\tilde{\Omega}(v) = \hat{Z}_1^{(0)} d\Omega(v) \ , \qquad (5.20)$$

and corresponds to the one in (9.2.21) for the case of photon states.

The *quantum gravitational pregeometry fibre* $\mathcal{Z}_{\boldsymbol{u}}$ over each affine coframe $\boldsymbol{u} \in \mathbf{A(S)}$ can be constructed from the graviton fibre $\mathbf{Z}_{\boldsymbol{u}}$ in the manner customary in the construction of Fock spaces for bosons – which was adapted in Sec. 7.4 to the GS framework. Thus we first attach to each $\boldsymbol{u} \in \mathbf{A(S)}$ a vacuum state vector $\Psi_{0;\boldsymbol{u}}$, which spans the one-dimensional local vacuum sector $\mathcal{Z}_{0;\boldsymbol{u}}$. We then add to $\mathcal{Z}_{0;\boldsymbol{u}}$ the J-direct sums of symmetrized tensor products of the graviton pregeometry fibre $\mathbf{Z}_{\boldsymbol{u}}$, so that:

$$\mathcal{Z}_{\boldsymbol{u}} = \bigoplus_{J}{}_{n=0}^{\infty} \mathcal{Z}_{n;\boldsymbol{u}} \ , \qquad \mathcal{Z}_{n;\boldsymbol{u}} = \mathbf{Z}_{\boldsymbol{u}} \otimes_S \cdots \otimes_S \mathbf{Z}_{\boldsymbol{u}} \ . \qquad (5.21)$$

Graviton annihilation operators can be defined by their action on n-graviton states:

$$\left(g_{ij}^{(-)}(\zeta)\Psi_{n;\boldsymbol{u}}\right)_{n-1}(\zeta_1, i_1 j_1, \ldots, \zeta_{n-1}, i_{n-1} j_{n-1})$$
$$= \sqrt{n}\, \Psi_{n;\boldsymbol{u}}(\zeta, ij, \zeta_1, i_1 j_1, \ldots, \zeta_{n-1}, i_{n-1} j_{n-1}) \ , \qquad \zeta = \boldsymbol{a}(x) + (q^j + iv^j)\boldsymbol{e}_j(x) \ . \ (5.22)$$

This definition gives rise to unbounded but J-closable [PQ] operators in $\mathcal{Z}_{\boldsymbol{u}}$, with well-defined adjoints and J-adjoints (Bognár, 1974, p. 122), which can be used to define graviton creation operators. These creation operators then act on n-graviton states as follows:

$$\left(g_{ij}^{(+)}(\zeta)\Psi_{n;\boldsymbol{u}}\right)_{n+1}(\zeta_1, i_1 j_1, \ldots, \zeta_{n+1}, i_{n+1} j_{n+1})$$
$$= i(n+1)^{-1/2} \sum_{r=1}^{n+1} \eta_{ii_r} \eta_{jj_r} D_x^{(+)}(\zeta_r; \zeta)\, \Psi_{n;\boldsymbol{u}}(\zeta_1, i_1 j_1, \ldots, \widehat{\zeta_r}, \widehat{i_r}\widehat{j_r}, \ldots, \zeta_{n+1}, i_{n+1} j_{n+1}) . \ (5.23)$$

We note that the two-point function in (5.23) is the one given in (9.3.13), and that the hats indicate that the variables under them should be omitted. It is then easily checked that

$$\left\langle g_{ij}^{(+)}(\zeta)\Psi \middle| \Psi' \right\rangle = \left\langle \Psi \middle| g_{ij}^{(-)}(\zeta)\Psi' \right\rangle \ , \qquad \Psi, \Psi' \in \mathcal{Z}_{\boldsymbol{u}} \ , \qquad (5.24)$$

for all vectors Ψ and Ψ' from the respective dense domains of definition of these operators. Hence, the *quantum gravitational pregeometry frame field*

$$g(\boldsymbol{u};\zeta) = g_{ij}(\zeta)\boldsymbol{\theta}^i(x)\otimes\boldsymbol{\theta}^j(x) , \qquad g_{ij}(\zeta) = g_{ij}^{(+)}(\zeta) + g_{ij}^{(-)}(\zeta) , \tag{5.25}$$

is given by operators that are selfadjoint in a Krein-space sense (Bognár, 1974, p. 133).

As a consequence of (5.22) and (5.23) we obtain at all $\boldsymbol{u} \in \mathbf{A(S)}$ the following *internal canonical commutation relations* :

$$\left[g_{ij}^{(-)}(\zeta), g_{i'j'}^{(+)}(\zeta')\right] = i\eta_{ii'}\,\eta_{jj'}\,D_x^{(+)}(\zeta;\zeta') , \qquad \left[g_{ij}^{(\pm)}(\zeta), g_{i'j'}^{(\pm)}(\zeta')\right] = 0 . \tag{5.26}$$

We can therefore define *quantum gravitational pregeometry frames*

$$\mathrm{GRF}(\boldsymbol{u}) = \left\{\boldsymbol{\Phi}_f \middle| \, \boldsymbol{f} \in \mathbf{Z}_{\boldsymbol{u}}\right\} , \qquad \boldsymbol{u} \in \mathbf{A(S)} , \tag{5.27}$$

at each $\boldsymbol{u} \in \mathbf{A(S)}$, by setting for all single-graviton local states $\boldsymbol{f}$

$$\boldsymbol{\Phi}_f = \exp\left[g^{(+)}(\boldsymbol{f}) - g^{(-)}(\boldsymbol{f})\right] \Psi_{0;\boldsymbol{u}} , \qquad \boldsymbol{f} \in \mathbf{Z}_{\boldsymbol{u}} , \tag{5.28a}$$

$$g^{(+)}(\boldsymbol{f}) = g^{(-)}(\boldsymbol{f})^{\dagger} := \int f^{ij}(\zeta) g_{ij}^{(+)}(\zeta)\, d\tilde{\Sigma}(\zeta) . \tag{5.28b}$$

It is easily seen that the considerations in Sec. 9.3, that led to (9.3.26) and (9.3.28), can be adapted with no problems to the present situation. Hence, the elements of such quantum gravitational pregeometry frames are eigenvectors of the graviton annihilation operators,

$$g_{ij}^{(-)}(\zeta)\,\boldsymbol{\Phi}_f = f_{ij\zeta}\,\boldsymbol{\Phi}_f , \qquad \boldsymbol{f} = f_{ij}\,\boldsymbol{\theta}^i(x)\otimes\boldsymbol{\theta}^j(x) \in \mathbf{Z}_{\boldsymbol{u}} , \tag{5.29}$$

and the family GRF($\boldsymbol{u}$) in (5.27) provides a continuous resolution of the identity $\mathbf{1}_{\boldsymbol{u}}$ in the quantum gravitational pregeometry fibre $\mathcal{Z}_{\boldsymbol{u}}$, when the functional integration is carried out with respect to the J-inner product of the majorant topology (Bognár, 1974) in that fibre:

$$\int_{\mathbf{Z}_{\boldsymbol{u}}} \left|\boldsymbol{\Phi}_f\right)\, \mathbf{d}\boldsymbol{f}\,\mathbf{d}\bar{\boldsymbol{f}}\, \left(\boldsymbol{\Phi}_f\right| = \mathbf{1}_{\boldsymbol{u}} . \tag{5.30}$$

In this manner we have recovered a great deal of the basic internal structure that characterizes the first and second quantized bundles of the preceding six chapters in general, and the ones for gauge fields in the preceding two chapters in particular. However, two fundamental distinctions emerge between the present case of a quantum pregeometry, and the previously treated cases of quantum geometries:

(I). In the manifold $\mathbf{S}$, called a base-segment for a quantum pregeometry, no particular Lorentzian metric is singled out, so that *no Poincaré gauge covariance* is yet in effect.

(II). The graviton pregeometry fibres $\mathbf{Z}_{\boldsymbol{u}}$ and the quantum gravitational pregeometry fibres $\mathcal{Z}_{\boldsymbol{u}}$ are *not soldered* by generalized soldering maps, such as those in (9.3.12), to the base-segment $\mathbf{S}$, so that quantum pregeometry bundles have as a base manifold the bundle-segment $\mathbf{A(S)}$ of affine coframes $\boldsymbol{u}$, rather than $\mathbf{S}$ itself.

In view of (I), and as indicated by the term "pregeometry", the fibre-bundle structure of a quantum pregeometry is not as yet associated to any principal bundle of physical interest, whereas, as indicated by the term "quantum", it already has quantum kinematics embedded into it. In view of (II), the dependence of the graviton states in (5.6) and of the quantum gravitational frame field in (5.25) on $x \in \mathbf{S}$ is only indirect, and it is due exclusively to the fact that the elements of the coframes $\boldsymbol{u} \in \mathbf{A}(\mathbf{S})$ assume values in the cotangent spaces of $\mathbf{S}$. On the other hand, as explained in Sec. 11.3, on account of the presence of the canonical form $\boldsymbol{\theta}$ in (2.6.4), these coframes are soldered to the base-segment $\mathbf{S}$, so that they are left invariant by the diffeomorphism group $\mathbf{Diff\,S}$. Hence, the pregeometry framework is diffeomorphism-invariant on each base-segment $\mathbf{S}$ – a feature that will be inherited by the Lorenz quantum gravitational bundles of the next section. In fact, this diffeomorphism-invariance extends into the general diffeomorphism-invariance given by

$$\psi : \mathbf{S} \to \mathbf{S}' \ , \qquad \Pi \to \Pi' = \psi \circ \Pi \ , \qquad \boldsymbol{\theta} \to \boldsymbol{\theta}' = \boldsymbol{\theta} \ , \tag{5.31}$$

whereby the original base-segment itself is exchanged for a diffeomorphic one – a situation which in the classical context would correspond to a change of segmentation of the classical spacetime manifold $\mathbf{M}$, resulting from a change of foliation of $\mathbf{M}$. If the structure group, the soldering form, and the total spaces of the physically relevant bundles are regarded as "absolute elements" (Trautman, 1981) of the theory, which are preserved by such transformations, then invariance under them will result if quantum events, like their classical general-relativistic counterparts, are represented by maps from those fibre bundles into their base manifolds, and not by elements in the base manifolds themselves.

11.6. Lorenz Quantum Gravitational Geometries

In the preceding section the process of metrization of a quantum bundle-segment was described, both mathematically as well as at a physical and operational level, as a process of *natural selection* from amongst all inertial quantum frames of a subset of Lorentz (or, more generally, Poincaré) quantum frames best suited to the inflow of quantum information.

Mathematically, the transformation of the quantum pregeometries in the preceding section into fibre bundles which have the orthochronous Poincaré group $\mathrm{ISO}^{\uparrow}(3,1)$ as a structure group (i.e., as gauge group of the first kind – cf. Sec. 11.3) can be achieved by soldering the graviton pregeometry fibres $\mathbf{Z}_{\boldsymbol{u}}$ to the base-segment $\mathbf{S}$ corresponding to all affine coframes $\boldsymbol{u} \in \mathbf{A}(\mathbf{S}, \boldsymbol{g}^{\mathrm{M}})$ that emerge, as described in the preceding section, from a metrization of a base-segment $\mathbf{S}$. Physically, such a metrization takes place as a result of an inflow into $\mathbf{S}$ of quantum matter and radiation in the course of their GS propagation – whose mechanism will be described in Sec. 11.11, once all the necessary mathematical tools have been developed. However, it is clear already at this stage that only those metrizations of a base-segment $\mathbf{S}$ can be deemed physical for which its boundary $\partial\mathbf{S}$ emerges as spacelike, and for which the time-orientation of the coframes in $\mathbf{A}(\mathbf{S}, \boldsymbol{g}^{\mathrm{M}})$ is such that the outflow surface $\boldsymbol{\Sigma}''$ lies in the future of the inflow surface $\boldsymbol{\Sigma}'$ – i.e. for which the causal futures of all points in $\boldsymbol{\Sigma}'$ intersect $\boldsymbol{\Sigma}''$, whereas the causal pasts of all points in $\boldsymbol{\Sigma}''$ intersect $\boldsymbol{\Sigma}'$, rather than the other way around. This is the essence of GS quantum gravitational causality in a strong as well as in a weak setting (cf. Chapter 7).

In studies of the topology of classical spacetimes, two 3-manifolds $\mathbf{\Sigma}'$ and $\mathbf{\Sigma}''$ are called *cobordant* if there is a 4-manifold $\mathbf{S}$ whose boundary is the disjoint union of $\mathbf{\Sigma}'$ and $\mathbf{\Sigma}''$. Two such cobordant manifolds are called *Lorentz cobordant* if there is a Lorentzian metric on $\mathbf{S}$ so that $\mathbf{\Sigma}'$ and $\mathbf{\Sigma}''$ are both spacelike. A theorem by Lickorish (1963) asserts that any two 3-manifolds are cobordant, and its extension by Reinhart (1963) asserts that any two 3-manifolds are actually Lorentz cobordant[46]. Thus, without any loss of generality, we can assume that in the case of any initial-data base-segment $\mathbf{S}_0$ its inflow and outflow surfaces $\mathbf{\Sigma}_0'$ and $\mathbf{\Sigma}_0'' = \mathbf{\Sigma}_1'$ are Lorentz cobordant, and that the base-segment $\mathbf{S}_0$ is the 4-manifold interpolated between them which carries one of the Lorentzian metrics that makes them both spacelike. For all other base-segments $\mathbf{S}$ this feature will then follow in the course of formulating quantum gravitational GS propagation in Secs. 11.10 and 11.11.

As we explained in the preceding section, the metrization of a base-segment $\mathbf{S}$ can be operationally achieved by singling out from all the affine frames $\{(\boldsymbol{a}(x),\boldsymbol{\theta}^i(x))\} \in \mathbf{A}(\mathbf{S})$ those which belong to the Poincaré bundle-segment $\mathbf{A}(\mathbf{S},\boldsymbol{g}^{\mathrm{M}})$ for the actually prevailing mean metric $\boldsymbol{g}^{\mathrm{M}}$. The mathematical embodiment of this procedure consists of specifying within such a base-segment $\mathbf{S}$ a vierbein field $\lambda^i{}_\mu$, which determines a cross-section $\boldsymbol{s}_0$ of $\mathbf{A}(\mathbf{S},\boldsymbol{g}^{\mathrm{M}})$, so that, in accordance with (2.3.2)-(2.3.4) as well as (5.2),

$$\boldsymbol{g}^{\mathrm{M}} = \eta_{ij}\,\boldsymbol{\theta}^i \otimes \boldsymbol{\theta}^j \quad , \qquad \boldsymbol{\theta}^i = \lambda^i{}_\mu\, \boldsymbol{d}x^\mu \in \boldsymbol{s}_0 \subset \mathbf{A}(\mathbf{S},\boldsymbol{g}^{\mathrm{M}}) \ . \tag{6.1}$$

The transition to other cross-sections of the Poincaré bundle-segment $\mathbf{A}(\mathbf{S},\boldsymbol{g}^{\mathrm{M}})$ can be then effected by means of the procedure in (5.3), whereby $\mathbf{A}(\mathbf{S},\boldsymbol{g}^{\mathrm{M}})$ is converted into a Poincaré frame bundle with typical fibre equal to the orthochronous Poincaré group $\mathrm{ISO}^{\uparrow}(3,1)$.

The fibres $\mathbf{Z}_{\boldsymbol{u}}$ corresponding to all coframes $\boldsymbol{u} \in \mathbf{A}(\mathbf{S},\boldsymbol{g}^{\mathrm{M}})$ above each given $x \in \mathbf{S}$ can be now soldered to the base-segment $\mathbf{S}$ by means of the generalized soldering maps

$$\sigma_x^{\boldsymbol{u}} : \quad \boldsymbol{f} = f_{ij\zeta}\boldsymbol{\theta}^{ij\zeta} \mapsto f = \left\{f_{ij\zeta}\right\} \in \mathbf{Z} \quad , \qquad \boldsymbol{f} \in \mathbf{Z}_{\boldsymbol{u}} \ . \tag{6.2}$$

In this manner a *Poincaré graviton bundle* $\mathbf{E}(\mathbf{S},\boldsymbol{g}^{\mathrm{M}})$ over $\mathbf{S}$ can be produced, which is associated to the principal frame bundle $\mathbf{A}(\mathbf{S},\boldsymbol{g}^{\mathrm{M}})$. This is achieved by means of an identification of all fibres $\mathbf{Z}_{\boldsymbol{u}}$ which correspond to all Poincaré coframes $\boldsymbol{u} \in \mathbf{A}(\mathbf{S},\boldsymbol{g}^{\mathrm{M}})$ above each given point x, resulting in a single fibre $\mathbf{Z}_x$ by the imposition of the equivalence relation

$$\left\{f_{ij\zeta}\right\} \sim \left\{f'_{ij\zeta}\right\} \;\Leftrightarrow\; \boldsymbol{f} \in \mathbf{Z}_{\boldsymbol{u}} \sim \boldsymbol{f}' \in \mathbf{Z}_{\boldsymbol{u}'} \ , \quad \boldsymbol{u},\boldsymbol{u}' \in \Pi^{-1}(x) \subset \mathbf{A}(\mathbf{S},\boldsymbol{g}^{\mathrm{M}}) \ , \tag{6.3}$$

between each pair of their respective elements whose coordinate wave functions in (6.3) are related by the following representation of $\mathrm{ISO}^{\uparrow}(3,1)$:

$$U(a,\Lambda) \ : \ f_{ij}(q+iv) \ \mapsto \ f'_{ij}(q+iv) = \Lambda_i{}^{i'}\Lambda_j{}^{j'} f_{i'j'}(\Lambda^{-1}(q-a)+i\Lambda^{-1}v) \ . \tag{6.4}$$

The formulation and study of the above representation of the orthochronous Poincaré group can be carried out very much as in the case of that for photons, presented in (9.2.24), so that we shall mention only its distinctive main points.

Let us proceed as in Sec. 9.1, and consider first the following counterpart of (6.4),

$$\tilde{U}(a,\Lambda)\ :\ \tilde{f}_{ij}(k)\ \mapsto\ \tilde{f}'_{ij}(k)\ = \exp(ia\cdot k)\Lambda_i^{\ i'}\Lambda_j^{\ j'}\tilde{f}_{i'j'}(\Lambda^{-1}k)\ , \tag{6.5}$$

in the momentum representation. If we impose the Lorenz gauge conditions

$$k_i\tilde{f}^{ij}(k) := k_i\eta^{ik}\eta^{jl}\tilde{f}_{kl}(k) = 0\ , \qquad j = 0,1,2,3\ , \tag{6.6}$$

the representation in (6.5) reduces to one which is equal to the symmetrized tensor product of two of the representations in (9.1.13). As such, it would in general consist of a direct sum of a spin-0 and a spin-2 representation. However, the spin-0 component was eliminated from the outset by imposing the additional trace-zero condition (cf. Note 45):

$$\mathrm{Tr}\,\tilde{f}(k) := \tilde{f}_i^{\ i}(k) = \eta^{ij}\tilde{f}_{ij}(k) = 0\ . \tag{6.7}$$

When the representation in (6.4) is restricted to a *Lorenz graviton fibre*

$$\mathbf{Z}_x^{(0)} = \left\{ \boldsymbol{f} \in \mathbf{Z}_x \middle|\ \mathrm{Tr}\,\boldsymbol{f} = f_i^{\ i} \equiv 0\ ,\ \ \partial_i f^{ij\zeta} = 0\ ,\ j = 0,1,2,3 \right\}\ , \tag{6.8}$$

it leaves invariant the subspace consisting of the null elements in $\mathbf{Z}_x$, namely (cf. 7.4))

$$\mathbf{N}_x^{(0)} = \left\{ \boldsymbol{f} \in \mathbf{Z}_x^{(0)} \middle| \langle \boldsymbol{f} | \boldsymbol{f} \rangle = 0 \right\} = \left\{ \boldsymbol{f} \in \mathbf{Z}_x^{(0)} \middle|\ f_{ij} = \partial_j\lambda_j + \partial_i\lambda_j\ ,\ i,j = 0,1,2,3 \right\}\ . \tag{6.9}$$

However, the analysis in Sec. 9.1 tells us that the representation in (6.5), which acts on the wave functions that satisfy (6.6) and (6.7), becomes equivalent to a unitary irreducible Wigner-type representation of the orthochronous Poincaré group $\mathrm{ISO}^{\uparrow}(3,1)$ for spin-2 and zero mass if it is deemed to act on classes of such functions, which are equivalent modulo elements of the set of null elements. Hence, that representation gives rise to a corresponding spin-2 and mass-0 irreducible representation of $\mathrm{ISO}^{\uparrow}(3,1)$, when defined on the quotient of the spaces in (6.8) and in (6.9).

We shall refer to the subbundle $\mathbf{E}^L(\mathbf{S},\boldsymbol{g}^{\mathrm{M}})$ of the Poincaré graviton bundle $\mathbf{E}(\mathbf{S},\boldsymbol{g}^{\mathrm{M}})$ over $\mathbf{S}$, whose fibres are given by (6.8), as a *Lorenz graviton bundle*. From it we can construct the *Lorenz gravitational bundle* $\mathcal{E}^L(\mathbf{S},\boldsymbol{g}^{\mathrm{M}})$, with fibres

$$\mathcal{Z}_x^{(0)} = \bigoplus_{a\ n=0}^{\infty} \mathcal{Z}_{n;x}^{(0)}\ , \qquad \mathcal{Z}_{n;x}^{(0)} = \mathbf{Z}_x^{(0)} \underset{S}{\otimes} \cdots \underset{S}{\otimes} \mathbf{Z}_x^{(0)}\ . \tag{6.10}$$

Problems with the noninvariance of typical fibres constructed by taking J-direct sums under the J-closure of

$$\mathbf{U}(a,\Lambda) = \bigoplus_{a\ n=0}^{\infty} U(a,\Lambda)^{\otimes n}\ , \qquad (a,\Lambda) \in \mathrm{ISO}^{\uparrow}(3,1)\ , \tag{6.11}$$

which are totally analogous to those encountered in Sec. 9.3, force us to restrict ourselves to algebraic direct sums in (6.10), in order to have well-defined generalized soldering maps

$$\sigma_x^u : \ \Psi_{n;x}(\zeta_1,i_1j_1,\ldots,\zeta_n,i_nj_n) \mapsto \Psi_{n;x}(\zeta_1,i_1j_1,\ldots,\zeta_n,i_nj_n) \ , \quad \Psi_{n;x} \in Z_{n;x}^{(0)} \ . \tag{6.12}$$

The remedies offered to this situation could be the same as the ones in Chapter 9. However, as we shall see in the next section, the physical nature of present problems makes a route based on the considerations in Chapter 10 more advantageous. Hence, for the time being we limit ourselves to defining exclusively within the core gravitational bundle $\mathcal{E}^L(\mathbf{S},\boldsymbol{g}^{\mathrm{M}})$ a *semiclassical connection* whose covariant derivatives are given by

$$\overline{\boldsymbol{\nabla}}_{\boldsymbol{X}}\, \Psi_x = [\partial_{\boldsymbol{X}} - \overline{\boldsymbol{\theta}}^j(\boldsymbol{X})\overline{\mathbf{P}}_{j;\boldsymbol{u}} + \tfrac{i}{2}\overline{\boldsymbol{\omega}}_{kl}(\boldsymbol{X})\overline{\mathbf{M}}_{\boldsymbol{u}}^{kl}]\, \Psi_x \ , \tag{6.13}$$

for sections $\boldsymbol{s}$ of the extended Lorentz bundle $L^{\uparrow}\mathbf{M}$ corresponding to the mean metric $\boldsymbol{g}^{\mathrm{M}}$. The infinitesimal generators in (6.13) are those corresponding to the representation in (6.11), whereas the connection coefficients are those corresponding, in accordance with (2.6.19) and (2.7.6), to the classical connection compatible with $\boldsymbol{g}^{\mathrm{M}}$. Their values can be therefore deduced directly from those for vierbein fields (Drechsler, 1984, p. 455):

$$\overline{\boldsymbol{\omega}}_{kl} = \overline{\boldsymbol{\theta}}^j\left(\Omega_{klj} - \Omega_{jkl} - \Omega_{ljk}\right) \ , \qquad \Omega_{jkl} = \tfrac{1}{2}\lambda_j{}^{\mu}\lambda_k{}^{\nu}\left(\partial_\mu \lambda^i{}_\nu - \partial_\nu \lambda^i{}_\mu\right)\eta_{il} \ . \tag{6.14}$$

If we introduce now at each point $x \in \mathbf{S}$ the *quantum-gravitational metric field*

$$\boldsymbol{g}^{\mathrm{Q}}(x;\zeta) = \eta_{ij}\, \boldsymbol{\theta}^i(x)\otimes\boldsymbol{\theta}^j(x) + \boldsymbol{g}(\boldsymbol{u};\zeta) \ , \qquad \boldsymbol{u} = \boldsymbol{s}(x) \ , \tag{6.15}$$

then the quantum gravitational frame field in (5.25) can be viewed as the source of quantum fluctuations around the mean value

$$\overline{\boldsymbol{g}}(x) := \left\langle \Psi \middle| \boldsymbol{g}^{\mathrm{Q}}(x;\zeta)\Psi \right\rangle \big/ \left\langle \Psi \middle| \Psi \right\rangle = \boldsymbol{g}^{\mathrm{M}}(x) \ , \qquad \Psi \in Z_{0;x} \ , \tag{6.16}$$

supplied by the vierbein field in (5.6b). Thus, the tangent space $T_x\mathbf{S}$ plays in the present context a role somewhat analogous to the one which the background Minkowski space plays in the most common formulation of covariant quantum gravity – but this role is in GS quantum gravity purely local, as opposed to being global, as in the latter case.

We also see that, in the same manner in which the Levi-Civita connection of any specific Lorentzian manifold is compatible with its metric structure, the semiclassical connection on the Lorenz graviton bundle $\mathbf{E}^L(\mathbf{S},\boldsymbol{g}^{\mathrm{M}})$ and the Lorenz gravitational bundle $\mathcal{E}^L(\mathbf{S},\boldsymbol{g}^{\mathrm{M}})$, that are associated in the above described manner with $\mathbf{P}(\mathbf{S},\mathrm{ISO}^{\uparrow}(3,1))$, are compatible with the metric structures determined by the respective indefinite inner products defined in the preceding section in the fibres of these quantum Lorenz bundles. Therefore, we shall refer to these bundles as instances of *Lorenz quantum geometries*. In fact, it should be kept in mind that the soldering of the fibres of both $\mathbf{E}^L(\mathbf{S},\boldsymbol{g}^{\mathrm{M}})$ and $\mathcal{E}^L(\mathbf{S},\boldsymbol{g}^{\mathrm{M}})$ depends in a crucial manner on the choice of mean metric $\boldsymbol{g}^{\mathrm{M}}$ in $\mathbf{S}$, since it is reflected in the choice of vierbein fields.

It is of interest to point out already at this stage that the framework for GS quantum gravity, which is hereby gradually emerging, has some common features with both the canonical as well as with the covariant approach to the quantization of gravity – although it is in other respects completely dissimilar from both these approaches. Thus, the prequantum geometry stage, described in the preceding section, has at its disposal *all* the possible Lorentzian metrics over the base segment **S**. If we restrict ourselves to those metrics which are space-compatible with the boundary of **S**, in the sense of making that boundary spacelike, then those Lorentzian metrics $\boldsymbol{g}^{\mathrm{L}}$ will give rise to all possible Riemannian metrics $\boldsymbol{\gamma}^{\mathrm{R}}$ along the inflow boundary $\boldsymbol{\Sigma}'$ of **S** – with the components of all $\boldsymbol{g}^{\mathrm{L}}$ in the timelike directions orthogonal to $\boldsymbol{\Sigma}'$ supplying, in accordance with (1.20), all the various possible extrinsic curvature data in the limit of an "infinitely-thin" base-segment. Hence, as far as the element of choice is concerned, GS quantum gravity has at its disposal at least as many possibilities for metric fluctuations as the canonical approach. The crucial difference is that, amongst all these possibilities, the GS approach first selects, on the basis of a quantum information inflow, a mean metric $\boldsymbol{g}^{\mathrm{M}}$, and then it envisages those fluctuations as taking place *locally* and *stochastically* – rather than globally and coherently, as it is the case in the canonical approach. As we saw in (6.15), in this respect the mean metric $\boldsymbol{g}^{\mathrm{M}}$ displays some similarities with the background metric $\boldsymbol{g}^{\mathrm{B}}$ introduced via (2.6) in the covariant approach – except that the latter is globally defined over a *given* manifold **M**, whereas $\boldsymbol{g}^{\mathrm{M}}$ provides only one of the *stages* $(\mathbf{S}, \boldsymbol{g}^{\mathrm{M}})$ in the never-ending "creation" of a Lorentzian manifold $(\mathbf{M}, \boldsymbol{g}^{\mathrm{M}})$.

The present GS approach is therefore epistemologically more general than either the canonical or the covariant approaches to quantum gravity, in the sense that it does not make any physically unwarranted assumptions, which are difficult to justify from a nondeterministic quantum point of view. Thus, as we discussed in Sec. 11.3, in the face of quantum indeterminism and of the mutual interaction of quantum matter and general-relativistic geometry, it is impossible to reconcile the *assumption* of the global spacetime manifold **M** of covariant approaches with any quantum ontology. Furthermore, even *if* we assume that a sensible physical interpretation can be assigned to the "wave function" $\Psi(\gamma,\phi)$ of the canonical approach in (2.1)-(2.4), it is equally difficult to reconcile the quantum type of stochasticity, in which quantum fluctuations occur randomly, *anywhere* in spacetime, with "fluctuations" of globally defined "spatial" and *smooth* Riemannian geometries, which somehow manifest themselves "simultaneously" across the entire reach of our Universe. Indeed, not only is such an epistemic hypothesis impossible to verify empirically in the absence of a global "time" observable that would provide the operational instructions as to *when* the measurement of a given $\boldsymbol{\gamma}$ should be performed, but it also contradicts the most basic aspects of any form of stochasticity at both the classical and the quantum level: such a *presumed* smoothness represents as sound an assumption as the one that *all* paths of a Brownian motion should be deemed to be smooth, on account of the fact that the trajectories of particles that obey Newton's second law have to be C^2-smooth. Thus, if we use as a guide all past experience with classical stochastic processes, as well as with path integrals, and recall that in CGR the splitting of spacetime into space and time is not absolute, then it appears more sensible to assume that smooth classical geometries would turn out to be the ones of "measure-zero" *if* a mathematically correct probabilistic interpretation of $|\Psi(\gamma,\phi)|^2$ will be eventually discovered. In that case it would be the "stochastic" metrics, including the "distributional" ones, which would carry the true quantum content of such canonical models of quantum gravity.

On the other hand, such an epistemologically sound approach to canonical quantum gravity would have to overcome very difficult mathematical obstacles, since "it is not at all clear what is meant by a 'distributional' metric, and how this would affect the ideas of

Riemannian geometry" (Isham, 1991, p. 366). Moreover, even *if* those technical problems were somehow solved, the physical problems of interpretation would still have to be faced – assuming that one is not willing to postulate fictitious "parallel universes", populated by equally fictitious observers, capable of carrying out the "simultaneous" measurements of various $\boldsymbol{\gamma}$ for the purpose of empirically verifying $|\Psi(\boldsymbol{\gamma},\boldsymbol{\phi})|^2$ (cf. Sec. 11.12).

The GS approach to quantum gravity totally dispenses with the need for any such theoretical fiction, by envisaging quantum metric fluctuations which occur only *locally*. Such *GS metric fluctuations are therefore locally measurable*, since they do not rely on wave functions involving information about physical data along an entire maximal hypersurface of any base-segment $\mathbf{S}$. The method of their measurement is therefore physically realistic, and (in principle) it can be deduced from the operational procedure of verification, described in the preceding section, for ascertaining which quantum frames constructed from quantum test bodies actually represent orthonormal frames. In particular, this implies that no "many-worlds" hypothesis is required for the operational interpretation of GS quantum metric fluctuation probabilities.

11.7. Internal Graviton Gauges and Linear Polarizations

The fibres of the Poincaré graviton bundle $\mathbf{E}(\mathbf{S},\boldsymbol{g}^{\mathrm{M}})$ over $\mathbf{S}$ contain, in addition to the elements of the subfibres in (6.8) that correspond to the Lorenz graviton bundle $\mathbf{E}^L(\mathbf{S},\boldsymbol{g}^{\mathrm{M}})$, also elements that satisfy the *generalized Lorenz gauge conditions*

$$\partial^i f_{ij\zeta} = b_j(\zeta) \ , \quad \partial^i = \eta^{ij}\partial_j \ , \quad \partial_j = \partial/\partial q^j \ , \quad j = 0,1,2,3 \ , \tag{7.1}$$

for local graviton states. These internal graviton gauges obviously correspond to various choices of 4-tuples of functions $b_j(\zeta)$, which under changes of Poincaré frames transform as the components of a covector:

$$U(a,\Lambda) \ : \ b_j(q+iv) \ \mapsto \ b'_j(q+iv) = \Lambda_j{}^k b_k(\Lambda^{-1}(q-a)+i\Lambda^{-1}v) \ . \tag{7.2}$$

They give rise to fibres which can be constructed as in (6.3) from the pregeometry fibres

$$\mathbf{Z}_{\boldsymbol{u}}^{(b)} = \left\{ \boldsymbol{f} \in \mathbf{Z}_{\boldsymbol{u}} \,\middle|\, \mathrm{Tr}\,\boldsymbol{f} = 0 \, , \ \ \partial^i f_{ij} = b_j \ , \ j = 0,1,2,3 \right\} \subset \mathbf{Z}_{\boldsymbol{u}} \ . \tag{7.3}$$

The latter are left invariant by the *internal Lorenz gauge transformations*

$$\boldsymbol{f} = f_{ij}\,\boldsymbol{\theta}^i(x)\otimes\boldsymbol{\theta}^j(x) \ \mapsto \ \boldsymbol{f}' = f'_{ij}\,\boldsymbol{\theta}^i(x)\otimes\boldsymbol{\theta}^j(x) \, , \qquad \boldsymbol{f},\boldsymbol{f}' \in \mathbf{Z}_{\boldsymbol{u}} \ , \tag{7.4a}$$

$$f_{ij}'(\zeta) = f_{ij}(\zeta) + \partial_i\lambda_j(\zeta) + \partial_j\lambda_i(\zeta) \ , \qquad \partial^i\partial_i\lambda_j(\zeta) = 0 \ . \tag{7.4b}$$

Indeed, the zero-trace property of the local graviton states $\boldsymbol{f}$ and $\boldsymbol{f}'$ implies that

$$\partial^i\lambda_i(\zeta) = \tfrac{1}{2}\mathrm{Tr}[\boldsymbol{f}'(\boldsymbol{u};\zeta) - \boldsymbol{f}(\boldsymbol{u};\zeta)] = 0 \ , \tag{7.5}$$

so that the gauge transformations in (7.4) leave invariant each one of the generalized Lorenz gauge conditions specified by (7.1). On account of (6.4), we must also have

$$U(a,\Lambda) : \lambda_j(q+iv) \mapsto \lambda'_j(q+iv) = \Lambda_j{}^k\lambda_k(\Lambda^{-1}(q-a)+i\Lambda^{-1}v) \qquad (7.6)$$

under changes of local Poincaré frame.

The physical significance of the internal gauge transformations in (7.4) can be easily understood if one considers the following changes of coordinates,

$$y^\mu \mapsto y^\mu - \psi^\mu , \qquad y \in \mathcal{N}_x \subset \mathbf{S} , \qquad \mu = 0,1,2,3 , \qquad (7.7)$$

within some "infinitesimal" neighborhood of the point $x \in \mathbf{S}$ in (7.4a). A straightforward computation shows that under such an "infinitesimal" coordinate transformation the components of the mean metric tensor $\boldsymbol{g}^{\mathrm{M}}$ change at the considered point $x \in \mathbf{S}$ as follows (cf. Wheeler, 1979, p. 434):

$$g^{\mathrm{M}}_{\mu\nu} \mapsto g^{\mathrm{M}}_{\mu\nu} + \psi_{\mu;\nu} + \psi_{\nu;\mu} , \quad \psi_\mu = g^{\mathrm{M}}_{\mu\nu}\psi^\nu, \ \psi_{\mu;\nu} = \partial\psi_\mu/\partial x^\nu - \bar{\Gamma}^\kappa{}_{\mu\nu}\psi_\kappa , \qquad (7.8a)$$

$$\bar{\Gamma}^\kappa{}_{\mu\nu} = \tfrac{1}{2}\bar{g}^{\kappa\lambda}\left(\bar{g}_{\lambda\mu,\nu} + \bar{g}_{\lambda\nu,\mu} - \bar{g}_{\mu\nu,\lambda}\right) , \quad \bar{g}^{\kappa\lambda}g^{\mathrm{M}}_{\lambda\mu} = \delta^\kappa_\mu , \ \bar{g}_{\lambda\mu,\nu} = \partial g^{\mathrm{M}}_{\lambda\mu}/\partial x^\nu . \qquad (7.8b)$$

Since the local graviton states in (6.12) depend on the frame but not on the choice of the coordinate charts in $\mathbf{S}$, such a transformation affects the mean metric, but not the quantum part $\boldsymbol{g}(x;\zeta)$ of the quantum-gravitational metric field in (6.14). Upon introducing in the considered neighborhood the Riemann normal coordinates at x whose coordinates lines have the vectors in the tetrad $\{\boldsymbol{e}_i(x)\}$ as tangents (cf. [M], p. 285), (7.8) assumes the form:

$$\bar{g}_{ij} \mapsto \bar{g}_{ij} + \psi_{i,j} + \psi_{j,i} , \qquad \bar{g}_{ij} = g^{\mathrm{M}}_{ij} = \eta_{ij} , \quad \bar{\Gamma}^i{}_{jk} = 0 . \qquad (7.9)$$

On the other hand, if we adopt an "active" instead of a "passive" point of view (cf. Sec. 10.2), then we can view the transformation in (7.9) as one that corresponds to a diffeomorphism ψ in $\mathbf{S}$, which shifts the location of the base point x within the considered neighborhood, while giving rise to an isometric transformation of the mean metric. The outcome of such a transformation can be transferred to the graviton states by carrying out, in accordance with (7.6), the following approximate identifications:

$$\lambda_j^{(\psi)}(q+iv) = (\Lambda_v)_j{}^k\left[\psi_k(\Lambda_v^{-1}q(x)) + \left(q^i + a^i\right)\psi_{k,i}(\Lambda_v^{-1}q(x))\right] + O(|q+a|^2), \qquad (7.10a)$$

$$\psi_{k,i} = \partial\psi_k/\partial y^i , \quad y = \exp_x[(q^i + a^i)\boldsymbol{e}_i(x)] \in \mathcal{N}_x , \quad |q|^2 = \textstyle\sum_{i=0}^3 (q^i)^2 . \qquad (7.10b)$$

This gives rise, in the quantum-gravitational metric field in (6.14), to an *internal quantum gravitational gauge transformation*. Hence, if we keep (7.10) in mind, it is clear that such a transformation represents an "infinitesimal" version of the type of "classical" diffeomorphisms ψ: $\mathbf{S} \to \mathbf{S}$ of $\mathbf{S}$ onto itself that were discussed in Sec. 11.3, and which were presented there as being the constituents of gauge transformations of the second kind in CGR.

In order to find an algebraically convenient gauge-fixing condition that singles out a graviton state vector from each gauge orbit of such quantum gravitational gauge transformations, let us introduce in each quantum pregeometry fibre $\mathbf{Z}_{\boldsymbol{u}}$, corresponding to some $\boldsymbol{u} \in \mathbf{A}(\mathbf{S}, \boldsymbol{g}^{\mathrm{M}})$, the *graviton linear polarization frames* whose elements are given by

$$\boldsymbol{\varepsilon}^{\mu\nu\zeta}(\boldsymbol{u};\zeta') = \int_{\mathbf{V}_0^+} \exp[i(\zeta' - \bar{\zeta}) \cdot k]\, \boldsymbol{\varepsilon}^{(\mu)}(\boldsymbol{u};k) \otimes \boldsymbol{\varepsilon}^{(\nu)}(\boldsymbol{u};k)\, d\Omega_0(k) \,, \tag{7.11}$$

where the above k-dependent polarization coframes are defined by (5.10)-(5.12) – except for a set of Ω_0-measure zero along the forward light-cone. Then we can replace (5.18) by

$$\boldsymbol{f} = \int d\tilde{\Sigma}(\zeta) f_{\mu\nu}(\zeta) \boldsymbol{\varepsilon}^{\mu\nu\zeta} \in \mathbf{Z}_{\boldsymbol{u}} \,, \qquad f_{\kappa\lambda}(\zeta) = \eta_{\kappa\mu}\eta_{\lambda\nu} \left\langle \boldsymbol{\varepsilon}^{\mu\nu\zeta} \middle| \boldsymbol{f} \right\rangle \,, \tag{7.12}$$

so that the inner products in (5.16) and (5.17) assume the following respective forms:

$$\langle \boldsymbol{f} | \boldsymbol{f}' \rangle = \int \bar{f}^{\mu\nu}(\zeta)\, f'_{\mu\nu}(\zeta)\, d\tilde{\Sigma}(\zeta) \,, \qquad f^{\mu\nu}(\zeta) = \left\langle \boldsymbol{\varepsilon}^{\mu\nu\zeta} \middle| \boldsymbol{f} \right\rangle \,, \tag{7.13}$$

$$(\boldsymbol{f} | \boldsymbol{f}')_J = \sum_{\mu,\nu=0}^{3} \int \bar{f}_{\mu\nu}(\zeta)\, f'_{\mu\nu}(\zeta)\, d\tilde{\Sigma}(\zeta) \,, \qquad \boldsymbol{f}, \boldsymbol{f}' \in \mathbf{Z}_{\boldsymbol{u}} \,. \tag{7.14}$$

In this decomposition into polarization modes the symmetry of the local graviton state vector (5.18) in the ij-indices becomes equivalent to

$$f_{\mu\nu}(\zeta) \equiv f_{\nu\mu}(\zeta) \,, \qquad \mu, \nu = 0,1,2,3 \,, \qquad \boldsymbol{f} \in \mathbf{Z}_{\boldsymbol{u}} \,, \tag{7.15}$$

and the four equations

$$f_{\mu 0}(\zeta) \equiv f_{\mu 3}(\zeta) \,, \qquad \mu = 0,1,2,3 \,, \qquad \boldsymbol{f} \in \mathbf{Z}_{\boldsymbol{u}}^{(0)} \,, \tag{7.16}$$

provide necessary as well as sufficient conditions for their belonging to the Lorenz graviton fibre in (6.8). Hence, using the analogy with the *TT (transverse traceless) gauge* in the linearized theory of classical gravity [M,W], we impose the additional gauge condition

$$f_{\mu 0}(\zeta) \equiv f_{\mu 3}(\zeta) \equiv 0 \,, \qquad \mu = 0,1,2,3 \,, \qquad \boldsymbol{f} \in \mathbf{Z}_{\boldsymbol{u}}^{\mathrm{TT}} \subset \mathbf{Z}_{\boldsymbol{u}}^{(0)} \,, \tag{7.17}$$

which singles out a TT-subspace in each graviton pregeometry fibre. Since $\boldsymbol{f}$ is traceless, for each choice of coframe $\boldsymbol{u}$ corresponding to an element of the Poincaré coframe bundle $\mathbf{A}(\mathbf{S}, \boldsymbol{g}^{\mathrm{M}})$, this singles out at the point $x \in \mathbf{S}$ above which that coframe lies the following two independent modes of local graviton state vector linear polarization:

$$\boldsymbol{f}^{(1)}(\zeta) = f_{11}(\zeta)\left(\boldsymbol{\varepsilon}^{11\zeta} - \boldsymbol{\varepsilon}^{22\zeta}\right) \,, \qquad \boldsymbol{f}^{(1)} \in \mathbf{Z}_{\boldsymbol{u}}^{\mathrm{TT}} \,, \tag{7.18a}$$

$$\boldsymbol{f}^{(2)}(\zeta) = f_{12}(\zeta)\left(\boldsymbol{\varepsilon}^{12\zeta} + \boldsymbol{\varepsilon}^{21\zeta}\right) \,, \qquad \boldsymbol{f}^{(2)} \in \mathbf{Z}_{\boldsymbol{u}}^{\mathrm{TT}} \,. \tag{7.18b}$$

Hence the *TT-subspace* of each graviton pregeometry fibre can be expressed in the form

$$\mathbf{Z}_{\boldsymbol{u}}^{\mathrm{TT}} = \left\{ \boldsymbol{f}^{(1)} + \boldsymbol{f}^{(2)} \middle| \boldsymbol{f} \in \mathbf{Z}_{\boldsymbol{u}}^{(0)} \right\} , \tag{7.19}$$

and any element in that fibre can be unambiguously decomposed as follows:

$$\boldsymbol{f} = \boldsymbol{f}^{\mathrm{TT}} + \boldsymbol{f}^{\perp} \in \mathbf{Z}_{\boldsymbol{u}}^{(0)} , \qquad \boldsymbol{f}^{\mathrm{TT}} = \boldsymbol{f}^{(1)} + \boldsymbol{f}^{(2)} \in \mathbf{Z}_{\boldsymbol{u}}^{\mathrm{TT}} , \qquad \boldsymbol{f}^{\perp} \in \mathbf{N}_{\boldsymbol{u}} , \tag{7.20a}$$

$$\langle \boldsymbol{f} | \boldsymbol{f} \rangle = \langle \boldsymbol{f}^{\mathrm{TT}} | \boldsymbol{f}^{\mathrm{TT}} \rangle , \qquad \langle \boldsymbol{f}^{\perp} | \boldsymbol{f}^{\perp} \rangle = \langle \boldsymbol{f}^{\perp} | \boldsymbol{f}^{\mathrm{TT}} \rangle = 0 . \tag{7.20b}$$

It should be emphasized that the above considerations were carried out on the Lorenz subfibres of the graviton pregeometry bundle of Sec. 11.5, rather than of the Lorenz graviton bundle of Sec. 11.6, since the TT-gauge condition in (7.17) is not left invariant by Lorentz boosts, so that the above decomposition is not Lorentz-invariant. On the other hand, the graviton pregeometry subfibres

$$\mathbf{N}_{\boldsymbol{u}}^{(0)} = \left\{ \boldsymbol{f} \in \mathbf{Z}_{\boldsymbol{u}}^{(0)} | \langle \boldsymbol{f} | \boldsymbol{f} \rangle = 0 \right\} , \tag{7.21}$$

consisting of graviton null vectors, are Lorentz-invariant, and can be therefore identified with those in (6.9) for the appropriate choice of $\boldsymbol{x}$. This suggests a restructuring of each graviton pregeometry fibre $\mathbf{Z}_{\boldsymbol{u}}$, based on the following decomposition:

$$\boldsymbol{f} = \boldsymbol{f}^{\mathrm{TT}} + \boldsymbol{f}^{\perp} + \boldsymbol{f}^{\mathrm{B}} , \qquad \boldsymbol{f}^{\mathrm{TT}} = (\boldsymbol{f}^{L})^{\mathrm{TT}} \in \mathbf{Z}_{\boldsymbol{u}}^{\mathrm{TT}} , \qquad \boldsymbol{f}^{\perp} = (\boldsymbol{f}^{L})^{\perp} \in \mathbf{N}_{\boldsymbol{u}}^{(0)} , \tag{7.22a}$$

$$\boldsymbol{f}^{\mathrm{B}} = \boldsymbol{f} - \boldsymbol{f}^{L} , \qquad \partial^{i} f_{ij}^{\mathrm{B}} = \partial^{i} f_{ij} = b_{j} , \qquad f_{11}^{\mathrm{B}} = f_{12}^{\mathrm{B}} = f_{21}^{\mathrm{B}} = f_{22}^{\mathrm{B}} = 0 . \tag{7.22b}$$

Such a decomposition can be regarded as the outcome of a "gauge-fixing", resulting from a specification of the generalized Lorenz class (7.3) to which $\boldsymbol{f}$ belongs, and the pinpointing of its Lorenz-gauge representative $\boldsymbol{f}^{L}$, to which the decomposition in (7.20) is then applied.

This Lorenz-gauge representative can be obtained by using the projection operator

$$\tilde{\mathrm{P}}_{\boldsymbol{u}} : \tilde{\boldsymbol{f}}(\boldsymbol{u};k) \;\mapsto\; \left(\delta_{i}^{\,k} - \bar{k}_{i} k_{m} \eta^{mk}\right)\left(\delta_{j}^{\,l} - \bar{k}_{j} k_{n} \eta^{nl}\right) \tilde{f}_{kl}(k) \boldsymbol{\theta}^{i}(x) \otimes \boldsymbol{\theta}^{j}(x) , \tag{7.23a}$$

$$\boldsymbol{k} = k_{(0)}\left(\boldsymbol{\varepsilon}^{(0)}(\boldsymbol{u};k) + \boldsymbol{\varepsilon}^{(3)}(\boldsymbol{u};k)\right) , \quad \bar{\boldsymbol{k}} = (2k_{(0)})^{-1}\left(\boldsymbol{\varepsilon}^{(0)}(\boldsymbol{u};k) - \boldsymbol{\varepsilon}^{(3)}(\boldsymbol{u};k)\right) . \tag{7.23b}$$

It is easy to check that, on account of the fact that $\boldsymbol{k}$ is a null covector, the map in (7.23a) produces a wave function which satisfies the Lorenz gauge condition, and that a second application leaves that function unchanged. Consequently, the counterpart of this operator in the fibre $\mathbf{Z}_{\boldsymbol{u}}$ is the operator whose action in $\mathbf{Z}_{\boldsymbol{u}}$ is that which follows from (5.6),

$$\mathrm{P}_{\boldsymbol{u}} : \boldsymbol{f}(\boldsymbol{u};\zeta) \;\mapsto\; \left(\delta_{i}^{\,k} + \bar{\partial}_{i} \partial_{m} \eta^{mk}\right)\left(\delta_{j}^{\,l} + \bar{\partial}_{j} \partial_{n} \eta^{nl}\right) f_{kl}(\zeta) \boldsymbol{\theta}^{i}(x) \otimes \boldsymbol{\theta}^{j}(x) , \tag{7.24}$$

so that multiplication with k_j is replaced by i times differentiation with respect to q^j. Hence

$$f^L = \mathrm{P}_u f \in \mathbf{Z}_u^{(0)} \ , \qquad f \in \mathbf{Z}_u \ , \tag{7.25}$$

provides the desired decomposition in (7.22).

11.8. Null Polarization Tetrads and Graviton Polarization Frames

A restructuring of the Lorenz graviton fibre in (6.8) which is suitable for the later introduction of BRST transformations related to changes in internal graviton gauges can be achieved by introducing in the complexification of the tangent space $T_x\mathbf{S}$ the following *circular polarization vectors*,

$$\boldsymbol{\varepsilon}^{[\pm1]}(\boldsymbol{u};k) = R(\hat{\mathbf{k}})\boldsymbol{\varepsilon}^{[\pm1]}(\boldsymbol{k}_u) \ , \qquad \boldsymbol{\varepsilon}^{[\pm1]}(\boldsymbol{k}_u) = \mp\tfrac{1}{\sqrt{2}}\left(\boldsymbol{\theta}^1(x) \pm i\boldsymbol{\theta}^2(x)\right) \ , \tag{8.1a}$$

$$\hat{\mathbf{k}} = \mathbf{k}/|\mathbf{k}| \ , \quad \mathbf{k} = k_1\boldsymbol{\theta}^1(x) + k_2\boldsymbol{\theta}^2(x) + k_3\boldsymbol{\theta}^3(x) \ , \quad \boldsymbol{k}_u = \boldsymbol{\theta}^0(x) + \boldsymbol{\theta}^3(x) \ , \tag{8.1b}$$

where the above rotation is the one which takes $\boldsymbol{\theta}^3(x)$ into $\hat{\mathbf{k}}$, and takes place around the axis which is orthogonal to these two covectors. The Lorentz boost which takes place in the direction of $\boldsymbol{e}_3(x)$, and is given by

$$\Lambda^{(3)}_{\vartheta(k)}\boldsymbol{\varepsilon}^{[\pm1]}(\boldsymbol{k}_u) = \boldsymbol{\varepsilon}^{[\pm1]}(\boldsymbol{k}_u) \ , \qquad \Lambda^{(3)}_{\vartheta(k)}\boldsymbol{k}_u = e^{\vartheta(k)}\boldsymbol{k}_u \ , \qquad e^{\vartheta(k)} = k_0 \ , \tag{8.2}$$

coincides with the one in (5.2.21). It then follows that

$$\hat{\Lambda}_{\boldsymbol{k}}\boldsymbol{k}_u = \boldsymbol{k} \ , \qquad \hat{\Lambda}_{\boldsymbol{k}} := R(\hat{\mathbf{k}})\Lambda^{(3)}_{\vartheta(k)} \ , \qquad \boldsymbol{k} \in \mathbf{V}^+_{0,u} \ , \tag{8.3}$$

so that, upon introducing in accordance with (7.23) the null vectors

$$\boldsymbol{\varepsilon}^{[+2]}(\boldsymbol{u};k) = \boldsymbol{k} \ , \qquad \boldsymbol{\varepsilon}^{[-2]}(\boldsymbol{u};k) = \overline{\boldsymbol{k}} = k_0^{-1}\boldsymbol{\theta}^0(x) - \tfrac{1}{2}k_0^{-2}\boldsymbol{k} \ , \tag{8.4}$$

we obtain the following *null polarization tetrads*[47]

$$\boldsymbol{\varepsilon}^{[r]}(\boldsymbol{u};k) = \hat{\Lambda}_{\boldsymbol{k}}\boldsymbol{\varepsilon}^{[r]}(\boldsymbol{k}_u) \ , \qquad k \in \mathbf{V}_0^+ \ , \qquad r = \pm1, \pm2 \ . \tag{8.5}$$

Indeed, it is easily verified that for any covector $\boldsymbol{k}$ lying in the forward light cone these four covectors coincide with the ones defined by (8.1) and (8.4), and that they obey the following relations:

$$\boldsymbol{\varepsilon}^{[r]}(\boldsymbol{u};k)\cdot\boldsymbol{\varepsilon}^{[r]}(\boldsymbol{u};k) = 0 \ , \qquad r = \pm1, \pm2 \ , \tag{8.6a}$$

$$\boldsymbol{\varepsilon}^{[r]}(\boldsymbol{u};k)\cdot\boldsymbol{\varepsilon}^{[s]}(\boldsymbol{u};k) = 0 \ , \qquad r = \pm1, \quad s = \pm2 \ , \tag{8.6b}$$

$$\boldsymbol{\varepsilon}^{[\pm1]}(\boldsymbol{u};k)\cdot\boldsymbol{\varepsilon}^{[\mp1]}(\boldsymbol{u};k) = \boldsymbol{\varepsilon}^{[\pm2]}(\boldsymbol{u};k)\cdot\boldsymbol{\varepsilon}^{[\mp2]}(\boldsymbol{u};k) = 1 \ . \tag{8.6c}$$

For any given proper Lorentz transformation Λ the following Lorentz transformation

$$D(\Lambda,\boldsymbol{k}) = \hat{\Lambda}_{\boldsymbol{k}}^{-1}\Lambda\,\hat{\Lambda}_{\Lambda^{-1}\boldsymbol{k}} \ , \qquad \Lambda\in \mathrm{SO}_0(3,1) \ , \qquad \boldsymbol{k}\in \mathbf{V}_{0,\boldsymbol{u}}^{+} \ , \tag{8.7}$$

belongs to the little group of $\boldsymbol{k}_{\boldsymbol{u}}$ for all covectors $\boldsymbol{k}$ within the forward light cone (cf. Note 6 to Chapter 9). From the representation of this little group in terms of matrix elements with respect to linear polarization tetrads, we can compute that[48]

$$D(\Lambda,\boldsymbol{k})\boldsymbol{\varepsilon}^{[+2]}(\boldsymbol{k}_{\boldsymbol{u}}) = \boldsymbol{\varepsilon}^{[+2]}(\boldsymbol{k}_{\boldsymbol{u}}) \ , \quad \theta=\theta(\Lambda,\boldsymbol{k})\in\mathbf{R}^1 \ , \quad z=z(\Lambda,\boldsymbol{k})\in\mathbf{C}^1 \ , \tag{8.8a}$$

$$D(\Lambda,\boldsymbol{k})\boldsymbol{\varepsilon}^{[+1]}(\boldsymbol{k}_{\boldsymbol{u}}) = e^{i\theta}\boldsymbol{\varepsilon}^{[+1]}(\boldsymbol{k}_{\boldsymbol{u}}) + z\boldsymbol{\varepsilon}^{[+2]}(\boldsymbol{k}_{\boldsymbol{u}}) \ , \tag{8.8b}$$

$$D(\Lambda,\boldsymbol{k})\boldsymbol{\varepsilon}^{[-1]}(\boldsymbol{k}_{\boldsymbol{u}}) = e^{-i\theta}\boldsymbol{\varepsilon}^{[-1]}(\boldsymbol{k}_{\boldsymbol{u}}) - \bar{z}\boldsymbol{\varepsilon}^{[+2]}(\boldsymbol{k}_{\boldsymbol{u}}) \ , \tag{8.8c}$$

$$D(\Lambda,\boldsymbol{k})\boldsymbol{\varepsilon}^{[-2]}(\boldsymbol{k}_{\boldsymbol{u}}) = \boldsymbol{\varepsilon}^{[-2]}(\boldsymbol{k}_{\boldsymbol{u}}) + ze^{-i\theta}\boldsymbol{\varepsilon}^{[-1]}(\boldsymbol{k}_{\boldsymbol{u}}) - \bar{z}e^{i\theta}\boldsymbol{\varepsilon}^{[+1]}(\boldsymbol{k}_{\boldsymbol{u}}) + |z|^2\boldsymbol{\varepsilon}^{[+2]}(\boldsymbol{k}_{\boldsymbol{u}}). \tag{8.8d}$$

Consequently, upon taking (8.5) and (8.7) into account, we find that for $\Lambda\in\mathrm{SO}_0(3,1)$

$$\Lambda\,\boldsymbol{\varepsilon}^{[+2]}(\Lambda^{-1}k) = \boldsymbol{\varepsilon}^{[+2]}(k) \ , \qquad \boldsymbol{\varepsilon}^{[r]}(k) := \boldsymbol{\varepsilon}^{[r]}(\boldsymbol{u};k) \ , \quad r=\pm1,\pm2 \ , \tag{8.9a}$$

$$\Lambda\,\boldsymbol{\varepsilon}^{[+1]}(\Lambda^{-1}k) = e^{i\theta}\boldsymbol{\varepsilon}^{[+1]}(k) + z\boldsymbol{\varepsilon}^{[+2]}(k) \ , \tag{8.9b}$$

$$\Lambda\,\boldsymbol{\varepsilon}^{[-1]}(\Lambda^{-1}k) = e^{-i\theta}\boldsymbol{\varepsilon}^{[-1]}(k) - \bar{z}\,\boldsymbol{\varepsilon}^{[+2]}(k) \ , \tag{8.9c}$$

$$\Lambda\,\boldsymbol{\varepsilon}^{[-2]}(\Lambda^{-1}k) = \boldsymbol{\varepsilon}^{[-2]}(k) + ze^{-i\theta}\boldsymbol{\varepsilon}^{[-1]}(k) - \bar{z}e^{i\theta}\boldsymbol{\varepsilon}^{[+1]}(k) + |z|^2\boldsymbol{\varepsilon}^{[+2]}(k) \ . \tag{8.9d}$$

Let us now re-express the Poincaré group representation in (6.5) in the form:

$$\tilde{U}(a,\Lambda):\ \tilde{\boldsymbol{f}} \mapsto \tilde{\boldsymbol{f}}' = \exp(ia\cdot k)\,\tilde{f}_{[rs]}(\Lambda^{-1}k)\,\Lambda\boldsymbol{\varepsilon}^{[r]}(\boldsymbol{u};\Lambda^{-1}k)\otimes\Lambda\boldsymbol{\varepsilon}^{[s]}(\boldsymbol{u};\Lambda^{-1}k), \tag{8.10a}$$

$$\tilde{\boldsymbol{f}} = \tilde{f}_{ij}(k)\boldsymbol{\theta}^i(x)\otimes\boldsymbol{\theta}^j(x) = \tilde{f}_{[rs]}(k)\boldsymbol{\varepsilon}^{[r]}(\boldsymbol{u};k)\otimes\boldsymbol{\varepsilon}^{[s]}(\boldsymbol{u};k)\in\mathbf{Z}_{\boldsymbol{u}}^{(0)},\ r,s=\pm1,\pm2. \tag{8.10b}$$

On account of the above relation, as well as of (5.6) and (8.9), the subfibre of $\mathbf{Z}_{\boldsymbol{u}}$ corresponding to circular polarization modes in (8.1a) and to "good" null polarization modes[49], namely to $r,s = +2,+1,-1$, is left invariant by the representation in (6.4), and in fact

$$\mathbf{Z}_{\boldsymbol{u}}^{(0)} = \left\{ \boldsymbol{f} \,\middle|\, \boldsymbol{f} = \sum_{r,s=-1}^{+2}\int_{V_0^+}\exp\left(-i\bar{\zeta}\cdot k\right)\tilde{f}_{[rs]}(k)\boldsymbol{\varepsilon}^{[r]}(\boldsymbol{u};k)\otimes\boldsymbol{\varepsilon}^{[s]}(\boldsymbol{u};k)\,d\Omega_0(k)\right\}, \tag{8.11}$$

Indeed, upon making in (7.11)-(7.12) the transition to the *graviton null polarization frames* with elements

$$\boldsymbol{\varepsilon}^{[rs]\zeta}(\boldsymbol{u};\zeta') = \int_{V_0^+}\exp[i(\zeta'-\bar{\zeta})\cdot k]\,\boldsymbol{\varepsilon}^{[r]}(\boldsymbol{u};k)\otimes\boldsymbol{\varepsilon}^{[s]}(\boldsymbol{u};k)\,d\Omega_0(k) \ , \tag{8.12}$$

this subfibre can be expressed in the form:

$$\mathbf{Z}_{\boldsymbol{u}}^{(0)} = \left\{ \boldsymbol{f} \,\middle|\, \boldsymbol{f} = \sum_{r,s=-1}^{+2} \int d\tilde{\Sigma}(\zeta) f_{[rs]}(\zeta) \boldsymbol{\varepsilon}^{[rs]\zeta} \right\} \ . \tag{8.13}$$

It is then easily established that

$$\mathbf{Z}_{\boldsymbol{u}}^{\mathrm{TT}} = \left\{ \boldsymbol{f} \,\middle|\, \boldsymbol{f} = \sum_{\rho=-1}^{+1} \int d\tilde{\Sigma}(\zeta) f_{[\rho\rho]}(\zeta) \boldsymbol{\varepsilon}^{[\rho\rho]\zeta} \ , \ f_{[\pm1\pm1]} = f_{11} \mp i f_{12} \right\} \subset \mathbf{Z}_{\boldsymbol{u}}^{(0)} \ , \tag{8.14}$$

$$\mathbf{Z}_{\boldsymbol{u}}^{\mathrm{C}} = \left\{ \boldsymbol{f} \,\middle|\, \boldsymbol{f} = \sum_{r=-1}^{+2} \int d\tilde{\Sigma}(\zeta) f_{[r2]}(\zeta) \boldsymbol{\varepsilon}^{[r2]\zeta} + \sum_{s=-1}^{+1} \int d\tilde{\Sigma}(\zeta) f_{[2s]}(\zeta) \boldsymbol{\varepsilon}^{[2s]\zeta} \right\} = \mathbf{N}_{\boldsymbol{u}}^{(0)} \ . \tag{8.15}$$

As with any structure group, the Poincaré group acts from the right[50] on the elements $\boldsymbol{u}$ of the Poincaré bundle-segment $\mathbf{A}(\mathbf{S},\boldsymbol{g}^{\mathrm{M}})$. Using (8.1)-(8.4) we find that for any proper Lorentz transformation Λ we have,

$$\Lambda\, \boldsymbol{\varepsilon}^{[\pm1]}(\boldsymbol{u};\Lambda^{-1}k) = e^{\pm i\theta(\Lambda,k)}\, \boldsymbol{\varepsilon}^{[\pm1]}(\boldsymbol{u}\cdot\Lambda^{\mathrm{T}};k), \quad \Lambda\, \boldsymbol{\varepsilon}^{[\pm2]}(\boldsymbol{u};\Lambda^{-1}k) = \boldsymbol{\varepsilon}^{[\pm2]}(\boldsymbol{u}\cdot\Lambda^{\mathrm{T}};k), \tag{8.16}$$

where the above action from the right of Λ on $\boldsymbol{u}$ coincides with that of $(0,\Lambda)$. All this suggests that the decompositions in (7.22) should be re-expressed in the form of the following direct sums:

$$\mathbf{Z}_{\boldsymbol{u}} = \mathbf{Z}_{\boldsymbol{u}}^{\mathrm{TT}} \oplus \mathbf{Z}_{\boldsymbol{u}}^{\mathrm{B}} \oplus \mathbf{Z}_{\boldsymbol{u}}^{\mathrm{C}} \ , \qquad \mathbf{Z}_{\boldsymbol{u}}^{(0)} = \mathbf{Z}_{\boldsymbol{u}}^{\mathrm{TT}} \oplus \mathbf{Z}_{\boldsymbol{u}}^{\mathrm{C}} \ . \tag{8.17}$$

In order to associate this decomposition with the elements of a quantum bundle over the base-segment $\mathbf{S}$ rather than over the Poincaré bundle-segment $\mathbf{A}(\mathbf{S},\boldsymbol{g}^{\mathrm{M}})$, let us introduce the *graviton polarization frame bundle*

$$\mathrm{Grp}(\mathbf{S},\boldsymbol{g}^{\mathrm{M}}) = \bigcup_{\boldsymbol{u}\in\mathbf{A}(\mathbf{S},\boldsymbol{g}^{\mathrm{M}})} \mathrm{Grp}(\boldsymbol{u}) \ , \tag{8.18a}$$

$$\mathrm{Grp}(\boldsymbol{u}) = \left\{ \boldsymbol{\varepsilon}^{[rs]\zeta} \,\middle|\, r,s = 0,1,2,3, \ \ \zeta = q + iv \in \mathbf{R}^4 \times \mathbf{V}^+ \right\} \ , \tag{8.18b}$$

We then define the action from the right of the orthochronous Poincaré group $\mathrm{ISO}^{\uparrow}(3,1)$ on its elements, given by (8.12), as follows:

$$\begin{aligned}(a,\Lambda):\ \boldsymbol{\varepsilon}^{[rs]\zeta}(\boldsymbol{u};\zeta') \ &\mapsto\ \boldsymbol{\varepsilon}'^{[rs]\zeta}(\boldsymbol{u}\cdot(a,\Lambda);\zeta') \\ &= \int_{V_0^+} \exp[i(\zeta' - \bar{\zeta} - a)\cdot k]\ \boldsymbol{\varepsilon}^{[r]}(\boldsymbol{u}\cdot\Lambda;k) \otimes \boldsymbol{\varepsilon}^{[s]}(\boldsymbol{u}\cdot\Lambda;k)\, d\Omega_0(k) \ .\end{aligned} \tag{8.19}$$

This action is consistent[51] with group multiplication for the Poincaré group on account of the fact that, upon inserting $\Lambda\Lambda'$ into (8.9b) and (8.9c) and comparing coefficients after executing the group multiplications, we obtain

$$\exp[\pm i\theta(\Lambda\Lambda',k)] = \exp[\pm i\theta(\Lambda,k)]\, \exp\!\left[\pm i\theta(\Lambda',\Lambda^{-1}k)\right], \quad \Lambda,\Lambda' \in \mathrm{SO}_0(3,1). \tag{8.20}$$

We can construct now a graviton polarization bundle with typical fibre $\mathbf{A}\oplus\mathbf{N}^{\mathrm{B}}\oplus\mathbf{N}^{\mathrm{C}}$ which consists of Hilbert spaces of functions

$$\alpha(\zeta) = \bigoplus_{\rho=-1}^{+1} \alpha_{[\rho\rho]}(\zeta) \ , \qquad \beta(\zeta) = 2\bigoplus_{r=-2}^{+2} \beta_{[-2r]}(\zeta) \ , \tag{8.21a}$$

$$\gamma(\zeta) = \left(\bigoplus_{r=-1}^{+2} \gamma_{[r2]}(\zeta)\right) \oplus \left(\bigoplus_{s=-1}^{+1} \gamma_{[2s]}(\zeta)\right) \ , \tag{8.21b}$$

which carry inner products similar to those in (5.17):

$$\langle\alpha|\alpha'\rangle = \sum_{\rho=-1}^{+1} \int \overline{\alpha}_{[\rho\rho]}(\zeta)\, \alpha'_{[\rho\rho]}(\zeta) d\tilde{\Sigma}(\zeta) \ , \qquad \alpha,\alpha' \in \mathbf{A} \ , \tag{8.22a}$$

$$\langle\beta|\beta'\rangle = 4\sum_{r=-2}^{+2} \int \overline{\beta}_{[-2r]}(\zeta)\, \beta'_{[-2r]}(\zeta) d\tilde{\Sigma}(\zeta) \ , \qquad \beta,\beta' \in \mathbf{N}^{\mathrm{B}} \ , \tag{8.22b}$$

$$\langle\gamma|\gamma'\rangle = \sum_{r=-1}^{+2} \int \overline{\gamma}_{[r2]}(\zeta)\, \gamma'_{[r2]}(\zeta) d\tilde{\Sigma}(\zeta) + \sum_{s=-1}^{+2} \int \overline{\gamma}_{[2s]}(\zeta)\, \gamma'_{[2s]}(\zeta) d\tilde{\Sigma}(\zeta) \ . \tag{8.22c}$$

The action of the orthochronous Poincaré group in the typical fibre $\mathbf{A}\oplus\mathbf{N}^{\mathrm{B}}\oplus\mathbf{N}^{\mathrm{C}}$ is given by means of the unitary representations

$$U^{\mathrm{A}}(a,\Lambda) \ : \ \alpha_{[\pm1\pm1]}(q+iv) \ \mapsto \ e^{\pm 2i\theta(\Lambda,i\partial)}\, \alpha_{[\pm1\pm1]}(\Lambda^{-1}(q-a)+i\Lambda^{-1}v) \ , \tag{8.23a}$$

$$U^{\mathrm{B}}(a,\Lambda) \ : \ \beta_{[-2r]}(q+iv) \ \mapsto \ e^{\pm i\delta_{\pm1r}\theta(\Lambda,i\partial)} \beta_{[-2r]}(\Lambda^{-1}(q-a)+i\Lambda^{-1}v) \ , \tag{8.23b}$$

$$U^{\mathrm{C}}(a,\Lambda) \ : \ \gamma_{[rs]}(q+iv) \ \mapsto \ e^{i(\pm\delta_{\pm1r}\pm\delta_{\pm1s})\theta(\Lambda,i\partial)} \gamma_{[rs]}(\Lambda^{-1}(q-a)+i\Lambda^{-1}v) \ . \tag{8.23c}$$

We can therefore define the following *graviton polarization bundle*

$$\mathrm{Gpb}(\mathbf{S},\boldsymbol{g}^{\mathrm{M}}) = \mathrm{Grp}(\mathbf{S},\boldsymbol{g}^{\mathrm{M}}) \times_{\mathrm{ISO}^{\uparrow}(3,1)} \mathbf{A}\oplus\mathbf{N}^{\mathrm{B}}\oplus\mathbf{N}^{\mathrm{C}} \ , \tag{8.24}$$

as a bundle associated to the graviton polarization frame bundle in (8.18) by the action of the orthochronous Poincaré group $\mathrm{ISO}^{\uparrow}(3,1)$. We can then embed the elements of the fibre in (8.17) into those of this graviton polarization bundle in the following natural manner,

$$\boldsymbol{f} \mapsto \boldsymbol{\alpha}+\boldsymbol{\beta}+\boldsymbol{\gamma}, \ \ \boldsymbol{f} \in \mathbf{Z}_{\boldsymbol{u}} \ , \ \ \boldsymbol{\alpha} \in \mathbf{A}_x \ , \ \ \boldsymbol{\beta} \in \mathbf{N}_x^{\mathrm{B}} \ , \ \ \boldsymbol{\gamma} \in \mathbf{N}_x^{\mathrm{C}} \ , \ \ \boldsymbol{u} = \Pi^{-1}(x) \ , \tag{8.25a}$$

$$f_{[\rho\rho]}^{\mathrm{TT}} = \alpha_{[\rho\rho]} \ , \ \ f_{[-2r]}^{\mathrm{B}} = \beta_{[-2r]} \ , \ \ f_{[rs]}^{\perp} = \gamma_{[rs]} \ , \ \ \alpha\oplus\beta\oplus\gamma = \boldsymbol{\sigma}_x^{\boldsymbol{u}}(\boldsymbol{\alpha}\oplus\boldsymbol{\beta}\oplus\boldsymbol{\gamma}), \tag{8.25b}$$

where the generalized soldering map underlying (8.25) is, naturally, the one determined by the $\mathrm{ISO}^{\uparrow}(3,1)$-product in (8.24). In this manner Poincaré gauge covariance can be enforced not only in the Lorenz quantum gravitational geometries of Sec. 11.6, but also in the generalized Lorenz quantum gravitational geometries based on the pregeometry single-graviton fibres in (7.3), as well as in their multi-graviton counterparts constructed in accordance with (6.10). However, in these generalized gravitational bundles the action of the orthochronous Poincaré group as a structure group is not determined by the representation in (6.11), but rather form its counterpart constructed from the representations in (8.23).

11.9. Quantum Gravitational Faddeev-Popov Fields and Gauge Groups

The Poincaré group acting in the role of structure group can be extended into that of a *graviton structure group* defined by the following semi-direct product:

$$\mathbf{G} = \mathbf{G}_0 \wedge \mathrm{ISO}^{\uparrow}(3,1) \ , \quad \mathbf{G}_0 = \left\{(b,\lambda) \,\middle|\, \{b_i\} \in \mathbf{B}, \{\partial_i\lambda_j + \partial_j\lambda_i\} \in \mathbf{N}^{\mathrm{C}}\right\} , \tag{9.1a}$$

$$(b,\lambda) \ : \ \left(\beta_{[-2r]}, \gamma_{ij}\right) \ \mapsto \ \left(\beta_{[-2r]} + b_{[r]}, \ \gamma_{ij}(\zeta) + \partial_i\lambda_j + \partial_j\lambda_i\right) \ . \tag{9.1b}$$

The group multiplication within the above *internal graviton structure group* $\mathbf{G}_0$ is defined by the pointwise summation of the 4-tuples b and λ of functions, representing the internal gauge transformations (9.1b), which can be also written in the following form:

$$(b,\lambda) : \ (\boldsymbol{\beta},\boldsymbol{\gamma}) \ \mapsto \ (\boldsymbol{\beta}',\boldsymbol{\gamma}') \ , \qquad \partial^i(\beta'_{ij} - \beta_{ij}) = b_j \ , \quad \gamma'_{ij} - \gamma_{ij} = \partial_i\lambda_j + \partial_j\lambda_i \ . \tag{9.2}$$

In view of (8.23), the semi-direct product in (9.1a) operates under the following rules for group multiplication,

$$((b,\lambda),(a,\Lambda))((b',\lambda'),(a',\Lambda')) = ((b,\lambda) + (a,\Lambda)\cdot(b',\lambda'),(a,\Lambda)(a',\Lambda')) \ , \tag{9.3a}$$

$$(a,\Lambda)\cdot(b',\lambda') \equiv \left(\Lambda_j{}^k b'_k(\Lambda^{-1}(q-a) + i\Lambda^{-1}v), \Lambda_j{}^k\lambda'_k(\Lambda^{-1}(q-a) + i\Lambda^{-1}v)\right) . \tag{9.3b}$$

with Poincaré group multiplication defined as in (3.3.2).

To describe such internal gauge transformations in mathematically precise terms, we require a suitable parametrization of the internal graviton structure group $\mathbf{G}_0$, which is obviously an infinite-dimensional Lie group. This group is Abelian, since its group multiplication is supplied by vector addition. In fact $\mathbf{G}_0$ is identifiable with a direct sum,

$$\mathbf{G}_0 = \mathbf{B} \oplus \mathbf{C} \ , \qquad \mathbf{C} = \left\{\lambda \,\middle|\, \{\partial_i\lambda_j + \partial_j\lambda_i\} \in \mathbf{N}^{\mathrm{C}}\right\} \ , \tag{9.4}$$

which constitutes a Krein space, whose above two subspaces carry the inner products:

$$\langle b|b'\rangle = \int \overline{b}^i(\zeta) b'_i(\zeta)\, d\tilde{\Sigma}(\zeta) \ , \qquad (b|b')_J = \sum_{i=0}^{3} \int \overline{b}_i(\zeta) b'_i(\zeta)\, d\tilde{\Sigma}(\zeta) \ , \tag{9.5a}$$

$$\langle \lambda|\lambda'\rangle = \int \overline{\lambda}^i(\zeta) \lambda'_i(\zeta)\, d\tilde{\Sigma}(\zeta) \ , \qquad (\lambda|\lambda')_J = \sum_{i=0}^{3} \int \overline{\lambda}_i(\zeta) \lambda'_i(\zeta)\, d\tilde{\Sigma}(\zeta) \ . \tag{9.5b}$$

We note that $\mathbf{G}_0$ has the topological structure of the Krein-Maxwell space described in Sec. 9.2, and on account of (7.5), $\mathbf{C}$ is a subspace of the Lorenz space in (9.3.20).

Let us now introduce in these spaces, regarded as typical fibres of a principal *internal graviton gauge bundle* with structure group $\mathbf{G}$, a family of conveniently labeled state vectors constituting bases

$$\{b^\alpha | \ \alpha = 1,2,3,\ldots\} \subset \mathbf{B} \ , \qquad \{\lambda^\alpha | \ \alpha = 1,2,3,\ldots\} \subset \mathbf{C} \ , \tag{9.6}$$

which are orthonormal in the respective J-inner products in (9.5). Since $\mathbf{G}_0$ is Abelian[52], it can be identified with its own Lie algebra $L(\mathbf{G}_0)$, so that these bases provide also a basis in $L(\mathbf{G}_0)$. However, this feature is certainly not retained by the full graviton structure group $\mathbf{G}$ in (9.1), on account of the presence of Lie algebra elements for the Poincaré group.

To each cross-section $\boldsymbol{s}$ of the Poincaré bundle-segment $\mathbf{A}(\mathbf{S}, \boldsymbol{g}^{\mathrm{M}})$ we can now associate a corresponding fundamental field (cf. Sec. 2.5) of *graviton internal gauge frames*

$$\left\{b^{s;\alpha} \middle| \alpha = 1,2,3,\ldots\right\} \subset \mathbf{B}_x \ , \qquad \left\{\lambda^{s;\alpha} \middle| \alpha = 1,2,3,\ldots\right\} \subset \mathbf{C}_x \ , \tag{9.7}$$

in terms of which we can expand any element (b,λ) in the corresponding graviton internal gauge fibres:

$$b = z_\alpha b^{s;\alpha} := \sum\nolimits_{\alpha=1}^{\infty} z_\alpha b^{s;\alpha} \in \mathbf{B}_x \ , \qquad z_\alpha = \left(b^{s;\alpha} \middle| b\right)_J \ , \tag{9.8a}$$

$$\lambda = z_\alpha \lambda^{s;\alpha} := \sum\nolimits_{\alpha=1}^{\infty} z_\alpha \lambda^{s;\alpha} \in \mathbf{C}_x \ , \qquad z_\alpha = \left(\lambda^{s;\alpha} \middle| \lambda\right)_J \ . \tag{9.8b}$$

If we extend the structure group $\mathrm{ISO}^{\uparrow}(3,1)$ of the graviton polarization bundle in (8.24) into the group $\mathbf{G}$, then the resulting *physical, antighost* and *ghost graviton polarization bundles* are, respectively:

$$\mathbf{G}^{\mathrm{A}}(\mathbf{S}, \boldsymbol{g}^{\mathrm{M}}) = \mathrm{Grp}(\mathbf{S}, \boldsymbol{g}^{\mathrm{M}}) \times_{\mathbf{G}} \mathbf{A} \ , \tag{9.9a}$$

$$\mathbf{G}^{\mathrm{B}}(\mathbf{S}, \boldsymbol{g}^{\mathrm{M}}) = \mathrm{Grp}(\mathbf{S}, \boldsymbol{g}^{\mathrm{M}}) \times_{\mathbf{G}} \mathbf{N}^{\mathrm{B}} \ , \quad \mathbf{G}^{\mathrm{C}}(\mathbf{S}, \boldsymbol{g}^{\mathrm{M}}) = \mathrm{Grp}(\mathbf{S}, \boldsymbol{g}^{\mathrm{M}}) \times_{\mathbf{G}} \mathbf{N}^{\mathrm{C}} \ . \tag{9.9b}$$

As such, they consist, respectively, of purely *physical graviton polarization modes*, corresponding to $\boldsymbol{\alpha}$ in (8.25), and of *graviton antighost* and *ghost polarization modes* corresponding to $\boldsymbol{\beta}$ and $\boldsymbol{\gamma}$ states in (8.25).

The physical significance of the internal graviton structure group $\mathbf{G}_0$ emerges from the identification in (7.10) of local infinitesimal diffeomorphisms with internal gauge transformations. Indeed, the transformations $(0,\lambda) \in \mathbf{G}_0$ do not correspond to a general type of infinitesimal diffeomorphism in (7.8), since according to (7.4) the components of λ have to satisfy the wave equation. Hence, when their values, as well as the values of their first partial derivatives in timelike directions are given along a hyperplane σ in $T_x\mathbf{S}$, then their values everywhere else in $T_x\mathbf{S}$ are thereby determined. This means that the prescription of each component of λ has to be supplemented by a function that takes care of these degrees of freedom in timelike directions, which from the passive point of view reside in a generic coordinate transformation in (7.7), and from the active point of view in its corresponding local diffeomorphism. Of course, those additional functions are supplied by the four components of $b \in \mathbf{B}$. Thus, *the diffeomorphism invariance in CGR, which reflects the principle of general covariance from an active point of view, is transposed into QGR in the form of invariance under gauge transformations of the second kind related to the internal graviton structure group* $\mathbf{G}_0$.

To arrive at such a reformulation of diffeomorphism invariance, let us consider a family of diffeomorphisms $\mathrm{D}_\xi(\varepsilon) : x \mapsto x'$ of $\mathbf{S}$ onto itself, determined by the integral curves[53] of a globally defined vector field ξ, so that they can be expressed as follows,

$$D_\xi(\varepsilon): x^\mu \mapsto x'^\mu = x^\mu + \varepsilon\xi^\mu(x) + O(\varepsilon^2), \quad x \in S_\alpha \subset \mathbf{S}, \quad \mu = 0,1,2,3, \tag{9.10}$$

in any atlas of coordinate charts (cf. Sec. 2.1) on **S**. We can relate the corresponding infinitesimal diffeomorphism (Isham and Kuchař, 1985)

$$D_{\varepsilon\xi} : x^\mu \mapsto x^\mu + \varepsilon\xi^\mu(x), \qquad x \in S_\alpha \subset \mathbf{S}, \qquad \mu = 0,1,2,3, \tag{9.11}$$

in a unique manner to the following operator acting on scalar fields in **S** :

$$\boldsymbol{X}_\xi : f \mapsto f', \quad f'(x) = \lim_{\varepsilon\to 0} \varepsilon^{-1}\left[f(D_\xi^{-1}(\varepsilon)(x)) - f(x)\right]. \tag{9.12}$$

Thus, we are clearly dealing with an associated vector field over **S**, assuming at all points $x \in \mathbf{S}$ the following values:

$$\boldsymbol{X}_\xi(x) = -\xi^\mu(x)\boldsymbol{\partial}_\mu \in T_x\mathbf{S}, \qquad \boldsymbol{\partial}_\mu = \partial/\partial x^\mu. \tag{9.13}$$

As infinitesimal generators of gauge transformations of the second kind, represented by elements of the diffeomorphism group **Diff S**, such fields can be deemed to represent the elements ξ of the Lie algebra L(**DiffS**) of this group, which then carries the Lie product[54]:

$$\boldsymbol{X}_{\xi\times\eta} = -\left[\boldsymbol{X}_\xi, \boldsymbol{X}_\eta\right], \qquad \boldsymbol{X}_\xi = -\xi^\mu\boldsymbol{\partial}_\mu, \qquad \boldsymbol{X}_\eta = -\eta^\mu\boldsymbol{\partial}_\mu, \tag{9.14a}$$

$$\boldsymbol{X}_{\xi\times\eta} = -(\xi\times\eta)^\mu\boldsymbol{\partial}_\mu, \qquad (\xi\times\eta)^\mu = \xi^\nu\partial_\nu\eta^\mu - \eta^\nu\partial_\nu\xi^\mu. \tag{9.14b}$$

As formally defined in the original physics literature[55] on the subject, the action of BRST transformations upon a generic tensor field $\boldsymbol{T}$ is obtained by considering the infinitesimal gauge transformation

$$\delta_\xi : \boldsymbol{T} \mapsto \boldsymbol{T}', \quad \boldsymbol{T}'(x) = \lim_{\varepsilon\to 0}\varepsilon^{-1}\left[D_\xi(\varepsilon)_*\boldsymbol{T}(D_\xi^{-1}(\varepsilon)(x)) - \boldsymbol{T}(x)\right], \tag{9.15}$$

which is obviously intimately related to the Lie derivative in the direction of the vector field $\boldsymbol{X}_\xi$ [C,I], and formally replacing ξ with a "ghost" field c. Such a transformation provides the change of value of the tensor field under the infinitesimal diffeomorphism in (9.11). When applied to the mean Lorentzian metric $\boldsymbol{g}^{\mathrm{M}}$ it gives rise to

$$\delta_\xi : \bar{g}_{\mu\nu} \mapsto -\bar{g}_{\lambda\nu}\boldsymbol{\partial}_\mu\xi^\lambda - \bar{g}_{\mu\lambda}\boldsymbol{\partial}_\nu\xi^\lambda - \xi^\lambda\boldsymbol{\partial}_\lambda\bar{g}_{\mu\nu}, \qquad \bar{g}_{\mu\nu} = g^{\mathrm{M}}_{\mu\nu}, \tag{9.16}$$

whereas, when applied to the vierbein field in (6.1), it gives rise to

$$\delta_\xi : \boldsymbol{\theta}^i = \lambda^i{}_\mu\boldsymbol{dx}^\mu \mapsto -(\lambda^i{}_\nu\partial_\mu\xi^\nu + \xi^\nu\partial_\nu\lambda^i{}_\mu)\boldsymbol{dx}^\mu. \tag{9.17}$$

We note that the transformations in (9.15) mediate the transference of the action of diffeomorphisms in the base manifolds of the respective tensor bundles into action along their fibres, giving rise to vertical bundle automorphisms.

To transfer such action to local graviton states, and thus arrive at gravitational gauge transformations of the second kind, we shall transpose the "ghost" and "antighost" components of those states into the following geometrized *FP (Faddeev-Popov) graviton states* that correspond to the decomposition in (7.22),

$$\boldsymbol{f}^{\perp} \mapsto \boldsymbol{c} \in \mathbf{H}_x\,, \qquad f_{ij}^{\perp} = \partial_i c_j + \partial_j c_i\,, \quad \partial_j \partial^j c_i = 0\,, \quad \partial^i c_i = 0\,, \tag{9.18a}$$

$$\boldsymbol{f}^{\mathrm{B}} \mapsto \bar{\boldsymbol{c}} \in \overline{\mathbf{H}}_x\,, \qquad f_{ij}^{\mathrm{B}} = \partial_i \bar{c}_j + \partial_j \bar{c}_i\,, \quad \partial_j \partial^j \bar{c}_i = 0\,, \quad \partial_j \partial^i \bar{c}_i = b_j\,, \tag{9.18b}$$

whose components are identified with the respective fibre elements in (8.25). Note should be taken of the fact that the bars in (9.18b) do not in general refer to complex conjugation; rather, they represent an adaptation of the standard notation for ghost and antighost states, which will be related to Grassmannian variables of the type encountered in Sec. 8.3.

For infinitesimal diffeomorphisms (9.11) which are *sufficiently local* in the sense that the exponential map at all points x within their supports is one-to-one, the components of the vector field ξ determining that diffeomorphism can be transferred in a natural manner to the tangent spaces of the base-segment $\mathbf{S}$ by setting

$$\xi_x^i(q) = \xi^i\big(\exp_x[(q^i + a^i)\boldsymbol{e}_i(x)]\big)\,, \qquad x \in \operatorname{supp}\xi \subset \mathbf{S}\,, \quad q = q^i \boldsymbol{e}_i(x) \in T_x\mathbf{S}. \tag{9.19}$$

The identification of such an infinitesimal diffeomorphism with a pair of fields

$$C(x;c^{\boldsymbol{s}}) = C^{(+)}(x;c^{\boldsymbol{s}}) + C^{(-)}(x;c^{\boldsymbol{s}})\,, \quad \overline{C}(x;\bar{c}^{\boldsymbol{s}}) = \overline{C}^{(+)}(x;\bar{c}^{\boldsymbol{s}}) + \overline{C}^{(-)}(x;\bar{c}^{\boldsymbol{s}})\,, \tag{9.20}$$

consisting of a Faddeev-Popov (FP) *gravitational polarization ghost field* and a *gravitational polarization antighost field*, respectively, can be then effected by first carrying out a suitable identification of initial conditions in the tangent spaces $T_x\mathbf{S}$ with respect to each affine coframe in the given section $\boldsymbol{s}$ of $\mathbf{A}(\mathbf{S}, \boldsymbol{g}^{\mathrm{M}})$,

$$\bar{c}_i^{\boldsymbol{s}}(x;\hat{q} + i\hat{v}) + c_i^{\boldsymbol{s}}(x;\hat{q} + i\hat{v}) = \eta_{ij}\xi_x^j(\hat{q})\,, \qquad \hat{q}^0 = -a^0\,, \qquad \hat{v}^0 = 1\,, \tag{9.21a}$$

$$\partial_0 \bar{c}_i^{\boldsymbol{s}}(x;\hat{q} + i\hat{v}) + \partial_0 c_i^{\boldsymbol{s}}(x;\hat{q} + i\hat{v}) = \eta_{ij}\partial_0\xi_x^j(\hat{q})\,, \qquad \hat{q}^0 = -a^0\,, \qquad \hat{v}^0 = 1\,. \tag{9.21b}$$

The so determined values of ghost and antighost graviton local states in relation to the "rest" frame components of a massive quantum frame can be then extended, in accordance with (7.10) and (8.23), to the components of generic mean stochastic frame 4-velocity v:

$$\bar{c}_i^{\boldsymbol{s}}(x;q + iv) = (\Lambda_v)_i^{\ k}\, \bar{c}_k^{\boldsymbol{s}}(x;\Lambda_v^{-1}q + i\hat{v}), \quad c_i^{\boldsymbol{s}}(x;q + iv) = (\Lambda_v)_i^{\ k}\, c_k^{\boldsymbol{s}}(x;\Lambda_v^{-1}q + i\hat{v}). \tag{9.22}$$

The FP fields in (9.20) are then the outcome of the "smearing"[56] of the following *FP gravitational polarization frame* and *coframe fields* with the thus obtained FP graviton states,

$$C(x;\zeta) = C^{(+)}(x;\zeta) + C^{(-)}(x;\zeta) \ , \qquad C^{(+)}(x;\zeta) = C^{(-)\dagger}(x;\zeta) \ , \tag{9.23a}$$

$$\overline{C}(x;\zeta) = \overline{C}^{(+)}(x;\zeta) + \overline{C}^{(-)}(x;\zeta) \ , \qquad \overline{C}^{(+)}(x;\zeta) = -\overline{C}^{(-)\dagger}(x;\zeta) \ , \tag{9.23b}$$

where, for example, the action of the "antighost" annihilation operator on local quantum gravitational states with m antighosts is given by the following expression (cf. Sec. 10.6):

$$\begin{aligned}&\left(\overline{C}_i^{(-)}(x;\zeta)\Psi_{k;m;n;u}\right)_{m-1}(\zeta_1,i_1j_1,\ldots,\zeta_k,i_kj_k;\zeta_{k+1},i_{k+1},\ldots,\zeta_{k+m+n},i_{k+m+n})\\ &\quad = i\sqrt{m}\ \Psi_{k;m;n;u}(\zeta_1,i_1j_1,\ldots,\zeta_k,i_kj_k;\zeta,i,\zeta_{k+1},i_{k+1},\ldots,\zeta_{k+m+n},i_{k+m+n})\ .\end{aligned} \tag{9.24}$$

The *quantum gravitational polarization bundle* in which these operators act can be constructed by first considering the following Whitney sums and products of the graviton polarization bundles in (9.9):

$$\mathcal{G}(\mathbf{S},\boldsymbol{g}^{\mathrm{M}}) = \mathcal{G}^{\mathrm{A}}(\mathbf{S},\boldsymbol{g}^{\mathrm{M}})\otimes\left(\mathcal{G}^{\mathrm{B}}(\mathbf{S},\boldsymbol{g}^{\mathrm{M}})\underset{A}{\otimes}\mathcal{G}^{\mathrm{C}}(\mathbf{S},\boldsymbol{g}^{\mathrm{M}})\right) \ , \tag{9.25a}$$

$$\mathcal{G}^{I}(\mathbf{S},\boldsymbol{g}^{\mathrm{M}}) = \bigoplus_{n=0}^{\infty} \mathcal{G}_n^{I}(\mathbf{S},\boldsymbol{g}^{\mathrm{M}}) \ , \qquad I = \mathrm{A,B,C} \ , \tag{9.25b}$$

$$\mathcal{G}_n^{\mathrm{A}} = \mathbf{G}^{\mathrm{A}}\underset{S}{\otimes}\cdots\underset{S}{\otimes}\mathbf{G}^{\mathrm{A}} \ , \qquad \mathcal{G}_n^{\mathrm{B}} = \mathbf{G}^{\mathrm{B}}\underset{A}{\otimes}\cdots\underset{A}{\otimes}\mathbf{G}^{\mathrm{B}} \ , \qquad \mathcal{G}_n^{\mathrm{C}} = \mathbf{G}^{\mathrm{C}}\underset{A}{\otimes}\cdots\underset{A}{\otimes}\mathbf{G}^{\mathrm{C}}. \tag{9.25c}$$

Its state vectors are related in accordance with (9.18) to those on which the operators in (9.23) act directly, and which belong to the *FP gravitational polarization bundle*

$$\mathcal{G}^{\mathrm{FP}}(\mathbf{S},\boldsymbol{g}^{\mathrm{M}}) = \mathcal{G}^{\mathrm{A}}(\mathbf{S},\overline{\boldsymbol{g}})\otimes\left(\overline{\mathcal{H}}(\mathbf{S},\overline{\boldsymbol{g}})\underset{A}{\otimes}\mathcal{H}(\mathbf{S},\overline{\boldsymbol{g}})\right) \ , \qquad \overline{\boldsymbol{g}} = \boldsymbol{g}^{\mathrm{M}} \ , \tag{9.26a}$$

$$\overline{\mathcal{H}}(\mathbf{S},\overline{\boldsymbol{g}}) = \bigoplus_{n=0}^{\infty} \overline{\mathcal{H}}_n(\mathbf{S},\overline{\boldsymbol{g}}), \qquad \overline{\mathcal{H}}_n(\mathbf{S},\overline{\boldsymbol{g}})) = \overline{\mathbf{H}}(\mathbf{S},\overline{\boldsymbol{g}})\underset{A}{\otimes}\cdots\underset{A}{\otimes}\overline{\mathbf{H}}(\mathbf{S},\overline{\boldsymbol{g}}) \ , \tag{9.26b}$$

$$\mathcal{H}(\mathbf{S},\overline{\boldsymbol{g}}) = \bigoplus_{n=0}^{\infty} \mathcal{H}_n(\mathbf{S},\overline{\boldsymbol{g}}), \qquad \mathcal{H}_n(\mathbf{S},\overline{\boldsymbol{g}})) = \mathbf{H}(\mathbf{S},\overline{\boldsymbol{g}})\underset{A}{\otimes}\cdots\underset{A}{\otimes}\mathbf{H}(\mathbf{S},\overline{\boldsymbol{g}}) \ , \tag{9.26c}$$

whose single ghost and antighost fibres are those in (9.18a) and (9.18b), respectively.

In the next section we shall introduce the *quantum gravitational gauge supergroup*

$$Q\mathcal{G}(\mathbf{S},\boldsymbol{g}^{\mathrm{M}}) = \mathcal{G}_0(\mathrm{Diff}\mathbf{S})\wedge\mathcal{G}\left(\mathbf{A}(\mathbf{S},\boldsymbol{g}^{\mathrm{M}})\right) \ , \tag{9.27}$$

associated with the Poincaré bundle-segment $\mathbf{A}(\mathbf{S},\boldsymbol{g}^{\mathrm{M}})$ as a gauge group of the second kind which combines, in the form of a semi-direct product, segmental Poincaré gauge covariance with covariance under vertical automorphisms resulting from the embeddings, given by (9.21) and (9.22), of locally infinitesimal diffeomorphisms into the internal graviton gauge group. Hence, these embeddings give rise to a Lie algebra structure of the diffeomorphism-generated subgroup of this gauge group. Thus, diffeomorphism gauge invariance is seen as a necessary, but certainly not as a sufficient feature of the GS formulation of quantum gravity, whereas Poincaré gauge invariance retains[57] its key role.

In fact, as we emphasized in Sec. 11.3, it is Poincaré gauge covariance, rather than diffeomorphism invariance, which provides, via the equivalence principle, actual physical input into general relativity, at the classical as well as at the present GSQG level. Indeed, as first discussed in Sec. 4.1 of [P], the former has actual operational meaning, so that it can be directly confirmed (or refuted) by experiments; whereas the latter merely reflects a mathematical feature of CGR, first implicitly revealed by Einstein's "hole" argument, which showed that a classical spacetime *cannot* be represented by a *unique* Lorentzian manifold $(\mathbf{M},\boldsymbol{g})$, but rather by an equivalence class of such manifolds related by the diffeomorphisms in (3.2)-(3.3). At the present GS quantum level that equivalence relation is transposed from its "horizontal" formulation by means of mappings *within* a base-segment $\mathbf{S}$, into a "vertical" formulation by means of mappings along quantum fibres *above* $\mathbf{S}$.

We note that the unitarity of the representations in (8.23) of the orthochronous Poincaré group $\mathrm{ISO}^{\uparrow}(3,1)$ removes the technical problem to which the pseudo-unitarity of their indefinite-metric counterparts used in (6.4) gave rise to: instead of the algebraic direct sum in (6.10), a Hilbert direct sum [PQ] can be used in (9.25b) and (9.26b)-(9.26c). This was achieved by viewing these bundles as associated to quantum rather than to classical principal polarization frame bundles. Thus, if it were not for the coupling of quantum gravitational fields to massive matter fields, one could dispense with FP "ghost" and "antighost" fields altogether, and work only with physical polarization modes in (9.9a). It is this coupling, however, that constitutes the basis of Einstein's equations, and necessitates their retention in the construction of the quantum gravitational bundle in the next section.

11.10. Quantum Gravitational BRST Symmetries and Connections

In this section we shall adapt to GS quantum gravity the geometric approach to BRST gauge symmetries presented[58] in Secs. 10.2-10.4. The basic idea consists of second-quantizing the connection in (6.13) in the TT-gauge, in which no ghost or antighost states appear, and then extending it in the vertical directions of the fibres of a quantum gravitational bundle, so that it yields a quantum connection over the base manifold of that bundle, which can serve in defining a notion of parallel transport that is invariant under the action of the quantum gravitational gauge supergroup in (9.27).

The first step in the second-quantization of the semi-classical connection in (6.13) is achieved by choosing any cross-section $\boldsymbol{s}$ of $\mathbf{A}(\mathbf{S},\boldsymbol{g}^{\mathrm{M}})$, and then proceeding as in previous chapters, namely by introducing the following Poincaré-covariant operator-valued connection form associated with that section:

$$\boldsymbol{\nabla}^{\mathrm{TT}} = \boldsymbol{d} + \mathcal{P}^{\mathrm{TT}}_{\boldsymbol{u}}\tilde{\boldsymbol{\Gamma}}\mathcal{P}^{\mathrm{TT}}_{\boldsymbol{u}} \quad , \qquad \tilde{\boldsymbol{\Gamma}} = -i\bar{\boldsymbol{\theta}}^{j}\tilde{\boldsymbol{P}}_{j;\boldsymbol{u}} + \tfrac{i}{2}\bar{\boldsymbol{\omega}}_{kl}\tilde{\boldsymbol{M}}^{kl}_{\boldsymbol{u}} \quad , \tag{10.1a}$$

$$\boldsymbol{d} = \boldsymbol{\theta}^{i}\boldsymbol{e}_{i} \quad , \qquad \boldsymbol{u} = (\boldsymbol{a}(x),\boldsymbol{\theta}^{i}(x)) \in \boldsymbol{s} \subset \mathbf{A}(\mathbf{S},\boldsymbol{g}^{\mathrm{M}}) \quad , \tag{10.1b}$$

$$\mathcal{P}^{\mathrm{TT}}_{\boldsymbol{u}} = \bigoplus_{J}{}_{n=0}^{\infty}\left(\mathbf{P}^{\mathrm{TT}}_{\boldsymbol{u}}\otimes\cdots\otimes\mathbf{P}^{\mathrm{TT}}_{\boldsymbol{u}}\right)_{n}, \quad \mathbf{P}^{\mathrm{TT}}_{\boldsymbol{u}} = \sum_{\rho=-1}^{+1}\int\left|\boldsymbol{\varepsilon}^{[\rho\rho]\zeta}\right\rangle d\tilde{\Sigma}(\zeta)\left\langle\boldsymbol{\varepsilon}^{[\rho\rho]\zeta}\right| \; . \tag{10.1c}$$

The infinitesimal generators in (10.1a) are those of the restriction of the representation in (6.11) to TT internal graviton gauge modes, so that for the generators of spacetime translations we have

$$\tilde{P}_{j;u} = i\int \eta^{ik}\eta^{i'k'} g^{(+)}_{ii'}(\zeta)\,\partial_j g^{(-)}_{kk'}(\zeta)\,d\tilde{\Sigma}(\zeta)\,, \qquad \partial_j = \partial/\partial q^j\ . \tag{10.2}$$

The derivation leading in the electromagnetic case from (9.5.1) to (9.5.11) can be adapted to this internal graviton gauge in a straightforward manner, by taking advantage of the form (5.20) of the inner product in graviton fibres. Thus, we obtain:

$$\tilde{P}_{j;u} = \int :\tilde{T}_{jk}[g(u;\zeta)]:\, d\sigma^k(q)\, d\tilde{\Omega}(v)\ , \quad j = 0,1,2,3\ , \tag{10.3a}$$

$$\tilde{T}_{ij}[g] = \tfrac{1}{2}\eta_{ij} g_{kk',l} g^{kk',l} - g_{kl,i} g^{kl}{}_{,j}\ , \qquad g_{ik,j} = \partial g_{ik}/\partial q^j\ . \tag{10.3b}$$

In turn, this yields the following expressions for the generators of Lorentz transformations,

$$\tilde{M}^{jk}_u = \int :\tilde{M}^{jkl}_u[g(u;\zeta)] : d\sigma_l(q) d\tilde{\Omega}(v)\ , \tag{10.4a}$$

$$\tilde{M}^{jkl}_u[g] = Q^j_u \tilde{T}^{kl}[g] - Q^k_u \tilde{T}^{jl}[g] + g^\dagger \tilde{S}^{jkl}_u g\ , \tag{10.4b}$$

in complete analogy with (9.5.12).

In order to extend the operator-valued connection form in (10.1) into one that incorporates the diffeomorphism-invariance embedded in the quantum gravitational gauge group (9.27), let us reconsider the quantum pregeometries of Sec. 11.5 from the point of view of such gauge invariance. Hence, let us denote by $\mathbf{Riem_M S}$ the family of all physically acceptable Lorentzian mean metrics in $\mathbf{S}$, namely all those Lorentzian metrics that can be the outcome of a metrization resulting from physical information coming through the inflow surface $\mathbf{\Sigma'}$ of a base-segment $\mathbf{S}$. Clearly $\mathbf{Riem_M S}$ does not coincide with the family $\mathbf{Riem_L S}$ of all Lorentzian metrics in $\mathbf{S}$, since, first of all, only orientable and time-orientable (cf. Sec. 2.3) metrics for which $\mathbf{\Sigma'}$ and $\mathbf{\Sigma''}$ are spacelike, and for which $\mathbf{\Sigma''}$ is in the future of $\mathbf{\Sigma'}$, are acceptable; second, in view of Geroch's theorem, in order to be able to accommodate also quantum information related to half-integer spin states, the Poincaré frame bundle corresponding to any physically acceptable mean metric should be trivial[59].

According to the fundamental lemma of (pseudo-)Riemannian geometry, each mean metric g^{M} gives rise to a unique Levi-Civita connection, which in turn gives rise to a unique operator-valued connection form in (10.1). The diffeomorphism group $\mathbf{DiffS}$ casts all such connection forms into equivalence classes, which in view of the aforementioned one-to-one correspondence between these connection forms and mean metrics, can be identified with the elements of the base manifold of the principal bundle given by (cf. (3.1))

$$\mathbf{DiffS} \to \mathbf{Riem_M S} \to \mathbf{Riem_M S/DiffS}\ , \tag{10.5}$$

which represent gauge orbits (cf. Sec. 10.4) of the diffeomorphism group over $\mathbf{S}$.

To map these orbits in the vertical directions of a quantum gravitational bundle, let us introduce in the typical fibres of the single-ghost and single-antighost FP gravitational polarization bundles in (9.26) the respective orthonormal bases

$$\{C_\alpha | \alpha = 1,2,3,\ldots\} \subset \mathbf{H}\ , \qquad \{\overline{C}_\alpha | \alpha = 1,2,3,\ldots\} \subset \overline{\mathbf{H}}\ , \tag{10.6}$$

in terms of which we can expand the generic elements of those typical fibres:

$$c = \theta^\alpha(c) C_\alpha \in \mathbf{H}\,,\ \theta^\alpha(c) = \left(C_\alpha | c\right)_{J_0},\quad \bar{c} = \bar{\theta}^\alpha(\bar{c})\bar{C}_\alpha \in \overline{\mathbf{H}}\,,\ \bar{\theta}^\alpha(\bar{c}) = \left(\bar{C}_\alpha | \bar{c}\right)_{J_0}. \tag{10.7}$$

In accordance with (9.14) and the identifications in (9.21)-(9.22), we shall introduce in the direct sum of these two typical fibres the following Lie bracket,

$$[h,k] = [h,k]' \oplus [h,k]'' \in \mathbf{H} \oplus \overline{\mathbf{H}}\ , \qquad h,k \in \mathbf{H} \oplus \overline{\mathbf{H}}\ , \tag{10.8a}$$

$$[h,k]' := -\sum_{\alpha=1}^{\infty} \left[\left(C_\alpha | h_i \partial^i k - k_i \partial^i h\right)_{J_0} C_\alpha + \left(\bar{C}_\alpha | h_i \partial^i k - k_i \partial^i h\right)_{J_0} \bar{C}_\alpha \right]\ , \tag{10.8b}$$

which leaves them invariant, so that they constitute an infinite-dimensional Lie algebra $L_{\mathbf{H}}$. As a consequence of this definition, we shall be able to embed those elements of the Lie algebra $L(\mathbf{Diff\,S})$ of the diffeomorphism group $\mathbf{Diff\,S}$, which are "sufficiently local" in the sense defined in the preceding section (namely for which the exponential map at each point x within their supports is one-to-one), into the family of cross-sections of the bundle

$$\boldsymbol{L}(\mathbf{S}) = \mathbf{A}(\mathbf{S}) \times (\mathbf{H} \oplus \overline{\mathbf{H}})\ . \tag{10.9}$$

Indeed, in accordance with (10.3.1), the elements of the above bundle are $L_{\mathbf{H}}$-valued one-forms on the manifold constituting the total space of the principal bundle $\mathbf{A}(\mathbf{S})$. On the other hand, to each field $\xi \in L(\mathbf{Diff\,S})$ which is sufficiently local, the relations in (9.21)-(9.22) assign non-zero elements at all $\boldsymbol{u} \in \mathbf{A}(\mathbf{S})$ above points x within its support, namely

$$\xi \ \mapsto\ -\left(c(\boldsymbol{u};\xi), \bar{c}(\boldsymbol{u};\xi)\right) \ \leftrightarrow\ -\left(\theta^\alpha(c(\boldsymbol{u};\xi)), \bar{\theta}^\alpha(\bar{c}(\boldsymbol{u};\xi))\right)\,, \tag{10.10a}$$

$$\bar{c}(\boldsymbol{u};\xi) = \bar{\theta}^\alpha(\bar{c}(\boldsymbol{u};\xi))\bar{C}_\alpha \in \overline{\mathbf{H}}\ , \qquad \bar{\theta}^\alpha(\bar{c}(\boldsymbol{u};\xi)) = \left(\bar{C}_\alpha | \bar{c}(\boldsymbol{u};\xi)\right)_{J_0}\,, \tag{10.10b}$$

$$c(\boldsymbol{u};\xi) = \theta^\alpha(c(\boldsymbol{u};\xi)) C_\alpha \in \mathbf{H}\ , \qquad \theta^\alpha(c(\boldsymbol{u};\xi)) = \left(C_\alpha | c(\boldsymbol{u};\xi)\right)_{J_0}\ , \tag{10.10c}$$

and they assign zero at x outside that support. This results in the desired embedding into $\Gamma(\boldsymbol{L}(\mathbf{S}))$ of the Lie algebra of sufficiently local infinitesimal diffeomorphisms in $\mathbf{S}$.

Our next task is to construct a diffeomorphism supergroup $\mathcal{G}_{\boldsymbol{s}}(\mathbf{S}, \boldsymbol{g}^{\mathrm{M}})$ which incorporates the Lie algebra $L_0(\mathbf{Diff\,S})$ of all the infinitesimal generators of $\mathbf{Diff\,S}$ which are sufficiently local. By definition, a *Lie supergroup* [BI] (or a *super Lie group* [D]) is a supermanifold whose group operations of multiplication and inversion are given by Grassmann analytic functions in any coordinate chart. In turn, a *supermanifold* is a manifold M together with a sheaf[60] of algebras of functions which supply the coordinate charts for that supermanifold, and which assume values in a Grassmann algebra (cf. Sec. 8.3).

In order to realize such a construction in the present context, let us proceed as in Sec. 3.8, namely by viewing for all $\boldsymbol{u} = \mathbf{s}(x)$ the linear functionals

$$\theta^\alpha_{\boldsymbol{s}} : \boldsymbol{c} \mapsto \left(C_\alpha | \boldsymbol{\sigma}^{\boldsymbol{u}}_x(\boldsymbol{c})\right)_{J_0} = \left(C^{\boldsymbol{s}(x)}_\alpha | \boldsymbol{c}\right)_J\,,\quad C^{\boldsymbol{s}(x)}_\alpha = (\boldsymbol{\sigma}^{\boldsymbol{u}}_x)^{-1}(C_\alpha)\,,\quad \boldsymbol{c} \in \mathbf{H}_x\,, \tag{10.11a}$$

$$\overline{\theta}^{\alpha}_{s} : \overline{c} \mapsto \left(\overline{C}_{\alpha}\middle|\sigma^{u}_{x}(\overline{c})\right)_{J_0} = \left(\overline{C}^{s(x)}_{\alpha}\middle|\overline{c}\right)_{J} , \quad \overline{C}^{s(x)}_{\alpha} = (\sigma^{u}_{x})^{-1}(\overline{C}_{\alpha}) , \quad \overline{c} \in \overline{\mathbf{H}}_{x} ,\tag{10.11b}$$

as representatives of 1-forms in a given cross-section s of the principal bundle $\mathbf{A}(\mathbf{S},g^{\mathrm{M}})$. As such, the collection of these one-forms provides all the generators of an infinite dimensional Grassmann algebra Λ_s if their wedge-product is taken to represent their Grassmann product in (8.3.1). Hence, the corresponding family of Grassmann-valued functions

$$F(x,\overline{\theta}_s,\theta_s) = F_{0,0}(x) + \sum\nolimits_{m,n=1}^{\infty} F_{m,n}(x,\overline{\theta}_s,\theta_s) \ , \qquad x \in \mathbf{S} \ , \tag{10.12a}$$

$$F_{m,n}(x,\overline{\theta}_s,\theta_s) = \sum\nolimits_{m,n=1}^{\infty} \frac{1}{m!n!} F_{\beta_1,\dots,\beta_m;\gamma_1,\dots,\gamma_n}(x)\overline{\theta}^{\beta_1}_s \cdots \overline{\theta}^{\beta_m}_s \theta^{\gamma_1}_s \cdots \theta^{\gamma_n}_s \ , \tag{10.12b}$$

constitutes a supermanifold. In this context it should be recalled that, in accordance with (8.3.1), the coefficients in (10.12b) are antisymmetric under the permutation of their indices, so that as the summation is taken over all repeated indices, the same outcome is obtained when any two of those indices are interchanged, and that the Grassmann product in (10.12b) is non-zero only when all the β-indices as well as all the γ-indices are distinct.

A sub-supermanifold[61] $M(\mathbf{S},g^{\mathrm{M}})$ is obtained when those functions (10.12a) are singled out whose coefficients in (10.12b) can be identified with the coefficients of the purely ghost-antighost states of sections Ψ of the FP gravitational bundle in (9.26):

$$M(\mathbf{S},g^{\mathrm{M}}) = \left\{ F^{\Psi} \middle| \Psi \in \Gamma(\mathcal{G}^{\mathrm{A}}_{0}(\mathbf{S},\overline{g}) \otimes \overline{\mathcal{H}}(\mathbf{S},\overline{g}) \otimes_{A} \mathcal{H}(\mathbf{S},\overline{g})) \right\} , \tag{10.13a}$$

$$F^{\Psi}_{\beta_1,\dots,\beta_m;\gamma_1,\dots,\gamma_n}(x) = \left\langle \Psi_{0;m;n;u} \middle| C^{s}_{\beta_1,\dots,\beta_m;\gamma_1,\dots,\gamma_n}(x) \right\rangle \ , \qquad u = s(x) \ , \tag{10.13b}$$

$$C^{s}_{\beta_1,\dots,\beta_m;\gamma_1,\dots,\gamma_n}(x) = (\sigma^{u}_{x})^{-1}(\overline{C}_{\beta_1}) \otimes_{A} \cdots \otimes_{A} (\sigma^{u}_{x})^{-1}(C_{\gamma_n}) \ . \tag{10.13c}$$

On account of (9.25) and (9.26), the structure group of the *FP ghost-antighost bundle*

$$\mathcal{G}^{\mathrm{G}}(\mathbf{S},g^{\mathrm{M}}) = \mathcal{G}^{\mathrm{A}}_{0}(\mathbf{S},\overline{g}) \otimes \overline{\mathcal{H}}(\mathbf{S},\overline{g}) \otimes_{A} \mathcal{H}(\mathbf{S},\overline{g}) \tag{10.14}$$

is given by the representation of $\mathrm{ISO}^{\uparrow}(3,1)$ consisting of the operators

$$\mathbf{U}^{\mathrm{G}}(a,\Lambda) = \bigoplus\nolimits_{m,n=0}^{\infty} U^{\mathrm{BG}}(a,\Lambda)^{\otimes m} \otimes U^{\mathrm{CG}}(a,\Lambda)^{\otimes n} \ . \tag{10.15}$$

Since the action of the Poincaré group elements upon each component of a single-ghost or a single-antighost local state is that corresponding to the representations in (8.23), which are unitary with respect to the inner products in (8.22), it follows that the above direct sum is independent of the choice of section s. Consequently, the same is true of the prequantum gravitational supermanifold $M(\mathbf{S},g^{\mathrm{M}})$ in (10.13). Hence, $M(\mathbf{S},g^{\mathrm{M}})$ will be able to serve in the construction of a base quantum gravitational supermanifold for a quantum gravitational superbundle, to whose realization we turn next.

In view of the fact that a Grassmann algebra Λ_s has emerged naturally, we can apply the procedure in Sec. 8.3 to construct a *Berezin-Faddeev-Popov (BFP) superfibre bundle*

$$\mathcal{B}^{\mathrm{FP}}(\mathbf{S},\boldsymbol{g}^{\mathrm{M}}) = \mathcal{G}_0^{\mathrm{A}}(\mathbf{S},\overline{\boldsymbol{g}}) \otimes \overline{\mathcal{B}}(\mathbf{S},\overline{\boldsymbol{g}}) \underset{A}{\otimes} \mathcal{B}(\mathbf{S},\overline{\boldsymbol{g}}) \ , \tag{10.16a}$$

whose cross-sections coincide with the elements of the quantum gravitational supermanifold $M(\mathbf{S},\boldsymbol{g}^{\mathrm{M}})$:

$$M(\mathbf{S},\boldsymbol{g}^{\mathrm{M}}) = \left\{ \boldsymbol{\Psi} \middle| \boldsymbol{\Psi} \in \Gamma(\mathcal{B}^{\mathrm{FP}}(\mathbf{S},\boldsymbol{g}^{\mathrm{M}})) \right\} . \tag{10.16b}$$

On the other hand, in the BFP bundle we can introduce the Berezin coherent states

$$\boldsymbol{\Phi}_{\boldsymbol{c}}^{s} = \exp\left(-\tfrac{1}{2}\overline{\boldsymbol{\theta}}_s \cdot \boldsymbol{\theta}_s\right) \exp\left(\boldsymbol{\theta}_s \cdot C^{(+)}(\boldsymbol{c})\right) \boldsymbol{\Psi}_{0,0,0;x} \ , \tag{10.17a}$$

$$\overline{\boldsymbol{\theta}}_s \cdot \boldsymbol{\theta}_s = \sum\nolimits_{\alpha=1}^{\infty} \overline{\theta}_s^{\alpha} \theta_s^{\alpha} \ , \qquad \boldsymbol{\theta}_s \cdot C^{(+)}(\boldsymbol{c}) = \sum\nolimits_{\alpha=1}^{\infty} \theta_s^{\alpha} \left(C_{\alpha}^{s} | \boldsymbol{c}\right)_J C^{(+)}\left(C_{\alpha}^{s}\right) \ , \tag{10.17b}$$

$$\boldsymbol{\Phi}_{\overline{\boldsymbol{c}}}^{s} = \exp\left(-\tfrac{1}{2}\boldsymbol{\theta}_s \cdot \overline{\boldsymbol{\theta}}_s\right) \exp\left(\overline{\boldsymbol{\theta}}_s \cdot \overline{C}^{(+)}(\overline{\boldsymbol{c}})\right) \boldsymbol{\Psi}_{0,0,0;x} \ , \tag{10.17c}$$

$$\boldsymbol{\theta}_s \cdot \overline{\boldsymbol{\theta}}_s = \sum\nolimits_{\alpha=1}^{\infty} \theta_s^{\alpha} \overline{\theta}_s^{\alpha} , \qquad \overline{\boldsymbol{\theta}}_s \cdot \overline{C}^{(+)}(\overline{\boldsymbol{c}}) = \sum\nolimits_{\alpha=1}^{\infty} \overline{\theta}_s^{\alpha} \left(\overline{C}_{\alpha}^{s} | \overline{\boldsymbol{c}}\right)_J \overline{C}^{(+)}\left(\overline{C}_{\alpha}^{s}\right) \ . \tag{10.17d}$$

According to the type of arguments leading to (8.3.19), such states are eigenvectors of the ghost and antighost annihilation operators,

$$C_{j\zeta}^{(-)}(x)\boldsymbol{\Phi}_{\boldsymbol{c}}^{s} = c_{j\zeta}(x,\boldsymbol{\theta}_s)\boldsymbol{\Phi}_{\boldsymbol{c}}^{s} \ , \qquad c_{j\zeta}(x,\boldsymbol{\theta}_s) = \sum\nolimits_{\alpha=1}^{\infty} \theta_s^{\alpha} \left(C_{\alpha}^{s(x)} | \boldsymbol{c}_x\right)_J C_{\alpha j\zeta}^{s(x)} \ , \tag{10.18a}$$

$$\overline{C}_{j\zeta}^{(-)}(x)\boldsymbol{\Phi}_{\overline{\boldsymbol{c}}}^{s} = i\overline{c}_{j\zeta}(x,\overline{\boldsymbol{\theta}}_s)\boldsymbol{\Phi}_{\overline{\boldsymbol{c}}}^{s} \ , \qquad \overline{c}_{j\zeta}(x,\overline{\boldsymbol{\theta}}_s) = \sum\nolimits_{\alpha=1}^{\infty} \overline{\theta}_s^{\alpha} \left(\overline{C}_{\alpha}^{s(x)} | \overline{\boldsymbol{c}}_x\right)_J \overline{C}_{\alpha j\zeta}^{s(x)} \ . \tag{10.18b}$$

Furthermore, together with their duals, they provide the elements of *BFP quantum gravitational superframes*, since, in accordance with (8.3.17)-(8.3.18), we can decompose as follows the identity operators in the fibres of the FP ghost or antighost bundle:

$$\int_{\hat{\mathbf{H}}_x} \left|\boldsymbol{\Phi}_{\boldsymbol{c}}^{s}\right\rangle d\overline{\boldsymbol{c}}\, d\boldsymbol{c} \left\langle \boldsymbol{\Phi}_{\boldsymbol{c}}^{s}\right| = \mathbf{1}_{\mathcal{H}_x} \ , \qquad \int_{\hat{\overline{\mathbf{H}}}_x} \left|\boldsymbol{\Phi}_{\overline{\boldsymbol{c}}}^{s}\right\rangle d\boldsymbol{c}\, d\overline{\boldsymbol{c}} \left\langle \boldsymbol{\Phi}_{\overline{\boldsymbol{c}}}^{s}\right| = \mathbf{1}_{\overline{\mathcal{H}}_x} \ . \tag{10.19}$$

We can now adapt to the present situation the superfield approach to BRST symmetries developed in the context of Yang-Mills fields by Bonora *et al.* (1981b, 1982a), Delbourgo and Jarvis (1982), Hoyos *et al.* (1982), and others[62]. Thus, we shall introduce the following supergroups of operators acting in the superfibres of the BFP bundle,

$$\boldsymbol{U}(\boldsymbol{\xi},\overline{\boldsymbol{\theta}},\boldsymbol{\theta}) = \exp\left[\boldsymbol{\theta}\overline{\boldsymbol{C}}(\overline{\boldsymbol{c}}) + \overline{\boldsymbol{\theta}}\boldsymbol{C}(\boldsymbol{c}) + \boldsymbol{\theta}\overline{\boldsymbol{\theta}}\left(\boldsymbol{B}(\boldsymbol{b}) + \tfrac{1}{2}\boldsymbol{C}(\boldsymbol{c}) \times \overline{\boldsymbol{C}}(\overline{\boldsymbol{c}})\right)\right], \tag{10.20}$$

$$\boldsymbol{B}(\boldsymbol{b}) = B^{(+)j\zeta}(x)b_{j\zeta}(x) + B^{(-)j\zeta}(x)b_{j\zeta}^{\dagger}(x) \ , \qquad b_i = \partial^j \partial_j \overline{c}_i \ , \tag{10.21a}$$

$$\boldsymbol{C}=C^{(+)j\zeta}(x)c_{j\zeta}(x)+C^{(-)j\zeta}(x)c^{\dagger}_{j\zeta}(x),\ \ \overline{\boldsymbol{C}}=\overline{C}^{(+)j\zeta}(x)\overline{c}_{j\zeta}(x)-\overline{C}^{(-)j\zeta}(x)\overline{c}^{\dagger}_{j\zeta}(x), \quad (10.21b)$$

$$\boldsymbol{C}(\boldsymbol{c})\times\overline{\boldsymbol{C}}(\overline{\boldsymbol{c}})=\sum_{\alpha,\beta}\left[\boldsymbol{C}([C_\alpha^{\boldsymbol{s}(x)},\overline{C}_\beta^{\boldsymbol{s}(x)}]')+\overline{\boldsymbol{C}}([C_\alpha^{\boldsymbol{s}(x)},\overline{C}_\beta^{\boldsymbol{s}(x)}]'')\right]$$
$$\times\,\theta_{\boldsymbol{s}}^{\alpha}\left(C_\alpha^{\boldsymbol{s}(x)}|\boldsymbol{c}\right)_J\overline{\theta}_{\boldsymbol{s}}^{\beta}\left(\overline{C}_\beta^{\boldsymbol{s}(x)}|\overline{\boldsymbol{c}}\right)_J\ ,\quad \xi=\boldsymbol{c}\oplus\overline{\boldsymbol{c}}\in\Gamma(\mathbf{H}(\mathbf{S},\overline{\boldsymbol{g}})\oplus\overline{\mathbf{H}}(\mathbf{S},\overline{\boldsymbol{g}}))\ , \quad (10.21c)$$

where $\boldsymbol{b}$ is the quantum gauge-fixing field corresponding to b in (9.8a) and (9.18b).

To each one-parameter local group of diffeomorphisms defined, in accordance with (9.13), for sufficiently small values of the real parameter τ by the maps

$$\mathrm{D}_\xi(\tau)=\exp(\xi\tau):\ \ x\mapsto\exp[-\boldsymbol{X}_\xi(x)\tau]\in\mathbf{S}\ ,\qquad x\in\mathbf{S}\ , \quad (10.22)$$

we can assign, for each choice of section $\boldsymbol{s}$ of $\mathbf{A}(\mathbf{S},\boldsymbol{g}^{\mathrm{M}})$, the following family of ghost and antighost states determined by (10.10):

$$\tau\boldsymbol{c}_\xi^{\boldsymbol{s}(x)}=c_i(\boldsymbol{s}(x);\xi\tau)\boldsymbol{\theta}^i(x)\ ,\qquad \tau\overline{\boldsymbol{c}}_\xi^{\boldsymbol{s}(x)}=\overline{c}_i(\boldsymbol{s}(x);\xi\tau)\boldsymbol{\theta}^i(x)\ ,\qquad \tau\in\mathbf{R}^1\ . \quad (10.23)$$

In turn, to this family of states we can assign the following supergroup in the parameter τ,

$$\boldsymbol{U}(\tau\xi,\overline{\boldsymbol{\theta}}_{\boldsymbol{s}},\boldsymbol{\theta}_{\boldsymbol{s}})=\mathbf{1}+\tau\boldsymbol{\theta}_{\boldsymbol{s}}\overline{\boldsymbol{C}}(x;\overline{\boldsymbol{c}}_\xi^{\boldsymbol{s}})+\tau\overline{\boldsymbol{\theta}}_{\boldsymbol{s}}\boldsymbol{C}(x;\boldsymbol{c}_\xi^{\boldsymbol{s}})$$
$$+\tau\boldsymbol{\theta}_{\boldsymbol{s}}\overline{\boldsymbol{\theta}}_{\boldsymbol{s}}\left(\boldsymbol{B}(x;\overline{\boldsymbol{c}}_\xi^{\boldsymbol{s}})+\tfrac{1}{2}\boldsymbol{C}(x;\boldsymbol{c}_\xi^{\boldsymbol{s}})\times\overline{\boldsymbol{C}}(x;\overline{\boldsymbol{c}}_\xi^{\boldsymbol{s}})\right)\ , \quad (10.24)$$

which are locally isomorphic to the local group of diffeomorphisms in (10.22). The *diffeomorphism supergroup* $\mathcal{G}_{\boldsymbol{s}}(\mathbf{S},\boldsymbol{g}^{\mathrm{M}})$ is then obtained by taking products of the elements of the supergroup images under these maps corresponding to all one-parameter local groups of diffeomorphisms $\exp(s\xi)$ generated by sufficiently local fields $\xi\in L_0(\mathrm{Diff}\,\mathbf{S})$.

The *quantum gravitational gauge supergroup* $Q\mathcal{G}(\mathbf{S},\boldsymbol{g}^{\mathrm{M}})$ itself results upon letting $\boldsymbol{s}$ vary over all the sections of $\mathbf{A}(\mathbf{S},\boldsymbol{g}^{\mathrm{M}})$, and noting the existence of the following one-to-one maps between Grassmann algebra generators,

$$\theta_{\boldsymbol{s}}^{\alpha}\mapsto\theta_{\boldsymbol{s}'}^{\alpha}\ ,\qquad \overline{\theta}_{\boldsymbol{s}}^{\alpha}\mapsto\overline{\theta}_{\boldsymbol{s}'}^{\alpha}\ ,\qquad \boldsymbol{s}'=\boldsymbol{s}\cdot g\ ,\quad g\in\mathcal{G}(\mathbf{A}(\mathbf{S},\boldsymbol{g}^{\mathrm{M}}))\ . \quad (10.25)$$

These maps give rise to smooth mappings of the prequantum gravitational supermanifold in (10.16) onto itself, and supply a representation of the Poincaré gauge group $\mathcal{G}(\mathbf{A}(\mathbf{S},\boldsymbol{g}^{\mathrm{M}}))$:

$$\mathbf{U}(g):\ \ \Psi(x,\overline{\boldsymbol{\theta}}_{\boldsymbol{s}},\boldsymbol{\theta}_{\boldsymbol{s}})\mapsto\Psi(x,\overline{\boldsymbol{\theta}}_{\boldsymbol{s}'},\boldsymbol{\theta}_{\boldsymbol{s}'})\ ,\qquad g\in\mathcal{G}(\mathbf{A}(\mathbf{S},\boldsymbol{g}^{\mathrm{M}}))\ , \quad (10.26a)$$

$$\Psi(x,\overline{\boldsymbol{\theta}}_{\boldsymbol{s}},\boldsymbol{\theta}_{\boldsymbol{s}})=\Psi_{0,0}(x)+\sum_{m,n=1}^{\infty}\Psi_{m,n}(x,\overline{\boldsymbol{\theta}}_{\boldsymbol{s}},\boldsymbol{\theta}_{\boldsymbol{s}})\ ,\qquad \Psi\in M(\mathbf{S},\boldsymbol{g}^{\mathrm{M}})\ . \quad (10.26b)$$

The *quantum gravitational superfibre bundle* can be now defined as the $\mathcal{G}$-product (cf. Secs. 10.2-10.4) resulting from the adjoint action of $Q\mathcal{G}(\mathbf{S},\boldsymbol{g}^{\mathrm{M}})$ on $M(\mathbf{S},\boldsymbol{g}^{\mathrm{M}})$:

$$\mathcal{GS}(\mathbf{S},\boldsymbol{g}^{\mathrm{M}}) = \mathbf{P}(\mathcal{M},\mathcal{G}) \times_{\mathcal{G}} \Gamma(\mathcal{E}^{\mathrm{TT}}(\mathbf{S},\boldsymbol{g}^{\mathrm{M}})) \ , \tag{10.27a}$$

$$\mathcal{G} = \mathcal{QG}(\mathbf{S},\boldsymbol{g}^{\mathrm{M}}) \ , \qquad \mathcal{M}(\mathbf{S},\boldsymbol{g}^{\mathrm{M}}) = M(\mathbf{S},\boldsymbol{g}^{\mathrm{M}})/\mathcal{QG}(\mathbf{S},\boldsymbol{g}^{\mathrm{M}}) \ . \tag{10.27b}$$

Hence, its typical fibre consists of all the cross-sections of the TT-subbundle $\mathcal{E}^{\mathrm{TT}}(\mathbf{S},\boldsymbol{g}^{\mathrm{M}})$ of the Lorenz gravitational bundle $\mathcal{E}^{L}(\mathbf{S},\boldsymbol{g}^{\mathrm{M}})$ defined in Sec. 11.6.

The treatment afforded to Yang-Mills theories in the preceding chapter, as well as in the earlier cited literature, can be now applied to extend the connection form in (10.1) in the vertical directions of the above superfibre bundle (namely along the superfibres in which the quantum gravitational gauge group acts), so as to assume the form

$$\boldsymbol{\nabla} = \mathbf{d} + \boldsymbol{\Gamma} + \boldsymbol{C} + \overline{\boldsymbol{C}} \ , \qquad\qquad \mathbf{d} = \boldsymbol{d} + \boldsymbol{\delta} + \overline{\boldsymbol{\delta}} \ , \tag{10.28}$$

which incorporates the following geometrized version of *quantum gravitational BRST and anti-BRST operators*:

$$\boldsymbol{\delta} = \boldsymbol{d}\overline{\boldsymbol{\theta}}\, \partial/\partial\overline{\boldsymbol{\theta}} \ , \qquad\qquad \overline{\boldsymbol{\delta}} = \boldsymbol{d}\boldsymbol{\theta}\, \partial/\partial\boldsymbol{\theta} \ . \tag{10.29}$$

Indeed, in accordance with (10.4.3)-(10.4.5) and (10.4.10), upon setting $\Gamma_\mu = \tilde{\boldsymbol{\Gamma}}_\mu$ we get

$$\boldsymbol{\nabla}_\mu = \boldsymbol{\partial}_\mu + \boldsymbol{\Gamma}_\mu \ , \qquad\qquad \Gamma_\mu = \mathrm{Ad}_{U^{-1}(g)}\, \Gamma_\mu + U^{-1}(g)\, \boldsymbol{\partial}_\mu U(g) \ , \tag{10.30}$$

where the elements $g(x)$ are the representatives of the various gravitational gauge supergroup transformations $g \in \mathcal{QG}(\mathbf{S},\boldsymbol{g}^{\mathrm{M}})$ in the fibres above the various points x in the base-segment $\mathbf{S}$. If we take note of the fact that according to (10.20)

$$U(g(x)) = \exp\left[\boldsymbol{\theta}\overline{C}(x) + \overline{\boldsymbol{\theta}}C(x) + \boldsymbol{\theta}\overline{\boldsymbol{\theta}}\left(B(x) + \tfrac{1}{2}C(x)\times\overline{C}(x)\right)\right] , \tag{10.31}$$

$$U^{-1}(g(x)) = \exp\left[-\boldsymbol{\theta}\overline{C}(x) - \overline{\boldsymbol{\theta}}C(x) - \boldsymbol{\theta}\overline{\boldsymbol{\theta}}\left(B(x) + \tfrac{1}{2}C(x)\times\overline{C}(x)\right)\right], \tag{10.32}$$

then straightforward algebra yields (Bonora *et al.*, 1981b, 1982a; Hoyos *et al.*, 1982) that

$$\Gamma_\mu(x,\overline{\boldsymbol{\theta}},\boldsymbol{\theta}) = \Gamma_\mu(x) + \boldsymbol{\theta}D_\mu\overline{C}(x) + \overline{\boldsymbol{\theta}}D_\mu C(x) + \boldsymbol{\theta}\overline{\boldsymbol{\theta}}\left(D_\mu B(x) + D_\mu C(x)\times\overline{C}(x)\right), \tag{10.33a}$$

$$B(x) = \boldsymbol{B}(\boldsymbol{b}(x)) \ , \ C(x) = \boldsymbol{C}(\boldsymbol{c}(x)) \ , \ \overline{C}(x) = \overline{\boldsymbol{C}}(\overline{\boldsymbol{c}}(x)) \ , \ D_\mu = \partial_\mu + \left[\Gamma_\mu(x), \cdot\right] , \tag{10.33b}$$

and that the Maurer-Cartan form in (10.26), generically given in the case of ordinary gauge groups of the second kind by (10.2.12)-(10.2.13), is now represented by:

$$\boldsymbol{C}(x,\overline{\boldsymbol{\theta}},\boldsymbol{\theta}) = C(x) + \boldsymbol{\theta}\overline{B}(x) - \tfrac{1}{2}\overline{\boldsymbol{\theta}}C(x)\times C(x) + \boldsymbol{\theta}\overline{\boldsymbol{\theta}}\left[\overline{B}(x), C(x)\right] \ , \tag{10.34a}$$

$$\overline{\boldsymbol{C}}(x,\overline{\boldsymbol{\theta}},\boldsymbol{\theta}) = \overline{C}(x) + \overline{\boldsymbol{\theta}}B(x) - \tfrac{1}{2}\boldsymbol{\theta}\overline{C}(x)\times\overline{C}(x) - \boldsymbol{\theta}\overline{\boldsymbol{\theta}}\left[B(x), \overline{C}(x)\right] \ , \tag{10.34b}$$

$$\overline{B}(x) = -B(x) - C(x) \times \overline{C}(x) \quad . \tag{10.34c}$$

On the other hand, we see from (10.22) and (10.23) that any change linear in an infinitesimal diffeomorphism ξ induces a corresponding change in the single-ghost and single-antighost fields:

$$\xi \mapsto \xi + \delta\xi \qquad \Rightarrow \qquad \overline{\boldsymbol{c}} \mapsto \overline{\boldsymbol{c}} + \delta\overline{\boldsymbol{c}} \quad , \qquad \boldsymbol{c} \mapsto \boldsymbol{c} + \delta\boldsymbol{c} \quad . \tag{10.35}$$

In turn, due to the linearity of the operator-valued one-forms in (10.20), and in view of (10.24), each one of these changes can be interpreted as one of the following respective translations

$$\boldsymbol{\theta} \mapsto \boldsymbol{\theta} + \delta\boldsymbol{\theta} \quad , \qquad \overline{\boldsymbol{\theta}} \mapsto \overline{\boldsymbol{\theta}} + \delta\overline{\boldsymbol{\theta}} \quad , \tag{10.36}$$

in the values assumed by the Grassmannian variables in (10.20). Hence, infinitesimal transformations of this nature, as well as those corresponding to (10.24) and (10.25), give rise to covariant differentiation operators represented by the following *quantum gravitational covariant BRST* and *anti-BRST connection forms*:

$$\boldsymbol{S} = \boldsymbol{\delta} + U^{-1}(g)\partial_{\overline{\boldsymbol{\theta}}} U(g)\,\boldsymbol{d}\overline{\boldsymbol{\theta}} \quad , \qquad \partial_{\overline{\boldsymbol{\theta}}} = \partial/\partial\overline{\boldsymbol{\theta}} \quad , \tag{10.37a}$$

$$\overline{\boldsymbol{S}} = \overline{\boldsymbol{\delta}} + U^{-1}(g)\partial_{\boldsymbol{\theta}} U(g)\,\boldsymbol{d}\overline{\boldsymbol{\theta}} \,, \qquad \partial_{\boldsymbol{\theta}} = \partial/\partial\boldsymbol{\theta} \quad . \tag{10.37b}$$

It then follows[63] from (10.33) and (10.34) that the translations in (10.36) are associated with the following counterparts of the BRST and anti-BRST transformations in (10.6.19):

$$\boldsymbol{\delta}\Gamma_\mu(x) = \delta\overline{\boldsymbol{\theta}}\, D_\mu C(x) \quad , \qquad \boldsymbol{\delta}B(x) = \boldsymbol{0} \quad , \tag{10.38a}$$

$$\boldsymbol{\delta}C(x) = -\tfrac{1}{2}\delta\overline{\boldsymbol{\theta}}\, C(x) \times C(x) \,, \qquad \boldsymbol{\delta}\overline{C}(x) = \delta\overline{\boldsymbol{\theta}} B(x) \quad , \tag{10.38b}$$

$$\overline{\boldsymbol{\delta}}\Gamma_\mu(x) = \delta\boldsymbol{\theta}\, D_\mu \overline{C}(x) \quad , \qquad \overline{\boldsymbol{\delta}}\,\overline{B}(x) = \boldsymbol{0} \quad , \tag{10.39a}$$

$$\overline{\boldsymbol{\delta}}\overline{C}(x) = -\tfrac{1}{2}\delta\boldsymbol{\theta}\,\overline{C}(x) \times \overline{C}(x) \,, \qquad \overline{\boldsymbol{\delta}}C(x) = \delta\boldsymbol{\theta}\overline{B}(x) \quad . \tag{10.39b}$$

We observe that the parallel transport to which the quantum gravitational connection in (10.24) gives rise can be carried out in arbitrary directions. In particular, when it is carried out in the directions of the superfibres of the quantum gravitational base supermanifold in (10.23b), then the covariant derivatives in those directions are given by the covariant BRST and anti-BRST connection forms in (10.37), which can be expressed as follows[64]:

$$\boldsymbol{S} = \boldsymbol{\delta} + \boldsymbol{C} \quad , \qquad \overline{\boldsymbol{S}} = \overline{\boldsymbol{\delta}} + \overline{\boldsymbol{C}} \quad . \tag{10.40}$$

At a general differential-geometric level, these operators are the geometric counterparts of the nilpotent BRST and anti-BRST charge operators that were originally used by Curci and Ferrari (1976c), Kugo and Ojima (1978, 1979), and others, to define physical states as being those states which belong to the kernel of such operators (Henneaux, 1985, 1986).

11.11. Principles of GS Propagation in Quantum Gravitational Bundles

With the quantum gravitational supergroup and connection derived in the preceding section, the stage is set for the formulation of the main ideas of GS propagation in quantum gravitational bundles which incorporate quantum gravitational fields, as well as quantum fields describing matter and nongravitational radiation. The basic idea of such GS propagation is rooted in the iterative formulation of the classical geometrodynamic evolution described in Sec. 11.1: initial data are prescribed within an initial-data segment $\mathbf{S}_0$, which provide the metrization of $\mathbf{S}_0$, and implicitly serve to determine the quantum gravitational supermanifold $\mathcal{M}(\mathbf{S}_0, \boldsymbol{g}^{\mathrm{M}})$, as well as the quantum states of matter and radiation fields (including quantum gravitational radiation) within $\mathcal{M}(\mathbf{S}_0, \boldsymbol{g}^{\mathrm{M}})$. The GS evolution of these states then sequentially metrizes the ensuing *subsequent* segments $\mathbf{S}_n$, $n = 1,2,3,....$, thus *creating* the corresponding quantum gravitational supermanifolds $\mathcal{M}(\mathbf{S}_n, \boldsymbol{g}^{\mathrm{M}})$, as well as the quantum states of matter and radiation fields (including quantum gravitational radiation) within these supermanifolds. Indeed, since by its very nature GS propagation is *not* time-reversible (as it was evident already in the cases with external gravitational fields treated in the preceding six chapters), it imposes a "time arrow" (Penrose, 1987, 1989; Zeh, 1989) on quantum evolution, so that the "present" does not unambiguously determine the "past", but can only be claimed to be compatible or non-compatible with any specified set of data in the "past ".

The data required along the inflow surface $\boldsymbol{\Sigma}_0'$ of a would-be initial-data segment $\mathbf{S}_0$ consist, in analogy with the classical case, of the specification, at all points $x \in \boldsymbol{\Sigma}_0'$, of a mean 3-metric $\boldsymbol{\gamma}_0$ and the mean extrinsic curvature $\boldsymbol{K}_0$, of the quantum gravitational states $\boldsymbol{\Psi}_x$ describing the quantum fluctuations around those mean values, as well as of an informationally complete set of initial data for all the matter and nongravitational radiation quantum fields, in accordance with the GS schemes for such fields discussed in Chapters 7-10. These initial data can be operationally most easily specified upon choosing in $\mathbf{S}_0$ a mean metric representative $\boldsymbol{g}^{\mathrm{M}}$ for which $\boldsymbol{\Sigma}_0'$ is a synchronous inflow surface, so that, in a suitably chosen atlas of Gaussian normal coordinates, it will assume the form of (1.7), namely:

$$\boldsymbol{g}^{\mathrm{M}} = d\tau \otimes d\tau - \gamma^{\mathrm{M}}_{ab}\, dx^a \otimes dx^b \ , \qquad x \in \mathbf{S}_0 \ . \tag{11.1}$$

All other mean metrics amongst all the Lorentzian metrics in $\mathbf{S}_0$ diffeomorphically equivalent to $\boldsymbol{g}^{\mathrm{M}}$, in the physically acceptable manner discussed in the preceding section, must have the generic form corresponding to (1.4), namely can be given by

$$\bar{\boldsymbol{g}} = \left(\bar{N}^2 - \bar{N}_a \bar{N}_b\, \gamma^{ab}\right) dx^0 \otimes dx^0 + \bar{N}_a \left(dx^0 \otimes dx^a + dx^a \otimes dx^0\right) - \bar{\gamma}_{ab}\, dx^a \otimes dx^b \tag{11.2}$$

in terms of lapse and shift functions. The transition to such metrics can be then effected by using the relationships (10.22)-(10.24), which map the diffeomorphisms between such metrics into elements of the diffeomorphism supergroup $\mathcal{G}_s(\mathbf{S}_0, \boldsymbol{g}^{\mathrm{M}})$, and thus make the transition to other locations in the quantum gravitational supermanifold $\mathcal{M}(\mathbf{S}_0, \boldsymbol{g}^{\mathrm{M}})$.

The specification of quantum gravitational initial data can be carried out in the context of the quantum pregeometry over $\mathbf{S}_0$ (cf. Sec. 11.5), by selecting a cross-section

$$\mathbf{s}_0 = \{\ \boldsymbol{u}(x) = (\mathbf{0}, \boldsymbol{\theta}^i(x)) \in \mathbf{A}(\Sigma_0') \mid x \in \Sigma_0'\ \} \tag{11.3}$$

of the restriction $\mathbf{A}(\mathbf{\Sigma}_0')$ of the bundle-segment $\mathbf{A}(\mathbf{S}_0)$ to $\mathbf{\Sigma}_0'$, for which, in accordance with (11.1) and (1.20), we can impose the initial conditions[65]

$$g^{\mathrm{M}}_{00} = 1 \ , \quad g^{\mathrm{M}}_{a0} = g^{\mathrm{M}}_{0a} = 0 \ , \quad g^{\mathrm{M}}_{ab} = -\gamma_{ab} \ , \quad \dot{g}^{\mathrm{M}}_{ab} = -2K_{ab} \tag{11.4}$$

along $\mathbf{\Sigma}_0'$. To treat these objects geometro-stochastically, we shall adopt the approach to the vacuum Einstein equations in (1.10) advocated in (Deser; 1970; Boulware and Deser, 1975), whereby those equations are viewed as belonging to a gauge theory with self-coupling. In this approach the Hilbert-Palatini classical action in (1.11a), which we can re-write in a form involving a Lagrangian density [M,W] for the classical metric $\mathbf{g}$,

$$S_{\mathrm{HP}} = \int d^4x\, \mathcal{L}_{\mathrm{HP}}[\mathbf{g},\overline{\nabla}] \quad , \quad \mathcal{L}_{\mathrm{HP}}[\mathbf{g},\overline{\nabla}] = (1/16\pi)\sqrt{-\mathrm{g}}\, \mathrm{g}^{ij} R_{ij} \ , \tag{11.5}$$

is transformed into one with a Lagrangian density which is the sum of two terms:

$$S_{\mathrm{D}} = \int d^4x \left\{ \mathcal{L}_{\mathrm{L}}[\overline{\boldsymbol{h}},\overline{\nabla}] + \overline{h}^{ij}\left(\Gamma^l{}_{ij}\Gamma^k{}_{kl} - \Gamma^l{}_{ki}\Gamma^k{}_{jl}\right)\right\} \ , \tag{11.6a}$$

$$\mathcal{L}_{\mathrm{L}}[\overline{\boldsymbol{h}},\overline{\nabla}] = \overline{h}^{ij}\left(\Gamma^k{}_{ij,k} - \Gamma^k{}_{ki,j} - \lambda_{kj}{}^l\Gamma^k{}_{li}\right) + \eta^{ij}\left(\Gamma^l{}_{ij}\Gamma^k{}_{kl} - \Gamma^l{}_{ki}\Gamma^k{}_{jl} - \lambda_{kj}{}^l\Gamma^k{}_{li}\right) \ , \tag{11.6b}$$

$$\sqrt{-\mathrm{g}}\, \mathrm{g}^{ij} = \eta^{ij} - \overline{h}^{ij} \ , \quad \mathrm{g}_{ij} = \eta_{ij} + h_{ij} \ , \quad \mathrm{g}^{ij}\mathrm{g}_{jk} = \delta^i{}_k \ . \tag{11.6c}$$

The connection coefficients in (11.6) are those of the Levi-Civita connection belonging to the Lorentzian metric $\mathbf{g}$. Hence, according to (2.6.23)-(2.6.26), the components of that metric in an arbitrary global Cartan moving frame over a segment $\mathbf{S}$, namely in a cross-section of the linear frame bundle $GL\mathbf{S}$ which is identifiable with the $GA\mathbf{S}$ cross-section

$$\boldsymbol{s} = \left\{ (\mathbf{0}, \boldsymbol{e}_i(x)) \,\middle|\, \boldsymbol{e}_i(x) = \lambda_i{}^\mu(x)\partial_\mu \ , \ x \in \mathbf{S} \right\} \subset GA\mathbf{S} \ , \tag{11.7}$$

are related to the connection coefficients and to the Riemann curvature tensor components in that moving frame as follows[66]:

$$\Gamma^i{}_{jk} = \mathrm{g}^{il}\Gamma_{ljk} \ , \quad \Gamma_{ijk} = \tfrac{1}{2}\left(h_{ij,k} + h_{ik,j} - h_{jk,i} + \mathrm{g}_{il}\lambda_{jk}{}^l + \mathrm{g}_{jl}\lambda_{ik}{}^l + \mathrm{g}_{kl}\lambda_{ij}{}^l\right) \ , \tag{11.8a}$$

$$R^i{}_{jkl} = \Gamma^i{}_{lj,k} - \Gamma^i{}_{kj,l} + \Gamma^i{}_{km}\Gamma^m{}_{lj} - \Gamma^i{}_{lm}\Gamma^m{}_{kj} - \lambda_{kl}{}^m\Gamma^i{}_{mj} \ , \tag{11.8b}$$

$$h_{ij,k}(x) = \boldsymbol{e}_k(x)h_{ij}(x) = \lambda_k{}^\mu(x)\partial_\mu h_{ij}(x) \ , \quad \Gamma^i{}_{jk,l}(x) = \boldsymbol{e}_l(x)\Gamma^i{}_{jk}(x) \ , \tag{11.8c}$$

$$\lambda_{ij}{}^k = \left(\lambda_i{}^\mu\partial_\mu\lambda_j{}^\nu - \lambda_j{}^\mu\partial_\mu\lambda_i{}^\nu\right)\lambda^k{}_\nu \ , \quad \boldsymbol{\theta}^k = \lambda^k{}_\nu\, \boldsymbol{dx}^\nu \ . \tag{11.8d}$$

The Lagrangian density in (11.6b) corresponds to the approximation that emerges from the linearized theory of gravity [M,W], and which in holonomic frames can be expressed as

$$\mathcal{L}_{\mathrm{L}}[\boldsymbol{h}] = -\tfrac{1}{2}\left(h_{\mu\nu,\lambda}h^{\mu\nu,\lambda} - 2h_\mu h^\mu + h_\mu h^{,\mu} - h_{,\mu}h^{,\mu}\right) \ , \quad h := \eta^{\mu\nu}h_{\mu\nu} \ , \tag{11.9}$$

where the repeated indices stand for contractions in which the Minkowski metric is used – cf. (Boulware and Deser, 1975), p. 206. Hence, by a standard use of variational principles one obtains (Gupta, 1952; Kraichnan, 1955) the following well-known equations for the linearized theory of gravity (cf. [M], Eq. (18.5); Boulware and Deser, 1975, Eq. (26)):

$$2G^{\mathrm{L}}_{\mu\nu} := h_{\mu\nu,\lambda}{}^{\lambda} - h_{\mu\lambda,\nu}{}^{\lambda} - h_{\nu\lambda,\mu}{}^{\lambda} + h_{,\mu\nu} + \eta_{\mu\nu}\left(h_{\kappa\lambda}{}^{,\kappa\lambda} - h_{,\lambda}{}^{\lambda}\right) = 0 \ . \tag{11.10}$$

In the TT-gauge these equations reduce to ten independent D'Alembert wave equations:

$$2G^{\mathrm{TT}}_{\mu\nu} := h_{\mu\nu,\lambda}{}^{\lambda} = \eta^{\kappa\lambda}\partial_\kappa\partial_\lambda h_{\mu\nu} = 0 \ , \qquad \mu,\nu = 0,1,2,3 \ . \tag{11.11}$$

Deser's (1970) key observation was that, upon using in (11.6) the expansion

$$\bar{h}_{ij} = h_{ij} - \tfrac{1}{2}\eta_{ij}h + O(k^2) \ , \qquad \bar{h}_{ij} = g_{ik}g_{jl}\bar{h}^{kl} \ , \tag{11.12}$$

which is formally indicative (from the point of view of the momentum representation) of a gravitational radiation low-frequency approximation, one should neglect in the stress-energy tensor that corresponds to the quadratic Lagrangian density in (11.9) (Gupta, 1952; Kraichnan, 1955) all the divergence terms in it, and hence set it equal to

$$T^{\mathrm{L}}_{ij} = (1/16\pi)\left(\Gamma^{l}{}_{ij}\Gamma^{k}{}_{kl} - \Gamma^{l}{}_{ki}\Gamma^{k}{}_{jl}\right) \ . \tag{11.13}$$

Then one can reproduce the Einstein equations in the following form:

$$G^{\mathrm{L}}_{\mu\nu} = -8\pi\left(T^{\mathrm{L}}_{\mu\nu} + T^{\mathrm{mat}}_{\mu\nu}\right) \ , \tag{11.14}$$

where the second term on the right-hand side of (11.14) stands for the stress-energy tensor of "matter" – into which, however, nongravitational radiation is also included.

By taking advantage of this observation in their study of the tree graph structure of graviton-graviton and graviton-matter interaction in the covariant formulation of quantum gravity (cf. Sec. 11.2) with a Minkowski background metric, Boulware and Deser arrive at the following conclusion (italicized as in the original): "*a quantum particle description of local (noncosmological) gravitational phenomena necessarily leads to a classical limit which is just a metric theory of gravity. As long as only helicity ±2 gravitons are included, the theory is precisely Einstein's general relativity*" (Boulware and Deser, 1975, p. 195).

In the GS approach this idea is modified by, first of all, replacing the Minkowski background metric with generic mean metrics (11.2) over base-segments. Within the initial-data segment $\mathbf{S}_0$ a mean metric (11.1) is generated by the vacuum Einstein equations from the initial data given along $\mathbf{\Sigma}_0'$ in (11.4). The quantum evolution obtained from the subsequently described GS reformulation of the equations in (11.14) will then lead to a quantum metric field whose expectation values along the outflow surface $\mathbf{\Sigma}_0''$ of $\mathbf{S}_0$ will supply the mean metric data along the inflow surface $\mathbf{\Sigma}_1' = \mathbf{\Sigma}_0''$ of the next base-segment $\mathbf{S}_1$. In accordance with (6.16) and (10.25)-(10.27), the required expectation values are given by

$$\bar{g}(x) = \left\langle \Psi \middle| g^{Q}(X;\tilde{\zeta})\Psi \right\rangle / \langle \Psi | \Psi \rangle , \quad \tilde{\zeta} = (-a^{j}(x) + in^{j}(x))e_{j}(x) , \tag{11.15a}$$

$$\dot{\bar{g}}(x) = \left\langle \Psi \middle| n^{j}(x)\partial_{j} g(X;\tilde{\zeta})\Psi \right\rangle / \langle \Psi | \Psi \rangle , \quad \Psi \in \mathcal{GS}(\mathbf{S}_{0}, g^{M}) , \tag{11.15b}$$

$$g^{Q}(X;\zeta) = \eta_{ij}\, \theta^{i}(x) \otimes \theta^{j}(x) + g(X;\zeta) , \quad X = (x, \bar{\theta}, \theta) , \tag{11.15c}$$

$$\theta^{i}(x;e_{j}) = \delta^{i}{}_{j} , \quad u = (a(x), \theta^{i}(x)) \in \mathbf{A}(\mathbf{S}_{0}, g^{M}) \subset \mathbf{A}(\mathbf{S}_{0}) , \tag{11.15d}$$

where $n(x)$ denotes the future-pointing normal at $x \in \Sigma_0''$. These data will serve as initial data for the vacuum Einstein equations in $\mathbf{S}_1$, thus producing the mean metric g^M in that next base-segment. This process can be then repeated iteratively for base-segments $\mathbf{S}_n$ whose outflow surfaces Σ_n'' have the maximum g^M-*mean proper* time (i.e., proper time determined by the mean metric g^M) separation ε from *any* base point x in their inflow surfaces Σ_n'. From a mathematical point of view, the limit $\varepsilon \to +0$ has to be ultimately considered. However, on account of the Planck length being *the* fundamental length in GSQM, maximum proper time distances ε of orders of magnitude smaller than one in Planck units do not appear to be physically essential, whereas from a computational point of view maximum proper time distances significantly larger than one might be more realistically implementable in many computer-generated algorithmic schemes[67].

The basic ideas of the formulation of the GS propagation of the quantum gravitational and matter fields in any base-segment $\mathbf{S}$ follow rather directly when the considerations in the preceding four chapters are combined with the superfield formalism of the preceding section. Foremost in this approach is the geometric notion of connection and associated parallel transport, which, for reasons explained already in Secs. 7.5 and 9.5, in the quantum field theoretical regime should take precedence over Lagrangian principles – despite the great appeal of such variational principles at the classical level. Indeed, as can be easily seen from many good treatments[68] of Noether's theorems on which these principles rely, their range of applicability extends, strictly speaking, only to classical field theory, since domain problems with quantum fields, their ordering in the Lagrangian, as well as the very notion of variation of a quantum Lagrangian are ill-defined. Thus, it is in fact only the heuristic path-integral notion of superposition over "field histories" that provides the bridge between classical variational principles and quantum field propagation. That idea is embedded in GSQG in a purely geometric manner via the notion of superposition of outcomes of parallel transport of quantum field theoretical states along all physically allowed paths.

The path γ of the propagator for parallel transport, within the superspace $\mathcal{M}(\mathbf{S}, g^M)$, of the quantum states of matter and (gravitational or nongravitational) radiation,

$$K_{\gamma}(\mathbf{f}'';\mathbf{f}') = \left\langle \Phi_{\mathbf{f}''} \middle| \mathcal{T}_{\gamma}(X'',X')\Phi_{\mathbf{f}'} \right\rangle , \quad X' = (x', \bar{\theta}', \theta') , \quad X'' = (x'', \bar{\theta}'', \theta'') , \tag{11.16}$$

cannot be arbitrary in order to qualify as "physical". Thus, first of all, in terms of the BRST and anti-BRST connection forms in (10.40) it must satisfy the subsidiary conditions

$$S_{Y}\Psi = \bar{S}_{Y}\Psi = 0 , \quad Y = \dot{x}^{\mu}\partial_{\mu} + \dot{\bar{\theta}}\partial_{\bar{\theta}} + \dot{\theta}\partial_{\theta} \in T_{X}\mathcal{M} , \quad X = (x, \bar{\theta}, \theta) \in \gamma , \tag{11.17}$$

where Y denotes the field of tangents along the smooth curve γ within that superspace. As we mentioned in the preceding section, in the conventional quantum field theory of gauge fields this corresponds to the Kugo-Ojima (1979) condition that physical states of such fields must belong to the kernel of BRST and anti-BRST operators. On the other hand, the physical meaning of this GSQG condition is very transparent in its geometric form (11.17): *physical* parallel transport leading to GSQG propagation has to take place along paths $\boldsymbol{\gamma}$ which are the horizontal lifts of paths in $\mathbf{S}$ (cf. also Sec. 9.4), for any fixed choice of gauge for the diffeomorphism group, reflected in a choice of mean metric from the equivalence class of diffeomorphically equivalent mean metrics in (11.2), and for any fixed Poincaré gauge, corresponding to any choice $\boldsymbol{s}$ of section of the bundle-segment for that metric.

The subsidiary conditions in (11.17) ensure that the *physical* parallel transport is governed, in accordance with (10.28), (10.30) and (10.33), by the equation

$$\boldsymbol{\nabla}_{\mathbf{Y}}\boldsymbol{\Psi} = \dot{x}^\mu\left(\partial_\mu + \tilde{\boldsymbol{\Gamma}}_\mu + \boldsymbol{\theta} D_\mu \bar{C} + \bar{\boldsymbol{\theta}} D_\mu C + \boldsymbol{\theta}\bar{\boldsymbol{\theta}}(D_\mu B + D_\mu C \times \bar{C})\right)\boldsymbol{\Psi} = \mathbf{0} \ , \qquad (11.18a)$$

where, as it follows from (10.1), at a fixed choice $\boldsymbol{s}$ of section of $\mathbf{A}(\mathbf{S},\boldsymbol{g}^{\mathrm{M}})$ we have:

$$\dot{x}^\mu(t)\tilde{\boldsymbol{\Gamma}}_\mu(x(t)) = -i\lambda^j{}_\mu(x(t))\dot{x}^\mu(t)\tilde{\boldsymbol{P}}_{j;\boldsymbol{s}(x(t))} + \tfrac{i}{2}\overline{\boldsymbol{\omega}}_{kl}(\dot{x}^\mu(t)\partial_\mu)\tilde{\boldsymbol{M}}^{kl}_{\boldsymbol{s}(x(t))} \ . \qquad (11.18b)$$

The computation of the propagator for physical parallel transport in (11.16) can then proceed in accordance with the guidelines established in Secs. 7.5, 8.4 and 9.5. Thus, we first adopt the TT-gauge and isolate from the quantum gravitational field in (11.15c) its TT-component in the form:

$$\boldsymbol{g}^{\mathrm{TT}}(x;\zeta) = g^{\mathrm{TT}}_{ij}(x;\zeta)\,\boldsymbol{\theta}^i(x)\otimes\boldsymbol{\theta}^j(x) = \mathcal{P}^{\mathrm{TT}}_{\boldsymbol{u}}\,\boldsymbol{g}(\boldsymbol{u};\zeta)\mathcal{P}^{\mathrm{TT}}_{\boldsymbol{u}} \ , \qquad \boldsymbol{u} = \boldsymbol{s}(x) \ . \qquad (11.19)$$

We then construct from the quantum gravitational pregeometry in (5.27) the *TT-quantum gravitational frames*

$$\boldsymbol{\Phi}^{\boldsymbol{s}}_{\boldsymbol{f}^{\mathrm{TT}}} = \mathcal{P}^{\mathrm{TT}}_{\boldsymbol{u}}\boldsymbol{\Phi}_{\boldsymbol{f}} \ , \qquad \boldsymbol{f}^{\mathrm{TT}} = \mathbf{P}^{\mathrm{TT}}_{\boldsymbol{u}}\boldsymbol{f} \ , \quad \boldsymbol{f}\in\mathbf{Z}_{\boldsymbol{u}} \ , \quad \boldsymbol{u} = \mathbf{s}(x) \ , \qquad (11.20a)$$

so that we obtain the following resolution of the identity in each TT-quantum gravitational fibre of the quantum gravitational superfibre bundle in (10.27):

$$\int_{\mathbf{Z}^{\mathrm{TT}}_{\boldsymbol{u}}} \left|\boldsymbol{\Phi}^{\boldsymbol{s}}_{\boldsymbol{f}^{\mathrm{TT}}}\right\rangle \mathbf{d}\boldsymbol{f}^{\mathrm{TT}}\mathbf{d}\bar{\boldsymbol{f}}^{\mathrm{TT}} \left\langle\boldsymbol{\Phi}^{\boldsymbol{s}}_{\boldsymbol{f}^{\mathrm{TT}}}\right| = \mathbf{1}^{\mathrm{TT}}_{\boldsymbol{u}} \ . \qquad (11.20b)$$

According to (5.29) we then have:

$$g^{\mathrm{TT}(-)}_{ij}(x;\zeta)\boldsymbol{\Phi}^{\boldsymbol{s}}_{\boldsymbol{f}^{\mathrm{TT}}} = f^{\mathrm{TT}}_{ij\zeta}\,\boldsymbol{\Phi}^{\boldsymbol{s}}_{\boldsymbol{f}^{\mathrm{TT}}} \ , \qquad \boldsymbol{f} = f^{\mathrm{TT}}_{ij}\,\boldsymbol{\theta}^i(x)\otimes\boldsymbol{\theta}^j(x)\in\mathbf{Z}^{\mathrm{TT}}_{\boldsymbol{u}} \ . \qquad (11.21)$$

We have already introduced within the fibres of the BFP ghost and antighost bundle the BFP quantum gravitational superframes whose elements are the Berezin coherent states in (10.17), respectively, and which give rise to resolutions of the identity in (10.19). We can therefore combine these three types of frames into the following *quantum gravitational superframes*,

$$\Phi_g^s = \Phi_{f^{TT}}^s \otimes \Phi_{\bar{c}}^s \otimes \Phi_c^s \ , \qquad g = f^{TT} \otimes \bar{c} \otimes c \ , \tag{11.22}$$

which are eigenstates of the quantum gravitational annihilation operators in the TT-gauge modes, as well as in ghost and antighost modes. Consequently, for the *physical quantum gravitational propagator for parallel transport*,

$$K_\gamma(g(X''); g(X')) = \left\langle \Phi_{g(X'')}^s \middle| \mathcal{T}_\gamma(X'', X') \Phi_{g(X')}^s \right\rangle \ , \tag{11.23}$$

we obtain, by combining the methods in Sec. 8.4 and 9.5, that

$$K_\gamma(g(X''); g(X')) =$$

$$= \lim_{\varepsilon \to +0} \int K_\gamma(\, g(X_N); g(X_{N-1})) \prod_{n=N-1}^{1} \ K_\gamma(\, g(X_n); g(X_{n-1})) \mathcal{D}[g(X_n)], \tag{11.24}$$

where the functional integration is carried out for TT-components by the methods of Sec. 9.5, and for the ghost and the antighost components by the methods of Sec. 8.4. In turn, using (10.18) we can express the each factor of the integrand of (11.24) as follows:

$$\begin{aligned} K_\gamma(g(X_n); g(X_{n-1})) = \Big\{ & \left(1 - i\delta x_n^j P_j + i\bar{\omega}_{kl}(\delta x_n) M^{kl}\right)(f^{TT}(x_n); f^{TT}(x_{n-1})) \\ & + \delta x_n^\mu \left(\theta_n D_\mu \bar{C} + \bar{\theta}_n D_\mu C + \theta_n \bar{\theta}_n (D_\mu B + D_\mu C \times \bar{C})\right)(\xi(x_n); \xi(x_{n-1}))\Big\} \\ & \times \left\langle \Phi_{g_n(X_n)} \middle| \Phi_{g_{n-1}(X_n)} \right\rangle + O\left((\delta t_n)^2\right) \ . \end{aligned} \tag{11.25}$$

The expressions on the first line of the above equation are those already familiar from the treatment of the propagators for the parallel transport of the electromagnetic field in Sec. 9.5, so that, for example,

$$P_j(f^{TT}(x_n); f^{TT}(x_{n-1})) = \int \tilde{T}_{jk}[f^{TT}(x_n; \zeta) + f^{TT}(x_{n-1}; \zeta)] \ d\sigma^k(q) d\tilde{\Omega}(v) \ , \tag{11.26}$$

with the stress-energy tensor provided by (10.3b). In view of the expression in (10.33b) of the covariant derivatives of the ghost and antighost fields, as well as of the gauge-fixing fields, we see that the expressions on the second line of the equation (11.25) correspond to the single-ghost and single-antighost states created, in accordance with (8.7)-(8.10) and (9.18), by the parallel transport from x_{n-1} to x_n, along the projection in any base segment $\mathbf{S}$ of the path $\boldsymbol{\gamma}$. Hence, using the expression for inner products exemplified by (5.20), we can write, for instance,

$$D_\mu \bar{C}(\xi(x_n); \xi(x_{n-1})) = i \int \bar{c}^\dagger(x_n; \zeta) \overleftrightarrow{\partial}_k \cdot D_\mu \bar{c}(x_{n-1}; \zeta) \ d\sigma^k(q) d\tilde{\Omega}(v) \ . \tag{11.27}$$

Finally, the treatment of the inner product on the last line of (11.25) can proceed as in (9.5.19) and in (8.4.11) for the TT modes, as well as for the ghost and antighost modes, respectively. Thus, in the end, an action integral will result for the propagator for physical quantum gravitational parallel transport,

$$K_\gamma(\boldsymbol{g}(X'');\boldsymbol{g}(X')) = \int \mathcal{D}g \exp\left(iS_\gamma[g]\right), \quad \mathcal{D}g = \prod\nolimits_{t''>t\geq t'} \mathcal{D}[g(X(t))] \ , \quad (11.28)$$

in very much the same manner (but with a considerably increased degree of algebraic complexity – cf. Note 68) as on the previous occasions considered in Chapters 8 and 9.

The GS propagation of the quantum gravitational field proceeds by treating both terms on the right hand side of (11.14) as source terms, in keeping with the treatment of interactions and self-interactions presented[69] in Sec. 7.8. Thus, the source term for GS quantum gravitational radiation will be

$$\hat{T}_{ij}[\boldsymbol{g}] = (1/16\pi)\left(\hat{\boldsymbol{\Gamma}}^l{}_{ij}\hat{\boldsymbol{\Gamma}}^k{}_{kl} - \hat{\boldsymbol{\Gamma}}^l{}_{ki}\hat{\boldsymbol{\Gamma}}^k{}_{jl}\right), \quad \hat{\boldsymbol{\Gamma}}^i{}_{jk} = \tfrac{1}{2}\eta^{il}\left(\boldsymbol{g}_{lj,k} + \boldsymbol{g}_{lk,j} - \boldsymbol{g}_{jk,l}\right). \quad (11.29)$$

On the other hand, the source terms of the quantum fields for matter and nongravitational radiation are obtained from the stress-energy tensors computed in the preceding four chapters by executing two extrapolations: first, the domains of definition of these fields are to be extended to the quantum base supermanifolds $\mathcal{M}(\mathbf{S}, \boldsymbol{g}^{\mathrm{M}})$ over each base-segment $\mathbf{S}$, and, second, all the Poincaré-covariant derivatives in them, which in those chapters served to define their parallel transport, are to be extended into $\mathcal{GS}(\mathbf{S}, \boldsymbol{g}^{\mathrm{M}})$-covariant derivatives.

The first step can be carried out by extending all "matter" fields over any given base-segment $\mathbf{S}$ (specified by means of sections of Poincaré-covariant bundles $\mathcal{R}(\mathbf{S}, \boldsymbol{g}^{\mathrm{M}})$ constructed as in the preceding four chapters) into fields belonging to the associated bundles[70]

$$\mathcal{GR}(\mathbf{S}, \boldsymbol{g}^{\mathrm{M}}) = \mathbf{P}(\mathcal{M}, \mathcal{G}) \times_{\mathcal{G}} \Gamma(\mathcal{E}^{\mathrm{TT}}(\mathbf{S}, \boldsymbol{g}^{\mathrm{M}}) \otimes \mathcal{R}(\mathbf{S}, \boldsymbol{g}^{\mathrm{M}})) \ . \quad (11.30)$$

The resulting matter superfield $\boldsymbol{\Psi}^{\mathrm{mat}}$ can be then deemed to be a section of a superbundle whose elements can be represented by functions of the type

$$\boldsymbol{\Psi}^{\mathrm{mat}}(x, \bar{\boldsymbol{\theta}}, \boldsymbol{\theta}) = \boldsymbol{U}^{-1}(\boldsymbol{c}(x) \oplus \bar{\boldsymbol{c}}(x), \bar{\boldsymbol{\theta}}, \boldsymbol{\theta})\, \boldsymbol{\Psi}^{\mathrm{TT}}(x) \otimes \boldsymbol{\Psi}^{\mathrm{mat}}(x) \ , \quad (11.31)$$

so that they satisfy the matter counterparts of the subsidiary conditions in (10.40):

$$(\boldsymbol{\delta} + \boldsymbol{C})\, \boldsymbol{\Psi}^{\mathrm{mat}}(x, \bar{\boldsymbol{\theta}}, \boldsymbol{\theta}) = \left(\bar{\boldsymbol{\delta}} + \bar{\boldsymbol{C}}\right) \boldsymbol{\Psi}^{\mathrm{mat}}(x, \bar{\boldsymbol{\theta}}, \boldsymbol{\theta}) = \mathbf{0} \ . \quad (11.32)$$

This immediately determines the action of the GS quantum gravitational connection on such fields, and consequently also the parallel transport governed by that connection, thus completing the second step of the construction.

Propagators for the parallel transport of matter fields under the action of the GS quantum gravitational connection can be defined, and their action integral representations can be determined by using the same procedure as the one outlined in this section for the quantum gravitational field. In turn, this determines their strongly causal as well as their weakly causal modes of GS propagation, whose principal features were discussed and compared in Sec. 7.7.

11.12. Foundational Aspects of GS Quantum Cosmology

The 1980s bore witness to the effective merger[71] of elementary particle theory and quantum cosmology. This merger eventually led to the evolution of the idea of "wormholes" into that of "baby universes" – described by its most recent reviewer as a "rapidly developing ... subject based on the non-existent (even by physicist's standards) Euclidean formulation of quantum gravity" (Strominger, 1991, p. 272). However, this latest theoretical development has highlighted a process that had begun with the advent of the renormalization program soon after the end of World War II, and accelerated considerably after the "social and conceptual diversity of theoretical practice [in high-energy physics] fostered an instrumentalist rather than a realist view of quarks." (Pickering, 1984, p. 115). It is therefore imperative that we survey these recent developments in quantum cosmology, before we outline an alternative approach to this subject, based on GS theory.

The idea of "wormholes" resulting from topological fluctuations in spacetime is due to Wheeler (1957, 1964, 1968), so that it has been around for quite a while – cf. [M], p. 1200. However, recent developments relate it to topologically nontrivial instantons in the Euclidean regime. The principal mathematical justification for that, as well as its drawbacks, are as follows: "To begin with, the spatial cross sections of a classical spacetime cannot change topology without metric singularities occurring. One can, of course, attempt to get around this difficulty by rotating the time axis in the complex plane and 'going Euclidean'. However, most interesting metrics, above all those that contain gravitational wave packets, are not analytic and cannot be smoothly continued to a Euclidean region. Moreover, most metrics fail to remain real after the time axis has been rotated. Finally, one loses any sense of a 'process' occurring and thereby any intuitive understanding of topological transitions. This is particularly true in the functional integral definition of transition amplitudes [cf. Eq. (2.5) in Sec. 11.2] where one sees that the Euclidean approach is fraudulent in at least one basic aspect: there is no homotopy relationship between topologies which allow one to determine the relative phases with which amplitudes for different topologies should appear together." (DeWitt, 1984b, p. 448).

Indeed, Wheeler (1957, 1964) speculated about the possible existence of "Lorentzian wormholes" creating a "spacetime foam" in a general relativistic spacetime manifold depicted by a Lorentzian manifold. The transition to "Euclidean wormholes" in Riemannian manifolds was motivated not by physical considerations, but by purely mathematical ones, centered around the fact that the "path integral" in (2.5) would not converge in the real-time regime of Lorentzian manifolds (Gibbons *et al.* 1978; Hartle and Hawking, 1983). A very simple example of a "Euclidean wormhole", which represents a "baby universe" is a 3-sphere $x^\mu x^\nu \delta_{\mu\nu} = b^2$ that joins the two asymptotically flat 4-spaces $x^\mu x^\nu \delta_{\mu\nu} > b^2$ and $x^\mu x^\nu \delta_{\mu\nu} < b^2$, which carry the Riemannian (i.e., "Euclidean") metrics (Hawking, 1987):

$$g^{\mathrm{E}} = \left(1 + b^2/x^\kappa \delta_{\kappa\lambda} x^\lambda\right) \delta_{\mu\nu}\, dx^\mu \otimes dx^\nu \quad . \tag{12.1}$$

Of course, in complete concordance with the above cited remarks, this metric does not satisfy Einstein's equations even when it is continued to the Lorentzian regime. In view of the tendency of advocates of "baby universes" to dismiss such observations as "not very relevant", it is of some significance that other noted researchers into quantum cosmology have pointed out that "topology changes do not have to be allowed in quantum gravity in order to construct a consistent theory." (Banks, 1985, p. 358).

A clear presentation of the issues involved can be found in (Klebanov *et al.*, 1989). However, an enlightening explanation of some of the basic motivation and prospects of this subject can be found in a highly acclaimed article by S. Coleman, which has given considerable impetus to the idea of "baby universes": "Although I find this theory [that if wormholes exist, they have the effect of making the cosmological constant vanish] in many ways very attractive, I must in honesty stress its speculative character. It rests on wormhole dynamics and the euclidean formulation of quantum gravity. Thus it is doubly a house built on sand. Wormholes may not exist, or, if they do exist, their effects may be overwhelmed by those of some more exotic configurations. Likewise, the euclidean formulation of gravity is not a subject with firm foundations and clear rules of procedure; indeed, it is more like a trackless swamp." (Coleman, 1988, p. 647).

Thus, whereas for some cosmologists the motivation for merging their theories with those of some particle physicists is to solve some major technical problems with the Big Bang model (Guth and Steinhardt, 1984; Rothman and Ellis, 1987), on the part of the particle physicists it is to explain one of the most glaring inconsistencies to which the conventional renormalization program has by now given rise (very much as predicted by Dirac – cf. the introduction to Chapter 7): on one hand, spontaneous symmetry breaking leading to Higgs bosons *had* to be introduced in order to make the electroweak model as well as QCD "perturbatively" renormalizable, but on the other hand, not only have such "particles" *not* been observed (Harari, 1983; Veltman, 1986; Weinberg, 1987) despite extensive searches after them, but their *assumed* existence gives rise to an enormous cosmological constant, whose value is in flagrant contradiction with the very existence of our entire Universe!

Indeed, as one of the well-known contributors to the "mathematical theory line" for these models points out, "the only legitimate reason for introducing the Higgs boson is to make ... the theory renormalizable" (Veltman, 1986, pp. 76 and 81). On the other hand: "The biggest drawback of the Higgs boson is that so far no evidence of its existence has been found. Instead, a fair amount of indirect evidence already suggests that this elusive particle does not exist." (*ibid.*, p. 76). In fact, one of two main progenitors of the electroweak model acknowledges the following: "We don't know where the Higgs is; we're not even really sure that we should regard the Higgs as an elementary particle, but we believe that something must show up by the time we get up to available energies of a TeV: Higgs, or new kinds of strong interactions, or entirely new physics." (Weinberg, 1987, p. 8).

Hence, there is absolutely no observational evidence in favor of the Higgs "particle" introduced solely for the purpose of making QCD "renormalizable". In fact, the theory cannot even predict its mass (Veltman, 1986, p. 84), so that "one worries that the next particle to be found with a mass between a few giga-electron-volts and a tera-electron-volt will be christened the Higgs boson by fiat." (Oldershaw, 1988, p. 1077). But that is not all: "The coupling of the graviton to the Higgs field – ever present in all space – would generate a huge 'cosmological constant': it would curve the universe into an object roughly the size of a football." (Veltman, 1986, p. 78).

Since we obviously do not live in one of the "parallel universes" (cf. Barrow and Tipler, 1986, as well as Notes 15 to 21) of the size of a football, a great variety of *ad hoc* theoretical devices have been invented in recent times to deal with this "cosmological constant problem" (Weinberg, 1989). The very latest are the aforementioned "baby universes", which obviously represent a more acceptable variant of the "parallel universe" idea.

The following explanation was offered in a recent review article by Weinberg (1989):

"Hawking (1987, 1988) has suggested that since baby universes are unobservable, their effect is an effective loss of quantum coherence. Recently Coleman (1988) has argued (convincingly, in my view) for a different interpretation, [so that] ... [i]f all we want is to explain why the cosmological constant is not enormous, then our task is essentially done." (Weinberg, 1989, p. 18). That task has been completed because, as the theory goes, we all happen to live in a baby universe where the *effective* cosmological constant $\lambda_{\rm eff}$ is probably zero, or in any event negligible, so that we have all been rescued by one of the "anthropic principles" advocated by Barrow and Tipler (1986). Indeed: "Now, generic baby-universe states $|B\rangle$ will have components $|\alpha\rangle$ for which the [effective cosmological constant] $\lambda_{\rm eff}(\alpha)$ is very small, as well as others for which it is enormous. The anthropic considerations of Sec. VI [cf. Barrow and Tipler, 1986] tell us that any scientist who asks about the value of the cosmological constant can only be living in components $|\alpha\rangle$ for which $\lambda_{\rm eff}$ is quite small, for otherwise galaxies and stars would never have formed (for $\lambda_{\rm eff} > 0$), or else there would not be time to evolve (for $\lambda_{\rm eff} < 0$)." (Weinberg, 1989, p. 18).

As one reads about "solutions" which have to invoke some form of the "anthropic principle" in order to rescue a fashionable theory from gross disagreements with empirical facts, one cannot but be reminded of Heisenberg's observations that, in high-energy physics, "it is only a slight exaggeration to say that good physics has at times been spoiled by poor philosophy" (Heisenberg, 1976, p. 32). Trying to reinstate agreement by means of the predictive power of "anthropic principles" reminds one of another one of his notable statements, namely that "if predictive power were indeed the only criterion for truth, Ptolemy's astronomy would be no worse than Newton's" (Heisenberg, 1971, p. 212). Indeed, not only was Ptolemy's astronomy the product of an anthropocentric philosophy, that considered man to be the central and most significant part of the Universe, but its technique of adding epicycles upon epicycles in order to secure agreement with astronomic observations can be viewed as a methodological prototype for many developments in particle physics during the past three decades (cf. Sec 12.3), since quark theory made its first appearance. In fact, as has been noted by some, "so many new 'fundamental' particles, force-carrying particles, and theoretical devices have been tacked on to the original quark model that the new physics is beginning to make the Ptolomaic universe look rather svelte." (Oldershaw, 1988, pp. 1077-8).

One is therefore left to ponder over the following assessment of the "baby universe" solution of the cosmological constant problem: "Has the cosmological constant problem been solved? Perhaps so, but there are things to worry about in Coleman's approach, as also in the earlier work of Hawking. Here is a short list of qualms. (1) Does Euclidean quantum cosmology have anything to do with the real world? ... (2) What are the boundary conditions? ... (3) Are wormholes real? ... (4) What about the other terms in the effective action [in 'baby-universe' quantum cosmology]?" (Weinberg, 1989, p. 19).

In addition to these serious "qualms", there is the ubiquitous problem of divergences, which the conventional renormalization program was supposed to have "resolved", but which eventually led to the above described cosmological constant problem[72]. Indeed,

Coleman's (1988) method of dealing with this problem "is based on a nonexistent path integral quantization" (Fischler *et al.*, 1989), due to the unboundedness of the action

$$S_{\mathrm{E}} = \int d^4x\sqrt{g^{\mathrm{E}}}\left(\Lambda - (16\pi)^{-1}R^{\mathrm{E}}\right) \ , \tag{12.2}$$

that enters a Euclidean "path integral" over geometries with the various kinds of topologies containing "wormholes". The essence of the purported "solution" to the cosmological constant problem resides in the argument that such a "path integral" is dominated by special topological configurations which have the appearance of large "spheres" connected by wormholes (cf. (11.1)). However, as pointed out by Susskind (1991) in a recent critique of Coleman's theory, the formal "summation" over such configurations gives rise to the divergent integral

$$\int d\alpha \exp(-D\alpha^2/2)\left[\exp[3/8\Lambda(\alpha)]\exp(\exp[3/8\Lambda(\alpha)])\right] \ , \tag{12.3}$$

over the wormhole parameter α mentioned in the earlier quotation from (Weinberg, 1989). Thus, the entire argument relies[73] on the fact that the "weight" of the integral in (12.3) is "concentrated at vanishing effective cosmological constant" (Susskind, 1991, p. 348). On the other hand, the "path integral" itself is "violently divergent", whereas its use "as a measure of probability" is totally unwarranted.

Alternative approaches have been proposed based on "a completely nonlocal theory ... [of a] bifurcating universe ... in which the universe splits into two disconnected closed pieces" (Banks, 1988, p. 494). Mathematically, this leads to "a 'third quantization' which replaces the Wheeler-DeWitt wave function by an operator which acts on a Fock space of universes" (Fischler *et al.*, p. 158). From the ontological point of view, the assumptions involved are rather drastic: "We imagine that 'we' live in one of the disconnected universes. Everything that goes on into the other one is lost to us forever. Thus, any measurement in our universe is an 'inclusive' one: it is the probability for some process to occur accompanied by something going on in the other universe, summed over all things which disappear from our ken." (Banks, 1988, p. 494).

It thus appears that all the most recent developments in quantum cosmology are aimed primarily at the resolution of the earlier described "cosmological constant problem". This problem has emerged at the developmental end of a long string of theories based on "working rules" and other *ad hoc* theoretical devices, which started with the advent of conventional renormalization theory at the beginning of the instrumentalistically-dominated post-World War II era in quantum theory (cf. Secs. 12.2 and 12.7). Its resolution seems to require the invention of new cosmological theories which deal with the very nature of physical reality in a manner that is radically different from that advocated by any of the traditional philosophies of science. Indeed, ever since the time of Galileo, all the principal approaches to science have, on one hand, treated any form of the anthropocentrism with great skepticism; whereas, on the other hand, they have all shied away from empirically unwarranted speculations, and encouraged a scientific objectivity rooted in observations whose explanations did not have to invoke any postulates about the existence of "other worlds", in the form of ontological presuppositions inaccessible to empirical verification on account of the very nature of those postulates.

The quantum geometry framework presented in this monograph views conventional renormalization theory as nothing more than a once useful computational technique, which has, however, outlived its usefulness by the time it has given rise to such physically paradoxical situations as those inherent in the "cosmological constant problem". It also notes that such physically unacceptable "scenarios" have arisen from the indiscriminate applications of "working rules" for which, as pointed out by Dirac (cf. Note 23 to Chapter 9), Pauli (cf. Note 43 to Chapter 12), and others, there was no good foundational justification in the first place. Hence, it advances a nonperturbative approach to quantum field theory, in which no "cosmological constant problem" occurs because, from the outset, no *ad hoc* renormalization rules are invoked. Thus, by contrast with the above described types of approaches to fundamental ontological questions about the physical nature of our universe, the ontology of GS quantum cosmology is rooted in the traditional philosophy of science. As such, GS ontology is based on the premise that there *is* a *quantum* reality, whose essence is totally independent of human conventions. A *quantum* spacetime whose physical manifestations are all around us is deemed to be an integral part of that reality, and the search for a suitable mathematical description of that quantum spacetime is considered to be one of the main tasks of quantum physics.

Indeed, as discussed in Sec. 11.5, and as further elaborated in Sec. 12.4, a *natural* choice, independent of all conventions, *exists* for the specification and measurement of spatio-temporal relationships. That choice is, *de facto, inscribed in every single bit of matter in existence* since, on account of its rest mass m, each massive elementary quantum "object" represents (de Broglie, 1923; Penrose, 1968) a *natural* clock with period $T = 2\pi/m$ in Planck *natural* units; moreover, each such "object" also provides a "natural rod" of length $L = 2\pi/m$ in Planck natural units – as witnessed by the interference fringes produced by a beam of such "objects". Hence, the universal constancy of the ratios of the observed rest masses of elementary particles guarantees that (strongly or weakly causal) GS propagation can take place under *locally well-specified* spatio-temporal conditions.

Indeed, de Broglie has pointed out the following in his recollections of the fundamental ideas that gave birth to wave mechanics: "When, in 1922-1923, I had my first ideas about wave mechanics, I ... noticed that if the particle was regarded as containing the rest energy $M_0c^2 = h\nu_0$, it was natural to compare it with a small clock of proper frequency ν_0 " (de Broglie, 1979, p. 7). In view of the subsequent undisputed success of the modern quantum mechanics that has emerged from this fundamental idea, this feature of all massive elementary "objects" (regardless of whether they are called "particles", "excitons", or something else) can be regarded as the most fundamental and the empirically best established *fact* in all of quantum theory, demonstrated each time an interference or diffraction pattern is being observed. Hence, contrary to the prevailing post-World War II instrumentalist trends in quantum physics (cf. Sec. 12.2), but very much in agreement with the philosophies of the founders of relativity and quantum mechanics (who all opposed such instrumentalist attitudes[74]), the GS epistemology does not expect Nature to conform to any of the fashionable theoretical ideas that relegate such basic physical facts, as the existence of spacetime, to the status of fiction – while proclaiming formalistic (and, as shown in Sec. 12.3, as well as elsewhere, mathematically often poorly grounded) theoretical fiction as valid physical theory. Rather, the GS epistemology is to *search* for the *right* theoretical ideas to fit well-established *physical facts*.

Epistemologically, therefore, the GS attitude is not to question whether "time" is a valid notion in QGR, but rather to *find* the correct formalism to describe *time* as it *exists*.

As a fundamental epistemic guide, it adopts Bohr's attitude that the basic concepts of classical physics should be retained at the measurement theoretical level in quantum physics – albeit, as discussed in Secs. 1.3 and 12.6, it rejects Bohr's artificially imposed dichotomy in the treatment of "system" and "apparatus". The GS epistemic idea is, therefore, that the CGR conceptualization of spacetime in terms of a Lorentzian manifold is not to be discarded even in GS quantum gravity, since that would imply the complete renunciation of all the physically well-grounded intuitive ideas that we all possess about spacetime, and would make, as Bohr contended, *unambiguous* and *meaningful* communication unfeasible[75]. Consequently, classical spacetime *has* to be incorporated into quantum spacetime explicitly and from the outset, and not added as an afterthought, as it is the case in some of the contemporary cosmological "scenarios" discussed in Secs. 11.2 and 12.3.

On the other hand, if we accept basic quantum epistemology, a viable framework for quantum gravity should incorporate additional, purely stochastic degrees of freedom. Those degrees of freedom should be related to the nondeterministic behavior in a *quantum* spacetime of wave functions resulting from nongravitational sources, which, however, in any gravitational setting *can* influence, via wave function reduction (Pearle, 1986; Penrose, 1986), the geometry of that spacetime. Hence, they should allow for the possibility that measurement acts *can* affect that spacetime geometry in an *active* manner, which could not have been possibly "prescribed" at the "instant of creation" of the Universe if there is any freedom for the *experimenter* to even locally *alter* and *adjust* initial conditions of the quantum phenomena under observation – as opposed to merely "shifting the focus of his attention" in an *absolutely*[76] passive role of a disembodied "observer".

Indeed, in a strictly deterministic theory, such as CGR, there appears to be no freedom for the experimenter – who can act only as a passive observer. Thus, the entire history of a *classical* spacetime, depicted by a Lorentzian manifold $(\mathbf{M}, \boldsymbol{g})$, was determined for "eternity" by the initial conditions following the "instant" of the Big Bang within an "infinitesimal" period (since there is no natural role for the Planck time in a classical theory!). Of course, one might not *know*, or even hope to ever know, what those conditions were like *exactly*, but the "naïve" realism which underlies the epistemology of classical physics, postulates that even if there is no *anthropic* possibility to ever acquire such complete *knowledge*, such exact conditions did *exist*. Thus, in a totally deterministic universe every phenomenon which is part of *physical* reality (as opposed to being merely a conception in someone's mind) was intrinsically *caused* to happen at the "instant" of cosmic creation[77]. In particular, since in CGR the behavior of *all* matter, and the "evolution" of the entire *real* spacetime (and not just some particular "model"), are inextricably intertwined by Einstein's equations, no change from the predetermined course can take place in one without it affecting and producing a change in the other. *Ergo*, the freedom to even *locally* change initial conditions in order to perform experiments is only illusory. Therefore, the freedom to influence and change the behavior of matter, or of any of the "gauge" fields partaking in any observation of such "experiments", is also illusory: everything was already determined an "infinitesimal instant" after the Big Bang.

This means that such very fundamental questions[78] are intimately related to deep philosophical problems concerning "free will" vs "determinism", which have preoccupied such great philosophers of the past as Descartes, Hume and Kant, as well as such outstanding philosophers of this century as Russell (1945, 1948) and Popper (1963, 1982). In fact, Popper had dedicated an entire volume to arguments in favor of a fundamentally indeterministic conceptualization of the world around us. In that context he describes how he "tried to persuade [Einstein] to give up his determinism, which amounted to a view that the world is a four-dimensional Parmenidean block universe in which change was a human illusion", and how "by 1954 Einstein appears to have changed his mind fundamentally with regard to determinism" (Popper, 1982b, p. 2).

Clearly, the phenomenon of the "reduction" of any universal wave function, meant to describe our Universe in its entirety, cannot be fitted within any theoretical framework based on a deterministic interpretation of that same framework and on mathematically well-posed initial-value problems (i.e., problems that produce *unique* solutions for any *complete* specification of initial data – a feature to which all successful theories aspire). Moreover, even in frameworks that are based on a statistical interpretation, similar problems are encountered if one tries to derive the "reduction of the wave packet" within the framework itself, by incorporating the "system" as well as the "apparatus" into a single "large system", while not allowing for the possibility of a new even "larger apparatus". Indeed, the latter possibility would merely represent the beginning of an infinite recession of "larger and larger systems" and "larger and larger apparatuses". Such infinite recessions are certainly not feasible in *cosmological* frameworks, which are meant to describe our entire Universe, so that there is no room left for an *external* "apparatus" for "observing *It* ".

Of course, at this point one should pause and contemplate the epistemic advisability of even considering the introduction of a "universal wave function" that results from the quantum gravity framework of this chapter by *imagining* that the quantum fields for *all* matter and radiation *in existence* have been included in it, in a manner which is conceptually (albeit not technically) similar to the well-known "wave function of the Universe" proposal of Hartle and Hawking (1983). Indeed, if one seriously ponders the feasibility of acquiring *complete information* of such a function even in some "asymptotic" sense of "arbitrarily good accuracy", then not only do the problems of limitations inherent in the limited capacity of the human mind come to fore, but the problem of information storage gives rise to paradoxes reminiscent of Russell's well-known paradox about the "set of all sets" (cf., e.g., Kline, 1980). Indeed, limitations on the "compressibility of information" (cf., e.g., Zurek, 1990) indicate that in order to store, process and retrieve, even in an "asymptotic" sense, the *complete* information about the states of *all* matter and radiation in existence, we would require "memories" whose capacity exceeds those of the states of all matter already "in existence", which we are actually trying to describe. Hence, the question of "universal wave function" can be treated only in an *ontological* sense, by arguing that such an entity must *exist*, since the Universe exists, but that there are absolute limitations in acquiring and processing "arbitrarily accurate" information about such a universal "physical state". As we shall see shortly, this observation will help us resolve the apparent paradox that such a *universal* wave function might not undergo any "reduction" whatsoever, but only "decoherence" induced by an anthropically *willful* act of data simplification.

Indeed, when our entire Universe is deemed to be the system in the formulation of GS quantum propagation in the preceding section, then quantum frames of reference become integral parts of the resulting *universal wave function* embodying that propagation. Since then there is no longer any external agent that can *cause* the reduction of the universal wave function, such reduction has to assume the form of *decoherence*, viewed as a natural process that can take place without direct or indirect human intervention (cf., e.g., Gell-Mann and Hartle, 1990). On the other hand, in order to effectively deal with a most basic feature of quantum theory – namely the fact that the *intervention* of the "observer" by means of experiments gives rise to changes of quantum states that would not have taken place if such observations were not performed – it might appear that in quantum cosmology one is *forced* to adopt either a "many-worlds" approach (Barrow and Tipler, 1986), or an approach based on the "reduction in the mind of the observer" idea of von Neumann (1932, 1955), which was subsequently discussed by Wigner (1962).

A "many-worlds" type of "solution" is, however, at best illusory, since it is not able to provide an *empirically meaningful* interpretation that would satisfy even the most minimal of *scientific* standards of verifiability (cf. Note 18). Indeed, since in the MWI ("many worlds") framework one *postulates* an ensemble of *distinct* "universes"[79], then there are only two logical possibilities left: 1) It is postulated from the outset that there *cannot* be any type of physical communication between "observers" within different universes belonging to the envisaged superspace of "universes" – in which case a merely *metaphysical* framework results that does not comply to even the most basic standards of empirical verifiability required of any theoretical idea that can call itself "scientific"[80]. 2) It is postulated from the outset that there *can* be some types of physical communication between the member universes, so that the postulated ensemble of universes plays the role of a "super-universe" – in which case, from the epistemological point of view, one is faced again with the very problem which one is purportedly trying to solve. Indeed, if the second possibility were taken seriously, one would have to entertain the question of *where* to place a "super-apparatus" for "measuring" various quantities in this "super-universe". Hence, an infinite regression of "super-super-universes" and "super-super-apparatuses" would result.

Thus, we are left with the von Neumann-Wigner alternative. On the surface, this is an alternative that would be unpalatable to any form of philosophical realism. Nevertheless, we shall argue that this alternative is capable of providing an epistemologically satisfactory solution *after* one rejects all of its metaphysical undertones.

To achieve a reconciliation between realism and such a conceptualization, which, as pointed out by Wigner (1962), is seemingly in tune only with philosophical idealism, and borders on solipsism, one has to recognize the basic fact that any physical theory can provide, at best, a *description* of some part of "reality", rather any kind of faithful mirror of that "reality" in all its various aspects. Indeed, as indicated by such fundamental theorems as Gödel's (1931) undecidability theorem, any claim that a given theory is capable of "mirroring reality" would have to contend with the question as to what is the status of those statements within its framework (i.e., its "language") which have empirical content, but for which neither truth nor falsehood can be established *within* that framework: do they therefore "reflect reality" or not ?

Once it is accepted that the correspondence between theoretical constructs *describing* an underlying quantum reality and actual observational acts does not have to be *absolutely* unambiguous, but that a certain interpretational lack of "sharpness" might be present (intrinsically embedded in the notion of "unsharp stochastic value" – cf. Sec 1.3), then one can provide a sensible and realistic resolution of the above dilemma within the GS framework – albeit not within the conventional one. This resolution consists of envisaging that, in the course of the geometro-stochastic evolution of the universal wave function, that describes *all* matter and gauge fields in existence, *no* reduction ever takes place, but that such reduction occurs only "in the mind of the observer". As such, it certainly does not produce any of the "slightly imperfect copies of [that observer], all constantly splitting into further copies, which ultimately become unrecognizable" (DeWitt and Graham, 1973, p. 161) – as claimed in "many-worlds" scenarios. In other words, the process of the "reduction" of *localized parts* of such a GS "universal wave function" is merely a procedure of *simplification*, whereby on account of *empirically* acquired information about the prevalence of GS *decoherence* at the *macro*scopic level[81], a very complex but, in principle, totally accurate description of the localized *part* of the quantum reality under observation is replaced by a less accurate, but computationally more manageable description of that same reality. Indeed, such a simplified *description* better suits the *macroscopic* (and therefore classically biased) mode of conceptualization of the observational process that was favored by Bohr, and many others.

On the other hand, it might appear that such a proposal flatly contradicts the orthodox interpretation of quantum mechanics in even such simple measurements of quantum observables as the one of spin components performed by the Stern-Gerlach experiment (Messiah, 1962; Bohm, 1986). However, that is not at all the case. First of all, as pointed out in Sec. 1.3, already within the context of orthodox theory, the Wigner-Araki-Yanase theorem prohibits the possibility of arbitrarily sharp measurements of spin components even with respect to an axis of a *classical* frame. But that is not the most essential point. Rather, the main point is that such a description is based on the (linear) Schrödinger equation, in which the actual Stern-Gerlach apparatus is represented in the idealized form of an external magnetic field, whereas the *quantum* reality is that of a source, a detector and an electromagnet, which together form a complex system of quantum constituents, that produce a quantum electromagnetic field whose photons interact *non*linearly with the "system" – all of which is, in turn, only a (small) part of the "universal wave packet". The GS description of the larger system consisting of the Stern-Gerlach apparatus and the original "system" should be in terms of local quantum state vectors that are part of this "universal wave packet" – but for all practical purposes a quantum *statistical* mechanics description would have to be substituted for the "apparatus" part.

On the other hand, as discussed in Sec. 1.4, some recent theories of quantum measurement point out that not every "very large" system of "micro-objects" can be deemed to be a (macroscopic) detector, but rather that some very specific conditions have to be met (Namiki, 1988; Namiki and Pascazio, 1991); whereas some other theories take into account the nonlinearity of realistically formulated quantum interactions (shared by all interactions in GS quantum field theories), when studying the interaction of a detecting apparatus and the micro-system which it is supposed to detect (Ghirardi *et al.*, 1990). Such theories can be incorporated into the GS framework upon interpreting "large" *local* probability amplitudes as propensities for macroscopic manifestation of excitons above the base points of those amplitudes, and representing a detecting apparatus by a GS wave function (cf. Sec. 5.5) for which such "large" probability amplitudes are localized in world-tubes of macroscopic dimensions within the base Lorentzian manifold of a quantum spacetime. However, what "large" actually means in a particular context is a matter of human *judgement* – a fact that remains unchanged even if the entire observational procedure is computerized, so that no human "observer" is required. Hence, from the point of view of *quantum* reality, it is the macroscopic world that is the "shadow" world, in which boundaries between "objects" become "fuzzy" – rather than the other way around. In particular, in accordance with Bell's (1987) ideas, from a microscopic point of view the boundary between "system" and "apparatus" is "fuzzy", and it is only through a procedure of conceptual and computational simplification, meant to recover the tremendously over-simplified conventional treatment of the Stern-Gerlach experiment, that the transition can be made from the GS *cosmological* conceptualization of a "universal wave function" to an orthodox type of treatment, in which wave functions can undergo "reductions" during certain observational processes.

Thus, the GS "universal wave function" represents only an ontological entity, of which no *complete* knowledge can ever be acquired even in principle. Its noncomputability is of the same general nature as that of the "wave function of the universe", discussed by Geroch and Hartle (1986). Of course, an orthodox-minded operationalist would object to the introduction of such a concept in the first place, on the purely utilitarian grounds that it is computationally "useless". It has to be therefore pointed out that, until the day when such deep mathematical issues as the role of noncomputable numbers and the validity of *reductio ad absurdum* existence proofs are clarified, the "existence" of this kind of entities cannot be dismissed as illegitimate.

In spirit, albeit not in the technical nature of the proposed solutions, the GS approach to quantum cosmology partially agrees with some other recently proposed modifications of the orthodox interpretation of quantum theory, such as those based on the following kind of general assessment of the situation: "The founders of quantum mechanics were right in pointing out that something external to the framework of wave function and Schrödinger equation *is* needed to interpret the theory. But it is not a postulated classical world to which quantum mechanics does not apply. Rather it is the initial condition of the universe that, together with the action function of the elementary particles and the throw of quantum dice from the beginning, explains the origin of *quasiclassical domain*(s) within quantum theory itself." (Gell-Mann and Hartle, 1990a, p. 455).

The part of the above quotation with which the GS approach fundamentally disagrees is the contention that "the initial condition of the universe" is the decisive factor that "explains the origin of quasiclassical domain(s)". Indeed, not only do we know *far too little* about those initial conditions to speculate reliably about their "present effects", but such an epistemic approach sets the cart in front of the horse: it is not "quasiclassical domain(s)" that require "explanation"; as Bohr has consistently maintained, they are part and parcel of the world as we *perceive* it, and an essential component of human communication – so that they have been embedded in GS quantum gravity form the outset (cf. Sec. 11.5). It is rather the presently popular "canonical route" of conventional quantum cosmology that is in dire need of "explanation", in view of its many conceptual as well as mathematical deficiencies, that were discussed in Secs. 11.2-11.4, and in some of the notes to this chapter.

Thus, in the end the GS approach to quantum gravity and cosmology recognizes that many valuable lessons can be learned from epistemological and historical factors that have shaped, and then gradually altered the orthodox interpretation. In fact, those lessons were of paramount importance in the development of the present approach to QGR. We therefore turn to them in the next, and last, chapter of this monograph.

Notes to Chapter 11

[1] Cf. (Einstein and Grossmann, 1913, 1914) for the original "hole" argument. Norton (1987, 1989) and Stachel (1989) have provided insightful contemporary historical as well as epistemologies analyses, which dispel misconceptions about the "hole" argument resulting from a "trivial mathematical error" on the part of Einstein and Grossmann – cf. Sec. 11.4 for some further details.

[2] To the extent that B. S. DeWitt, who has made fundamental contributions to both basic types of approaches, namely to the "canonical" as well as to the "covariant" approach (cf. Sec. 11.2), has remarked remarked in one of his key papers (DeWitt, 1967) that "... no rigorous mathematical link has thus far been established between the canonical and covariant theories. In the case of infinite worlds it is believed that the two theories are merely two versions of the same theory, expressed in different languages, but no one knows for sure." In essence, DeWitt's 1967 assessment has remained valid to the present day.

[3] In the Euclidean path integral approach to quantum cosmology (cf. Sec. 11.2), "baby universes" are connected to the "parent universe" as "wormholes" – cf. (Coleman *et al.*, 1991) for recent review articles, and Sec. 11.12 for a discussion of these subjects. A very readable account of Ashtekar's formulation of general relativity, and of the ensuing loop-space nonperturbative approach to quantum gravity, was recently published by Rovelli (1991d) – cf. the subsequent Secs. 11.3 and 11.4 for a discussion of some of its basic aspects. For electromagnetism, a relation between loop representations and coherent state representations, of the type used in in Chapter 9, is provided by the functional integral transform in Eq. (3.35) of (Rovelli, 1991d). However, its non-Abelian analogue in Eq. (3.40) "is far from being well defined because we do not have at our disposal a well defined gauge invariant measure $\mu[A]$" (*ibid.*, p. 1644). Furthermore, as will be amply illustrated in this chapter, the epistemic attitude towards quantum gravity exhibited in the GS approach is very distinct from that in the "loop-space" approach (cf. Note 23). Indeed, the GS approach is firmly rooted in a consistent measurement-theoretical scheme from the outset, so that what are its *basic* "physical observables" has been established from the beginning (cf. Sec 11.4), exactly as it was the case when CGR was developed; whereas, in the "loop-space" approach: "Major problems are open. Among these, the construction of the physically observable quantities and the definition of the inner product." (*ibid.*, p. 1654).

[4] This can be probably achieved in an uncountable infinity of fashions. The recent review by C.J. Isham suggests, of necessity, only a small finite number, but wisely concludes with the recommendation that "one should be cautious therefore about taking too literally ideas like 'virtual black holes', 'Planck length wormholes,' 'topology on $\tau(X)$' and similar ideas." (Isham, 1991, p. 392).

[5] Cf. Sec. 11.2 as well as the eight review articles in the section *The Issue of Time in Quantum Gravity*, in (Ashtekar and Stachel, 1991) – many of which contain extensive lists of original references. In addition, Sec. 4 in the review article by Ashtekar (1991), published in the same volume, is devoted to this issue, and two recent papers by Rovelli (1990, 1991a) also deal with this problem. At the classical level, a canonical approach to gravity that advocates the existence of an "intrinsic time" in CGR has been developed systematically in four papers by Kuchař (1976, 1977), in which the CGR "field dynamics is not properly described as taking place in spacetime, or along a single foliation of hypersurfaces drawn in spacetime, but in hyperspace. Heuristically, hyperspace is the (infinitely dimensional) manifold of all spacelike hypersurfaces drawn in a given Riemannian spacetime." (Kuchař, p. 777). Some of the principal "philosophical" aspects of the issue of "time" and its "arrow" vis-à-vis classical and quantum theory have been reviewed and discussed by Reichenbach (1956). A review of the most popular attitudes towards these issues, that includes the most fashionable lines of thought amongst leading contemporary cosmologists, was recently provided by Zeh (1989).

[6] A *globally hyperbolic* Lorentzian manifold is one which possesses a Cauchy surface Σ_0. By definition, a *Cauchy surface* is a spacelike *slice* (i.e., a closed achronal set without an edge – cf. [W], p. 201) across **M**, which is intersected by *any* causal smooth curve which can no further be extended into a causal curve in the past, or in the future. Also by definition, a *causal curve* is one whose tangent at every single point is null or timelike. These definitions imply that in a globally hyperbolic classical spacetime there are no naked singularities (cf. [W], p. 201). This basically means that only singularities which might occur in such a spacetime as a result of gravitational collapse are black holes. Hence, any causal path that encounters such a singularity ends in it, so that there is no breakdown in predictability of its future behavior. The so-called *strong cosmic censorship principle* (Penrose, 1969) "forbids" the occurrence of naked singularities, and it is equivalent to requiring global hyperbolicity of any physically acceptable classical model of spacetime.

[7] Cf. Sec. 2.1 for the definition of an atlas of charts. Coordinate-independent formulations of canonical gravity and the initial-value problem can be found in the review articles by Fisher and Marsden (1979), Isenberg and Nester (1980), and Choquet-Bruhat and York (1980).

[8] Regge and Teitelboim (1974) have argued, however, that in some cases, such as for spatially non-compact and asymptotically flat spacetimes, these boundary terms cannot be dropped, since then there are no solutions to Einstein's equations for those situations. Hence, in such cases the boundary terms must be retained in the CGR Hamiltonian in (1.16) for the Einstein vacuum equations.

[9] In purely geometric terms, these constraint equations are basically the Gauss-Codacci equations, relating the extrinsic curvature and the 3-metric via the Levi-Civita connection and the curvature tensor for that metric – cf. [M], Eqs. (21.75)-(21.76), or [W], Eqs. (10.2.23)-(10.2.24). A straightforward method of deriving them, which takes advantage of Gaussian normal coordinates, can be found in [C], pp. 315-316.

[10] These are couplings for which the Lagrangian for the source fields can be written in a form that does not contain a term involving the connection. All the practically important cases (Maxwell, Yang-Mills, scalar and spin-1/2 massive fields) belong to this category.

[11] Cf. (Choquet-Bruhat and York, 1980), Secs. 1.2, 3.3 and 3.4 for a review. In such schemes it is more efficient to choose coordinate suitable charts in which $N \neq 1$ and $N_a \neq 0$, since the Gaussian normal ones eventually develop caustics, namely coordinate singularities. However, such caustics constitute manifolds of lower dimensionality (Lifschitz and Khalatnikov, 1963), which in the quantum regime give rise to sets of measure zero. Hence, although they give rise to computational difficulties, from the point of view of basic principles they are not bothersome. In any event, the outcome of the described algorithmic procedure is in essence independent of the choice of coordinates, and Gaussian normal ones merely make the physical meaning of the initial conditions in (1.20) more transparent.

[12] Cf. Thm. 10.2.2 in [W] for a mathematically more precise statement, and (Hawking and Ellis, 1973) for proofs and for all the technical details.

[13] Cf., e.g., (Isham, 1975), p. 54, (Bergmann and Komar, 1980), p. 244, or (Isham, 1991), p. 354. The use of the notation $\delta(\mathrm{x}-\mathrm{y})$ in (1.22b) was customary in all earlier physics literature on the subject, but it is not justified when the manifold Σ_0 is not linear. Hence, Eq. (1.18) in (Isham, 1991) exhibits the correct notation for a generic Σ_0.

[14] As known from conventional quantum field theory in Minkowski space, "local" canonical commutation relations, such as the ones in (1.22), cannot be satisfied by operator-valued functions, and require the use of operator-valued distributions. This makes any mathematical steps that involves the pointwise multiplication of such "operators" (as it is the case in (1.23) and (1.24)) mathematically ill-defined. In conventional quantum field theory in Minkowski space this problem can be by-passed in the free case by using Fock space methods. These are, however, unavailable in the present context, due to the absence of a globally defined vacuum state – a problem which we already encountered in Sec. 7.1. The various "cutoff" methods devised for interacting "local" quantum fields in Minkowski space are also bound to fail, on account of the reasons pointed out by Dirac in the quotation at the beginning of Sec. 11.1. In the subsequent derivation we ignore all these mathematical difficulties, and concentrate instead on purely formal manipulations.

[15] One of the professionally trained popularizers of contemporary physics, who bases his most recent book on the contemporary ideas in quantum cosmology that have captured the fancy of some outstanding theoretical physicists as much as that of dedicated sci-fi readers, since they are based on the apparently very intriguing Many-World-Interpretation (MWI) of quantum mechanics, writes the following: "Parallel universes have existed in the fantasies of science-fiction writers ever since the genre began. Apparently science-fiction writers have a firm grasp of this rather startling idea. Yet when I try to explain scientifically just what a parallel universe *is*, I find myself stumbling over words. Perhaps the reason is that these universes which are 'old hat' to science-fiction writers are rather new to physicists." (Wolf, 1990, p. 26). However, when it comes to quantum cosmology, some physicists do not appear to lag behind science-fiction writers, as implicitly noted in the following quotation: "In general, when the proponents of the new physics apply their art to cosmological problems they usually invoke untestable physics that is supposed to have taken place in the unobservable past in order to explain current observations. Are they solving problems or hiding them behind a curtain of sophistry? Without adequate testability, how are we to decide this question? ... It seems hard to avoid the conclusion that the new physics and its applications to cosmology have begun to transcend the limits of physical science, as traditionally defined." (Oldershaw, 1988, p. 1080).

[16] Cf. Note 27 to Chapter 7. This measurement theoretical fact, incidentally, also suggests that the great concern in conventional quantum gravity with "pre-Planckian times" (Alvarez, 1989, p. 599), namely with "the first 10^{-43} sec following the Big Bang singularity", might be misplaced: no quantum theory can *meaningfully* deal with such spacetime intervals. The following perceptive remarks explain some of the basic reasons: "The real numbers are used in quantum theory because they model the results of making measurements, and all measurements involve ultimately measurements in space, or intervals of time. Hence, it could be argued, conventional quantum theory is valid in regimes where the representation of space-time by a differential manifold is valid. In particular, one must keep well away from the Planck length, in which case to talk of 'quantum' effects, topological or otherwise, at that scale would be totally misguided." (Isham, 1991, p. 383).

17 Cf. (Banks, 1985), Eq. (3) on p. 5. The need of corrections to the "measure" in (2.4) is also mentioned. It is almost needless to add that the mathematical existence of such "measures" over infinite-dimensional manifolds gives rise to very difficult problems – as can be deduced form the simpler case of the Berezin (1964) methods of integration, used from Chapter 7 onwards. The tendency of researchers in this field is, however, to concentrate on "toy models", in which these difficult mathematical questions are by-passed by limiting the number of *infinite* degrees of freedom to a small *finite* number with the imposition of various symmetries. The dangers lurking in the assumption that what is true for a finite number of degrees of freedom remains true for an infinite number of degrees of freedom are well-known not only to mathematicians, but also to all mathematical physicists who have to contend with the break-down in quantum field theory [BL] of von Neumann's theorem on canonical commutation relations for a finite number of degrees of freedom [PQ], and with a plethora of other similar problems (cf. also Sec. 12.3).

18 Cf. their articles in (DeWitt and Graham, 1973). Wheeler later changed his mind, and in acknowledging that publicly, he also pointed out some of the key weaknesses of MWI: "Imaginative Everett's thesis is, and instructive, we agree. We once subscribed to it. In retrospect, however, it looks like the wrong track. First, this formulation of quantum mechanics denigrates the quantum. It denies from the start that the quantum character of Nature is any clue to the plan of physics. ... Second, its infinitely many unobservable worlds make a heavy load of metaphysical baggage. They would seem to defy Mendeléev's demand of any proper scientific theory, that it should 'expose itself to destruction'." (Wheeler, 1979, p. 397). The second section of (Vilenkin, 1989) contains a review of the various proposals for an interpretation of the "wave function of the universe", including that by DeWitt (1967), and points out their technical weaknesses. Vilenkin then presents his own proposal for an "approximate interpretation", but frankly admits that his "paper involves a number of simplifying assumptions, [of which] the most important is the restriction to homogeneous superspace models". He subsequently concludes his paper with the following epistemologically very revealing observation: "Finally, I should mention the semiphilosophical issues arising when one attempts to apply a probabilistic theory to the Universe, of which one has only a single copy. Here I made no attempt to deal with these issues and took a simple-minded approach that the theory describes an ensemble of Universes." (Vilenkin, 1989, p. 1121).

19 Everett and many of his followers tend to ignore the fact that observables with a continuous spectrum are the rule rather than the exception in quantum mechanics (*viz* position, momentum, energy for scattering states, etc.), so that one has to contend with infinities of the cardinality of the continuum, i.e., certainly not smaller than $\aleph_1$. For example, DeWitt writes, only figuratively feigning shock: "The idea of 10^{100+} slightly imperfect copies of oneself all constantly splitting into further copies, which ultimately become unrecognizable, is not easy to reconcile with common sense." (DeWitt and Graham, 1973, p. 161). Since this, and other similar statements, are not tongue-in-cheek, but part of a serious review article reprinted from *Physics Today*, it should be pointed out that in view of the basic fact that "oneself" has a location in space, and since the position observable is supposed to have a purely continuous spectrum with an uncountable number of points of cardinality $\aleph_1$ (if the continuum hypothesis is correct), there is one conspicuous mathematical misrepresentation in the above statement: the *finite* number 10^{100+} should be replaced by the "MWI cardinal number" $\aleph_1{}^{\aleph_1+}$. Unfortunately, there might be some difficulty with determining what such a cardinal number actually *means*, since the plus sign in the exponent obviously has to be interpreted as $\aleph_1$ raised to the $\aleph_1$ power, raised to the $\aleph_1$ power, etc., etc., where the "number" of etceteras is itself equal to $\aleph_1$ – if we assume, as it is customary, that the "time" variable can certainly cover a nondegenerate interval in $\mathbf{R}^1$.

20 Some of the possibilities originally considered by Wheeler are described by Anderson and DeWitt (1986), but two insurmountable "difficulties" are subsequently pointed out. The final conclusion reached is: "Because [the Hamiltonian and momentum canonical gravity] constraints are local it is obvious that the topology of 3-space is dynamically preserved even if the space is noncompact and its volume is infinite." (Anderson and DeWitt, 1986, p. 105) – cf. also Note 3.

21 "It is difficult to solve differential equations on an infinite-dimensional manifold. Attention has therefore been concentrated on finite-dimensional approximations to [a superspace] W, called 'minisuperspaces'. In other words, one restricts the number of gravitational and matter degrees of freedom to a finite number and then solves the Wheeler-DeWitt equations on a finite-dimensional manifold with boundary conditions that reflect the fact that the wave function is given by a path integral over compact four-metrics." (Halliwell and Hawking, 1985, p. 1778). However, in addition to all the mathematical dangers inherent (cf. Note 17) in such a radical "approximation", whereby an uncountable infinity of degrees of freedom

is replaced by a very small finite number, the question of the physical interpretation of this minisuperspace formalism, and of its probability measure, still remains "problematic" (Halliwell, 1991, p. 179).

22 Strictly speaking, at a mathematically rigorous level, such quantum corrections should be treated as operator-valued distributions. However, in a highly nonlinear theory, which any form of quantum gravity has to represent, that entails all the technical problems that have not been successfully dealt with even in the conventional theory of interacting quantum fields in Minkowski space. Therefore, the covariant quantum theory has been always considered only perturbatively, namely as an algorithm-generating formalism for a formal perturbation expansion of an S-"matrix" involving graviton scattering.

23 Cf., e.g., (Ashtekar, 1990) and (Rovelli, 1990, 1991), as well as representative articles in (Ashtekar and Stachel, 1991). A typical comment, which illustrates how much the epistemology of this approach is at odds with Einstein's epistemology followed in this chapter, is the following: "From the mathematical point of view, *time* is a structure on the set of observables. From a physical point of view, time is an *experimental fact* that, in nature as we see it, meaningful observables are always constructed out of two partial observables." (Rovelli, 1991c, pp. 129-30). By contrast, Einstein's (1916) epistemology in developing CGR was precisely the opposite one: spacetime coincidences *are* deemed as being *the primary* physical entities, and *all* other observables are *secondary constructs, to be derived from purely spacetime* notions. For example, the "reading of a clock" with hands moving on a dial simply represents the "observation" of a *spatio-temporal* coincidence between those hands and numerals on the dial. On the other hand, that does not imply in the least that that in either relativity or quantum theory "instead of asking about the probabilities of sequences of observations in time, we [can] ask about the probability distribution of experimental records, bubble chambers photographs and counter positions" (Coleman, 1988, p. 652). To this instrumentalist claim the obvious retort is: When and where should such a distribution of "records" be ascertained? And, if terrestrial spacetime locations are deemed to be favored on account of some of the several types of "anthropic principles" (Barrow and Tipler, 1986; Weinberg, 1989), are we to peruse *all* the past "experimental records, bubble chambers photographs and counter positions"? Then, why not include also the yet nonexisting ones, which are, however, in the planning stages of various contemporary experiments? However, what about unanticipated experiments? Hence in Secs. 11.4 and 12.4 we shall contrast such instrumentalist claims, with display a strong anthropocentric flavor, with an epistemology based on a natural *local* time rooted in *quantum* reality, which, as such, is encoded in *all* matter, and which is independent of whatever *conventions* happen to be prevalent amongst theoretical physicists at any given *historical* (in an anthropic sense!) time.

24 As is well-known, superstring theory was dubbed the "theory of everything", and was greeted by some with such "euphoric statements" as: "It is a miracle; it is the theory of the world" (Wali, 1986, p. 390). As seen from the following assessment, conditions for such a reception seem to have been very auspicious in the early 1980s: "Hawking (1980) has claimed that 'the end of theoretical physics is in sight.' He refers to the promising progress made in unification, and the possibility that the 'theory of everything' might be around the corner. Although many physicists flatly reject this, it may nevertheless be correct. As Feynman has remarked, we cannot go on making discoveries in physics at the present rate for ever. ... Suppose that [this] optimistic view is correct ... Then it will be the case that a very limited period of mathematical development (300 or 3000 years, depending on where you start) will have proved sufficient to encapsulate the ultimate laws of the cosmos." (Davies, 1990, pp. 64-65). However, this enthusiasm for string theory was not unanimous amongst leading theoretical physicists: Feynman remained a sceptic until his death in early 1988 (cf. Davies and Brown, 1989, p. 192), whereas Glashow described superstrings as "an interesting sociological example of the tendency of physicists to jump on a theoretical bandwagon", and predicted in 1986, when their popularity was still at its peak, a "half-life of two years" – a remarkably accurate prediction (cf. Horwitz, 1991, p. 419).

25 Cf. (Hartle, 1991), pp. 184-185. It is, however, acknowledged in the same publication that "in the real world we never measure a three-geometry, and certainly we do not carry out observations over any but extraordinarily small regions of a spacelike surface" (*ibid.*, p. 189).

26 It should be noted that if, on account of Geroch's theorem (cf. Sec. 6.3), we concentrate in CGR on classical spacetimes for which the Poincaré frame bundles possess global sections, then such spacetimes can be identified with Poincaré subbundles of the general affine bundle GAM. This fact will be exploited in Sec. 11.5, in constructing quantum pregeometries. In Sec. 11.9 Poincaré gauge invariance of the first kind will be combined with diffeomorphism-group invariance of the second kind into a gauge group for quantum gravity.

27 According to Hartle (1991, p. 176), "the fact that, as observed on all accessible scales, over the whole accessible universe, spacetime has a classical geometry" is a consequence of some "particular initial conditions of our universe", that prevailed soon after the Big Bang. However, until quantum cosmology becomes firmly grounded in a *consistent* QGR theory, it appears more prudent to view this fact as an *a priori given* theoretical input, justified by the results of direct observations of the universe around us, rather than as a consequence of some still questionable cosmological models. The fundamental need, in the treatment of any gravitional phenomena, for a classical metric that gives rise to Poincaré gauge degrees of freedom of the first kind was emphasized on many occasions by Drechsler (1981, 1982, 1984). On the other hand, in his recent work, Drechsler (1991) discounts the need for quantizing gravity, and advocates instead certain new types of currents, which act as sources for bundle curvature via direct coupling to a connection, rather than via a second-quantized stress-energy tensor interacting with a second-quantized metric field.

28 Dirac arrived at the following conclusions in the course of his well-known studies of the Hamiltonian method in CGR: "The exact Hamiltonian for the theory of gravitation ... turns out to be simpler than one might have expected ... *but it can be obtained only at the expense of giving up four-dimensional symmetry*. I am inclined to believe that four-dimensional symmetry is not a fundamental property of the physical world." (Dirac, 1958, p. 343). In discussing the quantization of gravity Dirac later stated that "my own belief is that it will not be possible to dispense entirely with the Hamiltonian method. The Hamiltonian method dominates mechanics from the classical point of view. It may be that our method of passing from classical to quantum mechanics is not yet correct. I still think that in any future quantum theory there will be something corresponding to Hamiltonian theory, even if it is not in the same form as at present." (Dirac, 1964, p. 86). In GS quantum gravity that "something corresponding to Hamiltonian theory" is the "segmentation" of the base-manifold of a quantum spacetime (cf. Sec 11.5) – which is in keeping with the Hamiltonian method – and the conspicuous presence of "local Hamiltonians" that occur in the form of generators for spacetime translations in timelike directions.

29 We do *not* intend to invoke here any of the popular forms of "anthropic cosmological principles", and imply that "intelligent life has some global cosmological significance, [i.e.] that life will someday begin to transform and continue to transform the Universe on a cosmological scale." (Barrow and Tipler, 1986, p. 615). Clearly, after only several millennia in the development of "civilization", and only three centuries in the development of "exact science", such opinions can reflect at best wishful thinking. On the other hand, when discussing from a cosmological perspective the process of "state preparation" and "observation-cum-registration", it is impossible to avoid ancient philosophical questions concerning the "freedom-of-will" and "mind-body" problem, briefly touched upon in Secs. 11.12 and 12.6. Such philosophically most fundamental questions should not be confused with dubious associations of "states of consciousness" with Everett's MWI and other "airy-fairy metaphysics" (Deutsch, 1986).

30 Cf. (Stachel, 1991), p. 38. Stachel also goes on to say: "So we are pulling a swindle when we are telling students ... that you first pick the manifold and then solve the field equations on it." (*ibid.*, p.38). The ultimate effects of such "swindles" is discussed in a wider historical context in Sec. 12.3.

31 Cf., e.g., the review articles in the section *Approaches to the Quantization of Gravity* in (Ashtekar and Stachel, 1991): they cite all the main recent references on this subject that follow the most fashionable contemporary trends. In particular, the recent papers by Rovelli (1990, 1991a,b) contain extensive discussions of foundational issues from the "loop-space" point of view.

32 In fact, the following has been acknowledged with regard to this approach: "Two crucial steps are missing for the definition of a complete theory of quantum gravity. The first is construction of the physical inner product. For a closed universe this problem is related to the meaning of time and probability in quantum cosmology. The other is the definition of the gauge-invariant observables. Although the invariants of knot theory may be used to construct a large class of operators on the physical state space, the physical meaning of these operators is unclear, as no explicit physical observables are known for classical general relativity for a compact universe in the absence of matter." (Rovelli and Smolin, 1988, p. 1158). The main significance of knot theory (Kauffman, 1983, 1987; Burde and Zieschang, 1985; Booss and Bleeker, 1985; Atiyah, 1990) to this approach lies in the fact that the gauge orbits (cf. Sec 10.4) of the diffeomorphism group happen to represent knot classes, since obviously the manner in which knots are "knotted" is left invariant by diffeomorphisms. On the physical side, the main justification for considering "loops tied into knots" as being of paramount significance to quantum gravity seems to be the following: "*the loops of the loop representation are precisely the quantum version of*

Faraday's 'force lines', which historically gave birth to gauge theories." (Rovelli, 1991d, p. 1643) – italics as in the original. And indeed, the origins of knot theory can be traced to the nineteenth century physics of the aether (Whittaker, 1951), namely to the idea originally put forward in 1987 Lord Kelvin that "atoms were knotted vortex tubes of ether" (Atiyah, 1990, p. 6).

33 Cf., e.g., Kronheimer and Penrose (1967), Castagnino (1971), Ehlers *et al.* (1972), Ehlers (1973), Ehlers and Schild (1973), Woodhouse (1973).

34 These problems are pointed out in a footnote on p. 305 of (Rovelli, 1991b), and are closely related to those pointed out in Note 25 to Chapter 2. Indeed, as stressed by Ehlers (1987), all attempts to *derive* the geodesic postulate from the equations of motion in CGR have thus far proved unsuccessful at a mathematically rigorous level, so that this postulate has to be adopted as fundamental not only in QGR, but in CGR as well.

35 Cf. (Rovelli, 1991b), p. 300. However, knowledgeable researchers carefully qualify in this context the use of the term "observable"; to wit: "In the Dirac theory of constraints, one *sometimes calls 'observables'* real even phase functions $A_0(z)$ which have weakly vanishing Poisson brackets with the constraints ... . This property expresses that is unaffected by a gauge transformation. Hence, the terminology 'observable.' " (Henneaux, 1985, p. 20) – emphasis added. At the end of Rovelli's article the following is, however, correctly pointed out: "Gravitational physics cannot be properly understood unless one takes into account the physical nature and the gravitational interactions of the bodies that form the reference system. In the classical theory one can always work in the approximation in which the effects and the dynamics of the material reference systems are neglected; but in the quantum theory one has to take into account the quantum properties of the objects that form the reference system. It is in this sense that we have a 'quantized spacetime'." (Rovelli, 1991b, p. 329).

36 In the text of Dirac's key papers on the subject (Dirac 1950, 1958, 1959), the term "observable" occurs only once, namely in the following passage: "To pass over to the quantum theory [of gravitation], we must make our dynamical variables into operators satisfying commutation relations corresponding to the new P.b.'s [i.e., Poisson brackets]. We must then pick out a complete set of commuting observables." (Dirac, 1959, p. 930). However, Dirac goes on to point out the fundamental difficulties with this procedure, and ends his last key paper on the subject with the following observation: "*The gravitational treatment of point particles thus brings in one further difficulty, in addition of the usual ones in the quantum theory.*" (*ibid.*, p. 930) – emphasis added. This observation is obviously related to the one made by him ten years earlier, and cited in Note 11 to Chapter 3.

37 This conclusion does not contradict the quotation in Note 1 to Chapter 12, which deals with the period of *growth* of a physical theory, in the course of which new observables might be discovered, as its foundational structure becomes better understood. During that developmental stage, it can be indeed "really dangerous that one should *only* speak about observable quantities. Every reasonable theory will, besides all things which one can observe directly, also give the possibility of observing things more indirectly."

38 This is the type of formalism that is obviously regarded as "standard" in the paper "Loop Representation of Quantum Gravity" (in which "states, observables and constraints may be naturally represented in terms of functionals over the space of piecewise-smooth loops") when it is claimed the "results [of this representation] are obtained without any additional physical input or assumptions besides general relativity and standard quantum mechanics" (Rovelli, 1991, pp. 429 and 438). For, not only does not *standard* CGR recognize "loops in three dimensions" as being "natural", but *standard* quantum mechanics is formulated in Hilbert space (von Neumann, 1932), so that *its starting point is an inner product.* It is, therefore, only proper when it is pointed out later on in the same article that the "loop representation of quantum gravity" is "still far from being a complete theory of quantum gravity since two elements are missing: the Hilbert structure on the physical states, and the definition of physical, gauge-invariant operators" (*ibid.*, p. 438). That means, however, that the key *standard* ingredients of *standard* CGR as well as of *standard* quantum mechanics are *totally* missing in this still nonstandard framework for quantum gravity, which is "based on functionals over a set of loops in a three-dimensional manifold Σ" (Smolin, 1991, p. 440).

39 The persistence of this belief, despite the clear-cut analyses of Wigner (1963, 1981) and others, showing its untenability in light of the progress in understanding the quantum theory of measurement, is merely a reflection of the prevalent post-World-War II instrumentalist attitude, to the effect that, if something "works", then it must be "true", combined with the conventionalistic attitude, described in Secs. 12.2 and 12.3, according to which something "works" if there is a "consensus" that it works. This kind of re-

liance on "conventional wisdom" allows no room for inquiring into the *meaning* of such statements as that, in conventional quantum mechanics, "any Hermitian operator A is an observable", but simply reduces their significance to the computation of "predictions", and to checks on their purported "agreement with observational data". However, without blindly relying on the majority opinion amongst leading contemporary physicists, how does one *prove* that, e.g., such an operator as $A = QP^2 + P^2Q$ is, or is not, an observable? Clearly, the first step in that direction would be to spell out exactly what one *means* by the term "observable", in classical as well as in quantum physics – a step which conventionalists much prefer to avoid undertaking.

40 Some instrumentalistically-minded physicists maintain that the expectation value of any "Hermitian" operator A can be "measured" by first "measuring" the state vector Ψ of the system, and then computing the "expectation value" of A . A moment's reflection reveals, however, that if such purported "indirect measurements" really deserved the title of "measurement", then the "expectation value" of any operator in $\mathcal{H}$ could be "measured" – so that then, in fact, the term "measurement" loses its operational meaning, as well as all its distinctions from the term "computation".

41 Methods for measuring spatio-temporal distances with microscopic apparatuses were actually previously presented by Salecker and Wigner (1958). In particular, the question of minimum mass required for a given degree of accuracy was posed and studied in that context. The limitations of "quantum clocks" of the type they proposed was later reconsidered by Peres (1980), and their use in defining an "internal clock time" in quantum field theory and quantum gravity was studied by Page and Wootters (1983).

42 Throughout this section we set $\ell = 1$, since the consideration of the sharp-point limit of GS quantum gravity is totally unnecessary: by general consensus, there is no conventional quantum field theoretical approach, based on sharply localized fields, that has come even close to providing a viable method for quantizing gravity.

43 We make this assumption for the sake of simplicity, as well as on account of Geroch's theorem (cf. Sec. 6.3). The arguments could be extended, however, to the generic case by using local sections of $\boldsymbol{QPS}$, as in (10.2.15).

44 Such a definition was first explicitly provided in its present form in (Prugovečki, 1981d), and then elaborated in Sec. 4.3 of [P]. The basic idea was implicit, however, already in the treatment of informational completeness and stochastic phase space presented in (Prugovečki, 1977b, 1978a,b). A description of *quantum* gravimeters based on the behavior of charged particles, such as electrons, in gravitational fields is presented in (Anandan, 1986). In particular, the phase shift for electron interference in the presence of a gravitational field is derived in (Anandan and Stodolski, 1983) and in (Anandan, 1984).

45 An anti-symmetric component would carry spin-1, which would give rise, as in the electromagnetic case, to repulsive forces. The trace-zero condition eliminates spin-0 gravitons, whose existence would contradict the observed bending of light (Seielstad *et al.*, 1970; Muckelman *et al.*, 1970; Sramek, 1971). A non-zero graviton mass would have the same undesired effect even for spin-2 gravitons (van Dam and Veltman, 1970). Various indefinite-metric representations of the Poincaré group for spin-2 and mass-0, as well as the associated "Fock spaces", are presented in (Bracci and Strocchi, 1972, 1975), and in (Bertrand, 1978). A formal treatment of the scattering of gravitons, which underlines the similarities with that of photons, is presented in (Weinberg, 1964, 1965). However, the later work of Strocchi (1968) revealed, at a mathematically rigorous level, basic technical difficulties in combining the Einstein equations for weak gravitational fields with conventional formulations of quantum field theory.

46 A further generalization of this result to the case of n-dimensional manifolds $\boldsymbol{\Sigma}'$ and $\boldsymbol{\Sigma}''$ has been stated and proved by Sorkin (1986) for n even, and by Friedman (1991) for n odd – cf. also (Milnor and Stasheff, 1974).

47 Cf. Sec. 9.1, and in particular (9.1.29). Tetrads which consist of null vectors are called *null tetrads* (cf. Penrose and Rindler, 1984, p. 119). In this context note that the circular polarization covectors in (8.1a) are null in the bilinear (as opposed to the sesquilinear – cf. Note 9 to Chapter 7) extension of the Lorentzian metric to the complexified cotangent space at x, which is implicitly used in (8.6). The representations of the little group [BR] to which the null tetrads in (8.5) lead were first applied by Weinberg (1964) to the case of photons and gravitons, and were recently extended by Warlow (1992) into a systematic approach to indefinite-metric representations for mass-0 particles which is applicable to arbitrary integer-spin values.

48 Cf., e.g., (Kim and Noz, 1986), p. 167, where the linear polarization tetrad defined at $\boldsymbol{k}_{\boldsymbol{u}}$ by (5.10) is implicitly used. Upon setting $\alpha = \theta, z = u + iv$, one can make the transition (Warlow, 1992) to the corresponding null tetrad in (8.5) in order to arrive at (8.8).

49 Photon polarization states which are multiples of the null elements in (8.1b) have been referred to in literature (Dürr and Rudolph, 1969) as "good ghosts" if $r= +2$, and as "bad ghosts" if $r = -2$. In some recent studies (Nishijima, 1984) of BRST theories applicable to Yang-Mills fields the respective terms "daughter" states and "parent" states are used instead – cf. (Horuzhy and Voronin, 1989), p. 682.

50 In action from the left the Lorentz transformation Λ acts upon a polarization frame by acting upon its coordinate components with respect to a Lorentzian coframe $\{\boldsymbol{\theta}^i(x)\}$, whereas in action from the right upon a polarization frame it acts from the right on the elements of that coframe, while the aforementioned components remain fixed. Hence, the matrices executing these two actions are the inverses of each other (cf. Notes 11 and 15 to Chapter 2).

51 The proof can be most easily carried out in the momentum representation, and reveals that the entire procedure is applicable equally well to any integer-spin value (Warlow, 1992).

52 According to Nakanishi (1983, p. 174): "Though it is widely believed that quantum gravity is similar to Yang-Mills theory, I emphasize that *quantum gravity is more similar to quantum electrodynamics.*" The present treatment of GS quantum gravity brings out this "abelian nature" (Nakanishi and Ojima, 1990, p. 298), but only to a restricted degree of an internal graviton gauge group, which approximates only infinitesimal diffeomorphisms that are so local that heuristically they can be viewed as limits of deformations at a single point. The subsequent transformation of $\mathbf{G}_0$ into a Lie algebra will be required in order to deal with the generic case.

53 Cf., e.g., (Abraham *et al.*, 1988), §4.1, for a systemetic study of diffeomorphisms generated by a vector field X in any given manifold M. In those cases where the vector field X is *complete*, in the sense that all its integral curves (i.e., curves parametrized by a real-valued parameter t, whose tangents belong to X) are defined for all real values of t, then the diffeomorphisms it generates constitute a one-parameter group, called a *flow* (cf. Abraham *et al.*, 1988, p. 249). For the present purposes, the existence of a flow is an unnecessary as well as too restrictive an assumption, since we are concerned with local properties.

54 Cf. (Isham and Kuchař, 1985), p. 291, for an explanation of the origins of the sign difference between "Lie bracket" (namely what we call "Lie product") and the "commutator" of the corresponding vector fields. This same type of Lie algebra structure has been considered by Capper and Medrano (1974), as well as by Nishijima and Okawa (1978), from the "passive" point of view of the "infinitesimal" types of coordinate transformations that occur in (7.7). However, in this latter case one is not dealing, strictly speaking, with a Lie algebra, since coordinate transformations do *not* constitute a group, in view of the fact that generically they do not possess the same domain of definition.

55 Cf., e.g., the gauge-group treatments of quantum gravity by Delbourgo and Medrano (1976), Townsend and van Nieuwenhuizen (1977), Nakanishi (1978, 1979), Kugo and Ojima (1978b), Nishijima and Okawa (1978), which are all based on formal analogies with the quantization of Yang-Mills fields. The basic mathematical ideas are most lucidly presented in the last reference, from whose definitions in Eqs. (2.19)-(2.23) the relationship to Lie derivatives emerges most clearly. The present treatment of GSQG, originally presented in (Prugovečki, 1989b), dictates the use of the internal graviton structure group $\mathbf{G}_0$ as the carrier space of a Lie algebra in the subsequent formulation of the quantum gravitational gauge supergroup.

56 Note that such a "smearing" procedure defines the FP fields as linear operator-valued functionals of single-ghost and single-antighost states. It has to be carried out by first computing the components of those single-ghost and single-antighost states, as well as of the FP gravitational polarization frame and coframe fields in (9.23), with respect to ghost and antighost polarization frames corresponding to the respective modes in the graviton polarizaton frame bundle in (8.18); whereupon the "smearing" is carried out in the form of the same types of integrations as in (8.22b) and (8.22c).

57 Although Poincaré gauge invariance of the second kind is totally ignored in the most recent approaches to quantum gravity discussed in Secs. 11.3 and 11.4 (obviously due to the fact that these approaches rely on a canonical procedure in which such invariance cannot be embedded), earlier work on quantum gravity (Delbourgo *et al.*, 1982; Pasti and Tonin, 1982; Baulieu and Thierry-Mieg, 1984; Falck and Hirshfeld, 1984) displayed full awareness of the need for incorporating Poincaré as well as diffeomorphism gauge invariance into any *consistent* theory of quantum gravity. However, when using the type of superfield

formalism employed in the next section, this earlier work treated Grassmannian variables in a purely formal manner, rather than *deriving* them (as it is done in the next section) from a consistent supersymmetry reformulation of diffeomorphism invariance. Sometimes this work also replaced Poincaré with Lorentz gauge invariance, as well as diffeomorphism gauge invariance with "coordinate gauge invariance" (albeit coordinate transformations do not constitute a group), thus arriving at the claim that "gravitation theories expressed in terms of vierbein fields are invariant under two local groups: the general co-ordinate transformation group and the internal Lorentz group." (Pasti and Tonin, 1982), p. 99.

58 This approach was apparently initiated by Thierry-Mieg (1980), and subsequently extensively developed by Baulieu *et al.* (1982-1991) in the context of conventional, stochastic (Parisi and Wu, 1981), and, most recently, topological (Witten, 1988) quantum field theories. In fact, Baulieu *et al.* (1990), as well as Baulieu and Singer (1991), deal with topological gravity in two dimensions.

59 This dispenses with Gribov-Singer ambiguites, that would occur if in (10.5) $\mathrm{Riem}_M S$ were replaced with $\mathrm{Riem}_L S$ – cf., e.g., [I], pp. 132-134.

60 All the basic definitions of superanalysis required in this section can be found in the Introduction of [BI], whose terminology we shall follow. The same reference also contains the mathematically rigorous proofs of all the basic results on supermanifolds, Lie superalgebras and supergroups, whereas [D] presents an approach to the same subject which is aimed primarily at physicists.

61 Cf. [D], p. 53, for a formal definition of *sub-supermanifold*. In the present context the underlying gravitational polarization bundle structure can be used as a source of all the required basic geometric notions.

62 However, the mathematical as well as the physical aspects of all these treatments have remained very formal, for reasons that clearly emerge from the following observations: "The elegance and efficiency of the unified [superfield BRST] formalism are certainly *consequences of a geometrical structure that is still unknown*. One may hope that a superfield formalism exists in which the full theory ... would only be expressed in terms of a single generalized gauge field containing classical and ghost components, leading thereby to a true unification of ghost and classical fields in field theory. Unfortunately no one has been able to build such a superformalism, in spite of several attempts. The main problem is of course a deeper understanding of a quantum field theory with a dependence on unphysical variables in θ, $\bar{\theta}$." (Baulieu, 1985), p. 18 – emphasis added. It is therefore worth noting that this "unknown geometrical structure" reveals itself in (10.21)-(10.24) as actually due to the quantum gravitational counterpart of the diffeomorphism gauge invariance of CGR gravitational theories.

63 The proof is straightforward, and it is presented in Sec. IV of (Hoyos *et al.*, 1982). It has to be observed, however, that we the notation we use is more in keeping with that of Baulieu *et al.* (1982-1991) – cf. the next note.

64 We follow the notation systematically favored by Baulieu *et al.* (1982-1991) in the treatment of BRST symmetries. Indeed, if "vielbeins" are introduced consisting of the vierbeins that are the integral parts of Poincaré moving coframes and of the one-forms in (10.11), then a framework formally very similar to the one proposed by Baulieu and Thierry-Mieg (1984) for the formulation of the algebraic structure of quantum gravity results. On the other hand, as seen from Sec. 3 in (Baulieu, 1984), the introduction of "a pair of Grassmann coordinates" has decided advantages. The ghost and antighost fields for BRST symmetries not based on a superfield formalism, such as those in (10.4.12)- (10.4.13), or in (10.6.19), then formally result upon setting the Grassmann variables equal to zero.

65 Note that such initial conditions on $\boldsymbol{g}^{\mathrm{M}}$ are compatible with the mean metric conditions of the type that occurs in (11.15) only if the physical part of the quantum gravitational state $\boldsymbol{\Psi}_x$ at each $x \in \boldsymbol{\Sigma}_0$"contains an exact and finite number of gravitons. That, however, can be expected to be the outcome of an actual measurement procedure. On the other hand, if that turns out not to be the case, then the properly normalized actual expectation value for graviton states has to be added to the mean 3-metric and mean extrinsic curvature.

66 The original considerations in (Deser; 1970) and (Boulware and Deser, 1975) were carried out in the coordinate-dependent fashion which is necessitated by the linearized gravity approximation, which breaks general covariance, by selecting a preferred family of coordinate charts. However, the decomposition of such expressions as those in (11.6) can be carried out in arbitrary Cartan moving frames, once the S-coordinate independent approach of Chapter 2 is adopted in CGR .

67 One possible method for performing nonperturbative computations emerges from numerical techniques developed in lattice gauge theory (Li *et al.*, 1987; Satz *et al*, 1987). For example, the transition from the

continuum to the lattice formulation of gauge theories (Ktorides and Mavromatos, 1985) in SQM-based methods (Ktorides and Papaloucas, 1989) can be supplemented with Monte Carlo techniques (Creutz, 1983; Sabelfeld, 1991) founded on guided-random-walk algorithms (Barnes and Kotchan, 1987).

68 Cf., e.g., (Bogolubov and Shirkov, 1959), or (Nakanishi and Ojima, 1990). The latter reference provides in its §5.2.4 also a treatment of Noether's theorems in the context of CGR, and points out that some of its outcomes are "*valid only in the classical theory*" (Nakanishi and Ojima, 1990), p. 296. A partial illustration of the ambiguities of the Lagrangian approach to quantum gravity can be found in Table 1 of Pasti and Tonin (1982), which list a total of twenty-seven variants of "gauge-breaking" Lagrangian terms for their superfield approach to quantum gravity, and only six relations for reducing the number of possibilities. By contrast, the "Lagrangian" terms in (11.28) can be derived uniquely from (11.25) – although they are not explicitly displayed because of their apparent complexity. On the other hand, the Maurer-Cartan-based relations in (10.38) might prove effective in reducing the algebraic complexity of the most straightforwardly computable explicit expression for the action integral in (11.28).

69 Strictly speaking, the application of the method of Sec. 7.8 requires taking the Naimark extension of the fibres for all matter and radiation fields. However, eventually the projections onto the original fibres are applied (cf. (7.8.11)) to the interaction terms playing the role of sources. Since in the GS path integral approach such terms are placed between quantum frame elements belonging to the original fibres, this projection is automatically carried out as part of the required functional integration – cf. (9.8.12).

70 Such a definition is in agreement with that introduced in Sec. 4 of Bonora *et al.* (1982a) for matter fields parallel transported by means of Yang-Mills connections. Furthermore, the subsidiary conditions in (11.32) coincide with those in Eq. (2.41b) of Baulieu (1985), developed in that same context.

71 Apparently, the original reasons for the "marriage" between high-energy physics and cosmology were as follows: "There was a two-standard-deviation between the [GUT] theory [of the electroweak mixing angle] and experiment. As usual [in high-energy physics], this discrepancy was treated as an important result rather than a serious problem." (Pickering, 1984, p. 387). Hence, again as usual, the proposed solution was to built more complicated models. The problem was, however, that in order to "test" them proton synchrotrons with a radius comparable to that of the solar system would have been required. Hence (*ibid.*): "One response to the astronomical size of the unification mass was to go where the energy was – the Big Bang at the start of the universe. This led to a social and conceptual unification of high-energy physics and cosmology." According to faithful "believers" in this type of approach to the solution of fundamental difficulties, "the 'marriage' of cosmology and particle physics heralded the birth of a complete understanding of nature" (Oldershaw, 1988, p. 1079). However, according to some "skeptics" (*ibid.*): "A much smaller group of scientists worried that it looked more like an incestuous affair with high probability of yielding unsound progeny." The extent to which such rumors might not have been totally unfounded – and in fact vindicated Dirac's consistently skeptical attitude towards the conventional renormalization program, together with all the theoretical "dead wood" that followed from it – will be discussed in Sec. 12.3.

72 As we shall discuss in Sec. 12.3, this development totally vindicates Dirac's consistently critical attitude towards that programme.

73 Of course, the divergence of (12.3) is indicative of an inconsistency, and, as is well-known in any form of mathematical logic, from an inconsistent theory one can deduce *any* conclusion one likes.

74 As substantiated in Chapter 12, the belief in an underlying quantum reality was present not only in Einstein's, but (contrary to some contemporary opinions) also in Bohr's philosophy of physics (cf., e.g., Folse, 1985, p. 224).

75 The objection might be raised at this point that the present-day debates about the "existence of time" in quantum cosmology demonstrate that unambiguous and meaningful communication is feasible even after such a most fundamental notion of classical physics as "time" is relegated to the status of fiction. We will let, however, the reader decide, upon consulting the references which we have provided, whether there is truly meaningful communication between the participants in these debates.

76 As discussed in Sec. 12.6, this issue is related to the deep mind-body issue in philosophy – or, what von Neumann (1932, 1955) has called "psycho-physical parallelism". Indeed, "shifting the focus of one's attention" is a *mental* act, and *not* the physical act of "focusing one's eyes", or other sense organs. It was realized since the time of Descartes that the manner in which 'mind' influences 'matter' – including the matter constituting our own bodies, is a deep philosophical issue (cf., e.g., Russell, 1945). As further discussed in Sec. 12.6, these issues are related to the question of "free will" since, in a totally

deterministic universe, there does not appear to be any other solution than to envisage that freedom of action is only illusory, and to presuppose, with Descartes, that the train of all our thoughts, which *apparently* produce changes in our environment (including the state of motion of our own bodies), merely are in tune with an already predetermined course of events.

77 The presence of "chaos" (cf., e.g., Gutzwiller, 1990) in deterministic theories might make the *computability* of such phenomena *practically* unfeasible, but it would not change the conclusion from the point of view of basic principles. An entertaining and yet perceptive analysis of the foundational issues involved can be found in Chapters 2 to 5 of (Penrose, 1989).

78 The deep relevance of these issues to formulations of cosmology have also surfaced in a recent valiant attempt of a group of researchers to allow, on account of certain features of the earlier describe wormhole dynamics, for the possibility of closed timelike curves in general relativity, by postulating "a *principle of self-consistency*, which states *the only solutions to the laws of physics that can occur locally in the real Universe are those which are globally self-consistent.*" (Friedman *et al.*, 1990, p. 1916). This "principle of self-consistency by fiat forbids changing the past ... [i.e. the possibility] that a system that, after travelling around a nearly lost timelike world line, can interact with its younger self (e.g., a person who tries to kill his younger self)." (*ibid.*, pp. 1917-8). In a strictly deterministic universe there is no danger of such "anomalies", since any *apparently* free action would merely be one that reproduces itself again and again along any timelike closed worldline.

79 The replacement of the term "many worlds" with "many histories" has been proposed as an alternative (Gell-Mann and Hartle, 1990, p. 430). We shall continue using the term "universes" not only because it is used by the strongest proponents of MWI (Barrow and Tipler, 1986), but also because the term "history" might make the whole issue of the "existence" of "many worlds" more palatable to some, but in essence it does not change the following key issues: there is only *one* Universe that *we* live in, there is only *one* history of mankind, including its records of *all* observational procedures, and every single observation can have only *one* observed outcome in those cases where clear-cut alternatives exist. Hence, if another *potential* outcome is not observed in *this* Universe, but it is nevertheless *realized* in the manner *postulated* by MWI, that "realization" must happen in the "history" some other "universe".

80 Cf Wheeler's (1979) verdict, cited in Note 18. Of course, one can always postulate (Wolf, 1990) a purely "psychic" mode of communication. However, in that case the entire "many worlds" framework can be, at best, relegated to the realm of "unexplainable psychic phenomena", capable of producing, at most, entertaining science-fiction – as it apparently has been doing very successfully for quite a while (cf. Note 15).

81 Indeed, the following was recently noted in the context of developing a theory of measurement for quantum cosmology: "Defining a measurement situation solely as the existence of correlations in a quasiclassical domain, if suitable general definitions of maximality and classicity can be found, would have the advantages of clarity, economy, and generality. Measurement situations occur throughout the universe and without the necessary intervention of anything as sophisticated as an 'observer'. Thus, by this definition, the production of fission tracks in mica deep in the earth by decay of a uranium nucleus leads to a measurement situation in a quasiclassical domain in which the tracks directions decohere, whether or not these tracks are ever registered by an 'observer'." (Gell-Mann and Hartle, 1990a, p. 453).

Chapter 12

Historical and Epistemological Perspectives on Developments in Relativity and Quantum Theory

What is number? What are space and time? What is mind, and what is matter? I do not say that we can here and now give definitive answers to all those ancient questions, but I do say that a method has been discovered by which . . . we can make successive approximations to the truth, in which each new stage results from an improvement, not a rejection, of what has gone before.

In the welter of conflicting fanaticisms, one of the unifying forces is scientific truthfulness, by which I mean the habit of basing our beliefs upon observations and inferences as impersonal, and as much divested of local and temperamental bias, as is possible for human beings.

Bertrand Russell (1945)

The founders of relativity theory and of quantum mechanics were as concerned with the epistemological aspects and mathematical consistency of these theories, as they were with their empirical accuracy as reflected by experimental tests. In fact, some of them gave to epistemological scope and soundness preference over immediately *apparent* agreement with experiment, since they were acutely aware that all raw empirical data are submitted to a considerable amount of theoretical analysis and interpretation, before they are eventually released for publication. Of necessity, all such interpretations reflect the experimentalists' conscious or subconscious biases. Hence, the outcome is prone to various kinds of errors, ranging from systematic ones, due to the faulty design of apparatus or erroneous analysis of the raw data, to the subtle ones, due to misinterpretation or unwarranted extrapolation.

Nowhere is the setting of priority on sound epistemology ahead of the immediate agreement with experiment better illustrated than in Einstein's (1907) response to Kaufmann's (1905, 1906) negative experimental verdict on Einstein's (1905) at-that-time-just formulated special theory of relativity, and to the claim that the just-acquired experimental evidence provided indubitable verification of Abraham's (1902, 1903) theory of the electron. G. Holton describes that situation as follows: "We know what Einstein did when he heard about Kaufmann's results – one of the foremost experimentalists in Europe disproving this unknown person's work. Einstein did not respond for nearly two years. Finally, ... [in 1907] Einstein wrote that he had not found any obvious errors in Kaufmann's article, but that the theory that was being proved by Kaufmann's data was a theory of so much smaller generality than his own, and therefore so much less probable, that he would prefer for the time being to stay with it. Actually, it took until 1916 for a fault in Kaufmann's experimental equipment to be discovered." (Holton, 1980, p. 92).

Einstein himself made clear[1] the reasons for his primary concern with epistemological questions when he wrote: "[A physical] theory must not contradict empirical facts. However evident this demand may in the first place appear, its application turns out to be quite delicate. For it is often, perhaps even always, possible to adhere to a general theoretical foundation by securing the adaptation of the theory to the facts by means of additional artificial assumptions." (Einstein, 1949, p. 23).

This fundamental concern with sound epistemology, as reflected by the internal consistency and "elegance" of the advocated theoretical ideas, was exhibited in equal measure by the main founders of quantum theory – as amply witnessed in the writings of Bohr (1934, 1955, 1961), Born, Dirac and Heisenberg. For example, in a paper entitled "Why We Believe in Einstein's Theory?", Dirac (1980) asserts that the real basis for that belief does not lie merely in the experimental evidence itself; rather: "It is the essential beauty of the theory which, I feel, is the real reason for believing in it." And, in a similar vein, Heisenberg (1971) comments: "If predictive power were the only criterion of truth, Ptolomy's astronomy would be no worse than Newton's."

Unfortunately, after the Second World War this attitude towards epistemology and foundational issues in quantum physics became reversed[2], as leading physicists of the post-war generation obviously decided that, contrary to the opinions of their great predecessors, it was legitimate to secure "the adaptation of the theory to the facts by means of additional artificial assumptions". Thus, soon after the "triumph" of renormalization theory, Dirac (1951) felt compelled to point out in print that: "Recent work by Lamb, Schwinger and Feynman and others has been very successful . . . but the resulting theory is an ugly and incomplete one." He reiterated and expanded on this theme on many occasions. For example, in a 1968 lecture entitled "Methods in Theoretical Physics", in which he explained the methodology and epistemology of his approach to developing new physical theories, he stated[3]: "The difficulty with divergencies proved to be a very bad one. No progress was made for twenty years. Then a development came, initiated by Lamb's discovery and explanation of the Lamb shift, which *fundamentally changed the character of theoretical physics*. It involved setting up rules for discarding the infinities, rules which are precise, so as to leave well-defined residues that can be compared with experiment. But still one is using working rules and not regular mathematics. Most theoretical physicists nowadays appear to be satisfied with this situation, but I am not. I believe that *theoretical physics has gone on the wrong track with such developments and one should not be complacent about it*." In the end, true to his initial verdict[4], in his very last paper he concluded: "I want to emphasize that many of these modern quantum field theories are not reliable at all, even though many people are working on them and their work sometimes gets detailed results." (Dirac, 1987, p. 196).

Although, unfortunately, the many admonitions that were publicly pronounced by Dirac from 1951 until his death in 1984 have remained largely unheeded, the past decade has witnessed a gradual revival of interest in foundational questions. It is hoped that the present monograph will contribute to that revival in a constructive manner, which would reestablish the high standards for mathematical truth and epistemological soundness in science, to which the founders of twentieth century physics devoted their professional lives. Consequently, it is fitting, now that all the basic technicalities implicit in the formulation of quantum geometries have been presented in the preceding ten chapters, to devote this concluding chapter[5] to a clearly stated analysis of the epistemological meaning and significance of the physical ideas underlying the present mathematical framework for quantum geometries, framed against the historical background that has shaped those ideas.

We shall start, therefore, by reviewing the clash of philosophies that marked the rather turbulent early development of quantum theory in the pre-World War II years. We shall then describe the radical shift in values that characterized the post-World War II developments in quantum physics. These historical and sociological factors will help set into the proper perspective the multitude of glaring inconsistencies in conventional relativistic quantum theories, that have been simply glossed over, or even totally ignored, during the past four decades. After that, we shall analyze the most essential epistemological aspects that underlie the mathematical framework described in Chapters 3-11 of this monograph, keeping those historical perspectives in mind.

12.1. Positivism vs. Realism in Relativity Theory and Quantum Mechanics

The advent of the orthodox interpretation of quantum mechanics in the mid-1920s gave rise to what one of the leading contemporary philosophers of science, K.R. Popper, has called a "schism" in twentieth century physics: "The two greatest physicists, Einstein and Bohr, perhaps the two greatest thinkers of the twentieth century, disagreed with one another. And their disagreement was as complete at the time of Einstein's death in 1955 as it had been at the Solvay meeting in 1927." (Popper, 1976, p. 91).

During the 1920s and 1930s this schism was manifested as a sharp division of the leading physicists of the first half of this century into two camps: the cohesive Copenhagen school, led by Bohr, which included Heisenberg and Pauli as its other two leading proponents, with Born and Dirac as sympathizers, and a disunited opposition to that school, whose most outspoken representative was Einstein, but which also included such distinguished physicists as Planck and Schrödinger, and which was eventually also joined by de Broglie and Landé. To this day, there are many myths and misconceptions about the positions held by the main protagonists of the various public debates to which this schism has given rise – of which the Bohr-Einstein debate is the best known. This is closely connected to the still prevailing misconceptions about the degree of success which Bohr had in solving the basic epistemological issues confronting quantum theory. Through no fault of Bohr, the myths and misconceptions are in this regard so widespread that M. Gell-Mann once felt compelled to remark that: "Niels Bohr brainwashed a whole generation of physicists into thinking that the job [of an adequate interpretation of quantum mechanics] was done 50 years ago" (Gell-Mann, 1979, p. 29).

A critical examination of the main textbooks on quantum mechanics, which have shaped the beliefs held by most physicists since the thirties, seems to support this bluntly stated charge. Fortunately, in recent years, such publications as those by Popper (1982), MacKinnon (1982), Folse (1985), Murdoch (1989), Selleri (1990), and others, are beginning to set the record straight, by depicting and analyzing, amongst other things, the reasons behind the misconception that Bohr was the "winner" in the Bohr-Einstein debate. On the other hand, these and other similar studies are primarily written by scientific realists, and therefore sometimes tend to give the over-simplified impression of a clash between realism and positivism, with Einstein being cast as the "realist", and Bohr as the "positivist". For the more detached observer, who sees merit in both these most important streams in twentieth century philosophy, the situation appears to be considerably more complex.

First of all, from a broader historical perspective (Mehra and Rechenberg, 1982; Pais, 1982), the above classification of the philosophical beliefs held by Einstein is very much a function of the time period in his life which one chooses to examine. Indeed, in their heyday logical positivists were proud to point out that both special and general relativity were the outgrowth of a positivistic epistemology (cf., e.g., Ayer, 1946), which can be traced to Mach. Even a cursory reading of Einstein's main papers on these subjects confirms their judgment. In fact, if the operationalist attitude is expurgated from Einstein's 1905 paper, which launched special relativity, much of its basic motivation disappears. Indeed, Einstein was not the one to discover the Lorentz transformations; rather, he was the one to give to Lorentz transformations a straightforward operational interpretation, which did not rely on preconceived ideas about the nature of physical reality, in general, and about the intrinsic properties of the electron, in particular – i.e., the type of ideas which Lorentz was advocating at that time. Eventually, that simple and elegant *operationalistic* approach gave rise to far-reaching consequences, that would have been inconceivable without it. Similarly, Einstein's 1916 paper, in which classical general relativity was formulated in its final form, is operationally motivated and founded, even to the extent that it contains such extreme anti-realist statements as that the "requirement of general co-variance takes away from space and time the last remnant of physical objectivity" (Einstein, 1916, p. 117).

Thus, in some of his writings Popper had to admit that: "It is an interesting fact that Einstein himself was for years a dogmatic positivist and operationalist." (Popper, 1976, p. 96). But then he hastened to add that Einstein "later rejected this interpretation: he told me in 1950 that he regretted no mistake he ever made as much as this mistake." (*ibid.*, p. 97).

Regardless of whether Einstein's recantation was as extreme as all that, it remains a historical fact that, on one hand, by 1920 Einstein started to embrace the cause of realism; but, on the other hand, after that time he never came even close to matching any of the great achievements of his 1905 – 1916 period, during which his entire mode of thinking was heavily influenced by operational considerations. This perhaps contributed[6] to the fate endured for a long time by Einstein's crown achievement, namely his classical theory of general relativity (CGR). One of the most prominent historians of the subject, J. Stachel, has recently described that fate as follows:

"From the late 1920s until the late 1950s, general relativity was considered by most physicists a detour well off the main highway of physics, which ran through quantum theory. ... The low estimate of general relativity was not unconnected with the prevalence of a pragmatic attitude toward physics among its practitioners. Only the calculation of a testable number counted as valid theoretical physics. This attitude often was associated with an uncritical acceptance of a positivistic and operationalistic outlook on science. ... In recent years the situation has changed ... Difficulties encountered by the quantum field theory program made theorists more sympathetic to such explorations [as the relationship between general relativity and quantum theory]. Suggestions that the foundations of quantum mechanics might be subject to critical scrutiny and alteration were no longer taken as signs of mental incompetence." (Stachel, 1989, pp. 1-2).

Indeed, Bohr's attitude is often depicted as being staunchly positivistic, so that, historically speaking, it is fair to identify the uncritical acceptance of his ideas with an "uncritical acceptance of positivistic and operationalistic outlook on science". However, the mode of thinking which led Bohr to his complementarity principle was influenced by philosophical ideas which *fundamentally* transcended the tenets of any form of positivism. In fact, Jammer (1966) seems to have been the first historian of twentieth-century physics to point out the influence of Kierkegaard's existentialistic and irrationalistic philosophy on Bohr. More recently, Folse (1985), Murdoch (1989) and Selleri (1990) have documented this influence via Bohr's father and via his mentor, the Danish philosopher Harald Høffding. However, the fact that Bohr went well beyond a merely "positivistic and operationalistic outlook on science" in his writings should be evident to anyone familiar with logical positivism. Thus, Bohr's insistence that *either* only sharp position *or* only sharp momentum can be measured, which has influenced the thinking of entire generations of physicists, has not been so much an outgrowth of operationalism, as much as a reflection of "the impossibility of overcoming the conflict between thesis and antithesis – with a consequent existential pessimism – [which] was one of the cardinal features of existentialist philosophy" (Selleri, 1990, p. 348) that subconsciously influenced Bohr's thinking (Folse, 1985). Indeed, as pointed out in Chapter 1, and as further discussed later in this chapter, this "either-or" stance towards measurement outcomes runs counter to what is *actually* operationally feasible in practice, where, on one hand, truly sharp values of position or momentum can *never* be measured, and, on the other hand, information about unsharp simultaneous values of both position and momentum is *always* available to those *willing* to look for it.

Thus, the influence of positivism on Bohr, and, in turn, Bohr's direct or indirect influence on entire generations of physicists, is no doubt responsible for the following verdict, which still reflects an opinion widely held amongst quantum physicists, and especially those elementary particle physicists who have wholeheartedly embraced the instrumentalist doctrine discussed in the next two sections: "As every physicist knows, or is supposed to have been taught, physics does not deal with physical reality. Physics deals with mathematically describable patterns in our observations. It is only these patterns in our observations that can be tested empirically." (Stapp, 1991, p. 1). Indeed, a very close collaborator of Bohr confirms the following: "When asked whether the algorithm of quantum mechanics could be considered as somehow mirroring an underlying quantum world, Bohr would answer, 'There is no quantum world. There is only an abstract quantum physical description. It is wrong to think that the task of physics is to find out how nature is. Physics concerns what we can say about nature'." (Peterson, 1985, p. 305). On the other hand, Bohr made many statements to the opposite effect, so that a recent analyst of his philosophy has arrived at the following over-all appraisal: "Just as to the religious apologist it is never God's existence which is really at issue, but His nature that needs defense and elaboration, so to Bohr it was never the existence of the objects of quantum mechanical description which was in question, but only how to understand that description." (Folse, 1985, p. 224). And another recent analyst of his philosophy describes it by the (intentionally) contradictory terms of "instrumentalistic realism", claiming that "the realist component and the instrumentalist component are, so to speak, complementary sides to the phenomenon that is Bohr" (Murdoch, 1989, p. 222), but emphasizing that "it would be quite wrong to describe Bohr as a weak instrumentalist" – least of all of the kind that bears any relationship to the brand of strong *conventionalistic* instrumentalism discussed in the next two sections.

However, as we shall discuss in Sec. 12.3, Bohr's insistence that "the results of observations must be expressed in unambiguous language with suitable application of the terminology of classical physics" (Bohr, 1961, p. 39), rather than with a suitable application of the terminology of some new language, specifically designed for quantum theory, has nothing to do with positivism, operationalism, or any form of empiricism. Perhaps, the dictum that Kant's "inability to conceive of another geometry convinced him that there could be no other" (Kline, 1980, p. 76) could be also applied to Bohr vis-à-vis the possibility of introducing new quantum geometries – or purely quantum languages, in general. Indeed, although Bohr never referred to Kant in his writings, and never even acknowledged any influence of Kant's philosophy on his own, there is a certain parallelism between their epistemic stances: "Bohr's claim that the classical concepts are necessary for an objective description of experience may seem similar to Kant's view that the concepts of space, time, and causality can be known to apply to experienced phenomena *a priori*. Furthermore his view that these concepts apply only to phenomenal objects and cannot be used to characterize an independent physical reality seems to parallel Kant's ban on the application of these concepts to a transphenomenal reality." (Folse, 1985, p. 217).

In any event, one can speculate that the hidden influence of Kierkegaard's existentialist philosophy might have removed from Bohr any incentive to look into new *non*classical possibilities with regard to the geometries adopted in quantum mechanics. For the success of such an enterprise would have threatened to resolve the "conflict between thesis and antithesis", to which Bohr was exposed during his formative years. In Bohr's mind, such a "conflict" might have very well taken the form of the conflict between *sharp* simultaneous measurements of position and momentum, of various spin components, etc. And the existentialist side of Bohr might have been predisposed to see this "conflict" as a manifestation of a "complementarity principle", rather than allow for the possibility of realistic, and therefore necessarily *un*sharp values for those quantities to be incorporated into new mathematical frameworks, designed *specifically* for the needs of quantum physics.

In Secs. 12.4 and 12.6 we shall argue that it is not only possible[7], but even *necessary*, to combine the mutually consistent aspects of the "classical realism" (Folse, 1985) advocated by Einstein in his later years with the "existentialistic positivism" of Bohr, in order to arrive at an epistemology capable of consistently dealing with relativistic quantum theory. Indeed, the fact that both protagonists in this great debate went to unwarranted extremes becomes evident as soon as we take a second look at the epistemology of *classical* general relativity. Thus, while Bohr kept insisting that the language of "classical" physics was absolutely essential to describe all experimental data, he arbitrarily restricted himself to *nonrelativistic* classical physics, even when discussing essentially relativistic phenomena, such as the purportedly instantaneous propagation of measurement effects in the EPR paradox. On the other hand, as we have seen in Secs. 11.1 and 11.4, the epistemological questions in CGR concerning what is observable are by no means straightforward when viewed through the lense of nonrelativistic physics – as the confusion surrounding Einstein's "hole argument" vividly illustrates (Stachel, 1989).

An *implicit assumption* of Bohr's epistemology was that the basic language of physics, required for the communication of experimental data, is *static* at the historical level. Thus, he asserted that: "Strictly speaking, the mathematical formalism of quantum theory and electrodynamics merely offers rules of calculation for the deduction of expectations about observations obtained under well-defined experimental conditions specified by classical physical concepts". From his published debate with Einstein and his other writings (Bohr, 1955, 1962), it is clear that the "classical physical concepts" he had in mind were steeped in Newtonian classical physics, rather than reflecting those of classical general relativity – where point coincidences represent the most fundamental reflection of physical reality. Furthermore, it is an obvious and basic fact that every language, including our "everyday" language, constantly *grows* as it incorporates *new* concepts, that not only were not conceived, but might have been even unimaginable to earlier generations speaking that language[8]. It is therefore not unfair to conclude that it is dogmatic to insist that the language of Newton's classical mechanics, taken in conjunction with the "everyday" language of any given era in human history, is the one and only language capable of describing all conceivable "experimental conditions".

As pointed out in Sec. 1.3, the other leading proponents of the Copenhagen school were more than willing to look well beyond Bohr's "terminology of classical physics" in the search for solutions to the *new* problems raised by quantum theory, that were not shared by classical physics. Indeed, Heisenberg was one of the first proponents of the introduction of a fundamental length in quantum physics, whereas Born's maxims cited in Sec. 1.1 actually paved the way for the introduction of fundamentally indeterminate values of quantum observables, which underlie Principle 1 in Sec. 1.3.

12.2. Conventionalistic Instrumentalism in Contemporary Quantum Physics

While advocating, as the undisputed leader of the Copenhagen school, his peculiar mixture of positivism, realism, and existentialism, Bohr unfortunately did not anticipate the long-range effects of his teachings on all those in the future generations of physicists who lacked the philosophical training or the sophistication required to distinguish between subtle philosophical nuances (Murdoch, 1990, Chapter 10) and their gross over-simplifications. Such physicists condensed Bohr's entire philosophy into simplified enunciations[9] of the principles of complementarity, wave-particle duality and the purportedly "classical nature" of the "apparatus", and simply ignored the rest. Indeed, what Karl Popper calls the "third group of physicists", who emerged right after World War II, and soon became the overwhelming majority, is described by him as follows: "It consists of those who have turned away from discussions [concerning the confrontation between positivism and realism in quantum physics] because they regard them, rightly, as philosophical, and because they believe, wrongly, that philosophical discussions are unimportant for physics. To this group belong many younger physicists who have grown up in a period of over-specialization, and in the newly developing cult of narrowness, and the contempt for the non-specialist older generation: a tradition which may easily lead to the end of science and its replacement by technology." (Popper, 1982a, p. 100). Upon labeling the attitude of this "third group of physicists" a form of instrumentalism, Popper goes on to say: "But this instrumentalism, this fashionable attitude of being tough and not standing for any nonsense – is itself an old philosophical theory, however modern it may seem to us. For a long time the Church used the instrumentalist view of science as a weapon against a rising science ... [as can be seen in the] argument with which Cardinal Bellarmino opposed Galileo's teachings of the Copernican system, and with which Bishop Berkeley opposed Newton. ... Thus instrumentalism only revives a philosophy of considerable antiquity. But modern instrumentalists are, of course, unaware that they are philosophizing. Accordingly, they are unaware of even the possibility that their fashionable philosophy may in fact be uncritical, irrational, and objectionable – as I am convinced that it is." (*ibid.*, pp. 102-103).

One does not have to subscribe to the tenets of Popper's realism – or, for that matter, of any of the various coexisting brands of philosophical "realism" (d'Espagnat, 1989) – to agree with these assessments. In fact, some of his observations not only receive support from the statements of the founders of quantum theory (Dirac, Heisenberg, Born, etc.), cited earlier in this monograph, but were unwittingly echoed by one of the most outstanding members of the "third group of physicists" in the following statement: "The post-war developments of quantum electrodynamics have been largely dominated by questions of formalism and technique, and do not contain any fundamental improvement in the physical foundations of the theory." (Schwinger, 1958, p. xv). Unfortunately, this and other similar statements by one of the most outstanding and talented theoretical physicists of the post-World War II era, have not had any deeper impact on those of his contemporaries who belonged to the group of "younger physicists who have grown up in a period of over-specialization". In fact, one cannot help but agree with Popper as he arrives at the following pessimistic assessment of the post-World War II developments in quantum physics:

"A very serious situation has arisen. The general anti-rationalist atmosphere which has become a major menace of our time, and which to combat is the duty of every thinker who cares for the traditions of our civilization, has led to a most serious deterioration of the standards of scientific discussion. It is all connected with the difficulties of the theory – or rather, not so much with the difficulties of the theory itself as with the difficulties of the new techniques which threaten to engulf the theory. It started with brilliant young physicists who gloried in their mastery of the tools and look down upon us amateurs who struggle to understand what they are doing and saying. It became a menace when this attitude hardened into a kind of professional etiquette. But the greatest among the contemporary physicists never adopted such an attitude. This holds for Einstein and Schrödinger, and also for Bohr. They never gloried in their formalism, but always remained seekers, only too conscious of the vastness of their ignorance." (Popper, 1982a, p. 156).

Historically, this "very serious situation" began with the wholehearted acceptance by the new post-World War II generation of physicists of an algorithmic scheme for removing "infinities" from the perturbation expansion for the S-matrix in quantum electrodynamics (QED) – the same QED that was founded by Dirac (1927), but in whose formulation he began to publicly express doubts already in the mid-1930s (cf. Sec. 9.6). Indeed, after coming upon certain experimental discrepancies, in his habitual forthright and decisively uncompromising manner, which he used even with regard to his own theories, Dirac stated the following: "The only important part [of theoretical physics] that we have to give up is quantum electrodynamics ... We may give it up without regrets ... ; in fact, on account of its extreme complexity, most physicists will be glad to see the end of it." (Dirac, 1936). However, Dirac invested an additional ten years of hard work aimed at trying to come to grips with the infinities in QED by studying classical electrodynamics, only to eventually come "to the view that the infinities are a mathematical artifact resulting from expansions in [the coupling constant] α that are actually invalid (Dirac, 1946)." (Pais, 1987, p. 106).

Consequently, as opposed to the new post-World War II generation of physicists, Dirac remained totally unimpressed by the numerical successes of the renormalization programme in QED. As mentioned in the introductory remarks to this chapter, he declared from the outset: "Recent work by Lamb, Schwinger and Feynman and others has been very successful . . . but the resulting theory is an ugly and incomplete one." (Dirac, 1951). And, as seen from the many quotations of Dirac's words in this monograph, and as extensively documented in his recent biography (Kragh, 1990), throughout the remainder of his life he never wavered in the verdict that "these [renormalization] rules, even though they may lead to results in agreement with observations, are artificial rules, and I just cannot accept that the present foundations [of relativistic quantum theory] are correct." (Dirac 1978a, p. 20).

That this verdict is a fair and correct one was confirmed by one of the main founders of the conventional renormalization programme, when he stated: "The observational basis of quantum electrodynamics is self-contradictory . . . . We conclude that a convergent theory cannot be formulated consistently within the framework of present space-time concepts." (Schwinger, 1958, pp. xv-xvi).

Indeed, how can one possibly arrive at any other verdict if one is *rationally* considering the following plain facts: 1) QED, as well as all other "renormalizable" conventional quantum field theories, are formulated in terms of quantum field operators which *do not exist* as functions of points in Minkowski space (cf. [BL], Sec. 10.4); 2) a list of renormalization *rules* are derived, however, *as if* those quantum fields at a point did have a mathematical meaning (cf. [IQ], Sec. 6-1); 3) the practitioner of the conventional renormalization is asked to implement them, without concern for mathematical consistency or epistemological validity, but by using them as algorithmic *rules* for subtracting divergencies from originally *meaningless* integrals[10]; 4) those finite expressions are then claimed to provide the terms of a perturbation "expansion" – but after more than forty years there is no *proof* that the objects which one "expands", namely the S-matrix elements for various processes, actually exist in any well-defined mathematical sense; 5) in fact, not only does the "perturbation series" not converge, but the generally accepted *conjecture* in QED, as well as in other conventional quantum field theoretical models, is that this "perturbation expansion" is an asymptotic series (Dyson, 1952); 6) however, in the absence of a *proof* of the existence of an S-matrix – i.e., of functions in relation to which such series are supposedly asymptotic – the concept of "asymptotic series" is itself mathematically meaningless[11]; 7) after a protracted effort of more than twenty years, constructive quantum field theoretical attempts at imparting rigorous mathematical meaning to the conventional renormalization procedure has resulted merely in the conclusion that the S-"matrix" in QED, as well as in other "renormalizable" conventional quantum field theories in four spacetime dimensions, is most probably trivial, i.e., equal to the identity matrix (Glimm and Jaffe, 1987, p. 120).

On purely rational grounds, it might have been expected that even before this last bit of distressing information became available in the 1980s, Dirac's public admonishments and Schwinger's remarks would have already been taken to heart in the early 1950s, and a concerted effort would have been mounted to investigate the *foundations* of relativistic quantum mechanics in general, and of quantum field theory in particular. However, as is well-known, that is not at all what took place. Rather, in the mid-fifties a parade of changing fashions began to unfold in elementary particle physics, and is still continuing unabated to the present day[12]. During most of this period the prevailing *belief* was that these developments led to predictions which were "in agreement with experiment" – which, in a gross oversimplification and distortion of Bohr's teachings, was viewed as the ultimate arbiter of the validity of the various (and many transiently) fashionable theories. However, not only has it been repeatedly demonstrated[13] that the analysis of experimental results can be wrong, theoretical computation can be incorrect, and the very comparison between theory and experiment can be faulty, but as Heisenberg acerbicly pointed out on one occasion, "if predictive power were indeed the only criterion for truth, Ptolemy's astronomy would be no worse than Newton's" (Heisenberg, 1971, p. 212).

Indeed, in addition to Dirac, the only other founding father of quantum theory who lived to see these developments expressed his dismay and disapproval in an article which was published at the very same time that his death was announced to the professional world of physics. The second paragraph of this article contains the following declaration: "I believe that certain erroneous developments in particle theory – and I am afraid that such developments do exist – are caused by a misconception by some physicists that it is possible to avoid philosophical arguments altogether. Starting with poor philosophy, they pose the wrong questions. It is only a slight exaggeration to say that good physics has at times been spoiled by poor philosophy." (Heisenberg, 1976, p. 32).

As we mentioned earlier, Karl Popper very appropriately classified this type of "poor philosophy" as a form of instrumentalism, and described it as "the view that there is nothing to be understood in a [scientific theory]: that we can do no more than *master the mathematical formalism*, and then learn *how to apply it.*" (Popper, 1982a, p. 101).

However, the cavalier manner in which mathematics itself has been treated after the inception of the renormalization programme, indicates that instrumentalism *per se* is not actually solely responsible for this considerable decline in the standards for establishing *truth* in science. Indeed, ever since the advent of renormalization theory in QED, in quantum theory (as opposed to CGR) "mastering a mathematical formalism" has meant developing the *computational* skills to *algorithmically* derive from the "theory" numerical "predictions". Such practices require the *uncritical* acceptance of a series of computational "working rules", or, at the very least, the acceptance of the most *subjective* types of criteria for their mathematical validity – even when those criteria run counter to all deductive norms accepted in contemporary mathematics. In Dirac's words, these practices represent "a drastic departure from ... logical deduction to a mere setting up of working rules." (Dirac, 1965, p. 685). In fact, as we shall see from the examples cited in the next section, ever since the advent of renormalization theory, the prevailing attitude became to ignore *objective* mathematical criteria of truth and consistency, and to substitute instead *conventionally* acceptable mathematical procedures, i.e., *formal* computational rules *conventionally* deemed to produce valid results as long as they were declared as acceptable by those whom Dyson (1983) describes as the "mandarins" of the post-World War II generation of physicists[14].

It therefore seems appropriate to categorize this kind of approach to science by the more precise label of *conventionalistic instrumentalism.* This label is intended to reflect the fact that its general practices ignore or dismiss not only all the truth-values which scientific realism expects to be fulfilled by physical theories (cf. Murdoch, 1989, pp. 200-201), but even the most basic forms of *mathematical* truth – replacing them with mere conventions. As reflected by the activities of the "mainstream" in quantum theory, such conventions are primarily based on the consensus prevailing amongst the leading physicists of the present instrumentalist period in quantum physics as to what types of computational procedures are "acceptable". As such, the conventionalistic aspects of this form of instrumentalism should be strictly distinguished from Poincaré's type of conventionalism, mentioned in Sec. 1.2 (which indeed viewed the choice of geometries suitable for the description of the physical world as being a matter of convention, but otherwise reflected a deep respect for objective mathematical truth, and a love for mathematical beauty on a par with that displayed by Dirac[15]), or from contemporary forms of conventionalism in philosophy, which view all statements in logic and mathematics as being purely analytic, and as such emerging from *linguistic* conventions (Quine, 1949). It should also be distinguished from logical positivism, especially in view of the fact that when some contemporary authors discuss the foundations of quantum theory, they tend to identify the term "instrumentalism" with the so-called "positivism of physicists" (d'Espagnat, 1989, p. 28). Indeed, although when viewed as general philosophies, as well as working philosophies applied to science, logical positivism and instrumentalism share some common points, they are fundamentally distinct in many aspects – as will become apparent from the considerations in the next section.

Is conventionalistic instrumentalism an intrinsically unavoidable feature of contemporary quantum physics?

The preceding nine chapters of this monograph are meant to *prove* that *it is not.* On the other hand, some other critics of instrumentalism in contemporary quantum physics seem very anxious to affirm that the brand of instrumentalism which has become "the foundation stone of contemporary physics" in the second half of this century, "has been astonishingly successful in various fields, from elementary particle physics to astrophysics, quantum optics, solid state physics" (d'Espagnat, 1989, p. 28). It is therefore argued that "it is hardly surprising that physicists should see this foundation stone as being very solid and providing a basis for objective reality, [so that merely] a number of philosophers of science are not of the same opinion. ... The truth here is perhaps that since there is a high level of instrumentalist technical sophistication which science apparently cannot legitimately avoid, there is a gap of some kind between the theoretical physicist's activities and his thinking. *Either he thinks or he develops physics.*" (*ibid*, pp. 28-31) – emphasis added.

The above type of rationalization of the prevailing instrumentalist attitudes in quantum physics seems, however, to ignore a fact amply demonstrated by the founders of relativity and quantum theory, namely that a theoretical physicist can *both* think *and* develop *outstanding* physics – and, in fact, that the first activity is *necessary* for the second. This monograph is dedicated to the memory of P. A. M. Dirac, since he was the most outspoken and persistent of the critics of the values and practices of conventionalistic instrumentalism in quantum field theory. However, he most certainly was not alone in his critical attitude towards these types of developments in post-World War II physics (cf. Note 23 to Chapter 9). Indeed, it is an acknowledged historical fact that "the workers of the 1930s, particularly Bohr and Dirac, had sought solutions to the problems [of quantum field theory] in terms of revolutionary departures. ... The solution advanced by Feynman, Schwinger, and Dyson was at its core conservative: it asked to take seriously the received formulation of quantum mechanics and special relativity and to explore the content of [their] synthesis. A generational conflict manifested itself in the contrast between the revolutionary and conservative stances of the pre- and post-World War II theoreticians." (Schweber, 1986, p. 299).

In Chapters 3, 5, 7, 9 and 11 we have provided extensive evidence that no consistent "synthesis" of these two fields was *ever* achieved in the context of conventional theory – albeit a public relations campaign was launched after the advent of conventionalistic renormalization theory, meant to convince everybody that such a "synthesis" had already become *fait accompli*. In the next section we shall demonstrate that the problems that have been left open by this "renormalization" theory are deep rooted. Until recently this PR campaign had, however, by and large succeeded to gloss them over with a glittering veneer of formal manipulations, protected from closer scrutiny by the nurturing of a cavalier attitude towards all the basic tenets of mathematical truth and deductive validity. Indeed, amongst many "mainstream" quantum physicists, it only very recently became true that "suggestions that the foundations of quantum mechanics might be subject to critical scrutiny and alteration [are] no longer taken as signs of mental incompetence" (Stachel, 1989, p. 2). In the meantime, "old" *unsolved* problems remained deeply entrenched, but were left untouched, due to a systematic neglect of the foundations of quantum physics. That neglect can be clearly perceived (Bell, 1990) in the mainstream textbooks on quantum mechanics and quantum field theory. In particular, as will be illustrated in the next section, it is especially evident in the manner in which much of the required mathematics is treated in them.

12.3. Inadequacies of Conventionalistic Instrumentalism in Quantum Physics

In contemporary philosophy, the term "instrumentalism" is primarily applied to the theory about the nature of truth and falsehood advocated by John Dewey, which emerged on the North American continent as a natural extrapolation of the pragmatism of C.S. Peirce and

William James (1970) – cf. (Mackay, 1961). As seen by a contemporary elementary particle physicist: "James argued at length for a certain conception of what it means for an idea to be true. This conception was, in brief, that an idea is true if it works." (Stapp, 1972, p. 1103). In turn, John Dewey adapted this pragmatic criterion for truth in philosophy and science, as well as in everyday life, as being that which "works satisfactorily in the widest sense of the word", and based his instrumentalist concept of "truth" on the achievement of consensus. Thus, in scientific applications: "The significance of this viewpoint for science is its negation of the idea that the aim of science is to construct a mental or mathematical image of the world itself. According to the pragmatist view, the proper goal of science is to augment and order our experience. A scientific theory should be judged on how well it serves to extend the range of our experience and reduce it to order." (*ibid.*, p. 1104).

Such a *principal* criterion for *judging* a scientific theory can have some rather undesirable *social* consequences. Indeed, in his "History of Western Philosophy" Bertrand Russell writes that Dewey "quotes with approval Peirce's definition: 'Truth' is 'the opinion which is fated to be ultimately agreed to by all who investigate'." (Russell, 1945, p. 824). Then, upon demonstrating the *logical* untenability of the criterion that "an idea is 'true' so long as to believe it is profitable to our lives"[16], he concludes the chapter on the philosophy of John Dewey with the following critical observations: "The concept of 'truth' as something dependent upon facts largely outside human control has been one of the ways in which philosophy hitherto has inculcated the necessary element of humility. When this check upon pride is removed, a further step is taken on the road towards a certain kind of madness – the intoxication of power which invaded philosophy with Fichte, and to which modern men, whether philosophers or not, are prone[17]. I am persuaded that this intoxication is the greatest danger of our time, and that any philosophy which, however unintentionally, contributes to it is increasing the danger of vast social disaster." (Russell, 1945, p. 1828).

Thus, the emergence of conventionalistic instrumentalism as the officially undeclared, but functionally prevalent philosophy amongst quantum physicists of the post-World War II generation, might indeed represent a manifestation[18] of the "general anti-rationalist atmosphere which has become a major menace of our time" (Popper, 1982a, p. 156). And that in the eyes not only of such advocates of realism as Popper (1983), but also of those who accept the standard criteria of truth and deductive validity in mathematics[19], and yet believe that quantum mechanics and quantum field theory are very important and fundamental theories in science, in which the *traditional* standards of Truth should be preserved.

Indeed, the initial indifference of the undeclared adherents to conventionalistic instrumentalism towards the criticisms from Dirac, Heisenberg, and other leading physicists of the pre-World War II generation (i.e., from the very *founders* of quantum mechanics and quantum field theory), ultimately proved to be only a preamble to the eventually prevailing institutional intolerance in the most active areas of quantum physics towards anything that was out of step with the *prevailing* instrumentalist *conventions*. This intolerance manifested itself most clearly in the new criteria for acceptance of papers in major physics journals – which began to favor those based on sheer formal computations at the expense of those emphasizing mathematically and conceptually sound arguments – as well as by the cavalier manner in which relevant mathematics was treated in the most popular textbooks on quantum theory. It also manifested itself as a breakdown of the close contact and communication[20] between physicists and mathematicians, which, from Newton's era to Einstein's time, has been underlying all significant progress in theoretical physics[21]. In fact, it is only in the course of the 1980s that new channels of communication have re-opened between some of the leading physicists of the younger generation and some leading mathematicians – cf., e.g., (Witten, 1988), (Atiyah, 1990), (Nahm *et al.*, 1991). On the other hand, in addition to exhibiting foundational weaknesses (Bell, 1990), many of the mathematical standards exhibited by conventionally oriented quantum theoretical textbooks and practices are still rather distant from those acceptable in contemporary mathematics[22].

The most serious breaches of basic mathematical standards of consistency occur in relativistic quantum theory. However, telltale signs are already apparent in the nonrelativistic context. Since, in some of the preceding chapters, we have extensively discussed and analyzed the main failings of *conventional* relativistic quantum theory, let us now focus our attention for a while on the deficiencies of the conventionalistic approach to nonrelativistic quantum mechanics – illustrating in the process how, by violating the laws of standard mathematics, even some rather basic and crucial physics can be misrepresented.

We shall devote most of that attention to the deficiencies exhibited by the treatment which this subject receives in mainstream textbooks. Indeed, such textbooks not only reflect prevailing standards, but also shape and instill them in the minds of new generations of physicists. We shall strive to provide by means of readily comprehensible, and therefore of necessity elementary examples, a demonstration of the fact that the indiscriminate use, in professional practice, of the instrumentalist idea of "truth" can lead to a poor understanding of fundamental issues. In everyday practice, such a *mis*understanding is then maintained by institutionally reinforcing conformity (namely what Feynman (1954) colorfully called the "pack effect") by a variety of means – ranging from the criteria used in the refereeing of research papers submitted for publication in leading professional journals, to the standards applied during the allocation of research grants and other forms of financial support[23]. Naturally, with such means of "persuasion", the criterion that "truth" is "the opinion which is fated to be ultimately agreed to by all who investigate" is certainly "destined" to prevail.

Two years after Dirac published his justly famous textbook entitled "Principles of Quantum Mechanics", the German original of the "Mathematical Foundations of Quantum Mechanics" by von Neumann (1932) made its appearance. In it, von Neumann provided rigorous mathematical justification for many of the heuristic procedures used by Dirac – who, naturally, as a physicist totally involved with the various very rapidly expanding fields of quantum theory, was in no position to follow developments in functional analysis, which was emerging at that time as a new and separate discipline in mathematics. It might have been expected, however, that once the period of rapid growth in nonrelativistic quantum theory had came to an end – as it most certainly did by the end of the 1940s – all the subsequently written and published textbooks in quantum mechanics would begin to reflect at least the main lessons that could be learned from von Neumann's outstanding monograph – whose translation in English was eventually published in 1955.

That, however, did not take place at that time – and has still not taken place even in the most recent mainstream textbooks on nonrelativistic quantum theory[24]. This clearly demonstrates how the instrumentalistic identification of mathematical and other forms of "truth" with "generally held opinion" and "professional consensus" can act as a bulwark against *true progress* in the understanding of the basic structure of quantum theories.

An elementary but notable example of the deficient mathematical standards prevalent in mainstream textbooks is the treatment of those quantum mechanical observables which are represented by *unbounded* self-adjoint operators – such as is the case with the majority of important observables, namely energy, position, momentum, (external) angular momentum, etc. According to a theorem by Hellinger and Toeplitz[25], no such operators can be defined on the *entire* Hilbert space of a quantum system, which as a rule is separable but *not* finite-dimensional. However, not only is this *most basic mathematical fact*, which was very clearly emphasized already by von Neumann (1932, 1955), *not* mentioned *at all* in any of the mainstream textbooks on quantum mechanics, but the student of quantum theory is as a rule left with the *false* impression that every state vector of the quantum system *is* in the domain of definition of these operators.

While the failings of the conventionalistic approach to this type of problem might be deemed innocuous – as it rarely gives rise *directly* to physically incorrect conclusions – we shall see that there are other closely related problems which lead to physically questionable, and even to false *physical* conclusions. In fact, one of the sources of the foundational difficulties encountered by conventional relativistic quantum mechanics can be traced to its purely conventionalistic treatment of eigenfunction expansions for position and momentum operators in nonrelativistic quantum mechanics, which ignores some very essential mathematical as well as physical points. Let us therefore first examine the key aspects of this treatment on a few simple examples.

As is well-known, in the configuration representation the elements of *eigenfunction expansions* for position and momentum are given by δ-"functions" and plane waves, respectively. Thus, in the simple case of a single nonrelativistic quantum particle of zero spin, one *conventionally* writes:

$$\langle \mathbf{x}' | \mathbf{x} \rangle = \delta^3(\mathbf{x}-\mathbf{x}') , \qquad \langle \mathbf{x}' | \mathbf{k} \rangle = (2\pi)^{-3/2} \exp(i\mathbf{k}\cdot\mathbf{x}') , \qquad \mathbf{x},\mathbf{x}',\mathbf{k} \in \mathbf{R}^3 . \tag{3.1}$$

It is clear, however, that neither the δ-"functions", nor the plane waves, are Lebesgue square-integrable functions [PQ], so that they do not belong to the Hilbert space with the inner product defined in (3.1.1). For that reason, von Neumann (1932) avoided the use of δ-"functions". Eventually their mathematical nature was, however, totally clarified by L. Schwartz (1945). The mathematically *correct* general treatment of the objects in (3.1) was subsequently supplied by the theory of rigged Hilbert spaces (Gel'fand *et al.*, 1964, 1968), as well as that of equipped Hilbert spaces (Berezanskii, 1968, 1978). These mathematical frameworks pinpoint the objects in (3.1) as elements of eigen*function* expansions – and *not* as eigen*vectors* of Hilbert space operators. Adaptations of both these general frameworks to the needs of quantum physics have actually been in existence for quite a while (cf., e.g., Antoine, 1969, 1980; Prugovečki, 1973). Regardless of which one of these particular frameworks one adopts, they all underline the fact that

$$(\mathbf{x}|,(\mathbf{k}| \in \mathcal{H}_- \supset \mathcal{H} \supset \mathcal{H}_+ , \qquad (\mathbf{x}|,(\mathbf{k}| \notin \mathcal{H} = L^2(\mathbf{R}^3) \cong \mathcal{H}^* , \qquad \mathcal{H}_+^\dagger = \mathcal{H}_- , \tag{3.2}$$

where $\mathcal{H}_-$ is, in general, a topological vector space which provides an *extension* of the Hilbert space $\mathcal{H}$ of state vectors. The space $\mathcal{H}_+$ is dense in $\mathcal{H}$ in the norm topology of $\mathcal{H}$, and it is equipped with a topology that is finer than the norm topology of $\mathcal{H}$, and which makes $\mathcal{H}_-$ equal to the dual of $\mathcal{H}_+$ (whereas $\mathcal{H}$ can be identified with its own dual $\mathcal{H}^*$).

The key point, that had become clear a couple of decades after the appearance in 1930 of Dirac's famous textbook, is that these eigenfunctions do *not* provide resolutions of the identity operator $\mathbf{1}$ in the Hilbert space $\mathcal{H}$ of state vectors, but, strictly speaking[26], *only* of the identity operator $\mathbf{1}_+$ in $\mathcal{H}_+$, i.e.,

$$\int_{\mathbf{R}^3} |\mathbf{x})\, d^3\mathbf{x}\, (\mathbf{x}| = 1_+ \quad , \qquad\qquad \int_{\mathbf{R}^3} |\mathbf{k})\, d^3\mathbf{k}\, (\mathbf{k}| = 1_+ \ . \tag{3.3}$$

Furthermore, the choice of $\mathcal{H}_+$ is generally dictated by mathematical convenience, rather than by general physical principles. The use of the round brackets in (3.3) is, therefore, meant to emphasize that, although the theory of equipped Hilbert spaces allows us to write

$$(\mathbf{x}|\psi\rangle = \psi(\mathbf{x}) \ , \qquad\qquad \forall \psi \in \mathcal{H}_+ \subset \mathcal{H} = L^2(\mathbf{R}^3) \ , \tag{3.4}$$

the sesquilinear form on the left-hand side of the above relation is *not* an inner product. In fact, the domain of definition for the variable ψ on its right-hand side *cannot* be extended to the entire Hilbert space $\mathcal{H}$ – as is the custom in all conventional literature which adopts an instrumentalist attitude towards *mathematical* truth. However, that this feature of the sesquilinear form in (3.4) is an *unavoidable mathematical fact* follows from another basic mathematical fact: the generic element of $\mathcal{H}$ is not a single function, but rather an equivalence class of almost everywhere (in the Lebesgue sense [PQ]) equal functions, which are such that one can change the value of any one of these functions $\psi(\mathbf{x})$ at any given point $\mathbf{x}$ without leaving that equivalence class – namely, in physical terms, without changing the quantum state vector. Upon restricting oneself to mathematically convenient[27] dense subspaces $\mathcal{H}_+$, one can choose representative functions for which (3.4) holds true – but that is *not* possible globally on $\mathcal{H}$. Thus, strictly speaking, one can speak of the probabilities (3.1.7) for sharp position measurement outcomes within Borel regions B in configuration space, but *not* of probability densities for *arbitrary* wave functions at *single* points in configuration space. For that reason, von Neumann concentrated on the probability measures in (3.1.7), rather than on the probability densities in (3.5.1).

This seemingly innocuous mathematical point has significant physical repercussions. Thus, although the conventionalistic custom is to refer to $|\mathbf{x}\rangle$ as an "eigenvector" of the nonrelativistic position operators, and to consider the left-hand side of (3.4) a "transition probability" purportedly corresponding to a *sharp* measurement of position, we see that actually these "transition probabilities" are not generically well-defined at the mathematical level. Does that mean that they are not well-defined also operationally, at a physical level?

That does not immediately follow, but the above points indicate that caution should be exercised even in nonrelativistic quantum mechanics, and that *one should regard sharp localization as a limit of realistic measurement procedures*, which *necessarily* entail only unsharp localizations. In fact, the adaptation to position measurements of the Wigner-Araki-Yanase (1952, 1960) arguments on the impossibility of arbitrarily precise measurements of quantities which do not commute with an additive conserved quantity (i.e., with momentum, in the case of position measurements), shows that sharp localization is *un*achievable not only in practice, but also in principle, even in the context of the nonrelativistic quantum theory of measurement (Busch, 1985b). Hence, the fundamental impossibility of *sharp* relativistic localization of quantum systems, discussed in Secs. 3.3 and 3.5, has its roots in nonrelativistic quantum mechanics – but that fact is *conventionally* ignored.

It might be believed that these rather elementary observations are of no deeper consequence, since the conventionalistically predisposed quantum theorist can in practice easily avoid all the ensuing pitfalls. We shall, therefore, now present two elementary examples which demonstrate that this is not always the case.

First, it should be recalled that the EPR paradox was originally formulated (Einstein *et al.*, 1935) in the language of sharp position and momentum measurements, based on the above interpretation of the quantities in (3.1) and (3.4) as *bona fide* transition probabilities, and that it was only later adapted by Bohm (1951) to measurements of spin – but with the original epistemic assumption of (an arbitrarily close) realizability of sharp measurement outcomes retained. This led to Bell's inequalities, whose first experimental tests were performed in the 1970s. However, it was only with the experiments of Aspect *et al.* (1981, 1982) that the basic issue of nonexistence of local hidden variables was settled in favor of quantum mechanics. On the other hand, the discussion of the consequences of those experiments for the concept of locality is still going on unabated *as if* the macroscopic concept of arbitrary precise localization could be transferred without major revisions to the micro-domain, so that microscopic localizability could be identified with macroscopic separability (Selleri, 1990, p. 202). However, in Chapters 1 and 3 we reviewed conclusive evidence to the effect that such a transference leads to definite contradictions with the concept of Einstein causality – which is the hub of the ongoing disputes (van der Merwe *et al.*, 1988; Tarozzi and van der Merwe, 1988; Kafatos, 1989) about the significance of the EPR paradox. Once the impossibility of such transference is generally acknowledged, the focus of these debates could be shifted to posing the EPR problem in an epistemologically correct manner – namely as a natural by-product of the need for using at the microlevel geometries specifically designed to take the fundamental *quantum* features of localizability into account from the outset, and dispense with the interpretation of (3.4) as a *literal* representation of a *transition* probability amplitude for "observing" a "quantum particle at $\mathbf{x}$".

A second illustration of physical misconceptions that have resulted from the same type of interpretation of elements of eigenfunction expansions as "transition probability amplitudes" is provided by the conventionalistic derivation of such a most basic formula as that for the differential cross-section in two-body nonrelativistic scattering theory.

First of all, it should be noted that the conventionalistic approach tends to favor the stationary, i.e., time-independent formulation[28], despite the fact that the time-dependent approach comes much closer to reflecting physical reality by treating the scattering operator S as related to an *idealization* of a scattering process – namely as a process which evolves in Newtonian time t, but entails the *physically unachievable* limits $t \to \pm\infty$. This preference of stationary methods is, however, not accidental, since the S-matrix program of the 1960s (cf. Notes 35-36) was headed by elementary particle physicists whose advocacy of instrumentalist standards in physics eventually led to the conjecture that the entire concept of spacetime might be just a macroscopic "illusion" (Kaplunowski and Weinstein, 1985).

In keeping with such attitudes (which for a while threatened to prevail in all of quantum physics), in mainstream textbooks on quantum mechanics one typically begins the derivation of the aforementioned differential cross-section by considering the asymptotic expansion (cf., e.g., Messiah, 1961, p. 371)

$$\Phi_{\mathbf{k}}^{(+)}(\mathbf{x}) = (2\pi)^{-3/2}\left(e^{i\mathbf{k}\cdot\mathbf{x}} + \left[f_{\mathbf{k}}(\theta,\varphi)e^{ikr}/r\right]\right) + O(r^{-1-\varepsilon}) \quad , \qquad r = |\mathbf{x}| \to +\infty \quad , \qquad (3.5)$$

of an incoming distorted plane wave, which represents an eigenfunction (in the extension to $\mathcal{H}_-$) of the total internal Hamiltonian of the two-body system (cf. [PQ], pp. 425-436 and 553-556). One then conventionalistically interprets the plane wave on the right-hand side of (3.5) as a "probability amplitude" that gives rise, in accordance with (3.5.7), to a current density **k**/m. This current density is again conventionalistically interpreted as representing the incident flux of an incoming beam; whereas the term between square brackets is similarly interpreted as a probability amplitude of an outgoing (scattered) spherical wave. Then, treating, again by convention, the plane wave and the spherical wave *as if* they were *not* superimposed, and hence neglecting the cross term resulting from that superposition – typically on grounds that it "oscillates very rapidly as a function of r as r becomes large" (Joachain, 1975, p. 51) – one arrives at the well-known formula

$$\sigma_{\mathbf{k}}(\theta,\varphi) = |f_{\mathbf{k}}(\theta,\varphi)|^2 \ , \qquad 0 < \theta < \pi \ , \qquad 0 \leq \varphi < 2\pi \ , \tag{3.6a}$$

$$f_{\mathbf{k}}(\theta,\varphi) = -(2\pi)^2 m \langle \mathbf{k}' | T | \mathbf{k} \rangle \ , \qquad |\mathbf{k}| = |\mathbf{k}'| \ , \qquad \mathbf{k} \cdot \mathbf{k}' = |\mathbf{k}|^2 \cos\theta \ , \tag{3.6b}$$

for the differential scattering cross section in the "center-of-mass reference frame" of the two-body system – where the expression on the right-hand side of the first equation in (3.6b) is the so-called T-"matrix".

The physical meaning of the "center-of-mass reference frame" is not questioned in such derivations, as it is taken for granted that "somehow" classical concepts still apply. When some fundamental difficulties with this type of conventionalistic derivation of (3.6) were pointed out by Band and Park (1978), it was, however, acknowledged by the author of one of the leading mainstream textbooks on quantum scattering theory that: "The traditional derivation (as given, for example, by Goldberger and Watson, 1964, or by Newton, 1966) involves a bit of fakery that hides the issue of pure states *versus* mixed states. A correct derivation uses a beam represented as a mixed state of packets with different impact parameters. Such a derivation (Taylor, 1972) is analogous to the classical one, in which it is also necessary to assume that the incident beam consists of particles whose impact parameters are uniformly distributed." (Newton, 1979, pp. 929-930).

The response of Band and Park to the above statement was: "Newton's revelation of 'fakery' in orthodox pure-state collision theory and admission of an analogy with the coarse-graining device used classically to suspend basic mechanical laws are welcome confirmations of our main contention, that, if collision theory is followed consistently with quantum mechanical unitary evolution, it is impossible to explain thereby the approach to equilibrium in a gas." (Band and Park, 1979, p. 938).

It turns out, however, that an alternative to the "suspension of basic mechanical laws" is possible, on account of the existence[29] of *single-target* differential cross-section, whose derivation does not involve coarse-graining. This type of cross-section is therefore given by a formula that is distinct from (3.6), since it involves a T-"supermatrix" (rather than a T-"matrix"), as well as the confidence function in (3.5.3) (cf. [PQ], p. 518; [P], p. 170):

$$\tilde{\sigma}_{\mathbf{k}}(\theta,\varphi) = \int_0^\infty dk k^2 \int_{\mathbf{R}^3} d^3\mathbf{k}' \, \tilde{\chi}_{\mathbf{k}}(\mathbf{k}')(m/k') \langle \mathbf{k}' | \boldsymbol{T} | \mathbf{k} \rangle \ , \quad k = |\mathbf{k}| \ , \quad k' = |\mathbf{k}'| \ . \tag{3.7}$$

Indeed, it is *not* true that any of the rigorous derivations of (3.6), namely those based on wave packets, rather than on plane waves and spherical waves (cf. [PQ], pp. 430-436; [Messiah, 1961], Ch. X, §§5-6; Taylor, 1972], Sec. 3-e; [Newton, 1979]), are "analogous to the classical" derivation. In fact, in the classical context it is *not* at all necessary to assume that the "incident beam consists of particles whose impact parameters are uniformly distributed" in order to derive the classical scattering differential cross-section formula in its most basic form, namely in the form (cf., e.g., Balescu, 1975)

$$\sigma_{\rm cl}(\theta,\varphi) = (b\,db\,d\varphi)/d\Omega \quad , \qquad \Omega \leftrightarrow (\theta,\varphi) \quad . \tag{3.8}$$

On the other hand, if one does make the transition from classical mechanics to classical statistical mechanics, one obtains from (3.8) a formula which is the equivalent of (3.7), and *not* of (3.6). This was actually proved by developing a common framework for classical as well as quantum statistical mechanics (Prugovečki, 1978a,b), in which it is possible to derive (3.7) and its classical counterpart within the same Liouville superspace. Under reasonable assumptions on the orders of magnitude of basic parameters in a scattering experiment, (3.6a) and (3.7) appear to be numerically very close, but they certainly are not equal!

The above elementary example illustrates how "theory selection" is *actually* effected in the purely pragmatic approach to quantum theory, which has become the trademark of post-World War II conventionalistic instrumentalism in quantum physics. The type of attitude it reflects is aptly described in the following quotation (which, in its original context, concentrated on the *modus operandi* of the "new physics" from the 1960s to the present): "Having decided upon how the natural world really is, those data which supported that image were granted the status of natural facts, and the theories which constituted the chosen world-view were presented as intrinsically plausible." (Pickering, 1984, p. 404).

Thus, instead of relying on the uncovering of scientific *truth* based *exclusively* on analytic and rigorously formulated thought, combined with impartial observations vis-à-vis fashionable theories, post-World War II instrumentalism identifies "truth" with "consensus", which, in turn, becomes a matter of institutionally enforced[30] "convention". Over the past four decades such practices have provided dramatic illustrations of the reasons for Heisenberg's deep concern (which we cited already in Sec. 1.5) about the "erroneous developments ... [that] are caused by a misconception by some physicists that it is possible to avoid philosophical arguments altogether". That concern added to Dirac's deep distress about the "complacency" of contemporary "theoretical physicists [who are satisfied with the use of] working rules and not regular mathematics". Clearly, in relativistic quantum field theory, *both* these concerns have to be addressed simultaneously – as demonstrated by the failure of the constructive quantum field theory program to establish the consistency of QED after more than a quarter century of effort (cf. Secs. 1.2 and 7.8, as well as Note 33 to Chapter 7). The lesson that might be learned from that failure is that it is not sufficient to try to impart mathematical respectability to the algorithms of the conventional approach in order to arrive at a mathematically consistent and yet physically nontrivial framework for relativistic quantum field theory. Rather, an epistemological analysis of its fundamental concepts is also required, and the implemented mathematically sound techniques have to reflect that analysis. In other words, "one must seek a *new* relativistic quantum mechanics *and* one's prime concern must be to base it on *sound* mathematics" (Dirac, 1978b, p. 6) – emphases added.

We have already documented in appropriate sections of the preceding chapters many of the failings of the conventionalistic outlook on relativistic quantum theory. Hence, we shall only very briefly review the principal ones in the remainder of this section, and then indicate how the existence of the "cosmological constant problem" described in Sec. 11.12 totally vindicates Dirac's steadfastly critical attitude towards all the developments in the post-World War II renormalization program.

Perhaps the most striking instance of a claim made in conventionalistic literature, which has been rigorously *proved* (Gerlach *et al.*, 1967) *to be false*, is the assertion that the timelike component $j^0(x)$ of the Klein-Gordon current in (3.3.9) is positive definite if one restricts oneself to positive-energy solutions of the Klein-Gordon equation [SI]. This and other similar claims in otherwise respectable conventionalistic textbooks have influenced the thinking of generations of physicists, since they left them with the impression that "old" problems concerning relativistic quantum particle localizability have been "solved" by conventional relativistic quantum theory a long time ago, when actually the opposite is the case: not only have those problems *not* been solved, but proofs exist (Hegerfeldt, 1974, 1985, 1989) that they are not solvable within the conventionalistic framework – namely that all formulations of quantum particle localizability based on classical geometries give rise to violations of relativistic Einstein causality, albeit the opposite is maintained.

To some of those predisposed to favor either the conventionalistic instrumentalism of the contemporary mainstreams in quantum theoretical physics, or the formal instrumentalism of the dominant contemporary school in quantum mathematical physics, the answer to this type of insurmountable difficulty with conventional concepts for particle localization appears to lie in the substitution of quantum field localization for quantum particle localization. However, not only does this substitution replace one set of difficulties with another – namely with the still unresolved fundamental problem of a mathematically cogent concept of (interacting) quantum fields, that can be *mathematically* localized in arbitrarily small regions of classical spacetimes, e.g., by using test functions of arbitrarily small supports in the Wightman formalism [BL] – but the following physical question is then not asked and answered: how does one *operationally* localize a classical or a quantum field?

If, however, the above question is asked, then the only answer available is: by the use of massive test bodies. In their well-known papers on this subject, Bohr and Rosenfeld (1933, 1950) employed an analysis of the behavior of such *classical* test bodies, which therefore necessarily have to occupy macroscopic domains. Indeed, once regions of atomic and subatomic size are reached, the "consideration of the atomistic structure of measuring instruments", whose need they emphasized in their work, becomes unavoidable, so that one has come full circle: a consistent theory of localization of material quantum objects is needed in order to be able to formulate, in a *physically* meaningful manner, the concept of quantum field localization.

Until the last decade, conventionalistic instrumentalism tended to ignore such foundational questions on the pragmatic grounds that the agreement of *its* theoretical predictions with experimental results is *all* that matters. However, it has been demonstrated in a number of recent studies (Cushing, 1990; Franklin, 1986, 1990; Pickering, 1984, 1989) that experimental technique is itself highly conditioned by theoretical outlook. Furthermore, as illustrated in an extensively documented sociological history of post-1960 developments in high-energy physics, "the idea that experiment produces unequivocal fact is deeply problematic. ... [Actual experiments] are better regarded as being performed upon 'open', imperfectly understood systems, and therefore experimental reports are *fallible*." (Pickering, 1984, p. 6). Therefore, fundamental faults in theory can give rise to fundamental deficiencies in experimental design and technique, thus creating a vicious circle of feedbacks. In fact, as we have seen already in Sec. 9.6, when we discussed Dirac's critical attitude towards the experimental confirmation of QED predictions that are very highly acclaimed in conventional literature, in the absence of a mathematically *sound* theory it becomes a matter of *subjective belief* whether such apparent agreement represents confirmation of a theory intrinsically based on conventional "working rules", or just mere coincidence.

This becomes especially evident when closer scrutiny reveals that some such "coincidences" could be ascribed to fortuitous theoretical manipulation, since conventionalistic instrumentalism has facilitated the fine-tuning of theoretical computations to fit the experimental results by simply ignoring or discarding what is undesired, under the heading of such typical rationalizations as that it might be "naive", or "irrelevant", or "renormalizable", or "compactifiable", etc., etc. For instance, in the earlier cited carefully documented study of the development of the "new physics" in the 1960s and 1970s, we are provided with example after example of the following *sociological* high-energy phenomenon: "Discrepancies between prediction and data were taken as important results rather than serious problems: topics for further work rather than objections to the model." (*ibid.*, p. 266). Moreover, "fine-tuning" in such "further work" was greatly facilitated by the fact that *theoretical error bounds were intrinsically unavailable* in the computation of the "predicted" values of fundamental physical quantities, such as the S-matrix elements of conventional quantum field theories. Indeed, what would be the possible use and *meaning* of such traditional theoretical tools to the theorist who deals with theoretical constructs whose very mathematical *existence* is not at all assured? Or to the theorist who can conveniently stop the summation of a "perturbation" series, for constructs of undecided mathematical existence, as soon as the *desired* agreement with experimental data is achieved? On the other hand, it might be asked: What if its summation *were* continued? And, in view of the *presumed* "asymptotic" nature (Dyson, 1952) of all "renormalized perturbation series": *Where* should one *stop* the summation, from an *objective* point of view?

With regard to measurements of spatio-temporal relationships at the microlevel, even the reliability of experimental results as a direct guide to the validity of fashionable theories deserves closer scrutiny. Indeed, as discussed and documented by Hacking (1983), Cartwright (1983), Ackerman (1985), Galison (1987), Franklin (1986, 1990), and others, contemporary experimental procedures are heavily theory-dependent. Hence, just as with Kaufmann's (1905, 1906) negative experimental verdicts on Einstein's special relativity, cited in the introduction to this chapter, and other similar historically well-documented cases, some experimental results might have to be critically reevaluated if Dirac's often repeated urgings for the use of "sound mathematics" in relativistic quantum physics are eventually heeded, and a mathematically *sound*[31] reappraisal of some key theories is undertaken.

The fundamental inadequacies of the conventionalistic outlook emerge with full force when quantum fields in curved classical spacetimes are considered: as described in Secs. 7.2 and 7.3, not only do the fundamental mathematical difficulties of the conventionalistic approach to quantum field theory become then more pronounced, but even old and very well established *physical* principles are sacrificed in order to maintain some particularly favored *conventionalistic* scheme. Thus, as can be seen from the review and analysis of conventional quantum field theory in curved classical spacetime presented in Secs. 7.1-7.3, some of the adherents to conventionalistic instrumentalism transform even the law of local conservation of energy and momentum into a matter of mere convention, which can be violated in order to save the formal aspects of conventional quantum field theories in curved spacetime. These aspects, in turn, are disregarded at the level of quantum gravity and cosmology, where concern with unitarity of the S-matrix seems to take precedence over formulating a concept of physical time based on a consistent theory of measurement. On the other hand, the existence of a unitary S-matrix solution for *any* realistic quantum theory of interacting relativistic fields has *never* been proved[32] even in Minkowski space (cf. Sec. 7.6 as well as Note 31 to Chapter 9) – not to mention in any kind of curved spacetime. Thus, whereas conventionalistic instrumentalism has failed to meet in quantum physics even its own most basic criteria during the span of close to half-a-century of intense computational activities, its preoccupation with those criteria has derailed it on a sidetrack, where some of the most sensible and best established physical principles of quantum theory in the pre-instrumentalist era are ultimately ignored, or even violated.

As if all these distressing inadequacies were not enough, the developments in particle physics and quantum cosmology over the past three decades indicate "a blurring of distinction between physical science and mathematical abstraction ... [reflecting] a growing tendency to accept, and in some cases ignore, serious testability problems" (Oldershaw, 1988, p. 1076). Thus, no less than *twenty* major *effectively untestable* problems are listed in (Oldershaw, 1988) – each one of which is of the type that would have been deemed a serious cause for concern in the *pre*-instrumentalist era. In view of Dirac's steadfast opposition to the renormalization program, from the time of its inception in the late 1940s until his death (cf. the introduction to Chapter 7), we shall discuss only one of those twenty issues. It is the one which shows that his criticism of the *ad hoc* nature of that program, and of the fact that it does *not* provide "a correct mathematical theory at all", has been completely vindicated by some of the developments which took place after his death.

First of all, let us remind the reader that one of the two main progenitors of the renormalization program has recognized from the outset that "the observational basis of quantum electrodynamics is self-contradictory", and that "a convergent theory cannot be formulated consistently within the framework of present space-time concepts" (Schwinger, 1958, pp. xv-xvi); whereas, the second one eventually acknowledged that "it's also possible that electrodynamics [namely conventional QED] is not a consistent theory" (Feynman, 1989, p. 199). Furthermore, in this regard, to the end of his life Dirac's main point had been the following: "Just because the results [of the conventional renormalization procedures in quantum field theory] happen to be in agreement with observation does not prove that one's theory is correct." (Dirac, 1987, p. 196).

The glaring *observational* inconsistencies(cf. Sec 11.12), to which the introduction of the Higgs boson in the offspring of conventional QED (namely in electroweak theory and in QCD) has led, have proven Dirac absolutely right in *all* respects, *including* the observational ones. Indeed, on one hand, despite a wide-spread search (cf., e.g., Harari, 1983; Weinberg, 1987) there is *absolutely no observational evidence* in favor of such a Higgs "particle", introduced *solely* for the purpose of making QCD "renormalizable"; on the other hand, its *assumed* existence gives rise to an *enormous* cosmological constant – in blatant contradiction to the most basic observational facts. Of course, many "solutions" to the "cosmological constant problem" have been proposed (cf., e.g., the review article by Weinberg, 1989), but in the end one has to concur with the opinion that: "None of [these] attempts has succeeded. If anything matters have grown worse because theorists keep dumping more particles and fields into the vacuum." (Veltman, 1986, p. 78).

In fact, ever since the advent of quarks, which after the failure to be observed were simply declared to be permanently "confined" (with no indubitable proof of confinement yet in existence), there has been such a proliferation of *ad hoc* theoretical devices, designed solely to remove flagrant disagreements between conventional theories and experimental facts, that the above cited leading researcher in the theory of quantum Yang-Mills fields figuratively exclaimed in a tone of utter exasperation: "Indeed, modern theoretical physics is constantly filling the vacuum with so many contraptions such as the Higgs boson that it is amazing a person can see the stars on a clear night!" (*ibid.*, p. 76). The following are just a few examples of the "contraptions" that have highlighted the "progress" from the 1960s to the late 1980s: "Instead of one photon we have 12; three of them have acquired masses from spontaneous symmetry breaking, and eight of them are trapped. Instead of one electron, we have a whole menu of quarks and leptons defined by their representations with respect to the weak and strong gauge groups, and this menu is replicated three times: There are three generations." (Weinberg, 1987, p. 7). It is therefore of no surprise that when faced with such a cornucopia of offerings from particle physicists, a noted astrophysicist felt compelled to remark: "Indeed I sometimes have the feeling of taking part in a vaudeville skit: '... You want massive weakly interacting particles? We have a full rack. You want an effective potential for inflation with a shallow slope? We have several possibilities.' This is a lot of activity to be fed by the thin gruel of theory and negative observational results, with no prediction and experimental verification of the sort that, according to the usual rules of evidence in physics, would lead us to think we are on the right track of the physics of the universe at [a redshift epoch] $z > 10^{10}$." (Peebles, 1987, p. 236).

So, in the end one can ask, who was proven right by all these developments: Dirac, or the multitude of "dynamically acquiescent" (Pickering, 1984, p. 272) theorists, whom Dirac often described (cf. Sec. 9.6) as being too "complacent about the faults" of the renormalization programme instituted after World War II ?

Keeping all of the above points in mind, we can summarize the situation by saying that, at the foundational level, contemporary conventionalistic instrumentalism is confronted with two fundamental types of problems.

1) Mathematically, there is the one of *logical consistency*: as is well-known, from an inconsistent set of statements *any* other statement can be in principle derived. Thus, the deductive power of the scientific method can be in practice unwittingly undermined by *ad hoc* manipulations that are not dictated by *logical* necessity, but rather by the desire to achieve agreement with experiment – not to mention professional recognition. This was obviously central to Dirac's often expressed concern that the laws of "regular", "sound" and "sensible" mathematics be followed in contemporary relativistic quantum field theory.

2) Physically, there is Heisenberg's concern with posing the *epistemologically correct* questions: the use of *formal* analogies can lead to the introduction and development of concepts in a new context where such concepts no longer have a legitimate physical meaning, and lead to physically meaningless "scenarios". Perhaps the most extreme example of this type is provided by the *ex nihilo* "scenario" of the creation of our Universe. Indeed, the concept of a wave function, representing a quantum particle, "tunneling through" the potential barrier to which another system of *existing* quantum particles gives rise, is operationally well-defined, and it makes physical sense; however, what is the possible physical *meaning*[33] of Nothing tunneling through a potential barrier produced by Nothing, in order to "create" our Universe in some present-day cosmological "scenarios"? Even though such a "phenomenon" can be *formally* described (Tryon, 1973; Vilenkin, 1982, 1988), and certain features of inflationary cosmological models that are currently in fashion can be then reproduced by the *mathematics* employed, does that *physically* validate such a "scenario"? The fact that there are some features of the inflationary model that can be "deduced" from such a "scenario" *cannot* establish its *physical meaning* and *validity* any more than the existence of Santa Claus can be established by the mock argument of Bertrand Russell, cited in Note 16, which was aimed at demonstrating the utter fallacy of the principal instrumentalist

criterion of "truth" for a hypothesis – namely that "an idea is true if it works" (Stapp, 1972, p. 1103). Indeed, if that were so, then as Bertrand Russell pointed out with refined irony, the application of this most basic instrumentalist doctrine would allow us to infer that "Santa Claus exists" from the obviously correct statement that "the hypothesis [of the existence of] Santa Claus 'works satisfactorily in the widest sense of the word'"!

It would appear that one of the basic methodologies of conventionalistic instrumentalism is to pick fundamental techniques and results from a domain of quantum physics, where those results have a consistent and well-defined physical and mathematical meaning, and then transfer them to some new area of quantum physics, where both those types of meanings might be lost, and where only entrenched *conventionalism* provides the thread that holds together a thus newly created theoretical framework. Of course, as long as "truth" is to be found in the "wide acceptance of a theoretical idea", which can be secured by a variety of means (such as skillful promotional techniques, which in pre-instrumentalist times would have been more characteristic of practices in business and commerce, rather than in science), then there is nothing wrong with such an approach.

On the other hand, we have seen from the numerous quotations presented in this monograph, that Dirac and Heisenberg have criticized in print many of the post-World War II developments in conventional relativistic quantum theory which, as we approach the end of this century, have become entrenched in "pragmatic" attitudes towards what constitutes "truth" in many key areas of what Pickering (1984) and others have described as the "new physics". Popper ascribes such attitudes to "a tradition which may easily lead to the end of science and its replacement by technology"[34], and which is based on a "fashionable philosophy [which] may in fact be uncritical, irrational, and objectionable" (Popper, 1982a, pp. 100-103).

These are unequivocal and strong statements. They have to be weighed, however, against the fact that the protracted and practically unchallenged dominance of conventionalistic instrumentalism in quantum theory has given rise to a situation without exact precedent in the history of science. One commentator, who finds some of the latest manifestations of this phenomenon to be "a cause for concern", rhetorically asks: "If the empirical foundation of the new physics is so insecure, and if it is still an axiom of science that without an empirical foundation a paradigm is dangerously adrift in a sea of abstraction, then why is there an unquestioned faith in the new physics? How can we understand the remarkable optimism and credulity demonstrated by theorists, experimentalists, peer reviewers, editors, and science popularizers?" (Oldershaw, 1988, p. 1080).

As illustrated in this section, and as demonstrated in some other specific instances discussed in appropriate previous sections of this monograph, to this "insecure empirical foundation" has to be added the fact that the mathematical and epistemological foundations of this "new physics" are at least as "insecure". So, instead of answering the above two questions, let us merely pose a counter-question: Sociologically speaking, what else can be expected when traditional standards of epistemological soundness and mathematical truth have been uprooted, and replaced by purely instrumentalist standards of "truth" which encourage, and in many key institutional settings even *enforce*, the type of conformity whose manifestations Feynman (1954) has so colorfully described as the "pack effect"?

As witnessed by the earlier cited *public* statements of Dirac, Einstein, Heisenberg, Popper, Russell, and many other outstanding physicists and philosophers of this century, those men of vision have given proper and timely warnings as to what *can* be expected to happen. And what they foresaw and feared *has been happening* with increasing frequency and intensity ever since "World War II altered the character of science in a fundamental and irreversible way" (Schweber, 1989 – cf. also Note 47).

Perhaps it is time that those warnings were heeded.

12.4. General Epistemological Aspects of Quantum Geometries

The quantum geometry framework described in the present monograph grew out of a systematic effort at trying to see whether the *numerical* successes of the conventional approach to relativistic quantum theory could be explained from a mathematically and physically cogent point of view. It appeared obvious from the beginning that, at the epistemological level, such a point of view would have to reexamine the very foundations of relativity and quantum theory. It was also clear that, in so doing, it would have to reconcile Einstein's "realism" with Bohr's "positivism", by concentrating on the epistemological issues that united those two giants of twentieth century physics, and possibly ignoring the others – or, if absolutely necessary, even contradicting them on those issues that separated their distinct but not at all totally irreconcilable points of view.

Indeed, it was pointed out in Sec. 12.1 that the basically operationalist attitude of Bohr was very much shared by Einstein during the period when he created special as well as general relativity. On the other hand, it should be obvious to readers who have read most of Chapters 3-11, that the operationalism of Bohr, as well as that of the pre-1920 Einstein, is retained in the formulation of the quantum geometries studied in those chapters. The concept of frame of reference, already so crucial to Einstein in the formulation of special relativity, and of "event", defined as a spacetime coincidence, and viewed as the fundamental building block of all our observational constructs, namely all measurable physical quantities, were instrumental in those formulations. Such formulations are, therefore, also in agreement with Bohr's point of view – except that Bohr might have insisted on a classical description of all frames of reference.

On the other hand, a form of *quantum realism* decidedly manifests itself in the present framework in the form of the, until now, implicit premise that there *is* a physical reality, which is independent of any operational or linguistic conventions which any group of individuals happen to adopt. In other words the present work is founded on the belief that there is a *single* reality, which is *quantum* in its manifestations at the most fundamental level, and totally independent of any theoretical or experimental *conventions*. Hence, the quantum geometry framework presented in this monograph strives to remove the artificial dividing line which Bohr imposed between "system" and "apparatus": there is only one reality, and that reality is quantum; *ergo*, any apparatus should be described at the most fundamental level in purely quantum terms. In particular, that conclusion is applied to frames of reference, which are viewed as *quantum* "objects". However, as we have seen in Secs. 3.7 and 3.9, that does not preclude in some such frames the possibility of *approximations* of classical behavior: as we discussed in Sec. 3.9, such behavior is indeed manifested by sufficiently massive quantum frames. Thus, Bohr's teachings on the significance of classical concepts in the quantum theory of measurement are not ignored, but rather modified.

Bohr's insistence on the importance and the role of language is not ignored either. In this respect the present approach is at odds with Popper's (1976, 1982, 1983) type of classical realism, which downgrades that role. However, there is absolutely no contradiction in maintaining that, on one hand, there *is* a *micro*reality, and that the purpose of quantum theory is to reflect that reality *as closely as possible*, but that, on the other hand, in so doing it *should employ the type of language best suited for that task,* by incorporating all essential aspects of microreality, and at the same time avoiding, in accordance with Born's second maxim cited in Sec. 1.1, the introduction of redundant theoretical notions with no empirical counterpart. Consequently, the fundamental stance of *quantum* realism is epistemologically totally opposed to that of a "microrealism, according to which entities such as electrons, quarks, and the like, to which the name 'particle' is ascribed, are deemed to have a specific position at all times (and in terms of this conception, should also have, 'for reasons of symmetry', a specific velocity)" (d'Espagnat, 1989, p. 83).

Indeed, the type of "microrealism" defined by d'Espagnat tries to understand the behavior of such "objects" as molecules, atoms, elementary particles, etc. exclusively in terms of concepts that have grown out of the fertile soil of our experiences with the *macro*scopic world, which we routinely encounter in our everyday lives. Of course, such concepts are perpetually nurtured by those experiences, so that they are our principal source of physical intuition – as rightly emphasized by Bohr. On the other hand, that does not mean that they have to remain our *only* source of such intuition, and that the human mind cannot grasp concepts and relationships that transcend the most immediate types of sense-impressions that reach it. Hence, the quantum realism underlying the present work tries to understand the microworld on its own terms, by developing the conceptual, linguistic and mathematical tools best suited for that task – irrespective of whether or not they are in accordance with the commonsensical ideas rooted in our everyday experiences.

It could be said that as a conceptual and mathematical *framework*, rather than as a family of quantum theories, the purpose of quantum geometry is to supply a precise *operationally-based* mathematical language, as well as a metalanguage, for the description of quantum phenomena in purely quantum mechanical terms. In this context, the concept of informational completeness (cf. Sec. 3.7) emerges as fundamental, and it supersedes the EPR-type of *classical* realism, as applied to the quantum domain: a quantum theoretical description is not considered complete "if, without in any way disturbing the system, we can predict with certainty (i.e., with probability equal to unity) the value of a physical quantity" (Einstein *et al.*, 1935, p. 777); on the contrary, at the most fundamental quantum level, Wigner-Araki-Yanase types of arguments (discovered long after the advent of the EPR paper) indicate that in quantum theory there is no place for *sharp* stochastic values (i.e., for values predictable "with probability equal to unity"), so that the EPR formulation cannot possibly lead to valid criteria of completeness for the theoretical description of any *quantum* reality. Thus, since even in principle, and not only in practice, *all* values of physical quantities are unsharp at the quantum level, one of the basic principles adopted in quantum geometry is that of informational completeness (cf. Principle 2 in Sec. 1.3) at the *local* level, i.e., in the quantum fibres *above* the points of a base spacetime manifold (cf. Principle 3 in Sec. 1.3). In other words, *any* quantum state in those fibres is completely determined by the measurement of its Fubini-Study distance from the elements of an informationally complete quantum frame in that fibre, which in turn is given in terms of operationally directly measurable "transition" probabilities – cf. Eqs. (3.7.10) to (3.7.15).

This fundamental feature also dispenses with the need for von Neumann's questionable postulate about the identifiability of the set of quantum observables with the set of *all* self-adjoint operators in a Hilbert space (cf. Note 27 to Chapter 7). Furthermore, in the presence of quantum frame analyticity, only measurements of Fubini-Study distances of the local quantum state of a system to frame elements within arbitrarily small neighborhoods of the point of contact between tangent space and base manifold are required for the complete determination of that state. Therefore, such measurements are in principle implementable in the presence of arbitrarily strong gravitational fields. Thus, quantum realism is operationally based only in the context of measurement theoretical concepts (cf. Sec. 12.5).

On the other hand, by introducing the concepts of *proper* quantum state vector and of *quantum* frame as fundamental, it clearly recognizes that not all basic elements in its theoretical superstructure can possess *direct* operational counterparts, which, as such, would be simply groupings of our sense-experiences. Indeed: "In order to be able to consider a logical system as a physical theory it is not necessary to demand that all of its assertions can be independently interpreted and 'tested' 'operationally'; *de facto* this has never yet been achieved by any theory and can not at all be achieved." (Einstein, 1949, p. 679,). Rather: "Although [theoretical] conceptual systems are logically entirely arbitrary, they are bound by the aim to permit the *most nearly possible certain (intuitive) and complete co-ordination with the totality of sense-experiences*; secondly they aim at the *greatest possible sparsity of their logically independent elements* (basic concepts and axioms), i.e., undefined concepts and underived (postulated) propositions." (Einstein, 1949, p. 13) – emphasis added.

The fundamental role played by measurement theoretical aspects brings, however, to the fore the question of where the present quantum geometry framework stands in the ongoing realism–anti-realism dispute over the ontological status of the measured quantities. The following quotation succinctly reviews the issues in question:

"Anti-realism with respect to measurement can assume a variety of forms. The simplest is an austere operationalism [expressed by the idea that measurable quantities] derive their meaning entirely from our measurement practices. ... This outlook is a species of a more general and widespread view, according to which the fundamental facts about measurement are grounded in *conventions* ... . A much more sophisticated conventionalism ... [is the] carefully qualified development of the idea that measurement operations can be *said* to measure the same thing if they give rise to the same ordering of objects under the same conditions. By contrast, I take realism with respect to measurement to be the view that in many cases measurement can give information about objective features of phenomena that is tinged with interesting elements of convention. ... The realist's thesis is that there are objective facts about what the length of something is, facts that are – within precisely specifiable limits – independent of our linguistic and scientific conventions, the particular theories we happen to accept, and the beliefs we happen to hold. Length can be measured on a ratio scale, and that means once a unit (e.g., the meter) is conventionally selected, there will be an objective fact as to how many meters long any given object is (since this will just be a fact about the ratio of its length to that of the meter bar). The realism–anti-realism dispute over measurement is not usually cast in terms of semantic issues, but it is important to realize that they are just there beneath the surface." (Swoyer, 1987).

This and other publications (Bergmann, 1960; Reichenbach, 1961) on these issues in contemporary philosophy of science reveal that "semantic issues" are indeed at stake in much of the ongoing polemic. For those concerned with more substantive issues, there are merits and demerits in both the operationalist as well as in the realist points of view. It is, therefore, tempting for a scientist to completely ignore such polemics, and dismiss them as totally "irrelevant" to the actual practice of science.

The history of science teaches us, however, that utterly erroneous opinions were sometimes held because certain beliefs as to *what* is actually measurable, and *how* it is to be measured, were uncritically held in the face of existing strong evidence to the contrary. For example, since quantum geometry is applicable, amongst other fields, to quantum cosmology, the following comments might be of interest: "On various occasions in the history of cosmology the subject has been dominated by the bandwagon effect, that is, strongly held beliefs have been widely held because *they were unquestioned or fashionable, rather than because they were supported by evidence.* As a result, particular theories have sometimes dominated the discussion while more convincing explanations were missed or neglected for a substantial time, even though the basis for their understanding was already present." (Ellis, 1989, p. 367) – emphasis added.

Thus, "strongly held beliefs" can color[35] one's *perception* as to what observational evidence supports and what it does not, and, in fact, even shape one's *beliefs* as to what is observable and what is not. For example, in the heyday of *S*-matrix theory in the 1960s the opinion that the description of quantum phenomena did not require any concept of spacetime was not only widely held in elementary particle circles, but became thoroughly institutionalized[36]. In fact, opinions to this effect were heralded at international conferences and in review articles as the only acceptable approach to the physics of fundamental quantum phenomena[37] – without such basic questions being asked and answered, as to how such a *belief* could be reconciled with the fact that a spacetime background was essential to the rest of physics. In fact, even nowadays, residues of that belief condition research in conventional quantum gravity and instrumentalistically motivated quantum cosmology, where the question of "renormalizability" of the so-called "perturbation" series for the *S*-"matrix" still occupy center stage. However, in such cosmological pursuits, the following elementary question is not asked: what is the possible *literal* physical meaning of the concept of *S*-matrix in the *real* universe in which we live, namely in a universe in which, according to all evidence, asymptotic flatness of spacetime is certainly not present in the "cosmic" past, and, by all accounts, will never become realized in the "cosmic" future.

This is not to say that, if one subscribes to the point of view of quantum realism, according to which spatio-temporal relationships have an *objective* existence, which is independent of prevailing theories and operational procedures, then those operational procedures are automatically provided by the quantum reality. Rather, the opposite is true in practice: operational procedures are heavily theory-dependent, even to the extent that modifications of the underlying theories entail radical modifications in the measured values.

Consequently, one of the key questions from the point of view of a *quantum* realist, concerned with *empirical* reality (rather than with so-called "intrinsic reality" – cf. d'Espagnat, 1989), is what are the truly fundamental units for the measurement of spacetime separations in Nature. In other words, special relativity was grounded in an operationalist attitude, which stipulated that spatial distances are to be measured with "rigid" rods, and temporal separations with "standard" clocks (Einstein, 1905). Although the concept of strictly rigid rod is actually untenable in relativity (Stachel, 1980), that of standard clock suffices under the assumption of the constancy of the speed of light with respect to all Lorentz frames. That raises the question, however, as to what choice of clock should be made for that standard; and, even more importantly, why would Nature abide even at the *micro*level by any particular choice of macroscopic clock, made on technological or other anthropic grounds? In other words, except if *real* (as opposed to operational) time is somehow an intrinsic property of *all* matter in existence, it would be unrealistic to expect that Nature would abide by any purely *conventional* (Jammer, 1979) choice at all of its levels of magnitude, from the very smallest subnuclear processes, to the large-scale structure of our

Universe. Indeed, in practice, totally different units and operational procedures are used at the two ends of this scale of magnitudes, as well as at many particular stages in between.

The present quantum geometry framework is based on the premise that *a fundamental choice, independent of all conventions, does exist for the specification and measurement of spatio-temporal relationships, and that, therefore, it has to be inscribed in every single bit of matter in existence.* That *natural* choice can be found by simply tracing the origins of de Broglie's idea, which heralded the emergence of quantum mechanics[38]: namely that, on account of its rest mass m, each massive elementary quantum object represents a *natural* clock with period $T = 2\pi/m$ in Planck *natural* units. The universal constancy of the ratios of the observed rest masses of elementary particles vouches that *all the elementary particles in Nature keep the same local time*, so that any geometro-stochastic propagation can take place under well-specified spatio-temporal conditions. Without *that* assumption, the proposed idea of any quantum geometry would make no sense at all as a candidate for a *physical* geometry. But, without the hypothesis of cosmic constancy of the ratios of the rest masses of all "elementary particles", elementary particle physics would not make any sense either!

So, those in elementary particle circles who argue that at a fundamental level the concept of spacetime might not be meaningful (Chew and Stapp, 1988), or that it might be a mere illusion (Kaplunowski and Weinstein, 1985), are simply ignoring the most fundamental evidence in their own field: the existence of quantum entities which conventional terminology has labelled as "elementary particles". The fact that it might eventually turn out that all of these massive "objects" are neither "elementary" nor "particles" is irrelevant: the main point is that they *do* possess rest masses, and therefore they *are* localizable in reality, and that they *do* keep their own proper time. It is, therefore, a matter for theoreticians to display enough imagination in the creation of theories which *properly* reflect these *quantum facts*. In particular, this intrinsically fundamental *physical* significance of the concept of spacetime has to reflect the measurement-theoretical limitations imposed by the existence of the Planck length and of the Planck time.

For this very reason, these basic constants are embedded, in the form of the fundamental length ℓ (= 1 in Planck natural units), into the very structure of the fibres of quantum geometries. This is very much in keeping with Einstein's epistemology (albeit it would not have been in keeping with his predilection for *classical* realism):

"The relations between the concepts and propositions [of a theoretical framework] are of a logical nature, and the business of logical thinking is strictly limited to the achievement of the connection between concepts and propositions among each other according to *firmly laid down rules, which are the concern of logic*. The concepts and propositions get 'meaning', viz., 'content', only through their connection with sense experiences. The connection of the latter with the former is purely intuitive, not itself of a logical nature. The *degree of certainty* with which this connection, viz., intuitive combination, can be undertaken, *and nothing else, differentiates empty phantasy from scientific 'truth'*." (Einstein, 1949, pp. 11-13) – emphases added.

Finally, the retention of the equivalence principle in the relativistic quantum regime is the last, but certainly not the least, of the epistemological cornerstones in the formulation of the quantum geometries in the preceding seven chapters. In fact, the simplest type of experimental test, helping to choose between theories formulated within the present quantum geometry framework and those based on conventional frameworks (cf. Secs. 7.2 and 7.3), lies in the verification of this very principle in the quantum regime: is there, or is there not, *actual* (as opposed to conventionally agreed upon) Rindler particle production in Nature? Is there spontaneous particle production *ex nihilo* in Nature, that as such can be *observed* by *inertial* observers under very different free-fall conditions? Is there, therefore, local energy-momentum violation that such observers *can* witness?

The answer of the present GS framework, based on the application to general relativity of ideas anchored in the epistemology of *quantum* realism, is a firm: No! Some of the papers cited in Secs. 7.2 and 7.3 (cf., e.g., Unruh, 1976; Unruh and Wald, 1984), based on conventional instrumentalist conceptualizations of relativistic quantum theory in curved spacetime, claim: Yes! Hence, this is a very clear-cut case where experiments, performed under carefully and properly controlled conditions (cf. p. 203), should decide the issue.

12.5. The Concept of Point and Form Factor in Quantum Geometry

At the most fundamental epistemological level, the distinction between classical geometries and the quantum geometries treated in this monograph lies in the treatment of the concept of "point". From a purely mathematical perspective, the distinction does not appear that great: the points of classical geometries belong to finite-dimensional manifolds; whereas, those of quantum geometries belong to fibre bundles which constitute infinite-dimensional manifolds or supermanifolds. However, physically, the distinction is considerably greater. It can be described by saying that the points of classical geometries are "sharp" and "structureless"; whereas, those of quantum geometries are "unsharp" and *can* possess an *internal* structure. In the quantum geometries that describe quantum spacetimes, that structure is embedded in their quantum spacetime form factors. It therefore seems mandatory to single out a *fundamental* quantum spacetime form factor, which distinguishes itself by an outstanding simplicity of its internal structure, as well as some very special physical characteristics vis-à-vis some model of universal significance in quantum physics.

At the very foundations of quantum physics lie the canonical commutation relations between position and momentum. The harmonic oscillator is the simplest as well as the most fundamental physical model that embeds the constituents of those canonical relations into the eigenvalue equation for its energy spectrum. In the case of the relativistic harmonic oscillator that equation assumes the form

$$D^2 f = -\lambda f \ , \qquad \lambda \in \mathbf{R}^1 \ , \tag{5.1a}$$

$$D^2 = Q^\mu Q_\mu + P^\mu P_\mu \ , \qquad Q^\mu = \eta^{\mu\nu} Q_\nu \ , \quad P_\mu = \eta_{\mu\nu} P^\nu \ , \tag{5.1b}$$

into which the Minkowski metric enters intrinsically, and into which the relativistic canonical commutation relations are also intrinsically embedded:

$$[Q_\mu, P_\nu] = -i\eta_{\mu\nu} \ , \qquad [Q_\mu, Q_\nu] = [P_\mu, P_\nu] = 0 \ . \tag{5.2}$$

For that reason, as well as on account of the formal symmetry played in (5.1) by the Q's and the P's, Born (1949) adopted (5.1a) as the basic eigenvalue equation for his *quantum metric operator*.

Naturally, as they stand, (5.1) and (5.2) do not constitute a well-posed eigenvalue problem without the stipulation of boundary conditions on the eigenfunctions. Such boundary conditions can be imposed in the traditional manner by the requirement that the eigenfunctions be square-integrable in $\mathbf{R}^8$ with respect to the Lebesgue measure. However, such a stipulation cannot be justified from the point of view of a *relativistic* "quantum metric operator", since it is obviously related to the Euclidean regime[39], and, moreover, it leads to an eigenvalue spectrum which is unbounded from below. On the other hand, if (5.1a) is interpreted as an eigenvalue equation for quantum metric fluctuation amplitudes which result in *local* exciton propagators (cf. Sec. 7.4), then it turns out (cf. [P], Sec. 4.5) that its spectrum consists of eigenvalues bounded from below by a unique minimum eigenvalue, which corresponds to the fundamental quantum spacetime form factor f_ℓ in (5.5.5).

In view of the close connection between oscillator states and the realizations of Virasoro algebras emerging from some of the older treatments of string quantization (Green *et al.*., 1987, Sec. 2.2), a treatment of the eigenvalue problem in (5.1) can be devised which results in an entire family of "stringlike" quantum metric fluctuation amplitudes. Of course, although such possibilities of interpretation of excited states of the quantum metric operator in (5.1b) are intriguing, they are not particularly compelling, since the conjectures that excited string states might have occurred only during the "Planck era after the Big Bang" represent sheer speculation, which is unlikely to receive any direct experimental support in the foreseeable future. Nevertheless, in view of some still prevailing popularity of string theories, we shall briefly review them, before turning in the last part of this section towards the much firmer ground which underlies the choice of *fundamental* quantum spacetime form factor in this monograph. Hence, this review is intended primarily as an illustration of the fact that, although there are many other technical as well as conceptual differences between string theory and the present geometro-stochastic framework, there is also a certain underlying affinity of heuristic physical ideas, which could be used to establish closer theoretical links.

The incorporation[40] of massless oscillatory exciton states into a previous adaptation (Prugovečki, 1981b) of Born's (1949) quantum metric operator to GS quantum theory leads to a quantum relativistic harmonic oscillator, whose eigenstates display some of the features of string modes that are present in the fibres of a prequantum bundle over a ten-dimensional base space embedded in the bundle $T^*\mathbf{M} \oplus T^*\mathbf{M}$ over the Lorentzian manifold $\mathbf{M}$. In such a model for GS excitons the proper wave function for a graviton at any base location $x \in \mathbf{M}$ can be identified with the spin-2 ground state of the quantum metric operator $\boldsymbol{D}^2(x) = \boldsymbol{Q}^2(x) + \boldsymbol{P}^2(x)$ at that location.

From a semiclassical point of view, this treatment envisages a stringlike GS exciton at $x \in \mathbf{M}$ to be an excited eigenstate of a relativistic harmonic oscillator at that location. At such a heuristic level, a GS exciton above the base location $x \in \mathbf{M}$ can be visualized as a string of points $q \in T_x\mathbf{M}$ executing, in general, vibratory as well as rotational motions with respect to a local Lorentz frame $\{\boldsymbol{e}_i(x)\}$. The ground modes of such stringlike GS excitons would correspond to stochastic vibrations in the direction of motion specified by its 3-momentum $\mathbf{k}$, transversal oscillations in the polarization planes orthogonal to $\mathbf{k}$, and rotations around the direction in which $\mathbf{k}$ points. As a result of all these motions, its suitably renormalized probability wave amplitudes

$$f_{\mathrm{A}}(\breve{q},\breve{v}) = \lim_{\mu_{\mathrm{A}}\to+0} N_{\mu_{\mathrm{A}}} \int \exp(-\mathrm{i}\breve{q}\cdot\breve{k})\, \tilde{f}_{\mathrm{A}}(\Lambda_{\breve{v}}^{-1}\breve{k})\, \tilde{f}_{\mathrm{A}}(\breve{k})\; \delta(\breve{k}^2 - \mu_{\mathrm{A}}^2)\, d^2\breve{k} \quad , \tag{5.3}$$

satisfy the string equation

$$\left(\partial^2/\partial\sigma^2 - \partial^2/\partial\tau^2\right) f_{\rm A}(\breve{q},\breve{v}) = 0 \ , \qquad \sigma = \breve{q}^{(0)} \ , \qquad \tau = \breve{q}^{(3)} \ , \tag{5.4}$$

in the frontal localization frame (Prugovečki, 1978c) determined in $T_x{}^*\mathbf{M}$ by $(k_0,\mathbf{0})$ and $(0,\mathbf{k})$. In general we can also expect, however, more complex internal motions, involving additional rotational degrees of freedom that are not around the axis provided by their direction of motion **k**. If it is assumed that all GS exciton transition amplitudes (cf. [P], Sec. 4.5) to excited modes for such motions are eigenfunctions of Born's quantum metric operator, and that they satisfy the equation proposed in the context of Born's reciprocity theory by Yukawa (1953), then the proper state vectors $f_{\rm B,A}$ describing these higher exciton states satisfy the relativistic harmonic oscillator equation

$$\left(-\partial^2/\partial u_i \partial u^i + u_i u^i\right) f_{\rm B}(u) = \lambda_{\rm B,A}\, f_{\rm B}(u) \ , \tag{5.5}$$

in the variables $u_i = p_i - k_i$, representing relative internal 4-momentum components with respect to the dual $\{\boldsymbol{\theta}^i(x)\}$ of the local Lorentz frame $\{\boldsymbol{e}_i(x)\}$. The rest masses $m_{\rm B,A}$ carried by these excited modes $f_{\rm B,A}$ are then given, in Planck natural units, by the following equation,

$$m^2_{\rm B,A} = \lambda_{\rm B,A} - 4 \ , \tag{5.6}$$

relating them to the eigenvalues in (5.5), whose explicit values will be provided in (5.12).

Indeed, the eigenvalues and eigenstates of the relativistic harmonic oscillator equation in (5.5) can be computed by the standard use of raising and lowering operators, provided in the present context by the following expressions:

$$a_j^\dagger = 2^{-1/2} i\left(u_j + \partial/\partial u^j\right) \ , \qquad a_j = 2^{-1/2} i\left(-u_j + \partial/\partial u^j\right) \ . \tag{5.7}$$

In the present context these operators satisfy relativistic canonical commutation relations that are equivalent to those in (5.2):

$$\left[a_i^\dagger, a_j\right] = \eta_{ij} \ , \qquad \left[a_i^\dagger, a_j^\dagger\right] = \left[a_i, a_j\right] = 0 \ , \qquad i,j = 0,1,2,3 \ . \tag{5.8}$$

However, the ground state is degenerate since it corresponds to zero mass, so that various polarization modes exist that give rise to a great variety of internal gauges – as exemplified in Chapter 11 in the case of the graviton. Indeed, these ground states display invariance under the SO(2) group of motions that leaves **k** invariant. Consequently, they can be factorized as follows

$$f_{\rm A}(u; s_{\rm A}) = Z_{\rm A}(s_{\rm A})\, f_{\rm A}(u) \ , \tag{5.9}$$

where each $Z(s_{\rm A})$ is constructed from polarization frames, such as those in Chapters 9 and 11, so that they can be grouped into sets $\{Z(s_{\rm A})\}$ providing integer-spin frames. The spin

$s_A = 1$ and $s_A = 2$ cases provide ground exciton states that are capable of representing photons and gravitons, respectively.

All ground GS exciton states share the common form factor

$$f_A(u) = \exp(-u_i u^i/2) \ , \tag{5.10}$$

reflecting a string length $\ell_A = 2$ in Planck units, and supplying the fundamental quantum spacetime form factor in (9.2.14) upon setting $u_i = v_i - k_i$, and then renormalizing as $\mu_A \to +0$. The higher exciton modes can be obtained from the solutions for the eigenstates in (5.5) in the following simple manner (cf. [P], p. 204):

$$f_{B(\hat{n})} = a_0^{\dagger n_0} a_1^{\dagger n_1} a_2^{\dagger n_2} a_3^{\dagger n_3} f_A \ , \quad \hat{n} = (n_0, n_1, n_2, n_3), \ \ n_0, n_1, n_2, n_3 = 0,1,2,\ldots \tag{5.11}$$

Since by (5.6) these states are massive, they reflect a breaking of the SO(2) symmetry that left **k** invariant. However, in order to be *physical* GS exciton modes, they have to display an SO(3) invariance that reflects the presence of specific internal spin value. Thus, they correspond to the following eigenvalues of the quantum metric operator $\boldsymbol{D}^2(x)$ at each $x \in \mathbf{M}$ (cf. [P], p. 205):

$$\lambda_{B,A} = 2\left(2 + n_0 + 2n + s_{B,A}\right) \ , \qquad n_0, n, s_{B,A} = 0,1,2,\ldots \ . \tag{5.12}$$

The proper state vectors describing their internal stochastic motion with respect to the local Lorentz frame $\{e_i(x)\}$ can be then computed as in (Brooke and Prugovečki, 1984).

As mentioned earlier, much more compelling than the above string-motivated type of heuristics is the adoption of the quantum spacetime form factor f_ℓ in (5.5.5) as fundamental to any model of quantum spacetime – regardless of whether it manifests itself as the ground state of a quantum metric operator, or simply as the *only* quantum spacetime form factor in existence. Indeed, as we pointed out in Sec. 1.5, quantum geometries do not require the existence of physical "objects" and test "bodies" which *exactly* "fit" into their points, any more than classical geometries require truly pointlike test particles that exactly fit into theirs: in either case, the concept of point can be viewed as an abstraction, suggested by an empirical reality which is quantum in the former case, and classical in the latter, but without necessarily faithfully reflecting those respective realities. On the other hand, the adoption of f_ℓ as *the* quantum spacetime form factor can be justified purely on grounds of mathematical simplicity and aesthetics, combined with the fact that, as demonstrated in Sec. 11.4, it assures the informational completeness of the ensuing quantum frames.

Indeed, it is well known that, as a methodological guide to uncovering new physical laws and features of Nature, the principle of mathematical simplicity was already advocated by Newton, and that Einstein championed it throughout his life. The idea of mathematical beauty as methodological guide had its recent advocates in Poincaré and Weyl, and perhaps its strongest champion in Dirac: "For Dirac the principle of mathematical beauty was partly a methodological moral and partly a postulate about nature's qualities. It was clearly inspired by the theory of relativity, the general theory in particular, and also by the development of quantum mechanics." (Kragh, 1990, p. 277).

Of course, both these principles should be used only sparingly and judiciously, as they have been (justifiably) criticized on the basis that not all mathematicians or physicists share the same idea of either mathematical simplicity or beauty. In other words, mathematical beauty as well as simplicity might exist only "in the *mind* of the beholder". But then, we have seen in many previous examples that, to a certain extent, the same can be said even of the *appraisals* of the degree of support received by a very popular theory from various experiments. In fact, there are cases in which a compelling simplicity and beauty can be even more universally "obvious" in a theory than its purported agreement with experiment, since in the latter case, one often merely tries "to make sense of the mass of data provided by the experimentalists" (cf. Note 28); whereas, the former might almost be "able to speak for itself", on account of elegant features in its appearance as well as in its underlying ideas – as, most certainly, is the case with the Dirac equation. Hence, it is not at all surprising that Dirac "asserted that mathematical-aesthetic considerations should (sometimes) have priority over experimental facts and in this way act as criteria of truth" (Kragh, 1990, p. 284).

The adoption of the quantum spacetime form factor f_ℓ in (5.5.5) as *fundamental* embodies the criterion of mathematical simplicity in a most direct and evident form. It also incorporates one of Dirac's favorite paradigms of mathematical beauty – namely the theory of functions of one or more complex variables. Indeed, upon adopting f_ℓ as being *the* fundamental quantum spacetime form factor, the following straightforward substitution can be carried out in all local quantum fluctuation amplitudes (cf., e.g., (9.2.22), or (9.6.3) and (9.6.4)), whereby real Poincaré gauge variables are replaced with complex ones:

$$(q,v) \mapsto \zeta = q + iv \ , \qquad q \in \mathbf{R}^4 \ , \quad v \in \boldsymbol{V}^+ \ . \tag{5.13}$$

It thus solves one of the "many problems left over concerning particles other than those that come into electrodynamics: ... how to introduce the fundamental length to physics in some natural way" (Dirac, 1963, p. 50). It also mediates in a most *natural* way the strongly-advocated-by-Dirac replacement in quantum theory of real with complex variables. Indeed:

"As an interesting mathematical theory that fulfilled his criteria of mathematical beauty, Dirac emphasized in 1939 the theory of functions of a complex variable. He found this field to be of 'exceptional beauty' and hence likely to lead to deep physical insight. In quantum mechanics the state of a system is usually represented by a function of real variables, the domains of which are the eigenvalues of certain observables. In 1937, Dirac suggested that the condition of realness be dropped and the variables be considered as complex quantities so that the representatives of dynamical variables could be worked out with the powerful mathematical machinery belonging to the theory of complex functions. If dynamical variables are treated as complex quantities, they can no longer be associated with physical observables. Dirac admitted this loss of physical understanding but did not regard the increased level of abstraction as a disadvantage. ... Dirac never gave up his idea of mathematical beauty, to which he referred in numerous publications, technical as well as nontechnical." (Kragh, 1990, pp. 282-283).

The GS interpretation of the components of the complex variables in (5.13) not only removes any possibility of some "loss of physical understanding", but it also harmonizes very well with Born's (1938, 1949) reciprocity ideas about the symmetric role played in nature by the position and momentum variables. At the same time, the introduction of the complex variables in (5.13), mediated in a most natural manner by the choice of the fundamental quantum spacetime form factor f_ℓ in (5.5.5), also ensures that the GS quantum fluctuation amplitudes (i.e., local GS propagators such as $\Delta^{(\pm)}$ and $S^{(\pm)}$ in Secs. 7.4 and 8.1, respectively) are analytic extensions (in the sense of distributions) of their conventional counterparts. In view of the status of contemporary experimental high energy technology, which is still far from being able to probe energies and distances of "Planckian" orders of magnitude, this feature is bound to secure numerical agreement at the formal perturbative level, and within the domains experimentally reached thus far, between conventional quantum field theoretical models and their GS counterparts that are based on the fundamental quantum spacetime form factor f_ℓ. Hence, the choice between conventional models and their GS counterparts is not one that could be made, at the present technological level, on the basis of experiment alone. Rather, it is one that involves criteria for mathematical and epistemic soundness, which reflect a long-range view of the role of a quantum theory that incorporates gravity, rather than the immediate gratification of some simple-minded instrumentalist criterion of "agreement with experiment".

12.6. The Physical Significance of Quantum Geometries

The framework for quantum geometries presented in this monograph enables the embedding of fundamental measurement-theoretical limitations directly into the very structure of relativistic quantum field theories formulated in terms of such geometries. We have pointed out in the last section of Chapter 9 that the formal manipulations characteristic of conventional quantum field theoretical models can be duplicated in the context of GS models, and their "perturbation expansions" could be then recovered term by term in the Minkowski regime by taking the limit $\ell \to +0$ in the fundamental length ℓ. There appears to be no point, however, in such formal manipulations, except as paradigms in the study of the fundamental question of relativistic microcausality.

The central observation here is that, in the absence of a *proof* of the existence of the S-matrix in the quantum field theoretical models, from QED to QCD, that are currently in vogue in elementary particle physics, *no* test of the formulation of microcausality based on "local" (anti)commutativity can be said to have been performed thus far. Furthermore, *even if we grant* the existence of the S-matrix in such quantum field theoretical models, the fact that certain well-known properties of the S-matrix can be formally derived (cf., e.g., Blokhintsev, 1973) by the use of "local" (anti)commutativity does not prove that such (anti)commutativity is a *necessary* (and not just sufficient) condition for those properties to hold. For example, the violations of "local" commutativity for asymptotic fields in QED (Fröhlich *et al.*, 1979) provide one of the many indications that no such necessity is, in fact, present even within the conventional quantum field theoretical framework. Furthermore, as discussed in Sec. 7.6, the mere *postulation* of "algebras of observables" which purportedly satisfy "local" commutativity neither proves their mathematical existence for physically nontrivial conventional models, nor does it settle any fundamental measurement-theoretical questions as to the operational feasibility of associating actual observables with arbitrarily sharply delineated domains in classical spacetime manifolds.

In fact, in Secs. 7.6 and 9.6 we have pointed out that the conventionalistic identification of "microcausality" with "local" (anti)commutativity has no bearing on the GS approach, since such (anti)commutativity has no physically truly meaningful relationship to the question of Einstein causality any more than it would in classical relativistic theory.

Indeed, in classical *special* relativistic theory, the commutativity of all observables is trivially satisfied, since all classical fields and their observables commute. On the other hand, in a classical *general* relativistic theory such commutativity for non-scalar fields is *undefined* at distinct spacetime points. Of course, in the special relativistic regime, the concept of locality that emerges from the "naïve" realism predating modern quantum theory makes such a concept "plausible". However, there has never been any serious attempt in the literature to rigorously *prove* that the identification of "local" (anti)commutativity with some form of Einstein causality follows from any cogent quantum theory of measurement. Rather, from the earliest days this idea was introduced by *postulation* in the LSZ formulation (cf. Note 31 to Chapter 9), as well as in axiomatic quantum field theory (Streater and Wightman, 1964).

On the other hand, in the GS approach microcausality is *directly* related to the mode of propagation, i.e., to the realistically posed question as to which stochastic paths are followed in GS propagation: are only those paths allowed which can be approximated by piecewise smooth curves, whose smooth segments are strictly causal in the classical sense, as in strongly causal GS propagation, or are certain types of noncausal smooth arcs also allowed, as is the case in weakly causal GS propagation?

In developing a framework within which such questions can be *meaningfully* posed, the quantum geometry framework assigns total priority to geometric over variational principles. This is in contradistinction to Feynman's path-integral formulation of quantum propagation, which assigns the most prominent role to Lagrangians, and underplays the fact that each "sum-over-paths" is fundamentally a geometric concept, which can be formulated in a Lagrangian-independent manner. Hence, in the GS approach the entities of *direct* physical significance are the GS propagators themselves, which describe propagation between base spacetime points along causally ordered 3-manifolds, rather than being the conventional "propagators" in momentum space representations, whose introduction is motivated by the computational expediency imposed by conventional "perturbation" theories.

The ultimate question of choice between strongly and weakly causal GS propagation will have to be obviously answered by experiments based on properly formulated theoretical predictions of measurable effects that can distinguish between these two modes of propagation. Such predictions will have to take advantage of the fundamentally *non*perturbative formulation of GS propagation. Indeed, clearly specified error bounds would have to be computed at those base spacetime points where probability transition amplitudes for the two modes might be observationally distinguishable by means of present-day technology[41].

The fundamentally *non*perturbative nature of GS propagation is a reflection of the fact that the quantum reality envisaged by the GS approach is based on *quantum* stochasticity. The manifestations of this kind of stochasticity are in their most essential aspects totally different from those assumed in classical physics. This fundamental distinction emerges from the fact that *in quantum GS formulations the concept of probability measure for quantum stochastic paths does not exist*[42]. Hence, of necessity, GS propagation has to be formulated in terms of probability *amplitudes* over broken paths, with a subsequent specification of limits – the same type of limits as in Riemannian integration – rather than in terms of probability measures over stochastic paths that employ Lebesgue integration, as is the case in the theory of classical stochastic processes.

These GS probability amplitudes are superimposed in a *coherent* manner, due to the intrinsic proper time kept by proper state vectors, represented by *local* coherent states, as they propagate along such paths. As discussed in Sec. 1.4, the process of observation corresponds to decoherence, so that the "classical path" would be the most likely one to be "observed" in the sense that it might provide the best fit for the *discrete* set of base spacetime locations where actual macroscopic registrations have taken place. On the other hand, the existence of *proper* state vectors permits, by the application of the superposition principle, the possibility of *weak* relativistic GS microcausality – a concept that makes absolutely no sense for point particles whose behavior is governed by classical diffusion processes.

The existence in the GS approach of *proper* state vectors also enables the formulation of new types of quantum models based on the adopted structures of quantum spacetime form factors – such as those briefly mentioned in Sec. 1.5. Thus, in the GS context the problem of strong interactions can be approached from two very distinct angles: 1) with an *external dynamics* perspective in mind, which would lead to a GS counterpart of QCD, and in which the fundamental quantum spacetime form factor f_ℓ in (5.5.5) would be the only quantum spacetime form factor, while the interactions between quantum fields creating and annihilating quarks would take place by means of the external exchange of gluons; 2) from an *internal statics* point of view, whereby new quantum spacetime form factors would be "shaped" either by the presence of a quantum metric operator (such as the one discussed in the preceding section, or by an internal "Hamiltonian" based on fundamental oscillator and rotator models – cf. Bohm *et al.*, 1988), or on account of having a ground exciton "trapped" in some internal geometry – such as in the de Sitter types of quantum geometries adopted in (Drechsler and Prugovečki, 1991) and in (Drechsler, 1991).

The latter type of approach based on "internal statics" has an essential bearing on the epistemological significance of the concept of *congruence* in physical geometries. As discussed in (Jammer, 1969), pp. 208-211, various conceptions of geometric congruence were advanced in this century by Russell, Whitehead, Eddington, Bridgeman and others, and their significance to the empirical role of metric in CGR was debated by Einstein, Reichenbach and Robertson in (Einstein, 1949). The possibility of the existence of a *quantum* metric operator that "shapes" the points of quantum spacetimes obviously opens new possible perspectives on these "old" issues, by showing that the foundations of the *physical* geometries used in the description of spacetime do not reside in any kind of geochronometric conventionalism, such as that advocated by Reichenbach and Grünbaum, but rather in the intrinsic *quantum* features of spacetime.

In general, it can be said that quantum geometries throw new light on some of the "old" problems, that were raised in earlier times in the context of classical geometries, and that at the same time they give rise to new physical concepts of a geometric nature, whose very meaning would be nonexistent in their absence.

The quantum geometry framework also opens new possibilities in the theory of quantum measurement. In fact, as presented in Chapters 3–10, the GS approach has been in this respect quite conservative: the development of the GS theory of measurement was based on a very gradual and very careful extrapolation of the orthodox approach – so as to avoid any of the needless epistemological excesses encountered in some other non-orthodox approaches to the quantum theory of measurement (DeWitt and Graham, 1973; Barrow and Tipler, 1986). Thus, as described in Chapter 3, at the measurement theoretical level the GS program began with the assumption that the existence of (previously unsuspected) Galilei and Poincaré covariant and conserved *probability* currents, such as the ones in (3.5.1)–(3.5.9) and (3.5.13)–(3.5.15), respectively, were not due to sheer coincidence. The validity of this conjecture was reinforced by the striking similarity in external appearance of those currents, and by the fact that the nonrelativistic ones merge in the sharp-point limit into the conventional ones in (3.5.7). As we have seen in Chapter 3, in the nonrelativistic regime, this assured the possibility of a *gradual transition* from the orthodox to the SQM theory of measurement; whereas, in the special relativistic regime it enabled the *straightforward extrapolation* of the basic *formal* aspects of the conventional theory, and the avoidance of the difficulties created in that theory by the absence of *bona fide* probability currents. In the presence of gravity, an interpretation that was still very close to the orthodox one was adopted in the semiclassical approximation, in which the gravitational field is treated only as an external field (cf. Sec. 5.5).

The application of GS quantum gravity in Chapter 11 to quantum cosmology led, however, to the introduction of a "universal GS wave function", which represents a GS counterpart of the "wave function of the universe" (Hartle and Hawking, 1983; Barrow and Tipler, 1986), since it is meant to describe all the matter and gauge fields in existence. This necessitated the consideration of very fundamental epistemological questions, traceable in the history of philosophy to the mind-body problem and to the question of free will (cf., e.g., Weyl, 1949). Historically, these questions have impinged upon the epistemology of quantum mechanics in the form of von Neumann's (1932, 1955) "psycho-physical parallelism", and Wigner's (1962) subsequent analysis of the thesis that the "reduction of the wave packet" might take place in the mind of the "observer".

Whereas the empirical significance of such a thesis in ordinary quantum mechanics is very much open to debate, the general questions that it implicitly raises in cosmology are related to the issue of the freedom of the experimenter to locally change *physical* conditions, rather than act as merely a passive "observer". Indeed, such measurement-producing *actions* can give rise to "reductions of the universal wave function" that would not have occurred otherwise. Hence, in any theory describing a *single* universe (as opposed to "scenarios" based on any form of "parallel universes" – cf. Barrow and Tipler, 1986), they give rise to profound questions concerning the nature of *fundamental* causality – namely of the forms of causality in the traditional philosophic sense (Weyl, 1949; Bunge, 1970), some of which predate by millennia the notions of microcausality and of Einstein causality, which we discussed earlier.

Thus, as we saw in Chapter 11, the quantum geometry framework based on a GS conceptualization of *quantum* reality reverses Bohr's epistemic outlook, and asks us to *envisage how macroscopic phenomena appear from a microscopic point of view*. In other words, it poses from a *micro*scopic perspective the questions: What *is* an "observation"? What *is* an "apparatus" ? All of this provided, of course, that we grant as a basic *methodological* feature that the latter must be, in some *unambiguously* prescribed sense, a macroscopic object whose behavior can be *approximately* described in classical terms.

In this type of GS conceptualization, any phrase providing the "probability of detection of a GS exciton within a region $B \subset \mathbf{M}$" is merely a short-hand for the descriptively more accurate, but cumbersome and tediously long phrase, asserting the provision of the "probability of a *macro*scopic manifestation, within a region $B \subset \mathbf{M}$, of a given form of perturbation in a particular type of conglomeration of local state vectors, that constitute a GS wave function primarily localized in some vicinity of B". Furthermore, any such provision has to be supplemented by an unambiguous and detailed description of all the "well-defined experimental conditions specified by *quantum* physical concepts" under which such manifestations are to be observed.

In paraphrasing, in this last stipulation, one of Bohr's principal dictums (cf. Sec. 12.1) by the simple expedient of replacing in the original text the term "classical" with the term "quantum", we wish to underline the fact that much of the essence of Bohr's philosophical outlook can, and must, be retained in the future developments of any quantum theory of measurement. It is only what might be termed "epistemological dead wood" that has to be trimmed away, in order to arrive at a better understanding of the foundational issues whose study was initiated by the Copenhagen school, as well as by many of its outstanding contemporaries in the "opposing camp".

12.7. Summary and Conclusions

As we have seen in the first section of this chapter, from the perspective of philosophy of science, the development of quantum theory during the first half of this century was marked by the confrontation between classical realism and logical positivism. This confrontation was personified by Einstein and Bohr, respectively – although neither of them fully and consistently embraced the philosophies they were supposed to represent.

During the second half of this century the arena of such confrontations changed radically. As we discussed in the second and third section, most of the basic issues which were the focus of attention in the historic confrontation between Bohr and Einstein became either irrelevant or largely forgotten soon after World War II, while a new philosophy took over, and has dominated, practically unchallenged, the world of quantum physics for the last four decades. That publicly unacknowledged, but in everyday practice of quantum physics all-pervasive, philosophy was identified by Popper (1976, 1982), as well as other authors, to be a form of instrumentalism. This label was further qualified in this chapter as *conventionalistic instrumentalism*, in order to descriptively incorporate and characterize by it also the new attitudes towards mathematical standards of truth and deductive validity that emerged in quantum physics during the first decade of the post-World War II era.

Bohr, Dirac, Heisenberg, Pauli[43], Popper and many others of the pre-World War II "older generation" of physicists and philosophers of science have reacted with distinct disapproval towards the most prominent aspects and practices of this tacitly but widely accepted philosophy of the new generation of physicists – especially towards its computational opportunism, and its lack of commitment to the *rational and objective* mathematical and/or epistemological standards, which in previous eras represented the traditional hallmark of the scientific outlook. In fact, disapproval became a general sounding of the alarm when Popper described the professional atmosphere created by the functionally unconditional adoption of instrumentalism in contemporary quantum physics as "a major menace of our time", and when he stated that "to combat it is the duty of every thinker who cares for the traditions of our civilization".

This concern is understandable: the loss of dedication to a fundamental notion of truth in science – namely *truth* which stands above any of the temporary fashions reflected in whatever "conventional wisdom" might prevail in a given era in the history of mankind – is a very serious matter to anybody who believes that basic science is one of the very last bastions of rationality and integrity in contemporary civilization. Indeed, in similar cautionary words aimed at instrumentalism in general, Bertrand Russell pointed out that "the intoxication of power [reflected by the advocacy of an instrumentalist notion of 'truth'], which invaded philosophy with Fichte, and to which modern men, whether philosophers or not, are prone ... is the greatest danger of our time, and any philosophy which, however unintentionally, contributes to it is increasing the danger of vast social disaster" (Russell, 1945,

p. 1828). Hence, one does not have to subscribe to all, or even to some, of the tenets of Popper's form of realism in order to share his misgivings about the uncontested prevalence of an instrumentalist attitude in contemporary quantum physics.

The reasons for this practically unchallenged dominance of instrumentalistic philosophy in quantum physics cannot be attributed to either its use of carefully and rigorously reasoned arguments, or to the revelation of some deeper and previously unsuspected fundamental truths. The development of the S-matrix program, which enjoyed overwhelming popularity in the 1960s, only to fall out of favor in the 1970s, provides a good illustration:

"The dispersion-theory and S-matrix theory programs of the late 1950s and early 1960s had great appeal initially because they *worked* (i.e., they successfully related many directly measurable experimental quantities to each other). Of course, some of this success was 'arranged' (or greatly aided) since needed results (such as dispersion relations for massive particles and for nonforward directions, Regge asymptotic behavior, etc.) were assumed long before they could be proved (and many never were). ... These programs were characterized by a desire to 'get on with things', to 'do something'. Cini (1980) and Pickering (1989a) have stressed the pragmatic aspect of these approaches and Schweber (1989) has suggested that this was a hallmark of much of theoretical physics after the Second World War (as contrasted with the period before the War)." (Cushing, 1990, p. 214).

Thus, the trademark of conventionalistic instrumentalism was, and in a large measure still is, computational facility based on formal manipulations that disregard deeper physical, mathematical or epistemological questions[44]. Since its basic appeal is not to critical faculties, or to a sense of mathematical beauty, or to the desire to *truly* understand the workings of Nature, the reasons for its dominance must be *primarily*[45] sociological. Indeed, a pragmatism reflecting primarily the desire to "get on with things", even at the price of ignoring foundational issues, would not have surfaced to such an overwhelming degree in a science based on a quantum theory founded by individuals, such as Bohr, Born, Heisenberg, Pauli and Schrödinger[46], with a deep concern with fundamental philosophical issues, were it not for a specific type of change in the social climate brought about in science, as well as in other spheres of human activity, by the Second World War[47]. Clearly, this social change has shaped a new generation of quantum physicists with a strong predisposition to conform, and "to follow the very latest fashion". This was reinforced to a considerable extent by a tight control of institutional powers[48], and by the exercise of those powers to shape mental attitudes and professional opinions, in a manner which systematically rewarded conformity and discouraged critical appraisals of the *status quo*.

Some sociologists of science have documented these features in their studies of the "big science" that emerged after the Second World War[49]. However, since this sociological phenomenon lies outside the scope of the present monograph, it was pursued in the present chapter only in the context of specific instances, which dealt with the history and development of the pertinent ideas in quantum theory. Readers interested in it at a general level are referred to the work of Mitroff (1974), Pickering (1984), Savan (1988), and other sociologists of science cited in Sec. 12.3, who have written and published on this subject.

On the other hand, there are various publications which try to *rationalize* the reasons and origins of the domination exercised by conventionalistic instrumentalism on contemporary quantum theory by presenting them as the natural outgrowth of the philosophy of the Copenhagen school. We hope that the brief historical survey in this chapter has convinced the reader that, although the Copenhagen school may have unwittingly created a fertile soil for the seeds of such ideas, the post-World War II conventionalistic instrumentalism is not in the least its brainchild – as witnessed by the publicly stated opposition[50] of Dirac and Heisenberg, as well as others (cf. the quotation at the end of Sec. 12.2), to some of its practices, since its inception in the second half of the 1940s.

Of course, some instrumentalists might argue that Bohr was on their side throughout his life. However, as pointed out in the most recent expository analysis of Niels Bohr's philosophy of physics, "it would be quite wrong to describe Bohr as a weak instrumentalist, because for the latter the truth, as distinct from empirical adequacy, of a physical theory is of no concern whatever." (Murdoch, 1989, p. 222). Another recent analyst of Bohr's philosophical ideas has, independently, arrived at the same conclusion: "As there are various forms of realism, so there are different forms of anti-realism. The dominant one during Bohr's career was that of 'instrumentalism', the view that theoretical terms serve only as constructs enabling correct inferences to predictions concerning phenomena observed in specified circumstances. Many defenders of anti-realism also hold the view of 'phenomenalism', the assertion that the only reality of which we can form an idea with any content is that of phenomena, and that therefore statements about a reality behind phenomena are meaningless. *Both of these views have been imputed to Bohr quite incorrectly*" (Folse, 1985, p. 195) – emphasis added. Indeed, some key correspondence between Bohr and Born is reproduced in (Folse, 1985), p. 248, which conclusively demonstrates that both these great physicists and founders of quantum mechanics were very decidedly opposed to the "instrumentalist standpoint".

There also are hundreds of publications, ranging from textbooks to popularizations of quantum theory in general, which are aimed at convincing their readers that giant strides were made by post-world War II physics not only in the realm of technology (which is indisputable), but also in the realm of fundamental ideas in quantum physics. Explicitly or implicitly, these publications ascribe all those purported successes to the conventionalistic outlook. The fact is, however, that if one leaves aside various extreme ideas in quantum cosmology[51], then Schwinger's 1958 assessment of post-World War II developments in relativistic quantum physics can be, by and large, extrapolated to the present time[52]: at a fundamental level all post-World War II developments "have been largely dominated by questions of formalism and technique, and do not contain any fundamental improvement in the physical foundations of the theory" (Schwinger, 1958, p. xv).

As discussed in Secs. 9.6 and 12.2, other physicists and historians of science, who took a careful look at those developments, have arrived at similar conclusions[53]. In particular, Dirac believed that the type of renormalization theory that became fashionable soon after the end of World War II represents "a drastic departure from logic. It changes the whole character of the theory, from logical deduction to a mere setting up of working rules." (Dirac, 1965, p. 685).

Thus, from an informed and purely rational point of view, the case in favor of adopting conventionalistic instrumentalism as a valid and fruitful philosophy for quantum theory rests exclusively with its systematically and widely advertised successes in the production of *numerical* predictions, which are purportedly in good to excellent agreement with the experiment results. When this claim is assessed, it should be recalled, however, that instead of deeming them as clear-cut confirmations of the advocated theories, Dirac suggested[54] that even the agreements between the numerically most successful of models in quantum field theory (namely conventional QED) with the experimental results *might* be due to coincidence, and backed this observation with similar previous occurrences that took place in Bohr's semiclassical quantum theory of the 1910s.

Indeed, when the theoretical manipulations are based on simply "discarding" undesired terms, and on "asymptotic" series in which the summation is carried out only as far as it is necessary for "agreement with observation", the possibility of repeated occurrences of "coincidences" is not that easy to rule out. Furthermore, as discussed and documented in Sec. 9.6, as well as in this chapter, the analysis of the raw experimental data is prone to various types of systematic errors, whose likelihood increases dramatically once a strong predisposition exists to confirm a highly acclaimed theory (cf. Sec. 12.3, as well as Note 6). Healthy skepticism is therefore called for until the theoretical underpinnings of present-day fashionable theoretical models in high-energy physics are considerably strengthened, and the basic mathematical standards are fundamentally improved. It is only when all such theories become founded on *sound* mathematics – namely mathematics based on well-established canons of *logical* deduction, rather than on the "mere setting up of working rules" – that those believing in the rationality of science can attain the confidence that such theories provide a reliable account of *quantum* reality. And even for those who do not believe that there is a quantum reality, but that quantum theories are mere "*instruments*, which enable us, on the basis of the observed facts, to predict either with certainty or probabilistically the results of observations" (d'Espagnat, 1989, p. 27), such mathematical legitimacy can still provide the needed assurance of anthropic *objectivity* and *reliability*.

The present quantum geometry framework has been formulated during the span of many years, with the above type of healthy skepticism in mind, but with the otherwise constructive and progressive type of attitude that is suggested by the quotation of Bertrand Russell heading this chapter. Thus, as opposed to other types of stochastic approaches to quantum theory (cf. Note 2 to Chapter 1), it was never the intention of the GS program to try to turn back the clock of history, and impose in quantum theory values derived from some kind of "physical realism" (Bunge, 1967; d'Espagnat, 1989) with its roots in classical physics. Rather, the challenge met was to try to understand the numerical successes of post-War War II relativistic quantum theory by developing mathematically sound methods, that would enable "successive approximations to the truth, [and] in which each new stage results from an improvement, not a rejection, of what has gone before". On the other hand, another one of the principal aims of the program that eventually matured into the framework presented in this monograph, was to systematically reapply to quantum physics the traditional[55] pre-World War II criteria of "scientific truthfulness" (Russell, 1948), rather than to rely exclusively on instrumentalist criteria based on "conventional wisdom" and on "general consensus", that have become entrenched in the conventional relativistic quantum mechanics and quantum field theory of the post-World War II era.

In order to have any chance at achieving such a goal, it became mandatory to dig deep into the foundations of relativity and quantum theory in general, and to appeal not only to physical insights and intuition, but also to a wide range of ideas and techniques of contemporary mathematics, as well as to carefully formulated epistemological studies of those foundations. The central conclusions reached in this manner, and which pertain primarily to foundations, were discussed in Secs. 12.4–12.6. Those sections also contain, sometimes in an explicit form, but mostly implicitly, the main tenets of a *quantum* realism which is distinct from both classical realism as well as from logical positivism, and yet incorporates key epistemological ideas from both these very fundamental philosophies of the twentieth century. Naturally, the acceptance of a philosophy that envisages a quantum reality which exists independently of whether we "observe" it or not, is not necessary for the understanding and application of the present quantum geometry framework any more than the comprehension and adoption of the philosophy of the Copenhagen school is necessary for acquiring a working knowledge of nonrelativistic quantum mechanics. However, as Heisenberg has emphasized in his last (1976) paper, in the long run, philosophical assumptions can play a decisive role in the formation and development of physical theories.

During the course of most of the post-World War II developments in relativistic quantum physics, concentrating one's attention on anything but the *conventional* formalism of quantum field theory has been very unfashionable, not only amongst theoretical physicists, but also in the dominant mathematical physics circles. Fortunately, the last decade has witnessed, at least amongst certain types of theoretical physicists and mathematicians, a growth of interest in deeper mathematical questions, that call for the development of advanced nonperturbative mathematical tools in relativistic quantum theory. It has also witnessed, amongst a relatively small number of yet another type of physicist, a gradual revival of professional concern with the deeper epistemological questions pertaining to the foundations of relativity and quantum physics. As a result: "Physics finds itself in recent years in an exciting and revolutionary phase of development: after a long intermission – and despite practical successes – critical questions about the proper foundations are being asked, and far-reaching attempts are being made to gain a deeper understanding of the whole structure of the theory of our time." (Bleuler, 1991, p. 304).

It is hoped that the epistemological ideas and mathematical techniques expounded in the present monograph will contribute to the future merging of the above mentioned two very healthy trends in contemporary relativistic quantum theory, and to their joint subsequent development.

Notes to Chapter 12

[1] As related by Heisenberg in a 1968 talk delivered at ICTP in Trieste, during the early stages in the development of quantum mechanics, he himself thought that the most important philosophical idea was that of "introducing only observable quantities". Then Heisenberg went on to say: "But when I had to give a talk about quantum mechanics in Berlin in 1926, Einstein listened to the talk and corrected this view. ... He said 'whether you can observe a thing or not depends on the theory which you use. It is the theory which decides what can be observed'. His argument was like this: 'Observation means that we construct some connection between a phenomenon and our realization of the phenomenon. There is something happening in the atom, the light is emitted, the light hits the photographic plate, we see the photographic plate and so on and so on. In the whole course of events between the atom and your eye and your consciousness you must assume that everything works as in the old physics. If you change the theory concerning this sequence of events then the course of observation would be altered'. ... Einstein had pointed out to me that it is really dangerous that one should only speak about observable quantities. Every reasonable theory will, besides all things which one can observe directly, also give the possibility of observing things more indirectly. ... I should also add that when one has invented a new scheme which concerns observable quantities, the decisive question is: which of the old concepts can you really abandon?" (cf. Salam, 1990, pp. 98-101).

[2] As will be discussed in Sec. 12.3, this reversal was basically unrelated to any deeper epistemological considerations – as is illustrated most dramatically by the rapidly changing "fashions" in the elementary particle physics of the post-World War II era. Thus, the reasons for this rather dramatic change in basic attitudes were purely social and sociological, and closely related to the global aftereffects of World War II, whereby the focus of advanced research in basic science was shifted to societies in which pragmatic and instrumentalist attitudes towards science were already generally entrenched (cf. Note 12).

[3] The text of this talk (cf. Note 36 to Chapter 7) has been reprinted in (Salam, 1990), pp. 125-143, and the present quote (to which italics have been added for emphasis) can be found on pp. 137-138.

[4] An amusing and yet enlightening anecdote that illustrates Feynman's reaction to Dirac's verdict on renormalization theory is reported by A. D. Krisch (1987). It ends with the following observation: "What I concluded from this incident was that either Feynman shared Dirac's concerns [that an inelegant theory, such as QED, could not possibly be correct] or that there may be levels in the Theoretical 'pecking order' that are not easily observable to an experimenter." (Krisch, 1987, p. 50).

[5] Most of the sections in this chapter provide, primarily in the form of additional notes, an elaboration and further documentation of those previously presented in an article entitled "Realism, Positivism, Instrumentalism and Quantum Geometry", due to appear in an issue of *Foundations of Physics* dedicated to the ninetieth birthday of K. R. Popper. Much of this documentation is in the form of quotations from historical and sociological studies. In particular, pertinent quotations are provided from a recent scientific biography of Dirac by H. S. Kragh (1990), from a sociological study of theory selection in contemporary physics by J. T. Cushing (1990), and from a historical study of developments in particle physics in the 1950s by S. S. Schweber (1989). We would like to thank Cambridge University Press, which holds the exclusive copyrights to these publications, for the permission to extensively quote from them.

[6] The somewhat shaky experimental status of CGR might have been a contributing factor, as witnessed by the following observations: "Before [Eddington's famous account of the eclipse observations in] 1919 no one claimed to have obtained spectral shifts of the [by CGR] required size; but within one year of the announcement of the eclipse results several researchers reported finding the Einstein effect. The red-shift was confirmed because *reputable people agreed to throw out a good part of their observations* [emphasis added]. They did so in part because they believed in the theory; and they believed in the theory, again at least in part, because they believed that the British eclipse expeditions had confirmed it. Now the eclipse expeditions confirmed the theory only if part of the observations were thrown out and the discrepancies in the remainder ignored; Dyson and Eddington, who presented the results to the scientific world, threw out a good part of the data and ignored the discrepancies." (Earman and Glymour, 1980, p. 85). Indeed, from the 1920s to the present time, a great many observations, as well as disputes over the validity of a number of experimental results that were originally *claimed* to support Einstein's CGR and its basic principles, seemed at times to almost invalidate it. It was only in the summary of the 1989 General Relativity and Gravitation conference that it could be finally stated with confidence that: "In view of the now quite manifold and accurate empirical evidence it seems, then, that there is no reason, at least in the macro-domain, to look for an alternative to Einstein's theory." (Ehlers, 1990, p. 493).

[7] In fact, a recent analyst of Bohr's philosophy of physics has arrived at the following conclusions: "Thus Bohr was indeed a foe of the realistic understanding of particle and wave as viewed from within the classical framework. He was, in other words, against the realism that Einstein seemingly wanted to defend, what might be called 'classical realism'. However, to conclude from this fact that he embraced an anti-realist understanding of science would require to assume that there is no other interpretation of science other than that which operates from the viewpoint of the classical framework. ... The reason for the common misreading of Bohr as an anti-realist lies not only in his attack against classical realism but also in his lack of any criticism of such an interpretation in quantum physics. But Bohr never attacked anti-realism not because he embraced this view but simply because he considered it foreign to the basic presuppositions of natural philosophy. ... He took it as empirically demonstrated that atomic systems were real objects which it is the goal of acceptable atomic theory to describe. At least as Bohr understood it, the debate was joined over the nature of the framework within which the description of such objects is to be understood." (Folse, 1985, p. 22).

[8] Naturally, Bohr was aware of this fact, but chose to underestimate its significance. Thus, according to Petersen (1985, p. 305), Bohr once said jokingly: "Of course, it may be that when, in a thousand years, the electronic computers begin to talk, they will speak a language completely different from ours and lock us in asylums because they cannot communicate with us." In the same spirit, it should be noted that instead of a thousand years, computers have started to "talk" less than fifty years after this statement might have been made, so that Bohr was wrong on *that* count; hopefully, he will also turn out to have been wrong on the rest of his prediction!

[9] For example, an expert on the complementarity principle has the following to say about this topic: "In spite of [the] dominance [of the complementarity principle] during this period of the awesome growth of

atomic physics, the 'textbook' presentations of complementarity which introduced most physics students to Bohr's views hardly could be considered to do the subject justice." (Folse, 1985, p. 27).

10 Because of the extensive use in this context of the term 'regularization', it might be thought that these procedures can be made mathematically legitimate by the use of the theory of distributions (Schwartz, 1945) and of generalized functions (Gel'fand *et al.*, 1964-68). However, a *distribution* or *generalized function* is a continuous *linear* functional, so that the theory of distributions cannot handle *nonlinear* expressions, which are characteristic of interacting fields. This is precisely the difficulty encountered by the constructive quantum field theory program, whose *ad hoc* methods of trying to by-pass the need for defining directly the nonlinear terms in interactions of quantum fields represented by operator-valued distributions have eventually led to the conclusion that the interactions of primary interest in physics lead to trivial quantum field theories – cf. (Glimm and Jaffe, 1987), p. 120.

11 As opposed to a convergent series, whose partial sums approach in the limit well-defined values in its domain of convergence (so that they can be used to define a function in that domain as equal to the sum of that series), that is not the case for an asymptotic series. Consequently, in rigorous mathematics, an asymptotic series is always a series for an analytic function $f(z)$, which has to be defined independently of that series. Indeed, in general, for a given value of z, the partial sums $s_n(z)$ of such a series approach with increasing n the value of $f(z)$ up to a given optimum distance, reached for some $n_0(z)$, but then they get to be further and further away from $f(z)$ as n becomes larger and larger. Borel summability can be used to reconstruct a function from a divergent power series (Hardy, 1949; Sokai, 1980), but the preconditions for its applicability are not satisfied in QED.

12 According to Dyson, who was one of the principal contributors of the very first of those fashions, namely renormalization theory, this is a general and necessary phenomenon in science. He says of his own experience: "When I first came [to the Institute of Advanced Studies in Princeton] as a visiting member 34 years ago, the ruling mandarin was Robert Oppenheimer. Oppenheimer decided which areas of physics were worth pursuing. His tastes always coincided with the most recent fashions. Being then young and ambitious, I came to him with a quick piece of work dealing with a fashionable problem and was duly awarded with a permanent appointment. ... I am now, after 30 years, one of the mandarins. I try in a vague and feeble way to encourage young physicists to work outside the fashionable areas, ... [but] the young people are compelled today to follow fashion by forces stronger than wording of contracts and the authority of mandarins." (Dyson, 1983, p. 48). On a more general level, he states: "It has always been true, and it is true now more than ever, that the path of wisdom for a young scientist of mediocre talent is to follow the prevailing fashion. ... To find and keep a job you have to be competent in an area of science which the mandarins who control the job market find interesting. ... Anybody doing fundamental work in mathematical physics is almost certain to be unfashionable." (*ibid.*, pp. 47 and 53). However, while these assessments are most certainly very accurate for the post-World War II era, their degree of accuracy diminishes very rapidly when we look at the past history of science. The main work of most outstanding mathematicians and physicists in the preceding two centuries (e.g., Euler, Laplace, Lagrange, Gauss, Maxwell, etc.) was in mathematical physics, and yet they enjoyed status and recognition in their own times, whereas that would not be the case in the post-World War II era – cf. also (Popper, 1982a), as well as the next note.

13 Cf. Note 5 as well as the introductory paragraphs to this chapter. Another interesting example is provided by the developments in cosmology in the late 1920s, when "it was taken for granted that the universe must be static – *despite data being available that would shortly be taken to prove the contrary* [emphasis added], with at least three published papers proposing the idea of an expanding universe" (Ellis, 1989, p. 379). More recent examples, related to the theory of parity violations as well as to electroweak interactions, are extensively documented by Franklin (1986, 1990). The philosophy behind the development of the *S*-matrix program, which enjoyed overwhelming popularity in the 1960s, has been recently analyzed by Cushing (1990).

14 The most significant influence on the developments of post-World War II physics was the shift of central focus from Europe to North-America, and the concurrent entrance into the era of "big science". As Bertrand Russell pointed out as early as 1945, "it is natural that the strongest appeal of [John Dewey] should be to Americans [since all men's views] are influenced by [their] social environment." It is therefore not surprising that, as documented by many studies in the sociology of science, such as those by Mitroff (1974) and Savan (1988), one of the corollaries of shift development was, on one hand, the introduction of Madison Avenue techniques in the promotion of scientific ideas, and, on the other hand,

of the elimination of ideas that challenged whatever fashionable orthodoxy the "mandarins" (cf. Note 10) chose to enforce at a given time. This was achieved *not* by means of public debates, as in the pre-World War II era, but rather by the tight control exerted by the North American scientific establishment over the publication and dissemination of scientific ideas. "I can't find any fundamental difference between the scientific method and the procedures for making progress in business and the arts" says one of the North American physicists interviewed by Mitroff (1974, p. 65). As a natural consequence, the "selling of ideas" (Polkinghorne, 1985) became a generally accepted practice in much of theoretical and mathematical physics: "Some [people] are very successful in pure science but it really isn't pure; nobody is pure. ... People want to sell their point of view, beat down the other guy because it means more glory, more ego satisfaction, more money" says another of the physicists interviewed by Mitroff (1974, p. 70).

15 Cf. (Kragh, 1990), p. 278. On the other hand, there is no doubt that as far as mathematical *rigor* is concerned, Dirac's standards were rather lax, as witnessed by the critical remarks of Birkhoff that are cited in (Kragh, 1990), pp. 279-280. However, traditionally the standards of mathematical argumentation and proof used in physics, even in the pre-instrumentalist era, were not as demanding as those in pure mathematics – and Dirac was certainly no exception to that rule. Hence, he insisted only on mathematically *sound* arguments, by which he obviously meant arguments in which not all details are carefully spelled out (as is, in fact, the case with many arguments in this monograph, as well as in [P], as opposed to those in [PQ]); or even arguments in which the entire deductive approach requires modification in order to produce a *rigorous* version, but which are at least *capable* of such modifications. Such mathematical "soundness", rather than rigor, was exhibited by many of Dirac's own mathematical concepts and arguments, which eventually received a mathematically rigorous treatment – the theory of distributions of Schwartz (1945), that emerged from Dirac's introduction and use of the δ-"function", being the best known example. In contradistinction, in conventional renormalization theory the Feynman *rules* reflect only a physical heuristics whose end products (namely the "perturbation series" for various processes in QED and in other conventional quantum field theoretical models) are not *capable* of receiving mathematically rigorous justification – as was pointed out in Sec. 9.6. In fact, an acknowledgedly asymptotic series cannot serve as the basis for any kind of *reliable* computation, in the absence of an *independent proof* of existence of the function whose expansion it is supposed to represent, capable of producing *independent* estimates for the values of that function. For example, "Dyson estimates in quantum electrodynamics the terms of the [perturbation] series will decrease to a minimum and then increase again without limit" ([SI], p. 644). However, how does one *know* that, first of all, the S-matrix exists in QED (according to Glimm and Jaffe (1987), it *probably* exists, but it is trivial!); and second, even if an independent existence proof is produced, how does one *estimate* how far a "perturbative" partial sum for a given process is from the *actual* total S-matrix theoretical prediction for that same process?

16 Cf. (Russell, 1945), p. 816. With characteristic wit, Russell also writes: "With James's definition [of truth], it might happen that '*A* exists' is true although in fact *A* does not exist. I have found always that the hypothesis of Santa Claus 'works satisfactorily in the widest sense of the word'; therefore 'Santa Claus exists' is true, although Santa Claus does not exist." (Russell, 1945, pp. 817-818). On a more serious note, the following is a common reaction amongst those philosophers who are critical of the instrumentalist criteria for truth in everyday life, as well as in mathematics, science and philosophy: "To say that the truth of a belief or judgement depends on its practical consequences was to debase truth to considerations of personal profit or to other mercenary aims, while the attempt to enlarge the scope of the practical so as to include the abstract results of mathematical analysis or theoretical conclusions in pure physics was to deprive the word, *practice*, of any distinctive meaning." (Mackay, 1961, p. 393). On a loftier plane, a systematic case against instrumentalism in science is made in (Popper, 1983), pp. 111-131, where the positions of some other well-known philosophers with instrumentalist leanings are analyzed and criticized.

17 One might be prepared to believe that modern scientists are immune to the effects of such "intoxication of power", even if they subscribe, either explicitly and openly, or only in the manner in which they conduct their professional activities, to instrumentalist doctrines. A few nagging doubts might surface, however, in one's mind as one reads some of the over-confident claims following the emergence in the 1980s of superstring theory as the "Theory of Everything" – claims totally opposite in spirit to the humility exhibited by Newton, Einstein, Dirac and many other truly great physicists, as they contemplated the limitations of their own theories, when confronted with the mysteries of Nature. For example, quotations in Note 24 to Chapter 11 might be compared with the following prophetic quotation from the article "The Evolution of the Physicist's Picture of Nature" by P.A.M. Dirac: "There are a good many

problems left over concerning particles other than those that come into electrodynamics: ... how to introduce the fundamental length to physics in some natural way, how to explain the ratio of masses of the elementary particles and how to explain their other properties. I believe that separate ideas will be needed to solve these distinct problems and that they will be solved one at a time through successive stages in the future evolution of physics. At this point I find myself in disagreement with most physicists. They are inclined to think one master idea will be discovered that will solve all these problems together." (Dirac, 1963, p. 50).

18 The reader who desires illustrations of such an "anti-rationalist atmosphere which has become a major menace of our time" can easily find many examples even amongst the references cited in this monograph. Unfortunately for the future of some areas of quantum physics, this "menace of our time" represented by the practice and *imposition* of unadulterated instrumentalism is not only figurative, but a very real threat to all researchers who oppose this trend. Indeed, after taking control of funding agencies in many countries that are in the foreground of pure research (cf., e.g., Note 47), instrumentalists have in some instances prevailed on the bureaucracies in those agencies to support their own research at the expense of research dedicated to traditional values in science. As a rule, the argument offered is that traditionally-oriented research, based on goals and values that used to be the hallmark of all basic research in the pre-instrumentalist era, is no longer "in the mainstream". It is largely due to such practices that contemporary instrumentalists have succeeded to eliminate in almost all areas of theoretical and mathematical physics the very last traces of significant opposition to their doctrines, which set "belonging to the mainstream" and "following the general consensus" ahead of the search for *actual* truth, and a deeper understanding of Nature. Of course, by the systematic use of such means of "persuasion", the assertion that "truth is the opinion which is fated to be ultimately agreed to by all who investigate" regrettably becomes mere self-fulfilling and self-serving prophecy. And, unfortunately, by rewarding conformity, such professional practices thwart initiative and stultify the spirit of free inquiry in science.

19 Admittedly, it cannot be said that everything is satisfactory in the world of contemporary mathematics for those in search of objective truth, rather than solutions to fashionable problems. A trend decreeing, under the banner of "mathematics for mathematics' sake", that the ultimate arbiter of what is valuable in mathematics lies exclusively in the opinion of "leading mathematicians", rather than in its potential of solving problems of the *real* world around us, had emerged and became dominant in this century soon after the confrontation between Hilbert and Brouwer in the late 1920's. As most mathematicians uncritically sided with Hilbert (van Dalen, 1990), the aftermath practically destroyed the intuitionistic school, and set the cause of constructivism in mathematics back by decades – albeit such a great mathematician as H. Weyl, generally deemed to be the successor of Hilbert in both depth and stature, remained predisposed to the intuitionistic as well as the applied point of view ("Mathematics with Brouwer gains its highest intuitive clarity. ... [I]t is the function of mathematics to be at the service of the natural sciences." – cf. Weyl, 1949, pp. 54 and 61). This unfortunate state of affairs was compounded by the difficulties which Hilbert's ambitious program (Weyl, 1949; Reid, 1986), aimed at establishing the consistency of all the major areas of mathematics, encountered after the discovery of Gödel's (1931) incompleteness theorem – cf. (Kline, 1980), pp. 260-264. On the other hand, in contemporary mathematics the pursuit of fashions is considerably more subdued than in quantum physics in general, and in elementary particle physics in particular. Furthermore, although even such a fundamental question as the consistency of arithmetic remains unresolved, at least as long as that consistency is accepted together with the law of excluded middle, an *objective* state of affairs prevails with regard to the criteria for mathematical *truth*. Hence, although what is at present deemed to be "deep" mathematics is very much a function of fashions dictated by the prevailing circles of "mandarins" (cf. Notes 10, 16, 18, 20 and 22), at least what is deemed to be *valid* and *valuable* mathematics is not exclusively a function of their pronouncements.

20 The well-documented (Kline, 1980) isolation of modern pure mathematics from all applications was most definitely a contributing factor to this counterproductive breakdown. Courant is cited to have remarked as early as 1927: "The predominant characteristic of American mathematicians seemed to be a tendency to favor abstract and the so-called areas of pure mathematics. ... Applied mathematics was treated as a stepchild in America." (Reid, 1986, p. 382). With the post-World War II shift of focus in mainstream research from Europe to North America, this fact no doubt became one of the major factors that enlarged the chasm which emerged between the physics and mathematics communities during the second half of this century – a chasm which has begun to be bridged to a significant degree only in the course of the last decade.

21 Indeed, *all* physical theories, from Newtonian mechanics onwards, were developed by individuals who either simultaneously introduced the required mathematical tools at a level commensurate with the prevailing mathematical standards of their generation, or worked in close contact and collaboration with competent mathematicians, who helped steer them away from deductive mathematical errors that might have affected the physical content and predictions of their theories. In recent history, the best example is the 1913-1915 period in the development of general relativity by Einstein (Norton, 1989). That period started with Einstein's collaboration with his mathematician-friend M. Grossmann, and culminated in the triumphant final version of classical general relativity systematically presented by Einstein in 1916. The painstaking historical research by Norton (1987, 1989), Stachel (1980, 1989), and others, vividly illustrates how Einstein's superb physical intuition for once led him astray, so that in his *Entwurf* paper (Einstein and Grossmann, 1913) he discarded, on the basis of *physical* misconceptions, the requirement of general covariance. Then, for two full years he expressed in public as well as in private satisfaction with *non*-generally covariant equations – cf. (Cattanin and De Maria, 1989), p. 179. The constructive criticism and suggestions of such outstanding mathematicians as Hilbert and Levi-Civita emerges as a major, and perhaps even as the decisive factor (cf. Cattanin and De Maria, 1989, p. 185), which eventually enabled Einstein to publicly present to the Berlin Academy the *correct* field equations, in their final form (2.7.3), on November 25, 1915 – namely five days after, unbeknownst to him, Hilbert had already presented the same equations to the Göttingen Academy (Norton, 1989, p. 150).

22 Some of the examples which we shall provide in this chapter lead to straight *contradictions*, so that they would be unacceptable even to those mathematicians who accept the following evaluation: "There is no rigorous definition of rigor. A proof is accepted if it obtains the endorsement of the leading specialists of the time or employs the principles that are fashionable at the moment. But no standard is universally accepted today." (Kline, 1980, p. 315). In this context, it should be noted that in this monograph the term "mathematically rigorous" is used as a short version for "mathematically acceptable by generally agreed upon contemporary standards in the mainstream areas of mathematics". The present author is highly sympathetic to the constructivist school in mathematics, and hopes that one day it will prevail and supply constructive proofs for all theorems of relevance in applications – but that day seems to be still far off in the future. For the present, however, existence proofs relying on the law of the excluded middle are certainly preferable to no proofs at all, and to the blind application of "working rules" for arriving at "theoretical results in agreement with experiments" (cf. Note 6 on this last score).

23 Cf. (Savan, 1988) for a general analysis and documentation of this phenomenon in modern science. As for the situation in high-energy physics specifically, S. S. Schweber asks the following pertinent questions, and then provides some answers: "Did the involvement of many of the leading American high-energy theorists in defense matters reinforce a particular kind of theoretical orientation – pragmatic, phenomenological, with '*S* matrix theory' as its most impressive statement – to the exclusion of others? Did it affect developments in theoretical high-energy physics? I would suggest that the fragmentation of interests by these leading theorists, stemming from their consulting and their involvement in defense matters, hindered – and to a certain extent prevented – their maintaining a sustained effort on fundamental theory. Also, in their capacity as reviewers of research proposals, and by virtue of their dominance in the funding process, they tended to reinforce their dominant view." (Schweber, 1989), p. 681.

24 Naturally, a number of textbooks aimed at would-be "mathematical physicists" eventually made their appearance – of which the original 1971 edition of [PQ] represented perhaps the first attempt at rederiving *all* the major results presented in a typical "mainstream" textbook on nonrelativistic quantum mechanics in a mathematically acceptable manner. Unfortunately, although by the 1980s a score of very good textbooks and monographs published by various mathematical physicists made it easy for potential authors of "mainstream" textbooks on quantum theory to raise their mathematical standards to an acceptable level, that opportunity was ignored – and it is still being by and large ignored. Clearly, the underlying feeling must be that, since "truth" can be identified with the "professional consensus as to what works", and since that "consensus" was reached and firmly established within the profession by the working practices of the leading physicists of the post-World War II generation (namely Dyson's "mandarins" – cf. Note 12), which "work well" due to that very same consensus, there is no room left for further doubts, or for any critical reconsiderations of those practices.

25 Cf. [PQ], p. 195. The Hellinger-Toeplitz theorem is a special case of the (to mathematicians) very well-known closed-graph theorem (which is stated and proved on p. 210 of [PQ] for the case of closed operators in Hilbert space), but which holds for much more general cases of topological vector spaces.

[26] Naturally, *after* the resolutions in (3.3) are applied to all pairs of vectors from $\mathcal{H}_+$, the resulting Lebesgue integrals can be extended to all of $\mathcal{H}$ by virtue of the fact that $\mathcal{H}_+$ is dense in $\mathcal{H}$. But such a procedure merely reproduces the formula (3.1.1) for the inner product in $\mathcal{H}$, and still leaves open the question as to how to specify $\psi(x)$ at every single $x \in \mathbf{R}^3$ so that an extension to $\psi \notin \mathcal{H}_+$ of the formal inner product in (3.4) would be achieved. It should be noted that, for the sake of notational simplicity, in all these considerations Berezanskii's (1968) notation for equipped Hilbert spaces is being used, but all the presented arguments apply equally well to rigged Hilbert spaces.

[27] Such as to Schwartz $\mathcal{S}$-spaces in the rigged Hilbert space approaches (Gel'fand *et al.*, 1964; Antoine, 1969). In the equipped Hilbert space approaches (Berezanskii, 1968; Prugovečki, 1973) $\mathcal{H}_+$ and $\mathcal{H}_-$ are both Hilbert spaces, with $\mathcal{H}_+$ being the domain of an unbounded equipping operator, and $\mathcal{H}_-$ the range of its extension to $\mathcal{H}$.

[28] In Chapter V of [PQ] it is shown how one can start, in a physically legitimate and meaningful manner, from the time-dependent approach to scattering theory, and then derive from it in a mathematically rigorous manner all the main results of the stationary approach that are formalistically derived in typical textbooks on conventional quantum scattering theory exclusively in the stationary context.

[29] This formula was originally derived in (Prugovečki, 1978b), pp. 240-247, in the context of developing a quantum mechanical counterpart of the well-known Boltzmann equation – cf., e.g., (Balescu, 1975). An alternative derivation was subsequently provided by Turner and Snider (1980), which corresponds to applying the sharp-momentum limit in (3.5.5) to (3.7). However, the physical significance of such a limit is rather questionable, and a problem of mathematical existence also emerges.

[30] Cf., e.g., Note 25 to Chapter 9. The documentation of such cases could easily fill volumes, of the same length as the present monograph – which is, however, chiefly concerned with the *constructive* task of setting the *worthwhile* aspects of present-day conventional theories on a solid foundation, and only incidentally (as well as rather reluctantly) with their critical evaluation. The fact that not many such volumes are in existence is not primarily due to lack of material, or even to a total absence of dedicated individuals willing to engage in such a thankless task. The chief explanation is that, from the point of view of the individual researcher striving for professional survival – not to mention professional recognition – it can be a professionally self-destructive enterprise to collect and document fallacies of institutionally strongly sponsored points of view, in the face of the multifarious devices for pressure and control (Savan, 1988) that are exerted in the modern era of "big science" in order to enforce and preserve conformity. Of course, such professional pressures would have been ineffective against such outstanding and well-known physicists as Dirac and Heisenberg, but their use against those who have heeded their open and justified criticisms has until recently made that criticism virtually counterproductive. Hence, as will be pointed out in Sec. 12.7, it was only in the course of the last few years that open and critical inquiry into the foundations of quantum mechanics and quantum field theory has begun to reassert itself at the international level.

[31] We use Dirac's own stipulation of "mathematical *soundness*", rather than the stronger condition of mathematical *rigor* (cf. Note 12). *Eventual* mathematically rigorous justification, rather than mere reliance on "agreement with experiment" is, however, especially essential at the present technological frontier of measurements in the microdomain, where the *independent* and *repeated* verifiability of experimental results becomes ever more questionable. Indeed, in contemporary elementary particle physics, very expensive experiments are carried out with costly experimental equipment, which, of necessity (Yaes, 1974), has become the monopoly of a handful of teams of experimentalists, who work in close contact with leading proponents of the theories they are verifying. Although such contact is in many respects desirable, it can be also conducive to erroneous analyses of experimental results, produced under various types of conscious or unconscious professional influence. A senior elementary particle physicist, who clearly perceives only the positive aspects of such close contact, and wholeheartedly approves of it, describes its effects as follows: "Constructing modern theories also means constructing new concepts and abandoning old ones ... [as it] would be obvious to all if all had a chance to experience life in a great research center in fundamental physics. In such places ... a permanent exchange of views is observed to take place between the two teams of people [namely experimentalists and theorists]; they seem both to understand and to need each other. When we see all this going on, it is not hard to appreciate that *in order to make sense of the mass of data provided by the experimentalists, the theorists have to create new concepts.*" (d'Espagnat, 1987, p. 40) – emphasis added. Thus, in practice, the empirical verification

of a theory is not all a mere matter of comparing "theoretical predictions" with "experimental results". And, once officially sanctioned by being accepted for publication in professional journals, such results tend to become uncritically accepted as unconditionally valid – with occasionally published retractions proving the existence of occurrences of faulty analyses of data, but not giving any ideas as to the frequencies of such occurrences. Hence, the impossibility of their *routine* reproducibility in practice, and therefore of *plentiful and independent* verification, makes the acceptance of the remaining ones to a considerable degree a matter of *subjective* faith. Relevant examples of well-documented cases where that faith might have been misplaced can be found in Note 25 to Chapter 9.

32 The fact that such proofs are absolutely necessary is indicated already by mathematically rigorous results in nonrelativistic quantum scattering theory. Thus, contrary to *assertions* made in some textbooks on quantum mechanics that the unitarity of the S-"matrix" (i.e., of the scattering operator $S = \Omega_-^* \Omega_+$) is a consequence of the "conservation of probability" (i.e., of the fact that the time-evolution governed by the Schrödinger equation is represented by a family of unitary operators, which, as such, preserve all transition probabilities), such assertions are actually *false*: a mathematically rigorous theorem (cf. Thm. 2.5 in [PQ], p. 443) shows that, even if the initial domains, $\mathbf{M}_+$ and $\mathbf{M}_-$, of the partial isometries Ω_+ and Ω_- (representing Møller wave operators) are equal, a necessary as well as sufficient condition for the unitarity of S is that the ranges $\mathbf{R}_+$ and $\mathbf{R}_-$ of these wave operators be equal. An early but physically artificial model by Kato and Kuroda (1959) has shown that it can happen that $\mathbf{R}_+ \neq \mathbf{R}_-$ even when time-evolution is unitary, whereas later Pearson (1975) rigorously demonstrated that it *can* happen that $\mathbf{R}_+ \neq \mathbf{R}_-$ even in physically acceptable models. Pearson's model employs a local potential that oscillates ever more rapidly as one approaches the origin of the center-of-mass reference system, so that a quantum particle in certain asymptotically free incoming states becomes forever trapped in that potential, and does not produce corresponding asymptotically free outgoing states.

33 J. Gribbin recounts the amusing circumstances of Tryon's creative spark, which triggered his "creation *ex nihilo*" idea, whereby at one of Sciama' seminars, "Tryon blurted out, to his own surprise as much as everyone else's, 'maybe the Universe is a vacuum fluctuation!'" (Gribbin, 1986, p. 376). The sober evaluation of its *meaning* and *significance*, outside the realm of religion or science fiction, actually does not encounter any problems with the superficial appearance that such a "concept of the universe being created from nothing is a crazy one" (Vilenkin, 1982, p. 26). Indeed, the underlying mathematics is rudimentary and well understood by any student of quantum mechanics (Vilenkin, 1982, 1988); whereas, at first sight, Einstein's ideas on relativity theory appeared much "crazier" to some of his contemporaries. The crucial difference is that Einstein's ideas were operationally well-founded, empirically directly testable, and ontologically sensible. But *what* is it that is supposedly "tunneling", and through a barrier of *what* does that purported "tunneling" take place in the Tryon-Vilenkin "scenario"? The fact that such basic questions are *not even* asked (not to mention answered) by conventionally-minded instrumentalists in quantum cosmology forcefully illustrates the general grounds for the type of concern voiced in Heisenberg's last paper (cf. Sec. 1.5). Indeed, instead of being provided with such questions and answers, we are simply authoritatively told that "the only relevant question seems to be whether or not the spontaneous creation of universes is possible", and that, after all, "obviously, we must live in one of the rare universes which tunneled to the symmetric vacuum state" (*ibid*., p. 27). Of course, the most conventionalistic of instrumentalists argue that we are to judge any "theory" exclusively by its "observational consequences". Does that mean, however, that if ancient Greek mythology, present-day religious fables, or even ordinary fairy tales make "predictions" which are indeed in accordance with *certain* "observations", then we are to accept them as serious contenders for "valid" scientific theories?

34 Perhaps Dirac's determination to publicly condemn the replacement of science by technology can be traced to his experiences as a student in an engineering college: "In [the] engineering courses [which he then attended] the emphasis was on mathematical rules with the help of which problems could be solved, without giving strict proofs or asking how or why the rules worked. Dirac remembered that there always remained a kind of magic about these rules, and frequently he had a strange feeling about how he ever got answers out of them." (Mehra and Rechenberg, 1982, vol. 4, p. 11). However, in engineering those rules were at least deduced from an underlying consistent physical theory based on sound mathematics, but that is not the case in conventional renormalization theory. Hence, six decades later he was to say: "Working with the present foundations [of conventional quantum field theory], people have done an awful lot of work in making applications in which they find rules for discarding the infinities. But

these rules, even though they may lead to results in agreement with observations, are artificial rules, and I just cannot accept that the present foundations are correct." (Dirac 1978a, p. 20).

35 A well-documented example of systematic propagation of errors in the analysis of the raw experimental data in measurements of large-scale spatio-temporal separations is provided by the changes in the estimates of Cepheid distances by a factor of 2.6 in 1952, and by another 2.2 factor in 1958 (Ellis, 1989, p. 391). This led to radical changes in estimates of the age of our universe, as well as in estimates of all intergalactic distances.

36 In a recent monograph, Cushing (1990) provides an exhaustive account and analysis of the rise and fall of the *S*-matrix program in elementary particle physics. It began as follows: "Cosmic ray showers (or 'explosions') and the divergence of cross-sections beyond a certain energy in a classical (nonlinear) field theory version of Fermi's β-decay formalism were taken by Heisenberg (1936, 1938a) to indicate the existence of a fundamental length and the need for a profound revision of elementary particle dynamics. Not knowing what the future theory might be, he proposed the *S*-matrix theory *as an interim program.*' (Cushing, 1990, p. 33) – emphasis added. Later on Landau (1955) expressed his conviction that the only *directly* observable physical quantities were those associated with asymptotically free particles, such as their initial or final momenta before and after a scattering process. He therefore concluded that any quantum fields interpolated between asymptotic states were physically meaningless, and advocated a break with quantum field theory, while supporting a program similar to the one of Heisenberg. However, it was Chew who ultimately "made a radical break with quantum field theory" (*ibid.*, p. 167). The role of sociological factors in the adoption of theories in high-energy physics receives separate attention in Sec. 10.2 of (Cushing, 1990), where it is pointed out that "the very nature of scientific practice has changed significantly with the advent of 'big science' after the Second World War."

37 For example, in his talk delivered at the Twelfth Solvay Conference in Physics, M. L. Goldberger said: "My own feeling is that we have learned a great deal from field theory as we shall see, even dispersion theory came from it; that I am quite happy to discard it as an old but friendly mistress, who I would even be willing to recognize on the street if I should encounter her again. ... It is perhaps correct to say that much of the deeper philosophy of the *S*-matrix approach held by some of us, in particular Chew, who believe that there are no elementary particles, and that there are no undetermined dimensionless constants in the theory, has not yet been put to a test." (Goldberger, 1961, pp. 179-180). According to G. F. Chew: "The capacity for experimental predictions is the only reliable measure of a physical theory. ... No suggestion is being made that space and time do not continue to be the basis of *macroscopic* physics; ... Does this mean that there can be no continuous connection between the microscopic and macroscopic worlds? The situation is no more uncomfortable than it has always been for quantum theory, where the conventional explanation of the relation between the classical observer and quantum laws leaves most people feeling queasy." (Chew, 1963, pp. 533, 538). Of course, after the revival of interest in quantum field theories in the 1970s, this "uncomfortable situation" was forgotten: "The development of *S*-matrix theory was characterized by a certain degree of sectarian strife. ... It was not so much a question of its being expedient to be on the mass-shell as of its being sinful to be anywhere else. In particular, [the advocates of the *S*-matrix program] proclaimed the demise of quantum field theory. ... [However] the *S*-matrix endeavor looks a good deal less beguiling [now, in the late 1970s] than it did in those brave early days [namely the 1960s]." (Polkinghorne, 1979, p. 87).

38 Louis de Broglie recalls this emergence as follows: "When, in 1922-1923, I had my first ideas about wave mechanics, I was guided by the vision of constructing a true physical synthesis, resting upon precise concepts, of the coexistence of waves and particles. I never questioned then the nature of the physical reality of waves and particles. ... I also noticed that if the particle was regarded as containing the rest energy $M_0c^2 = h\nu_0$, it was natural to compare it with a small clock of proper frequency ν_0." (de Broglie, 1979, p. 7).

39 Cf. (Kim and Noz, 1986), Chapter V, where such a mathematical treatment is provided in the context of a quark model for mesons. However, a wave function that is square integrable in space *and* in time is of questionable physical significance even in the nonrelativistic regime, since it suggests that whatever entity is described by it spontaneously disappears from existence the further we look into the distant past or into the distant future.

40 The mathematical heuristics of this procedure was presented in (Prugovečki, 1988a), and further elaborated in Appendix A of (Prugovečki, 1989b), which the present review basically reproduces. However, as mentioned in Sec. 1.5, originally such considerations were used in the treatment of hadrons

as GS excitons. That led to a mass formula (Prugovečki, 1981b), which produced Regge trajectories that were found (Brooke and Guz, 1984) to be in otherwise good agreement with the experimental data available at that time.

41 The very rough estimates in (Greenwood and Prugovečki, 1984) do not indicate that the prospects are very good. However, the intrinsic non-linearity of GS models for interacting quantum fields might produce new and unexpected results if numerical computations are performed even by the use of existing lattice approximation methods.

42 It is the neglect of this most crucial aspect of quantum propagation that causes Popper's (1967, 1982, 1988) "propensity interpretation" of quantum mechanics not to come even close to an adequate depiction of quantum phenomena – as persuasively demonstrated by a number of critics (cf. Jammer, 1974, pp. 448-453). A similar neglect also makes Nelson's (1986) "stochastic mechanics" depiction of the two-slit experiment to be totally at odds with *quantum* reality.

43 Due to lack of space, in this chapter we have concentrated most of the attention on the principal protagonists in the historical drama in the quantum physics of this century, that began with the confrontation between Einstein and Bohr in the 1920s, and after World War II developed into a historically most paradoxical situation, in which the usual stereotypes about the conservatism of "older" generations vs. the radicalism of "younger" generations in all walks of life are totally reversed: during the second half of this century, the "older generation" of theoretical physicists, incorporating all those who founded quantum theory, *remained* "revolutionary" in its outlook; whereas the "younger generation" turned out to be deeply "conservative", as well as strikingly conformist in all the principal facets of its regular professional practice. For example, in a recent article entitled "Wolfgang Pauli: His Scientific Work and His Ideas on the Foundations of Physics", the following was pointed out by Bleuler (1991, pp. 306-307): "I would like to conjecture that Pauli himself already had, in an early stage of the development of quantum field theory, profound doubts as to its adequacy and mathematical consistency as a general theory of elementary particles. He refused to be 'renormalized', as he expressed himself in relation to that famous principle of renormalisation, thus setting himself (together, however, with Dirac, Bohr and others) for a time into strict opposition with an enthusiastic, and at first very successful, younger generation. A (partial?) concession in this respect came only after Pauli's death, firstly in connection with the (unsuccessful?) attempts to 'save' local quantum field theory by seeking recourse to 'strings', and secondly in the recent profound and far-reaching attempts at a 'non-commutative geometry' of A. Connes, D. Kastler and others. It seems to me that his most recent (as yet not generally recognized) development is perhaps the fulfilment of a 'vision', which Pauli expressed in a long and unforgettable discussion a few weeks before his death: 'For a real solution of the problem of singularities (i.e. the question of renormalization) a step of the same size and significance as that which was taken once before in the twenties might be necessary'."

44 Ironically, it appears that only when a given instrumentalistically motivated theory begins to "fall out of fashion" that the concern of its still faithful adherents starts to shift to this type of questions. For example: "At the same time that the predictive fertility of the *S*-matrix program waned, it continued to have considerable philosophical appeal (Cushing, 1985).... The duality and superstring models also became theory-driven, having little contact with experiment (Schwarz, 1975, p. 67; 1982, p. 7). Consistency, potential scope, and hoped-for contact (in a limited regime) with an empirically adequate theory (such as QCD) remained the major motivations for pursuit." (Cushing, 1990, p. 215).

45 We italicized "primarily" since, of course, sociological factors have always played a role in science: "Pickering (1984) has presented the process of choice (or judgment) as a largely social exercise. In the tradition of the radical relativist-constructivist program in the sociology of knowledge, he has attempted to show that not only the *form*, but even the very *content*, of scientific knowledge is sociologically determined. ... However, he (1989b) has recently argued for a pragmatic realism in which not just anything goes. ... Galison (1987), in his study of the change in experimental practice in high-energy physics during the twentieth century, argues convincingly that it is not by deductive reasoning alone that scientists pass from the raw data of an experiment to a conviction that an effect has been seen." (Cushing, 1990, p. 217). Thus, it is a matter of the *degree* to which sociological factors have become predominant in contemporary quantum physics.

46 A vividly drawn portrait of Schrödinger can be found in (Bernstein, 1991, pp. 32-33, where the following is pointed out: "All the inventors of the quantum theory, as it happened, were men of broad culture, perhaps attributable to their European *gymnasium* educations, but even in this group

Schrödinger stood out." This "broad culture" is in sharp contrast to the "newly developing cult of narrowness" exhibited by the "many younger physicists who have grown up in [the] period of over-specialization" (Popper, 1982a, p. 100) of the post-World War II era.

47 This change in social climate is described and documented by Schweber (1989, 1991), who states the following: "World War II altered the character of science in a fundamental and irreversible way: the importance and magnitude of the contribution of scientists and engineers, particularly physicists, to the American war effort changed the relationship between the scientists and the military, industry, and government. The Department of Defense, realizing that the security of the nation depended on the strength and creativity of the scientific community, invested heavily in both their support and control." (Schweber, 1989, p. 670). Later on he describes how widespread this phenomenon soon became: "This 'American' style of doing physics was characteristic of the great wartime laboratories: the Radiation Laboratory at MIT, the Met Lab in Chicago, and Los Alamos. It was in these wartime laboratories that many of the outstanding theoreticians of the 1950s were molded: Feynman, Goldberger, Chew, Robert Marshak. It is a style that became institutionalized at all the leading departments during the fifties and became the national norm. ... At the leading high-energy centers, [the] fortunes and future [of talented young people] were often determined by their skill in explaining experimental results, and more generally by their usefulness to their experimental colleagues; the latter had invested enormous energy, skills, and government resources in building their high-energy machines." (*ibid.*, p. 672). In the end this phenomenon became "hegemonic worldwide": "The defense connection during the 1950s reinforced the pragmatic, utilitarian, instrumental style so characteristic of theoretical physics in the United States. The successes of this mode of doing theoretical physics help explain its diffusion to Europe and elsewhere. The pragmatic ideal of American physics that had been visible from early on now became not only the national norm but in fact hegemonic worldwide." (*ibid.*, p. 673).

48 Indeed, "the structure of the scientific community is that of a pyramid, the apex being occupied by the relatively few creative people [cf. the description of Dyson's 'mandarins' in Note 13] who can invent and sustain successful theories. They ultimately make the rules of the game." (Cushing, 1990, p. 253). Another source points out the following: "It is as though most of the members of the community consider it worthwhile to work out the approach suggested by the intellectual leader of the moment (e.g., Gell-Mann, Mandelstam, Chew) than work on their own ideas or on longer-range programs of research. As early as 1951 Feynman called it the 'pack' effect. The work on dispersion relations after Gell-Mann's and Goldberger's initial papers is an example of this phenomenon; the almost wholesale adoption of Chew's *S*-matrix program is another. The community at one time or another seems to be dominated by a single individual. Gell-Mann, Goldberger, Lee, Yang, and Chew were the dominant figures from the mid-1950s to the mid-1960s, a role Steven Weinberg assumed in the late sixties." (Schweber, 1989, p. 673).

49 Cf. Schweber (1989, 1990). The controversial writings of Kuhn (1970) and Feyerabend (1975) might make it appear that such sociological phenomena are universal in the history of science, being part and parcel of its very methodology. This view is disputed, with ample documentation, by Franklin (1986), and implicitly also by Cushing (1990) – cf., e.g., the last quotation in Note 25 to Chapter 9.

50 As a recent biographer of Dirac has stated: "Dirac firmly believed that a new revolution was needed. His lack of sympathy for the new quantum electrodynamics involved a lack of appreciation for the values of the new [i.e., post-World War II] generation of physicists." (Kragh, 1990, p. 184). A similar "lack of appreciation" was displayed by Heisenberg (1976).

51 We refer here to such "predictions" as those discussed in Secs. 7.2-7.3, to the effect that particles are created *ex nihilo* in violation of energy conservation laws. Even more "daring" is the idea that our entire universe was created by a tunneling of Nothing through a (potential?) barrier of Nothing (Vilenkin, 1982, 1988) – which might indeed appear very innovative to physicists, but perhaps not so to scholars of various religious scriptures dealing with the creation of our world, or to science fiction writers specializing in "alternate realities" and "parallel universes" (Wolf, 1990).

52 This applies even to the idea of gauge fields, which was initiated by Weyl in 1918, and extended by him to quantum mechanics in 1929. The idea of supersymmetry was new, but in addition to not receiving any experimental confirmation, it was also "dominated by questions of formalism and technique". The theory of superstrings (which at the present time is suffering a sharp decline on the popularity scale) would have been an exception, had it been originally derived from clearly stated first principles, rather than in a manner which left "the fundamental physical and geometric principles that lie at [its]

foundation . . . still unknown" (Kaku, 1988, p. viii). However, the most recent developments in topological quantum field theory are of definite mathematical interest. They are especially intriguing since they are based on the study of knot theory, which has its roots in the nineteenth century physics of the ether – cf. (Atiyah, 1990), Sec. 1.3. Perhaps, after all, the history of science does move in circles!

53 For example the following quotation of M. Moravczik can be found on pp. 279 and 280 of (Cushing, 1990): "[A]s far as strong interactions are concerned, we have not made any substantial physical progress since Yukawa in 1935.... [O]ld fashioned field theory, then dispersion theory, then Regge poles, and now QCD are simply reincarnations of the same Yukawa idea."

54 Dirac made many public statements to this effect. Those statements quoted in Sec. 9.6 are amongst the most representative. It should be also observed that Note 25 in that chapter provides additional circumstantial evidence supporting Dirac's point of view.

55 It is in this respect that the motivating factors behind the present work are consonant with the spirit of Popper (1963, 1968, 1982, 1983) and his followers (Lakatos, 1976; Watkins, 1984; etc.), who maintain that there *is* rationality in science, and that the goal of science is the pursuit of an *objective* truth that is independent of transitory fashions and other social factors. Such Truth can be arrived at by the traditional methods characteristic of the science of Newton, Maxwell, Einstein, Dirac, and other *truly* great physicists. The attitude reflected in this monograph is therefore very much at odds with the view of science advocated by some contemporary sociologists and historians of science, according to whom: "Although philosophy of science may once have set truth as a goal to which science aspires (Watkins, 1984, p. 155), ... closer examination of the historical record of actual scientific practice has shown that things are not as simple as we might hope them to be. ... A relativist or irrationalist (perhaps better, *a*rationalist in contrast to the rationalist) school sees scientific knowledge as contingent, being determined by social and historical factors, so that the specific laws of science become arbitrary." (Cushing, 1990, pp. 282-283). All that a practising scientist, who *is* dedicated to Truth in science, can provide in the way of a retort to such an assertion is the following: Whereas it might be, unfortunately, true that, during the present instrumentalist era, many of the *advocated* "laws" of "big science" have become indeed rather arbitrary, that is a phenomenon which characterizes the activities of many, and perhaps even most *scientists* in some areas of contemporary science – but not *Science* itself. As with any *social* phenomenon, this particular one *will* turn out to be only transitory *if* there is enough determination and dedication to reverse the trend amongst all those who take the opposite point of view as to what makes Science worth pursuing: *the concerted search for Truth, and ultimately its revelation.*

References

Key references. *A number of titles have been singled out from the following list of references, since they contain material essential for the understanding of the mathematical framework presented in this monograph, and therefore are frequently cited. These citations are primarily in the form of the first letter of the surname of their (first) author set in between square brackets. In those cases where the same first letter occurs in the surnames of the authors of two or more of these key references, the first letter in the surname of the second author, or the first letter in the title proper (i.e., ignoring definite or indefinite articles) is appended to avoid confusion – as it would be the case with: [B], [BD], [BG], [BI], [BL] and [BR]; [N] and [NT]; [P] and [PQ]; [SC], [SI] and [ST]; [W] and [WQ]. Furthermore, a number of these key references overlap in some of their subject matter. Hence, for most of the considerations in the present monograph, it is sufficient to choose and consult only one of the references from each of the following six groups: [BI] and [D]; [BD] and [N]; [C], [I] and [NT]; [IQ] and [SI]; [K] and [SC]; [M] and [W].*

Abraham, M.: 1902, *Phys. Z.* **4**, 57.

Abraham, M.: 1902, *Ann. d. Phys.* **10**, 105.

Abraham, R., and J. E. Marsden: 1978, *Foundations of Mechanics*, 2nd edition, Benjamin, Reading, Mass.

Abraham, R., J. E. Marsden, and T. Ratiu: 1988, *Manifolds, Tensor Analysis, and Applications*, 2nd edition,, Springer, New York.

Accardi, L.: 1988, in *The Nature of Quantum Paradoxes*, G. Tarozzi and A. van der Merwe, (eds.) Kluwer, Dordrecht.

Accardi, L., A. Frigerio, and V. Gorini (eds.): 1984, *Quantum Probability and Applications to the Theory of Irreversible Processes*, Springer Lecture Notes in Mathematics, vol. 1055, Berlin.

Acharya, R., and R. Sudarshan: 1960, *J. Math. Phys.* **1**, 532.

Ackermann, R.: 1985, *Data, Instruments, and Theory*, Princeton University Press, Princeton.

Adamowicz, W., and A. Trautman: 1975, *Bull. Acad. Pol. Sci., Ser. Sci. Math. Astro. Phys.* **21**, 345.

Adler, R., M. Bazin, and M. Schiffer: 1975, *Introduction to General Relativity*, 2nd edition, McGraw-Hill, New York.

Adler, S. L.: 1982, *Rev. Mod. Phys.* **54**, 729.

Aharonov, Y., D. Z. Albert, and C. K. Au: 1981, *Phys. Lett. A* **124**, 199.

Aharonov, Y., D. Z. Albert, A. Casher, and L. Vaidman: 1987, *Phys. Rev. Lett.* **47**, 1029.

Aharonov, Y., and J. Anandan: 1987, *Phys. Rev. Lett.* **58**, 1593.

Aharonov, Y., and T. Kaufherr: 1984, *Phys. Rev. D* **30**, 368.

Aharonov, Y., and M. Vardi: 1980, *Phys. Rev. D* **21**, 2235.

Albeverio, S. A., and R. Høegh-Krohn: 1976, *Mathematical Theory of Feynman Path Integrals*, Springer Lecture Notes in Mathematics, vol. 523, Berlin.

Alcalde, C., and D. Sternheimer: 1989, *Lett. Math. Phys.* **17**, 117.

Aldinger, R. R., *et al.*: 1983, *Phys. Rev. D* **28**, 3020.
Ali, S. T.: 1985, *Riv. Nuovo Cimento* **8**(11), 1.
Ali, S. T., and J.-P. Antoine: 1989, *Ann. Inst. H. Poincaré* **51**, 23.
Ali, S. T., J.-P. Antoine, and J.-P. Gazeau: 1990, *Ann. Inst. H. Poincaré* **52**, 83.
Ali, S. T., J. A. Brooke, P. Busch, R. Gagnon, and F. E. Schroeck, Jr.: 1988, *Can. J. Phys.* **66**, 238.
Ali, S. T., and H.-D. Doebner: 1990, *Phys. Rev. A* **41**, 1199.
Ali, S. T., R. Gagnon, and E. Prugovečki: 1981, *Can. J. Phys.* **59**, 807.
Ali, S. T., and E. Prugovečki: 1977a, *J. Math. Phys.* **18**, 219.
Ali, S. T., and E. Prugovečki: 1977b, *Physica A* **89**, 501.
Ali, S. T., and E. Prugovečki: 1977c, *Int. J. Theor. Phys.* **16**, 689.
Ali, S. T., and E. Prugovečki: 1981, *Nuovo Cimento A* **63**, 171.
Ali, S. T., and E. Prugovečki: 1986, *Acta Appl. Math.* **6**, 1, 19, 47.
Alvarez, E.: 1989, *Rev. Mod. Phys.* **61**, 561.
Anandan, J.: 1984, in *Proc. Int. Symp. on Foundations of Quantum Mechanics in the Light of New Technology, Tokyo, 1983*, S. Kamefuchi, H. Ezawa, Y. Murayama, M. Namiki and T. Yajima (eds.), Physical Society of Japan, Tokyo.
Anandan, J.: 1986, in *Quantum Concepts in Space and Time*, R. Penrose and C. J. Isham (eds.), Clarendon Press, Oxford.
Anandan, J.: 1988, *Phys. Lett. A* **129**, 201.
Anandan, J.: 1990, *Phys. Lett. A* **147**, 3.
Anandan, J., and Y. Aharonov: 1988, *Phys. Rev. D* **38**, 1863.
Anandan, J., and Y. Aharonov: 1990, *Phys. Rev. Lett.* **65**, 1697.
Anandan, J., and A. Pines: 1989, *Phys. Lett. A* **141**, 335.
Anandan, J., and L. Stodolski: 1983, *Phys. Rev. Lett.* **50**, 1730.
Anderson, A., and B. DeWitt: 1986, *Found. Phys.* **16**, 91.
Antoine, J.-P.: 1969, *J. Math. Phys.* **10**, 53.
Antoine, J.-P.: 1980, *J. Math. Phys.* **21**, 268, 2067.
Antoine, J.-P.: 1990, in *Wavelets, Time-Frequency Methods and Phase Space*, 2nd edition, J. M. Combes, A. Grossmann and Ph. Tchamitchian (eds.), Springer, Berlin.
Antoine, J.-P., and A. Grossmann: 1976, *J. Func. Anal.* **23**, 369, 379.
Antoine, J.-P., and A. Grossmann: 1978, *J. Math. Phys.* **19**, 329.
Applebaum, D.: 1988a, in *Proceedings 1987 Oberwolfach Conference on Quantum Stochastic Processes*, L. Accardi and W. von Walderfels (eds.), Springer Lecture Notes in Mathematics, vol. 1303, Berlin.
Applebaum, D.: 1988b, *Lett. Math. Phys.* **16**, 93.
Applebaum, D., and R. L. Hudson: 1984a, *J. Math. Phys.* **25**, 858.
Applebaum, D., and R. L. Hudson: 1984b, *Commun. Math. Phys.* **96**, 413.
Araki, H.: 1985, *Commun. Math. Phys.* **97**, 149.
Araki, H., and M. M. Yanase: 1960, *Phys. Rev.* **120**, 622.
Arnold, L.: 1974, *Stochastic Differential Equations: Theory and Applications*, Wiley, New York.
Arnowitt, R., S. Deser, and C. W. Misner: 1962, in *Gravitation: An Introduction to Current Research*, L. Witten (ed.), Wiley, New York.
Arshansky, R., L. P. Horwitz, and Y. Lavie: 1983, *Found. Phys.* **13**, 1167.
Arzeliés, H.: 1966, *Relativistic Kinematics*, Pergamon Press, Oxford.
Ashtekar, A.: 1984, in *General Relativity and Gravitation*, B. Bertotti, F. DeFelice and A. Pascolini (eds.), Reidel, Dordrecht.
Ashtekar, A.: 1987, *Phys. Rev. D* **36**, 1587.
Ashtekar, A. (ed.).: 1988, *Recent Developments in Canonical Gravity*, Bibliopolis, Naples.

Ashtekar, A.: 1990, in *General Relativity and Gravitation*, N. Ashby, D. F. Bartlett and W. Wyss (eds.), Cambridge University Press, Cambridge.
Ashtekar, A.: 1991, in *Conceptual Problems in Quantum Gravity*, A. Ashtekar and J. Stachel (eds.), Birkhäuser, Boston.
Ashtekar, A., and A. Magnon: 1975, *Proc. Roy. Soc. London A* **346**, 375.
Ashtekar, A. J., and J. Stachel (eds.): 1991, *Conceptual Problems in Quantum Gravity*, Birkhäuser, Boston.
Aspect, A., P. Grangier, and G. Rogers: 1981, *Phys. Rev. Lett.* **47**, 460.
Aspect, A., P. Grangier, and G. Rogers: 1982, *Phys. Rev. Lett.* **49**, 1804.
Atiyah, M., N. J. Hitchin, and I. M. Singer: 1978, *Proc. Roy. Soc. London A* **362**, 425.
Atiyah, M.: 1990, *The Geometry and Physics of Knots*, Cambridge University Press, Cambridge.
Avis, S. J., and C. J. Isham: 1980, *Commun. Math. Phys.* **72**, 103.
Ayer, A. J.: 1946, *Language, Truth and Logic*, Dover, New York.
Ayer, A. J. (ed.): 1966, *Logical Positivism*, Free Press, Glencoe, Illinois.
Azizov, T. Ya., and I. S. Iokhvidov: 1989, *Linear Operators in Spaces With an Indefinite Metric*, Wiley, New York.
Azizov, T. Ya., and S. S. Khoruzhii: 1990, *Theor. Math. Phys.* **80**, 671.
Babelon, O., and C. M. Viallet: 1981, *Commun. Math. Phys.* **81**, 515.
Bacry, H.: 1988, *Localizability and Space in Quantum Physics*, Springer Lecture Notes in Physics, vol. 308, Berlin.
Badurek, G., H. Rauch, and D. Tuppinger: 1986, *Phys. Rev. A* **34**, 2600.
Balescu, R.: 1975, *Equilibrium and Non-Equilibrium Statistical Mechanics*, Wiley, New York.
Ballentine, L. E.: 1970, *Rev. Mod. Phys.* **42**, 358.
Ballentine, L. E.: 1973, *Found. Phys.* **3**, 1973.
Ballentine, L. E.: 1990, *Found. Phys.* **20**, 1329.
Band, W., and J. L. Park: 1978, *Found. Phys.* **8**, 677.
Band, W., and J. L. Park: 1979, *Found. Phys.* **9**, 937.
Banks, T.: 1985, *Nucl. Phys. B* **249**, 332.
Banks, T.: 1988, *Nucl. Phys. B* **309**, 493.
Barber, D. P., *et al.*: 1979, *Phys. Rev. Lett.* **43**, 1915.
Barbour, J. B., and L. Smolin: 1988, *Can Quantum Mechanics be Sensibly Applied to the Universe as a Whole?*, Yale report (unpublished).
Bargmann, V.: 1947, *Ann. Math.* **48**, 568.
Bargmann, V.: 1954, *Ann. Math.* **59**, 1.
Bargmann, V.: 1961, *Commun. Pure Appl. Math.* **14**, 187.
Bargmann, V., and E. P. Wigner: 1948, *Proc. Nat. Acad. Sci.* **34**, 211.
Barnes, T., and D. Kotchan: 1987, *Phys. Rev. D* **35**, 1947.
Barrow, J. D., and F. J. Tipler: 1986, *The Anthropic Cosmological Principle*, Clarendon Press, Oxford.
Barut, A. O., and L. Girardello: 1971, *Commun. Math. Phys.* **21**, 41.
[BR] Barut, A. O., and R. Raczka: 1986, *Theory of Group Representations and Applications*, 2nd revised edition, World Scientific, Singapore.
Baulieu, L.: 1984, *Nucl. Phys. B* **241**, 557.
Baulieu, L.: 1985, *Phys. Reports* **129**, 1.
Baulieu, L.: 1986a, *Nucl. Phys. B* **270**, 507.
Baulieu, L.: 1986b, in *Quantum Mechanics of Fundamental Systems 1*, C. Teitelboim (ed.), Plenum, New York.
Baulieu, L.: 1989, *Phys. Lett. B* **232**, 473, 479.

Baulieu, L., A. Bilal, and M. Picco: 1990, *Nucl. Phys. B***346**, 507.

Baulieu, L., and B. Grossman: 1988, *Phys. Lett. B* **212**, 351; **214**, 223.

Baulieu, L., B. Grossman, and R. Stora: 1986, *Phys. Lett. B* **180**, 95.

Baulieu, L., and I. M. Singer: 1988, *Nucl. Phys. B (Proc. Suppl.)* **5**, 12.

Baulieu, L., and I. M. Singer: 1991, *Commun. Math. Phys.* **135**, 253.

Baulieu, L., and J. Thierry-Mieg: 1982, *Nucl. Phys. B***197**, 477.

Baulieu, L., and J. Thierry-Mieg: 1984, *Phys. Lett. B* **145**, 53.

Becchi, C., A. Rouet, and R. Stora: 1976, *Ann. Phys. (N.Y.)* **98**, 287.

Bell, J. S.: 1975, *Epist. Letters.*

Bell, J. S.: 1987a, in *Quantum Implications*, B.J. Hiley and F. D. Peat (eds.), Routledge and Kegan Paul, London.

Bell, J. S.: 1987b, *Speakable and Unspeakable in Quantum Mechanics*, Cambridge University Press, Cambridge.

Bell, J. S.: 1990, *Phys. World (August)*, 33.

Bell, J. S., and J. M. Leinaas: 1983, *Nucl. Phys. B* **12**, 131.

Belopolskaya, Ya. I., and Yu. L. Dalecky: 1990, *Stochastic Equations and Differential Geometry*, Kluwer, Dordrecht.

Benn, I. M., and R. W. Tucker: 1987, *An Introduction to Spinors and Geometry with Applications in Physics*, Adam Hilger, Bristol.

Berestetskii, V. B., E. M. Lifshitz, and L. P. Pitaevskii: 1982, *Quantum Electrodynamics*, 2nd edition, Pergamon Press, Oxford.

Berezanskii, Yu. M.: 1968, *Expansions in Eigenfunctions of Self-Adjoint Operators*, American Mathematical Society, Providence, Rhode Island.

Berezanskii, Yu. M.: 1978, *Self-Adjoint Operators in Spaces of Functions of Infinitely Many Variables*, in Russian, Naukova Dumkas, Kiev.

[B] Berezin, F. A.: 1966, *The Method of Second Quantization*, Academic Press, New York.

Berezin, F. A.: 1975, *Commun. Math. Phys.* **40**, 153.

[BI] Berezin, F. A.: 1987, *Introduction to Superanalysis*, Reidel, Dordrecht.

Bergmann, G.: 1960, *Meaning and Existence*, University of Wisconsin Press, Madison, Wisconsin.

Bergmann, P. G., and A. Komar: 1980, in *General Relativity and Gravitation*, vol. 1, A. Held (ed.), Plenum, New York.

Bernstein, J.: 1991, *Quantum Profiles*, Princeton University Press, Princeton.

Berry, M. V.: 1984, *Proc. Roy. Soc. London A* **392**, 45.

Bertrand, J.: 1971, *Nuovo Cimento A* **1**, 1.

Bertrand, J.: 1978, *Rep. Math. Phys.* **14**, 101.

Billingsley, P.: 1979, *Probability and Measure*, Wiley, New York.

[BD] Birrell, N. D., and P. C. W. Davies: 1982, 1986, *Quantum Fields in Curved Space*, Cambridge University Press, Cambridge.

Bjorken, J. D., and S. D. Drell: 1964, *Relativistic Quantum Mechanics*, McGraw-Hill, New York.

Blaizot, J. P., and H. Orland: 1980, *J. Phys. Lett.* **41**, 523.

Blaizot, J. P., and H. Orland: 1981, *Phys. Rev. C***24**, 1740.

[BG] Bleecker, D.: 1981, *Gauge Theory and Variational Principles*, Addison-Wesley, Reading, Mass.

Bleuler, K.: 1950, *Helv. Phys. Acta* **23**, 567.

Bleuler, K.: 1991, in *Geometry and Theoretical Physics*, J. Debrus and A. C. Hirshfeld (eds.), Springer, Berlin.

Bloch, F., and A. Nordsieck: 1937, *Phys. Rev.* **52**, 54.

Blokhintsev, D. I.: 1968, *The Philosophy of Quantum Mechanics*, Reidel, Dordrecht.

Blokhintsev, D. I.: 1973, *Space and Time in the Microworld*, Reidel, Dordrecht.

Bognár, J.: 1974, *Indefinite Inner Product Spaces*, Springer, Berlin.

Bogolubov, N. N., and D. V. Shirkov: 1959, *Introduction to the Theory of Quantized Fields*, Interscience, New York.

[BL] Bogolubov, N. N., A. A. Logunov, and I. T. Todorov: 1975, *Introduction to Axiomatic Quantum Field Theory*, Benjamin, Reading, Mass.

Bogolubov, N. N., A. A. Logunov, A. I. Oksak and I. T. Todorov: 1990, *General Principles of Quantum Field Theory*, Kluwer, Dordrecht.

Bohm, A.: 1986, *Quantum Mechanics: Foundations and Applications*, 2nd edition, Springer, New York.

Bohm, A., *et al.*: 1988, *Int. J. Mod. Phys.* **3**, 1103.

Bohm, D.: 1951, *Quantum Theory*, Prentice-Hall, Englewood Cliffs.

Bohr, N.: 1934, *Atomic Theory and the Description of Nature*, Cambridge University Press, Cambridge.

Bohr, N.: 1949, in *Albert Einstein: Philosopher Scientist*, P. A. Schilpp (ed.), Harper & Row, New York.

Bohr, N.: 1955, *The Unity of Knowledge*, Doubleday, New York.

Bohr, N.: 1961, *Atomic Physics and Human Knowledge*, Science Editions, New York.

Bohr, N.: 1963, *Essays 1958/1962 on Atomic Physics and Human Knowledge*, Wiley, New York.

Bohr, N.: 1985, in *Niels Bohr: A Centenary Volume*, A. P. French and P. J. Kennedy, Harvard University Press, Cambridge, Mass.

Bohr, N., and L. Rosenfeld: 1933, *Mat.-fys. Medd. Dan. Vid. Selsk.*, **12**, No. 8; translation reprinted in [WQ], pp. 479-522.

Bohr, N., and L. Rosenfeld: 1950, *Phys. Rev.* **78**, 794; reprinted in [WQ], pp. 523-534.

Bombelli, L., L. Lee, D. Meyer, and R. D. Sorkin: 1987, *Phys. Rev. Lett.* **59**, 521; **60**, 656.

Bonora, L., P. Pasti, and M. Tonin: 1981a, *Nuovo Cimento A* **63**, 353.

Bonora, L., P. Pasti, and M. Tonin: 1981b, *Nuovo Cimento A* **64**, 307, 378.

Bonora, L., P. Pasti, and M. Tonin: 1982a, *J. Math. Phys.* **23**, 839.

Bonora, L., P. Pasti, and M. Tonin: 1982b, *Ann. Phys. (N.Y.)* **144**, 15.

Bonora, L., and P. Cotta-Ramusino: 1983, *Commun. Math. Phys.* **87**, 589.

Booss, B., and D. D. Bleecker: 1985, *Topology and Analysis: The Atiyah-Singer Index Formula and Gauge-Theoretic Physics*, Springer, New York.

Born, M.: 1938, *Proc. Roy. Soc. London A* **165**, 291.

Born, M.: 1949, *Rev. Mod. Phys.* **21**, 463.

Born, M.: 1955, *Dan. Mat. Fys. Medd.* **30**, No. 2.

Born, M.: 1956, *Physics in My Generation*, Pergamon Press, London.

Born, M.: 1971, *The Born-Einstein Letters*, MacMillan, London.

Born, M., and P. Jordan: 1925, *Z. Phys.* **34**, 858.

Boulware, D. G., and S. Deser: 1975, , *Ann. Phys. (N.Y.)* **89**, 193.

Bracci, L., and F. Strocchi: 1972, *J. Math. Phys.* **13**, 1151.

Bracci, L., and F. Strocchi: 1975, *J. Math. Phys.* **16**, 2522.

Bracci, L., G. Morchio, and F. Strocchi: 1975, *Commun. Math. Phys.* **41**, 289.

Breit, G.: 1928, *Proc. Nat. Acad. Sci. U.S.A.* **14**, 553.

Brink, L., and M. Henneaux: 1988, *Principles of String Theory*, Plenum, New York.

Brooke, J. A., and W. Guz: 1984, *Nuovo Cimento A* **78**, 221.

Brooke, J. A., and E. Prugovečki: 1984, *Nuovo Cimento A* **79**, 237.

Brooke, J. A., and E. Prugovečki: 1985, *Nuovo Cimento A* **89**, 126.

Brooke, J. A., and F. E. Schroeck, Jr.: 1989, *Nucl. Phys. B (Proc. Suppl.)* **6**, 104.

Bruzzo, U., and R. Cianci: 1984, *Class. Quantum Grav.* **1**, 213.

Bunge, M.: 1967, *Rev. Mod. Phys.* **39**, 463.

Bunge, M.: 1970, *Causality*, World Publishing, Cleveland, Ohio.
Burde, G., and H. Zieschang: 1985, *Knots*, Walter der Gruyter, Berlin.
Busch, P.: 1985a, *Int. J. Theor. Phys.* **24**, 63.
Busch, P.: 1985b, *J. Phys. A: Math. Gen.* **18**, 3351.
Busch, P.: 1986, *Phys. Rev. D* **33**, 2253.
Busch, P.: 1987, *Found. Phys.* **17**, 905.
Busch, P.: 1988, *Phys. Lett. A* **130**, 323.
Busch, P.: 1990, *Found. Phys.* **20**, 1, 33.
Busch, P., G. Cassinelli, and P. J. Lahti: 1990, *Found. Phys.* **20**, 757.
Busch, P., M. Grabowski, and P. J. Lahti: 1989, *Found. Phys. Lett.* **2**, 331.
Busch, P., and P. J. Lahti: 1990, *Ann. Physik (Leipzig)* **47**, 369.
Busch, P., P. J. Lahti, and P. Mittelstaedt: 1991, *Quantum Theory of Measurement*, Springer Lecture Notes in Physics, Berlin.
Busch, P., T. P. Schonbeck, and F. E. Schroeck, Jr.: 1987, *J. Math. Phys.* **28**, 2866.
Busch, P., and F. E. Schroeck, Jr.: 1989, *Found. Phys.* **19**, 807.
Carmeli, M.: 1977, *Group Theory and General Relativity*, McGraw-Hill, New York.
Carmeli, M.: 1982, *Classical Fields: General Relativity and Gauge Theory*, Wiley, New York.
Cartan, E.: 1913, *Bull. Soc. Math. France* **41**, 53.
Cartan, E.: 1923, *Ann. Ec. Norm. Sup.* **40**, 325.
Cartan, E.: 1924, *Ann. Ec. Norm. Sup.* **41**, 1.
Cartan, E.: 1925, *Ann. Ec. Norm. Sup.* **42**, 17.
Cartan, E.: 1935, *La méthode du Repère Mobile, la Théorie des Groupes Continus et les Espaces Généralisées*, Act. Sci. et Ind., #194, Hermann, Paris.
Cartan, E.: 1966, *The Theory of Spinors*, MIT Press, Cambridge, Mass.
Cartwright, N.: 1983, *How the Laws of Physics Lie*, Oxford University Press, Oxford.
Cassinelli, G., G. Olivieri, P. Truini, and V. S. Varadarajan: 1989, *J. Math. Phys.* **30**, 2692.
Cassinelli, G., P. Truini, and V. S. Varadarajan: 1991, *J. Math. Phys.* **32**, 1076.
Castagnino, M.: 1971, *J. Math. Phys.* **12** , 2203.
Cattani, C., and M. De Maria: 1989, in *Einstein and the History of General Relativity*, D. Howard and J. Stachel (eds.), Birkhäuser, Boston.
Cheng, K. S. : 1972, *J. Math. Phys.* **13** , 1723.
Chew, G. F.: 1963, *Sci. Prog.* **51**, 529.
Chew, G. F., and H. P. Stapp: 1988, *Found. Phys.* **18**, 809.
Choquet-Bruhat, Y.: 1968, in *Battelle Rencontres*, C. DeWitt and J. A. Wheeler (eds.), Benjamin, New York.
Choquet-Bruhat, Y., and J. W. York: 1980, in *General Relativity and Gravitation*, vol. 1, A. Held (ed.), Plenum, New York.
[C] Choquet-Bruhat, Y., C. DeWitt-Morette, and M. Dillard-Bleick: 1987, *Analysis, Manifolds and Physics*, revised edition, North-Holland, Amsterdam.
Choquet-Bruhat, Y., and C. DeWitt-Morette: 1989, *Analysis, Manifolds and Physics*, vol. II, North-Holland, Amsterdam.
Christodoulakis, T., and J. Zanelli: 1987, *Class. Quantum Grav.* **4**, 851.
Cini, M.: 1980, *Fundamenta Scientiae* **1**, 157.
Cini, M., and M. Serva: 1990, *Found. Phys. Lett.* **3**, 129.
Close, F. E.: 1969, *An Introduction to Quarks and Partons*, Academic Press, New York.
Cohen, L.: 1966, *Phil. Sci.* **33**, 317.
Coleman, S.: 1988, *Nucl. Phys. B* **310**, 643.

Coleman, S., J. Hartle, T. Piran, and S. Weinberg (eds.): 1991, *Quantum Cosmology and Baby Universes*, World Scientific, Singapore.
Cornwell, J. F.: 1984, *Group Theory in Physics*, vol. 1, Academic Press, New York.
Creutz, M.: 1983, *Quarks, Gluons and Lattices*, Cambridge University Press, Cambridge.
Curci, G., and R. Ferrari: 1976a, *Phys. Lett. B* **63**, 91.
Curci, G., and R. Ferrari: 1976b, *Nuovo Cimento A* **32**, 151.
Curci, G., and R. Ferrari: 1976c, *Nuovo Cimento A* **35**, 273.
Currie, D. G., T. F. Jordan, and E. C. G. Sudarshan: 1963, *Rev. Mod. Phys.* **35**, 350.
Cushing, J. T.: 1985, *Stud. Hist. Phil. Sci.* **16**, 31.
Cushing, J. T.: 1990, *Theory Construction and Selection in Modern Physics*, Cambridge University Press, Cambridge.
Daletskii, Yu. L.: 1983, *Russian Math. Surveys* **38**, 97.
Damgaard, P. H., and H. Hüffel: 1987, *Phys. Rep.* **152**, 227.
Daniel, M., and C. M. Viallet, *Rev. Mod. Phys.* **52**, 175.
Das, A.: 1960, *Nuovo Cimento* **18**, 482.
Davies, E. B.: 1976, *Quantum Theory of Open Systems*, Academic Press, New York.
Davies, E. B., and J. T. Lewis: 1969, *Commun. Math. Phys.* **17**, 239.
Davies, P. C. W.: 1975, *J. Phys. A: Gen. Phys.* **8**, 609.
Davies, P. C. W.: 1980, in *General Relativity and Gravitation*, vol. 1, A. Held (ed.), Plenum, New York.
Davies, P. C. W.: 1984, in *Quantum Theory of Gravity* , S.M. Christensen (ed.), Adam Hilger, Bristol.
Davies, P. C. W.: 1990, in *Complexity, Entropy, and the Physics of Information*, W. H. Zurek (ed.), Addison-Wesley, Reading, Mass.
Davies, P. C. W., and J. Brown (eds.): 1989, *Superstrings: A Theory of Everything?* Cambridge University Press, Cambridge.
Debever, R. (ed.): 1979, *Elie Cartan - Albert Einstein Letters on Absolute Parallelism 1929-1932*, Princeton University Press, Princeton.
De Bièvre, S.: 1989a, *Class. Quantum Grav.* **6**, 731.
De Bièvre, S.: 1989b, *J. Math. Phys.* **30**, 1401.
de Broglie, L.: 1923, *Comp. Rend.* **177**, 506, 548, 630.
de Broglie, L.: 1979, in *Perspectives in Quantum Theory*, W. Yougrau and A. van der Merwe (eds.), Dover, New York.
de Broglie, L.: 1990, *Heisenberg's Uncertainties and the Probabilistic Interpretation of Wave Mechanics*, Kluwer, Dordrecht.
Delbourgo, R., and M. R. Medrano: 1976, *Nucl. Phys. B* **110**, 467.
Delbourgo, R., and P. D. Jarvis: 1982, *J. Phys. A: Math. Gen.* **15**, 611.
Delbourgo, R., P. D. Jarvis, and G. Thompson: 1982a, *Phys. Lett. B* **109**, 25.
Delbourgo, R., P. D. Jarvis, and G. Thompson: 1982b, *Phys. Rev. D* **26**, 775.
Deser, S.: 1970, *Gen. Rel. Grav.* **1**, 9.
Deser, S.: 1975, in *Quantum Gravity: An Oxford Symposium*, C.J. Isham, R. Penrose and D.W. Sciama (eds.) Clarendon Press, Oxford.
d'Espagnat, B.: 1976, *Conceptual Foundations of Quantum Mechanics*, 2nd edition, Benjamin, Reading, Mass.
d'Espagnat, B.: 1989, *Reality and the Physicist*, Cambridge University Press, Cambridge.
Deutsch, D.: 1986, in *Quantum Concepts in Space and Time*, R. Penrose and C. J. Isham (eds.), Clarendon Press, Oxford.
DeWitt, B. S.: 1957, *Rev. Mod. Phys.* **29**, 377.
DeWitt, B. S.: 1962, in *Gravitation: An Introduction to Current Research*, L. Witten (ed.), Wiley, New York.
DeWitt, B. S.: 1967, *Phys. Rev.* **160**, 1113; **162**, 1195, 1239.

DeWitt, B. S.: 1975, *Phys. Reports* **19**, 295.

DeWitt, B. S.: 1979, in *General Relativity: An Einstein Centenary Survey*, S. W. Hawking and W. Israel (eds.), Cambridge University Press, Cambridge.

[D] DeWitt, B. S.: 1984a, *Supermanifolds* , Cambridge University Press, Cambridge.

DeWitt, B. S.: 1984b, in *General Relativity and Gravitation*, B. Bertotti, F. de Felice and A. Pascolini (eds.), Reidel, Dordrecht.

DeWitt, B. S., and N. Graham (eds.): 1973, *The Many-Worlds Interpretation of Quantum Mechanics*, Princeton University Press, Princeton.

DeWitt-Morette, C., A. Maheshwari, and B. Nelson: 1979, *Phys. Reports* **50**, 255.

Dicke, R. H.: 1964, *The Theoretical Significance of Experimental Relativity*, Gordon and Breach, New York.

Dimock, J.: 1980, *Commun. Math. Phys.* **77**, 219.

Dirac, P. A. M.: 1927, *Proc. Roy. Soc. London A* **114**, 243.

Dirac, P. A. M.: 1928, *Proc. Roy. Soc. London A* **117**, 610.

Dirac, P. A. M.: 1930, *The Principles of Quantum Mechanics*, Clarendon Press, London.

Dirac, P. A. M.: 1933, *Phys. Zeit. Sowjetunion* **3**, 64.

Dirac, P. A. M.: 1936, *Nature* **137**, 298.

Dirac, P. A. M.: 1942, *Proc. Roy. Soc. London A* **180**, 1.

Dirac, P. A. M.: 1945, *Rev. Mod. Phys.* **17**, 195.

Dirac, P. A. M.: 1946, *Commun. Dublin Inst. Adv. Studies A* **3**, 1.

Dirac, P. A. M.: 1947, *The Principles of Quantum Mechanics*, 3rd edition, Clarendon Press, London.

Dirac, P. A. M.: 1948, *Phys. Rev.* **73**, 1092.

Dirac, P. A. M.: 1949, *Rev. Mod. Phys.* **21**, 392.

Dirac, P. A. M.: 1950, *Can J. Math.* **2**, 129.

Dirac, P. A. M.: 1951, *Proc. Roy. Soc. London A* **209**, 251.

Dirac, P. A. M.: 1958, *Proc. Roy. Soc. London A* **246**, 326, 333.

Dirac, P. A. M.: 1959, *Phys. Rev.* **114**, 924.

Dirac, P. A. M.: 1962, *Proc. Roy. Soc. London A* **268**, 57.

Dirac, P. A. M.: 1963, *Sci. Am.* **208**(5), 45.

Dirac, P. A. M.: 1964, *Lectures on Quantum Mechanics*, Belfer Graduate School of Science, New York.

Dirac, P. A. M.: 1965, *Phys. Rev.* **139B**, 684.

Dirac, P. A. M.: 1968, in *Contemporary Physics*, vol. 1, Trieste Symposium (Vienna: IAEA).

Dirac, P. A. M.: 1973, in *The Past Decade in Particle Theory*, E. C. G. Sudarshan and Y. Ne'eman (eds.), Gordon and Breach, London.

Dirac, P. A. M.: 1977, in *Deeper Pathways in High-Energy Physics*, A. Perlmutter and L. F. Scott (eds.), Plenum, New York.

Dirac, P. A. M.: 1978a, in *Directions in Physics*, H. Hora and J. R. Shepanski (eds.), Wiley, New York.

Dirac, P. A. M.: 1978b, in *Mathematical Foundations of Quantum Mechanics*, A. R. Marlow (ed.), Academic Press, New York.

Dirac, P. A. M.: 1979, *Sov. Phys. Uspeki* **22**, 648.

Dirac, P. A. M.: 1987, in *Reminiscences About a Great Physicist: Paul Adrien Maurice Dirac* , B. N. Kursunoglu and E. P. Wigner (eds.), Cambridge University Press, Cambridge.

Dohrn, D., and F. Guerra: 1978, *Lett. Nuovo Cimento* **22**, 121.

Dombrowski, H. D., and K. Horneffer: 1964, *Math. Z.* **86**, 291.

Drechsler, W.: 1975, *Fortschr. Phys.* **23**, 607.

Drechsler, W.: 1977a, *Found. Phys.* **7**, 629.

Drechsler, W.: 1977b, *J. Math. Phys.* **18**, 1358.

Drechsler, W.: 1981, *Phys. Lett. B* **107**, 415.

Drechsler, W.: 1982, *Ann. Inst. H. Poincaré* **37**, 155.

Drechsler, W.: 1983, *Gen. Rel. Grav.* **15**, 703.

Drechsler, W.: 1984, *Fortschr. Phys.* **23**, 449.

Drechsler, W.: 1985, *J. Math. Phys.* **26**, 41.

Drechsler, W.: 1988, *Z. Phys. C* **41**, 197.

Drechsler, W.: 1989a, *Class. Quantum Grav.* **6**, 623.

Drechsler, W.: 1989b, *Found. Phys.* **19**, 1479.

Drechsler, W.: 1990, *Fortschr. Phys.* **38**, 63.

Drechsler, W.: 1991, "Quantized Fibre Dynamics for Extended Elementary Objects Involving Gravitation" *Found. Phys.* **22** (to appear).

Drechsler, W., and M. E. Mayer: 1977, *Fiber Bundle Techniques in Gauge Theories*, Springer Lecture Notes in Physics, vol. 67, Heidelberg.

Drechsler, W., and E. Prugovečki: 1991, *Found. Phys.* **21**, 513.

Drechsler, W., and R. Sasaki: 1978, *Nuovo Cimento A* **46**, 527.

Drechsler, W., and W. Thacker: 1987, *Class. Quantum Grav.* **4**, 291.

Duff, M. J.: 1975, in *Quantum Gravity* , C. J. Isham, R. Penrose and D.W. Sciama (eds.), Clarendon Press, Oxford.

Duff, M. J.: 1981, in *Quantum Gravity 2*, C. J. Isham, R. Penrose and D.W. Sciama (eds.), Clarendon Press, Oxford.

Dürr, H. P., and E. Rudolph: 1969, *Nuovo Cimento A* **62**, 411.

Duval C., and H. P. Künzle: 1977, *C. R. Acad. Sci., Paris* **285**, 813.

Duval C., and H. P. Künzle: 1984, *Gen. Rel. Grav.* **16**, 333.

Duval C., G. Burdet, H. P. Künzle, and M. Perrin: 1985, *Phys. Rev. D* **31**, 1841.

Dynkin, E. B.: 1968, *Soviet Math. Dokl.* **179**, 532.

Dyson, F. J.: 1949, *Phys. Rev.* **75**, 1736.

Dyson, F. J.: 1952, *Phys. Rev.* **85**, 631.

Dyson, F. J.: 1983, *Math. Intelligencer* **5**(3), 47.

Eberhard, P. H., and R. R. Ross: 1989, *Found. Phys. Lett.* **2**, 127.

Earman, J., and C. Glymour: 1980, *Hist. Studies Phys. Sci.* **11**, 49.

Efimov, G. V.: 1977, *Nonlocal Interactions of Quantized Fields*, Nauka, Moskow.

Ehlers, J.: 1971, in *General Relativity and Cosmology*, B.K. Sachs (ed.), Academic Press, New York.

Ehlers, J.: 1973, in *Relativity, Astrophysics and Cosmology*, W. Israel (ed.), Reidel, Dordrecht.

Ehlers, J.: 1987, in *General Relativity and Gravitation*, M.A.H. MacCallum (ed.), Cambridge University Press, Cambridge.

Ehlers, J.: 1990, in *General Relativity and Gravitation*, N. Ashby, D. F. Bartlett and W. Wyss (eds.), Cambridge University Press, Cambridge.

Ehlers, J., and A. Schild: 1973, *Commun. Math. Phys.* **32**, 119.

Ehlers, J., F.A.E. Pirani, and A. Schild: 1972, in *General Relativity*, L. O'Raifeartaigh (ed.), Clarendon Press, Oxford.

Einstein, A.: 1905, *Ann. Phys.* **17**, 891; translated in English by W. Perrett and G. B. Jeffery as "On the Electrodynamics of Moving Bodies", pp. 37-65 in *The Principle of Relativity* (Methuen, London, 1923; reprinted by Dover, New York, 1952).

Einstein, A.: 1907, *J. d. Radioaktivität.* **4**, 411.

Einstein, A.: 1916, *Ann. Phys.* **49**, 769; translated in English by W. Perrett and G. B. Jeffery as "The Foundation of the General Theory of Relativity", pp. 109-164 in *The Principle of Relativity* (Methuen, London, 1923; reprinted by Dover, New York, 1952).

Einstein, A.: 1918, *Ann. Physik* **55**, 241.
Einstein, A.: 1949, in *Albert Einstein: Philosopher Scientist*, P. A. Schilpp (ed.), Harper & Row, New York.
Einstein, A.: 1961, *Relativity: the Special and the General Theory*, 15th edition, Crown Publishers – Bonanza Books, New York.
Einstein, A., and J. Grommer: 1927, *Sitzber. Deut. Akad. Wiss. Berlin, Kl. Math. Phys. Tech.*, 2-13, 235-245.
Einstein, A., and M. Grossmann: 1913, *Entwurf einer verallgemeinerten Relativitätstheorie und einer Theorie der Gravitation*, Teubner, Leipzig. Reprinted in *Z. Math. u. Phys.* **62**, 225 (1914).
Einstein, A., and L. Infeld: 1938, *The Evolution of Physics*, Simon and Schuster, New York.
Einstein, A., B. Podolsky, and N. Rosen: 1935, *Phys. Rev.* **47**, 777.
Ellis, G. F. R.: 1989, in *Einstein and the History of General Relativity*, D. Howard and J. Stachel (eds.), Birkhäuser, Boston.
Emch, G. G.: 1972, *Algebraic Methods in Statistical Mechanics and Quantum Field Theory*, Wiley, New York.
Englert, F., and R. Brout: 1964, *Phys. Rev. Lett.* **13**, 321.
Evens, D., J. W. Moffat, G. Kleppe, and R. P. Woodward: 1991, *Phys. Rev. D* **43**, 499.
Everett, H.: 1957, *Rev. Mod. Phys.* **29**, 454.
Exner, P.: 1985, Open Quantum Systems and Feynman Integrals, Reidel, Dordrecht.
Faddeev, L. D., and V. N. Popov: 1967, *Phys. Lett. B* **25**, 29.
Faddeev, L. D., and A. A. Slavnov: 1980, *Gauge Fields*, Benjamin, Reading, Mass.
Falck, N. K., and A. C. Hirshfeld: 1984, *Phys. Lett. B* **138**, 52.
Fanchi, J. R.: 1988, *Phys. Rev. A* **37**, 3956.
Fanchi, J. R.: 1990, *Found. Phys.* **20**, 189.
Feigl, H., and M. Brodbeck (eds.): 1953, *Readings on the Philosophy of Science*, Appleton-Century-Crofts, New York.
Fell, J. M. G., and R. S. Doran: 1988, *Representations of *-Algebras, Locally Compact Groups, and Banach *-Algebraic Bundles*, vol. 1, Academic Press, Boston.
Fermi, E.: 1932, *Rev. Mod. Phys.* **4**, 87.
Ferrari, R., L. E. Picasso, and F. Strocchi: 1974, *Commun. Math. Phys.* **35**, 25.
Feyerabend, P.: 1975, *Against Method*, Humanities Press, London.
Feynman, R. P.: 1948, *Rev. Mod. Phys.* **20**, 367.
Feynman, R. P.: 1950, *Phys. Rev.* **80**, 440.
Feynman, R. P.: 1951, *Phys. Rev.* **84**, 108.
Feynman, R. P.: 1954, *Ann. da Acad. Brasileira de Ciencias* **26**, 51.
Feynman, R. P.: 1963, *Acta Phys. Polon.* **24**, 697.
Feynman, R. P.: 1985, *QED: The Strange Story of Light and Matter*, Princeton University Press, Princeton.
Feynman, R. P., and A. R. Hibbs: 1965, *Quantum Mechanics and Paths Integrals*, McGraw-Hill, New York.
Feynman, R. P.: 1989, in *Superstrings: A Theory of Everything?* P. C. W. Davies and J. Brown (eds.) Cambridge University Press, Cambridge.
Feynman, R. P., M. Kislinger, and F. Ravndal: 1971, *Phys. Rev. D* **3**, 2706.
Fine, A.: 1970, *Phys. Rev. D* **2**, 2783.
Finkelstein, D.: 1989, *Int. J. Theor. Phys.* **28**, 441, 1081.
Fisher, A. E., and J. E. Marsden: 1979, in *General Relativity: An Einstein Centenary Survey*, S. W. Hawking and W. Israel (eds.), Cambridge University Press, Cambridge.

Fischler, W., Klebanov, I., J. Polchinski, and L. Susskind: 1989, *Nucl. Phys. B* **327**, 157.
Fleming, G. N.: 1965, *Phys. Rev. B* **139**, 963.
Flint, H. T.: 1937, *Proc. Roy. Soc. London A* **159**, 45.
Flint, H. T.: 1948, *Phys. Rev.* **74**, 209.
Flint, H. T., and O. W. Richardson: 1928, *Proc. Roy. Soc. London A* **117**, 637.
Fock, V. A.: 1937, *Phys. Z. Sowjetunion* **12**, 404.
Foldy, L. L., and S. A. Wouthuysen: 1950, *Phys. Rev.* **78**, 29.
Folse, H. J.: 1985, *The Philosophy of Niels Bohr*, North-Holland, Amsterdam.
Frank, P.: 1909, *Sitzungsber. Kgl. Akad. Wiss. Wien, Math.-Naturw. Kl.* **118**(IIa), 373.
Franklin, A.: 1986, *The Neglect of Experiment*, Cambridge University Press, Cambridge.
Franklin, A.: 1990, *Experiment, Right or Wrong*, Cambridge University Press, Cambridge.
Freed, D. S.: 1985, in *Infinite Dimensional Groups with Applications*, V. Kac (ed.) Springer, New York.
Freed, D. S.: 1988, *J. Diff. Geom.* **28**, 223.
Friedman, J. L.: 1991, in *Conceptual Problems in Quantum Gravity*, A. Ashtekar and J. Stachel (eds.), Birkhäuser, Boston.
Friedman, J. L., M. S. Morris, I. D. Novikov, F. Echeverria, G. Klinkhammer, K. S. Thorne, and U. Yurtsever: 1990, *Phys. Rev. D.* **42**, 1915.
Friedman, M.: 1983, *Foundations of Space-Time Theories*, Princeton University Press, Princeton.
Fröhlich, J.: 1991, *Non-Perturbative Quantum Field Theory*, World Scientific, Singapore.
Fröhlich, J., G. Morchio, and F. Strocchi: 1979, *Ann. Phys. (N.Y.)* **119**, 241.
Fujimura, K., T. Kobayashi, and M. Namiki: 1970, *Prog. Theor. Phys.* **43**, 73.
Fujimura, K., T. Kobayashi, and M. Namiki: 1971, *Prog. Theor. Phys.* **44**, 193.
Fulling, S. A.: 1973, *Phys. Rev. D* **7**, 2850.
Fulling, S. A.: 1989, *Aspects of Quantum Field Theory in Curved Space-Time*, Cambridge University Press, Cambridge.
Fürth, R.: 1929, *Z. Phys.* **57**, 429.
Gagnon, R.: 1985, *J. Math. Phys.* **26**, 440.
Gale, W., E. Guth, and G. T. Trammell: 1968, *Phys. Rev.* **165**, 1434.
Galison, P.: 1987, *How Experiments End*, Chicago University Press, Chicago.
Gel'fand, I. M., and G. E. Shilov: 1964, *Generalized Functions*, vol. I, Academic Press, New York.
Gel'fand, I. M., and N. Vilenkin: 1968, *Generalized Functions*, vol. IV, Academic Press, New York.
Gel'fand, I. M., and A. M. Yaglom: 1960, *J. Math. Phys.* **1**, 48.
Gell-Mann, M.: 1979, in *The Nature of the Physical Universe: 1976 Nobel conference*, D. Huff and O. Prewett (eds.), Wiley, New York.
Gell-Mann, M., and J. B. Hartle: 1990a, in *Complexity, Entropy, and the Physics of Information*, W. H. Zurek (ed.), Addison-Wesley, Reading, Mass.
Gell-Mann, M., and J. B. Hartle: 1990b, in *Proceedings of the Third International Symposium on Foundations of Quantum Mechanics in the Light of New Technology*, H. Ezawa, S. Kobayashi, Y. Murayama and S. Nomura (eds.), Japan Physical Society, Tokyo.
Gerber, P.: 1898, *Z. Math. u. Phys. II* **43**, 93.
Gerlach, B., D. Gromes, and J. Petzold: 1967, *Z. Phys.* **202**, 401; **204**, 1.
Gerlach, B., D. Gromes, and J. Petzold: 1969, *Z. Phys.* **221**, 141.
Geroch, R. P.: 1968, *J. Math. Phys.* **9**, 1739.
Geroch, R. P., and J. B. Hartle: 1986, *Found. Phys.* **16**, 533.
Geroch, R. P., and G. T. Horowitz: 1979, in *General Relativity: An Einstein Centenary Survey*, S. W. Hawking and W. Israel (eds.), Cambridge University Press, Cambridge.
Ghirardi, G. C., R. Grassi, and P. Pearle: 1990a, *Found. Phys.* **20**, 1271.

Ghirardi, G. C., P. Pearle, and A. Rimini: 1990b, *Phys. Rev. A* **42**, 78.
Ghirardi, G. C., A. Rimini, and T. Weber: 1986, *Phys. Rev. D* **34**, 470.
Gibbons, G. W.: 1979, in *General Relativity: An Einstein Centenary Survey*, S. W. Hawking and W. Israel (eds.), Cambridge University Press, Cambridge.
Gibbons, G. W., S. W. Hawking, and M. J. Perry: 1978, *Nucl. Phys. B* **138**, 141.
Ginsparg, P.: 1989, *J. Phys. G: Nucl. Part. Phys.* **15**, 121.
Gitman, D. M., and I. V. Tyutin: 1990, *Quantization of Fields with Constraints*, Springer, Berlin.
Glaser, W., and K. Sitte: 1934, *Z. Phys.* **87**, 674.
Glashow, S. L.: 1980, *Rev. Mod. Phys.* **52**, 539.
Glauber, R. J.: 1963, *Phys. Rev.* **130**, 2529; **131**, 2766.
Glimm, J., and A. Jaffe: 1987, *Quantum Physics*, 2nd edition, Springer, New York.
Göckeler, M., and T. Schücker: 1987, *Differential Geometry, Gauge Theories, and Gravity*, Cambridge University Press, Cambridge.
Gödel, K.: 1931, *Monatshefte Math. Phys.* **38**, 173.
Goldberger, M. L.: 1961, in *La Theorie Quantique des Champs*, R. Stoops (ed.), Interscience, New York.
Goldberger, M. L., and K. M. Watson: 1964, *Collision Theory*, Wiley, New York.
Gol'fand, Yu. A.: 1963, *Sov. Phys. JETP* **16**, 184; **17**, 842.
Gomatam, J.: 1971, *Phys. Rev. D*3, 1292.
Goroff, M. H., and A. Sagnotti: 1986, *Nucl. Phys. B* **266**, 709.
Goto, T., and T. Imamura: 1955, *Prog. Theor. Phys.* **14**, 396.
Grabowski, M.: 1978, *Int. J. Theor. Phys.* **17**, 635.
Grabowski, M.: 1984, *Rep. Math. Phys.* **20**, 153.
Grabowski, M.: 1985, *Acta Phys. Polon. A* **67**, 503, 511.
Green, M. B.: 1989, in *Superstrings: A Theory of Everything?* P. C. W. Davies and J. Brown (eds.), Cambridge University Press, Cambridge.
Green, M. B., and J. H. Schwarz: 1984, *Phys. Lett. B* **149**, 117.
Green, M. B., J. H. Schwarz, and E. Witten: 1987, *Superstring Theory. 1*, Cambridge University Press, Cambridge.
Greenwood, D. P., and E. Prugovečki: 1984, *Found. Phys.* **14**, 883.
Greiner, W.: 1990, *Relativistic Quantum Mechanics*, Springer, Berlin.
Gribbin, J.: 1986, *In Search of the Big Bang*, Bantam Books, Toronto.
Gross, D. J., and V. Perival: 1988, *Phys. Rev. Lett.* **60**, 2105; **61**, 1517.
Gudder, S. P.: 1968, *J. Math. Mech.* **18**, 325.
Gudder, S. P.: 1979, *Stochastic Methods in Quantum Mechanics*, North-Holland, NewYork.
Gudder, S. P.: 1988a, *Quantum Probability*, Academic Press, New York.
Gudder, S. P.: 1988b, *Int. J. Theor. Phys.* **27**, 193.
Gudder, S. P.: 1989, *Found. Phys.* **19**, 293.
Guerra, F.: 1981, *Phys. Reports* **77**, 263.
Guerra, F., and P. Ruggiero: 1978, *Lett. Nuovo Cimento* **23**, 529.
Gupta, S. N.: 1950, *Proc. Phys. Soc. London A* **63**, 681.
Gupta, S. N.: 1952, *Proc. Phys. Soc. London A* **65**, 608.
Gupta, S. N.: 1959, *Prog. Theor. Phys.* **21**, 581.
Guth, A. H., and P. J. Steinhardt: 1984, *Sci. Am.* **250**(4), 116.
Gutzwiller, M. C.: 1990, *Chaos in Classical and Quantum Mechanics*, Springer, New York.
Haag, R., and D. Kastler: 1964, *J. Math. Phys.* **5**, 848.
Haag, R., H. Nahmhofer, and U. Stein: 1984, *Commun. Math. Phys.* **94**, 219.
Hacking, I.: 1983, *Representing and Intervening*, Cambridge University Press, Cambridge.

Hajra, K., and P. Bandyopadhyay: 1991, *Phys. Lett. A* **155**, 7.
Halliwell, J. J.: 1991, in *Quantum Cosmology and Baby Universes*, S. Coleman, J. Hartle, T. Piran, and S. Weinberg (eds.), World Scientific, Singapore.
Halliwell, J. J., and S. W. Hawking: 1985, *Phys. Rev. D* **31**, 1777.
Halzen, F., and A. D. Martin: 1984, *Quarks and Leptons*, Wiley, New York.
Hannay, J. H.: 1985, *J. Phys. A: Math. Gen.* **18**, 221.
Harary, H.: 1983, *Sci. Am.* **248**(4), 56.
Hardy, G. H.: 1949, *Divergent Series*, Oxford University Press, London.
Hartle, J. B.: 1988, *Phys. Rev D* **38**, 2985.
Hartle, J. B.: 1990, in *General Relativity and Gravitation*, N. Ashby, D. F. Bartlett and W. Wyss (eds.), Cambridge University Press, Cambridge.
Hartle, J. B.: 1991, in *Conceptual Problems in Quantum Gravity*, A. Ashtekar and J. Stachel (eds.), Birkhäuser, Boston.
Hartle, J. B., and S. W. Hawking: 1983, *Phys. Rev. D* **28**, 2960.
Hartle, J. B., and K. Kuchař: 1984, *J. Math. Phys.* **25**, 57.
Harvey, A.: 1976, *Gen. Rel. Grav.* **7**, 891.
Havas, P.: 1964, *Rev. Mod. Phys.* **36**, 938.
Havas, P.: 1989, in *Einstein and the History of General Relativity*, D. Howard and J. Stachel (eds.), Birkhäuser, Boston.
Hawking, S. W.: 1975a, *Commun. Math. Phys.* **43**, 199.
Hawking, S. W.: 1975b, in *Quantum Gravity*, C.J. Isham, R. Penrose and D.W. Sciama (eds.), Clarendon Press, Oxford.
Hawking, S. W.: 1977, *Sci. Am.* **236**, 34.
Hawking, S. W.: 1979, in *General Relativity: An Einstein Centenary Survey*, S. W. Hawking and W. Israel (eds.), Cambridge University Press, Cambridge.
Hawking, S. W.: 1980: *Is the End in Sight for Theoretical Physics?* 1979 Inaugural Lecture for the Lucasian Chair at the University of Cambridge, Cambridge University Press, Cambridge.
Hawking, S. W.: 1982, *Commun. Math. Phys.* **87**, 395.
Hawking, S. W.: 1984, *Nucl. Phys. B* **239**, 257.
Hawking, S. W.: 1985, *Phys. Rev. D* **32**, 2489.
Hawking, S. W.: 1987, *Phys. Lett. B* **195**, 337.
Hawking, S. W.: 1988, *Phys. Rev D* **37**, 904.
Hawking, S. W., and G.F.R. Ellis: 1973, *The Large Scale Structure of Space-Time*, Cambridge University Press, Cambridge.
Hawking, S. W., and I. G. Moss: 1982, *Phys. Lett. B* **110**, 35.
Hegerfeldt, G. C.: 1974, *Phys. Rev. D* **10**, 3320.
Hegerfeldt, G. C.: 1985, *Phys. Rev. Lett.* **54**, 2395.
Hegerfeldt, G. C.: 1989, *Nucl. Phys. B (Proc. Suppl.)* **6**, 231.
Hegerfeldt, G. C., K. Kraus, and E. P. Wigner: 1968, *J. Math. Phys.* **9**, 2029.
Hegerfeldt, G. C., and S. N. M. Ruijsenaars: 1980, *Phys. Rev. D* **22**, 377.
Hehl, F. W., P. von der Heyde, and G. D. Kerlick: 1976, *Rev. Mod. Phys.* **48**, 393.
Heisenberg, W.: 1927, *Z. Phys.* **43**, 172.
Heisenberg, W.: 1936, *Z. Phys.* **101**, 533.
Heisenberg, W.: 1938a, *Ann. Phys. (Leipzig)* **32**, 20.
Heisenberg, W.: 1938b, *Z. Phys.* **110**, 251.
Heisenberg, W.: 1943, *Z. Phys.* **120**, 513, 673.
Heisenberg, W.: 1958, *Physics and Philosophy*, Harper and Row, New York.

Heisenberg, W.: 1971, *Physics and Beyond*, Harper and Row, New York.

Heisenberg, W.: 1976, *Phys. Today* **29**(3), 32.

Helstrom, C.: 1976, *Quantum Detection and Estimation Theory*, Academic Press, New York.

Henneaux, M.: 1985, *Phys. Reports* **126**, 1.

Henneaux, M.: 1986, in *Quantum Mechanics of Fundamental Systems 1*, C. Teitelboim (ed.), Plenum, New York.

Higgs, P. W.: 1964, *Phys. Lett.* **12**, 132.

Hill, E. L.: 1950, *Phys. Rev.* **100**, 1780.

Hill, R. N.: 1967, *J. Math. Phys.* **8**, 1756.

Hille, E., and R. S. Phillips: 1968, *Functional Analysis and Semi-Groups*, revised edition, American Mathematical Society, Providence, Rhode Island.

Hofstadter, R.: 1956, *Rev. Mod. Phys.* **28**, 214.

Holevo, A. S.: 1972, *Probl. Peredachi Inform.* **8**, 63.

Holevo, A. S.: 1982, *Probabilistic and Statistical Aspects of Quantum Mechanics*, North Holland, Amsterdam.

Holevo, A. S.: 1984, in *Quantum Probability and Applications to the Theory of Irreversible Processes*, L. Accardi, A. Frigerio and V. Gorini (eds.), Springer Lecture Notes in Mathematics, vol. 1055, Berlin.

Holevo, A. S.: 1985, *Rep. Math. Phys.* **22**, 385.

Holevo, A. S.: 1987a, *Math. USSR Izvestiya*, **28**, 175.

Holevo, A. S.: 1987b, in *Advances in Statistical Signal Processing*, vol. 1, H. V. Poor (ed.), JAI Press.

Holton, G.: 1980, in *Some Strangeness in the Proportion*, H. Woolf (ed.), Addison-Wesley, Reading, Mass.

Horuzhy, S. S.: 1990, *Introduction to Algebraic Quantum Field Theory*, Kluwer, Dordrecht.

Horuzhy, S. S., and A. V. Voronin: 1989, *Commun. Math. Phys.* **123**, 677.

Horwitz, G. T.: 1990, in *General Relativity and Gravitation*, N. Ashby, D. F. Bartlett and W. Wyss (eds.), Cambridge University Press, Cambridge.

Horwitz, L. P.: 1984, *Found. Phys.* **10**, 1027.

Horwitz, L. P.: 1992, "On the Definition and Evolution of States in Relativistic Classical and Quantum Mechanics", *Found. Phys.* **22** (to appear).

Horwitz, L. P., R. I. Arshanski, and A. C. Elitzur: 1988, *Found. Phys.* **18**, 1159.

Horwitz, L. P., and C. Piron: 1973, *Helv. Phys. Acta* **46**, 316.

Horwitz, L. P., and M. Usher: 1991, *Found. Phys. Lett.* **4**, 289.

Howard, D., and J. Stachel (eds.): 1989, *Einstein and the History of General Relativity*, Birkhäuser, Boston.

Hoyos, J., M. Quirós, J. Ramirez Mittelbrunn, and F. J. de Urries: 1982, *J. Math. Phys.* **23**, 1504.

Hudson, R. L., and K.R. Parthasarathy: 1984a, in *Quantum Probability and Applications to the Theory of Irreversible Processes*, L. Accardi, A. Frigerio and V. Gorini (eds.), Springer Lecture Notes in Mathematics, vol. 1055, Berlin.

Hudson, R. L., and K.R. Parthasarathy: 1984b, *Commun. Math. Phys.* **93**, 301.

Hudson, R. L., and K.R. Parthasarathy: 1984c, *Acta Appl. Math.* **2**, 353.

Hudson, R. L., and K.R. Parthasarathy: 1986, *Commun. Math. Phys.* **104**, 457.

Hudson, R. L., and P. Robinson: 1988, *Lett. Math. Phys.* **15**, 47.

Hudson, R. L., and P. Robinson: 1989, in *Proceedings of the XIIth International Conference on Differential Geometry*, A. Solomon (ed.), World Scientific, Singapore.

Hurt, N. E.: 1983, *Geometric Quantization in Action*, Reidel, Dordrecht.

Inönü, E., and E. P. Wigner: 1952, *Nuovo Cimento* **9**, 705.

Inönü, E., and E. P. Wigner: 1953, *Proc. Nat. Acad. Sci. (USA)* **39**, 510.

Isenberg, J., and J. Nester: 1980, in *General Relativity and Gravitation*, A. Held (ed.), Plenum, New York.

Isham, C. J.: 1975, in *Quantum Gravity: An Oxford Symposium*, C.J. Isham, R. Penrose and D.W. Sciama (eds.) Clarendon Press, Oxford.

Isham, C. J.: 1978, in *Differential Geometrical Methods in Mathematical Physics*, K Bleuler, H. R. Petry and R. Reetz (eds.), Springer Lecture Notes in Mathematics, vol. 676, Berlin.

Isham, C. J.: 1981, in *Quantum Gravity 2*, C.J. Isham, R. Penrose and D.W. Sciama (eds.) Clarendon Press, Oxford.

Isham, C. J.: 1987, in *General Relativity and Gravitation*, M. A. H. MacCallum (ed.), Cambridge University Press, Cambridge.

[I] Isham, C. J.: 1989, *Modern Differential Geometry for Physicists*, World Scientific, Singapore.

Isham, C. J.: 1990, in *Proceedings of the Advanced Summer Institute on Physics, Geometry and Topology*, H. Lee (ed.), Plenum, New York.

Isham, C. J.: 1991, in *Conceptual Problems in Quantum Gravity*, A. Ashtekar and J. Stachel (eds.), Birkhäuser, Boston.

Isham, C. J., and A. C. Kakas: 1984, *Class. Quantum Grav.* **1**, 621, 633.

Isham, C. J., and K. V. Kuchař: 1985, *Ann. Phys. (N.Y.)* **164**, 288, 316.

Itô, K.: 1962, *Proc. Int. Congress Math. Stockholm*, pp. 536-539.

Itô, K.: 1975, in *Probabilistic Methods in Differential Equations*, M. A. Pinsky (ed.), Springer Lecture Notes in Mathematics, vol. 451, Berlin.

[IQ] Itzykson, C., and J.-B. Zuber: 1980, *Quantum Field Theory*, McGraw-Hill, New York.

Jaffe, A. M.: 1967, *Phys. Rev.* **158**, 1454.

James, W.: 1970, *The Meaning of Truth*, University of Michigan Press, Ann Arbor, Mich.

Jammer, M.: 1966, *The Conceptual Development of Quantum Mechanics*, McGraw-Hill, New York.

Jammer, M.: 1969, *Concepts of Space*, Harvard University Press, Cambridge, Mass.

Jammer, M.: 1974, *The Philosophy of Quantum Mechanics*, Wiley, New York.

Jammer, M.: 1979, in *Problems in the Foundations of Physics*, G. Toraldo de Francia (ed.), North-Holland, Amsterdam.

Jauch, J. M., and F. Rohrlich: 1976, *The Theory of Photons and Electrons*, 2nd expanded edition, Springer, New York.

Jaynes, E. T.: 1990, in *Complexity, Entropy, and the Physics of Information*, W. H. Zurek (ed.), Addison-Wesley, Reading, Mass.

Joachain, C. J.: 1975, *Quantum Collision Theory*, North-Holland, Amsterdam.

Kac, M.: 1959, *Probability and Related Topics in the Physical Sciences*, Interscience, New York.

Kac, M.: 1964, *Statistical Independence in Probability, Analysis and Number Theory*, 2nd printing, Wiley, New York.

Kadyshevskii, V. G.: 1962, *Sov. Phys. JETP* **14**, 1340.

Kadyshevskii, V. G.: 1963, *Sov. Phys. Doklady* **7**, 1013.

Kafatos, M. (ed.): 1989, *Bell's Theorem, Quantum Theory and Conceptions of the Universe*, Kluwer, Dordrecht.

Kaku, M.: 1988, *Introduction to Superstrings*, Springer, New York.

Kaku, M.: 1991, *Strings, Conformal Fields, and Topology*, Springer, New York.

Källen, G.: 1954, *Mat.-fys. Medd. Dan. Vid. Selsk.*, **27**, No. 12.

Kálnay, A. J.: 1971, in *Problems in the Foundations of Physics*, M. Bunge (ed.), Springer, New York.

Kant, I.: 1961, *Critique of Pure Reason*, 2nd revised edition of the translation by K.F. Müller, Doubleday, New York.

Kaplunowski, V., and M. Weinstein: 1985, *Phys. Rev. D* **31**, 1879.

Kato, T., and S. T. Kuroda: 1959, *Nuovo Cimento* **14**, 1102.

Kauffman, L. H.: 1983, *Formal Knot Theory*, Math. Notes 30, Princeton University Press, Princeton.

Kauffman, L. H.: 1987, *On Knots*, Math. Ann. Math. Studies 115, Princeton University Press, Princeton.
Kaufmann, W.: 1905, *Berl. Ber.* **45**, 949.
Kaufmann, W.: 1906, *Ann. d. Phys.* **19**, 487.
Kay, B. S.: 1978, *Commun. Math. Phys.* **62**, 55.
Kay, B. S.: 1980, *Commun. Math. Phys.* **71**, 29.
Kelly, R. L., *et al.* (Particle Data Group): 1980, *Rev. Mod. Phys.* **52**(2), S1.
Kerner, E. H.: 1965, *J. Math. Phys.* **6**, 1218.
Khrennikov, A. Yu.: 1988, *Russian Math. Surveys* **43**, 103.
Kibble ,T.W.B.: 1981, in *Quantum Gravity 2*, C.J. Isham, R. Penrose and D.W. Sciama (eds.) Clarendon Press, Oxford.
Kim, Y. S., and M. E. Noz: 1986, *Theory and Applications of the Poincaré Group*, Reidel, Dordrecht.
Kim, Y. S., and E. P. Wigner: 1987, *J. Math. Phys.* **28**, 1175.
Kim, Y. S., and E. P. Wigner: 1990, *J. Math. Phys.* **31**, 55.
Klebanov, I., L. Susskind, and T. Banks: 1989, *Nucl. Phys. B* **317**, 665.
Klein, O.: 1929, *Z. Phys.* **53**, 157.
Klein, O.: 1939, in *Le Magnétism*, Proceedings of the International Institute of Intellectual Cooperation, University of Strasbourg.
Kline, M.: 1980, *Mathematics: The Loss of Certainty*, Oxford University Press, New York.
Kobayashi, S.: 1956, *Can. J. Math.* **8**, 145.
Kobayashi, S.: 1957, *Annali di Math.* **43**, 119.
[K] Kobayashi, S., and K. Nomizu: 1963, *Foundations of Differential Geometry*, vol. I, Wiley, New York.
[KN] Kobayashi, S., and K. Nomizu: 1969, *Foundations of Differential Geometry*, vol. II, Wiley, New York.
Konisi, G., and T. Ogimoto: 1958, *Prog. Theor. Phys.* **20**, 868.
Kosinski, P., and P. Maslanka: 1990, *J. Math. Phys.* **31**, 1755.
Kostant, B.: 1970, *Quantization and Unitary Representations*, Springer Lecture Notes in Mathematics, vol. 170, New York.
Kostro, L., A. Posiewnik, J. Pykacz, and M. Zukowski (eds.): 1988, *Problems in Quantum Physics, Gdansk '87*, World Scientific, Singapore.
Kragh, H. S.: 1990, *Dirac: A Scientific Biography*, Cambridge University Press, Cambridge.
Kraichnan, R.: 1955, *Phys. Rev.* **98**, 1118.
Kraus, K.: 1983, *States, Effects and Operations*, Springer Lecture Notes in Physics, vol. 190, Berlin.
Kretschmann, E.: 1917, *Ann. Physik* **53**, 575.
Krisch, A. D.: 1987, in *Reminiscences About a Great Physicist: Paul Adrien Maurice Dirac*, B. N. Kursunoglu and E. P. Wigner (eds.), Cambridge University Press, Cambridge.
Kronheimer, E. H., and R. Penrose: 1967, *Proc. Cam. Phil. Soc.* **63**, 481.
Ktorides, C. N., and N. E. Mavromatos: 1985a, *Ann. Phys. (N.Y.)* **162**, 53.
Ktorides, C. N., and N. E. Mavromatos; 1985b, *Phys. Rev. D* **31**, 3187, 3193.
Ktorides, C. N., and L. C. Papaloucas; 1989, *Phys. Rev. A* **39**, 3310.
Kuchař, K.: 1971, *Phys. Rev. D* **4**, 4.
Kuchař, K.: 1973, in *Relativity, Astrophysics, and Cosmology*, W. Israel (ed.), Reidel, Dordrecht.
Kuchař, K.: 1976, *J. Math. Phys.* **17**, 777, 792, 801.
Kuchař, K.: 1977, *J. Math. Phys.* **18**, 1589.
Kuchař, K.: 1980, *Phys. Rev. D* **22**, 1285.
Kugo, T., and I. Ojima: 1978a, *Prog. Theor. Phys.* **60**, 1869.
Kugo, T., and I. Ojima: 1978b, *Nucl. Phys. B* **144**, 234.
Kugo, T., and I. Ojima: 1979, *Prog. Theor. Phys. Suppl.* **66**, 1.

Kuiper, N. H.: 1965, *Topology* **3**, 19.

Kuhn, T. S.: 1970, *The Structure of Scientific Revolutions*, University of Chicago Press, Chicago.

Künzle, H. P.: 1972, *Ann. Inst. H. Poincaré A* **17**, 337.

Künzle, H. P.: 1976, *Gen. Rel. Grav.* **7**, 445.

Künzle, H. P.: 1988, *Nuovo Cimento B* **101**, 721.

Künzle, H. P., and C. Duval: 1986, *Class. Quantum Grav.* **3**, 957.

Kursunoglu, B. N.: 1987, in *Reminiscences About a Great Physicist: Paul Adrien Maurice Dirac* , B. N. Kursunoglu and E. P. Wigner (eds.), Cambridge University Press, Cambridge.

Lahti, P., and P. Mittelstaedt (eds.): 1985, *Symposium on the Foundations of Modern Physics*, World Scientific, Singapore.

Lakatos, I.: 1976, in *Method and Appraisal in the Physical Sciences*, C. Howson (ed.), Cambridge University Press, Cambridge.

Landau, L. D.: 1955, in *Niels Bohr and the Development of Physics*, Pergamon Press, London.

Landau, L., and R. Peierls: 1931, Z. Phys. **69**, 56; reprinted in [WQ], pp. 465-476.

Landé, A.: 1939, *Phys. Rev.* **56**, 482, 486.

Lang, S.: 1972, *Differential Manifolds*, Addison-Wesley, Reading, Mass.

Latzin, H.: 1927, *Naturwiss.* **15**, 161.

Lautrup, B.: 1967, *Mat.-fys. Medd. Dan. Vid. Selsk.*, **12** (11), 1.

Lehmann, H., K. Symanzik, and W. Zimmermann: 1957, *Nuovo Cimento* **6**, 319.

Leibbrandt, G.: *Rev. Mod. Phys.* **59**, 1067.

Leinaas, J. M., and K. Olaussen: 1982, *Phys. Lett. B* **108**, 199.

Lévy-Leblond, J.-M.: 1963, *J. Math. Phys.* **4**, 776.

Lévy-Leblond, J.-M.: 1971, in *Group Theory and Its Applications*, vol. II, E. M. Loebl (ed.), Academic Press, New York.

Li, X., Z. Qiu, and H. Ren (eds.): 1987, *Lattice Gauge Theory Using Parallel Processors*, Gordon and Breach, New York.

Lifshitz, E. M., and I. M. Khalatnikov: 1963, *Adv. Phys.* **12**, 185.

Likorish, W. B. R.: 1963, *Proc. Phil. Soc. Cambridge* **59**, 307.

Logunov, A. A., and Yu. M. Loskutov: 1986, *Math. Theor. Phys.* **67**, 425.

Logunov, A. A., and Yu. M. Loskutov: 1988, *Math. Theor. Phys.* **74**, 215.

Logunov, A. A., Yu. M. Loskutov, and Yu. V. Chugreev: 1986, *Math. Theor. Phys.* **69**, 1179.

Lorenzen, P.: 1987, *Constructive Philosophy*, University of Massachusetts Press, Amherst.

Ludwig, G.: 1953, Z. *Phys.* **135**, 483.

Ludwig, G.: 1958, Z. *Phys.* **150**, 346; **152**, 98.

Ludwig, G.: 1983, *Foundations of Quantum Mechanics*, vol. 1, Springer, New York.

Ludwig, G.: 1985, *An Axiomatic Basis for Quantum Mechanics*, vol. 1, Springer, New York.

MacCallum, M. A. H. (ed.): 1987, *General Relativity and Gravitation*, Cambridge University Press, Cambridge.

Mach, E.: 1907, *The Science of Mechanics*, 3rd. edition, trans. by J. McCormack, Open Court, Chicago.

Machida, S., and M. Namiki: 1980, *Prog. Theor. Phys.* **63**, 1457, 1833.

Mackay, D. S.: 1961, in *A History of Philosophical Systems*, Littlefield, Adams & Co., Paterson, New Jersey.

Mackey, G. W.: 1951, *Amer. J. Math.* **73**, 576.

Mackey, G. W.: 1952, *Ann. Math.* **55**, 101.

Mackey, G. W.: 1953, *Ann. Math.* **58**, 193.

Mackey, G. W.: 1963, *The Mathematical Foundations of Quantum Mechanics*, Benjamin, New York.

Mackey, G. W.: 1968, *Induced Representations of Groups and Quantum Mechanics*, Benjamin, New York.

MacKinnon, E. M.: 1982, *Scientific Explanation and Atomic Physics*, University of Chicago Press, Chicago.
Mahoto, P., and P. Bandyopadhyay: 1987, *Nuovo Cimento B* **98**, 53.
Mandelbrot , B. B.: 1983, *The Fractal Geometry of Nature*, revised printing, Freeman, San Francisco.
March, A.: 1937, *Z. Phys.* **104**, 93, 161; **105**, 620; **106**, 49; **108**, 128.
Markov, M.: 1940, *Zh. Eksperim. Teor. Fiz.* **10,** 1311.
Markov, M. A., V. A. Berezin, and V.P. Frolov (eds.): 1988, *Quantum Gravity*, World Scientific, Singapore.
Martens, H., and W. M. de Muynck: 1990, *Found. Phys.* **20,** 255, 357.
Marzke, R. F., and J. A. Wheeler: 1964, in *Gravitation and Relativity*, H. Y. Chiu and W. F. Hoffmann (eds.), Benjamin, New York.
Mayer, M. E.: 1981, *Acta Phys. Austriaca Suppl.* **23**, 491.
Mayer, M. E.: 1983, *Phys. Lett. B* **120**, 355.
Mayer, M. E., and A. Trautman: 1981, *Acta Phys. Austriaca Suppl.* **23**, 433.
Maziarz, E. A., and T. Greenwood: 1968, *Greek Mathematical Philosophy*, Frederick Ungar, New York.
Mead, C. A.: 1964, *Phys. Rev. B* **135**, 849.
Mehra, J., and H. Rechenberg: 1982, *The Historical Development of Quantum Theory*, volumes 1-4, Springer, New York.
Messiah, A.: 1962, *Quantum Mechanics*, Wiley, New York.
Milnor, J., and J. O. Stasheff: 1974, *Characteristic Classes*, Princeton University Press, Princeton.
Mintchev, M.: 1980, *J. Phys. A: Math. Gen.* **13**, 1841.
[M] Misner, C. W., K. S. Thorne, and J. A. Wheeler: 1973, *Gravitation* , Freeman, San Francisco.
Mitroff, I. I.: 1974, *The Subjective Side of Science*, Elsevier, Amsterdam.
Mittelstaedt, P., A. Prieur, and R. Schieder: 1987, *Found. Phys.* **17**, 891.
Mitter, P. K., and C. M. Viallet: 1981, *Commun. Math. Phys.* **79**, 457.
Møller, C.: 1952, *The Theory of Relativity*, Clarendon Press, Oxford.
Montano, D., and J. Sonnenschein: 1989, *Nucl. Phys. B* **324**, 348.
Morchio, G., and F. Strocchi: 1980, *Ann. Inst. H. Poincaré A* **33**, 251.
Muckelman, D. O., R. d. Ehers,, and E. B. Fomalont: 1970, *Phys. Rev. Lett.* **24**, 1377.
Müller-Kirsten, H. J. W., and A. Wiedemann: 1987, *Supersymmetry*, World Scientific, Singapore.
Murayama, Y.: 1990, *Found. Phys. Lett.* **3**, 103.
Murdoch, D.: 1989, *Niels Bohr's Philosophy of Physics*, edition with corrections, Cambridge University Press, Cambridge.
Nachtmann, O.: 1990, *Elementary Particle Physics*, Springer, Berlin.
Nahm, W., S. Randjbar-Daemi, E. Sezgin, and E. Witten (eds.): 1991, *Topological Methods in Quantum Field Theories*, World Scientific, Singapore.
Naimark, M. A.: 1966, *Soviet Math. Dokl.* **7**, 1366.
Nakahara, M.: 1990, *Geometry, Topology and Physics*, Adam Hilger, Bristol.
Nakanishi, N.: 1966, *Prog. Theor. Phys.* **35**, 1111.
Nakanishi, N.: 1972, *Prog. Theor. Phys. Suppl.* **51**, 1.
Nakanishi, N.: 1978, *Prog. Theor. Phys.* **59**, 972; **60**, 1190, 1890.
Nakanishi, N.: 1979, *Prog. Theor. Phys.* **61**, 1536; **62**, 779, 1101, 1385.
Nakanishi, N.: 1983, in *Gauge Theory and Gravitation*, K. Kikkawa, N. Nakanishi and H. Nariai (eds.), Springer Lecture Notes in Physics, vol. 176, Berlin.
Nakanishi, N.: 1990, in *Quantum Electrodynamics*, T. Kinoshita (ed.), World Scientific, Singapore.
Nakanishi, N., and I. Ojima: 1990, *Covariant Operator Formalism of Gauge Theories and Quantum Gravity*, World Scientific, Singapore.

Namiki, M.: 1988a, *Found. Phys.* **18**, 29.

Namiki, M.: 1988b, in *Microphysical Reality and Quantum Formalism*, A. van der Merwe, F. Selleri and G. Tarozzi (eds.), Kluwer, Dordrecht.

Namiki, M., I. Ohba, K. Okano, and Y. Yamanaka: 1983, *Prog. Theor. Phys.* **69**, 1580.

Namiki, M., and S. Pascazio: 1991, *Found. Phys. Lett.* **4**, 203.

Namsrai, Kh.: 1986, *Nonlocal Quantum Field Theory and Stochastic Quantum Mechanics*, Reidel, Dordrecht.

Narain, K. S.: 1986, *Phys. Lett. B***169**, 41.

Narasimhan, M. S., and T. R. Ramadas: 1979, *Commun. Math. Phys.* **67**, 121.

[N] Narlikar, J. V., and T. Padmanabhan: 1986, *Gravity, Gauge Theories and Quantum Cosmology*, Reidel, Dordrecht.

Narlikar, J. V., and T. Padmanabhan: 1988, *Found. Phys.* **18**, 659.

Nash, C.: 1978, *Relativistic Quantum Fields*, Academic Press, London.

[NT] Nash, C., and S. Sen: 1983, *Topology and Geometry for Physicists*, Academic Press, London.

Ne'eman, Y.: 1986, *Found. Phys.* **16**, 361.

Nelson, E.: 1967, *Dynamical Theories of Brownian Motion*, Princeton University Press, Princeton.

Nelson, E.: 1986, *Quantum Fluctuations*, Princeton University Press, Princeton.

Newton, R. G.: 1966, *Scattering Theory of Waves and Particles*, McGraw-Hill, New York.

Newton, R. G.: 1979, *Found. Phys.* **9**, 929.

Newton, T. D., and E. P. Wigner: 1949, *Rev. Mod. Phys.* **21**, 400.

Nishijima, K.: 1984, *Nucl. Phys. B* **238**, 601.

Nishijima, K., and M. Okawa: 1978, *Prog. Theor. Phys.* **60**, 272.

Norton, J.: 1987, in *Measurement, Realism and Objectivity*, J. Forge (ed.), Reidel, Dordrecht.

Norton, J.: 1989, in *Einstein and the History of General Relativity*, D. Howard and J. Stachel (eds.), Birkhäuser, Boston.

Ohnuki, Y., and T. Kashiwa: 1978, *Prog. Theor. Phys.* **60**, 548.

Oldershaw, R. L.: 1988, *Am. J. Phys.* **56**, 1075.

Omnès, R.: 1990, in *Complexity, Entropy, and the Physics of Information*, W. H. Zurek (ed.), Addison-Wesley, Reading, Mass.

Osborn, H.: 1982, *Vector Bundles*, vol. 1, Academic Press, New York.

Page, D. N.: 1987, *Phys. Rev. A* **36**, 3479.

Page, D. N., and W. K. Wooters: 1983, *Phys. Rev. D* **27**, 2885.

Pais, A.: 1982, *Subtle Is the Lord: The Science and the Life of Albert Einstein*, Clarendon Press, Oxford.

Pais, A.: 1987, in *Reminiscences About a Great Physicist: Paul Adrien Maurice Dirac* , B. N. Kursunoglu and E. P. Wigner (eds.), Cambridge University Press, Cambridge.

Pap, A.: 1962, *An Introduction to the Philosophy of Science*, Free Press of Glencoe, New York.

Parisi, G., and Y.S. Wu: 1981, *Sci. Sinica* **24** , 483.

Park, J. L., and H. Margenau: 1968, *Int. J. Theor. Phys.* **1**, 211.

Parker, L.: 1966, *The Creation of Particles in Expanding Universes*, Ph. D. thesis, Harvard University.

Parker, L.: 1968, *Phys. Rev. Lett.* **21**, 562.

Parker, L.: 1969, *Phys. Rev.* **183**, 1057.

Parker, L.: 1971, *Phys. Rev. D* **3**, 346.

Parker, L.: 1977, in *Asymptotic Structure of Space-Time*, F. P. Esposito and L. Witten (eds.), Plenum, New York.

Parker, L.: 1983, in *Gauge Theory and Gravitation*, K. Kikkawa, N. Nakanishi and H. Nariai (eds.), Springer, Berlin.

Parthasarathy, K. R.: 1989, *Rev. Math. Phys.* **1**, 89.

Parvizi, D.: 1975, *Rep. Math. Phys.* **8**, 401.

Pasti, P., and M. Tonin: 1982, *Nuovo Cimento B* **69**, 97.

Pati, J. C., A. Salam, and J. Strathdee: 1975, *Phys. Lett. B* **59**, 265.

Pati, J. C., and A. Salam: 1984, *Nucl. Phys. B* **214**, 109.

Pearle. P.: 1986, in *Quantum Concepts in Space and Time*, R. Penrose and C. J. Isham (eds.), Clarendon Press, Oxford.

Pearson, D.: 1975, *Commun. Math. Phys.* **40**, 125.

Peebles, P. J. E.: 1987, *Science* **235**, 372.

Penrose, R.: 1968, in *Battelle Rencontres*, C. DeWitt and J. A. Wheeler (eds.), Benjamin, New York.

Penrose, R.: 1969, *Riv. Nuovo Cimento* **1**, 252.

Penrose, R.: 1979, in *General Relativity: An Einstein Centenary Survey*, S. W. Hawking and W. Israel (eds.), Cambridge University Press, Cambridge.

Penrose, R.: 1986, in *Quantum Concepts in Space and Time*, R. Penrose and C. J. Isham (eds.), Clarendon Press, Oxford.

Penrose, R.: 1987, in *Three Hundred Years of Gravitation*, S. W. Hawking and W. Israel (eds.), Cambridge University Press, Cambridge.

Penrose, R.: 1989, *The Emperor's New Mind*, Oxford University Press, New York.

Penrose, R., and R. Rindler: 1986, *Spinors and Space-Time*, vol. 1, printing with corrections, Cambridge University Press, Cambridge.

Perelomov, A. M.: 1972, *Commun. Math. Phys.* **26**, 22.

Perelomov, A. M.: 1986, *Generalized Coherent States and Their Applications*, Springer, Berlin.

Peres, A.: 1980, *Am. J. Phys.* **48**, 552.

Perez, J. F., and I. F. Wilde: 1977, *Phys. Rev. D* **16**, 315.

Petersen, A.: 1985, in *Niels Bohr: A Centenary Volume*, A. P. French and P. J. Kennedy (eds.), Harvard University Press, Cambridge, Mass.

Pierotti, D.: 1990, *J. Math. Phys.* **31**, 1862.

Pickering, A.: 1984, *Constructing Quarks – A Sociological History of Particle Physics*, University of Chicago Press, Chicago.

Pickering, A.: 1989a, in *Pions to Quarks: Particle Physics in the 1950's*, L. M. Brown, M. Dresden and L. Hoddeson (eds.), Cambridge University Press, Cambridge.

Pickering, A.: 1989b, in *The Uses of Experiment: Studies of Experimentation in the Natural Sciences*, D. Gooding, T. J. Pinch and S. Schaffer (eds.), Cambridge University Press, Cambridge.

Pokrowski, G. I.: 1928, *Z. Phys.* **51**, 737.

Polkinghorne, J. C.: 1979, *The Particle Play*, Freeman, Oxford.

Polkinghorne, J. C.: 1985, in *A Passion for Physics: Essays in Honor of Geoffrey Chew* , C. De Tar, J. Finkelstein and C. I. Tan (eds.), World Scientific, Singapore.

Popov, V. N.: 1983, *Functional Integrals in Quantum Field Theory and Statistical Physics*, Reidel, Dordrecht.

Popper, K. R.: 1963, *Conjectures and Refutations*, Harper& Row, New York.

Popper, K. R.: 1967, in *Quantum Theory and Reality*, M. Bunge (ed.), Springer, New York.

Popper, K. R.: 1968, *The Logic of Scientific Discovery*, revised edition, Hutchinson, London.

Popper, K. R.: 1976, *Unended Quest: An Intellectual Autobiography*, Open Court, La Salle, Illinois.

Popper, K. R.: 1982a, *Quantum Theory and the Schism in Physics*, Hutchinson, London.

Popper, K. R.: 1982b, *The Open Universe: An Argument for Indeterminism*, Rowman and Littlefield, Totowa, New Jersey.

Popper, K. R.: 1983, *Realism and the Aim of Science*, Rowman and Littlefield, Totowa, New Jersey.

Popper, K. R.: 1988, in *Microphysical Reality and Quantum Formalism*, A. van der Merwe, F. Selleri and G. Tarozzi (eds.), Kluwer, Dordrecht.

Prigogine, I.: 1980, *From Being to Becoming*, Freeman, San Francisco.

Prigogine, I., and Y. Elskens: 1989, in *Quantum Implications*, B. J. Hiley and F. D. Peat (eds.), Routledge & Kegan Paul, London.

Prugovečki, E.: 1964, *J. Math. Phys.* **5**, 442.

Prugovečki, E.: 1966, *J. Math. Phys.* **7**, 1054, 1070, 1680.

Prugovečki, E.: 1967, *Can. J. Phys.* **45**, 2173.

Prugovečki, E.: 1969a, *J. Math. Phys.* **10**, 933.

Prugovečki, E.: 1969b, *Can. J. Phys.* **47**,1083.

Prugovečki, E.: 1969c, *Can. J. Math.* **21**,158.

Prugovečki, E.: 1973, *J. Math. Phys.* **14**, 1410.

Prugovečki, E.: 1977a, *Int. J. Theor. Phys.* **16**, 321.

Prugovečki, E.: 1977b, *J. Phys. A: Math. Gen.* **10**, 543.

Prugovečki, E.: 1978a, *Ann. Phys. (N.Y.)* **110**, 102.

Prugovečki, E.: 1978b, *Physica A* **91**, 202, 229.

Prugovečki, E.: 1978c, *J. Math. Phys.* **19**, 2260, 2271.

Prugovečki, E.: 1978d, *Phys. Rev. D* **18**, 3655.

Prugovečki, E.: 1980, *Rep. Math. Phys.* **17**, 401.

[PQ] Prugovečki, E.: 1981, *Quantum Mechanics in Hilbert Space*, 2nd edition, Academic Press, New York.

Prugovečki, E.: 1981a, *Nuovo Cimento A* **61**, 85.

Prugovečki, E.: 1981b, *Lett. Nuovo Cimento* **32**, 277, 481; **33**, 480.

Prugovečki, E.: 1981c, *Nuovo Cimento B* **62**, 17.

Prugovečki, E.: 1981d, *Hadronic J.* **4**, 1018.

Prugovečki, E.: 1982a, *Found. Phys.* **12**, 555.

Prugovečki, E.: 1982b, *Phys. Rev. Lett.* **49**, 1065.

[P] Prugovečki, E.: 1984, *Stochastic Quantum Mechanics and Quantum Spacetime*, Reidel, Dordrecht; printing with corrections, Reidel, Dordrecht, 1986.

Prugovečki, E.: 1985, *Nuovo Cimento A* **89**, 105; **91**, 317.

Prugovečki, E.: 1987a, *Class. Quantum Grav.* **4**, 1659.

Prugovečki, E.: 1987b, *Nuovo Cimento A* **97**, 597, 837.

Prugovečki, E.: 1988a, *Nuovo Cimento A* **100**, 289.

Prugovečki, E.: 1988b, *Nuovo Cimento A* **100**, 827; **101**, 853.

Prugovečki, E.: 1989a, *Found. Phys. Lett.* **2**, 81, 163, 403.

Prugovečki, E.: 1989b, *Nuovo Cimento A* **102**, 881.

Prugovečki, E.: 1990, *Found. Phys. Lett.* **3**, 37.

Prugovečki, E.: 1991a, *Found. Phys.* **21**, 93.

Prugovečki, E.: 1991b, in *Group Theoretical Methods in Physics: Proceedings, Moscow 1990*, V. V. Dodonov and V. I. Man'ko (eds.), Springer Lecture Notes in Physics, vol. 382, Berlin.

Prugovečki, E.: 1991c, *Found. Phys. Lett.* **4**, 129.

Prugovečki, E., and S. Warlow: 1989a, *Found. Phys. Lett.* **2**, 409.

Prugovečki, E., and S. Warlow: 1989b, *Rep. Math. Phys.* **28**, 105.

Putnam, C. R.: 1967, *Commutation Properties of Hilbert Space Operators and Related Topics*, Springer, New York.

Quine, W. V.: 1949, in *Readings in Philosophical Analysis*, H. Feigle and W. Sellars (eds.), Apple-Century-Crofts, New York.

Quirós, M., F. J. de Urries, J. Hoyos, M. L. Mazón, and E. Rodriguez: 1981, *J. Math. Phys.* **22**, 1767.

Rauch, H.: 1987, *Contemp. Phys.* **27**, 345.
Rauch, H.: 1988, in *Microphysical Reality and Quantum Formalism*, A. van der Merwe, F. Selleri and G. Tarozzi (eds.), Kluwer, Dordrecht.
Recami, E.: 1977, in *Uncertainty Principle and Foundations of Quantum Mechanics*, W. C. Price and S. S. Chissic (eds.), Wiley, New York.
Recami, E.: 1987, *Found. Phys.* **17**, 239.
Regge, T., and C. Teitelboim: 1974, *Ann. Phys. (N.Y.)* **88**, 286.
Reich, K.: 1977, *Carl Friedrich Gauss 1777/1977*, Inter Nationes, Bonn-Bad Godesberg.
Reichenbach, H.: 1948, *Philosophic Foundations of Quantum Mechanics*, University of California Press, Berkeley.
Reichenbach, H.: 1956, *The Direction of Time*, University of California Press, Berkeley.
Reichenbach, H.: 1957, *The Philosophy of Space and Time*, Dover, New York.
Reichenbach, H.: 1961, *Experience and Prediction*, University of Chicago Press, Chicago.
Reid, C.: 1986, *Hilbert-Courant*, Springer, Berlin.
Reinhart, B. L.: *Topology* **2**, 173.
Rideau, G.: 1978, *J. Math. Phys.* **19**, 1627.
Riemann, B.: 1854, *Über die Hypothesen welche der Geometrie zu Grunde liegen*, Inaugural Lecture, translated in English as *On the Hypotheses Which Lie at the Foundations of Geometry*, in [SI], pp. 135-153.
Riesz, F., and B. Sz.-Nagy: 1990, *Functional Analyis*, Dover, New York.
Rindler, W.: 1969, *Essential Relativity*, Van Nostrand, New York.
Rivers, R. J.: 1987, *Path Integral Methods in Quantum Field Theory*, Cambridge University Press, Cambridge.
Rohrlich, F.: 1980, in *Foundations of Radiation Theory and Quantum Electrodynamics*, A. O. Barut (ed.), Plenum, New York.
Rosenfeld, L.: 1930a, *Ann. Phys. (Leipzig)* **5**, 113.
Rosenfeld, L.: 1930b, *Z. Phys.* **65**, 589.
Rosenfeld, L.: 1957, in *Report on the Conference on the Role of Gravitation in Physics*, C. M. DeWitt (ed.), Chapel Hill, North Carolina.
Rothman, T., and G. Ellis: 1987, *Astron. J.* **15**, 6.
Rovelli, C.: 1990, *Phys. Rev. D***42**, 2683.
Rovelli, C.: 1991a, *Phys. Rev. D***43**, 442.
Rovelli, C.: 1991b, *Class. Quantum Grav.* **8**, 297, 317.
Rovelli, C.: 1991c, in *Conceptual Problems in Quantum Gravity*, A. Ashtekar and J. Stachel (eds.), Birkhäuser, Boston.
Rovelli, C.: 1991d, *Class. Quantum Grav.* **8**, 1613.
Rovelli, C., and L. Smolin: 1988, *Phys. Rev. Lett.* **61**, 1155.
Rovelli, C., and L. Smolin: 1990, *Nucl. Phys. B* **331**, 80.
Ruark, A. E.: 1928, *Proc. Nat. Acad. Sci. Wash.* **14**, 322.
Russell, B.: 1945, *A History of Western Philosophy*, Simon and Shuster, New York.
Russell, B.: 1948, *Human Knowledge*, Simon and Shuster, New York.
Saad, D., L. P. Horwitz, and R. I. Arshansky: 1989, *Found. Phys.* **19**, 1125.
Sabelfeld, K. K.: 1991, *Monte Carlo Methods in Boundary Value Problems*, Springer, Berlin.
Salam, A.: 1987, in *Reminiscences About a Great Physicist: Paul Adrien Maurice Dirac*, B. N. Kursunoglu and E. P. Wigner (eds.), Cambridge University Press, Cambridge.
Salam, A.: 1990, *Unification of Fundamental Forces*, Cambridge University Press, Cambridge.
Salecker, H., and E.P. Wigner: 1958, *Phys. Rev.* **109**, 571.

Samuel, J., and R. Bhandari: 1988, *Phys. Rev. Lett.* **60**, 2339.

Satz, H., I. Harrity, and J. Potvin (eds.): 1987, *Lattice Gauge Theory '86*, Plenum, New York.

Savan, B.: 1988, *Science under Siege*, CBC Enterprises, Montréal.

Schames, L.: 1933, *Z. Phys.* **81**, 270.

Scharf, G.: 1989, *Finite Quantum Electrodynamics*, Springer, Berlin.

Schild, A.: 1948, *Phys. Rev.* **73**, 414.

Schilpp, A. (ed.): 1949, *Albert-Einstein: Philosopher-Scientist*, The Library of Living Philosophers, Evanston, Illinois.

Schrödinger, E.: 1930, *Sitzber. Preuss. Acad. Wiss. Berlin* **24**, 418.

Schrödinger, E.: 1939, *Physica* **6**, 899.

Schroeck, F. E., Jr.: 1981, *J. Math. Phys.* **22**, 2562.

Schroeck, F. E., Jr.: 1982a, *Found. Phys.* **12**, 479.

Schroeck, F. E., Jr.: 1982b, *Found. Phys.* **12**, 825.

Schroeck, F. E., Jr.: 1991, *Quantum Mechanics on Phase Space* (in preparation).

Schroeck, F. E., Jr., and D. J. Foulis: 1990, *Found. Phys.* **20**, 823.

[ST] Schulman, L. S.: 1981, *Techniques and Applications of Path Integration* , Wiley, New York.

Schwartz, L.: 1945, *Ann. de l'Univ. Grenoble* **21**, 57.

Schwarzschild, K.: 1916, *Sitzber. Deut. Akad. Berlin, Kl. Math.-Phys. Tech.* 189, 424.

[SI] Schweber, S. S.: 1961, *An Introduction to Relativistic Quantum Field Theory*, Row, Peterson and Company, Evanston, Illinois.

Schweber, S. S.: 1986, *Osiris* **2**, 265.

Schweber, S. S.: 1989, in *Pions to Quarks: Particle Physics in the 1950's*, L. M. Brown, M. Dresden and L. Hoddeson (eds.), Cambridge University Press, Cambridge.

Schweber, S. S.: 1991, *QED (1946-1950)*, Princeton University Press, Princeton (to appear).

Schwinger, J.: 1953, *Phys. Rev.* **91**, 728.

Schwinger, J. (ed.): 1958, *Quantum Electrodynamics*, Dover, New York.

Schwinger, J.: 1959a, *Proc. Nat. Acad. Sci.* **45**, 1542.

Schwinger, J.: 1959b, *Phys. Rev. Lett.* **3**, 296.

Seielstad, G. A., R. A. Sramek, and K. W. Weiler: 1970, *Phys. Rev. Lett.* **24**, 1373.

Selleri, F.: 1988, in *The Nature of Quantum Paradoxes*, G. Tarozzi and A. van der Merwe, (eds.), Kluwer, Dordrecht.

Selleri, F.: 1990, *Quantum Paradoxes and Physical Reality*, Kluwer, Dordrecht.

Shaw, R.: 1964, *Nuovo Cimento* **33**, 1074.

Shaw, R.: 1965, *Nuovo Cimento* **37**, 1086.

She, C. Y., and H. Hefner: 1966, *Phys. Rev.* **152**, 1103.

Shimony, A.: 1974, *Phys. Rev. D* **9**, 2321.

Shimony, A.: 1978, *Int. Phil Quart.* **18**, 3.

Simon, B.: 1983, *Phys. Rev. Lett.* **51**, 2176.

Skagerstam, B. K.: 1976, *Int. J. Theor. Phys.* **15**, 213.

Smolin, L.: 1986a, *Class. Quantum Grav.* **3**, 347.

Smolin, L.: 1986b, *Int. J. Theor. Phys.* **25**, 215.

Smolin, L.: 1991, in *Conceptual Problems in Quantum Gravity*, A. Ashtekar and J. Stachel (eds.), Birkhäuser, Boston.

Śniatycki, J.: 1980, *Geometric Quantization and Quantum Mechanics*, Springer, New York.

Socolovsky, M.: 1991, *J. Math. Phys.* **32**, 2522.

Sokai, A. D.: 1980, *J. Math. Phys.* **21**, 261.

Sorkin, R. D.: 1986a, *Phys. Rev. D* **33**, 978.

Sorkin, R. D.: 1986b, *Int. J. Theor. Phys.* **25**, 877.
Sorkin, R. D.: 1991, in *Conceptual Problems in Quantum Gravity*, A. Ashtekar and J. Stachel (eds.), Birkhäuser, Boston.
Souriau, J.-M.: 1970, *Structure des systèmes dynamiques*, Dunod, Paris.
[SC] Spivak, M.: 1979, *A Comprehensive Introduction to Differential Geometry*, vol. 2, 2nd edition, Publish or Perish, Wilmington, Delaware.
Sramek, R. A.: 1971, *Astrophys. J. Lett.* **167**, L55.
Stachel, J.: 1980, in *General Relativity and Gravitation*, vol. 1, A. Held (ed.), Plenum, New York.
Stachel, J.: 1987, in *General Relativity and Gravitation*, M. A. H. MacCallum (ed.), Cambridge University Press, Cambridge.
Stachel, J.: 1989, in *Einstein and the History of General Relativity and Gravitation*, D. Howard and J. Stachel (eds.), Birkhäuser, Boston.
Stachel, J.: 1991, in *Conceptual Problems in Quantum Gravity*, A. Ashtekar and J. Stachel (eds.), Birkhäuser, Boston.
Stapp, H. P.: 1972, *Am. J. Phys.* **40**, 1098.
Stapp, H. P.: 1991, *Found. Phys.* **21**, 1.
Stora, R.: 1976, in *Renormalization Theory*, G. Velo and A. S. Wightman (eds.), Reidel, Dordrecht.
Straumann, N.: 1984, *General Relativity and Relativistic Astrophysics*, Springer, Berlin.
Streater, R. F., and A. S. Wightman: 1964, *PCT, Spin and Statistics, and All That*, Benjamin, New York.
Strocchi, F: 1967, *Phys. Rev.* **162**, 1429.
Strocchi, F: 1968, *Phys. Rev.* **166**, 1302.
Strocchi, F: 1970, *Phys. Rev. D* **2**, 2334.
Strocchi, F: 1978, *Phys. Rev. D* **17**, 2010.
Strocchi, F., and A. S. Wightman: 1974, *J. Math. Phys.* **15**, 2198; **17**, 1930.
Streit, L.: 1969, *Nuovo Cimento* **62A**, 673.
Strominger, A.: 1991, in *Quantum Cosmology and Baby Universes*, S. Coleman, J. Hartle, T. Piran, and S. Weinberg (eds.), World Scientific, Singapore.
Strube, D.: 1990, *J. Math. Phys.* **31**, 2244.
Stuart, C. I. J. M.: 1991, *Found. Phys.* **21**, 591.
Stueckelberg, E. C. G.: 1941, *Helv. Phys. Acta* **14**, 372, 588.
Stulpe, W., and M. Singer: 1990, *Found. Phys. Lett.* **3**, 153.
Sunakawa, S.: 1958, *Prog. Theor. Phys.* **19**, 221.
Susskind, L.: 1991, in *Quantum Cosmology and Baby Universes*, S. Coleman, J. Hartle, T. Piran, and S. Weinberg (eds.), World Scientific, Singapore.
Suter, D., G. Chingas, R. A. Harris, and A. Pines: 1987, *Mol. Phys.* **61**, 1327.
Suter, D., K.T. Mueller, and A. Pines: 1987, *Phys. Rev. Lett.* **60**, 1218.
Swoyer, C.: 1987, in *Measurement, Realism and Objectivity*, J. Forge (ed.), Reidel, Dordrecht.
Sz.-Nagy, B.: 1960, *Appendix to: F. Riesz and B. Sz.-Nagy, Functional Analyis*, Frederick Ungar, New York; reprinted on pp. 457–490 of Riesz and Sz.-Nagy (1990).
Tarozzi, G., and A. van der Merwe (eds.): 1988, *The Nature of Quantum Paradoxes*, Kluwer, Dordrecht.
Taylor, J. R.: 1972, *Scattering Theory*, Wiley, New York.
Teitelboim, C.: 1982, *Phys. Rev. D* **25**, 3159.
Thaller, B., and S. Thaller: 1984, *Nuovo Cimento A* **82**, 222.
Thierry-Mieg, J.: 1980a, *J. Math. Phys.* **21**, 2834.
Thierry-Mieg, J.: 1980b, *Nuovo Cimento A* **56**, 396.
Tipler, F. J.: 1986, *Phys. Rep.* **137**, 231.
Townsend, P. K., and P. van Nieuwenhuizen: 1977, *Nucl. Phys. B* **120**, 301.

Trautman, A.: 1963, *C. R. Acad. Sci., Paris* **257**, 617.

Trautman, A.: 1966, in *Perspectives in Geometry and Relativity*, B. Hoffmann (ed.), Indiana University Press, Bloomington.

Trautman, A.: 1970, *Rep. Math. Phys.* **1**, 29.

Trautman, A.: 1972, *Bull. Acad. Pol. Sci., Sér. Sci. Math. Astro. Phys.* **20**, 185, 503, 895.

Trautman, A.: 1973a, *Bull. Acad. Pol. Sci., Sér. Sci. Math. Astro. Phys.* **21**, 345.

Trautman, A.: 1973b, *Symp. Math.* **12**, 139.

Trautman, A.: 1975, *Ann. N. Y. Acad. Sci.* **262**, 241.

Trautman, A.: 1976, *Rep. Math. Phys.* **10**, 297.

Trautman, A.: 1979, *Bull. Acad. Pol. Sci., Sér. Sci. Math. Astro. Phys.* **27**, 7.

Trautman, A.: 1980, in *General Relativity and Gravitation*, vol. 1, A. Held (ed.), Plenum, New York.

Trautman, A.: 1981, *Acta Phys. Austriaca Suppl.* **23**, 401.

Trautman, A.: 1982, in *Geometrical Techniques in Gauge Theories*, R. Martini and E.M. De Jager (eds.), Springer, Berlin.

Treder, H.-J., H.-H. von Borzeszkowski, A. van der Merwe, and W. Yourgrau: 1980, *Fundamental Principles of General Relativity Theories*, Plenum, New York.

Tryon, E. P.: 1973, *Nature* **246**, 396.

Turner, R. E., and R. F. Snider, 1980: *Can. J. Phys.* **58**, 1171.

Tycko, R.: 1987, *Phys. Rev. Lett.* **58**, 2281.

Tyutin, I. V.: 1975, Lebedev preprint, FIAN No. 39 (unpublished).

Unruh, W. G.: 1976, *Phys. Rev. D* **14**, 870.

Unruh, W. G.: 1989, *Int. J. Theor. Phys.* **28**, 1181.

Unruh, W. G., and R. M. Wald: 1984, *Phys. Rev. D* **29**, 1047.

Unruh, W. G., and R. M. Wald: 1989, *Phys. Rev. D* **40**, 2598.

Urbanik, K.: 1961, *Studia Math.* **21**, 117.

Utiyama, R.: 1956, *Phys. Rev.* **101**, 219, 1597.

van Dalen, D.: 1990, *Math. Intelligencer* **12**(4), 17.

van Dam, H., and M. Veltman: 1970, *Nucl. Phys. B* **22**, 397.

van der Merwe, A., F. Selleri, and G. Tarozzi (eds.): 1988, *Microphysical Reality and Quantum Formalism*, Kluwer, Dordrecht.

van Fraassen, B. C.: 1979, in *Problems in the Foundations of Physics*, G. Toraldo de Francia (ed.), North-Holland, Amsterdam.

van Kampen, N. G.: 1981, *Stochastic Processes in Physics and Chemistry*, North Holland, Amsterdam.

van Nieuwenhuizen, P: 1977, in *Proceedings of the First Marcel Grossmann Meeting on General Relativity*, North-Holland, Amsterdam.

Varadarajan, V. S.: 1962, *Commun. Pure. Appl. Math.* **15**, 189.

Vassiliadis, D. V.: 1989, *J. Math. Phys.* **30**, 2177.

Velo, G., and D. Zwanziger: 1969, *Phys. Rev.* **186**, 1337; **188**, 2218.

Velo, G., and D. Zwanziger: 1971, in *Troubles in the External Field Problem for Invariant Wave Equations*, M. D. Cin, G. J. Iverson and A Perlmutter (eds.), Gordon and Breach, New York.

Vilenkin, A.: 1982, *Phys. Lett. B* **117**, 26.

Vilenkin, A.: 1988, *Phys. Rev. D* **37**, 888.

Vilenkin, A.: 1989, *Phys. Rev. D* **39**, 1116.

Veltman, M. J. G.: 1986, *Sci. Am.* **255**(5), 76.

von Borzeszkowski, H.-H., and H.-J. Treder: 1988, *The Meaning of Quantum Gravity*, Reidel, Dordrecht.

von Neumann, J.: 1932, 1955, *Mathematische Grundlagen der Quantenmechanik*, Springer, Berlin, 1932; translated in English by R. T. Beyer as *Mathematical Foundations of Quantum Mechanics*, Princeton University Press, Princeton, 1955.

Wald, R. M.: 1975, *Commun. Math. Phys.* **45**, 9.

Wald, R. M.: 1977, *Commun. Math. Phys.* **54**, 1.

Wald, R. M.: 1978, *Phys. Rev. D* **17**, 1477.

[W] Wald, R. M.: 1984, *General Relativity*, University of Chicago Press, Chicago.

Wali, K. C.: 1986, *Prog. Theor. Phys. Suppl.* **86**, 387.

Wallstrom, T. C.: 1989, *Found. Phys. Lett.* **2**, 113.

Warlow, S.: 1992, *Krein Spaces and Fibre Bundles in the Quantum Theory of Gauge Fields*, Ph. D. thesis, University of Toronto.

Watkins, J. W. N.: 1984, *Science and Skepticism*, Princeton University Press, Princeton.

Weinberg, S.: 1964, *Phys. Rev.* **133**, B1049.

Weinberg, S.: 1965, *Phys. Rev.* **138**, B988.

Weinberg, S.: 1972, *Gravitation and Cosmology*, Wiley, New York.

Weinberg, S.: 1987, *Phys. Today* **40**(1), 7.

Weinberg, S.: 1989, *Rev. Mod. Phys.* **61**, 1.

Wess, J., and J. Bagger: 1983, *Supersymmetry and Supergravity*, Princeton University Press, Princeton.

West, P. C.: 1986, *Introduction to Supersymmetry and Supergravity*, World Scientific, Singapore.

Weyl, H.: 1923, *Raum-Zeit-Materie*, 5th edition, Springer, Berlin; 4th edition translated in English by H. L. Brose as: *Space-Time-Matter* , Methuen, London, 1921; reprinted by Dover, New York, 1952.

Weyl, H.: 1924, *Naturwissenschaften* **12**, 197.

Weyl, H.: 1929, *Z. Phys.* **56**, 330.

Weyl, H.: 1949, *Philosophy of Mathematics and Natural Science*, Princeton University Press, Princeton.

Wheeler, J. A.: 1957, *Ann. Phys. (N.Y.)* **2**, 604.

Wheeler, J. A.: 1962, *Geometrodynamics*, Academic Press, New York.

Wheeler, J. A.: 1964, in *Relativity, Groups and Topology*, C. DeWitt and B. S. DeWitt (eds.), Gordon and Breach, New York.

Wheeler, J. A.: 1968, in *Battelle Rencontres*, C. DeWitt and J. A. Wheeler (eds.), Benjamin, New York.

Wheeler, J. A.: 1979, in *Problems in the Foundations of Physics*, G. Toraldo de Francia (ed.), North-Holland, Amsterdam.

[WQ] Wheeler, J. A., and W. H. Zurek, (eds.): 1983, *Quantum Theory and Measurement*, Princeton University Press, Princeton.

Whittaker, E. T.: 1951, *A History of the Theory of Aether and Elasticity*, revised edition, Thomas Nelson, London.

Wick, G., A. S. Wightman, and E. P. Wigner: 1952, *Phys. Rev.* **88**, 101.

Wightman, A. S.: 1956, *Phys. Rev.* **101**, 860.

Wightman, A. S.: 1962, *Rev. Mod. Phys.* **34**, 845.

Wightman, A. S.: 1971, in *Troubles in the External Field Problem for Invariant Wave Equations*, M. D. Cin, G. J. Iverson and A. Perlmutter (eds.), Gordon and Breach, New York.

Wightman, A. S.: 1972, in *Aspects of Quantum Theory*, A. Salam and E. P. Wigner (eds.), Cambridge University Press, Cambridge.

Wightman, A. S.: 1973, in *Proceedings of the Summer Institute on Partial Differential Equations, Berkeley 1971*, D. Spencer (ed.), American Mathematical Society, Providence, R.I.

Wightman, A. S., and L. Gårding: 1964, *Arkiv för Fysik* **28**, 129.

Wightman, A. S., and S. S. Schweber: 1955, *Phys. Rev.* **98**, 812.

Wigner, E. P.: 1932, *Phys. Rev.* **40**, 749.

Wigner, E. P.: 1939, *Ann. Math* **40**, 149.

Wigner, E. P.: 1952, *Z. Phys.* **131**, 101.

Wigner, E. P.: 1959, *Group Theory and Its Application to the Quantum Mechanics of Atomic Spectra*, Academic Press, New York.

Wigner, E. P.: 1962, in *The Scientist Speculates*, I. G. Good (ed.), Basic Books, New York; reproduced in [WQ], pp. 168-181.

Wigner, E. P.: 1963, *Am. J. Phys.* **31**, 6; reproduced in [WQ], pp. 324-341.

Wigner, E. P.: 1964, *Monist* **48**, 248.

Wigner, E. P.: 1972, in *Aspects of Quantum Theory*, A. Salam and E. P. Wigner (eds.), Cambridge University Press, Cambridge.

Wigner, E. P.: 1976, 1981, *Interpretation of Quantum Mechanics*, Lectures originally given at Princeton University; reproduced in [WQ], pp. 260-314.

Wigner, E. P.: 1979, in *Perspectives in Quantum Theory*, W. Yourgrau and A. van der Merwe (eds.), Dover, New York.

Wilczek, F., and A. Zee: 1984, *Phys. Rev. Lett.* **52**, 2111.

Witten, E.: 1988, *Commun. Math. Phys.* **117**, 353.

Witten, E.: 1989, *Nucl. Phys. B* **311**, 46.

Witten, L. (ed.): 1962, *Gravitation: An Introduction to Current Research*,Wiley, New York.

Wolf, F. A.: 1990, *Parallel Universes*, Simon & Shuster, New York.

Woodhouse, N. M. J.: 1973, *J. Math. Phys.* **14**, 495.

Woodhouse, N. M. J.: 1980, *Geometric Quantization*, Clarendon Press, Oxford.

Wu, T. T., and C. N. Yang: 1975, *Phys. Rev. D* **12**, 3845.

Yaes, R.: 1974, *New Scientist* **63**, 462.

Yamamoto, Y., and H. A. Haus: 1986, *Rev. Mod. Phys.* **58**, 1001.

Yanase, M. M.: *Phys. Rev.* **72**, 874.

Yang, C. N.: 1947, *Phys. Rev.* **123**, 666.

Yang, C. N., and R. Mills: 1954, *Phys. Rev.* **96**, 191.

Yosida, K.: 1974, *Functional Analysis*, 4th edition, Springer, Berlin.

Yngvason, J.: 1977, *Rep. Math. Phys.* **12**, 57.

Yokoyama, K., and R. Kubo: 1990, in *Quantum Electrodynamics*, T. Kinoshita (ed.), World Scientific, Singapore.

Yukawa, H.: 1950, *Phys. Rev.* **77**, 219, 1047.

Yukawa, H.: 1953, *Phys. Rev.* **91**, 416.

Zardecki, A.: 1988, *J. Math. Phys.* **29**, 1661.

Zeh, H. D.: 1989, *The Physical Basis of the Direction of Time*, Springer, Berlin.

Zurek, W. H.: 1982, *Phys. Rev. D* **26**, 1862.

Zurek, W. H.: 1990a, in *Complexity, Entropy, and the Physics of Information*, W. H. Zurek (ed.), Addison-Wesley, Reading, Mass.

Zwanziger, D.: 1989, *Nucl.Phys. B* **323**, 513.

Index

Fundamental Theories of Physics

Series Editor: Alwyn van der Merwe, *University of Denver, USA*

1. M. Sachs: *General Relativity and Matter.* A Spinor Field Theory from Fermis to Light-Years. With a Foreword by C. Kilmister. 1982 ISBN 90-277-1381-2
2. G.H. Duffey: *A Development of Quantum Mechanics.* Based on Symmetry Considerations. 1985 ISBN 90-277-1587-4
3. S. Diner, D. Fargue, G. Lochak and F. Selleri (eds.): *The Wave-Particle Dualism.* A Tribute to Louis de Broglie on his 90th Birthday. 1984 ISBN 90-277-1664-1
4. E. Prugovečki: *Stochastic Quantum Mechanics and Quantum Species.* A Consistent Unification of Relativity and Quantum Theory based on Stochastic Spaces. 1984; 2nd printing 1986 ISBN 90-277-1617-X
5. D. Hestenes and G. Sobczyk: *Clifford Algebra to Geometric Calculus.* A Unified Language for Mathematics and Physics. 1984 ISBN 90-277-1673-0; Pb (1987) 90-277-2561-6
6. P. Exner: *Open Quantum Systems and Feynman Integrals.* 1985 ISBN 90-277-1678-1
7. L. Mayants: *The Enigma of Probability and Physics.* 1984 ISBN 90-277-1674-9
8. E. Tocaci: *Relativistic Mechanics, Time and Inertia.* Translated from Romanian. Edited and with a Foreword by C.W. Kilmister. 1985 ISBN 90-277-1769-9
9. B. Bertotti, F. de Felice and A. Pascolini (eds.): *General Relativity and Gravitation.* Proceedings of the 10th International Conference (Padova, Italy, 1983). 1984 ISBN 90-277-1819-9
10. G. Tarozzi and A. van der Merwe (eds.): *Open Questions in Quantum Physics.* 1985 ISBN 90-277-1853-9
11. J.V. Narlikar and T. Padmanabhan: *Gravity, Gauge Theories and Quantum Cosmology.* 1986 ISBN 90-277-1948-9
12. G.S. Asanov: *Finsler Geometry, Relativity and Gauge Theories.* 1985 ISBN 90-277-1960-8
13. K. Namsrai: *Nonlocal Quantum Field Theory and Stochastic Quantum Mechanics.* 1986 ISBN 90-277-2001-0
14. C. Ray Smith and W.T. Grandy, Jr. (eds.): *Maximum-Entropy and Bayesian Methods in Inverse Problems.* Proceedings of the 1st and 2nd International Workshop (Laramie, Wyoming, USA). 1985 ISBN 90-277-2074-6
15. D. Hestenes: *New Foundations for Classical Mechanics.* 1986 ISBN 90-277-2090-8; Pb (1987) 90-277-2526-8
16. S.J. Prokhovnik: *Light in Einstein's Universe.* The Role of Energy in Cosmology and Relativity. 1985 ISBN 90-277-2093-2
17. Y.S. Kim and M.E. Noz: *Theory and Applications of the Poincaré Group.* 1986 ISBN 90-277-2141-6
18. M. Sachs: *Quantum Mechanics from General Relativity.* An Approximation for a Theory of Inertia. 1986 ISBN 90-277-2247-1
19. W.T. Grandy, Jr.: *Foundations of Statistical Mechanics.* Vol. I: *Equilibrium Theory.* 1987 ISBN 90-277-2489-X
20. H.-H von Borzeszkowski and H.-J. Treder: *The Meaning of Quantum Gravity.* 1988 ISBN 90-277-2518-7
21. C. Ray Smith and G.J. Erickson (eds.): *Maximum-Entropy and Bayesian Spectral Analysis and Estimation Problems.* Proceedings of the 3rd International Workshop (Laramie, Wyoming, USA, 1983). 1987 ISBN 90-277-2579-9

22. A.O. Barut and A. van der Merwe (eds.): *Selected Scientific Papers of Alfred Landé. [1888-1975]*. 1988 ISBN 90-277-2594-2
23. W.T. Grandy, Jr.: *Foundations of Statistical Mechanics.* Vol. II: *Nonequilibrium Phenomena.* 1988 ISBN 90-277-2649-3
24. E.I. Bitsakis and C.A. Nicolaides (eds.): *The Concept of Probability.* Proceedings of the Delphi Conference (Delphi, Greece, 1987). 1989 ISBN 90-277-2679-5
25. A. van der Merwe, F. Selleri and G. Tarozzi (eds.): *Microphysical Reality and Quantum Formalism, Vol. 1.* Proceedings of the International Conference (Urbino, Italy, 1985). 1988 ISBN 90-277-2683-3
26. A. van der Merwe, F. Selleri and G. Tarozzi (eds.): *Microphysical Reality and Quantum Formalism, Vol. 2.* Proceedings of the International Conference (Urbino, Italy, 1985). 1988 ISBN 90-277-2684-1
27. I.D. Novikov and V.P. Frolov: *Physics of Black Holes.* 1989 ISBN 90-277-2685-X
28. G. Tarozzi and A. van der Merwe (eds.): *The Nature of Quantum Paradoxes.* Italian Studies in the Foundations and Philosophy of Modern Physics. 1988 ISBN 90-277-2703-1
29. B.R. Iyer, N. Mukunda and C.V. Vishveshwara (eds.): *Gravitation, Gauge Theories and the Early Universe.* 1989 ISBN 90-277-2710-4
30. H. Mark and L. Wood (eds.): *Energy in Physics, War and Peace.* A Festschrift celebrating Edward Teller's 80th Birthday. 1988 ISBN 90-277-2775-9
31. G.J. Erickson and C.R. Smith (eds.): *Maximum-Entropy and Bayesian Methods in Science and Engineering.* Vol. I: *Foundations.* 1988 ISBN 90-277-2793-7
32. G.J. Erickson and C.R. Smith (eds.): *Maximum-Entropy and Bayesian Methods in Science and Engineering.* Vol. II: *Applications.* 1988 ISBN 90-277-2794-5
33. M.E. Noz and Y.S. Kim (eds.): *Special Relativity and Quantum Theory.* A Collection of Papers on the Poincaré Group. 1988 ISBN 90-277-2799-6
34. I.Yu. Kobzarev and Yu.I. Manin: *Elementary Particles. Mathematics, Physics and Philosophy.* 1989 ISBN 0-7923-0098-X
35. F. Selleri: *Quantum Paradoxes and Physical Reality.* 1990 ISBN 0-7923-0253-2
36. J. Skilling (ed.): *Maximum-Entropy and Bayesian Methods.* Proceedings of the 8th International Workshop (Cambridge, UK, 1988). 1989 ISBN 0-7923-0224-9
37. M. Kafatos (ed.): *Bell's Theorem, Quantum Theory and Conceptions of the Universe.* 1989 ISBN 0-7923-0496-9
38. Yu.A. Izyumov and V.N. Syromyatnikov: *Phase Transitions and Crystal Symmetry.* 1990 ISBN 0-7923-0542-6
39. P.F. Fougère (ed.): *Maximum-Entropy and Bayesian Methods.* Proceedings of the 9th International Workshop (Dartmouth, Massachusetts, USA, 1989). 1990 ISBN 0-7923-0928-6
40. L. de Broglie: *Heisenberg's Uncertainties and the Probabilistic Interpretation of Wave Mechanics.* With Critical Notes of the Author. 1990 ISBN 0-7923-0929-4
41. W.T. Grandy, Jr.: *Relativistic Quantum Mechanics of Leptons and Fields.* 1991 ISBN 0-7923-1049-7
42. Yu.L. Klimontovich: *Turbulent Motion and the Structure of Chaos.* A New Approach to the Statistical Theory of Open Systems. 1991 ISBN 0-7923-1114-0

43. W.T. Grandy, Jr. and L.H. Schick (eds.): *Maximum-Entropy and Bayesian Methods.* Proceedings of the 10th International Workshop (Laramie, Wyoming, USA, 1990). 1991 ISBN 0-7923-1140-X
44. P.Pták and S. Pulmannová: *Orthomodular Structures as Quantum Logics.* Intrinsic Properties, State Space and Probabilistic Topics. 1991 ISBN 0-7923-1207-4
45. D. Hestenes and A. Weingartshofer (eds.): *The Electron.* New Theory and Experiment. 1991 ISBN 0-7923-1356-9
46. P.P.J.M. Schram: *Kinetic Theory of Gases and Plasmas.* 1991 ISBN 0-7923-1392-5
47. A. Micali, R. Boudet and J. Helmstetter (eds.): *Clifford Algebras and their Applications in Mathematical Physics.* 1992 ISBN 0-7923-1623-1
48. E. Prugovečki: *Quantum Geometry.* A Framework for Quantum General Relativity. 1992 ISBN 0-7923-1640-1

KLUWER ACADEMIC PUBLISHERS – DORDRECHT / BOSTON / LONDON

Batch number: 10809436

Printed by Printforce, the Netherlands